REAL ANALYSIS AND PROBABILITY

This much admired textbook, now reissued in paperback, offers a clear exposition of modern probability theory and of the interplay between the properties of metric spaces and probability measures.

The first half of the book gives an exposition of real analysis: basic set theory, general topology, measure theory, integration, an introduction to functional analysis in Banach and Hilbert spaces, convex sets and functions, and measure on topological spaces. The second half introduces probability based on measure theory, including laws of large numbers, ergodic theorems, the central limit theorem, conditional expectations, and martingale convergence. A chapter on stochastic processes introduces Brownian motion and the Brownian bridge.

The new edition has been made even more self-contained than before; it now includes early in the book a foundation of the real number system and the Stone-Weierstrass theorem on uniform approximation in algebras of functions. Several other sections have been revised and improved, and the extensive historical notes have been further amplified. A number of new exercises, and hints for solution of old and new ones, have been added.

R. M. Dudley is Professor of Mathematics at the Massachusetts Institute of Technology in Cambridge, Massachusetts.

T0297702

REAL ANALYSIS AND PROBABILITY

R. M. DUDLEY
Massachusetts Institute of Technology

CAMBRIDGE UNIVERSITY PRESS
Cambridge, New York, Melbourne, Madrid, Cape Town,
Singapore, São Paulo, Delhi, Tokyo, Mexico City

Cambridge University Press
The Edinburgh Building, Cambridge CB2 8RU, UK

Published in the United States of America by Cambridge University Press, New York

www.cambridge.org
Information on this title: www.cambridge.org/9780521007542

First published 1989 by Wadsworth, Inc.
Cambridge University Press edition published 2002

A catalogue record for this publication is available from the British Library

ISBN 978-0-521-80972-6 Hardback
ISBN 978-0-521-00754-2 Paperback

Contents

Preface to the Cambridge Edition

This is a text at the beginning graduate level. Some study of intermediate analysis in Euclidean spaces will provide helpful background, but in this edition such background is not a formal prerequisite. Efforts to make the book more self-contained include inserting material on the real number system into Chapter 1, adding a treatment of the Stone-Weierstrass theorem, and generally eliminating references for proofs to other books except at very few points, such as some complex variable theory in Appendix B.

Chapters 1 through 5 provide a one-semester course in real analysis. Following that, a one-semester course on probability can be based on Chapters 8 through 10 and parts of 11 and 12. Starred paragraphs and sections, such as those found in Chapter 6 and most of Chapter 7, are called on rarely, if at all, later in the book. They can be skipped, at least on first reading, or until needed.

Relatively few proofs of less vital facts have been left to the reader. I would be very glad to know of any substantial unintentional gaps or errors. Although I have worked and checked all the problems and hints, experience suggests that mistakes in problems, and hints that may mislead, are less obvious than errors in the text. So take hints with a grain of salt and perhaps make a first try at the problems without using the hints.

I looked for the best and shortest available proofs for the theorems. Short proofs that have appeared in journal articles, but in few if any other textbooks, are given for the completion of metric spaces, the strong law of large numbers, the ergodic theorem, the martingale convergence theorem, the subadditive ergodic theorem, and the Hartman-Wintner law of the iterated logarithm.

Around 1950, when Halmos' classic *Measure Theory* appeared, the more advanced parts of the subject headed toward measures on locally compact spaces, as in, for example, §7.3 of this book. Since then, much of the research in probability theory has moved more in the direction of metric spaces. Chapter 11 gives some facts connecting metrics and probabilities which follow the newer trend. Appendix E indicates what can go wrong with measures

on (locally) compact nonmetric spaces. These parts of the book may well not be reached in a typical one-year course but provide some distinctive material for present and future researchers.

Problems appear at the end of each section, generally increasing in difficulty as they go along. I have supplied hints to the solution of many of the problems. There are a lot of new or, I hope, improved hints in this edition.

I have also tried to trace back the history of the theorems to give credit where it is due. Historical notes and references, sometimes rather extensive, are given at the end of each chapter. Many of the notes have been augmented in this edition and some have been corrected. I don't claim, however, to give the last word on any part of the history.

The book evolved from courses given at M.I.T. since 1967 and in Aarhus, Denmark, in 1976. For valuable comments I am glad to thank Ken Alexander, Deborah Allinger, Laura Clemens, Ken Davidson, Don Davis, Persi Diaconis, Arnout Eikeboom, Sy Friedman, David Gillman, José Gonzalez, E. Griffor, Leonid Grinblat, Dominique Haughton, J. Hoffmann-Jørgensen, Arthur Mattuck, Jim Munkres, R. Proctor, Nick Reingold, Rae Shortt, Dorothy Maharam Stone, Evangelos Tabakis, Jin-Gen Yang, and other students and colleagues.

For helpful comments on the first edition I am thankful to Ken Brown, Justin Corvino, Charles Goldie, Charles Hadlock, Michael Jansson, Suman Majumdar, Rimas Norvaiša, Mark Pinsky, Andrew Rosalsky, the late Rae Shortt, and Dewey Tucker. I especially thank Andries Lenstra and Valentin Petrov for longer lists of suggestions. Major revisions have been made to §10.2 (regular conditional probabilities) and in Chapter 12 with regard to Markov times.

<div align="right">

R. M. Dudley

</div>

1

Foundations; Set Theory

In constructing a building, the builders may well use different techniques and materials to lay the foundation than they use in the rest of the building. Likewise, almost every field of mathematics can be built on a foundation of axiomatic set theory. This foundation is accepted by most logicians and mathematicians concerned with foundations, but only a minority of mathematicians have the time or inclination to learn axiomatic set theory in detail.

To make another analogy, higher-level computer languages and programs written in them are built on a foundation of computer hardware and systems programs. How much the people who write high-level programs need to know about the hardware and operating systems will depend on the problem at hand.

In modern real analysis, set-theoretic questions are somewhat more to the fore than they are in most work in algebra, complex analysis, geometry, and applied mathematics. A relatively recent line of development in real analysis, "nonstandard analysis," allows, for example, positive numbers that are infinitely small but not zero. Nonstandard analysis depends even more heavily on the specifics of set theory than earlier developments in real analysis did.

This chapter will give only enough of an introduction to set theory to define some notation and concepts used in the rest of the book. In other words, this chapter presents mainly "naive" (as opposed to axiomatic) set theory. Appendix A gives a more detailed development of set theory, including a listing of axioms, but even there, the book will not enter into nonstandard analysis or develop enough set theory for it.

Many of the concepts defined in this chapter are used throughout mathematics and will, I hope, be familiar to most readers.

1.1. Definitions for Set Theory and the Real Number System

Definitions can serve at least two purposes. First, as in an ordinary dictionary, a definition can try to give insight, to convey an idea, or to explain a less familiar idea in terms of a more familiar one, but with no attempt to specify or exhaust

1

completely the meaning of the word being defined. This kind of definition will be called *informal*. A *formal* definition, as in most of mathematics and parts of other sciences, may be quite precise, so that one can decide scientifically whether a statement about the term being defined is true or not. In a formal definition, a familiar term, such as a common unit of length or a number, may be defined in terms of a less familiar one. Most definitions in set theory are formal. Moreover, set theory aims to provide a coherent logical structure not only for itself but for just about all of mathematics. There is then a question of where to begin in giving definitions.

Informal dictionary definitions often consist of synonyms. Suppose, for example, that a dictionary simply defined "high" as "tall" and "tall" as "high." One of these definitions would be helpful to someone who knew one of the two words but not the other. But to an alien from outer space who was trying to learn English just by reading the dictionary, these definitions would be useless. This situation illustrates on the smallest scale the whole problem the alien would have, since all words in the dictionary are defined in terms of other words. To make a start, the alien would have to have some way of interpreting at least a few of the words in the dictionary other than by just looking them up.

In any case some words, such as the conjunctions "and," "or," and "but," are very familiar but hard to define as separate words. Instead, we might have rules that define the meanings of phrases containing conjunctions given the meanings of the words or subphrases connected by them.

At first thought, the most important of all definitions you might expect in set theory would be the definition of "set," but quite the contrary, just because the entire logical structure of mathematics reduces to or is defined in terms of this notion, it cannot necessarily be given a formal, precise definition. Instead, there are rules (axioms, rules of inference, etc.) which in effect provide the meaning of "set." A preliminary, informal definition of *set* would be "any collection of mathematical objects," but this notion will have to be clarified and adjusted as we go along.

The problem of defining *set* is similar in some ways to the problem of defining *number*. After several years of school, students "know" about the numbers 0, 1, 2, ..., in the sense that they know rules for operating with numbers. But many people might have a hard time saying exactly what a number is. Different people might give different definitions of the number 1, even though they completely agree on the rules of arithmetic.

In the late 19th century, mathematicians began to concern themselves with giving precise definitions of numbers. One approach is that beginning with 0, we can generate further integers by taking the "successor" or "next larger integer."

If 0 is defined, and a successor operation is defined, and the successor of any integer n is called n', then we have the sequence $0, 0', 0'', 0''', \ldots$. In terms of 0 and successors, we could then write down definitions of the usual integers. To do this I'll use an equals sign with a colon before it, ":=," to mean "equals by definition." For example, $1 := 0', 2 := 0'', 3 := 0''', 4 := 0''''$, and so on. These definitions are precise, as far as they go. One could produce a thick dictionary of numbers, equally precise (though not very useful) but still incomplete, since 0 and the successor operation are not formally defined. More of the structure of the number system can be provided by giving rules about 0 and successors. For example, one rule is that if $m' = n'$, then $m = n$.

Once there are enough rules to determine the structure of the nonnegative integers, then what is important is the structure rather than what the individual elements in the structure actually are.

In summary: if we want to be as precise as possible in building a rigorous logical structure for mathematics, then informal definitions cannot be part of the structure, although of course they can help to explain it. Instead, at least some basic notions must be left undefined. Axioms and other rules are given, and other notions are defined in terms of the basic ones.

Again, informally, a set is any collection of objects. In mathematics, the objects will be mathematical ones, such as numbers, points, vectors, or other sets. (In fact, from the set-theoretic viewpoint, all mathematical objects are sets of one kind or another.) If an object x is a member of a set y, this is written as "$x \in y$," sometimes also stated as "x belongs to y" or "x is in y." If S is a finite set, so that its members can be written as a finite list x_1, \ldots, x_n, then one writes $S = \{x_1, \ldots, x_n\}$. For example, $\{2, 3\}$ is the set whose only members are the numbers 2 and 3. The notion of membership, "\in," is also one of the few basic ones that are formally undefined.

A set can have just one member. Such a set, whose only member is x, is called $\{x\}$, read as "singleton x." In set theory a distinction is made between $\{x\}$ and x itself. For example if $x = \{1, 2\}$, then x has two members but $\{x\}$ only one.

A set A is *included* in a set B, or is a *subset* of B, written $A \subset B$, if and only if every member of A is also a member of B. An equivalent statement is that B *includes* A, written $B \supset A$. To say B *contains* x means $x \in B$. Many authors also say B contains A when $B \supset A$.

The phrase "if and only if" will sometimes be abbreviated "iff." For example, $A \subset B$ iff for all x, if $x \in A$, then $x \in B$.

One of the most important rules in set theory is called "extensionality." It says that if two sets A and B have the same members, so that for any object

$x, x \in A$ if and only if $x \in B$, or equivalently both $A \subset B$ and $B \subset A$, then the sets are equal, $A = B$. So, for example, $\{2, 3\} = \{3, 2\}$. The order in which the members happen to be listed makes no difference, as long as the members are the same. In a sense, extensionality is a definition of equality for sets. Another view, more common among set theorists, is that any two objects are equal if and only if they are identical. So "$\{2, 3\}$" and "$\{3, 2\}$" are two names of one and the same set.

Extensionality also contributes to an informal definition of *set*. A set is defined simply by what its members are—beyond that, structures and relationships between the members are irrelevant to the definition of the set.

Other than giving finite lists of members, the main way to define specific sets is to give a condition that the members satisfy. In notation, $\{x : \ldots\}$ means the set of all x such that. . . . For example, $\{x : (x - 4)^2 = 4\} = \{2, 6\} = \{6, 2\}$.

In line with a general usage that a slash through a symbol means "not," as in $a \neq b$, meaning "a is not equal to b," the symbol "\notin" means "is not a member of." So $x \notin y$ means x is not a member of y, as in $3 \notin \{1, 2\}$.

Defining sets via conditions can lead to contradictions if one is not careful. For example, let $r = \{x : x \notin x\}$. Then $r \notin r$ implies $r \in r$ and conversely (Bertrand Russell's paradox). This paradox can be avoided by limiting the condition to some set. Thus $\{x \in A : \ldots x \ldots\}$ means "the set of all x in A such that $\ldots x \ldots$." As long as this form of definition is used when A is already known to be a set, new sets can be defined this way, and it turns out that no contradictions arise.

It might seem peculiar, anyhow, for a set to be a member of itself. It will be shown in Appendix A (Theorem A.1.9), from the axioms of set theory listed there, that no set is a member of itself. In this sense, the collection r of sets named in Russell's paradox is the collection of all sets, sometimes called the "universe" in set theory. Here the informal notion of set as any collection of objects is indeed imprecise. The axioms in Appendix A provide conditions under which certain collections are or are not sets. For example, the universe is not a set.

Very often in mathematics, one is working for a while inside a fixed set y. Then an expression such as $\{x : \ldots x \ldots\}$ is used to mean $\{x \in y : \ldots x \ldots\}$.

Now several operations in set theory will be defined. In cases where it may not be obvious that the objects named are sets, there are axioms which imply that they are (Appendix A).

There is a set, called \emptyset, the "empty set," which has no members. That is, for all x, $x \notin \emptyset$. This set is unique, by extensionality. If B is any set, then 2^B, also called the "power set" of B, is the set of all subsets of B. For example, if B has 3 members, then 2^B has $2^3 = 8$ members. Also, $2^\emptyset = \{\emptyset\} \neq \emptyset$.

$A \cap B$, called the intersection of A and B, is defined by $A \cap B := \{x \in A : x \in B\}$. In other words, $A \cap B$ is the set of all x which belong to both A and B. $A \cup B$, called the union of A and B, is a set such that for any x, $x \in A \cup B$ if and only if $x \in A$ or $x \in B$ (or both). Also, $A \backslash B$ (read "A minus B") is the set of all x in A which are not in B, sometimes called the *relative complement* (of B in A). The *symmetric difference $A \bigtriangleup B$* is defined as $(A \backslash B) \cup (B \backslash A)$.

\mathbb{N} will denote the set of all nonnegative integers 0, 1, 2, (Formally, nonnegative integers are usually defined by defining 0 as the empty set \emptyset, 1 as $\{\emptyset\}$, and generally the successor operation mentioned above by $n' = n \cup \{n\}$, as is treated in more detail in Appendix A.)

Informally, an *ordered pair* consists of a pair of mathematical objects in a given order, such as $\langle x, y \rangle$, where x is called the "first member" and y the "second member" of the ordered pair $\langle x, y \rangle$. Ordered pairs satisfy the following axiom: for all x, y, u, and v, $\langle x, y \rangle = \langle u, v \rangle$ if and only if both $x = u$ and $y = v$. In an ordered pair $\langle x, y \rangle$ it may happen that $x = y$. Ordered pairs can be defined formally in terms of (unordered, ordinary) sets so that the axiom is satisfied; the usual way is to set $\langle x, y \rangle := \{\{x\}, \{x, y\}\}$ (as in Appendix A). Note that $\{\{x\}, \{x, y\}\} = \{\{y, x\}, \{x\}\}$ by extensionality.

One of the main ideas in all of mathematics is that of function. Informally, given sets D and E, a function f on D is defined by assigning to each x in D one (and only one!) member $f(x)$ of E. Formally, a *function* is defined as a set f of ordered pairs $\langle x, y \rangle$ such that for any x, y, and z, if $\langle x, y \rangle \in f$ and $\langle x, z \rangle \in f$, then $y = z$. For example, $\{\langle 2, 4 \rangle, \langle -2, 4 \rangle\}$ is a function, but $\{\langle 4, 2 \rangle, \langle 4, -2 \rangle\}$ is not a function. A set of ordered pairs which is (formally) a function is, informally, called the *graph* of the function (as in the case $D = E = \mathbb{R}$, the set of real numbers).

The *domain*, *dom f*, of a function f is the set of all x such that for some y, $\langle x, y \rangle \in f$. Then y is uniquely determined, by definition of function, and it is called $f(x)$. The *range*, *ran f*, of f is the set of all y such that $f(x) = y$ for some x.

A function f with domain A and range included in a set B is said to be *defined on A* or *from A into B*. If the range of f equals B, then f is said to be *onto B*.

The symbol "\mapsto" is sometimes used to describe or define a function. A function f is written as "$x \mapsto f(x)$." For example, "$x \mapsto x^3$" or "$f : x \mapsto x^3$" means a function f such that $f(x) = x^3$ for all x (in the domain of f). To specify the domain, a related notation in common use is, for example, "$f : A \mapsto B$," which together with a more specific definition of f indicates that it is defined from A into B (but does not mean that $f(A) = B$; to

distinguish the two related usages of \mapsto, A and B are written in capitals and members of them in small letters, such as x).

If X is any set and A any subset of X, the *indicator function* of A (on X) is the function defined by

$$1_A(x) := \begin{cases} 1 & \text{if } x \in A \\ 0 & \text{if } x \notin A. \end{cases}$$

(Many mathematicians call this the *characteristic* function of A. In probability theory, "characteristic function" happens to mean a Fourier transform, to be treated in Chapter 9.)

A *sequence* is a function whose domain is either \mathbb{N} or the set $\{1, 2, \ldots\}$ of all positive integers. A sequence f with $f(n) = x_n$ for all n is often written as $\{x_n\}_{n \geq 1}$ or the like.

Formally, every set is a set of sets (every member of a set is also a set). If a set is to be viewed, also informally, as consisting of sets, it is often called a family, class, or collection of sets. Let \mathcal{V} be a family of sets. Then the *union* of \mathcal{V} is defined by

$$\bigcup \mathcal{V} := \{x : x \in A \text{ for some } A \in \mathcal{V}\}.$$

Likewise, the *intersection* of a non-empty collection \mathcal{V} is defined by

$$\bigcap \mathcal{V} := \{x : x \in A \text{ for all } A \in \mathcal{V}\}.$$

So for any two sets A and B, $\bigcup\{A, B\} = A \cup B$ and $\bigcap\{A, B\} = A \cap B$. Notations such as $\bigcup \mathcal{V}$ and $\bigcap \mathcal{V}$ are most used within set theory itself. In the rest of mathematics, unions and intersections of more than two sets are more often written with indices. If $\{A_n\}_{n \geq 1}$ is a sequence of sets, their union is written as

$$\bigcup_n A_n := \bigcup_{n=1}^{\infty} A_n := \{x : x \in A_n \text{ for some } n\}.$$

Likewise, their intersection is written as

$$\bigcap_{n \geq 1} A_n := \bigcap_{n=1}^{\infty} A_n := \{x : x \in A_n \text{ for all } n\}.$$

The union of finitely many sets A_1, \ldots, A_n is written as

$$\bigcup_{1 \leq i \leq n} A_i := \bigcup_{i=1}^{n} A_i := \{x : x \in A_i \text{ for some } i = 1, \ldots, n\},$$

and for intersections instead of unions, replace "some" by "all."

More generally, let I be any set, and suppose A is a function defined on I whose values are sets $A_i := A(i)$. Then the union of all these sets A_i is written

$$\bigcup_i A_i := \bigcup_{i \in I} A_i := \{x : x \in A_i \text{ for some } i\}.$$

A set I in such a situation is called an *index set*. This just means that it is the domain of the function $i \mapsto A_i$. The index set I can be omitted from the notation, as in the first expression above, if it is clear from the context what I is. Likewise, the intersection is written as

$$\bigcap_i A_i := \bigcap_{i \in I} A_i := \{x : x \in A_i \text{ for all } i \in I\}.$$

Here, usually, I is a non-empty set. There is an exception when the sets under discussion are all subsets of one given set, say X. Suppose $t \notin I$ and let $A_t := X$. Then replacing I by $I \cup \{t\}$ does not change $\bigcap_{i \in I} A_i$ if I is non-empty. In case I is empty, one can set $\bigcap_{i \in \emptyset} A_i = X$.

Two more symbols from mathematical logic are sometimes useful as abbreviations: \forall means "for all" and \exists means "there exists." For example, $(\forall x \in A)(\exists y \in B)\ldots$ means that for all x in A, there is a y in B such that. . . .

Two sets A and B are called *disjoint* iff $A \cap B = \emptyset$. Sets A_i for $i \in I$ are called disjoint iff $A_i \cap A_j = \emptyset$ for all $i \neq j$ in I.

Next, some definitions will be given for different classes of numbers, leading up to a definition of real numbers. It is assumed that the reader is familiar with integers and rational numbers. A somewhat more detailed and formal development is given in Appendix A.4.

Recall that \mathbb{N} is the set of all nonnegative integers $0, 1, 2, \ldots, \mathbb{Z}$ denotes the set of all integers $0, \pm 1, \pm 2, \ldots$, and \mathbb{Q} is the set of all rational numbers m/n, where $m \in \mathbb{Z}, n \in \mathbb{Z}$, and $n \neq 0$.

Real numbers can be defined in different ways. A familiar way is through decimal expansions: x is a real number if and only if $x = \pm y$, where $y = n + \sum_{j=1}^{\infty} d_j / 10^j$, $n \in \mathbb{N}$, and each digit d_j is an integer from 0 to 9. But decimal expansions are not very convenient for proofs in analysis, and they are not unique for rational numbers of the form $m/10^k$ for $m \in \mathbb{Z}, m \neq 0$, and $k \in \mathbb{N}$. One can also define real numbers x in terms of more general sequences of rational numbers converging to x, as in the completion of metric spaces to be treated in §2.5.

The formal definition of real numbers to be used here will be by way of Dedekind cuts, as follows: A *cut* is a set $C \subset \mathbb{Q}$ such that $C \notin \emptyset; C \neq \mathbb{Q}$; whenever $q \in C$, if $r \in \mathbb{Q}$ and $r < q$ then $r \in C$, and there exists $s \in \mathbb{Q}$ with $s > q$ and $s \in C$.

Let \mathbb{R} be the set of all real numbers; thus, formally, \mathbb{R} is the set of all cuts. Informally, a one-to-one correspondence between real numbers x and cuts C, written $C = C_x$ or $x = x_C$, is given by $C_x = \{q \in \mathbb{Q}: q < x\}$.

The ordering $x \leq y$ for real numbers is defined simply in terms of cuts by $C_x \subset C_y$. A set E of real numbers is said to be *bounded above* with an *upper bound* y iff $x \leq y$ for all $x \in E$. Then y is called the *supremum* or *least upper bound* of E, written $y = \sup E$, iff it is an upper bound and $y \leq z$ for every upper bound z of E. A basic fact about \mathbb{R} is that for every non-empty set $E \subset \mathbb{R}$ such that E is bounded above, the supremum $y = \sup E$ exists. This is easily proved by cuts: C_y is the union of the cuts C_x for all $x \in E$, as is shown in Theorem A.4.1 of Appendix A.

Similarly, a set F of real numbers is *bounded below* with a *lower bound* v if $v \leq x$ for all $x \in F$, and v is the *infimum* of F, $v = \inf F$, iff $t \leq v$ for every lower bound t of F. Every non-empty set F which is bounded below has an infimum, namely, the supremum of the lower bounds of F (which are a non-empty set, bounded above).

The maximum and minimum of two real numbers are defined by $\min(x, y) = x$ and $\max(x, y) = y$ if $x \leq y$; otherwise, $\min(x, y) = y$ and $\max(x, y) = x$.

For any real numbers $a \leq b$, let $[a, b] := \{x \in \mathbb{R}: a \leq x \leq b\}$.

For any two sets X and Y, their *Cartesian product*, written $X \times Y$, is defined as the set of all ordered pairs $\langle x, y \rangle$ for x in X and y in Y. The basic example of a Cartesian product is $\mathbb{R} \times \mathbb{R}$, which is also written as \mathbb{R}^2 (pronounced r-two, not r-squared), and called the *plane*.

Problems

1. Let $A := \{3, 4, 5\}$ and $B := \{5, 6, 7\}$. Evaluate: (a) $A \cup B$. (b) $A \cap B$. (c) $A \backslash B$. (d) $A \triangle B$.

2. Show that $\emptyset \neq \{\emptyset\}$ and $\{\emptyset\} \neq \{\{\emptyset\}\}$.

3. Which of the following three sets are equal? (a) $\{\{2, 3\}, \{4\}\}$; (b) $\{\{4\}, \{2, 3\}\}$; (c) $\{\{4\}, \{3, 2\}\}$.

4. Which of the following are functions? Why?
 (a) $\{\langle 1, 2 \rangle, \langle 2, 3 \rangle, \langle 3, 1 \rangle\}$.
 (b) $\{\langle 1, 2 \rangle, \langle 2, 3 \rangle, \langle 2, 1 \rangle\}$.
 (c) $\{\langle 2, 1 \rangle, \langle 3, 1 \rangle, \langle 1, 2 \rangle\}$.
 (d) $\{\langle x, y \rangle \in \mathbb{R}^2: x = y^2\}$.
 (e) $\{\langle x, y \rangle \in \mathbb{R}^2: y = x^2\}$.

5. For any relation V (that is, any set of ordered pairs), define the *domain* of

V as $\{x: \langle x, y \rangle \in V$ for some $y\}$, and the *range* of V as $\{y: \langle x, y \rangle \in V$ for some $x\}$. Find the domain and range for each relation in the last problem (whether or not it is a function).

6. Let $A_{1j} := \mathbb{R} \times [j - 1, j]$ and $A_{2j} := [j - 1, j] \times \mathbb{R}$ for $j = 1, 2$. Let $B := \bigcup_{m=1}^{2} \bigcap_{n=1}^{2} A_{mn}$ and $C := \bigcap_{n=1}^{2} \bigcup_{m=1}^{2} A_{mn}$. Which of the following is true: $B \subset C$ and/or $C \subset B$? Why?

7. Let $f(x) := \sin x$ for all $x \in \mathbb{R}$. Of the following subsets of \mathbb{R}, which is f into, and which is it onto? (a) $[-2, 2]$. (b) $[0, 1]$. (c) $[-1, 1]$. (d) $[-\pi, \pi]$.

8. How is Problem 7 affected if x is measured in degrees rather than radians?

9. Of the following sets, which are included in others? $A := \{3, 4, 5\}$; $B := \{\{3, 4\}, 5\}$; $C := \{5, 4\}$; and $D := \{\{4, 5\}\}$. Assume that no nonobvious relations, such as $4 = \{3, 5\}$, are true. More specifically, you can assume that for any two sets x and y, at most one of the three relations holds: $x \in y$, $x = y$, or $y \in x$, and that each nonnegative integer k is a set with k members. Please *explain* why each inclusion does or does not hold. Sample: If $\{\{6, 7\}, \{5\}\} \subset \{3, 4\}$, then by extensionality $\{6, 7\} = 3$ or 4, but $\{6, 7\}$ has two members, not three or four.

10. Let $I := [0, 1]$. Evaluate $\bigcup_{x \in I} [x, 2]$ and $\bigcap_{x \in I} [x, 2]$.

11. "Closed half-lines" are subsets of \mathbb{R} of the form $\{x \in \mathbb{R}: x \le b\}$ or $\{x \in \mathbb{R}: x \ge b\}$ for real numbers b. A *polynomial of degree n* on \mathbb{R} is a function $x \mapsto a_n x^n + \cdots + a_1 x + a_0$ with $a_n \ne 0$. Show that the range of any polynomial of degree $n \ge 1$ is \mathbb{R} for n odd and a closed half-line for n even. *Hints:* Show that for large values of $|x|$, the polynomial has the same sign as its leading term $a_n x^n$ and its absolute value goes to ∞. Use the intermediate value theorem for a continuous function such as a polynomial (Problem 2.2.14(d) below).

12. A polynomial on \mathbb{R}^2 is a function of the form $\langle x, y \rangle \mapsto \sum_{0 \le i \le k, 0 \le j \le k} a_{ij} x^i y^j$. Show that the ranges of nonconstant polynomials on \mathbb{R}^2 are either all of \mathbb{R}, closed half-lines, or open half-lines $(b, \infty) := \{x \in \mathbb{R}: x > b\}$ or $(-\infty, b) := \{x \in \mathbb{R}: x < b\}$, where each open or closed half-line is the range of some polynomial. *Hint:* For one open half-line, try the polynomial $x^2 + (xy - 1)^2$.

1.2. Relations and Orderings

A *relation* is any set of ordered pairs. For any relation E, the inverse relation is defined by $E^{-1} := \{\langle y, x \rangle: \langle x, y \rangle \in E\}$. Thus, a function is a special kind

of relation. Its inverse f^{-1} is not necessarily a function. In fact, a function f is called 1–1 or *one-to-one* if and only if f^{-1} is also a function. Given a relation E, one often writes xEy instead of $\langle x, y \rangle \in E$ (this notation is used not for functions but for other relations, as will soon be explained). Given a set X, a relation $E \subset X \times X$ is called *reflexive* on X iff xEx for all $x \in X$. E is called *symmetric* iff $E = E^{-1}$. E is called *transitive* iff whenever xEy and yEz, we have xEz. Examples of transitive relations are orderings, such as $x \leq y$.

A relation $E \subset X \times X$ is called an *equivalence relation* iff it is reflexive on X, symmetric, and transitive. One example of an equivalence relation is equality. In general, an equivalence relation is like equality; two objects x and y satisfying an equivalence relation are equal in some way. For example, two integers m and n are said to be equal mod p iff $m - n$ is divisible by p. Being equal mod p is an equivalence relation. Or if f is a function, one can define an equivalence relation E_f by xE_fy iff $f(x) = f(y)$.

Given an equivalence relation E, an *equivalence class* is a set of the form $\{y \in X: yEx\}$ for any $x \in X$. It follows from the definition of equivalence relation that two equivalence classes are either disjoint or identical. Let $f(x) := \{y \in X: yEx\}$. Then f is a function and xEy if and only if $f(x) = f(y)$, so $E = E_f$, and every equivalence relation can be written in the form E_f.

A relation E is called *antisymmetric* iff whenever xEy and yEx, then $x = y$. Given a set X, a *partial ordering* is a transitive, antisymmetric relation $E \subset X \times X$. Then $\langle X, E \rangle$ is called a *partially ordered set*. For example, for any set Y, let $X = 2^Y$ (the set of all subsets of Y). Then $\langle 2^Y, \subset \rangle$, for the usual inclusion \subset, gives a partially ordered set. (*Note:* Many authors require that a partial ordering also be reflexive. The current definition is being used to allow not only relations '\leq' but also '$<$' to be partial orderings.) A partial ordering will be called *strict* if xEx does not hold for any x. So "strict" is the opposite of "reflexive." For any partial ordering E, define the relation \leq by $x \leq y$ iff (xEy or $x = y$). Then \leq is a reflexive partial ordering. Also, define the relation $<$ by $x < y$ iff (xEy and $x \neq y$). Then $<$ is a strict partial ordering. For example, the usual relations $<$ and \leq between real numbers are connected in the way just defined. A one-to-one correspondence between strict partial orderings E and reflexive partial orderings F on a set X is given by $F = E \cup D$ and $E = F \backslash D$, where D is the "diagonal," $D := \{\langle x, x \rangle: x \in X\}$. From here on, the partial orderings considered will be either reflexive, usually written \leq (or \geq), or strict, written $<$ (or $>$). Here, as usual, "$<$" is read "less than," and so forth.

Two partially ordered sets $\langle X, E \rangle$ and $\langle Y, G \rangle$ are said to be *order-isomorphic* iff there exists a 1–1 function f from X onto Y such that for any

u and x in X, uEx iff $f(u)Gf(x)$. Then f is called an *order-isomorphism*. For example, the intervals $[0, 1]$ and $[0, 2]$, with the usual ordering for real numbers, are order-isomorphic via the function $f(x) = 2x$. The interval $[0, 1]$ is not order-isomorphic to \mathbb{R}, which has no smallest element.

From here on, an ordered pair $\langle x, y \rangle$ will often be written as (x, y) (this is, of course, still different from the unordered pair $\{x, y\}$).

A *linear ordering* E of X is a partial ordering E of X such that for all x and $y \in X$, either xEy, yEx, or $x = y$. Then $\langle X, E \rangle$ is called a *linearly ordered set*. The classic example of a linearly ordered set is the real line \mathbb{R}, with its usual ordering. Actually, $(\mathbb{R}, <)$, (\mathbb{R}, \leq), $(\mathbb{R}, >)$, and (\mathbb{R}, \geq) are all linearly ordered sets.

If (X, E) is any partially ordered set and A is any subset of X, then $\{\langle x, y \rangle \in E : x \in A$ and $y \in A\}$ is also a partial ordering on A. Suppose we call it E_A. For most orderings, as on the real numbers, the orderings of subsets will be written with the same symbol as on the whole set. If (X, E) is linearly ordered and $A \subset X$, then (A, E_A) is also linearly ordered, as follows directly from the definitions.

Let W be a set with a reflexive linear ordering \leq. Then W is said to be *well-ordered* by \leq iff for every non-empty subset A of W there is a smallest $x \in A$, so that for all $y \in A$, $x \leq y$. The corresponding strict linear ordering $<$ will also be called a well-ordering. If X is a finite set, then any linear ordering of it is easily seen to be a well-ordering. The interval $[0, 1]$ is not well-ordered, although it has a smallest element 0, since it has subsets, such as $\{x : 0 < x \leq 1\}$, with no smallest element.

The method of proof by mathematical induction can be extended to well-ordered sets, as follows. Suppose $(X, <)$ is a well-ordered set and that we want to prove that some property holds for all elements of X. If it does not, then there is a smallest element for which the property fails. It suffices, then, to prove that for each $x \in X$, if the property holds for all $y < x$, then it holds for x. This "induction principle" will be treated in more detail in §1.3.

Problems

1. For any partial ordering E, show that E^{-1} is also a partial ordering.

2. For two partially ordered sets $\langle A, \leq \rangle$ and $\langle B, \leq \rangle$, the *lexicographical ordering* on the Cartesian product $A \times B$ is defined by $\langle a, b \rangle \leq \langle c, d \rangle$ iff $a < c$ or $(a = c$ and $b \leq d)$. (For example, if A and B are both an alphabet with the usual ordering, then we have the dictionary or "alphabetical" ordering of two-letter words or strings.) If the orderings on A and B are linear, show that the lexicographical ordering is linear on $A \times B$. If A

and B are well-ordered by the given relations, show that $A \times B$ is also well-ordered.

3. Instead, let $\langle a, b \rangle \leq \langle c, d \rangle$ iff both $a \leq c$ and $b \leq d$. Show that this is a partial ordering. If A and B each contain more than one element, show that $A \times B$ is never linearly ordered by such an ordering.

4. On \mathbb{R}^2 let $\langle x, y \rangle E \langle u, v \rangle$ iff $x + y = u + v$, $\langle x, y \rangle F \langle u, v \rangle$ iff $x + y \leq u + v$, and $\langle x, y \rangle G \langle u, v \rangle$ iff $x + u \leq y + v$. Which of E, F, and G is an equivalence relation, a partial ordering, or a linear ordering? Why?

5. For sequences $\{x_n\}$ of real numbers let $\{x_n\} E \{y_n\}$ iff $\lim_{n \to \infty} x_n - y_n = 0$ and $\{x_n\} F \{y_n\}$ iff $\lim_{n \to \infty} x_n - y_n = 1$. Which of E and F is an equivalence relation and/or a partial ordering? Why?

6. For any two relations E and F on the same set X, define a relation $G := E \circ F$ by xGz iff for some y, xEy and yFz. For each of the following properties, if E and F both have the property, prove, or disprove by an example, that G also has the property: (a) reflexive, (b) symmetric, (c) transitive.

7. Refer to Problem 6 and answer the same question in regard to the following properties: (d) antisymmetric, (e) equivalence relation, (f) function.

*1.3. Transfinite Induction and Recursion

Mathematical induction is a well-known and useful method of proving facts about nonnegative integers. If $F(n)$ represents a statement that one wants to prove for all $n \in \mathbb{N}$, and a direct proof is not apparent, one first proves $F(0)$. Then, in proving $F(n + 1)$, one can assume that $F(n)$ is true, which is often helpful. Or, if you prefer, you can assume that $F(0), F(1), \ldots, F(n)$ are all true. More generally, let $(X, <)$ be any partially ordered set. A subset $Y \subset X$ will be called *inductive* if, for every $x \in X$ such that $y \in Y$ for all $y \in X$ such that $y < x$, we have $x \in Y$. If X has a least element x, then there are no $y < x$, so x must belong to any inductive subset Y of X. In ordinary induction, Y is the set of all n for which $F(n)$ holds. Proving that Y is inductive gives a proof that $Y = \mathbb{N}$, so that $F(n)$ holds for all n. In \mathbb{R}, the set $(-\infty, 0)$ is inductive, but it is not all of \mathbb{R}. The set \mathbb{N} is well-ordered, but \mathbb{R} is not: the set $\{x \in \mathbb{R} : x > 1\}$ has no least element. One of the main advantages of well-orderings is that they allow the following extension of induction:

1.3.1. Induction Principle Let X be any set well-ordered by a relation $<$. Let Y be any inductive subset of X. Then $Y = X$.

Proof. If $X \setminus Y = \emptyset$, the conclusion holds. Otherwise, if not, let y be the least element of $X \setminus Y$. Then $x \in Y$ for all $x < y$ (perhaps vacuously, if y is the least element of X), so $y \in Y$, a contradiction. ☐

For any linearly ordered set (X, \leq), an *initial segment* is a subset $Y \subset X$ such that whenever $x < y$ and $y \in Y$, then also $x \in Y$. Then if (X, \leq) is the real line with usual ordering and Y an initial segment, then either $Y = X$ or, for some y, either $Y = \{x: x < y\}$ or $Y = \{x: x \leq y\}$.

In ordinary mathematical induction, the set $(X, <)$ is order-isomorphic to \mathbb{N}, the set of nonnegative integers, or to some initial segment of it (finite integer) with usual ordering. Transfinite induction refers to induction for an $(X, <)$ with a more complicated well-ordering. One example is "double induction." To prove a statement $F(m, n)$ for all nonnegative integers m and n, one can first prove $F(0, 0)$. Then in proving $F(m, n)$ one can assume that $F(j, k)$ is true for all $j < m$ and all $k \in \mathbb{N}$, and for $j = m$ and $k < n$. (In this case the well-ordering is the "lexicographical" ordering mentioned in Problem 2 of §1.2.) Other well-orderings of $\mathbb{N} \times \mathbb{N}$ may also be useful. Much of set theory is concerned with well-orderings more general than those of sequences, such as well-orderings of \mathbb{R}, although these are in a sense nonconstructive (well-ordering of general sets, and of \mathbb{R} in particular, depends on the axiom of choice, to be treated in §1.5).

Another very important method in mathematics, definition by recursion, will be developed next. In its classical form, a function f is defined by specifying $f(0)$, then defining $f(n)$ in terms of $f(n - 1)$ and possibly other values of $f(k)$ for $k < n$. Such recursive definitions will also be extended to well-ordered sets. For any function f and $A \subset \operatorname{dom} f$, the restriction of f to A is defined by $f \restriction A := \{\langle x, f(x)\rangle: x \in A\}$.

1.3.2. Recursion Principle Let $(X, <)$ be a well-ordered set and Y any set. For any $x \in X$, let $I(x) := \{u \in X: u < x\}$. Let g be a function whose domain is the set of all j such that for some $x \in X$, j is a function from $I(x)$ into Y, and such that $\operatorname{ran} g \subset Y$. Then there is a unique function f from X into Y such that for every $x \in X$, $f(x) = g(f \restriction I(x))$.

Note. If b is the least element of X and we want to define $f(b) = c$, then we set $g(\emptyset) = c$ and note that $I(b) = \emptyset$.

Proof. If $X = \emptyset$, then $f = \emptyset$ and the conclusion holds. So suppose X is non-empty and let b be its smallest element. Let $J(x) := \{u \in X: u \leq x\}$ for

each $x \in X$. Let T be the set of all $x \in X$ such that on $J(x)$, there is a function f such that $f(u) = g(f \restriction I(u))$ for all $u \in J(x)$. Let us show that if such an f exists, it is unique. Let h be another such function. Then $h(b) = g(\emptyset) = f(b)$. By induction (1.3.1), for each $u \in J(x)$, $h(u) = g(f \restriction I(u)) = f(u)$. So f is unique. If $x < u$ for some u in T and f as above is defined on $J(u)$, then $f \restriction J(x)$ has the desired properties and is the f for $J(x)$ by uniqueness. Thus T is an initial segment of X. The union of all the functions f for all $x \in T$ is a well-defined function, which will also be called f. If $T \neq X$, let u be the least element of $X \backslash T$. But then $T = I(u)$ and $f \cup \{\langle u, g(f) \rangle\}$ is a function on $J(u)$ with the desired properties, so $u \in T$, a contradiction. So f exists. As it is unique on each $J(x)$, it is unique. □

For any function f on a Cartesian product $A \times B$, one usually writes $f(a, b)$ rather than $f(\langle a, b \rangle)$. The classical recursion on the nonnegative integers can then be described as follows.

1.3.3. Corollary (Simple Recursion) Let Y be any set, $c \in Y$, and h any function from $\mathbb{N} \times Y$ into Y. Then there is a function f from \mathbb{N} into Y with $f(0) = c$ and for each $n \in \mathbb{N}$, $f(n + 1) = h(n, f(n))$.

Proof. To apply 1.3.2, let $g(\emptyset) = c$. Let j be any function from some non-empty $I(n)$ into Y. (Note that $I(n)$ is empty if and only if $n = 0$.) Then $n - 1$ is the largest member of $I(n)$. Let $g(j) = h(n - 1, j(n - 1))$. Then the function g is defined on all such functions j, and 1.3.2 applies to give a function f. Now $f(0) = c$, and for any $n \in \mathbb{N}$, $f(n + 1) = g(f \restriction I(n + 1)) = h(n, f(n))$. □

Example. Let t be a function with real values defined on \mathbb{N}. Let

$$f(n) = \sum_{j=0}^{n} t(j).$$

To obtain f by simple recursion (1.3.3), let $c = t(0)$ and $h(n, y) = t(n+1) + y$ for any $n \in \mathbb{N}$ and $y \in \mathbb{R}$. A computer program to compute f, given a program for t, could well be written along the lines of this recursion, which in a sense reduces the summation to simple addition.

Example. General recursion (1.3.2) can be used to define the function f such that for $n = 1, 2, \ldots$, $f(n)$ is the nth prime: $f(1) = 2$, $f(2) = 3$, $f(3) = 5$, $f(4) = 7$, $f(5) = 11$, and so on. On the empty function, g is defined as 2, and so $f(1) = 2$. Given j on $J(n) = \{1, 2, \ldots, n\} = I(n + 1)$, let $g(j)$

be the least $k > j(n)$ such that k is not divisible by any of $j(1), \ldots, j(n)$. Such a k always exists since there are infinitely many primes. We will have $f(1) < f(2) < \cdots$, and in constructing f, g will be applied only to functions j with $j(1) < j(2) < \cdots < j(n)$ (since each $j(i) = f(i)$). Note that all the values $j(1), \ldots, j(n)$, not only $j(n)$, are used in defining $f(n + 1) = g(j)$.

Problems

1. To define the factorial function $f(n) = n!$ by simple recursion, how can c and h in Corollary 1.3.3 be chosen? For h to serve this purpose, for which n and x (if any) are the values of $h(n, x)$ uniquely determined?

2. Let $(X, <)$ be a well-ordered set. Let $>$ be the reversed ordering as usual, that is, $x > y$ means $y < x$. Suppose that $(X, >)$ is also well-ordered. Let f be a function from \mathbb{N} onto X. Show that f cannot be one-to-one. *Hint:* If it is, then define h by recursion on \mathbb{N} such that $h(0)$ is the least element of X for $<$, $h(1)$ the next-least, and so on. The range of h has no largest element.

3. Given a partially ordered set (X, \leq), a subset $A \subset X$, and an element $x \in A$, x is called a *minimal* element of A iff there is no other $y \in A$ with $y < x$. Then (X, \leq) will be called *min-ordered* iff every non-empty subset of it has a minimal element. (Note that any well-ordered set is min-ordered.)
 (a) Show that the induction principle (1.3.1) holds with "min-ordered" in place of "well-ordered."
 (b) Do likewise for the recursion principle (1.3.2).

4. Refer to Problem 3. On $\mathbb{N} \times \mathbb{N}$, define an ordering by $(i, k) \leq (m, n)$ iff $i \leq m$ and $k \leq n$.
 (a) Show that then $(\mathbb{N} \times \mathbb{N}, \leq)$ is min-ordered.
 (b) If $A := \{(m, n) \in \mathbb{N} \times \mathbb{N} : m + n \geq 4\}$, what are the minimal elements of A? Show that $(\mathbb{N} \times \mathbb{N}, \leq)$ is not well-ordered.

 Note: When mathematics is being built on a set-theoretic foundation and \mathbb{N} has just been defined, the definitions of addition and then multiplication can be done by recursion either on a lexicographical ordering of $\mathbb{N} \times \mathbb{N}$ or by the min-ordering just defined (see Appendix A.3 and references there).

5. Refer again to Problem 3. Let (X, \leq) be a partially ordered set such that the only inductive subset of X is X (as in the induction principle). Show that (X, \leq) must be min-ordered.

6. Let (X, \leq) be a partially ordered set on which a function f is defined such that $f(x) < x$ for every $x \in X$. Prove by recursion that there exists a function g from \mathbb{N} into X such that $g(n + 1) < g(n)$ for all n.

1.4. Cardinality

Two sets X and Y are said to have the *same cardinality* iff there is a 1–1 function from X onto Y. We would like to say that two such sets have the "same number of elements," but for infinite sets, it is not necessarily clear what a "number" is. A set X is said to be *finite* iff it has the same cardinality as some $n \in \mathbb{N}$ (where n is represented as the set $\{0, 1, \ldots, n - 1\}$, so that it has n members). Otherwise, X is infinite. For example, \mathbb{N} is infinite. X is called *countable* iff there is a function f from \mathbb{N} onto X, *countably infinite* if X is also infinite. A set is *uncountable* iff it is not countable.

As an example, to see that \mathbb{N} and $\mathbb{N} \times \mathbb{N}$ have the same cardinality, let $f(m, n) := 2^m(2n + 1) - 1$. Then f is 1–1 from $\mathbb{N} \times \mathbb{N}$ onto \mathbb{N}.

X is said to have *smaller cardinality than Y* iff there is a 1–1 function from X into Y, but no such function onto Y. The following fact shows that this definition is coherent.

1.4.1. Equivalence Theorem *If A and B are sets, f is a 1–1 function from A into B, and g is a 1–1 function from B into A, then A and B have the same cardinality.*

Proof. For any function j and set X, let $j[X] := \{j(x): x \in X\} = \mathrm{ran}(j \restriction X)$. For any $X \subset A$, let $F(X) := A \backslash g[B \backslash f[X]]$. For any U such that $X \subset U \subset A$, we have $F(X) \subset F(U)$, since as X gets larger, $B \backslash f[X]$ gets smaller, so $F(X)$ gets larger. Let $W := \{X \subset A: X \subset F(X)\}$ and $C := \bigcup W$. For any $u \in C$, we have $u \in X$ for some $X \in W$, so $u \in X \subset F(X) \subset F(C)$. Thus $C \subset F(C)$ and $F(C) \subset F(F(C))$. So $F(C) \in W$, and by definition of C, $F(C) \subset C$. Thus $F(C) = C$. Then by definition of F, g is 1–1 from $B \backslash f[C]$ onto $A \backslash F(C) = A \backslash C$. In any case, f is 1–1 from C onto $f[C]$. Let $h(x) := f(x)$ if $x \in C$, $h(x) := g^{-1}(x)$ if $x \in A \backslash C$. Then h is 1–1 from A onto B. □

For any finite $n = 0, 1, 2, \ldots$, we have $n < 2^n$: for example, $0 < 1, 1 < 2, 2 < 4, 3 < 8$, and so forth. For a finite set X with n elements, the collection 2^X of all subsets of X has 2^n elements. The fact that 2^X is larger than X is true also for arbitrarily large (infinite) sets.

1.4.2. Theorem *For every set X, X has smaller cardinality than 2^X.*

Proof. Let $f(x) := \{x\}$. This gives a 1–1 function from X into 2^X. Suppose g is a function from X onto 2^X. Let $A := \{x \in X : x \notin g(x)\}$. Then $g(y) = A$ for some y. If $y \in A$, then $y \notin g(y) = A$, but if $y \notin A = g(y)$, then $y \in A$, a contradiction. \square

1.4.3. Corollary \mathbb{N} *has smaller cardinality than $2^\mathbb{N}$, so $2^\mathbb{N}$ is uncountable.*

The collection $2^\mathbb{N}$ of all subsets of a countably infinite set is said to have cardinality c, or the cardinality of the *continuum*, in view of the following fact.

1.4.4. Theorem *The set \mathbb{R} of real numbers and the interval $[0, 1] := \{x \in \mathbb{R} : 0 \le x \le 1\}$ both have cardinality c.*

Proof. See the problems below. \square

Are there cardinalities between those of \mathbb{N} and $2^\mathbb{N}$? This is known as the *continuum problem*. The conjecture that there are no such cardinalities, so that every uncountable set of real numbers has the same cardinality as all of \mathbb{R}, is called the *continuum hypothesis*. It turns out that the problem cannot be settled from the usual axioms of set theory (including the axiom of choice; see the notes to Appendix A.2). So one can assume either the continuum hypothesis or its negation without getting a contradiction. In this book, the negation is never assumed. When, occasionally, the continuum hypothesis is assumed, it will be pointed out (while the axiom of choice, in the next section, may sometimes be taken for granted).

Problems

1. Prove that for any countably infinite set S, there is a 1–1 function from \mathbb{N} onto S. *Hint:* Let h be a function from \mathbb{N} onto S. Define f using h and recursion. The problem is to show that some form of recursion actually applies (1.3.2).

2. To understand the situation in the equivalence theorem: if, in the notation of 1.4.1, $f(x) = y$ and $g(v) = u$, say x is an ancestor of y and v an ancestor of u. Also, say that any ancestor of an ancestor is an ancestor. If h is a 1–1 function from A onto B such that for all x in A, either $h(x) = f(x)$ or $h(x) = g^{-1}(x)$, show that:

(a) If x has a finite, even number of ancestors, say $2n$, then $h(x) = f(x)$. *Hints:* Use induction on n.

(b) If x has a finite, odd number of ancestors, then $h(x) = g^{-1}(x)$.

(c) If x has infinitely many ancestors, then how is $h(x)$ chosen according to the proof of the equivalence theorem (1.4.1)?

3. If X is uncountable and Y is a countable subset of X, show that $X \backslash Y$ has the same cardinality as X, assuming that \mathbb{N} has smaller cardinality than $X \backslash Y$ (which cannot be proved in general without the axiom of choice, to be treated in §1.5). *Hint:* Let B be a countably infinite subset of $X \backslash Y$. Then B and $B \cup Y$ have the same cardinality.

4. Prove Theorem 1.4.4. *Hints:* Use the equivalence theorem 1.4.1. To show that $[0, 1]$, the open interval $(0, 1)$, and \mathbb{R} have the same cardinality, define a 1–1 function from \mathbb{R} onto $(0, 1)$ of the form $a + b \cdot \tan^{-1}x$ for suitable constants a and b. If $A \subset \mathbb{N}$, let $x(A) := \sum_{n \in A} 1/2^{n+1}$ (binary expansion). This function is not quite 1–1, but use it and check that Problem 3 applies to show that $[0, 1]$ has cardinality c.

5. Let X be a non-empty set of cardinality less than or equal to c and let Y have cardinality c. Show that $X \times Y$ has cardinality c. *Hint:* Reduce to the case where X has cardinality c. Show that $2^{\mathbb{N}} \times 2^{\mathbb{N}}$ has the same cardinality as $2^{\mathbb{N}}$.

1.5. The Axiom of Choice and Its Equivalents

In Euclid's geometry, the "parallel postulate" was long recognized as having a weaker sense of intuitive rightness than most of the other axioms and postulates. Eventually, in the 19th century, "non-Euclidean geometry" was invented, in which the parallel postulate is no longer assumed. In set theory, the following axiom is, likewise, not always assumed; although it has been shown to be consistent with the other axioms, it is powerful, giving "existence" to structures that are less concrete, popular, or accessible to intuition than garden-variety sets.

Axiom of Choice (AC). Let A be any function on a set I such that $A(x) \neq \emptyset$ for all $x \in I$. Then there is a function f such that $f(x) \in A(x)$ for all $x \in I$.

The Cartesian product $\Pi_{x \in I} A(x)$ is defined as the set of all functions f on I such that $f(x) \in A(x)$ for all $x \in I$. Then AC says that the Cartesian product of non-empty sets is non-empty. Here I is often called an index set and f a

choice function. The following argument may seem to "prove" the axiom of choice: since for each x, $A(x)$ is non-empty, it contains some element, which we call $f(x)$. Then f is the desired choice function. This argument is actually valid if I is finite. If I is infinite, however, the problem is whether there is any systematic rule, which can be written down in finitely many symbols (as it always can if I is finite), for choosing a specific member of $A(x)$ for each x. In some cases the answer is just not clear, so one may have to choose an element $f(x)$ in some sense at random. The axiom of choice makes it legal for one to suppose an infinite set of such random choices can be made.

Here is another form of the axiom of choice:

AC'. Let X be any set and $I := 2^X \backslash \{\emptyset\}$, the set of all non-empty subsets of X. Then there is a function f from I into X such that $f(A) \in A$ for each $A \in I$.

It will be shown that AC and AC' are equivalent. Assuming AC, to prove AC', let A be the identity function on I, $A(B) = B$ for any $B \in I$. Then AC implies AC'. On the other hand, assuming AC', given any function A whose values are non-empty sets, take X as the union of the range of A. Then each value of A is a non-empty subset of X and AC follows from AC'.

The axiom of choice is widely used in other, equivalent forms. The equivalence will be proved for some of the best-known forms. A *chain* is a linearly ordered subset of a partially ordered set (with the same ordering). In a partially ordered set (X, \leq), an element z of X is called *maximal* iff there is no w with $z < w$. If X is not linearly ordered, it may have many maximal elements. For example, for the trivial partial ordering whose strict partial ordering $<$ is empty, every element is maximal. A *maximum* of X is a $y \in X$ such that $x \leq y$ for all $x \in X$. A maximum is a maximal element, but the converse is often not true. If an ordering is not specified, then inclusion is the intended ordering.

As an example of some of these notions, let X be the collection of all intervals $[a, b]$ in \mathbb{R} of length $b - a \leq 2$ (here $a \leq b$). These intervals are partially ordered by inclusion. Any interval of length equal to 2 is a maximal element. There is no maximum.

Here are three statements to be shown equivalent to AC:

Well-ordering principle (WO): Every set can be well-ordered.
Hausdorff's maximal principle (HMP): For every partially ordered set
 (P, \leq) there is a maximal chain $L \subset P$.

Zorn's Lemma (ZL): Let (P, \leq) be a partially ordered set such that for every chain $L \subset P$, there is an upper bound $y \in P$, that is, $x \leq y$ for all $x \in L$. Then P has a maximal element.

1.5.1. Theorem *AC, WO, HMP, and ZL are all equivalent.*

Note. Strictly speaking, the equivalence stated in Theorem 1.5.1 holds if we assume a system of axioms for set theory, such as the Zermelo-Fraenkel system, given in Appendix A. But the proof will proceed on the basis of the definitions given so far; it will not call on any axioms explicitly.

Proof. AC will be used in the form AC'. AC implies WO: given a set X, a choice function c which selects an element of each non-empty subset of X, a subset $A \subset X$, and an ordering $<$ on A, $(A, <)$ will be called *c-ordered* iff $<$ is a well-ordering on A such that for every $x \in A$, $x = c(X \backslash \{y \in A : y < x\})$. For any two c-ordered subsets (A, \leq) and (B, λ), it will be shown that one is an initial segment of the other (with the same ordering). If not, let x be the least element of A such that $\{u \in A : u \leq x\}$ is not an initial segment of B. Such an x exists, since any union of initial segments is an initial segment. Then $D := \{u \in A : u < x\}$ is an initial segment of both A and B on which the two orderings agree. Since A is c-ordered, $x = c(X \backslash D)$. If $B \backslash D \neq \emptyset$, let y be its least element for λ (see Figure 1.5). Then since B is c-ordered, $y = c(X \backslash D) = x$, contradicting the choice of x. Thus, $B = D$, an initial segment of A.

Let C be the union of all c-ordered sets and \leq the union of their orderings. Then (C, \leq) is c-ordered. If $C \neq X$, let $z = c(X \backslash C)$ and extend the definition of the ordering by setting $x \leq z$ for all $x \in C$ (and $z \leq z$). Then $(C \cup \{z\}, \leq)$ is c-ordered, a contradiction, so X is well-ordered.

WO implies HMP: Let (X, \leq) be a partially ordered set and let W be a well-ordering of X. Let t be the least element of X for W (for X empty there is no problem). Define a function f by recursion on (X, W) (1.3.2) such that $f(t) = t$ and

$$f(x) = \begin{cases} x & \text{if } \{x\} \cup \{f(y) : yWx, y \neq x\} \text{ is a chain for } \leq \\ t & \text{otherwise, so } x \text{ is not comparable for } \leq \text{ to some } y \\ & \text{with } yWx, \text{ and } f(y) = y. \end{cases}$$

Figure 1.5

Then the range of f is a maximal chain, as desired.

HMP implies ZL: Let (P, \le) be a partially ordered set in which every chain has an upper bound. By HMP take a maximal chain L. Let z be an upper bound for L. Then if for some w, $z < w$, $L \cup \{w\}$ is a chain strictly including L, a contradiction, so z is maximal.

ZL implies AC: Given a set X, let F be the set of all choice functions whose domains are subsets of $2^X \backslash \{\emptyset\}$. Then F is partially ordered by inclusion. Any chain in F has its union in F. Thus F has a maximal element f. If dom $f \ne 2^X \backslash \{\emptyset\}$, then for any A in the non-empty collection $(2^X \backslash \{\emptyset\}) \backslash \mathrm{dom}\, f$, and any x in such a (non-empty) set A, $f \cup \{\langle A, x \rangle\} \in F$, contradicting the maximality of f. Thus f is a choice function on all of $2^X \backslash \{\emptyset\}$. So AC$'$ holds, and thus AC. \square

Problems

1. Prove, without applying Theorem 1.5.1, that the well-ordering principle implies AC. *Caution:* Is it clear that the possibly very large family of sets $A(x)$ can all be well-ordered simultaneously, so that there is a function f on I such that for each $x \in I$, $f(x)$ is a well-ordering of $A(x)$? *Hint:* Use AC$'$.

2. Prove it is equivalent to AC that in every partially ordered set, every chain is included in a maximal chain.

3. In the partially ordered set $\mathbb{N} \times \mathbb{N}$ with the ordering $(j, k) \le (m, n)$ iff $j \le m$ and $k \le n$, consider the sequence (n, n), for $n = 0, 1, 2, \ldots$. Describe all the maximal chains that include the given sequence. Do any of these chains have upper bounds?

4. Find a partially ordered set (X, \le), with as few elements as possible, for which the hypothesis of Zorn's lemma holds (every chain has an upper bound) but which does not have a maximum.

5. Show, assuming AC, that any Cartesian product of finite sets is either finite or uncountable (can't be countably infinite).

6. Let X be the collection of all intervals $[a, b]$ of length $0 \le b - a \le 2$, with inclusion as partial ordering. Show that every chain in X has an upper bound.

7. In the application of the recursion principle 1.3.2 in the proof that WO implies HMP, how should g be defined? Show that the resulting argument does prove that WO implies HMP.

8. Let A and B be two sets such that there exists a function f from A onto B

and a function g from B onto A. Show, assuming AC, that A and B have the same cardinality. *Hint:* If $h(y) \in f^{-1}(\{y\})$ for all y in B, then h is 1–1 from B into A.

Notes

§§1.1–1.3 Notes on set theory are given at the end of Appendix A.

§1.4 Georg Cantor (1874) first proved the existence of an uncountable set, the set of all real numbers. Then, in 1892, he proved that every set X has smaller cardinality than its power set 2^X. The equivalence theorem came up as an open problem in Cantor's seminar in 1897. It was solved by Felix Bernstein, then a 19-year-old student. Bernstein's proof was given in the book of Borel (1898). Meanwhile, E. Schröder published another proof. The equivalence theorem has often been called the Schröder-Bernstein theorem. Korselt (1911), however, pointed out an error in Schröder's argument. In a letter quoted in Korselt's paper, Schröder gives full credit for the theorem to Bernstein. For further notes and references on the history and other proofs of these facts, see Fraenkel (1966, pp. 70–80). Thanks to Sy Friedman for telling me the proof of 1.4.1. Bernstein also worked in statistics (see Frewer, 1978) and in genetics; he showed that human blood groups A, B, and O are inherited through three alleles at one locus (Nathan, 1970; Boorman et al., 1977, pp. 41–43).

§1.5 Entire books have been published on the axiom of choice and its equivalents (Jech, 1973; Rubin and Rubin, 1985). Ernst Zermelo (1904, 1908a, 1908b) and Bertrand Russell (1906) formulated AC and used it to prove WO. Hausdorff (1914, pp. 140–141) gave his maximal principle (HMP) and showed that it follows from WO. Hausdorff (1937; 1978 transl., p. 5) says "most of the theory of ordered sets" from the first edition was left out of later editions. Zorn (1935) brought in a form of what is now known as Zorn's Lemma. (Zorn stated ZL in the case where the partial ordering is inclusion and the upper bound is the union.) He emphasized its usefulness in algebra. It is rather easily seen to be equivalent to HMP. Apparently Zorn was the first to point out that a statement such as ZL or HMP implies AC, although Zorn never published his proof of this (Rubin and Rubin, 1963, p. 11).

References

An asterisk identifies works I have found discussed in secondary sources but have not seen in the original.

Boorman, Kathleen E., Barbara E. Dodd, and Patrick J. Lincoln (1977). *Blood Group Serology*. 5th ed. Churchill Livingstone, Edinburgh.

*Borel, Émile (1898). *Leçons sur la théorie des fonctions*. Gauthier-Villars, Paris.

Cantor, Georg (1874). Ueber eine Eigenschaft des Inbegriffes aller reellen algebraischen Zahlen. *Journal für die reine u. angew. Math.* 77: 258–262.

——— (1892). Über eine elementare Frage der Mannigfaltigkeitslehre. *Jahresbericht der deutschen Mathematiker-Vereinigung 1*: 75–78.

Fraenkel, Abraham A. (1976). *Abstract Set Theory*. Rev. by Azriel Levy. 4th ed. North-Holland, Amsterdam.

Frewer, Magdalene (1978). Das wissenschaftliche Werk Felix Bernsteins. Diplomarbeit, Inst. f. Math. Statistik u. Wirtschaftsmath., Univ. Göttingen.

Hausdorff, Felix (1914). *Grundzüge der Mengenlehre.* Von Veit, Leipzig. 1st ed. repr. Chelsea, New York (1949). 2d ed., 1927. 3d ed., 1937, published in English as *Set Theory*, transl. J. R. Aumann et al., Chelsea, New York (1957; 2d Engl. ed., 1962; 3d Engl. ed., 1978).

Jech, Thomas (1973). *The Axiom of Choice*. North-Holland, Amsterdam.

Korselt, A. (1911). Über einen Beweis des Äquivalenzsatzes. *Math. Annalen 70*: 294–296.

Nathan, Henry (1970). Bernstein, Felix. In *Dictionary of Scientific Biography 2*: 58–59. Scribner's, New York.

Rubin, Herman, and Jean E. Rubin (1963). *Equivalents of the Axiom of Choice*. North-Holland, Amsterdam. 2d. ed. 1985.

Russell, Bertrand (1906). On some difficulties in the theory of transfinite numbers and order types. *Proc. London Math. Soc.* (Ser. 2) *4*: 29–53.

Zermelo, Ernst (1904). Beweis, dass jede Menge wohlgeordnet werden kann. *Math. Annalen 59*: 514–516.

——— (1908a). Neuer Beweis für die Möglichkeit einer Wohlordnung. *Math. Ann. 65*: 107–128.

——— (1908b). Untersuchungen über die Grundlagen der Mengenlehre I. *Math. Ann. 65*: 261–281.

Zorn, Max (1935). A remark on method in transfinite algebra. *Bull. Amer. Math. Soc. 41*: 667–670.

2

General Topology

General topology has to do with, among other things, notions of convergence. Given a sequence x_n of points in a set X, convergence of x_n to a point x can be defined in different ways. One of the main ways is by a metric, or distance d, which is nonnegative and real-valued, with $x_n \to x$ meaning $d(x_n, x) \to 0$. The usual metric for real numbers is $d(x, y) = |x - y|$. For the usual convergence of real numbers, a function f is called continuous if whenever $x_n \to x$ in its domain, we have $f(x_n) \to f(x)$.

On the other hand, some interesting kinds of convergence are not defined by metrics: if we define convergence of a sequence of functions f_n "pointwise," so that $f_n \to f$ means $f_n(x) \to f(x)$ for all x, it turns out that (for a large enough class of functions defined on an uncountable set) there may be no metric e such that $f_n \to f$ is equivalent to $e(f_n, f) \to 0$.

Given a sense of convergence, we can call a set F closed if whenever $x_i \in F$ for all i and $x_i \to x$ we have $x \in F$ also. Any closed interval $[a, b] := \{x : a \le x \le b\}$ is an example of a closed set. The properties of closed sets F and their complements $U := X \backslash F$, which are called open sets, turn out to provide the best and most accepted way of extending the notions of convergence, continuity, and so forth to nonmetric situations. This leads to kinds of convergence (nets, filters) more general than those of sequences. For example, pointwise convergence of functions is handled by "product topologies," to be treated in §2.2. Some readers may prefer to look at §2.3 and parts of §§2.4–2.5 on metrics, which may be more familiar, before taking up §§2.1–2.2.

2.1. Topologies, Metrics, and Continuity

From here on through the rest of the book, the axiom of choice will be assumed, without being mentioned in each instance.

Definition. Given a set X, a *topology* on X is a collection \mathcal{T} of subsets of X (in other words, $\mathcal{T} \subset 2^X$) such that

(1) $\emptyset \in \mathcal{T}$ and $X \in \mathcal{T}$.
(2) For every $U \in \mathcal{T}$ and $V \in \mathcal{T}$, we have $U \cap V \in \mathcal{T}$.
(3) For any $\mathcal{U} \subset \mathcal{T}$, we have $\bigcup \mathcal{U} \in \mathcal{T}$.

When a topology has been chosen, its members are called *open* sets. Their complements, $F = X \backslash U$, $U \in \mathcal{T}$, are called *closed* sets. The pair (X, \mathcal{T}) is called a *topological space*. Members of X are called *points*.

In \mathbb{R}, there is a "usual topology" for which the simplest examples of open sets are the open intervals $(a, b) := \{x : a < x < b\}$. Here (X, Y) is also a notation for the ordered pair $\langle X, Y \rangle$. These notations are both in general use in real analysis. The context should indicate which one is meant for X and Y real numbers.

General open sets in \mathbb{R} are arbitrary unions of open intervals (it will turn out later that any such union can be written as a countable union). Closed sets, which are the complements of the open sets, are also the sets which are closed in the sense defined in the introduction to this chapter, for the usual convergence of real numbers.

There are sets that are neither closed nor open, such as the "half-open intervals" $[a, b) := \{x : a \leq x < b\}$ or $(a, b] := \{x : a < x \leq b\}$ for $a < b$ in \mathbb{R}. On the other hand, in any topological space X, since \emptyset and X are both open and are complements of each other, they are also both closed.

Although in axiomatic set theory, as in Appendix A, all mathematical objects are represented by sets, it's easier to think of a point as just that, without any members or other internal structure. Such points will usually be denoted by small letters, such as x and y. Sets of points will usually be denoted by capital letters, such as A and B. Sets of such sets will often be called by other names, such as "collections" or "families" of sets, and denoted by script capitals, such as \mathcal{A} and \mathcal{B}. (Later, in functional analysis, functions will in turn be considered as points for some purposes.)

From (2) in the definition of topology and induction, any finite intersection of open sets (that is, an intersection of finitely many open sets) is open. An arbitrary union of open sets is open, according to (3).

For any set X, the power set 2^X is a topology, called the *discrete* topology on X. For this topology, all sets are open and all are closed. If X is a finite set, then a topology on X will be assumed to be discrete unless some other topology is provided.

When we are dealing for the time being with subsets of a set X, such as a topological space, then the *complement* of any set $A \subset X$ is defined as $X \backslash A$. Thus the complement of any open set is closed.

Definition. Given a set X, a *pseudometric* for X is a function d from $X \times X$ into $\{x \in \mathbb{R}: x \geq 0\}$ such that

(1) for all x, $d(x, x) = 0$,
(2) for all x and y, $d(x, y) = d(y, x)$ (symmetry), and
(3) for all x, y, and z, $d(x, z) \leq d(x, y) + d(y, z)$ (triangle inequality).
 If also
(4) $d(x, y) = 0$ implies $x = y$, then d is called a *metric*.

(X, d) is called a *(pseudo)metric space.*

A classic example of a metric space is \mathbb{R} with the "usual metric" $d(x, y) := |x - y|$. Also, the plane \mathbb{R}^2 with the usual Euclidean distance is a metric space, in which the triangle inequality says that the length of any side of a triangle is no more than the sum of the lengths of the other two sides.

Definition. A *base* for a topology \mathcal{T} is any collection $\mathcal{U} \subset \mathcal{T}$ such that for every $V \in \mathcal{T}$, $V = \bigcup \{U \in \mathcal{U}: U \subset V\}$. A *neighborhood* of a point x is any set N (open or not) such that $x \in U \subset N$ for some open U. A collection \mathcal{N} of neighborhoods of x is a *neighborhood-base* at x iff for every neighborhood V of x, $x \in N \subset V$ for some $N \in \mathcal{N}$.

The usual topology for \mathbb{R} has a base consisting of all the open intervals (a, b) for $a < b$. A neighborhood-base at any $x \in \mathbb{R}$ is the set of all intervals $(x - 1/n, x + 1/n)$ for $n = 1, 2, \ldots$. The intervals $[-1/3, 1/4]$ and $(-1/8, 1/7)$ are both neighborhoods of 0 in \mathbb{R}, while $[0, 1]$ is not.

The union of the empty set is empty (by definition), so that $\bigcup \emptyset = \bigcup \{\emptyset\} = \emptyset$. Thus, for $V = \emptyset$, V is always the union of those U in \mathcal{U} which it includes, whether or not $\emptyset \in \mathcal{U}$. If \mathcal{U} is a base, then $X = \bigcup \mathcal{U}$: every point of X must be in at least one set in \mathcal{U}.

Note that a set U is open if and only if it is a neighborhood of each of its points. Given a pseudometric space $\langle X, d \rangle$, $x \in X$, and $r > 0$, let $B(x, r) := \{y \in X: d(x, y) < r\}$. Then $B(x, r)$ is called the (open) ball with center at x and radius r.

Specifically, in \mathbb{R}, the ball $B(x, r)$ is the open interval $(x - r, x + r)$. Conversely, any open interval (a, b) for $a < b$ in \mathbb{R} can be written as $B(x, r)$ for $x = (a + b)/2$, $r = (b - a)/2$.

2.1.1. Theorem *For any pseudometric space $\langle X, d \rangle$, the collection of all sets $B(x, r)$, for $x \in X$ and $r > 0$, is a base for a topology on X. For fixed x it is a neighborhood-base at x for this topology.*

Proof. Suppose $x \in X$, $y \in X$, $r > 0$, and $s > 0$. Let $U = B(x, r) \cap B(y, s)$. Suppose $z \in U$. Let $t := \min(r - d(x, z), s - d(y, z))$. Then $t > 0$. To show $B(z, t) \subset U$, suppose $d(z, w) < t$. Then the triangle inequality gives $d(x, w) < d(x, z) + t < r$. Likewise, $d(y, w) < s$. So $w \in B(x, r)$ and $w \in B(y, s)$, so $B(z, t) \subset U$. Thus for every point z of U, an open ball around z is included in U, and U is the union of all open balls which it includes.

Let \mathcal{T} be the collection of all unions of open balls, so $U \in \mathcal{T}$. Suppose $V \in \mathcal{T}$ and $W \in \mathcal{T}$, so $V = \bigcup \mathcal{A}$ and $W = \bigcup \mathcal{B}$ where \mathcal{A} and \mathcal{B} are collections of open balls. Then

$$V \cap W = \bigcup \{A \cap B : A \in \mathcal{A}, B \in \mathcal{B}\}.$$

Thus $V \cap W \in \mathcal{T}$. The empty set is in \mathcal{T} (as an empty union), and X is the union of all balls. Clearly, any union of sets in \mathcal{T} is in \mathcal{T}. Thus \mathcal{T} is a topology. Also clearly, the balls form a base for it (and they are actually open, so that the terminology is consistent). Suppose $x \in U \in \mathcal{T}$. Then for some y and $r > 0$, $x \in B(y, r) \subset U$. Let $s := r - d(x, y)$. Then $s > 0$ and $B(x, s) \subset U$, so the set of all balls with center at x is a neighborhood-base at x. $\qquad \square$

The topology \mathcal{T} given by Theorem 2.1.1 is called a (pseudo)metric topology. If d is a metric, then \mathcal{T} is said to be *metrizable* and to be *metrized* by d. On \mathbb{R}, the topology metrized by the usual metric $d(x, y) := |x - y|$ is the usual topology on \mathbb{R}; namely, the topology with a base given by all open intervals (a, b).

If (X, \mathcal{T}) is any topological space and $Y \subset X$, then $\{U \cap Y : U \in \mathcal{T}\}$ is easily seen to be a topology on Y, called the *relative* topology.

Let f be a function from a set A into a set B. Then for any subset C of B, let $f^{-1}(C) := \{x \in A : f(x) \in C\}$. This $f^{-1}(C)$ is sometimes called the *inverse image* of C under f. (Note that f need not be 1–1, so f^{-1} need not be a function.) The inverse image preserves all unions and intersections: for any non-empty collection $\{B_i\}_{i \in I}$ of subsets of B, $f^{-1}(\bigcup_{i \in I} B_i) = \bigcup_{i \in I} f^{-1}(B_i)$ and $f^{-1}(\bigcap_{i \in I} B_i) = \bigcap_{i \in I} f^{-1}(B_i)$. When I is empty, the equation for union still holds, with both sides empty. If we define the intersection of an empty collection of subsets of a space X as equal to X (for $X = A$ or B), the equation for intersections is still true also.

Recall that a *sequence* is a function whose domain is \mathbb{N}, or the set $\{n \in \mathbb{N} : n > 0\}$ of all positive integers. A sequence x is usually written with subscripts, such as $\{x_n\}_{n \geq 0}$ or $\{x_n\}_{n \geq 1}$, setting $x_n := x(n)$. A sequence is said to be in some set X iff its range is included in X. Given a topological space (X, \mathcal{T}), we say a sequence x_n *converges* to a point x, written $x_n \to x$

(as $n \to \infty$), iff for every neighborhood A of x, there is an m such that $x_n \in A$ for all $n \geq m$.

The notion of continuous function on a metric space can be characterized in terms of converging sequences (if $x_n \to x$, then $f(x_n) \to f(x)$) or with ε's and δ's. It turns out that continuity (as opposed to, for example, uniform continuity) really depends only on topology and has the following simple form.

Definition. Given topological spaces (X, \mathcal{T}) and (Y, \mathcal{U}), a function f from X into Y is called *continuous* iff for all $U \in \mathcal{U}$, $f^{-1}(U) \in \mathcal{T}$.

Example. Consider the function f with $f(x) := x^2$ from \mathbb{R} into itself and let $U = (a, b)$. Then if $b \leq 0$, $f^{-1}(U) = \emptyset$; if $a < 0 < b$, $f^{-1}(U) = (-b^{1/2}, b^{1/2})$; or if $0 \leq a < b$, $f^{-1}(U) = (-b^{1/2}, -a^{1/2}) \cup (a^{1/2}, b^{1/2})$. So the inverse image of an open interval under f is not always an interval (in the last case, it is a union of two disjoint intervals) but it is always an open set, as stated in the definition of continuous function. In the other direction, $f((-1, 1)) := \{f(x): -1 < x < 1\} = [0, 1)$, which is not open.

If $n(\cdot)$ is an increasing function from the positive integers into the positive integers, so that $n(1) < n(2) < \cdots$, then for a sequence $\{x_n\}$, the sequence $k \mapsto x_{n(k)}$ will be called a *subsequence* of the sequence $\{x_n\}$. Here $n(k)$ is often written as n_k.

It is straightforward that if $x_n \to x$, then any subsequence $k \mapsto x_{n(k)}$ also converges to x. If \mathcal{T} is the topology defined by a pseudometric d, then it is easily seen that for any sequence x_n in X, $x_n \to x$ if and only if $d(x_n, x) \to 0$ (as $n \to \infty$).

Converging along a sequence is not the only way to converge. For example, one way to say that a function f is continuous at x is to say that $f(y) \to f(x)$ as $y \to x$. This implies that for *every* sequence such that $y_n \to x$, we have $f(y_n) \to f(x)$, but one can think of y moving continuously toward x, not just along various sequences. On the other hand, in some topological spaces, which are not metrizable, sequences are inadequate. It may happen, for example, that for every x and for every sequence $y_n \to x$, we have $f(y_n) \to f(x)$, but f is not continuous. There are two main convergence concepts, for "nets" and "filters," which do in general topological spaces what sequences do in metric spaces, as follows.

Definitions. A *directed* set is a partially ordered set (I, \leq) such that for any i and j in I, there is a k in I with $k \geq i$ (that is, $i \leq k$) and $k \geq j$. A *net* $\{x_i\}_{i \in I}$ is any function x whose domain is a directed set, written $x_i := x(i)$.

Let (X, \mathcal{T}) be a topological space. A net $\{x_i\}_{i \in I}$ *converges* to x in X, written $x_i \to x$, iff for every neighborhood A of x, there is a $j \in I$ such that $x_k \in A$ for all $k \geq j$.

Given a set X, a *filter base* in X is a non-empty collection \mathcal{F} of non-empty subsets of X such that for any F and G in \mathcal{F}, $F \cap G \supset H$ for some $H \in \mathcal{F}$. A filter base \mathcal{F} is called a *filter* iff whenever $F \in \mathcal{F}$ and $F \subset G \subset X$ then $G \in \mathcal{F}$. Equivalently, a filter \mathcal{F} is a non-empty collection of non-empty subsets of X such that (a) $F \in \mathcal{F}$ and $F \subset G \subset X$ imply $G \in \mathcal{F}$, and (b) if $F \in \mathcal{F}$ and $G \in \mathcal{F}$, then $F \cap G \in \mathcal{F}$.

Examples. (a) A classic example of a directed set is the set of positive integers with usual ordering. For it, a net is a sequence, so that sequences are a special case of nets.

(b) For another example, let I be the set of all finite subsets of \mathbb{N}, partially ordered by inclusion. Then if $\{x_n\}_{n \in \mathbb{N}}$ is a sequence of real numbers and $F \in I$, let $S(F)$ be the sum of the x_n for n in F. Then $\{S(F)\}_{F \in I}$ is a net. If it converges, the sum $\sum_n x_n$ is said to converge *unconditionally*. (You may recall that this is equivalent to absolute convergence, $\sum_n |x_n| < \infty$.)

(c) A major example of nets (although much older than the general concept of net) is the *Riemann integral*. Let a and b be real numbers with $a < b$ and let f be a function with real values defined on $[a, b]$. Let I be the set of all finite sequences $a = x_0 \leq y_1 \leq x_1 \leq y_2 \leq x_2 \cdots \leq y_n \leq x_n = b$, where n may be any positive integer. Such a sequence will be written $u := (x_j, y_j)_{j \leq n}$. If also $v \in I$, $v = (w_i, z_j)_{j \leq m}$, the ordering is defined by $v < u$ iff $m < n$ and for each $j \leq m$ there is an $i \leq n$ with $x_i = w_j$. (This relationship is often expressed by saying that the *partition* $\{x_0, \ldots, x_n\}$ of the interval $[a, b]$ is a *refinement* of the partition $\{w_0, \ldots, w_m\}$, keeping the w_j and inserting one or more additional points.) It is easy to check that this ordering makes I a directed set. The ordering does not involve the y_j. Now let $S(f, u) := \sum_{1 \leq j \leq n} f(y_j)(x_j - x_{j-1})$. This is a net. The *Riemann integral* of f from a to b is defined as the limit of this net iff it converges to some real number.

If \mathcal{F} is any filter base, then $\{G \subset X : F \subset G \text{ for some } F \in \mathcal{F}\}$ is a filter \mathcal{G}. \mathcal{F} is said to be a *base* of \mathcal{G}. The filter base \mathcal{F} is said to *converge* to a point x, written $\mathcal{F} \to x$, iff every neighborhood of x belongs to the filter. For example, the set of all neighborhoods of a point x is a filter converging to x. The set of all open neighborhoods of x is a filter base converging to x. If X is a set and f a function with $\operatorname{dom} f \supset X$, for each $A \subset X$ recall that $f[A] := \operatorname{ran}(f \restriction A) = \{f(x) : x \in A\}$. For any filter base \mathcal{F} in X let $f[[\mathcal{F}]] := \{f[A] : A \in \mathcal{F}\}$. Note that $f[[\mathcal{F}]]$ is also a filter base.

2.1.2. Theorem *Given topological spaces* (X, \mathcal{T}) *and* (Y, \mathcal{U}) *and a function* f *from* X *into* Y, *the following are equivalent (assuming AC, as usual):*

(1) f *is continuous.*
(2) For every convergent net $x_i \to x$ *in* X, $f(x_i) \to f(x)$ *in* Y.
(3) For every convergent filter base $\mathcal{F} \to x$ *in* X, $f[[\mathcal{F}]] \to f(x)$ *in* Y.

Proof. (1) implies (2): suppose $f(x) \in U \in \mathcal{U}$. Then $x \in f^{-1}(U)$, so for some j, $x_i \in f^{-1}(U)$ for all $i > j$. Then $f(x_i) \in U$, so $f(x_i) \to f(x)$.

(2) implies (3): let $\mathcal{F} \to x$. If $f[[\mathcal{F}]] \not\to f(x)$ (that is, $f[[\mathcal{F}]]$ does not converge to $f(x)$), take $f(x) \in U \in \mathcal{U}$ with $f[A] \not\subset U$ for all $A \in \mathcal{F}$. Define a partial ordering on \mathcal{F} by $A \leq B$ iff $A \supset B$ for A and B in \mathcal{F}. By definition of filter base, (\mathcal{F}, \leq) is then a directed set. Define a net (using AC) by choosing, for each $A \in \mathcal{F}$, an $x(A) \in A$ with $f(x(A)) \notin U$. Then the net $x(A) \to x$ but $f(x(A)) \not\to f(x)$, contradicting (2).

(3) implies (1): take any $U \in \mathcal{U}$ and $x \in f^{-1}(U)$. The filter \mathcal{F} of all neighborhoods of x converges to x, so $f[[\mathcal{F}]] \to f(x)$. For some neighborhood V of x, $f[V] \subset U$, so $V \subset f^{-1}(U)$, and $f^{-1}(U) \in \mathcal{T}$. \square

For another example of a filter base, given a continuous real function f on $[0, 1]$, let $t := \sup\{f(x): 0 \leq x \leq 1\}$. A sequence of intervals I_n will be defined recursively. Let $I_0 := [0, 1]$. Then the supremum of f on at least one of the two intervals $[0, 1/2]$ or $[1/2, 1]$ equals t. Let I_1 be such an interval of length $1/2$. Given a closed interval I_n of length $1/2^n$ on which f has the same supremum t as on all of $[0, 1]$, let I_{n+1} be a closed interval, either the left half or right half of I_n, with the same supremum. Then $\{I_n\}_{n \geq 0}$ is a filter base converging to a point x for which $f(x) = t$.

A topological space (X, \mathcal{T}) is called *Hausdorff*, or a *Hausdorff space*, iff for every two distinct points x and y in X, there are open sets U and V with $x \in U$, $y \in V$, and $U \cap V = \emptyset$. Thus a pseudometric space (X, d) is Hausdorff if and only if d is a metric. For any topological space (S, \mathcal{T}) and set $A \subset S$, the *interior* of A, or int A, is defined by int $A := \bigcup \{U \in \mathcal{T}: U \subset A\}$. It is clearly open and is the largest open set included in A. Also, the *closure* of A, called \overline{A}, is defined by $\overline{A} := \bigcap \{F \subset S: F \supset A$ and F is closed$\}$. It is easily seen that for any sets $U_i \subset S$, for i in an index set I, $S \setminus (\bigcup_{i \in I} U_i) = \bigcap_{i \in I}(S \setminus U_i)$. Since any union of open sets is open, it follows that any intersection of closed sets is closed. So \overline{A} is closed and is the smallest closed set including A.

Examples. If $a < b$ and A is any of the four intervals (a, b), $(a, b]$, $[a, b)$, or $[a, b]$, the closure \overline{A} is $[a, b]$ and the interior is int $A = (a, b)$.

Closure is related to convergent nets as follows.

2.1.3. Theorem *Let* (S, \mathcal{T}) *be any topological space. Then:*

(a) *For any* $A \subset S, \overline{A}$ *is the set of all* $x \in S$ *such that some net* $x_i \to x$ *with* $x_i \in A$ *for all* i.

(b) *A set* $F \subset S$ *is closed if and only if for every net* $x_i \to x$ *in* S *with* $x_i \in F$ *for all* i *we have* $x \in F$.

(c) *A set* $U \subset S$ *is open iff for every* $x \in U$ *and net* $x_i \to x$ *there is some* j *with* $x_i \in U$ *for all* $i \geq j$.

(d) *If* \mathcal{T} *is metrizable, nets can be replaced by sequences* $x_n \to x$ *in (a), (b), and (c).*

Proof. (a): If $x \notin \overline{A}$ and $x_i \to x$, then $x_i \notin A$ for some i. Conversely, if $x \in \overline{A}$, let \mathcal{F} be the filter of all neighborhoods of x. Then for each $N \in \mathcal{F}$, $N \cap A \neq \emptyset$. Choose (by AC) $x(N) \in N \cap A$. Then the net $x(N) \to x$ (where the set of neighborhoods is directed by reverse inclusion, as in the last proof).

(b): Note that F is closed if and only if $\overline{F} = F$, and apply (a).

(c): "Only if" follows from the definition of convergence of nets. "If": suppose a set B is not open. Then for some $x \in B$, by (b) there is a net $x_i \to x$ with $x_i \notin B$ for all i.

(d): In the proof of (a) we can take the filter base of neighborhoods $N = \{y : d(x, y) < 1/n\}$ to get a sequence $x_n \to x$. The rest follows. □

For any topological space (S, \mathcal{T}), a set $A \subset S$ is said to be *dense* in S iff the closure $\overline{A} = S$. Then (S, \mathcal{T}) is said to be *separable* iff S has a countable dense subset.

For example, the set \mathbb{Q} of all rational numbers is dense in the line \mathbb{R}, so \mathbb{R} is separable (for the usual metric).

(S, \mathcal{T}) is said to satisfy the first axiom of countability, or to be *first-countable*, iff there is a countable neighborhood-base at each point. For any pseudometric space (S, d), the topology is first-countable, since for each $x \in S$, the balls $B(x, 1/n) := \{y \in S : d(x, y) < 1/n\}, n = 1, 2, \ldots$, form a neighborhood-base at x. (In fact, there are practically no other examples of first-countable spaces in analysis.) A topological space (S, \mathcal{T}) is said to satisfy the second axiom of countability, or to be *second-countable*, iff \mathcal{T} has a countable base. Clearly any second-countable space is also first-countable.

2.1.4. Proposition *A metric space* (S, d) *is second-countable if and only if it is separable.*

Proof. Let A be countable and dense in S. Let \mathcal{U} be the set of all balls $B(x, 1/n)$ for x in A and $n = 1, 2, \ldots$. To show that \mathcal{U} is a base, let U be any open set and $y \in U$. Then for some m, $B(y, 1/m) \subset U$. Take $x \in A$ with $d(x, y) < 1/(2m)$. Then $y \in B(x, 1/(2m)) \subset B(y, 1/m) \subset U$, so U is the union of the elements of \mathcal{U} that it includes, and \mathcal{U} is a base, which is countable. Conversely, suppose there is a countable base \mathcal{V} for the topology, which we may assume consists of non-empty sets. By the axiom of choice, let f be a function on \mathbb{N} whose range contains at least one point of each set in \mathcal{V}. Then this range is dense. $\qquad\square$

Problems

1. On \mathbb{R}^2 let $d(\langle x, y \rangle, \langle u, v \rangle) := ((x - u)^2 + (y - v)^2)^{1/2}$ (usual metric), $e(x, y) := |x - u| + |y - v|$. Show that e is a metric and metrizes the same topology as d.

2. For any topological space (X, \mathcal{T}) and set $A \subset X$, the *boundary* of A is defined by $\partial A := \overline{A} \backslash \text{int } A$. Show that the boundary of A is closed and is the same as the boundary of $X \backslash A$. Show that for any two sets A and B in X, $\partial(A \cup B) \subset \partial A \cup \partial B$. Give an example where $\partial(A \cup B) \neq \partial A \cup \partial B$.

3. Let (X, d) and (Y, e) be pseudometric spaces with topologies \mathcal{T}_d and \mathcal{T}_e metrized by d and e respectively. Let f be a function from X into Y. Show that the following are equivalent (as stated in the first paragraph of this chapter):
 (a) f is continuous: $f^{-1}(U) \in \mathcal{T}_d$ for all $U \in \mathcal{T}_e$.
 (b) f is sequentially continuous: for every $x \in X$ and every sequence $x_n \to x$ for d, we have $f(x_n) \to f(x)$ for e.

4. Let (S, d) be a metric space and X a subset of S. Let the restriction of d to $X \times X$ also be called d. Show that the topology on X metrized by d is the same as the relative topology of the topology metrized by d on S.

5. Show that any subset of a separable metric space is also separable with its relative topology. *Hint:* Use the previous problem and Proposition 2.1.4.

6. Let $\{x_i\}_{i \in I}$ be a net in a topological space. Define a filter base \mathcal{F} such that for all x, $\mathcal{F} \to x$ if and only if $x_i \to x$.

7. A net $\{f_i\}_{i \in I}$ of functions on a set X is said to converge *pointwise* to a function f iff $f_i(x) \to f(x)$ for all x in X. The *indicator function* of a set A is defined by $1_A(x) = 1$ for $x \in A$ and $1_A(x) = 0$ for $x \in X \backslash A$. If X is uncountable, show that there is a net of indicator functions of finite sets converging to the constant function 1, but that the net cannot be replaced by a sequence.

8. (a) Let \mathbb{Q} be the set of rational numbers. Show that the Riemann integral of $1_{\mathbb{Q}}$ from 0 to 1 is undefined (the net in its definition does not converge). (\mathbb{Q} is countable and $[0, 1]$ is uncountable, so the integral "should be" 0, and will be for the Lebesgue integral, to be defined in Chapter 3.)

 (b) Show that for a sequence $1_{F(n)}$ of indicator functions of finite sets $F(n)$ converging pointwise to $1_{\mathbb{Q}}$, the Riemann integral of $1_{F(n)}$ is 0 for each n.

9. Let X be an infinite set. Let \mathcal{T} consist of the empty set and all complements of finite subsets of X. Show that \mathcal{T} is a topology in which every singleton $\{x\}$ is closed, but \mathcal{T} is not metrizable. *Hint:* A sequence of distinct points converges to every point.

10. Let S be any set and S^∞ the set of all sequences $\{x_n\}_{n\geq 1}$ with $x_n \in S$ for all n. Let C be a subset of the Cartesian product $S \times S^\infty$. Also, $S \times S^\infty$ is the set of all sequences $\{x_n\}_{n\geq 0}$ with $x_n \in S$ for all $n = 0, 1, \ldots$. Such a set C will be viewed as defining a sense of "convergence," so that $x_n \to_C x_0$ will be written in place of $\{x_n\}_{n\geq 0} \in C$. Here are some axioms: C will be called an *L-convergence* if it satisfies (1) to (3) below.

 (1) If $x_n = x$ for all n, then $x_n \to_C x$.
 (2) If $x_n \to_C x$, then any subsequence $x_{n(k)} \to_C x$.
 (3) If $x_n \to_C x$ and $x_n \to_C y$, then $x = y$.

 If C also satisfies (4), it is called an *L*-convergence*:

 (4) If for every subsequence $k \mapsto x_{n(k)}$ there is a further subsequence $j \mapsto y_j := x_{n(k(j))}$ with $y_j \to_C x$, then $x_n \to_C x$.

 (a) Prove that if T is a Hausdorff topology and $C(T)$ is convergence for T, then $C(T)$ is an L^*-convergence.

 (b) Let C be any L-convergence. Let $U \in T(C)$ iff whenever $x_n \to_C x$ and $x \in U$, there is an m such that $x_n \in U$ for all $n \geq m$. Prove that $T(C)$ is a topology.

 (c) Let X be the set of all sequences $\{x_n\}_{n\geq 0}$ of real numbers such that for some m, $x_n = 0$ for all $n \geq m$. If $y(m) = \{y(m)_n\}_{n\geq 0} \in X$ for all $m = 0, 1, \ldots$, say $y(m) \to_C y(0)$ if for some k, $y(m)_j = y(0)_j = 0$ for all $j \geq k$ and all m, and $y(m)_n \to y(0)_n$ as $m \to \infty$ for all n. Prove that \to_C is an L^*-convergence but that there is no metric e such that $y(m) \to_C y(0)$ is equivalent to $e(y(m), y(0)) \to 0$.

11. For any two real numbers u and v, $\max(u, v) := u$ iff $u \geq v$; otherwise, $\max(u, v) := v$. A metric space (S, d) is called an *ultrametric space* and d an *ultrametric* if $d(x, z) \leq \max(d(x, y), d(y, z))$ for all x, y, and z in S. Show that in an ultrametric space, any open ball $B(x, r)$ is also closed.

2.2. Compactness and Product Topologies

In the field of optimization, for example, where one is trying to maximize or minimize a function (often a function of several variables), it can be good to know that under some conditions a maximum or minimum does exist. As shown after Theorem 2.1.2 for $[0, 1]$, for any $a \leq b$ in \mathbb{R} and continuous function f from $[a, b]$ into \mathbb{R}, there is an $x \in [a, b]$ with $f(x) = \sup\{f(u): a \leq u \leq b\}$. Likewise there is a $y \in [a, b]$ with $f(y) = \inf\{f(v): a \leq v \leq b\}$. This property, that a continuous real-valued function is bounded and attains its supremum and infimum, extends to compact topological spaces, as will be defined. (See Problem 18.)

Compactness was defined for metric spaces before general topological spaces. In metric spaces it has several equivalent characterizations, to be given in §2.3. Among them, the following, called the "Heine-Borel property," is stated in terms of the topology, rather than a metric, so it has been taken as the definition of "compact" for general topological spaces. Although it perhaps has less immediate intuitive flavor and appeal than most definitions, it has proved quite successful mathematically.

Definition. A topological space (K, \mathcal{T}) is called *compact* iff whenever $\mathcal{U} \subset \mathcal{T}$ and $K = \bigcup \mathcal{U}$, there is a finite $\mathcal{V} \subset \mathcal{U}$ such that $K = \bigcup \mathcal{V}$.

Let X be a set and A a subset of X. A collection of sets whose union includes A is called a *cover* or *covering* of A. If it consists of open sets, it is called an *open cover*. If a subset A is not specified, then $A = X$ is intended. So the definition of compactness says that "every open cover has a finite subcover." The word "every" is crucial, since for any topological space, there always exist some open coverings with finite subcovers – in other words, there exist finite open covers, in fact open covers containing just one set, since the whole set X is always open.

For other examples, the open intervals $(-n, n)$ form an open cover of \mathbb{R} without a finite subcover. The intervals $(1/(n + 2), 1/n)$ for $n = 1, 2, \ldots$, form an open cover of $(0, 1)$ without a finite subcover. Thus, \mathbb{R} and $(0, 1)$ are not compact.

A subset K of a topological space X (that is, a set X where $\langle X, \mathcal{T} \rangle$ is a topological space) is called compact iff it is compact for its relative topology. Equivalently, K is compact if for any $\mathcal{U} \subset \mathcal{T}$ such that $K \subset \bigcup \mathcal{U}$, there is a finite $\mathcal{V} \subset \mathcal{U}$ such that $K \subset \bigcup \mathcal{V}$. We know that if a non-empty set A of real numbers has an upper bound b—so that $x \leq b$ for all $x \in A$—then A has a least upper bound, or *supremum* $c := \sup A$. That is, c is an upper bound

of A such that $c \leq b$ for any other upper bound b of A (as shown in §1.1 and Theorem A.4.1 of Appendix A.4). Likewise, a non-empty set D of real numbers with a lower bound has a greatest lower bound, or *infimum*, inf D. If a set A is unbounded above, let sup $A := +\infty$. If A is unbounded below, let inf $A := -\infty$.

2.2.1. Theorem *Any closed interval $[a, b]$ with its usual (relative) topology is compact.*

Proof. It will be enough to prove this for $a = 0$ and $b = 1$. Let \mathcal{U} be an open cover of $[0, 1]$. Let H be the set of all x in $[0, 1]$ for which $[0, x]$ can be covered by a union of finitely many sets in \mathcal{U}. Then since $0 \in V$ for some $V \in \mathcal{U}$, $[0, h] \subset H$ for some $h > 0$. If $H \neq [0, 1]$, let $y := \inf([0, 1] \backslash H)$. Then $y \in V$ for some $V \in \mathcal{U}$, so for some $c > 0$, $[y - c, y] \subset V$ and $y - c \in H$. Taking a finite open subcover of $[0, y - c]$ and adjoining V gives an open cover of $[0, y]$, so $y \in H$. If $y = 1$, we are done. Otherwise, for some $b > 0$, $[y, y + b] \subset V$, so $[0, y + b] \subset H$, contradicting the choice of y. \square

The next two proofs are rather easy:

2.2.2. Theorem *If (K, \mathcal{T}) is a compact topological space and F is a closed subset of K, then F is compact.*

Proof. Let \mathcal{U} be an open cover of F, where we may take $\mathcal{U} \subset \mathcal{T}$. Then $\mathcal{U} \cup \{K \backslash F\}$ is an open cover of K, so has a finite subcover \mathcal{V}. Then $\mathcal{V} \backslash \{K \backslash F\}$ is a finite cover of F, included in \mathcal{U}. \square

2.2.3. Theorem *If (K, \mathcal{T}) is compact and f is continuous from K onto another topological space L, then L is compact.*

Proof. Let \mathcal{U} be an open cover of L. Then $\{f^{-1}(U): U \in \mathcal{U}\}$ is an open cover of K, with a finite subcover $\{f^{-1}(U): U \in \mathcal{V}\}$ where \mathcal{V} is finite. Then \mathcal{V} is a finite subcover of L. \square

An example or corollary of Theorem 2.2.3 is that if f is a continuous real-valued function on a compact space K, then f is bounded, since any compact set in \mathbb{R} is bounded (consider the open cover by intervals $(-n, n)$).

Definition. A filter \mathcal{F} in a set X is called an *ultrafilter* iff for all $Y \subset X$, either $Y \in \mathcal{F}$ or $X \backslash Y \in \mathcal{F}$.

The simplest ultrafilters are of the form $\{A \subset X : x \in A\}$ for $x \in X$. These are called point ultrafilters. The existence of non-point ultrafilters depends on the axiom of choice. Some filters converging to a point x are included in the point ultrafilter of all sets containing x; but $(0, 1/n), n = 1, 2, \ldots$, for example, is a base of a filter converging to 0 in \mathbb{R}, where *no* set in the base contains 0.

The next two theorems provide an analogue of the fact that every sequence in a compact metric space has a convergent subsequence.

2.2.4. Theorem *Every filter \mathcal{F} in a set X is included in some ultrafilter. \mathcal{F} is an ultrafilter if and only if it is maximal for inclusion; that is, if $\mathcal{F} \subset \mathcal{G}$ and \mathcal{G} is a filter, then $\mathcal{F} = \mathcal{G}$.*

Proof. Let $Y \subset X$. If $F \in \mathcal{F}, G \in \mathcal{F}, F \subset Y$, and $G \cap Y = \emptyset$, then $F \cap G = \emptyset$, a contradiction. So in particular at most one of Y and $X \backslash Y$ belongs to \mathcal{F}, and among filters in X, any ultrafilter is maximal for inclusion. Either $G \cap Y \neq \emptyset$ for all $G \in \mathcal{F}$ or $F \backslash Y \neq \emptyset$ for all $F \in \mathcal{F}$. If $G \cap Y \neq \emptyset$ for all $G \in \mathcal{F}$, let $\mathcal{G} :=$ $\{H \subset X : \text{for some } G \in \mathcal{F}, H \supset G \cap Y\}$. Then clearly \mathcal{G} is a filter and $\mathcal{F} \subset \mathcal{G}$. Or, if $G \cap Y = \emptyset$ for some $G \in \mathcal{F}$, so $F \backslash Y \neq \emptyset$ for all $F \in \mathcal{F}$, define $\mathcal{G} := \{H \subset X : \text{for some } F \in \mathcal{F}, H \supset F \backslash Y\}$. Thus, always $\mathcal{F} \subset \mathcal{G}$ for a filter \mathcal{G} with either $Y \in \mathcal{G}$ or $X \backslash Y \in \mathcal{G}$. Hence a filter maximal for inclusion is an ultrafilter.

Next suppose \mathcal{C} is an inclusion-chain of filters in X and $\mathcal{U} = \bigcup \mathcal{C}$. If $F \subset G \subset X, F \in \mathcal{U}$, then for some $\mathcal{V} \in \mathcal{C}, F \in \mathcal{V}$ and $G \in \mathcal{V} \subset \mathcal{U}$. If $H \in \mathcal{U}, H \in \mathcal{H}$ for some $\mathcal{H} \in \mathcal{C}$. Either $\mathcal{H} \subset \mathcal{V}$ or $\mathcal{V} \subset \mathcal{H}$. By symmetry, say $\mathcal{V} \subset \mathcal{H}$. Then $F \cap H \in \mathcal{H} \subset \mathcal{U}$. Thus \mathcal{U} is a filter. Hence by Zorn's Lemma (1.5.1), any filter \mathcal{F} is included in some maximal filter, which is an ultrafilter. \square

In any infinite set, the set of all complements of finite subsets forms a filter \mathcal{F}. By Theorem 2.2.4, \mathcal{F} is included in some ultrafilter, which is not a point ultrafilter. The non-point ultrafilters are exactly those that include \mathcal{F}.

Here is a characterization of compactness in terms of ultrafilters, which is one reason ultrafilters are useful:

2.2.5. Theorem *A topological space (S, \mathcal{T}) is compact if and only if every ultrafilter in S converges.*

Proof. Let (S, \mathcal{T}) be compact and \mathcal{U} an ultrafilter. If \mathcal{U} is not convergent, then for all x take an open set $U(x)$ with $x \in U(x) \notin \mathcal{U}$. Then by compactness, there

is a finite $F \subset S$ such that $S = \bigcup \{U(x): x \in F\}$. Since finite intersections of sets in \mathcal{U} are in \mathcal{U}, we have $\emptyset = \bigcap \{S \backslash U(x): x \in F\} \in \mathcal{U}$, a contradiction. So every ultrafilter converges. Conversely, if \mathcal{V} is an open cover without a finite subcover, let \mathcal{W} be the set of all complements of finite unions of sets in \mathcal{V}. It is easily seen that \mathcal{W} is a filter base. It is included in some filter and thus in some ultrafilter by Theorem 2.2.4. This ultrafilter does not converge. □

Given a topological space (S, \mathcal{T}), a subcollection $\mathcal{U} \subset \mathcal{T}$ is called a *subbase* for \mathcal{T} iff the collection of all finite intersections of sets in \mathcal{U} is a base for \mathcal{T}. In \mathbb{R}, for example, a subbase of the usual topology is given by the open half-lines $(-\infty, b) := \{x: x < b\}$ and $(a, \infty) := \{x: x > a\}$, which do not form a base. Intersecting one of the latter with one of the former gives (a, b), and such intervals form a base.

2.2.6. Theorem *For any set X and collection \mathcal{U} of subsets of X, there is a smallest topology \mathcal{T} including \mathcal{U}, and \mathcal{U} is a subbase of \mathcal{T}. Given a topology \mathcal{T} and $\mathcal{U} \subset \mathcal{T}$, \mathcal{U} is a subbase for \mathcal{T} iff \mathcal{T} is the smallest topology including \mathcal{U}.*

Proof. Let \mathcal{B} be the collection of all finite intersections of members of \mathcal{U}. One member of \mathcal{B} is the intersection of *no* members of \mathcal{U}, which in this case is (hereby) defined to be X. Let \mathcal{T} be the collection of all (arbitrary) unions of members of \mathcal{B}. It will be shown that \mathcal{T} is a topology and \mathcal{B} is a base for it. First, $X \in \mathcal{B}$ gives $X \in \mathcal{T}$, and \emptyset, as the empty union, is also in \mathcal{T}.

Clearly, any union of sets in \mathcal{T} is in \mathcal{T}. So the problem is to show that the intersection of any two sets V and W in \mathcal{T} is also in \mathcal{T}. Now, V is the union of a collection \mathcal{V} and W is the union of a collection \mathcal{W}. Each set in $\mathcal{V} \cup \mathcal{W}$ is a finite intersection of sets in \mathcal{U}. The intersection $V \cap W$ is the union of all intersections $A \cap B$ for $A \in \mathcal{V}$ and $B \in \mathcal{W}$. But an intersection of two finite intersections is a finite intersection, so each such $A \cap B$ is in \mathcal{B}. It follows that $V \cap W \in \mathcal{T}$, so \mathcal{T} is a topology. Then, clearly, \mathcal{B} is a base for it, and \mathcal{U} is a subbase.

Any topology that includes \mathcal{U} must include \mathcal{B}, and then must include \mathcal{T}, by definition of topology. So \mathcal{T} is the smallest topology including \mathcal{U}. The subbase \mathcal{U} determines the base \mathcal{B} and then the topology \mathcal{T} uniquely, so \mathcal{U} is a subbase for \mathcal{T} if and only if \mathcal{T} is the smallest topology including \mathcal{U}. □

2.2.7. Corollary *(a) If (S, \mathcal{V}) and (X, \mathcal{T}) are topological spaces, \mathcal{U} is a subbase of \mathcal{T}, and f is a function from S into X, then f is continuous if and only if $f^{-1}(U) \in \mathcal{V}$ for each $U \in \mathcal{U}$.*

(b) If S and I are any sets, and for each $i \in I$, f_i is a function from S into X_i, where (X_i, \mathcal{T}_i) is a topological space, then there is a smallest topology \mathcal{T} on S for which every f_i is continuous. Here a subbase of \mathcal{T} is given by $\{f_i^{-1}(U): i \in I, U \in \mathcal{T}_i\}$, and a base by finite intersections of such sets for different values of i, where each \mathcal{T}_i can be replaced by a subbase of itself.

Proof. (a) This essentially follows from the fact that inverse under a function, $B \mapsto f^{-1}(B)$, preserves the set operations of (arbitrary) unions and intersections. Specifically, to prove the "if" part (the converse being obvious), for any finite set U_1, \ldots, U_n of members of \mathcal{U},

$$f^{-1}\left(\bigcap_{1 \leq i \leq n} U_i \right) = \bigcap_{1 \leq i \leq n} f^{-1}(U_i) \in \mathcal{V},$$

so $f^{-1}(A) \in \mathcal{V}$ for each A in a base \mathcal{B} of \mathcal{T}. Then for each $W \in \mathcal{T}$, W is the union of some collection $\mathcal{W} \subset \mathcal{B}$. So

$$f^{-1}(W) = f^{-1}\left(\bigcup \mathcal{W} \right) = \bigcup \{f^{-1}(B): B \in \mathcal{W}\} \in \mathcal{T},$$

proving (a). Part (b), through the subbase statement, is clear from Theorem 2.2.6. When we take finite intersections of sets $f_i^{-1}(U_i)$ to get a base, if we had more than one U_i for one value of i—say we had U_{ij} for $j = 1, \ldots, k$— then the intersection of the sets $f_i^{-1}(U_{ij})$ for $j = 1, \ldots, k$ equals $f_i^{-1}(U_i)$, where U_i is the intersection of the U_{ij} for $j = 1, \ldots, k$. Or, if the U_{ij} all belong to a base \mathcal{B}_i of \mathcal{T}_i, then their intersection U_i is the union of a collection \mathcal{U}_i included in \mathcal{B}_i. The intersection of the $f_i^{-1}(U_i)$ for i in a finite set G is the union of all the intersections of the $f_i^{-1}(V_i)$ for $i \in G$, where $V_i \in \mathcal{U}_i$ for each $i \in G$, so we get a base as stated. \square

Corollary 2.2.7(a) can simplify the proof that a function is continuous. For example, if f has real values, then, using the subbase for the topology of \mathbb{R} mentioned above, it is enough to show that $f^{-1}((a, \infty))$ and $f^{-1}((-\infty, b))$ are open for any real a, b.

Let (X_i, \mathcal{T}_i) be topological spaces for all i in a set I. Let X be the Cartesian product $X := \Pi_{i \in I} X_i$, in other words, the set of all indexed families $\{x_i\}_{i \in I}$, where $x_i \in X_i$ for all i. Let p_i be the projection from X onto the ith coordinate space X_i: $p_i(\{x_j\}_{j \in I}) := x_i$ for any $i \in I$. Then letting $f_i = p_i$ in Corollary 2.2.7(b) gives a topology \mathcal{T} on X, called the *product topology*, the smallest topology making all the coordinate projections continuous.

Let $\mathbb{R}^k := \{x = (x_1, \ldots, x_k): x_j \in \mathbb{R} \text{ for all } j\}$ be the Cartesian product of k copies of \mathbb{R}, with product topology. The ordered k-tuple (x_1, \ldots, x_k) can be

defined as a function from $\{1, 2, \ldots, k\}$ into \mathbb{R}. We also write $x = \{x_j\}_{1 \le j \le k} = \{x_j\}_{j=1}^k$. The product topology on \mathbb{R}^k is metrized by the Euclidean distance (Problem 16). For any real $M > 0$, the interval $[-M, M]$ is compact by Theorem 2.2.1. The cube in \mathbb{R}^k, $[-M, M]^k := \{\{x_j\}_{j=1}^k : |x_j| \le M, j = 1, \ldots, k\}$ is compact for the product topology, as a special case of the following general theorem.

2.2.8. Theorem (*Tychonoff's Theorem*) *Let* (K_i, \mathcal{T}_i) *be compact topological spaces for each* i *in a set* I. *Then the Cartesian product* $\Pi_i K_i$ *with product topology is compact.*

Proof. Let \mathcal{U} be an ultrafilter in $\Pi_i K_i$. Then for all i, $p_i[[\mathcal{U}]]$ is an ultrafilter in K_i, since for each set $A \subset K_i$, either $p_i^{-1}(A)$ or its complement $p_i^{-1}(K_i \backslash A)$ is in \mathcal{U}. So by Theorem 2.2.5, $p_i[[\mathcal{U}]]$ converges to some $x_i \in K_i$. For any neighborhood U of $x := \{x_i\}_{i \in I}$, by definition of product topology, there is a finite set $F \subset I$ and $U_i \in \mathcal{T}$ for $i \in F$ such that $x \in \bigcap \{p_i^{-1}(U_i) : i \in F\} \subset U$. For each $i \in F$, $p_i^{-1}(U_i) \in \mathcal{U}$, so $U \in \mathcal{U}$ and $\mathcal{U} \to x$. So every ultrafilter converges and by Theorem 2.2.5 again, $\Pi_i K_i$ is compact. $\qquad\square$

One of the main reasons for considering ultrafilters was to get the last proof; other proofs of Tychonoff's theorem seem to be longer.

Among compact spaces, those which are Hausdorff spaces have especially good properties and are the most studied. (A subset of a Hausdorff space with relative topology is clearly also Hausdorff.) Here is one advantage of the combined properties:

2.2.9. Proposition *Any compact set* K *in a Hausdorff space is closed.*

Proof. For any $x \in K$ and $y \notin K$ take open $U(x, y)$ and $V(x, y)$ with $x \in U(x, y)$, $y \in V(x, y)$, and $U(x, y) \cap V(x, y) = \emptyset$. For each fixed y, the set of all $U(x, y)$ forms an open cover of K with a finite subcover. The intersection of the corresponding finitely many $V(x, y)$ gives an open neighborhood $W(y)$ of y, where $W(y)$ is disjoint from K. The union of all such $W(y)$ is the complement $X \backslash K$ and is open. $\qquad\square$

On any set S, the *indiscrete* topology is the smallest topology, $\{\emptyset, S\}$. All subsets of S are compact, but only \emptyset and S are closed. This is the reverse of the usual situation in Hausdorff spaces.

If f is a function from X into Y and g a function from Y into Z, let $(g \circ f)(x) := g(f(x))$ for all $x \in X$. Then $g \circ f$ is a function from X into

Z, called the *composition* of g and f. For any set $A \subset Z, (g \circ f)^{-1}(A) = f^{-1}(g^{-1}(A))$. Thus we have:

2.2.10. Theorem *If (X, S), (Y, T), and (Z, U) are topological spaces, f is continuous from X into Y, and g is continuous from Y into Z, then $g \circ f$ is continuous from X into Z.*

Continuity of a composition of two continuous functions is also clear from the formulation of continuity in terms of convergent nets (Theorem 2.1.2): if $x_i \to x$, then $f(x_i) \to f(x)$, so $g(f(x_i)) \to g(f(x))$.

If (X, S) and (Y, T) are topological spaces, a *homeomorphism* of X onto Y is a 1–1 function f from X onto Y such that f and f^{-1} are continuous. If such an f exists, (X, S) and (Y, T) are called *homeomorphic*. For example, a finite, non-empty open interval (a, b) is homeomorphic to $(0, 1)$ by a linear transformation: let $f(x) := a + (b - a)x$. A bit more surprisingly, $(-1, 1)$ is homeomorphic to all of \mathbb{R}, letting $f(x) := \tan(\pi x / 2)$.

In general, if $f \circ h$ is continuous and h is continuous, f is not necessarily continuous. For example, h and so $f \circ h$ could be constants while f was an arbitrary function. Or, if T is the discrete topology 2^X on a set X, h is a function from X into a topological space Y, and f is a function from Y into another topological space, then h and $f \circ h$ are always continuous, but f need not be; in fact, it can be arbitrary. In the following situation, however, continuity of f will follow, providing another instance of how "compact" and "Hausdorff" work well together.

2.2.11. Theorem *Let h be a continuous function from a compact topological space T onto a Hausdorff topological space K. Then a set $A \subset K$ is open if and only if $h^{-1}(A)$ is open in T. If f is a function from K into another topological space S, then f is continuous if and only if $f \circ h$ is continuous. If h is 1–1, it is a homeomorphism.*

Proof. Note that K is compact by Theorem 2.2.3. Let $h^{-1}(A)$ be open. Then $T \backslash h^{-1}(A)$ is closed and hence compact by Theorem 2.2.2. Thus $h[T \backslash h^{-1}(A)] = K \backslash A$ is compact by Theorem 2.2.3, hence closed by Proposition 2.2.9, so A is open. If $f \circ h$ is continuous, then for any open $U \subset S, (f \circ h)^{-1}(U) = h^{-1}(f^{-1}(U))$ is open. So $f^{-1}(U)$ is open and f is continuous. The other implications are immediate from the definitions and Theorem 2.2.10. □

The power set 2^X, which is the collection of all subsets of a set X, can be viewed via indicator functions as the set of all functions from X into

$\{0, 1\}$. In other words, 2^X is a Cartesian product, indexed by X, of copies of $\{0, 1\}$. With the usual discrete topology on $\{0, 1\}$, the product topology on 2^X is compact.

The following somewhat special fact will not be needed until Chapters 12 and 13. Here $f(V) := \{f(x): x \in V\}$ and a bar denotes closure. $F_n \downarrow K$ means $F_n \supset F_{n+1}$ for all $n \in \mathbb{N}$ and $\bigcap_n F_n = K$.

***2.2.12. Theorem** *Let X and Y be Hausdorff topological spaces and f a continuous function from X into Y. Let F_n be closed sets in X with $F_n \downarrow K$ as $n \to \infty$ where K is compact. Suppose that either (a) for every open $U \supset K$, there is an n with $F_n \subset U$, or (b) F_1 is, and so all the F_n are, compact. Then*

$$f(K) = \bigcap_n f(F_n) = \bigcap_n f(F_n)^-.$$

Proof. Clearly,

$$f(K) \subset \bigcap_n f(F_n) \subset \bigcap_n f(F_n)^-.$$

For the converse, first assume (a). Take any $y \in \bigcap_n f(F_n)^-$. Suppose every x in K has an open neighborhood V_x with $y \notin f(V_x)^-$. Then the V_x form an open cover of K, having a finite subcover. The union of the V_x in the subcover gives an open set $U \supset K$ with $y \notin f(U)^-$, a contradiction since $F_n \subset U$ for n large. So take $x \in K$ with $y \in f(V)$ for every open V containing x. If $f(x) \neq y$, then take disjoint open neighborhoods W of $f(x)$ and T of y. Let $V = f^{-1}(W)$ to get a contradiction. So $f(x) = y$ and $y \in f(K)$, completing the chain of inclusions, finishing the (a) part.

Now, showing that (b) implies (a) will finish the proof. The sets F_n are all compact since they are closed subsets of the compact set F_1. Let $U \supset K$ where U is open. Then $F_n \backslash U$ is a decreasing sequence of compact sets with empty intersection, so for some n, $F_n \backslash U$ is empty (otherwise, U and the complements of the F_j would form an open cover of F_1 without a finite subcover), so $F_n \subset U$. □

Problems

1. If S_i are sets with discrete topologies, show that the product topology for finitely many such spaces is also discrete.

2. If there are infinitely many discrete S_i, each having more than one point, show that their product topology is not discrete.

3. Show that the product of countably many separable topological spaces, with product topology, is separable.

4. If (X, \mathcal{T}) and (Y, \mathcal{U}) are topological spaces, \mathcal{A} is a base for \mathcal{T} and \mathcal{B} is a base for \mathcal{U}, show that the collection of all sets $A \times B$ for $A \in \mathcal{A}$ and $B \in \mathcal{B}$ is a base for the product topology on $X \times Y$.

5. (a) Prove that any intersection of topologies on a set is a topology.
 (b) Prove that for any collection \mathcal{U} of subsets of a set X, there is a smallest topology on X including \mathcal{U}, using part (a) (rather than subbases).

6. (a) Let A_n be the set of all integers greater than n. Let \mathcal{B}_n be the collection of all subsets of $\{1, \ldots, n\}$. Let \mathcal{T}_n be the collection of sets of positive integers that are either in \mathcal{B}_n or of the form $A_n \cup B$ for some $B \in \mathcal{B}_n$. Prove that \mathcal{T}_n is a topology.
 (b) Show that \mathcal{T}_n for $n = 1, 2, \ldots$, is an inclusion-chain of topologies whose union is not a topology.
 (c) Describe the smallest topology which includes \mathcal{T}_n for all n.

7. Given a product $X = \Pi_{i \in I} X_i$ of topological spaces (X_i, \mathcal{T}), with product topology, and a directed set J, a net in X indexed by J is given by a doubly indexed family $\{x_{ji}\}_{j \in J, i \in I}$. Show that such a net converges for the product topology if and only if for every $i \in I$, the net $\{x_{ji}\}_{j \in J}$ converges in X_i for \mathcal{T}_i. (For this reason, the product topology is sometimes called the topology of "pointwise convergence": for each j, we have a function $i \mapsto x_{ij}$ on I, and convergence for the product topology is equivalent to convergence at each "point" $i \in I$. This situation comes up especially when the X_i are [copies of] the same space, such as \mathbb{R} with its usual topology. Then $\Pi_i X_i$ is the set of all functions from I into \mathbb{R}, often called \mathbb{R}^I.)

8. Let $I := [0, 1]$ with usual topology. Let I^I be the set of all functions from I into I with product topology.
 (a) Show that I^I is separable. *Hint:* Consider functions that are finite sums $\sum a_i 1_{J(i)}$ where the a_i are rational and the $J(i)$ are intervals with rational endpoints.
 (b) Show that I^I has a subset which is not separable with the relative topology.

9. (a) For any partially ordered set $(X, <)$, the collection $\{X\} \cup \{\{x : x < y\} : y \in X\} \cup \{\{x : x > z\} : z \in X\}$ is a subbase for a topology on X called the *interval topology*. For the usual linear ordering of the real numbers, show that the interval topology is the usual topology.
 (b) Assuming the axiom of choice, there is an uncountable well-ordered set (X, \leq). Show that there is such a set containing exactly one

element x such that $y < x$ for uncountably many values of y. Let $f(x) = 1$ and $f(y) = 0$ for all other values of $y \in X$. For the interval topology on X, show that f is not continuous, but for every sequence $u_n \to u$ in X, $f(u_n)$ converges to $f(u)$.

10. Let f be a bounded, real-valued function defined on a set X: for some $M < \infty$, $f[X] \subset [-M, M]$. Let \mathcal{U} be an ultrafilter in X. Show that $\{f[A]: A \in \mathcal{U}\}$ is a converging filter base.

11. What happens in Problem 10 if f is unbounded?

12. Show that Theorem 2.2.12 can fail without the hypothesis "for every open $U \supset K$, there is an n with $F_n \subset U$." *Hint:* Let $F_n = [n, \infty)$.

13. Show that Theorem 2.2.12 can fail, for an intersection of just two compact sets F_j, if neither is included in the other.

14. A topological space (S, \mathcal{T}) is called *connected* if S is not the union of two disjoint non-empty open sets.
 (a) Prove that if S is connected and f is a continuous function from S onto T, then T is also connected.
 (b) Prove that for any $a < b$ in \mathbb{R}, $[a, b]$ is connected. *Hint:* Suppose $[a, b] = U \cup V$ for disjoint, non-empty, relatively open sets U and V. Suppose $c \in U$ and $d \in V$ with $c < d$. Let $t := \sup(U \cap [c, d])$. Then $t \in U$ or $t \in V$ gives a contradiction.
 (c) If $S \subset \mathbb{R}$ is connected and $c < d$ are in S, show that $[c, d] \subset S$. *Hint:* Suppose $c < t < d$ and $t \notin S$. Consider $(-\infty, t) \cap S$ and $(t, \infty) \cap S$.
 (d) (Intermediate value theorem) Let $a < b$ in \mathbb{R} and let f be continuous from $[a, b]$ into \mathbb{R}. Show that f takes all values between $f(a)$ and $f(b)$. *Hint:* Apply parts (a), (b) and (c).

15. For x and y in \mathbb{R}^k, the *dot product* or *inner product* is defined by $x \cdot y := (x, y) := \sum_{j=1}^k x_j y_j$. The *length* of x is defined by $|x| := (x, x)^{1/2}$.
 (a) (Cauchy's inequality). Show that for any $x, y \in \mathbb{R}^k$, $(x, y)^2 \le |x|^2 |y|^2$. *Hint:* the quadratic $q(t) := |x + ty|^2$ must not have two distinct real roots.
 (b) Show that for any $x, y \in \mathbb{R}^k$, $|x + y| \le |x| + |y|$.
 (c) For $x, y \in \mathbb{R}^k$ let $d(x, y) := |x - y|$. Show that d is a metric on \mathbb{R}^k. It is called the *usual* or *Euclidean* metric.

16. Let d be as in the previous problem.
 (a) Show that d metrizes the product topology on \mathbb{R}^k. *Hint:* Show that any open ball $B(x, r)$ includes a product of open intervals $(x_i - u, x_i + u)$ for some $u > 0$, and conversely.
 (b) Show that any closed set F in \mathbb{R}^k, bounded (for d), meaning that

sup$\{d(x, y): x, y \in F\} < \infty$, is compact. *Hint:* It is a subset of a
product of closed intervals.

17. A topological space (S, \mathcal{T}) is called T_1 iff all singletons $\{x\}$, $x \in S$, are
closed. Let S be any set.
 (a) Show that the empty set and the collection of all complements of finite
 sets form a T_1 topology \mathcal{T} on S in which all subsets are compact.
 (b) If S is infinite with the topology in part (a), show that there exists a
 sequence of non-empty compact subsets $K_1 \supset K_2 \supset \cdots \supset K_n \supset \cdots$
 such that $\bigcap_{n=1}^{\infty} K_n = \emptyset$.
 (c) Show that the situation in part (b) cannot occur in a Hausdorff space.
 Hint: Use Proposition 2.2.9.

18. A real-valued function f on a topological space S is called *upper semi-
 continuous* iff for each $a \in \mathbb{R}$, $f^{-1}([a, \infty))$ is closed, or *lower semicon-
 tinuous* iff $-f$ is upper semicontinuous.
 (a) Show that f is upper semicontinuous if and only if for all $x \in S$,

 $$f(x) \geq \limsup_{y \to x} f(y) := \inf\{\sup\{f(y): y \in U, y \neq x\}: x \in U \text{ open}\},$$

 where $\sup \emptyset := -\infty$.
 (b) Show that f is continuous if and only if it is both upper and lower
 semicontinuous.
 (c) If f is upper semicontinuous on a compact space S, show that for
 some $t \in S$, $f(t) = \sup f := \sup\{f(x): x \in S\}$. *Hint:* Let $a_n \in \mathbb{R}$,
 $a_n \uparrow \sup f$. Consider $f^{-1}((-\infty, a_n))$, $n = 1, 2, \ldots$.

2.3. Complete and Compact Metric Spaces

A sequence $\{x_n\}$ in a space S with a (pseudo)metric d is called a *Cauchy
sequence* if $\lim_{n \to \infty} \sup_{m \geq n} d(x_m, x_n) = 0$. The pseudometric space (S, d)
is called *complete* iff every Cauchy sequence in it converges. A point x in
a topological space is called a *limit point* of a set E iff every neighborhood
of x contains points of E other than x. Recall that for any sequence $\{x_n\}$ a
subsequence is a sequence $k \mapsto x_{n(k)}$ where $k \mapsto n(k)$ is a strictly increasing
function from $\mathbb{N} \setminus \{0\}$ into itself. (Some authors require only that $n(k) \to \infty$ as
$k \to +\infty$.) As an example of a compact metric space, first consider the interval
$[0, 1]$. Every number x in $[0, 1]$ has a decimal expansion $x = 0.d_1 d_2 d_3 \ldots$,
meaning, as usual, $x = \sum_{j \geq 1} d_j / 10^j$. Here each $d_j = d_j(x)$ is an integer
and $0 \leq d_j \leq 9$ for all j. If a number x has an expansion with $d_j = 9$ for
all $j > m$ and $d_m < 9$ for some m, then the numbers d_j are not uniquely

determined, and

$$0.d_1d_2d_3 \ldots d_{m-1}d_m 9999 \ldots = 0.d_1d_2d_3 \ldots d_{m-1}(d_m + 1)0000. \ldots$$

In all other cases, the digits d_j are unique given x.

In practice, we work with only the first few digits of decimal expansions. For example, we use $\pi = 3.14$ or 3.1416 and very rarely need to know that $\pi = 3.14159265358979 \ldots$. This illustrates a very important property of numbers in $[0, 1]$: given any prescribed accuracy (specifically, given any $\varepsilon > 0$), there is a finite set F of numbers in $[0, 1]$ such that every number x in $[0, 1]$ can be represented by a number y in F to the desired accuracy, that is, $|x - y| < \varepsilon$. In fact, there is some n such that $1/10^n < \varepsilon$ and then we can let F be the set of all finite decimal expansions with n digits. There are exactly 10^n of these. For any x in $[0, 1]$ we have $|x - 0.x_1x_2 \ldots x_n| \leq 1/10^n < \varepsilon$ and $0.x_1x_2 \ldots x_n \in F$.

The above property extends to metric spaces as follows.

Definition. A metric space (S, d) is called *totally bounded* iff for every $\varepsilon > 0$ there is a finite set $F \subset S$ such that for every $x \in S$, there is some $y \in F$ with $d(x, y) < \varepsilon$.

Another convenient property of the decimal expansions of real numbers in $[0, 1]$ is that for any sequence x_1, x_2, x_3, \ldots of integers from 0 to 9, there is some real number $x \in [0, 1]$ such that $x = 0.x_1x_2x_3 \ldots$. In other words, the special Cauchy sequence $0.x_1, 0.x_1x_2, 0.x_1x_2x_3, \ldots, 0.x_1x_2x_3 \ldots x_n, \ldots$ actually converges to some limit x. This property of $[0, 1]$ is an example of completeness of a metric space (of course, not all Cauchy sequences in $[0, 1]$ are of the special type just indicated).

Now, here are some useful general characterizations of compact metric spaces.

2.3.1. Theorem *For any metric space (S, d), the following properties are equivalent (any one implies the other three):*

 (I) *(S, d) is compact: every open cover has a finite subcover.*
 (II) *(S, d) is both complete and totally bounded.*
 (III) *Every infinite subset of S has a limit point.*
 (IV) *Every sequence of points of S has a convergent subsequence.*

Proof. (I) implies (II): let (S, d) be compact. Given $r > 0$ and $x \in S$, recall that $B(x, r) := \{y \in S : d(x, y) < r\}$. Then for each r, the set of all such

neighborhoods, $\{B(x, r): x \in S\}$, is an open cover and must have a finite subcover. Thus (S, d) is totally bounded. Now let $\{x_n\}$ be any Cauchy sequence in S. Then for every positive integer m, there is some $n(m)$ such that $d(x_n, x_{n(m)}) < 1/m$ whenever $n > n(m)$. Let $U_m = \{x: d(x, x_{n(m)}) > 1/m\}$. Then U_m is an open set. (If $y \in U_m$ and $r := d(x_{n(m)}, y) - 1/m$, then $r > 0$ and $B(y, r) \subset U_m$.) Now $x_n \notin U_m$ for $n > n(m)$ by definition of $n(m)$. Thus $x_k \notin \bigcup\{U_m: 1 \le m < s\}$ if $k > \max\{n(m): m < s\}$. Since the U_m do not have a finite subcover, they cannot form an open cover of S. So there is some x with $x \notin U_m$ for all m. Thus $d(x, x_{n(m)}) \le 1/m$ for all m. Then by the triangle inequality, $d(x, x_n) < 2/m$ for $n > n(m)$. So $\lim_{n\to\infty} d(x, x_n) = 0$, and the sequence $\{x_n\}$ converges to x. Thus (S, d) is complete as well as totally bounded, and (I) does imply (II).

Next, assume (II) and let's prove (III). For each $n = 1, 2, \ldots$, let F_n be a finite subset of S such that for every $x \in S$, we have $d(x, y) < 1/n$ for some $y \in F_n$. Let A be any infinite subset of S. (If S is finite, then by the usual logic we say that (III) does hold.) Since the finitely many neighborhoods $B(y, 1)$ for $y \in F_1$ cover S, there must be some $x_1 \in F_1$ such that $A \cap B(x_1, 1)$ is infinite. Inductively, we choose $x_n \in F_n$ for all n such that $A \cap \bigcap\{B(x_m, 1/m): m = 1, \ldots, n\}$ is infinite for all positive integers n. This implies that $d(x_m, x_n) < 1/m + 1/n < 2/m$ when $m < n$ (there is some $y \in B(x_m, 1/m) \cap B(x_n, 1/n)$, and $d(x_m, x_n) < d(x_m, y) + d(x_n, y)$). Thus $\{x_n\}$ is a Cauchy sequence. Since (S, d) is complete, this sequence converges to some $x \in S$, and $d(x_n, x) < 2/n$ for all n. Thus $B(x, 3/n)$ includes $B(x_n, 1/n)$, which includes an infinite subset of A. Since $3/n \to 0$ as $n \to \infty$, x is a limit point of A. So (II) does imply (III).

Now assume (III). If $\{x_n\}$ is a sequence with infinite range, let x be a limit point of the range. Then there are $n(1) < n(2) < n(3) < \cdots$ such that $d(x_{n(k)}, x) < 1/k$ for all k, so $x_{n(k)}$ converges to x as $k \to \infty$. If $\{x_n\}$ has finite range, then there is some x such that $x_n = x$ for infinitely many values of n. Thus there is a subsequence $x_{n(k)}$ with $x_{n(k)} = x$ for all k, so $x_{n(k)} \to x$. Thus (III) implies (IV).

Last, let's prove that (IV) implies (I). Let \mathcal{U} be an open cover of S. For any $x \in S$, let

$$f(x) := \sup\{r: B(x, r) \subset U \text{ for some } U \in \mathcal{U}\}.$$

Then $f(x) > 0$ for every $x \in S$. A stronger fact will help:

2.3.2. Lemma $\mathrm{Inf}\{f(x): x \in S\} > 0$.

Proof. If not, there is a sequence $\{x_n\}$ in S such that $f(x_n) < 1/n$ for $n = 1, 2, \ldots$. Let $x_{n(k)}$ be a subsequence converging to some $x \in S$. Then for

some $U \in \mathcal{U}$ and $r > 0$, $B(x, r) \subset U$. Then for k large enough so that $d(x_{n(k)}, x) < r/2$, we have $f(x_{n(k)}) > r/2$, a contradiction for large k. □

Now continuing the proof that (IV) implies (I), let $c := \min(2, \inf\{f(x): x \in S\}) > 0$. Choose any $x_1 \in S$. Recursively, given x_1, \ldots, x_n, choose x_{n+1} if possible so that $d(x_{n+1}, x_j) > c/2$ for all $j = 1, \ldots, n$. If this were possible for all n, we would get a sequence $\{x_n\}$ with $d(x_m, x_n) > c/2$ whenever $m \neq n$. Such a sequence has no Cauchy subsequence and hence no convergent subsequence. So there is a finite n such that $S = \bigcup_{j \leq n} B(x_j, c/2)$. By the definitions of f and c, for each $j = 1, \ldots, n$ there is a $U_j \in \mathcal{U}$ such that $B(x_j, c/2) \subset U_j$. Then the union of these U_j is S, and \mathcal{U} has a finite subcover, finishing the proof of Theorem 2.3.1. □

For any metric space (S, d) and $A \subset S$, the *diameter* of A is defined as $\mathrm{diam}(A) := \sup\{d(x, y): x \in A, y \in A\}$. Then A is called *bounded* iff its diameter is finite.

Example. Let S be any infinite set. For $x \neq y$ in S, let $d(x, y) = 1$, and $d(x, x) = 0$. Then S is complete and bounded, but not totally bounded. The characterization of compact sets in Euclidean spaces as closed bounded sets thus does not extend to general complete metric spaces.

Totally bounded metric spaces can be compared as to how totally bounded they are in terms of the following quantities. Let (S, d) be a totally bounded metric space. Given $\varepsilon > 0$, let $N(\varepsilon, S)$ be the smallest n such that $S = \bigcup_{1 \leq i \leq n} A_i$ for some sets A_i with $\mathrm{diam}(A_i) \leq 2\varepsilon$ for $i = 1, \ldots, n$. Let $D(\varepsilon, S)$ be the largest number m of points x_i, $i = 1, \ldots, m$, such that $d(x_i, x_j) > \varepsilon$ whenever $i \neq j$.

Problems

1. Show that for any metric space (S, d) and $\varepsilon > 0$, $N(\varepsilon, S) \leq D(\varepsilon, S) \leq N(\varepsilon/2, S)$.

2. Let (S, d) be the unit interval $[0, 1]$ with the usual metric. Evaluate $N(\varepsilon, S)$ and $D(\varepsilon, S)$ for all $\varepsilon > 0$. *Hint:* Use the "ceiling function" $\lceil x \rceil :=$ least integer $\geq x$.

3. If S is the unit square $[0, 1] \times [0, 1]$ with the usual metric on \mathbb{R}^2, show that for some constant K, $N(\varepsilon, S) \leq K/\varepsilon^2$ for $0 < \varepsilon < 1$.

4. Give an open cover of the open unit square $(0, 1) \times (0, 1)$ which does not have a finite subcover.

5. Prove that any open cover of a separable metric space has a countable subcover *Hint:* Use Proposition 2.1.4.

6. Prove that a metric space (S, d) is compact if and only if every countable filter base is included in a convergent one.

7. For the covering of [0, 1] by intervals $(j/n, (j + 2)/n)$, $j = -1, 0, 1, \ldots, n - 1$, evaluate the infimum in Lemma 2.3.2.

8. Let (S, d) be a noncompact metric space, so that there is an infinite set A without a limit point. Show that the relative topology on A is discrete.

9. Show that a set with discrete relative topology may have a limit point.

10. A point x in a topological space is called *isolated* iff $\{x\}$ is open. A compact topological space is called *perfect* iff it has no isolated points. Show that:

 (a) Any compact metric space is a union of a countable set and a perfect set. *Hint:* Consider the set of points having a countable open neighborhood. Use Problem 5.

 (b) If (K, d) is perfect, then every non-empty open subset of K is uncountable.

11. Let $\{x_i, i \in I\}$ be a net where I is a directed set. For $J \subset I$, $\{x_i, i \in J\}$ will be called a *strict subnet* of $\{x_i, i \in I\}$ if J is *cofinal* in I, that is, for all $i \in I$, $i \le j$ for some $j \in J$.

 (a) Show that this implies J is a directed set with the ordering of I.

 (b) Show that in [0, 1] with its usual topology there exists a net having no convergent strict subnet (in contrast to Theorems 2.2.5 and 2.3.1). *Hint:* Let W be a well-ordering of [0, 1]. Let I be the set of all $y \in [0, 1]$ such that $\{t : t W y\}$ is countable. Show that I is uncountable and well-ordered by W. Let $x_y := y$ for all $y \in I$. Show that $\{x_y : y \in I\}$ has no convergent strict subnet.

 Compactness can be characterized in terms of convergent subnets (e.g. Kelley, 1955, Theorem 5.2), but only for nonstrict subnets; see also Kelley (1955, p. 70 and Problem 2.E).

2.4. Some Metrics for Function Spaces

First, here are three rather simple facts:

2.4.1. Proposition *For any metric space (S, d), if $\{x_n\}$ is a Cauchy sequence, then it is bounded (that is, its range is bounded). If it has a convergent*

subsequence $x_{n(k)} \to x$, *then* $x_n \to x$. *Any closed subset of a complete metric space is complete.*

Proof. If $d(x_m, x_n) < 1$ for $m > n$, then for all $m, d(x_m, x_n) < 1 + \max\{d(x_j, x_n): j < n\} < \infty$, so the sequence is bounded.

If $x_{n(k)} \to x$, then given $\varepsilon > 0$, take m such that if $n > m$, then $d(x_n, x_m) < \varepsilon/3$, and take k such that $n(k) > m$ and $d(x_{n(k)}, x) < \varepsilon/3$. Then $d(x_n, x) < d(x_n, x_m) + d(x_m, x_{n(k)}) + d(x_{n(k)}, x) < \varepsilon/3 + \varepsilon/3 + \varepsilon/3 = \varepsilon$, so $x_n \to x$. From Theorem 2.1.3(b) and (d), a closed subset of a complete space is complete. \square

A closed subset F of a noncomplete metric space X, for example $F = X$, is of course not necessarily complete. Here is a classic case of completeness:

2.4.2. Proposition \mathbb{R} *with its usual metric is complete.*

Proof. Let $\{x_n\}$ be a Cauchy sequence. By Proposition 2.4.1, it is bounded and thus included in some finite interval $[-M, M]$. This interval is compact (Theorem 2.2.1). Thus $\{x_n\}$ has a convergent subsequence (Theorem 2.3.1), so $\{x_n\}$ converges by Proposition 2.4.1. \square

Let (S, d) and (T, e) be any two metric spaces. It is easy to see that a function f from S into T is continuous if and only if for all $x \in S$ and $\varepsilon > 0$ there is a $\delta > 0$ such that whenever $d(x, y) < \delta$, we have $e(f(x), f(y)) < \varepsilon$. If this holds for a fixed x, we say f is *continuous at* x. If for every $\varepsilon > 0$ there is a $\delta > 0$ such that $d(x, y) < \delta$ implies $e(f(x), f(y)) < \varepsilon$ for all x and y in S, then f is said to be *uniformly continuous* from (S, d) to (T, e).

For example, the function $f(x) = x^2$ from \mathbb{R} into itself is continuous but not uniformly continuous (for a given ε, as $|x|$ gets larger, δ must get smaller).

Before taking countable Cartesian products it is useful to make metrics bounded, which can be done as follows. Here $[0, \infty) := \{x \in \mathbb{R}: x \geq 0\}$.

2.4.3. Proposition *Let f be any continuous function from $[0, \infty)$ into itself such that*

(1) f *is nondecreasing:* $f(x) \leq f(y)$ *whenever* $x \leq y$,
(2) f *is subadditive:* $f(x + y) \leq f(x) + f(y)$ *for all* $x \geq 0$ *and* $y \geq 0$, *and*
(3) $f(x) = 0$ *if and only if* $x = 0$.

Then for any metric space (S, d), $f \circ d$ is a metric, and the identity function $g(s) \equiv s$ from S to itself is uniformly continuous from (S, d) to $(S, f \circ d)$ and from $(S, f \circ d)$ to (S, d).

Proof. Clearly $0 \le f(d(x, y)) = f(d(y, x))$, which is 0 if and only if $d(x, y) = 0$, for all x and y in S. For the triangle inequality, $f(d(x, z)) \le f(d(x, y) + d(y, z)) \le f(d(x, y)) + f(d(y, z))$, so $f \circ d$ is a metric. Since $f(t) > 0$ for all $t > 0$, and f is continuous and nondecreasing, we have for every $\varepsilon > 0$ a $\delta > 0$ such that $f(t) < \varepsilon$ if $t < \delta$, and $t < \varepsilon$ if $f(t) < \lambda := f(\varepsilon)$. Thus we have uniform continuity in both directions. □

Suppose $f''(x) < 0$ for $x > 0$. Then f' is decreasing, so for any $x, y > 0$,

$$f(x + y) - f(x) = \int_0^y f'(x + t)\, dt < \int_0^y f'(t)\, dt = f(y) - f(0).$$

Thus if $f(0) = 0$, f is subadditive. There are bounded functions f satisfying the conditions of Proposition 2.4.3; for example, $f(x) := x/(1+x)$ or $f(x) := $ arc tan x.

2.4.4. Proposition *For any sequence (S_n, d_n) of metric spaces, $n = 1$, $2, \ldots$, the product $S := \Pi_n S_n$ with product topology is metrizable, by the metric $d(\{x_n\}, \{y_n\}) := \sum_n f(d_n(x_n, y_n))/2^n$, where $f(t) := t/(1 + t)$, $t > 0$.*

Proof. First, $f(x) = 1 - 1/(1 + x)$, so f is nondecreasing and $f''(x) = -2/(1 + x)^3$. Thus f satisfies all three conditions of Proposition 2.4.3, so $f \circ d_n$ is a metric on S_n for each n. To show that d is a metric, first let $e_n(x, y) := f(d_n(x_n, y_n))/2^n$. Then e_n is a pseudometric on S for each n. Since $f < 1$, $d(x, y) = \sum_n e_n(x, y) < 1$ for all x and y. Clearly, d is non-negative and symmetric. For any x, y, and z in S, $d(x, z) = \sum_n e_n(x, z) \le \sum_n e_n(x, y) + e_n(y, z) = d(x, y) + d(y, z)$ (on rearranging sums of nonneg-ative terms, see Appendix D). Thus d is a pseudometric. If $x \ne y$, then for some n, $x_n \ne y_n$, so $d_n(x_n, y_n) > 0$, $f(d_n(x_n, y_n)) > 0$, and $d(x, y) > 0$. So d is a metric. For any $x = \{x_n\} \in S$, the product topology has a neighborhood-base at x consisting of all sets $N(x, \delta, m) := \{y : d_j(x_j, y_j) < \delta$ for all $j = 1, \ldots, m\}$, for $\delta > 0$ and $m = 1, 2, \ldots$. Given $\varepsilon > 0$, for n large enough, $2^{-n} < \varepsilon/2$. Then since $(\sum_{1 \le j \le n} 2^{-j})\varepsilon/2 < \varepsilon/2$, noting that $f(x) < x$ for all x, and $\sum_{j > n} 2^{-j} < \varepsilon/2$, we have $N(x, \varepsilon/2, n) \subset B(x, \varepsilon) := \{y : d(x, y) < \varepsilon\}$ for each n.

Conversely, suppose given $0 < \delta < 1$ and n. Since $f(1) = 1/2$, $f(x) < 1/2$ implies $x < 1$. Then $x = (1+x)f(x) < 2f(x)$. Let $\gamma := 2^{-n-1}\delta$. If $d(x, y) < \gamma$, then for $j = 1, \ldots, n$, $f(d_j(x_j, y_j)) < 2^j\gamma < 1/2$, so $d_j(x_j, y_j) < 2^{j+1}\gamma \le \delta$. So we have $B(x, \gamma) \subset N(x, \delta, n)$. Thus neighborhoods of x for d are the same as for the product topology, so d metrizes the topology. $\qquad\square$

A product of uncountably many metric spaces (each with more than one point) is not metrizable. Consider, for example, a product of copies of $\{0, 1\}$ over an uncountable index set I; in other words, the set of all indicators of subsets of I. Let the finite subsets F of I be directed by inclusion. Then the net 1_F, for all finite F, converges to 1 for the product topology, but no sequence $1_{F(n)}$ of indicator functions of finite sets can converge to 1, since the union of the $F(n)$, being countable, is not all of I.

So, to get metrizable spaces of real functions on possibly uncountable sets, one needs to restrict the space of functions and/or consider a topology other than the product topology. Here is one space of functions: for any compact topological space K let $C(K)$ be the space of all continuous real-valued functions on K. For $f \in C(K)$, we have $\sup|f| := \sup\{|f(x)|: x \in K\} < \infty$ since $f[K]$ is compact in \mathbb{R}, by Theorem 2.2.3. It is easily seen that $d_{\sup}(f, g) := \sup|f - g|$ is a metric on $C(K)$.

A collection \mathcal{F} of continuous functions from a topological space S into X, where (X, d) is a metric space, is called *equicontinuous at* $x \in S$ iff for every $\varepsilon > 0$ there is a neighborhood U of x such that $d(f(x), f(y)) < \varepsilon$ for all $y \in U$ and all $f \in \mathcal{F}$. (Here U does not depend on f.) \mathcal{F} is called *equicontinuous* iff it is equicontinuous at every $x \in S$. If (S, e) is a metric space, and for every $\varepsilon > 0$ there is a $\delta > 0$ such that $e(x, y) < \delta$ implies $d(f(x), f(y)) < \varepsilon$ for all x and y in S and all f in \mathcal{F}, then \mathcal{F} is called *uniformly equicontinuous*. In terms of these notions, here is an extension of a better-known fact (Corollary 2.4.6 below):

2.4.5. Theorem *If (K, d) is a compact metric space and (Y, e) a metric space, then any equicontinuous family of functions from K into Y is uniformly equicontinuous.*

Proof. If not, there exist $\varepsilon > 0, x_n \in K, u_n \in K$, and $f_n \in \mathcal{F}$ such that $d(u_n, x_n) < 1/n$ and $e(f_n(u_n), f_n(x_n)) > \varepsilon$ for all n. Then since any sequence in K has a convergent subsequence (Theorem 2.3.1), we may assume $x_n \to x$ for some $x \in K$, so $u_n \to x$. By equicontinuity at x, for n large enough, $e(f_n(u_n), f_n(x)) < \varepsilon/2$ and $e(f_n(x_n), f_n(x)) < \varepsilon/2$, so $e(f_n(u_n), f_n(x_n)) < \varepsilon$, a contradiction. $\qquad\square$

2.4.6. Corollary *A continuous function from a compact metric space to any metric space is uniformly continuous.*

A collection \mathcal{F} of functions on a set X into \mathbb{R} is called *uniformly bounded* iff $\sup\{|f(x)|: f \in \mathcal{F}, x \in X\} < \infty$. On any collection of bounded real functions, just as on $C(K)$ for K compact, let $d_{\sup}(f, g) := \sup|f - g|$. Then d_{\sup} is a metric.

The sequence of functions $f_n(t) := t^n$ on $[0, 1]$ consists of continuous functions, and the sequence is uniformly bounded: $\sup_n \sup_t |f_n(t)| = 1$. Then $\{f_n\}$ is not equicontinuous at 1, so not totally bounded for d_{\sup}, by the following classic characterization:

2.4.7. Theorem (Arzelà-Ascoli) *Let (K, e) be a compact metric space and $\mathcal{F} \subset C(K)$. Then \mathcal{F} is totally bounded for d_{\sup} if and only if it is uniformly bounded and equicontinuous, thus uniformly equicontinuous.*

Proof. If \mathcal{F} is totally bounded and $\varepsilon > 0$, take $f_1, \ldots, f_n \in \mathcal{F}$ such that for all $f \in \mathcal{F}$, $\sup|f - f_j| < \varepsilon/3$ for some j. Each f_j is uniformly continuous (by Corollary 2.4.6). Thus the finite set $\{f_1, \ldots, f_n\}$ is uniformly equicontinuous. Take $\delta > 0$ such that $e(x, y) < \delta$ implies $|f_j(x) - f_j(y)| < \varepsilon/3$ for all $j = 1, \ldots, n$ and $x, y \in K$. Then $|f(x) - f(y)| < \varepsilon$ for all $f \in \mathcal{F}$, so \mathcal{F} is uniformly equicontinuous. In any metric space, a totally bounded set is bounded, which for d_{\sup} means uniformly bounded.

Conversely, let \mathcal{F} be uniformly bounded and equicontinuous, hence uniformly equicontinuous by Theorem 2.4.5. Let $|f(x)| \le M < \infty$ for all $f \in \mathcal{F}$ and $x \in K$. Then $[-M, M]$ is compact by Theorem 2.2.1. Let \mathcal{G} be the closure of \mathcal{F} in the product topology of \mathbb{R}^K. Then \mathcal{G} is compact by Tychonoff's theorem 2.2.8 and Theorem 2.2.2. For any $\varepsilon > 0$ and $x, y \in K$, $\{f \in \mathbb{R}^K: |f(x) - f(y)| \le \varepsilon\}$ is closed. So if $e(x, y) < \delta$ implies $|f(x) - f(y)| \le \varepsilon$ for all $f \in \mathcal{F}$, the same remains true for all $f \in \mathcal{G}$. Thus \mathcal{G} is also uniformly equicontinuous.

Let \mathcal{U} be any ultrafilter in \mathcal{G}. Then \mathcal{U} converges (for the product topology) to some $g \in \mathcal{G}$, by Theorem 2.2.5. Given $\varepsilon > 0$, take $\delta > 0$ such that whenever $e(x, y) < \delta$, $|f(x) - f(y)| \le \varepsilon/4 < \varepsilon/3$ for all $f \in \mathcal{G}$. Take a finite set $S \subset K$ such that for any $y \in K$, $e(x, y) < \delta$ for some $x \in S$. Let

$$U := \{f: |f(x) - g(x)| < \varepsilon/3 \text{ for all } x \in S\}.$$

Then U is open in \mathbb{R}^K, so $U \in \mathcal{U}$. If $f \in U$, then $|f(y) - g(y)| < \varepsilon$ for all $y \in K$, so $d_{\sup}(f, g) \le \varepsilon$. Thus $\mathcal{U} \to g$ for d_{\sup}. So \mathcal{G} is compact for d_{\sup} (by Theorem 2.2.5), hence \mathcal{F} is totally bounded (by Theorem 2.3.1). \square

For any topological space (S, \mathcal{T}) let $C_b(S) := C_b(S, \mathcal{T})$ be the set of all bounded, real-valued, continuous functions on S. The metric d_{sup} is defined on $C_b(S)$. Any sequence f_n that converges for d_{sup} is said to converge *uniformly*. Uniform convergence preserves boundedness (rather easily) and continuity:

2.4.8. Theorem *For any topological space* (S, \mathcal{T}), *if* $f_n \in C_b(S, \mathcal{T})$ *and* $f_n \to f$ *uniformly as* $n \to \infty$, *then* $f \in C_b(S, \mathcal{T})$.

Proof. For any $\varepsilon > 0$, take n such that $d_{\text{sup}}(f_n, f) < \varepsilon/3$. For any $x \in S$, take a neighborhood U of x such that for all $y \in U$, $|f_n(x) - f_n(y)| < \varepsilon/3$. Then

$$|f(x) - f(y)| \le |f(x) - f_n(x)| + |f_n(x) - f_n(y)| + |f_n(y) - f(y)|$$
$$< \varepsilon/3 + \varepsilon/3 + \varepsilon/3 = \varepsilon.$$

Thus f is continuous. It is bounded, since $d_{\text{sup}}(0, f) \le d_{\text{sup}}(0, f_n) + d_{\text{sup}}(f_n, f) < \infty$. □

2.4.9. Theorem *For any topological space* (S, \mathcal{T}), *the metric space* $(C_b(S, \mathcal{T}), d_{\text{sup}})$ *is complete.*

Proof. Let $\{f_n\}$ be a Cauchy sequence. Then for each x in S, $\{f_n(x)\}$ is a Cauchy sequence in \mathbb{R}, so it converges to some real number, call it $f(x)$. Then for each m and x, $|f(x) - f_m(x)| = \lim_{n\to\infty} |f_n(x) - f_m(x)| \le \lim \sup_{n\to\infty} d_{\text{sup}}(f_n, f_m) \to 0$ as $m \to \infty$, so $d_{\text{sup}}(f_m, f) \to 0$. Now $f \in C_b(S, \mathcal{T})$ by Theorem 2.4.8. □

We write $c_n \downarrow c$ for real numbers c_n iff $c_n \ge c_{n+1}$ for all n and $c_n \to c$ as $n \to \infty$. If f_n are real-valued functions on a set X, then $f_n \downarrow f$ means $f_n(x) \downarrow f(x)$ for all $x \in X$. We then have:

2.4.10. Dini's Theorem *If* (K, \mathcal{T}) *is a compact topological space,* f_n *are continuous real-valued functions on* K *for all* $n \in \mathbb{N}$, *and* $f_n \downarrow f_0$, *then* $f_n \to f_0$ *uniformly on* K.

Proof. For each n, $f_n - f_0 \ge 0$. Given $\varepsilon > 0$, let $U_n := \{x \in K : (f_n - f_0)(x) < \varepsilon\}$. Then the U_n are open and their union is all of K. So they have a finite subcover. Since the convergence is monotone, we have inclusions $U_n \subset U_{n+1} \subset \cdots$. Thus some U_n is all of K. Then for all $m \ge n$, $(f_m - f_0)(x) < \varepsilon$ for all $x \in K$. □

Examples. The functions $x^n \downarrow 0$ on $[0, 1)$ but not uniformly; $[0, 1)$ is not compact. On $[0, 1]$, which is compact, $x^n \downarrow 1_{\{1\}}$, not uniformly; here the limit function f_0 is not continuous. This shows why some of the hypotheses in Dini's theorem are needed.

A collection \mathcal{F} of real-valued functions on a set X forms a *vector space* iff for any $f, g \in \mathcal{F}$ and $c \in \mathbb{R}$ we have $cf + g \in \mathcal{F}$, where $(cf + g)(x) :=$ $cf(x) + g(x)$ for all x. If, in addition, $fg \in \mathcal{F}$ where $(fg)(x) := f(x)g(x)$ for all x, then \mathcal{F} is called an *algebra*. Next, \mathcal{F} is said to *separate points* of X if for all $x \neq y$ in X, we have $f(x) \neq f(y)$ for some $f \in \mathcal{F}$.

2.4.11. Stone-Weierstrass Theorem (M. H. Stone) *Let K be any compact Hausdorff space and let \mathcal{F} be an algebra included in $C(K)$ such that \mathcal{F} separates points and contains the constants. Then \mathcal{F} is dense in $C(K)$ for d_{\sup}.*

Theorem 2.4.11 has the following consequence:

2.4.12. Corollary (Weierstrass) *On any compact set $K \subset \mathbb{R}^d, d < \infty$, the set of all polynomials in d variables in dense in $C(K)$ for d_{\sup}.*

Proof of Theorem 2.4.11. A special case of the Weierstrass theorem will be useful. Define $\binom{x}{k} := x(x-1)\cdots(x-k+1)/k!$ for any $x \in \mathbb{R}$ and $k = 1, 2, \ldots$, with $\binom{x}{0} := 1$. The Taylor series of the function $t \mapsto (1-t)^{1/2}$ around $t = 0$ is the "binomial series"

$$(1-t)^{1/2} = \sum_{n=0}^{\infty} \binom{1/2}{n}(-t)^n.$$

For any $r < 1$, the series converges absolutely and uniformly to the function for $|t| \leq r$ (Appendix B, Example (c)). Thus for any $\varepsilon > 0$ the function

$$(1 + \varepsilon - t)^{1/2} = (1 + \varepsilon)^{1/2}(1 - t/[1 + \varepsilon])^{1/2}$$

has a Taylor series converging to it uniformly on $[0, 1]$. Letting $\varepsilon \downarrow 0$ we have

$$\sup_{0 \leq t \leq 1} \left[(1 + \varepsilon - t)^{1/2} - (1 - t)^{1/2} \right] \to 0,$$

so $(1-t)^{1/2}$ can be approximated uniformly by polynomials on $0 \leq t \leq 1$. Letting $t = 1 - s^2$, we get that the function $A(s) := |s|$ can be approximated uniformly by polynomials on $-1 \leq s \leq 1$. Let $(A \circ f)(x) := |f(x)|$ if $|f(x)| \leq 1$ for all x. If P is any polynomial and $f \in \mathcal{F}$ then $P \circ f \in \mathcal{F}$ where $(P \circ f)(x) := P(f(x))$ for all x.

Let $\overline{\mathcal{F}}$ be the closure of \mathcal{F} for d_{sup}. The closure equals the completion, by Proposition 2.4.1 and Theorem 2.4.9, and is also included in $C(K)$. It is easy to check that $\overline{\mathcal{F}}$ is also an algebra. For any $f \in \overline{\mathcal{F}}$ and $M > d_{\text{sup}}(0, f) = \sup |f|$ we have $|f| = M A \circ (f/M)$, so $|f| \in \overline{\mathcal{F}}$. Thus for any $f, g \in \overline{\mathcal{F}}$ we have

$$\max(f, g) = \tfrac{1}{2}(f + g) + \tfrac{1}{2}|f - g| \in \overline{\mathcal{F}},$$
$$\min(f, g) = \tfrac{1}{2}(f + g) - \tfrac{1}{2}|f - g| \in \overline{\mathcal{F}}.$$

Iterating, the maximum or minimum of finitely many functions in $\overline{\mathcal{F}}$ is in $\overline{\mathcal{F}}$.

For any $x \neq y$ in X take $f \in \mathcal{F}$ with $f(x) \neq f(y)$. Then for any real c, d there exist $a, b \in \mathbb{R}$ with $(af + b)(x) = c$ and $(af + b)(y) = d$, namely $a := (c - d)/(f(x) - f(y))$, $b := c - af(x)$. Note that $af + b \in \mathcal{F}$. Now take any $h \in C(K)$ and fix $x \in K$. For any $y \in K$ take $h_y \in \mathcal{F}$ with $h_y(x) = h(x)$ and $h_y(y) = h(y)$. Given $\varepsilon > 0$, there is an open neighborhood U_y of y such that $h_y(v) > h(v) - \varepsilon$ for all $v \in U_y$. The sets U_y form an open cover of K and have a finite subcover $U_{y(j)}$, $j = 1, \ldots, n$. Let $g_x := \max_{1 \leq j \leq n} h_y(j)$. Then $g_x \in \overline{\mathcal{F}}$, $g_x(x) = h(x)$, and $g_x(v) > h(v) - \varepsilon$ for all $v \in K$.

For each $x \in K$, there is an open neighborhood V_x of x such that $g_x(u) < h(u) + \varepsilon$ for all $u \in V_x$. The sets V_x have a finite subcover $V_{x(1)}, \ldots, V_{x(m)}$ of K. Let $g := \min_{1 \leq j \leq m} g_{x(j)}$. Then $g \in \overline{\mathcal{F}}$ and $d_{\text{sup}}(g, h) < \varepsilon$. Letting $\varepsilon \downarrow 0$ gives $h \in \mathcal{F}$, finishing the proof. \square

Complex numbers $z = x + iy$ are treated in Appendix B. The absolute value $|z| = \sqrt{x^2 + y^2}$ is defined, so we have a metric d_{sup} for bounded complex-valued functions. Here is a form of the Stone-Weierstrass theorem in the complex-valued case.

2.4.13. Corollary *Let (K, \mathcal{T}) be a compact Hausdorff space. Let \mathcal{A} be an algebra of continuous functions: $K \mapsto \mathbb{C}$, separating the points and containing the constants. Suppose also that \mathcal{A} is self-adjoint, in other words $\bar{f} = g - ih \in \mathcal{A}$ whenever $f = g + ih \in \mathcal{A}$ where g and h are real-valued functions. Then \mathcal{A} is dense in the space of all continuous complex-valued functions on K for d_{sup}.*

Proof. For any $f = g + ih \in \mathcal{A}$ with $g := \text{Re} f$ and $h := \text{Im} f$ real-valued, we have $g = (f + \bar{f})/2 \in \mathcal{A}$ and $h = (f - \bar{f})/(2i) \in \mathcal{A}$. Let \mathcal{C} be the set of real-valued functions in \mathcal{A}. Then \mathcal{C} is an algebra over \mathbb{R}. We also have $\mathcal{C} = \{\text{Re } f : f \in \mathcal{A}\}$ and $\mathcal{C} = \{\text{Im } f : f \in \mathcal{A}\}$. Thus \mathcal{C} separates the points of K. It contains the real constants. Thus \mathcal{C} is dense in the space of real-valued continuous functions on K by Theorem 2.4.11. Taking $g + ih$ for any $g, h \in \mathcal{C}$, the result follows. \square

Example. The hypothesis that \mathcal{A} is self-adjoint cannot be omitted. Let $T^1 :=$ $\{z \in \mathbb{C} : |z| = 1\}$, the unit circle, which is compact. Let \mathcal{A} be the set of all polynomials $z \mapsto \sum_{j=0}^{n} a_j z^j$, $a_j \in \mathbb{C}$, $n = 0, 1, \ldots$. Then \mathcal{A} is an algebra satisfying all conditions of Corollary 2.4.13 except self-adjointness. The function $f(z) := \bar{z} = 1/z$ on T^1 cannot be uniformly approximated by a polynomial $P_n \in \mathcal{A}$, as follows. For any such P_n,

$$\frac{1}{2\pi} \int_0^{2\pi} |f(e^{i\theta}) - P_n(e^{i\theta})|^2 \, d\theta \geq 1$$

because the "cross terms" $\int_0^{2\pi} e^{-i(k+1)\theta} \, d\theta = 0$ for $k = 0, 1, \ldots$, and likewise if $-i$ is replaced by i.

Problems

For any $M > 0$, $\alpha > 0$, and metric space (S, d), let $\mathrm{Lip}(\alpha, M)$ be the set of all real-valued functions f on S such that $|f(x) - f(y)| \leq M d(x, y)^\alpha$ for all $x, y \in S$. (For $\alpha = 1$, such functions are called Lipschitz functions. For $0 < \alpha < 1$ they are said to satisfy a Hölder condition of order α.)

1. If (K, d) is a compact metric space and $u \in K$, show that for any finite M and $0 < \alpha \leq 1$, $\{f \in \mathrm{Lip}(\alpha, M) : |f(u)| \leq M\}$ is compact for d_{\sup}.

2. If $S = [0, 1]$ with its usual metric and $\alpha > 1$, show that $\mathrm{Lip}(\alpha, 1)$ contains only constant functions. *Hint:* For $0 \leq x \leq x + h \leq 1$, $f(x+h) - f(x) = \sum_{1 \leq j \leq n} f(x + jh/n) - f(x + (j-1)h/n)$. Give an upper bound for the absolute value of the jth term of the right, sum over j, and let $n \to \infty$.

3. Find continuous functions f_n from $[0, 1]$ into itself where $f_n \to 0$ pointwise but not uniformly as $n \to \infty$. *Hint:* Let $f_n(1/n) = 1$, $f_n(0) \equiv f_n(2/n) \equiv 0$. (This shows why monotone convergence, $f_n \downarrow f_0$, is useful in Dini's theorem.)

4. Show that the functions $f_n(x) := x^n$ on $[0, 1]$ are not equicontinuous at 1, without applying any theorem from this section.

5. If (S_i, d_i) are metric spaces for $i \in I$, where I is a finite set, then on the Cartesian product $S = \Pi_{i \in I} S_i$ let $d(x, y) = \sum_i d_i(x_i, y_i)$.
 (a) Show that d is a metric.
 (b) Show that d metrizes the product of the d_i topologies.
 (c) Show that (S, d) is complete if and only if all the (S_i, d_i) are complete.

6. Prove that each of the following functions f has properties (1), (2), and (3) in Proposition 2.4.3: (a) $f(x) := x/(1+x)$; (b) $f(x) := \tan^{-1} x$; (c) $f(x) := \min(x, 1), 0 \le x < \infty$.

7. Show that the functions $x \mapsto \sin(nx)$ on $[0, 1]$ for $n = 1, 2, \ldots$, are not equicontinuous at 0.

8. A function f from a topological space (S, \mathcal{T}) into a metric space (Y, d) is called *bounded* iff its range is bounded. Let $C_b(S, Y, d)$ be the set of all bounded, continuous functions from S into Y. For f and g in $C_b(S, Y, d)$ let $d_{\sup}(f, g) := \sup\{d(f(x), g(x)): x \in S\}$. If (Y, d) is complete, show that $C_b(S, Y, d)$ is complete for d_{\sup}.

9. (Peano curves). Show that there is a continuous function f from the unit interval $[0, 1]$ onto the unit square $[0, 1] \times [0, 1]$. *Hints:* Let f be the limit of a sequence of functions f_n which will be piecewise linear. Let $f_1(t) \equiv (0, t)$. Let $f_2(t) = (2t, 0)$ for $0 \le t \le 1/4$, $f_2(t) = (1/2, 2t - 1/2)$ for $1/4 \le t \le 3/4$, and $f_2(t) = (2 - 2t, 1)$ for $3/4 \le t \le 1$. At the nth stage, the unit square is divided into $2^n \cdot 2^n = 4^n$ equal squares, where the graph of f_n runs along at least one edge of each of the small squares. Then at the next stage, on the interval where f_n ran along one such edge, f_{n+1} will first go halfway along a perpendicular edge, then along a bisector parallel to the original edge, then back to the final vertex, just as f_2 related to f_1. Show that this scheme can be carried through, with f_n converging uniformly to f.

10. Show that for $k = 2, 3, \ldots$, there is a continuous function $f^{(k)}$ from $[0, 1]$ onto the unit cube $[0, 1]^k$ in \mathbb{R}^k. *Hint:* Let $f^{(2)}(t) := (g(t), h(t)) := f(t)$ for $0 \le t \le 1$ from Problem 9. For any $(x, y, z) \in [0, 1]^3$, there are t and u in $[0, 1]$ with $f(u) = (y, z)$ and $f(t) = (x, u)$, so $f^{(3)}(t) := (g(t), g(h(t)), h(h(t))) = (x, y, z)$. Iterate this construction.

11. Show that there is a continuous function from $[0, 1]$ onto $\Pi_{n \ge 1}[0, 1]_n$, a countable product of copies of $[0, 1]$, with product topology. *Hint:* Take the sequence $f^{(k)}$ as in Problem 10. Let $F_k(t)_n := f^{(k)}(t)_n$ for $n \le k$, 0 for $n > k$. Show that F_k converge to the desired function as $k \to \infty$.

12. Let K be a compact Hausdorff space and suppose for some k there are k continuous functions f_1, \ldots, f_k from K into \mathbb{R} such that $x \mapsto (f_1(x), \ldots, f_k(x))$ is one-to-one from K into \mathbb{R}^k. Let \mathcal{F} be the smallest algebra of functions containing f_1, f_2, \ldots, f_k and 1. (a) Show that \mathcal{F} is dense in $C(K)$ for d_{\sup}. (b) Let $K := S^1 := \{(\cos\theta, \sin\theta): 0 \le \theta \le 2\pi\}$ be the unit circle in \mathbb{R}^2 with relative topology. Part (a) applies easily for $k = 2$. Show that it does not apply for $k = 1$: there is no 1–1 continuous

function f from S^1 into \mathbb{R}. *Hint:* Apply the intermediate value theorem, Problem 14(d) of Section 2.2. For θ consider the intervals $[0, \pi]$ and $[\pi, 2\pi]$.

13. Give a direct proof of the "if" part of the Arzelà-Ascoli theorem 2.4.7, without using the Tychonoff theorem or filters. *Hints:* Apply Theorem 2.4.5. Given $\varepsilon > 0$, take $\delta > 0$ for $\varepsilon/4$ and \mathcal{F}. Theorem 2.3.1 gives a finite δ-dense set in K, and $[-M, M]$ has a finite $\varepsilon/4$-dense set. Use these to get a finite ε-dense set in \mathcal{F} for d_{sup}.

2.5. Completion and Completeness of Metric Spaces

Let (S, d) and (T, e) be two metric spaces. A function f from S into T is called an *isometry* iff $e(f(x), f(y)) = d(x, y)$ for all x and y in S.

For example if $S = T = \mathbb{R}^2$, with metric the usual Euclidean distance (as in Problems 15–16 of Section 2.2), then isometries are found by taking $f(u) = u + v$ for a vector v (translations), by rotations (around any center), by reflection in any line, and compositions of these.

It will be shown that any metric space S is isometric to a dense subset of a complete one, T. In a classic example, S is the space \mathbb{Q} of rational numbers and $T = \mathbb{R}$. In fact, this has sometimes been used as a definition of \mathbb{R}.

2.5.1. Theorem *Let (S, d) be any metric space. Then there is a complete metric space (T, e) and an isometry f from S onto a dense subset of T.*

Remarks. Since f preserves the only given structure on S (the metric), we can consider S as a subset of T. T is called the *completion* of S.

Proof. Let $f_x(y) := d(x, y)$, $x, y \in S$. Choose a point $u \in S$ and let $F(S, d) := \{f_u + g : g \in C_b(S, d)\}$. On $F(S, d)$, let $e := d_{\mathrm{sup}}$. Although functions in $F(S, d)$ may be unbounded (if S is unbounded for d), their differences are bounded, and e is a well-defined metric on $F(S, d)$. For any x, y, and z in S, $|d(x, z) - d(y, z)| \leq d(x, y)$ by the triangle inequality, and equality is attained for $z = x$ or y. Thus f_z is continuous for any $z \in S$, and for any x, y, we have $f_y - f_x \in C_b(S, d)$ and $d_{\mathrm{sup}}(f_x, f_y) = d(x, y)$. Also, $f_y \in F(S, d)$, and $F(S, d)$ does not depend on the choice of u. It follows that the function $x \mapsto f_x$ from S into $F(S, d)$ is an isometry for d and e. Let T be the closure of the range of this function in $F(S, d)$. Since $(C_b(S, d), d_{\mathrm{sup}})$ is complete (Theorem 2.4.9), so is $F(S, d)$. Thus (T, e) is complete, so it serves as a completion of (S, d). $\qquad\square$

Let (T', e') and f' also satisfy the conclusion of Theorem 2.5.1 in place of (T, e) and f, respectively. Then on the range of f, $f' \circ f^{-1}$ is an isometry of a dense subset of T onto a dense subset of T'. This isometry extends naturally to an isometry of T onto T', since both (T, e) and (T', e') are complete. Thus (T, e) is uniquely determined up to isometry, and it makes sense to call it "the completion" of S.

If a space is complete to begin with, as \mathbb{R} is, then the completion does not add any new points, and the space equals (is isometric to) its completion.

A set A in a topological space S is called *nowhere dense* iff for every non-empty open set $U \subset S$ there is a non-empty open $V \subset U$ with $A \cap V = \emptyset$. Recall that a topological space (S, \mathcal{T}) is called *separable* iff S has a countable dense subset.

In $[0, 1]$, for example, any finite set is nowhere dense, and a countable union of finite sets is countable. The union may be dense, but it has dense complement. This is an instance of the following fact:

2.5.2. Category Theorem *Let (S, d) be any complete metric space. Let A_1, A_2, \ldots, be a sequence of nowhere dense subsets of S. Then their union $\bigcup_{n \geq 1} A_n$ has dense complement.*

Proof. If S is empty, the statement holds vacuously. Otherwise, choose $x_1 \in S$ and $0 < \varepsilon_1 < 1$. Recursively choose $x_n \in S$ and $\varepsilon_n > 0$, with $\varepsilon_n < 1/n$, such that for all n,

$$B(x_{n+1}, \varepsilon_{n+1}) \subset B(x_n, \varepsilon_n/2) \backslash A_n.$$

This is possible since A_n is nowhere dense. Then $d(x_m, x_n) < 1/n$ for all $m \geq n$, so $\{x_n\}$ is a Cauchy sequence. It converges to some x with $d(x_n, x) \leq \varepsilon_n/2$ for all n, so $d(x_{n+1}, x) \leq \varepsilon_{n+1}/2 < \varepsilon_{n+1}$ and $x \notin A_n$. Since $x_1 \in S$ and $\varepsilon_1 > 0$ were arbitrary, and the balls $B(x_1, \varepsilon_1)$ form a base for the topology, $S \backslash \bigcup_n A_n$ is dense. \square

A union of countably many nowhere dense sets is called a set of *first category*. Sets not of first category are said to be of *second category*. (This terminology is not related to other uses of the word "category" in mathematics, as in homological algebra.) The category theorem (2.5.2) then says that every complete metric space S is of second category. Also if A is of first category, then $S \backslash A$ is of second category. A metric space (S, d) is called *topologically complete* iff there is some metric e on S with the same topology as d such that (S, e) is complete. Since the conclusion of the category theorem is in terms of topology, not depending on the specific metric, the theorem also holds in

topologically complete spaces. For example, $(-1, 1)$ is not complete with its usual metric but is complete for the metric $e(x, y) := |f(x) - f(y)|$, where $f(x) := \tan(\pi x/2)$.

By definition of topology, any union of open sets, or the intersection of finitely many, is open. In general, an intersection of countably many open sets need not be open. Such a set is called a G_δ (from German *Gebiet-Durchschnitt*). The complement of a G_δ, that is, a union of countably many closed sets, is called an F_σ (from French *fermé-somme*).

For any metric space (S, d), $A \subset S$, and $x \in S$, let

$$d(x, A) := \inf\{d(x, y): y \in A\}.$$

For any x and z in S and y in A, from $d(x, y) \le d(x, z) + d(z, y)$, taking the infimum over y in A gives

$$d(x, A) \le d(x, z) + d(z, A)$$

and since x and z can be interchanged,

$$|d(x, A) - d(x, A)| \le d(x, z). \tag{2.5.3}$$

Here is a characterization of topologically complete metric spaces. It applies, for example, to the set of all irrational numbers in \mathbb{R}, which at first sight looks quite incomplete.

**2.5.4. Theorem* *A metric space (S, d) is topologically complete if and only if S is a G_δ in its completion for d.*

Proof. By the completion theorem (2.5.1) we can assume that S is a dense subset of T and (T, d) is complete.

To prove "if," suppose $S = \bigcap_n U_n$ with each U_n open in T. Let $f_n(x) := 1/d(x, T \backslash U_n)$ for each n and $x \in S$. Let $g(t) := t/(1 + t)$. As in Propositions 2.4.3 and 2.4.4 (metrization of countable products), let

$$e(x, y) := d(x, y) + \sum_n 2^{-n} g(|f_n(x) - f_n(y)|)$$

for any x and y in S. Then e is a metric on S.

Let $\{x_m\}$ be a Cauchy sequence in S for e. Then since $d \le e$, $\{x_m\}$ is also Cauchy for d and converges for d to some $x \in T$. For each n, $f_n(x_m)$ converges as $m \to \infty$ to some $a_n < \infty$. Thus $d(x_m, T \backslash U_n) \to 1/a_n > 0$ and $x \notin T \backslash U_n$ for all n, so $x \in S$.

For any set F, by (2.5.3) the function $d(\cdot, F)$ is continuous. So on S, all the f_n are continuous, and convergence for d implies convergence for e. Thus d

and e metrize the same topology, and $x_m \to x$ for e. So S is complete for e, as desired.

Conversely, let e be any metric on S with the same topology as d such that (S, e) is complete. For any $\varepsilon > 0$ let

$$U_n(\varepsilon) := \{x \in T : \mathrm{diam}\,(S \cap B_d(x, \varepsilon)) < 1/n\},$$

where diam denotes diameter with respect to e, and B_d denotes a ball with respect to d. Let $U_n := \bigcup_{\varepsilon > 0} U_n(\varepsilon)$. For any x and v, if $x \in U_n(\varepsilon)$ and $d(x, v) < \varepsilon/2$, then $v \in U_n(\varepsilon/2)$. Thus U_n is open in T.

Now $S \subset U_n$ for all n, since d and e have the same topology. If $x \in U_n$ for all n, take $x_m \in S$ with $d(x_m, x) \to 0$. Then $\{x_m\}$ is also Cauchy for e, by definition of the U_n and $U_n(\varepsilon)$. Thus $e(x_m, y) \to 0$ for some $y \in S$, so $x = y \in S$. □

In a metric space (X, d), if $x_n \to x$ and for each n, $x_{nm} \to x_n$ as $m \to \infty$, then for some $m(n)$, $x_{nm(n)} \to x$: we can choose $m(n)$ such that $d(x_{nm(n)}, x_n) < 1/n$. This iterated limit property fails, however, in some non-metrizable topological spaces, such as $2^{\mathbb{R}}$ with product topology. For example, there are finite $F(n)$ with $1_{F(n)} \to 1_{\mathbb{Q}}$ in $2^{\mathbb{R}}$, and for any finite F there are open $U(m)$ with $1_{U(m)} \to 1_F$. However:

***2.5.5. Proposition** *There is no sequence $U(1), U(2), \ldots$, of open sets in \mathbb{R} with $1_{U(m)} \to 1_{\mathbb{Q}}$ in $2^{\mathbb{R}}$ as $m \to \infty$.*

Proof. Suppose $1_{U(m)} \to 1_{\mathbb{Q}}$. Let $X := \bigcap_{m \geq 1} \bigcup_{n \geq m} U(n)$. Then X is an intersection of countably many dense open sets. Hence $\mathbb{R} \backslash X$ is of first category. But if $X = \mathbb{Q}$, then \mathbb{R} is of first category, contradicting the category theorem (2.5.2). □

This gives an example of a space that is not topologically complete:

***2.5.6. Corollary** \mathbb{Q} *is not a G_δ in \mathbb{R} and hence is not topologically complete.*

Next, (topological) completeness will be shown to be preserved by countable Cartesian products. This will probably not be surprising. For example, a product of a sequence of compact metric spaces is metrizable by Proposition 2.4.4 and compact by Tychonoff's theorem, and so complete by Theorem 2.3.1 (in this case for any metric metrizing its topology).

2.5.7. Theorem *Let (S_n, d_n) for $n = 1, 2, \ldots,$ be a sequence of complete metric spaces. Then the Cartesian product $\Pi_n S_n$, with product topology, is complete with the metric d of Proposition 2.4.4.*

Proof. A Cauchy sequence $\{x_m\}_{m \geq 1}$ in the product space is a sequence of sequences $\{\{x_{mn}\}_{n \geq 1}\}_{m \geq 1}$. For any fixed n, as m and $k \to \infty$, and since d is a sum of nonnegative terms, $f(d_n(x_{mn}, x_{kn}))/2^n \to 0$, so $d_n(x_{mn}, x_{kn}) \to 0$, and $\{x_{in}\}_{i \geq 1}$ is a Cauchy sequence for d_n, so it converges to some x_n in S_n. Since this holds for each n, the original sequence in the product space converges for the product topology, and so for d by Proposition 2.4.4. □

Problems

1. Show that the closure of a nowhere dense set is nowhere dense.

2. Let (S, d) and (V, e) be two metric spaces. On the Cartesian product $S \times V$ take the metric

$$\rho(\langle x, u \rangle, \langle y, v \rangle) = d(x, y) + e(u, v).$$

 Show that the completion of $S \times V$ is isometric to the product of the completions of S and of V.

3. Show that the intersection of the complement of a set of first category with a non-empty open set in a complete metric space is not only non-empty but uncountable. *Hint:* Are singleton sets $\{x\}$ nowhere dense?

4. Show that the set $\mathbb{R} \backslash \mathbb{Q}$ of irrational numbers, with usual topology (relative topology from \mathbb{R}), is topologically complete.

5. Define a complete metric for $\mathbb{R} \backslash \{0, 1\}$ with usual (relative) topology.

6. Define a complete metric for the usual (relative) topology on $\mathbb{R} \backslash \mathbb{Q}$.

7. (a) If (S, d) is a complete metric space, X is a G_δ subset of S, and for the relative topology on X, Y is a G_δ subset of X, show that Y is a G_δ in S.
 (b) Prove the same for a general topological space S.

8. Show that the plane \mathbb{R}^2 is not a countable union of lines (a line is a set $\{\langle x, y \rangle : ax + by = c\}$ where a and b are not both 0).

9. A C^1 curve is a function $t \mapsto (f(t), g(t))$ from \mathbb{R} into \mathbb{R}^2 where the derivatives $f'(t)$ and $g'(t)$ exist and are continuous for all t. Show that \mathbb{R}^2 is not a countable union of ranges of a C^1 curves. *Hint:* Show that the range of a C^1 curve on a finite interval is nowhere dense.

10. Let (S, d) be any noncompact metric space. Show that there exist bounded continuous functions f_n on S such that $f_n(x) \downarrow 0$ for all $x \in S$ but f_n do not converge to 0 uniformly. *Hint: S* is either not complete or not totally bounded.

11. Show that a metric space (S, d) is complete for every metric e metrizing its topology if and only if it is compact. *Hint:* Apply Theorem 2.3.1. Suppose $d(x_m, x_n) \geq \varepsilon > 0$ for all $m \neq n$ integers ≥ 1. For any integers $j, k \geq 1$ let

$$e_{jk}(x, y) := d(x, x_j) + |j^{-1} - k^{-1}|\varepsilon + d(y, x_k).$$

Let $e(x, y) := \min(d(x, y), \inf_{j,k} e_{jk}(x, y))$. To show that for any $j, k, r,$ and s, and any $x, y, z \in S$, $e_{js}(x, z) \leq e_{jk}(x, y) + e_{rs}(y, z)$, consider the cases $k = r$ and $k \neq r$.

*2.6. Extension of Continuous Functions

The problem here is, given a continuous real-valued function f defined on a subset F of a topological space S, when can f be extended to be continuous on all of S? Consider, for example, the set $\mathbb{R}\setminus\{0\} \subset \mathbb{R}$. The function $f(x) := 1/x$ is continuous on $\mathbb{R}\setminus\{0\}$ but cannot be extended to be continuous at 0. Likewise, the bounded function $\sin(1/x)$ is continuous except at 0. As these examples show, it is not possible generally to make the extension unless F is closed. If F is closed, the extension will be shown to be possible for metric spaces, compact Hausdorff spaces, and a class of spaces including both, called *normal* spaces, defined as follows.

Sets are called *disjoint* iff their intersection is empty. A topological space (S, \mathcal{T}) is called *normal* iff for any two disjoint closed sets E and F there are disjoint open sets U and V with $E \subset U$ and $F \subset V$. First it will be shown that some other general properties imply normality.

2.6.1. Theorem Every metric space (S, d) is normal.

Proof. For any set $A \subset S$ and $x \in S$ let $d(x, A) := \inf_{y \in A} d(x, y)$. Then $d(\cdot, A)$ is continuous, by (2.5.3). For any disjoint closed sets E and F, let $g(x) := d(x, E)/(d(x, E) + d(x, F))$. Since E is closed, $d(x, E) = 0$ if and only if $x \in E$, and likewise for F. Since E and F are disjoint, the denominator in the definition of g is never 0, so g is continuous. Now $0 \leq g(x) \leq 1$ for all x, with $g(x) = 0$ iff $x \in E$, and $g(x) = 1$ if and only if $x \in F$. Let $U := g^{-1}((-\infty, 1/3))$, $V := g^{-1}((2/3, \infty))$. Then clearly U and V have the desired properties. \square

2.6.2. Theorem *Every compact Hausdorff space is normal.*

Proof. Let E and F be disjoint and closed. For each $x \in E$ and $y \in F$, take open U_{xy} and V_{yx} with $x \in U_{xy}$, $y \in V_{yx}$, and $U_{xy} \cap V_{yx} = \emptyset$. For each fixed y, $\{U_{xy}\}_{x \in E}$ form an open cover of the closed, hence (by Theorem 2.2.2) compact set E. So there is a finite subcover, $\{U_{xy}\}_{x \in E(y)}$ for some finite subset $E(y) \subset E$. Let $U_y := \bigcup_{x \in E(y)} U_{xy}$, $V_y := \bigcap_{x \in E(y)} V_{yx}$. Then for each y, U_y and V_y are open, $E \subset U_y$, $y \in Y_y$, and $U_y \cap Y_y = \emptyset$. The V_y form an open cover of the compact set F and hence have an open subcover $\{V_y\}_{y \in G}$ for some finite $G \subset F$. Let $U := \bigcap_{y \in G} U_y$, $V := \bigcup_{y \in G} V_y$. Then U and V are open and disjoint, $E \subset U$, and $F \subset V$. □

The next fact will give an extension if the original continuous function has only two values, 0 and 1, as was done for a metric space in the proof of Theorem 2.6.1. This will then help in the proof of the more general extension theorem.

2.6.3. Urysohn's Lemma *For any normal topological space (X, \mathcal{T}) and disjoint closed sets E and F, there is a continuous real f on X with $f(x) = 0$ for all $x \in E$, $f(y) = 1$ for all $y \in F$, and $0 \le f \le 1$ everywhere on X.*

Proof. For each dyadic rational $q = m/2^n$, where $n = 0, 1, \ldots$, and $m = 0, 1, \ldots, 2^n$, so that $0 \le q \le 1$, first choose a unique representation such that m is odd or $m = n = 0$. For such q, m, and n, an open set $U_q := U_{mn}$ and a closed set $F_q = F_{mn}$ will be defined by recursion on n as follows. For $n = 0$, let $U_0 := \emptyset$, $F_0 := E$, $U_1 := X \backslash F$, and $F_1 := X$. Now suppose the U_{mj} and F_{mj} have been defined for $0 \le j \le n$, with $U_r \subset F_r \subset U_s \subset F_s$ for $r < s$. These inclusions do hold for $n = 0$. Let $q = (2k + 1)/2^{n+1}$. Then for $r = k/2^n$ and $s = (k + 1)/2^n$, $F_r \subset U_s$, so F_r is disjoint from the closed set $X \backslash U_s$. By normality take disjoint open sets U_q and V_q with $F_r \subset U_q$ and $X \backslash U_s \subset V_q$. Let $F_q := X \backslash V_q$. Then as desired, $F_r \subset U_q \subset F_q \subset U_s$, so all the F_q and U_q are defined recursively.

Let $f(x) := \inf\{q : x \in F_q\}$. Then $0 \le f(x) \le 1$ for all x, $f = 0$ on E, and $f = 1$ on F. For any $y \in [0, 1]$, $f(x) > y$ if and only if for some dyadic rational $q > y$, $x \in X \backslash F_q$. Thus $\{x : f(x) > y\}$ is a union of open sets and hence is open. Next, $f(x) < t$ if and only if for some dyadic rational $q < t$, $x \in F_q$, and so $x \in U_r$ for some dyadic rational r with $q < r < t$. So $\{x : f(x) < t\}$ is also a union of open sets and hence is open. So for any open interval (y, t), $f^{-1}((y, t))$ is open. Taking unions, it follows that f is continuous. □

Now here is the main result of this section:

2.6.4. Extension Theorem (Tietze-Urysohn) *Let (X, \mathcal{T}) be a normal topological space and F a closed subset of X. Then for any $c \geq 0$ and each of the following subsets S of \mathbb{R} with usual topology, every continuous function f from F into S can be extended to a continuous function g from X into S:*

(a) $S = [-c, c]$.
(b) $S = (-c, c)$.
(c) $S = \mathbb{R}$.

Proof. We can assume $c = 1$ (if $c = 0$ in (a), set $g = 0$). For (a), let $E :=$ $\{x \in F : f(x) \leq -1/3\}$ and $H := \{x \in F : f(x) \geq 1/3\}$. Since E and H are disjoint closed sets, by Urysohn's Lemma there is a continuous function h on X with $0 \leq h(x) \leq 1$ for all x, $h = 0$ on E, and $h = 1$ on H. Let $g_0 := 0$ and $g_1 := (2h - 1)/3$. Then g_1 is continuous on X, with $|g_1(x)| \leq 1/3$ for all x and $\sup_{x \in F} |f - g_1|(x) \leq 2/3$. Inductively, it will be shown that there are $g_n \in C_b(X, \mathcal{T})$ for $n = 1, 2, \ldots$, such that for each n,

$$\sup_{x \in F} |f - g_n|(x) \leq 2^n/3^n, \quad \text{and} \tag{2.6.5}$$

$$\sup_{x \in X} |g_{n-1} - g_n|(x) \leq 2^{n-1}/3^n. \tag{2.6.6}$$

Both inequalities hold for $n = 1$. Let g_1, \ldots, g_n be such that (2.6.5) and (2.6.6) hold for $j = 1, \ldots, n$. Apply the method of choice of g_1 to $(3/2)^n(f - g_n)$ in place of f, which can be done by (2.6.5). So there is an $f_n \in C_b(X, \mathcal{T})$ with $\sup_{x \in F} |(3/2)^n(f - g_n) - f_n|(x) \leq 2/3$ and $\sup_{x \in X} |f_n(x)| \leq 1/3$. Let $g_{n+1} := g_n + (2/3)^n f_n$. Then (2.6.5) and (2.6.6) hold with $n + 1$ in place of n, as desired.

Now g_n converge uniformly on X as $n \to \infty$ to a function g with $g = f$ on F and for all $x \in X$, $|g(x)| \leq \sum_{1 \leq n < \infty} 2^{n-1}/3^n \leq 1$. As each g_n is continuous, g is continuous (by Theorem 2.4.8), proving (a).

Now for (b), still with $c = 1$, first apply (a). Then let $G := |g|^{-1}(\{1\})$, a closed set disjoint from F. By Urysohn's Lemma take $h \in C_b(X, \mathcal{T})$ with $h = 0$ on G, $h = 1$ on F, and $0 \leq h \leq 1$ on X. Let $j := hg$. Then j is continuous, $j = f$ on F, and $|j(x)| < 1$ for all x, proving (b).

For (c), just apply (b) to the function $2(\arctan f)/\pi$, then take $\tan(\pi g/2)$ to complete the proof. \square

Problems

1. (a) Show that any open set U in \mathbb{R} is a union of countably many disjoint open intervals, one or two of which may be unbounded (the line or

half-lines). *Hint:* For each $x \in U$, find a largest open interval containing x and included in U.

(b) Let F be a closed set in \mathbb{R} whose complement is the union of disjoint open intervals (a_n, b_n) from part (a). Let f be a continuous real-valued function on F. Extend f to be linear on $[a_n, b_n]$, or if a_n or b_n is infinite, extend f to be constant on a half-line. Show that the resulting f is continuous on \mathbb{R}. *Note:* Be aware that subsequences of the a_n may converge.

2. Show that if (S, d) is any metric space which is not complete, then there exist disjoint closed sets E and F and points $x_n \in E$ and $y_n \in F$ with $d(x_n, y_n) \to 0$. Then the function f in Urysohn's Lemma cannot be uniformly continuous. *Hint:* Show that there is a nonconvergent Cauchy sequence $\{z_n\}$ with $z_m \neq z_n$ for $m \neq n$. Let $E = \{x_n\}$ and $F = \{y_n\}$ be subsequences of $\{z_n\}$.

3. Show that if $a < b$, the identity function from the two-point set $\{a, b\}$ onto itself cannot be extended to a continuous function from $[a, b]$ onto $\{a, b\}$. *Hint:* Use connectedness, see §2.2, Problem 14.

4. A topological space (S, \mathcal{T}) is called *perfectly normal* iff for every closed set F there is a continuous real function f on S with $F = f^{-1}(\{0\})$.
 (a) Show that every metric space is perfectly normal.
 (b) Show that every perfectly normal space is normal. *Hint:* Adapt the proof of Theorem 2.6.1.

5. Let (S, \mathcal{T}) be a perfectly normal space and F a closed subset of S. Here is another way to prove that (a) implies (b) in Theorem 2.6.4 in such spaces. Show that there is an $h \in C_b(S, \mathcal{T})$ such that for every continuous function f from F into $(-1, 1)$ and continuous function g from S into $[-1, 1]$ which equals f on F, hg is continuous from S into $(-1, 1)$ and equals f on F. *Hint:* Make $h = 1$ on F, $0 \leq h < 1$ elsewhere.

6. If (S, \mathcal{T}) is a normal space, F a closed subset of S, and f a continuous function from F into the subset Y of \mathbb{R}^k, with usual (relative) topology on Y, show that f can be extended to a continuous function from S into Y if:
 (a) Y is the closed unit ball $\{y \in \mathbb{R}^k : \sum_{j=1}^{k} y_j^2 \leq 1\}$. *Hint:* First extend each coordinate function f_j to g_j by Theorem 2.6.4(a) to get a function g. Let $h(x) := g(x)$ if $|g(x)| \leq 1$, $h(x) := g(x)/|g(x)|$ otherwise.
 (b) Y is the open unit ball $\{y \in \mathbb{R}^k : \sum_{j=1}^{k} y_j^2 < 1\}$. *Hint:* Use part (a) and see the proof of Theorem 2.6.4(b).

7. Let (S, d) be a metric space, F a closed set in S, and f a continuous, bounded, nonnegative function on F. For $x \in S$ let $g(x) :=$

$\sup_{t \in F} f(t)/(1 + d(t, x)^2)^{1/d(x,F)}$, $x \notin F$; $g(x) := f(x)$, $x \in F$. Prove g is bounded and continuous on S. *Hint:* Prove g is both upper and lower semicontinuous (as in §2.2, Problem 18), $\lim \sup_{y \to x} g(y) \le g(x) \le \lim \inf_{y \to x} g(y)$: (a) for $x \notin F$, and (b) for $x \in F$, $y \notin F$. In case (b), as $y = y_n \to x$ consider $t_n \in F$ with $d(y_n, t_n)/d(y_n, F) \to 1$ as $n \to \infty$.

8. A compact Hausdorff space is normal (Theorem 2.6.2), but a subset of such a space need not be: let (A, \le) be a countable well-ordered set such that there is just one $a \in A$ for which $x < a$ for infinitely many $x \in A$ (so a is the largest element of A). Let (B, \le) be an uncountable well-ordered set such that there is just one $b \in B$ for which $y < b$ for uncountably many $y \in B$ (so b is the largest element of B).
 (a) Show that with their interval topologies (Problem 9 in §2.2), A and B are compact, so $A \times B$ with product topology is compact and normal. (It is called the *Tychonoff plank*.)
 (b) Show that $A \times B$ with the upper "vertex" $\langle a, b \rangle$ deleted is not normal. *Hint:* Show that the closed sets $E := \{a\} \times (B\backslash\{b\})$ and $F := (A\backslash\{a\}) \times \{b\}$ cannot be separated. If U is an open set including F, show that for each $x < a$, $(x, y) \in U$ for all y not in some countable set B_x. Let C be the union of these B_x. For any $y \notin C$, $(x, y) \in U$ for all $x < a$. But an open set V including E must contain some of these points (x, y).

9. Let C_r be the circle $x^2 + y^2 = r^2$ and D the unit disk $x^2 + y^2 \le 1$. Show that the identity from C_1 onto itself cannot be extended to a continuous function g from D onto C_1. *Hint:* Take polar coordinates $(r(p), \theta(p))$ for any point $p \in D$, so $r \circ g \equiv 1$. Let (r, φ) be the coordinates of points in D, with $0 \le \varphi < 2\pi$. Show that $\theta(g(r, \varphi))$ can be defined to be continuous (not necessarily with values in $[0, 2\pi)$) as a function of φ for $r > 0$. Let $h(r) := \lim_{\varphi \to 2\pi} \theta(g(r, \varphi)) - \theta(g(r, 0))$. Show that h is continuous as a function of r, must always be a multiple of 2π, but has different values at $r = 0$ and $r = 1$.

*2.7. Uniformities and Uniform Spaces

Uniform spaces have some of the properties of metric spaces, so that uniform continuity of functions between such spaces can be defined. First, the following notion will be needed. Let A be a subset of a Cartesian product $X \times Y$ and B a subset of $Y \times Z$. Then $A \circ B$ is the set of all $\langle x, z \rangle$ in $X \times Z$ such that for some $y \in Y$, both $\langle x, y \rangle \in A$ and $\langle y, z \rangle \in B$. This is an extension of the usual notion of composition of functions, where y would be unique.

Definition. Given a set S, a *uniformity* on S is a filter \mathcal{U} in $S \times S$ with the following properties:

(a) Every $A \in \mathcal{U}$ includes the diagonal $D := \{\langle s, s \rangle : s \in S\}$.
(b) For each $A \in \mathcal{U}$, we have $A^{-1} := \{\langle y, x \rangle : \langle x, y \rangle \in A\} \in \mathcal{U}$.
(c) For each $A \in \mathcal{U}$, there is a $B \in \mathcal{U}$ with $B \circ B \subset A$.

The pair $\langle S, \mathcal{U} \rangle$ is called a *uniform space.*

A set $A \subset S \times S$ will be called *symmetric* iff $A = A^{-1}$.

Recall that a pseudometric d satisfies all conditions for a metric except possibly $d(x, y) = 0$ for some $x \neq y$. If (S, d) is a (pseudo)metric space, then the *(pseudo)metric uniformity* for d is the set \mathcal{U} of all subsets A of $S \times S$ such that for some $\delta > 0$, A includes $\{\langle x, y \rangle : d(x, y) < \delta\}$. It is easy to check that this is, in fact, a uniformity.

Let $\langle S, \mathcal{U} \rangle$ and $\langle T, \mathcal{V} \rangle$ be two uniform spaces. Then a function f from S into T is called *uniformly continuous* for these uniformities iff for each $B \in \mathcal{V}, \{\langle x, y \rangle \in S \times S : \langle f(x), f(y) \rangle \in B\} \in \mathcal{U}$. If T is the real line, then \mathcal{V} will be assumed (usually) to be the uniformity defined by the usual metric $d(x, y) := |x - y|$.

For any uniform space $\langle S, \mathcal{U} \rangle$, the *uniform topology* \mathcal{T} on S defined by \mathcal{U} is the collection of all sets $V \subset S$ such that for each $x \in V$, there is a $U \in \mathcal{U}$ such that $\{y : \langle x, y \rangle \in U\} \subset V$.

Since a uniformity is a filter, a *base* of the uniformity will just mean a base of the filter, as defined before Theorem 2.1.2. If (S, d) is a metric space, then clearly the topology defined by the metric uniformity is the usual topology defined by the metric. (Pseudo)metric uniformities are characterized neatly as follows:

2.7.1. Theorem *A uniformity \mathcal{U} for a space S is pseudometrizable (it is the pseudometric uniformity for some pseudometric d) if and only if \mathcal{U} has a countable base.*

Proof. "Only if": The sets $\{\langle x, y \rangle : d(x, y) < 1/n\}, n = 1, 2, \ldots$, clearly form a countable base for the uniformity of the pseudometric d.

"If": Let the uniformity \mathcal{U} have a countable base $\{U_n\}$. For any U in a uniformity \mathcal{U}, applying (c) twice, there is a $V \in \mathcal{U}$ with $(V \circ V) \circ (V \circ V) \subset U$. Recursively, let $V_0 := S \times S$. For each $n = 1, 2, \ldots$, let W_n be the intersection of U_n with a set $V \in \mathcal{U}$ satisfying $(V \circ V) \circ (V \circ V) \subset V_{n-1}$. Let $V_n := W_n \cap W_n^{-1} \in \mathcal{U}$. Then $\{V_n\}$ is a base for \mathcal{U}, consisting of symmetric sets, with $V_n \circ V_n \circ V_n \subset V_{n-1}$ for each $n \geq 1$. The next fact will yield a proof of Theorem 2.7.1:

2.7.2. Lemma *For V_n as just described, there is a pseudometric d on $S \times S$ such that*

$$V_{n+1} \subset \{\langle x, y \rangle : d(x, y) < 2^{-n}\} \subset V_n \quad \text{for all } n = 0, 1, \ldots.$$

Proof. Let $r(x, y) := 2^{-n}$ iff $\langle x, y \rangle \in V_n \backslash V_{n+1}$ for $n = 0, 1, \ldots,$ and $r(x, y) := 0$ iff $\langle x, y \rangle \in V_n$ for all n. Since each V_n is symmetric, so is $r: r(x, y) = r(y, x)$ for all x and y in S.

For each x and y in S, let $d(x, y)$ be the infimum of all sums $\sum_{0 \le i \le n} r(x_i, x_{i+1})$ over all $n = 0, 1, 2, \ldots,$ and sequences x_0, \ldots, x_{n+1} in S with $x_0 = x$ and $x_{n+1} = y$. Then d is nonnegative. Since r is symmetric, so is d. From its definition, d satisfies the triangle inequality, so d is a pseudometric. Since $d \le r$, clearly $V_{n+1} \subset \{\langle x, y \rangle : d(x, y) < 2^{-n}\}$. The next step is the following:

2.7.3. Lemma $r(x_0, x_{n+1}) \le 2 \sum_{0 \le i \le n} r(x_i, x_{i+1})$ *for any* x_0, \ldots, x_{n+1}.

Proof. We use induction on n. The lemma clearly holds for $n = 0$. Let $L(j, k) := \sum_{j \le i < k} r(x_i, x_{i+1})$ for $0 \le j < k \le n + 1$ and $L := L(0, n + 1)$. Let k be the largest $i \le n$ such that $L(0, i) \le L/2$. Then $L(k + 1, n + 1) = L - L(0, k + 1) \le L/2$ and $k < n + 1$. By induction hypothesis, $r(x_0, x_k)$ and $r(x_{k+1}, x_{n+1})$ are each at most $2(L/2) = L$, and clearly $r(x_k, x_{k+1}) \le L$. Let m be the smallest integer such that $2^{-m} \le L$. Then $\langle x_0, x_k \rangle, \langle x_k, x_{k+1} \rangle$, and $\langle x_{k+1}, x_{n+1} \rangle$ all belong to V_m. If $m = 0$, the lemma clearly holds, so suppose $m \ge 1$. Then by choice of the V_j, we have $\langle x_0, x_{n+1} \rangle \in V_{m-1}$, so $r(x_0, x_{n+1}) \le 2^{1-m} \le 2L$, proving Lemma 2.7.3. \square

Thus, if $d(x, y) < 2^{-n}$, then $r(x, y) < 2^{1-n}$, so in fact $r(x, y) \le 2^{-n}$ and $\langle x, y \rangle \in V_n$, finishing the proof of Lemma 2.7.2. Theorem 2.7.1 then follows. \square

Uniformities also come up naturally in the study of topological groups. A *group* is a set G together with a function $(x, y) \mapsto xy$ from $G \times G$ into G, which is associative: $x(yz) \equiv (xy)z$, for which there is an "identity" $e \in G$ such that $ex = xe = x$ for all $x \in G$, and such that for all $x \in G$, there is an inverse $y \in G$ such that $xy = yx = e$, written $y = x^{-1}$. A *topological group* is a group G together with a topology \mathcal{T} on G such that the multiplication $\langle x, y \rangle \mapsto xy$ is continuous from $G \times G$ (with product topology) onto G and the inverse operation $x \mapsto x^{-1}$ is continuous from G onto itself. For any topological group (G, \mathcal{T}), the collection of all sets $\{\langle x, y \rangle : xy^{-1} \in U\}$, for any neighborhood U of the identity in G, is a base for a uniformity,

called the *right uniformity*. Likewise the sets $\{\langle x, y \rangle: x^{-1}y \in U\}$, where U is any neighborhood of the identity, form a base of a uniformity called the *left uniformity*.

A group is called *Abelian* if $xy = yx$ for every x and y in G. For an Abelian group, clearly, the left and right uniformities are the same. The group operation is written as addition instead of multiplication, and $-x$ instead of x^{-1}.

Problems

1. If X, Y, and Z are uniform spaces, f is uniformly continuous from X into Y, and g is uniformly continuous from Y into Z, prove that $g \circ f$ is uniformly continuous from X into Z.

2. Prove that any uniformity has a base consisting of symmetric sets.

3. For the real line with usual topology, find two different uniformities giving the topology. *Hint:* Give two metrics for the topology such that the identity is not uniformly continuous from one metric to the other.

4. A uniform space (S, \mathcal{U}) is called *separated* iff for every $x \neq y$ in S, there is a $U \in \mathcal{U}$ with $\langle x, y \rangle \notin U$.
 (a) Show that the uniform topology of a separated uniformity is always Hausdorff.
 (b) Show that if S has more than one point, a separated uniformity never converges as a filter for its product topology.

5. Given a uniform space (S, \mathcal{U}), a net $\{x_\alpha\}_{\alpha \in I}$ in S is called a *Cauchy net* iff for any $V \in \mathcal{U}$ there is a $\gamma \in I$ such that $\langle x_\alpha, x_\beta \rangle \in V$ whenever $\alpha \geq \gamma$ and $\beta \geq \gamma$. The uniform space is called *complete* if every Cauchy net converges to some element of S (for the topology of \mathcal{U}). If a metric space (S, d) is complete, prove it is also complete as a uniform space.

6. Show that any compact uniform space S (compact for the topology of the uniformity) is complete. *Hint:* For a Cauchy net $\{x_\alpha\}$, the collection of all $\{x_\alpha : \alpha \geq \gamma\}$, $\gamma \in I$, is a filter base that extends to a filter and an ultrafilter (Theorem 2.2.4) which converges (Theorem 2.2.5). Show that the net x_α converges to the same limit.

7. If (S, \mathcal{U}) and (T, \mathcal{V}) are compact uniform spaces and f is a function from S into T, continuous for the topologies of the uniformities, show that it is uniformly continuous.

8. Referring to Problem 7, show that a compact Hausdorff space has a unique uniformity giving its topology.

9. Let G be the *affine group of the line*, namely, the set of all transformations $x \mapsto ax + b$ of \mathbb{R} onto itself where $a \neq 0$ and $b \in \mathbb{R}$. Let G have its relative topology as $\{\langle a, b\rangle : a \neq 0\}$ in \mathbb{R}^2. Show that G is a topological group for which the left and right uniformities are different.

10. A metric d on an Abelian group G will be called an *invariant* metric iff $d(x + z, y + z) = d(x, y)$ for all x, y, and z in G.
 (a) Show that a metrizable Abelian topological group can be metrized by an invariant metric d.
 (b) If an Abelian topological group G can be metrized by a metric for which it is complete, show that it is also complete for the invariant d from part (a). *Hint:* If not, take the completion of G for d and show it has an Abelian group structure extending that of G continuously; apply Theorem 2.5.4 and the category theorem (2.5.2).

*2.8. Compactification

In this section the question is, given a topological space (X, \mathcal{T}), is it homeomorphic to a subset of a compact Hausdorff space? If K is a compact Hausdorff space and f is a homeomorphism of X onto a dense subset of K, then K or (K, f) is called a *compactification* of X or (X, \mathcal{T}). There is an easy compactification for locally compact spaces (those in which every point has a compact neighborhood):

2.8.1. Theorem *If (X, \mathcal{T}) is a locally compact (but not compact) Hausdorff space, it has a compactification (K, f) where the range of f contains all but one point of K.*

Proof. Let ∞ be any point not in X and $K := X \cup \{\infty\}$. Let \mathcal{U} be the topology on K with a base given by the sets in \mathcal{T} and all sets $\{\infty\}\cup(X\setminus L)$ for all compact subsets L of X. The function f is the identity from X into K. It is easily seen that the topology \mathcal{U} is compact and that its relative topology on X is \mathcal{T}, so f is a homeomorphism. \square

The compactification $(X \cup \{\infty\}, \mathcal{U})$ of (X, \mathcal{T}) given by Theorem 2.8.1 is called the *one-point compactification* of (X, \mathcal{T}). For example, if $X = \mathbb{R}$ with usual topology, then it can be checked (this is left as a problem) that the one-point compactification of \mathbb{R} is homeomorphic to a circle; if we label the added point in this case as ω, then as $n \to \infty$ (in the usual sense of integers becoming large), then $n \to \omega$ (in the topology of the one-point compactification) and

also $-n \to \omega$. Actually, for \mathbb{R}, another compactification is more often useful: the two points $-\infty$ and $+\infty$ are adjoined to \mathbb{R}, where each neighborhood of $-\infty$ includes some set $[-\infty, -n)$ and each neighborhood of $+\infty$ includes some set $(n, \infty]$ for $n = 1, 2, \dots$. The result is the so-called set of extended real numbers, $[-\infty, \infty]$.

For metric spaces, one may ask whether there is a metrizable compactification. For \mathbb{R}, and many other spaces, the one-point compactification is metrizable. The one-point compactification works poorly for some other spaces, such as infinite-dimensional Banach and Hilbert spaces, defined in Chapter 5. Here is one example. Let ℓ^1 be the set of all sequences $\{x_n\}$ of real numbers such that $\sum_{n=1}^{\infty} |x_n| < \infty$. For two such sequences let $d(\{x_n\}, \{y_n\}) := \sum_n |x_n - y_n|$. Then (ℓ^1, d) is a metric space and is separable (using sequences of rational numbers r_n with $r_n = 0$ for n large enough). (ℓ^1, d) is not locally compact; for example, in any basic neighborhood U of 0, $U := \{\{y_n\}: \sum_n |y_n| < \delta\}$ for $\delta > 0$, let e_m be the sequence with $e_{mn} = \delta/2$ for $n = m$ and 0 for $n \neq m$. Then e_m has no convergent subsequence. The topology as defined above for the "one-point compactification" of ℓ^1 is in fact compact, but it is not Hausdorff.

Since compact metric spaces are separable (because totally bounded, by Theorem 2.3.1) and any subset of a separable metric space is separable (Problem 5 of Section 2.1), a necessary condition for a metrizable compactification of a metric space is separability. This condition is actually sufficient (so that, for example, ℓ^1 has a metrizable compactification):

2.8.2. Theorem *For any separable metric space (S, d), there is a totally bounded metrization. That is, there is a metric e on S, defining the same topology as d, such that (S, e) is totally bounded, so that the completion of S for e is a compact metric space and a compactification of S.*

Proof. Let $\{x_n\}_{n \geq 1}$ be dense in S. Let $f(t) := t/(1 + t)$, so that $f \circ d$ is a metric bounded by 1, with the same topology as d, as shown in Proposition 2.4.3. So we can assume $d < 1$. The Cartesian product $\Pi_{n=1}^{\infty} [0, 1]$ of copies of $[0, 1]$, with product topology, is compact by Tychonoff's theorem (2.2.8). A metric for the topology is, by Proposition 2.4.4,

$$\alpha(\{u_n\}, \{v_n\}) := \sum_n |u_n - v_n|/2^n.$$

So this metric is totally bounded (Theorem 2.3.1). Define a metric e on S by $e(x, y) := \alpha(\{d(x, x_n)\}, \{d(y, x_n)\})$. Then (S, e) is totally bounded. Now a

sequence $y_m \to y$ in S if and only if for all n, $\lim_{m \to \infty} d(y_m, x_n) = d(y, x_n)$: "only if" is clear, and "if" can be shown by taking x_n close to y. So $y_m \to y$ if and only if $e(y_m, y) \to 0$, as in Proposition 2.4.4. Thus e metrizes the d topology. The completion of (S, e) is still totally bounded, so it is compact by Theorem 2.3.1. $\qquad\qquad\square$

For a general Hausdorff space (X, \mathcal{T}), which may not be locally compact or metrizable, the existence of a compactification will be proved equivalent to the following: (X, \mathcal{T}) is called *completely regular* if for every closed set F in X and point p not in F there is a continuous real function f on X with $f(x) = 0$ for all $x \in F$ and $f(p) = 1$. Note that, for example, if we take a compact Hausdorff space K and delete one point q, the remaining space is easily shown to be completely regular: since $F \cup \{q\}$ is closed in K, by Theorem 2.6.2 and Urysohn's Lemma (2.6.3) there is a continuous f on K with $f(p) = 1$ and $f \equiv 0$ on $F \cup \{q\}$.

2.8.3. Theorem (Tychonoff) *A topological space (X, \mathcal{T}) is homeomorphic to a subset of a compact Hausdorff space if and only if (X, \mathcal{T}) is Hausdorff and completely regular.*

Remarks. A Hausdorff, completely regular space is called a $T_{3\frac{1}{2}}$ space, or a *Tychonoff* space. So Theorem 2.8.3 says that a space is homeomorphic to a subset of a compact Hausdorff space if and only if it is a Tychonoff space.

Proof. Let X be a subset of a compact Hausdorff space K. Let F be a (relatively) closed subset of X, and $p \in X \backslash F$. Let H be the closure of F in K. Then H is a closed subset of K and $p \notin H$. Also, $\{p\}$ is closed in K. So p can be separated from H by a continuous real function f by the Tietze-Urysohn theorems (2.6.3 and 2.6.4 in light of 2.6.2). Restricting f to X shows that X is a Tychonoff space.

Conversely, let X be a Tychonoff space. Let $G(X)$ be the set of all continuous functions from X into $[0, 1]$ with the usual topology on $[0, 1]$. Let K be the set of all functions from $G(X)$ into $[0, 1]$, with the product topology, which is compact Hausdorff (Tychonoff's theorem, 2.2.8). For each g in $G(X)$ and x in X let $f(x)(g) := g(x)$. Then f is a function from X into K. To show that f is continuous, it is enough (by Corollary 2.2.7) to check that $f^{-1}(U)$ is open in X for each U in the standard subbase of the product topology, that is, the collection of all sets $\{y \in K : y(g) \in V\}$ for each $g \in G(X)$ and open V in \mathbb{R}. For such a set, $f^{-1}(U) = g^{-1}(V)$ and is open since g is continuous. Next,

to show that f is a homeomorphism, first note that it is 1–1 since for any $x \neq y$ in X, by complete regularity, there is a $g \in G(X)$ with $g(x) \neq g(y)$ so $f(x) \neq f(y)$. Let W be any open set in X. To show that the direct image $f(W) := \{f(x): x \in W\}$ is relatively open in $f(X)$, let $x \in W$. By definition of completely regular space, take a continuous real function g on X with $g(x) = 1$ and $g(y) = 0$ for all $y \in X\backslash W$. We can assume that $0 \leq g \leq 1$, replacing g by $\max(0, \min(g, 1))$. Then $g \in G(X)$. Let U be the set of all $z \in K$ such that $z(g) > 0$. Then U is open. The intersection of U with $f(X)$ is included in $f(W)$ and contains $f(x)$. So $f(W)$ is relatively open, and f is a homeomorphism from X onto $f(X)$. □

The closure of the range $f(X)$ in K in the last proof is a compact Hausdorff space, which has been called the *Stone-Čech compactification* of X, although historically it might more accurately be called the *Tychonoff-Čech* compactification.

Another method of compactification applies to spaces Y^J where (Y, \mathcal{T}) is a Tychonoff topological space and Y^J is the set of all functions from J into Y, with product topology. Then if (K, \mathcal{U}) is a compactification of (Y, \mathcal{T}), it is easily seen that K^J, with product topology, is a compactification of Y^J. For example, if $Y = \mathbb{R}$, let $\overline{\mathbb{R}}$ be its two-point compactification $[-\infty, \infty]$. Then for any set J, the space \mathbb{R}^J of all real functions on J has a compactification $\overline{\mathbb{R}}^J$.

Problems

1. Prove that the one-point compactification of \mathbb{R}^k (with usual topology) is homeomorphic to the sphere $S^k := \{(x_1, \ldots, x_{k+1}) \in \mathbb{R}^{k+1}: x_1^2 + \cdots + x_{k+1}^2 = 1\}$ with its relative topology from \mathbb{R}^{k+1}:
 (a) for $k = 1$ (where S^1 is a circle),
 (b) for general k. *Hint:* In \mathbb{R}^{k+1} let S be the sphere with radius 1 and center $p = (0, \ldots, 0, 1)$, $S = \{y: |y - p| = 1\}$. For each $y \neq 2p$ in S the unique line through $2p$ and y intersects $\{x: x_{k+1} = 0\}$ at a unique point $g(y)$. Show that g gives a homeomorphism of $S\backslash\{2p\}$ onto \mathbb{R}^k.

2. Let (X, \mathcal{T}) have a compactification (K, f) where K contains only one point not in the range of f and K is a compact Hausdorff space. Prove that (X, \mathcal{T}) is locally compact.

3. Show that for any Tychonoff space X, any bounded, continuous, real-valued function on X can be extended to such a function on the Tychonoff-Čech compactification of X.

4. If (X, \mathcal{T}) is a locally compact Hausdorff space, show that X, as a subset of its Tychonoff-Čech compactification K, is open (if f is the homeomorphism of X into K given in the definition of Tychonoff-Čech compactification, the range of f is open). *Hint:* Given any $x \in X$, let U be a neighborhood of x with compact closure. Show that there is a continuous real function f on X, 0 at x, and 1 on $X \backslash U$. Use this function to show that x is not in the closure of $K \backslash U$.

5. Let K be the Tychonoff-Čech compactification of \mathbb{R}. Show that addition from $\mathbb{R} \times \mathbb{R}$ onto \mathbb{R} cannot be extended to a continuous function S from $K \times K$ into K. *Hint:* Let x_α be a net in \mathbb{R} converging in K to a point $x \in K \backslash \mathbb{R}$. Then $-x_\alpha$ converges to some point $y \in K \backslash \mathbb{R}$. If S exists, then $S(x, y) = 0 \in \mathbb{R}$. Then by Problem 4 there must be neighborhoods U of x and V of y in K such that $S(u, v) \in \mathbb{R}$ and $|S(u, v)| < 1$ for all $u \in U$ and $v \in V$. Show, however, that each of U and V contains real numbers of arbitrarily large absolute value, to get a contradiction.

6. Let (X, d) be a locally compact separable metric space. Show that its one-point compactification is metrizable.

7. Let X be any noncompact metric space, considered as a subset of its Tychonoff-Čech compactification K. Let $y \in K \backslash X$. Show that K is not metrizable by showing that there is no sequence $x_n \in X$ with $x_n \to y$ in K. *Hint:* If $x_n \to y$, by taking a subsequence, assume that the points x_n are all different. Then, $\{x_{2n}\}_{n \geq 1}$ and $\{x_{2n-1}\}_{n \geq 1}$ form two disjoint closed sets in X. Apply Urysohn's Lemma (2.6.3) to get a continuous function f on X with $f(x_{2n}) = 1$ and $f(x_{2n-1}) = 0$ for all n. So $\{x_n\}$ cannot converge in K to y.

8. Show that for any metric space S, if $A \subset S$ and $x \in \overline{A} \backslash A$, then there is a bounded, continuous real-valued function on A which cannot be extended to a function continuous on $A \cup \{x\}$. *Hint:* $f(t) := \sin(1/t)$ for $t > 0$ cannot be extended continuously to $t = 0$.

Notes

§2.1 According to Grattan-Guinness (1970, pp. 51–53, 76), priority for the definitions of limit and continuity for real functions of real variables belongs to Bolzano (1818), then Cauchy (1821).

For sets of real numbers, Cantor (1872) defined the notions of "neighborhood" and "accumulation point." A point x is an accumulation point of a set A iff every neighborhood of x contains points of A other than x. Cantor published a series of papers in which he developed the notion of "derived set" Y of a set X, where Y is the set of all accumulation points of X, also in several dimensions (Cantor, 1879–1883). The ideas of open and closed sets, interior, and closure are present, at least implicitly, in these papers. (The closure of a set is its union with its first derived set.) Maurice Fréchet

(1906, pp. 17, 30) began the study of metric spaces; Siegmund-Schultze (1982, Chap. 4) surveys the surrounding history. Fréchet also gave abstract formulations of convergence of sequences even without a metric. Hausdorff (1914, Chap. 9, §1), after defining what are now called Hausdorff spaces, gave the definition of continuous function in terms of open sets. Kuratowski (1958, pp. 20, 29) reviews these and other contributions to the definition of topological space, closure, etc. The concepts of nets and their convergence are due to E. H. Moore (1915), partly in joint work with H. L. Smith (Moore and Smith, 1922). Henri Cartan (1937) defined filters and ultrafilters. Earlier, Caratheodory (1913, p. 331) had worked with decreasing sequences of non-empty sets, which can be viewed as filter bases, and M. H. Stone (1936) had defined "dual ideals" which, if they do not contain \emptyset, are filters.

Hausdorff (1914) was the first book on general topology in not necessarily metric spaces. It also first proved some basic facts about metric spaces (see the notes to §§2.3 and 2.5). Felix Hausdorff lived from 1868 to 1942. As Eichhorn (1992) tells us, Hausdorff wrote several literary and philosophical works, including poems and a (produced) play, under the pseudonym Paul Mongré. Being Jewish, he encountered adversity under Nazi rule from 1933 on. In 1942 Hausdorff, his wife, and her sister all took their own lives to avoid being sent to a concentration camp.

Heine (1872, p. 186) proved that a continuous real function on a closed interval attains its maximum and minimum, by the successive bisection of the interval, as in the example after the proof of Theorem 2.1.2. Fréchet (1918) invented $L*$-spaces. Kisyński (1959–1960) proved that if C is an $L*$-convergence, then $C(T(C)) = C$.

Alexandroff and Fedorchuk (1978) survey the history of set-theoretic topology, giving 369 references. See also Arboleda (1979).

§2.2 The important book of Bourbaki (1953, p. 45) includes the Hausdorff separation condition ("*séparé*") in the definition of compact topological space. Most other authors prefer to write about "compact Hausdorff spaces." Several of the notions connected with compactness were first found in forms relating to sequences, countable open covers, etc., and only later put into more general forms. One of the first steps toward the notion of compactness was the statement by Bernard Bolzano (1781–1848) to the effect that every bounded infinite set of real numbers has an accumulation point. According to van Rootselaar (1970), no proof of this statement has been found in Bolzano's works, many of which remained in the form of unpublished manuscripts in the Austrian national library in Vienna. It appears that Bolzano made a number of errors on other points.

Borel (1895, pp. 51–52) showed that any covering of a bounded, closed interval in \mathbb{R} by a sequence of open intervals has a finite subcover. Lebesgue (1904, p. 117) extended the theorem to coverings by open "domaines" (homeomorphic to an open disk) of any set in \mathbb{R}^2 which is a continuous image of [0, 1], and specifically, via Peano curves, to any set homeomorphic to a closed square. Lebesgue (1907b) and Temple (1981) review the history of these so-called Heine-Borel or Heine-Borel-Lebesgue theorems. Borel (1895) gave the first explicit statement; its proof was implicit in a proof of Heine (1872).

In full generality, the current definition of compact space (then called "bicompact") was given by Alexandroff and Urysohn (1924). Alexandroff (1926, p. 561) showed that the range of a continuous function on a compact space is compact.

Fréchet (1906) proved that a countable product of copies of [0, 1] is compact. Tychonoff (1929–1930) actually proved that an arbitrary product of copies of [0, 1]

is compact. Čech (1937) proved the "Tychonoff" theorem that any Cartesian product of compact spaces is compact (2.2.8), which according to Kelley (1955, p. 143) is "probably the most important single theorem in general topology." Eduard Čech lived from 1893 to 1960. His papers on topology have been collected (Čech, 1968). The two books *Point Sets* (Čech 1936, 1966, 1969) and *Topological Spaces* (Čech 1959, 1966), on general topology, both posthumously translated into English, do not particularly address products of compact spaces. An introduction to Čech (1968) gives a 10-page scientific biography, "Life and work of Eduard Čech," by M. Katětov, J. Novák, and A. Švec. Čech contributed substantially to algebraic as well as general topology.

There were articles in honor of Tychonoff's fiftieth and sixtieth birthdays: Alexandroff et al. (1956, 1967). In fact, most of Tychonoff's work was in such fields as differential equations and mathematical physics.

H. Cartan (1937) defined ultrafilters and showed that a topological space is compact if and only if every ultrafilter converges (Theorem 2.2.5). He also showed that for any ultrafilter \mathcal{U} in a set X and function f from X to a set Y, the "direct image" of \mathcal{U}, defined as $\{B \subset Y : f^{-1}(B) \in \mathcal{U}\}$ is an ultrafilter. From these facts it is easy to get the ultrafilter proof of Tychonoff's theorem, although Cartan did not mention the Tychonoff theorem explicitly and the Čech general form was first published in the same year. Bourbaki (1940) gave a proof. Kelley (1955) gave two proofs. The second, referring to Bourbaki, is close to the ultrafilter proof but does not explicitly mention filters or ultrafilters. Chapter 2 of Kelley's book is on nets ("Moore-Smith convergence"), and filters are treated only in Problem L at the end of the chapter. See also Chernoff (1992) about proofs of Tychonoff's theorem. Feferman (1964, Theorem 4.12, p. 343) showed that without the axiom of choice, it is consistent with set theory that in \mathbb{N} the only ultrafilters are point ultrafilters.

Alexandroff (1926, p. 561) proved Theorem 2.2.11. Alexandroff also worked in algebraic topology, inventing, for example, the notion of exact sequence. Pontryagin and Mishchenko (1956), Kolmogoroff et al. (1966), and Arkhangelskii et al. (1976) wrote articles in honor of Alexandroff's sixtieth, seventieth, and eightieth birthdays.

§2.3 Hausdorff (1914, pp. 311–315) first defined the notion "totally bounded" and proved that a metric space is compact iff it is both totally bounded and complete.

§2.4 Cauchy, famous as the discoverer of, among other things, his integral theorem and integral formula in complex analysis, claimed mistakenly in 1823 that if a series of continuous functions converged at every point of an interval, the sum was continuous on that interval. Abel (1826) gave as a counterexample the sum $\sum_n (-1)^n (\sin nx)/n$, which converges to $x/2$ for $0 \le x < \pi$ and 0 at π. (Abel, for whom Abelian groups are named, lived from 1802 to 1829.) Cauchy (1833, pp. 55–56) did not notice, and repeated his error. The notion of uniform convergence began to appear in work of Abel (1826) and Gudermann (1838, pp. 251–252) in special cases. Manning (1975, p. 363) writes: "All of CAUCHY'S proofs prior to 1853 involving term-by-term integration of power series are invalid due to his failure to employ this concept [uniform convergence]." The theorem that a uniform limit of continuous functions is continuous (2.4.8, for functions of real variables) was proved independently by Seidel (1847–1849) and Stokes (1847–1848). Stokes mistakenly claimed that if a sequence of continuous functions converges pointwise, on a closed, bounded interval, to a continuous function, the convergence must be uniform. Seidel noted that he could *not* prove this converse. A leading mathematician of

a later era examined Stokes's and others' contributions (Hardy, 1916–1919). Eventually, Cauchy (1853) formulated the notion now called "Cauchy sequence" and showed the completeness not only of \mathbb{R} (2.4.2) but of C_b (2.4.9, for functions of real or complex variables on bounded sets).

Heine (1870, p. 361) apparently was the first to define uniform continuity of functions and (1872, p. 188) published a proof that any continuous real-valued function on a closed, bounded interval is uniformly continuous (the prototype of Cor. 2.4.6). Heine gave major credit to unpublished lectures and work of Weierstrass. Although much of Weierstrass's work in other fields was published after his death, apparently most of his work on real functions was still unpublished according to Biermann (1976). Heine, who was born Heinrich Eduard Heine in 1821, published under his middle name, perhaps to distinguish himself from the famous poet Heinrich Heine, 1797–1856. (A sister of Eduard's married a brother of the composer Felix Mendelssohn, who, among other composers, set to music some of the poet Heinrich Heine's poems.) According to Fréchet (1906, p. 36), who invented the d_{sup} metric (and metrics generally), Weierstrass was the first mathematician to make systematic use of uniform convergence (see Manning, 1975). The notion of equicontinuity is due to Arzelà (1882–1883) and Ascoli (1883–1884). The Arzelà-Ascoli theorem (2.4.7) is attributed to papers of Ascoli (1883–1884) and Arzelà (1889, 1895), although earlier Dini (1878) had proved a related result, as noted by Dunford and Schwartz (1958, pp. 382–383). Dini is best known, in real analysis, for his theorem on monotone convergence (2.4.10). He also did substantial work (21 papers) in differential geometry (Dini, 1953, vol. 1; Reich, 1973). Baire (1906) noticed that \mathbb{R} is homeomorphic to $(-1, 1)$. Hahn (1921) showed that any metric space is homeomorphic to a bounded one (2.4.3). Fréchet (1928) metrized countable products of metric spaces (2.4.4).

M. H. Stone (1947–48) proved the Stone-Weierstrass Theorem 2.4.11. Weierstrass (1885, pp. 5, 36) had proved polynomial approximation theorems (Corollary 2.4.12) for $d = 1$ and any finite d respectively. Weierstrass's convolution method seems to require the continuous function f to be defined in a neighborhood of the compact set K, as it could be, for example, by the Urysohn-Tietze extension theorem 2.6.4. On a bounded, closed interval in \mathbb{R}, an explicit approximation is given by Bernstein polynomials; see, for example, Bartle (1964, Theorem 17.6).

Peano (1890) defined a curve whose range is a square (Problem 2.4.9).

§2.5 Hausdorff (1914, pp. 315–316) proved that every metric space has a completion (2.5.1). The short proof given is due to Kuratowski (1935, p. 543), for bounded spaces. The extension to unbounded spaces is straightforward and was presumably noticed not long afterward. My thanks to L. Š. Grinblat for telling me the proof. The ideas in Kuratowski's proof are related to those of Fréchet (1910, pp. 159–161). Hausdorff's proof, given in many textbooks, is along the following lines. For any two Cauchy sequences $\{x_n\}$ and $\{y_n\}$ in S, let $e(\{x_n\}, \{y_n\}) := \lim_{n \to \infty} d(x_n, y_n)$. One proves that this limit always exists and that it defines a pseudometric on the set of all Cauchy sequences in S. Define a relation E by $x E y$ iff $e(x, y) = 0$. As with any pseudometric, this is an equivalence relation. On the set T of all equivalence classes for E, e defines a metric. Let f be the function from S into T such that for each x in S, $f(x)$ is the equivalence class of the Cauchy sequence $\{x_n\}$ with $x_n = x$ for all n. Then (T, e) is a completion of X. Although this proof is, in a way, natural and conceptually straightforward, there are more details involved in making it a full proof.

The category theorem (2.5.2) is often called the "Baire category theorem." Actually, Osgood (1897, pp. 171–173) proved it earlier in \mathbb{R}. Then Baire (1899, p. 65) proved it for \mathbb{R}^n in his thesis. Mazurkiewicz (1916) proved that every topologically complete space is a G_δ in its completion. Dugundji (1966) attributes the converse (and thus all of Theorem 2.5.4) to Mazurkiewicz. Alexandroff (1924) proved the converse, and Hausdorff (1924) gave a shorter proof.

§2.6 Lebesgue (1907a) proved the extension theorem (2.6.4) for $X = \mathbb{R}^2$ by a method that does not immediately extend beyond \mathbb{R}^k, at any rate. Tietze (1915, p. 14) first proved the theorem where f is bounded and X is a metric space. Borsuk (1934, p. 4) proved that if the closed subset F is separable in the metric space X, the extension of bounded continuous real functions, a mapping from $C_b(F)$ into $C_b(X)$, can be chosen to be linear. On the other hand, Dugundji (1951) extended Tietze's theorem to the case of more general, possibly infinite-dimensional range spaces in place of \mathbb{R}.

Tietze (1923, p. 301, axiom (h)) defined normal spaces. For normal spaces, Urysohn (1925, pp. 290–291), in a posthumous paper, proved his lemma (2.6.3), and then case (a) of the extension theorem (2.6.4), giving essentially the above proof (p. 293). Urysohn was born in 1898. Arkhangelskii et al. (1976) write that on a visit to Bonn in 1924, "every day Aleksandrov and Uryson swam across the Rhine—a feat that was far from being safe and provoked Hausdorff's displeasure ... on 17 August 1924, at the age of 26, Uryson drowned whilst bathing in the Atlantic." On Urysohn's life and career see Alexandroff (1950).

Alexandroff and Hopf (1935, p. 76) state Theorem 2.6.4 in general (case (c)) but actually prove only case (a). Caratheodory (1918, 1927, p. 619), for $X = \mathbb{R}^k$, noted that one can get from (a) to (b) by dividing g by $1 + d(x, F)$. This works in any metric space. For general normal spaces, the earliest reference I can give for the short but non-empty additional proof of the (b) case is Bourbaki (1948).

§2.7 André Weil (1937) began the theory of uniform spaces. For a more extensive exposition than that given here, see also, for example, Kelley (1955, Chap. 6). Kelley (p. 186) attributes the metrization Theorem 2.7.1 to Alexandroff and Urysohn (1923) and Chittenden (1927) and its current formulation and proof to Weil (1937), Frink (1937), and Bourbaki (1948).

§2.8 The one-point compactification is attributed to P. S. Alexandroff; it appears, for example, in Alexandroff and Hopf (1935, p. 93). A normal Hausdorff space is called a T_4 space. A topological space (X, \mathcal{T}) is called *regular* if for every point p not in a closed set F there are disjoint open sets U and V with $p \in U$ and $F \subset V$. A Hausdorff regular space is called a T_3 space. Every T_4 space is clearly a $T_{3.5}$ (= Tychonoff) space, and every Tychonoff space is T_3. Urysohn (1925, p. 292) used an assumption of complete regularity without naming it.

Tychonoff (1930) proved that every T_4 space has a compactification, defined completely regular spaces, gave examples of a T_3 space that is not $T_{3.5}$ and a $T_{3.5}$ space that is not T_4, and showed that a Hausdorff space has a (Hausdorff) compactification iff it is completely regular (Theorem 2.8.3).

The first paragraph of the paper Čech (1937), reprinted in Čech (1968), clearly states that Tychonoff (1930) had proved the existence, for any completely regular ($T_{3.5}$) space S of a compact Hausdorff space $\beta(S)$ such that (i) S is homeomorphic to a dense subset of $\beta(S)$ and (ii) every bounded continuous real function on S extends to such a function

on $\beta(S)$. Čech states "it is easily seen that $\beta(S)$ is uniquely defined by the two properties (i) and (ii). The aim of the present paper is chiefly the study of $\beta(S)$." Thus the "Stone-Čech" compactification $\beta(S)$ is due to Tychonoff and was developed by Čech. Stone (1937, pp. 455*ff.*, esp. 461–463) treats this compactification as one topic in a very long paper, citing Tychonoff in this connection only for the fact that the implications $T_4 \to T_{3.5} \to T_3$ cannot be reversed.

References

Lebesgue (1907b) criticizes Young and Young (1906) for a "*bibliographie trop copieuse*" of 300 references on point sets and topology, then called "analysis situs." Here are some 90 references. An asterisk identifies works I have found discussed in secondary sources but have not seen in the original.

Abel, Niels Henrik (1826). Untersuchungen über die Reihe $1 + \frac{m}{1}x + \frac{m(m-1)}{1\cdot2}x^2 + \frac{m(m-1)(m-2)}{1\cdot2\cdot3}x^2 + \cdots$, *J. reine angew. Math. 1*: 311–339. Also in *Oeuvres complètes*, ed. L. Sylow and S. Lie. Grøndahl, Kristiania [Oslo], 1881, I, pp. 219–250.

Alexandroff, Paul [Aleksandrov, Pavel Sergeevich] (1924). Sur les ensembles de la première classe et les espaces abstraits. *Comptes Rendus Acad. Sci. Paris 178*: 185–187.

——— (1926). Über stetige Abbildungen kompakter Räume. *Math. Annalen 96*: 555–571.

——— (1950). Pavel Samuilovich Urysohn. *Uspekhi Mat. Nauk 5*, no. 1: 196–202.

——— et al. (1967). Andrei Nikolaevich Tychonov (on his sixtieth birthday): On the works of A. N. Tychonov in *Uspekhi Mat. Nauk 22*, no. 2: 133–188 (in Russian); transl. in *Russian Math. Surveys 22*, no. 2: 109–161.

——— and V. V. Fedorchuk, with the assistance of V. I. Zaitsev (1978). The main aspects in the development of set-theoretical topology. *Russian Math. Surveys 33*, no. 3: 1–53. Transl. from *Uspekhi Mat. Nauk 33*, no. 3 (201): 3–48 (in Russian).

——— and Heinz Hopf (1935). *Topologie*, 1. Band. Springer, Berlin; repr. Chelsea, New York, 1965.

——— , A. Samarskii, and A. Sveshnikov (1956). Andrei Nikolaevich Tychonov (on his fiftieth birthday). *Uspekhi Mat. Nauk 11*, no. 6: 235–245 (in Russian).

——— and Paul Urysohn (1923). Une condition nécessaire et suffisante pour qu'une classe (\mathcal{L}) soit une classe (\mathcal{D}). *C. R. Acad. Sci Paris 177*: 1274–1276.

——— and——— (1924). Theorie der topologischen Räume. *Math. Annalen. 92*: 258–266.

Arboleda, L. C. (1979). Les débuts de l'école topologique soviétique: Notes sur les lettres de Paul S. Alexandroff et Paul S. Urysohn à Maurice Fréchet. *Arch. Hist. Exact Sci. 20*: 73–89.

Arkhangelskii, A. V., A. N. Kolmogorov, A. A. Maltsev, and O. A. Oleinik (1976). Pavel Sergeevich Aleksandrov (On his eightieth birthday). *Uspekhi Mat. Nauk 31*, no. 5: 3–15 (in Russian); transl. in *Russian Math. Surveys 31*, no. 5: 1–13.

*Arzelà, Cesare (1882–1883). Un'osservazione intorno alle serie di funzioni. *Rend. dell' Acad. R. delle Sci. dell'Istituto di Bologna*, pp. 142–159.

*———— (1889). Funzioni di linee. *Atti della R. Accad. dei Lincei, Rend. Cl. Sci. Fis. Mat. Nat.* (Ser. 4) *5*: 342–348.

*———— (1895). Sulle funzioni di linee. *Mem. Accad. Sci. Ist. Bologna Cl. Sci. Fis. Mat.* (Ser. 5) *5*: 55–74.

*Ascoli, G. (1883–1884). Le curve limiti di una varietà data di curve. *Atti della R. Accad. dei Lincei, Memorie della Cl. Sci. Fis. Mat. Nat.* (Ser. 3) *18*: 521–586.

Baire, René (1899). Sur les fonctions de variables réelles (Thèse). *Annali di Matematica Pura ed Applic.* (Ser. 3) *3*: 1–123.

———— (1906). Sur la représentation des fonctions continues. *Acta Math. 30*: 1–48.

Bartle, R. G. (1964). *The Elements of Real Analysis.* Wiley, New York.

Biermann, Kurt-R. (1976). Weierstrass, Karl Theodor Wilhelm. *Dictionary of Scientific Biography 14*: 219–224. Scribner's, New York.

*Bolzano, B. P. J. N. (1818). Rein analytischer Beweis des Lehrsatzes, dass zwischen je zwey Werthen, die ein entgegengesetztes Resultat gewähren, wenigstens eine reelle Wurzel der Gleichung liege. *Abh. Gesell. Wiss. Prague* (Ser. 3) *5*: 1–60.

Borel, Émile (1895). Sur quelques points de la théorie des fonctions. *Ann. Scient. Ecole Normale Sup.* (Ser. 3) *12*: 9–55.

Borsuk, Karol (1934). Über Isomorphie der Funktionalräume. *Bull. Acad Polon. Sci. Lett. Classe Sci. Sér. A Math.* 1933: 1–10.

Bourbaki, Nicolas [pseud.] (1940, 1948, 1953, 1961, 1971). *Topologie Générale, Chap. 9. Utilisation des nombres réels en topologie générale.* Hermann, Paris. English transl. *Elements of Mathematics. General Topology, Part 2.* Hermann, Paris; Addison-Wesley, Reading, Mass., 1966.

Cantor, Georg (1872). Ueber die Ausdehnung eines Satzes aus der Theorie der trigonometrischen Reihen. *Math. Annalen 5*: 123–132.

———— (1879–1883). Ueber unendliche, lineare Punktmannichfaltigkeiten. *Math. Annalen 15*: 1–7; *17*: 355–358; *20*: 113–121; *21*: 51–58.

Caratheodory, Constantin (1913). Über die Begrenzung einfach zusammenhängender Gebiete. *Math. Annalen 73*: 323–370.

———— (1918). *Vorlesungen über reelle Funktionen.* Teubner, Leipzig and Berlin. 2d ed., 1927.

Cartan, Henri (1937). Théorie des filtres: Filtres et ultrafiltres. *C. R. Acad. Sci. Paris 205*: 595–598, 777–779.

Cauchy, Augustin-Louis (1821). *Cours d'analyse de l'École Royale Polytechnique,* Paris.

———— (1823). *Résumés des leçons données à l'École Royale Polytechnique sur le calcul infinitésimal.* Debure, Paris; also in *Œuvres Complètes* (Ser. 2), IV, pp. 5–261, Gauthier-Villars, Paris, 1889.

*———— (1833). *Résumés analytiques.* Imprimerie Royale, Turin.

———— (1853). Note sur les séries convergentes dons les divers termes sont des functions continues d'une variable réelle ou imaginaire, entre des limites données. *Comptes Rendus Acad. Sci. Paris 36*: 454–459. Also in *Œuvres Complètes* (Ser. 1) XII, pp. 30–36. Gauthier-Villars, Paris, 1900.

Čech, Eduard (1936). *Point Sets.* In Czech, *Bodové Množiny.* 2d ed. 1966, Czech. Acad. Sci., Prague; *In English, transl. by Aleš Pultr, Academic Press, New York, 1969.

———— (1937). On bicompact spaces. *Ann. Math.* (Ser. 2) *38*: 823–844.

———— (1959). *Topological Spaces*. In Czech, *Topologické Prostory*. Rev. English Ed. (1966), Eds. Z. Frolík and M. Katětov; Czech. Acad. Sci., Prague; Wiley, London.
———— (1968). *Topological Papers of Eduard Čech*. Academia (Czech. Acad. Sci), Prague.
Chernoff, Paul R. (1992). A simple proof of Tychonoff's theorem via nets. *Amer. Math. Monthly 99*: 932–934.
Chittenden, E. W. (1927). On the metrization problem and related problems in the theory of abstract sets. *Bull. Amer. Math. Soc. 33*: 13–34.
*Dini, Ulisse (1878). *Fondamenti per la teorica delle funzioni di variabili reali*. Nistri, Pisa.
———— (1953–1959, posth.) *Opere. 5* vols. Ediz. Cremonese, Rome.
Dugundji, James (1951). An extension of Tietze's theorem. *Pacific J. Math. 1*: 352–367.
———— (1966). *Topology*. Allyn & Bacon, Boston.
Dunford, Nelson, and Jacob T. Schwartz, with the assistance of William G. Badé and Robert G. Bartle (1958). *Linear Operators, Part I: General Theory*. Interscience, New York.
Eichhorn, Eugen (1992). Felix Hausdorff—Paul Mongré. Some aspects of his life and the meaning of his death. In Gähler, W., Herrlich, H., and Preuß, G., eds., *Recent Developments of General Topology and its Applications; Internat. Conf. in Memory of Felix Hausdorff (1868–1942)*, Akademie Verlag, Berlin, pp. 85–117.
Feferman, Solomon (1964). Some applications of the notions of forcing and generic sets. *Fund. Math. 56*: 325–345.
Fréchet, Maurice (1906). Sur quelques points du calcul fonctionnel. *Rendiconti Circolo Mat. Palermo 22*: 1–74.
———— (1910). Les dimensions d'un ensemble abstrait. *Math. Annalen 68*: 145–168.
———— (1918). Sur la notion de voisinage dans les ensembles abstraits. *Bull. Sci. Math. 42*: 138–156.
———— (1928). *Les espaces abstraits*. Gauthier-Villars, Paris.
Frink, A. H. (1937). Distance functions and the metrization problem. *Bull. Amer. Math. Soc. 43*: 133–142.
Grattan-Guinness, I. (1970). *The Development of the Foundations of Analysis from Euler to Riemann*. MIT Press, Cambridge, MA.
Gudermann, Christof J. (1838). Theorie der Modular-Functionen und der Modular-Integrale, 4–5 Abschnitt. *J. reine angew. Math. 18*: 220–258.
Hahn, Hans (1921). *Theorie der reellen Funktionen 1*. Springer, Berlin.
Hardy, Godfrey H. (1916–1919). Sir George Stokes and the concept of uniform convergence. *Proc. Cambr. Phil. Soc. 19*: 148–156.
Hausdorff, Felix (1914). *Grundzüge der Mengenlehre*. Von Veit, Leipzig. (See References to Chap. 1 for later editions.)
———— (1924). Die Mengen G_δ in vollständigen Räumen. *Fund. Math. 6*: 146–148.
Heine, E. [Heinrich Eduard] (1870). Ueber trigonometrische Reihen. *Journal für die reine und angew. Math. 71*: 353–365.
———— (1872). Die Elemente der Functionenlehre. *Journal für die reine und angew. Math. 74*: 172–188.
Hoffman, Kenneth M. (1975). *Analysis in Euclidean Space*. Prentice-Hall, Englewood Cliffs, N.J.
Kelley, John L. (1955). *General Topology*. Van Nostrand, Princeton.

Kisyński, J. (1959–1960). Convergence du type L. *Colloq. Math.* 7: 205–211.

Kolmogorov, A. N., L. A. Lyusternik, Yu. M. Smirnov, A. N. Tychonov, and S. V. Fomin (1966). Pavel Sergeevich Alexandrov (On his seventieth birthday and the fiftieth anniversary of his scientific activity). *Uspekhi Mat. Nauk 21*, no. 4: 2–7 (in Russian); transl. in *Russian Math. Surveys 21*, no. 4: 2–6.

Kuratowski, Kazimierz (1935). Quelques problèmes concernant les espaces métriques non-séparables. *Fund. Math. 25*: 534–545.

——— (1958). *Topologie*, vol. 1. PWN, Warsaw; Hafner, New York.

Lebesgue, Henri (1904). *Leçons sur l'intégration et la recherche des fonctions primitives*. Gauthier-Villars, Paris.

——— (1907a). Sur le problème de Dirichlet. *Rend. Circolo Mat. Palermo 24*: 371–402.

——— (1907b). Review of Young and Young (1906). *Bull. Sci. Math.* (Ser. 2) *31*: 129–135.

Manning, Kenneth R. (1975). The emergence of the Weierstrassian approach to complex analysis. *Arch. Hist. Exact Sci. 14*: 297–383.

*Mazurkiewicz, Stefan (1916). Über Borelsche Mengen. *Bull. Acad. Cracovie 1916:* 490–494.

Moore, E. H. (1915). Definition of limit in general integral analysis. *Proc. Nat. Acad. Sci. USA 1*: 628–632.

——— and H. L. Smith (1922). A general theory of limits. *Amer. J. Math. 44*: 102–121.

Osgood, W. F. (1897). Non-uniform convergence and the integration of series term by term. *Amer. J. Math. 19*: 155–190.

Peano, G. (1890). Sur un courbe, qui remplit toute une aire plane. *Math. Ann. 36*: 157–160.

Pontryagin, L. S., and E. F. Mishchenko (1956). Pavel Sergeevich Aleksandrov (On his sixtieth birthday and fortieth year of scientific activity). *Uspekhi Mat. Nauk 11*, no. 4: 183–192 (in Russian).

Reich, Karin (1973). Die Geschichte der Differentialgeometrie von Gauss bis Riemann (1828–1868). *Arch. Hist. Exact Sci. 11*: 273–382.

van Rootselaar, B. (1970). Bolzano, Bernard. *Dictionary of Scientific Biography*, II, pp. 273–279. Scribner's, New York.

*Seidel, Phillip Ludwig von (1847–1849). Note über eine Eigenschaft der Reihen, welche discontinuierliche Functionen darstellen. *Abh. der Bayer. Akad. der Wiss.* (Munich) *5*: 379–393.

Siegmund-Schultze, Reinhard (1982). Die Anfänge der Funktionalanalysis und ihr Platz im Umwälzungsprozess der Mathematik um 1900. *Arch. Hist. Exact Sci. 26*: 13–71.

Stokes, George G. (1847–1848). On the critical values of periodic series. *Trans. Cambr. Phil. Soc. 8*: 533–583; *Mathematical and Physical Papers I*: 236–313.

Stone, Marshall Harvey (1936). The theory of representations for Boolean algebras. *Trans. Amer. Math. Soc. 40*: 37–111.

——— (1937). Applications of the theory of Boolean rings to general topology. *Trans. Amer. Math. Soc. 41*: 375–481.

——— (1947–48). The generalized Weierstrass approximation theorem. *Math. Mag. 21*: 167–184, 237–254. Repr. in *Studies in Modern Analysis 1*, ed. R. C. Buck, Math. Assoc. of Amer., 1962, pp. 30–87.

Temple, George (1981). *100 Years of Mathematics*. Springer, New York.

Tietze, Heinrich (1915). Über Funktionen, die auf einer abgeschlossenen Menge stetig sind. *J. reine angew. Math. 145*: 9–14.

——— (1923). Beiträge zur allgemeinen Topologie. I. Axiome für verschiedene Fassungen des Umgebungsbegriffs. *Math. Annalen 88*: 290–312.

Tychonoff, A. [Tikhonov, Andrei Nikolaevich] (1930). Über die topologische Erweiterung von Räumen. *Math. Ann. 102*: 544–561.

——— (1935). Über einen Funktionenraum. *Math. Ann. 111*: 762–766.

Urysohn, Paul [Urison, Pavel Samuilovich] (1925). Über die Mächtigkeit der zusammenhängenden Mengen. *Math. Ann. 94*: 262–295.

Weierstrass, K. (1885). Über die analytische Darstellbarkeit sogenannter willkürlicher Functionen reeller Argumente. *Sitzungsber. königl. preussischen Akad. Wissenschaften* 633–639, 789–805. *Mathematische Werke*, Mayer & Müller, Berlin, 1894–1927, vol. 3, pp. 1–37.

Weil, André (1937). *Sur les espaces à structure uniforme et sur la topologie générale. Actualités Scientifiques et Industrielles 551*, Paris.

Young, W. H., and Grace Chisholm Young (1906). *The Theory of Sets of Points.* Cambridge Univ. Press.

3

Measures

3.1. Introduction to Measures

A classical example of measure is the length of intervals. In the modern theory of measure, developed by Émile Borel and Henri Lebesgue around 1900, the first task is to extend the notion of "length" to very general subsets of the real line. In representing intervals as finite, disjoint unions of other intervals, it is convenient to use left open, right closed intervals. The length is denoted by $\lambda((a, b]) := b - a$ for $a \leq b$. Now, in the extended real number system $[-\infty, \infty] := \{-\infty\} \cup \mathbb{R} \cup \{+\infty\}$, $-\infty$ and $+\infty$ are two objects that are not real numbers. Often $+\infty$ is written simply as ∞. The linear ordering of real numbers is extended by setting $-\infty < x < \infty$ for any real number x. Convergence to $\pm\infty$ will be for the interval topology, as defined in §2.2; for example, $x_n \to +\infty$ iff for any $K < \infty$ there is an m with $x_n > K$ for all $n > m$. If a sequence or series of real numbers is called convergent, however, and the limit is not specified, then the limit is supposed to be in \mathbb{R}, not $\pm\infty$. For any real x, $x + (-\infty) := -\infty$ and $x + \infty := +\infty$, while $\infty - \infty$, or $\infty + (-\infty)$, is undefined, although of course it may happen that $a_n \to +\infty$ and $b_n \to -\infty$ while $a_n + b_n$ approaches a finite limit.

Let X be a set and \mathcal{C} a collection of subsets of X with $\emptyset \in \mathcal{C}$. Recall that sets $A_n, n = 1, 2, \ldots$, are said to be *disjoint* iff $A_i \cap A_j = \emptyset$ whenever $i \neq j$. A function μ from \mathcal{C} into $[-\infty, \infty]$ is said to be *finitely additive* iff $\mu(\emptyset) = 0$ and whenever A_i are disjoint, $A_i \in \mathcal{C}$ for $i = 1, \ldots, n$, and

$$A := \bigcup_{i=1}^{n} A_i \in \mathcal{C}, \quad \text{we have} \quad \mu(A) = \sum_{i=1}^{n} \mu(A_i).$$

(Thus, all such sums must be defined, so that there cannot be both $\mu(A_i) = -\infty$ and $\mu(A_j) = +\infty$ for some i and j.) If also whenever $A_n \in \mathcal{C}, n = 1, 2, \ldots, A_n$ are disjoint and $B := \bigcup_{n \geq 1} A_n \in \mathcal{C}$, we have $\mu(B) = \sum_{n \geq 1} \mu(A_n)$, then μ is called *countably additive*.

Recall that for any set X, the power set 2^X is the collection of all subsets of X.

Example. Let $p \neq q$ in a set X and let $m(A) = 1$ if A contains both p and q, and $m(A) = 0$ otherwise. Then m is *not* additive on 2^X.

Definitions. Given a set X, a collection $\mathcal{A} \subset 2^X$ is called a *ring* iff $\emptyset \in \mathcal{A}$ and for all A and B in \mathcal{A}, we have $A \cup B \in \mathcal{A}$ and $B \setminus A \in \mathcal{A}$. A ring \mathcal{A} is called an *algebra* iff $X \in \mathcal{A}$. An algebra \mathcal{A} is called a σ-*algebra* if for any sequence $\{A_n\}$ of sets in \mathcal{A}, $\bigcup_{n \geq 1} A_n \in \mathcal{A}$.

For example, in any set X, the collection of all finite sets is a ring, but it is not an algebra unless X is finite. The collection of all finite sets and their complements is an algebra but not a σ-algebra, unless, again, X is finite.

Note that for any A and B in ring \mathcal{R}, $A \cap B = A \setminus (A \setminus B) \in \mathcal{R}$. For any set X, 2^X is a σ-algebra of subsets of X. For any collection $\mathcal{C} \subset 2^X$, there is a smallest algebra including \mathcal{C}, namely, the intersection of all algebras including \mathcal{C}. Likewise, there is a smallest σ-algebra including \mathcal{C}. This algebra and σ-algebra are each said to be *generated* by \mathcal{C}.

For example, if \mathcal{A} is the collection of all singletons $\{x\}$ in a set X, the algebra generated by \mathcal{A} is the collection of all subsets A of X which are finite or have finite complement $X \setminus A$. The σ-algebra generated by \mathcal{A} is the collection of sets which are countable or have countable complement.

Here is a first criterion for being countably additive. For any sequence of sets A_1, A_2, \dots, $A_n \downarrow \emptyset$ means $A_n \supset A_{n+1}$ for all n and $\bigcap_n A_n = \emptyset$. For an infinite interval, such as $[c, \infty)$, with c finite, we have $\lambda([c, \infty)) := \infty$. Then for $A_n := [n, \infty)$, we have $A_n \downarrow \emptyset$ but $\lambda(A_n) = +\infty$ for all n, not converging to 0. This illustrates why, in the following statement, μ is required to have real (finite) values.

3.1.1. Theorem *Let μ be a finitely additive, real-valued function on an algebra \mathcal{A}. Then μ is countably additive if and only if μ is "continuous at \emptyset," that is, $\mu(A_n) \to 0$ whenever $A_n \downarrow \emptyset$ and $A_n \in \mathcal{A}$.*

Proof. First suppose μ is countably additive and $A_n \downarrow \emptyset$ with $A_n \in \mathcal{A}$. Then the sets $A_n \setminus A_{n+1}$ are disjoint for all n and their union is A_1. Also, their union for $n \geq m$ is A_m for each m. It follows that $\sum_{n \geq m} \mu(A_n \setminus A_{n+1}) = \mu(A_m)$ for each m. Since the series $\sum_{n \geq 1} \mu(A_n \setminus A_{n+1})$ converges, the sums for $n \geq m$ must approach 0, so μ is continuous at \emptyset.

Conversely, suppose μ is continuous at \emptyset, and the sets B_j are disjoint and in \mathcal{A} with $B := \bigcup_j B_j \in \mathcal{A}$. Let $A_n := B \setminus \bigcup_{j<n} B_j$. Then $A_n \in \mathcal{A}$ and $A_n \downarrow \emptyset$, so $\mu(A_n) \to 0$. By finite additivity,

$$\mu(B) = \mu(A_n) + \sum_{j<n} \mu(B_j) \qquad \text{for each } n.$$

Letting $n \to \infty$ gives $\mu(B) = \sum_j \mu(B_j)$. $\qquad\qquad\qquad\qquad\square$

Definition. A countably additive function μ from a σ-algebra \mathcal{S} of subsets of X into $[0, \infty]$ is called a *measure*. Then (X, \mathcal{S}, μ) is called a *measure space*.

For $A \subset X$ let $\mu(A)$ be the cardinality of A for A finite, and $+\infty$ for A infinite. Then μ is always a measure, called the *counting measure* on X.

To rearrange sums of nonnegative terms, the following will be of use. It is proved in Appendix D.

3.1.2. Lemma *If $a_{mn} \geq 0$ for all m and n in \mathbb{N} and $k \mapsto \langle m(k), n(k) \rangle$ is any $1{-}1$ function from \mathbb{N} onto $\mathbb{N} \times \mathbb{N}$, then*

$$\sum_{m=0}^{\infty} \sum_{n=0}^{\infty} a_{mn} = \sum_{n=0}^{\infty} \sum_{m=0}^{\infty} a_{mn} = \sum_{k=0}^{\infty} a_{m(k),n(k)}.$$

While showing that length is countably additive, and can be extended to be countably additive on a large collection of sets (a σ-algebra), it will be useful to show at the same time that countable additivity also holds if the length $b - a$ of the interval $(a, b]$ is replaced by $G(b) - G(a)$ for a suitable function G. A function G from \mathbb{R} into itself is called *nondecreasing* iff $G(x) \leq G(y)$ whenever $x \leq y$. Then for any x, the limit $G(x^+) := \lim_{y \downarrow x} G(y)$ exists. G is said to be *continuous from the right* iff $G(x^+) = G(x)$ for all x. Let G be a nondecreasing function which is continuous from the right. As $x \uparrow +\infty$, $G(x) \uparrow z$ for some $z \leq +\infty$. Set $G(+\infty) := z$. Likewise define $G(-\infty)$ so that $G(x) \downarrow G(-\infty)$ as $x \downarrow -\infty$.

For example, if G is the indicator function of the interval $(1, \infty)$, then G is nondecreasing, $G(1) = 0$, and $G(1^+) = 1$. So G is not continuous from the right (it is from the left).

Let $\mathcal{C} := \{(a, b] : -\infty < a \leq b < +\infty\}$. A function $\mu := \mu_G$ is defined on \mathcal{C} by $\mu((a, b]) := G(b) - G(a)$. If $a = b$, so that $(a, b] = \emptyset$, we have $\mu(\emptyset) = 0$. Then $0 \leq \mu(A) < +\infty$ for each $A \in \mathcal{C}$.

3.1.3. Theorem *If G is nondecreasing and continuous from the right, then on \mathcal{C}, μ is countably additive.*

Measures

Proof. Let us first show that μ is finitely additive. Suppose $(a, b] = \bigcup_{1 \leq i \leq n} (a_i, b_i]$ where the $(a_i, b_i]$ are disjoint and we may assume the intervals are non-empty. Without changing the sum of the $\mu((a_i, b_i])$, the intervals can be relabeled so that $b_{j-1} = a_j$ for $j = 2, \ldots, n$. Then

$$G(b) - G(a) = \sum_{1 \leq j \leq n} G(b_j) - G(a_j)$$

(by a telescoping sum), so μ is finitely additive on \mathcal{C}. The next step will be to show that μ is finitely subadditive: that is, if $a < b$, and $(a, b] \subset \bigcup_{i \leq j \leq n} (c_j, d_j]$, with $c_j < d_j$ for each j, then $G(b) - G(a) \leq \sum_{i \leq j \leq n} G(d_j) - G(c_j)$. This will be proved by induction on n. For $n = 1$ it is clear. In general, $c_j < b \leq d_j$ for some j (if $a = b$ there is no problem). By renumbering, let us take $j = n$. If $c_n \leq a$, we are done. If $c_n > a$, then $(a, c_n]$ is covered by the remaining $n - 1$ intervals, so by induction hypothesis

$$G(b) - G(a) = G(b) - G(c_n) + G(c_n) - G(a)$$

$$\leq G(b) - G(c_n) + \sum_{j=1}^{n-1} G(d_j) - G(c_j),$$

as desired. Now for countable additivity, suppose an interval $J := (c, d]$ is a union of countably many disjoint intervals $J_i := (c_i, d_i]$. For each finite n, J is a union of the J_i for $i = 1, \ldots, n$, and finitely many other left open, right closed intervals, disjoint from each other and the J_i. Specifically, relabel the intervals $(c_i, d_i], i = 1, \ldots, n$, as $(a_i, b_i], i = 1, \ldots, n$, where $c \leq a_1 < b_1 \leq a_2 < \cdots \leq a_n < b_n \leq d$. Then the "other" intervals are $(c, a_1], (b_j, a_{j+1}]$ for $j = 1, \ldots, n - 1$, and $(b_n, d]$. By finite additivity, for all n,

$$\mu(J) \geq \sum_{i=1}^{n} \mu(J_i), \quad \text{so} \quad \mu(J) \geq \sum_{i=1}^{\infty} \mu(J_i).$$

For the converse inequality, let $J := (c, d] \in \mathcal{C}$. Let $\varepsilon > 0$. For each n, using right continuity of G, there are $\delta_n > 0$ such that $G(d_n + \delta_n) < G(d_n) + \varepsilon/2^n$, and $\delta > 0$ such that $G(c + \delta) \leq G(c) + \varepsilon$. Now the compact closed interval $[c + \delta, d]$ is included in the union of countably many open intervals $I_n := (c_n, d_n + \delta_n)$. Thus there is a finite subcover. Hence by finite subadditivity,

$$G(d) - G(c) - \varepsilon \leq G(d) - G(c + \delta) \leq \sum_{n} G(d_n) - G(c_n) + \varepsilon/2^n,$$

and $\mu(J) \leq 2\varepsilon + \sum_{n} \mu(J_n)$. Letting $\varepsilon \downarrow 0$ completes the proof. $\qquad \square$

Returning now to the general case, we have the following extension property. The first main example of this will be where \mathcal{A} is the ring of all finite unions of left open, right closed intervals in \mathbb{R}. The σ-algebra generated by \mathcal{A} contains some quite complicated sets.

3.1.4. Theorem *For any set X and ring \mathcal{A} of subsets of X, any countably additive function μ from \mathcal{A} into $[0, \infty]$ extends to a measure on the σ-algebra \mathcal{S} generated by \mathcal{A}.*

Proof. For any set $E \subset X$ let

$$\mu^*(E) := \inf\left\{\sum_{1 \le n < \infty} \mu(A_n): A_n \in \mathcal{A}, E \subset \bigcup_n A_n\right\},$$

or $+\infty$ if no such A_n exist.

Then μ^* is called the *outer measure* defined by μ. Note that $\mu^*(\emptyset) = 0$ (let $A_n = \emptyset$ for all n). The proof will include four lemmas. The first gives countable subadditivity of μ^*:

3.1.5. Lemma *For any sets E and $E_n \subset X$, if $E \subset \bigcup_n E_n$, then $\mu^*(E) \le \sum_n \mu^*(E_n)$.*

Proof. If the latter sum is $+\infty$, there is no problem. Otherwise, given $\varepsilon > 0$, for each n take $A_{nm} \in \mathcal{A}$ such that $E_n \subset \bigcup_m A_{nm}$ and $\sum_m \mu(A_{nm}) < \mu^*(E_n) + \varepsilon/2^n$. Then E is included in the union over all m and j of A_{mj}, so (using Lemma 3.1.2) $\mu^*(E) \le \sum_{1 \le n < \infty} \mu^*(E_n) + \varepsilon/2^n \le \varepsilon + \sum_{1 \le n < \infty} \mu^*(E_n)$. Letting $\varepsilon \downarrow 0$ proves Lemma 3.1.5. \square

3.1.6. Lemma *For any $A \in \mathcal{A}$, $\mu^*(A) = \mu(A)$.*

Proof. If $A \subset \bigcup_n A_n$, $A_n \in \mathcal{A}$, let $B_n := A \cap A_n \setminus \bigcup_{j<n} A_j$. Then the B_n are disjoint and in \mathcal{A}, with union A, so by assumption $\mu(A) = \sum_n \mu(B_n) \le \sum_n \mu(A_n)$. Thus $\mu(A) \le \mu^*(A)$. Conversely, taking $A_1 = A$ and $A_n = \emptyset$ for $n > 1$ shows that $\mu^*(A) \le \mu(A)$, so $\mu^*(A) = \mu(A)$. \square

A set $F \subset X$ is called μ^*-*measurable*, $F \in M(\mu^*)$, iff for every set $E \subset X$, $\mu^*(E) = \mu^*(E \cap F) + \mu^*(E \setminus F)$. In other words, F splits all sets additively for μ^*. This is equivalent to

$$\mu^*(E) \ge \mu^*(E \cap F) + \mu^*(E \setminus F) \text{ whenever } \mu^*(E) < +\infty, \text{ hence always,}$$

and since the reverse inequality always holds (by Lemma 3.1.5).

3.1.7. Lemma $\mathcal{A} \subset \mathcal{M}(\mu^*)$ *(all sets in \mathcal{A} are μ^* measurable).*

Proof. Let $A \in \mathcal{A}$ and $E \subset X$ with $\mu^*(E) < \infty$. Given $\varepsilon > 0$, take $A_n \in \mathcal{A}$ with $E \subset \bigcup_n A_n$ and $\sum_n \mu(A_n) \le \mu^*(E) + \varepsilon$. Then $E \cap A \subset \bigcup_n (A \cap A_n)$, $E \setminus A \subset \bigcup_n (A_n \setminus A)$, and so $\mu^*(E \cap A) + \mu^*(E \setminus A) \le \sum_n \mu(A \cap A_n) + \mu(A_n \setminus A) = \sum_n \mu(A_n)$. Letting $\varepsilon \downarrow 0$ gives $\mu^*(E \cap A) + \mu^*(E \setminus A) \le \mu^*(E)$. \square

3.1.8. Lemma $\mathcal{M}(\mu^*)$ *is a σ-algebra and μ^* is a measure on it.*

Proof. Clearly $F \in \mathcal{M}(\mu^*)$ if and only if $X \setminus F \in \mathcal{M}(\mu^*)$. If $A, B \in \mathcal{M}(\mu^*)$, then for any $E \subset X$, note that $A \cup B = X \setminus [(X \setminus A) \cap (X \setminus B)]$, and

$$\mu^*(E) = \mu^*(E \cap A) + \mu^*(E \setminus A) \qquad\qquad \text{since } A \in \mathcal{M}(\mu^*)$$
$$= \mu^*(E \cap A \cap B) + \mu^*(E \cap A \setminus B) + \mu^*(E \setminus A) \quad \text{since } B \in \mathcal{M}(\mu^*)$$
$$= \mu^*(E \cap (A \cap B)) + \mu^*(E \setminus (A \cap B)) \qquad \text{since } A \in \mathcal{M}(\mu^*).$$

Thus $\mathcal{M}(\mu^*)$ is an algebra. Now let $E_n \in \mathcal{M}(\mu^*)$ for $n = 1, 2, \ldots$, $F := \bigcup_{1 \le j < \infty} E_j$, and $F_n := \bigcup_{1 \le j \le n} E_j \in \mathcal{M}(\mu^*)$. Since $E_n \setminus \bigcup_{j<n} E_j \in \mathcal{M}(\mu^*)$ for all n, we may assume E_n disjoint in proving F measurable. For any $E \subset X$ we have

$$\mu^*(E) = \mu^*(E \setminus F_n) + \mu^*(E \cap F_n) \qquad\qquad \text{since } F_n \in \mathcal{M}(\mu^*)$$

$$= \mu^*(E \setminus F_n) + \mu^*(E \cap E_n) + \mu^*\left(E \cap \bigcup_{j<n} E_j\right) \quad \text{since } E_n \in \mathcal{M}(\mu^*).$$

Thus by induction on n,

$$\mu^*(E) = \mu^*(E \setminus F_n) + \sum_{j=1}^n \mu^*(E \cap E_j) \ge \mu^*(E \setminus F) + \sum_{j=1}^n \mu^*(E \cap E_j).$$

Letting $n \to \infty$ gives

$$\mu^*(E) \ge \mu^*(E \setminus F) + \sum_{j=1}^\infty \mu^*(E \cap E_j) \ge \mu^*(E \setminus F) + \mu^*(E \cap F)$$

by Lemma 3.1.5. Thus $F \in \mathcal{M}(\mu^*)$, and by Lemma 3.1.5 again,

$$\mu^*(E) = \mu^*(E \setminus F) + \sum_{j \ge 1} \mu^*(E \cap E_j).$$

Letting $E = F$ shows μ^* is countably additive on $\mathcal{M}(\mu^*)$, proving Lemma 3.1.8 and Theorem 3.1.4. \square

Sets of outer measure 0 are always measurable:

3.1.9. Proposition *If* $\mu^*(E) = 0$, *then* $E \in \mathcal{M}(\mu^*)$.

Proof. For any $A \subset X$,

$$\mu^*(A) \geq \mu^*(A \backslash E) = \mu^*(A \backslash E) + \mu^*(E) \geq \mu^*(A \backslash E) + \mu^*(A \cap E).$$

\square

Recall that the *symmetric difference* of two sets A and B is defined by $A \bigtriangleup B := (A \backslash B) \cup (B \backslash A)$. So $A \bigtriangleup B$ is the set of points in one or the other, but not both, of A and B. For example, $[0, 2] \bigtriangleup [1, 3] = [0, 1) \cup (2, 3]$.

A function μ on a collection \mathcal{A} of subsets of a set X will be called σ-*finite* iff there is a sequence $\{A_n\} \subset \mathcal{A}$ with $|\mu(A_n)| < \infty$ for all n and $X = \bigcup_n A_n$. For example, if \mathcal{A} is the collection of all intervals $(a, b]$ for $a < b$ in \mathbb{R} with $\lambda((a, b]) := b - a$, then λ is σ-finite. Counting measure, defined on the collection of all subsets of a set X, is σ-finite if and only if X is countable. Or, if $\mu(A) = +\infty$ for all non-empty sets in some collection, and $\mu(\emptyset) = 0$, then μ is clearly not σ-finite. As another example, let \mathcal{A} be the algebra generated by all intervals $(a, b]$ and let $\mu(A) = +\infty$ for all non-empty sets in \mathcal{A}. Let \mathcal{B} be the σ-algebra generated by \mathcal{A}. One measure defined on \mathcal{B} which equals μ on \mathcal{A} is counting measure; another one assigns measure $+\infty$ to all non-empty sets in \mathcal{B}. This illustrates how σ-finiteness is useful in the following uniqueness theorem.

3.1.10. Theorem *Let* μ *be countably additive and nonnegative on an algebra* \mathcal{A}. *Let* α *be a measure on the* σ-*algebra* \mathcal{S} *generated by* \mathcal{A} *with* $\alpha = \mu$ *on* \mathcal{A}. *Then for any* $A \in \mathcal{S}$ *with* $\mu^*(A) < \infty$, $\alpha(A) = \mu^*(A)$. *If* μ *is* σ-*finite, then the extension of* μ *from the algebra* \mathcal{A} *to the* σ-*algebra* \mathcal{S} *it generates (given by Theorem 3.1.4) is unique, and the extension* α *equals* μ^* *on* \mathcal{S}.

Proof. For any $A \in \mathcal{S}$ and $A_n \in \mathcal{A}$ with $A \subset \bigcup_n A_n$, we have by the proof of Lemma 3.1.6 $\alpha(A) \leq \sum_n \alpha(A_n) = \sum_n \mu(A_n)$. Taking an infimum gives $\alpha(A) \leq \mu^*(A)$.

If $\mu^*(A) < \infty$, then given $\varepsilon > 0$, choose A_n with $\sum_n \mu(A_n) < \mu^*(A) + \varepsilon/2$. Let $B_k := \bigcup_{1 \leq m < k} A_m$. Then $B_k \in \mathcal{A}$ for k finite and $B_\infty \in \mathcal{S}$. As $A \subset B_\infty$,

$$\mu^*(A) \leq \mu^*(B_\infty) < \mu^*(A) + \varepsilon/2.$$

By Lemmas 3.1.7 and 3.1.8, which imply $\mathcal{S} \subset \mathcal{M}(\mu^*)$, and Theorem 3.1.1, for k large enough, $\mu^*(B_\infty \backslash B_k) < \varepsilon/2$, and $\mu^*(B_\infty \backslash A) = \mu^*(B_\infty) - \mu^*(A) <$

$\varepsilon/2$. Now $\alpha(B_\infty) \leq \mu^*(B_\infty) < \infty$, and since $\alpha \leq \mu^*$ on \mathcal{S},

$$\alpha(A) = \alpha(B_\infty) - \alpha(B_\infty \backslash A) \geq \alpha(B_k) - \mu^*(B_\infty \backslash A)$$
$$\geq \alpha(B_k) - \varepsilon/2 = \mu^*(B_k) - \varepsilon/2 \quad \text{(since } B_k \in \mathcal{A} \text{ and by Lemma 3.1.6)}$$
$$\geq \mu^*(B_\infty) - \varepsilon \geq \mu^*(A) - \varepsilon.$$

Letting $\varepsilon \downarrow 0$ gives $\alpha(A) \geq \mu^*(A)$, so $\alpha(A) = \mu^*(A)$. Or if A_n are disjoint with $\mu(A_n) < \infty$, then

$$\alpha(A) = \sum_n \alpha(A \cap A_n) = \sum_n \mu^*(A \cap A_n) = \mu^*(A)$$

by Lemma 3.1.8. Thus α is unique if μ is σ-finite. \square

Next it will be shown that an outer measure μ^* has a property of "continuity from below," even for possibly nonmeasurable sets.

***3.1.11. Theorem** *Let μ be countably additive and nonnegative on an algebra \mathcal{A} of subsets of a set X. Let μ^* be the outer measure defined by μ. Let A_n and A be any subsets of X (not necessarily measurable) such that $A_n \uparrow A$ as $n \to \infty$, meaning that $A_1 \subset A_2 \subset \cdots$ and $\bigcup_n A_n = A$. Then $\mu^*(A_n) \uparrow \mu^*(A)$.*

Example. Let B_1, \ldots, B_k be non-empty disjoint subsets whose union is X. Let \mathcal{A} be the algebra generated by $\{B_1, \ldots, B_k\}$. Then \mathcal{A} is a σ-algebra and there is a measure μ on \mathcal{A} with $\mu(B_j) = 1$ for $j = 1, \ldots, k$. For any $A \subset X$, $\mu^*(A)$ is the number of B_j which A intersects, in other words the number of values of j for which $A \cap B_j \neq \emptyset$. Under the conditions of Theorem 3.1.11, $\mu^*(A_n)$ will become equal to $\mu^*(A)$ for n large enough.

Proof. From the definition of μ^*, $\mu^*(B) \leq \mu^*(C)$ whenever $B \subset C$, so $\mu^*(A_n)$ is nondecreasing in n up to some limit $c \leq \mu^*(A)$. So we need to prove the converse inequality. The conclusion is clear if $\mu^*(A_n) \to +\infty$ as $n \to \infty$, so we can assume $c < \infty$. Given $\varepsilon > 0$, by definition of μ^* there are sets $A_{nj} \in \mathcal{A}$ for all j with $A_n \subset B_n := \bigcup_j A_{nj}$ and $\sum_j \mu(A_{nj}) < \mu^*(A_n) + \varepsilon$. Then if $B_{nk} := \bigcup_{1 \leq j < k} A_{nj}$, and the extension of μ to a measure on the σ-algebra \mathcal{L} generated by \mathcal{A}, given by Theorem 3.1.4, is also called μ, we have

$$\mu(B_n) = \mu(A_{n1}) + \sum_{k \geq 2} \mu(A_{nk} \backslash B_{nk}) \leq \sum_j \mu(A_{nj}) < \mu^*(A_n) + \varepsilon.$$

Let $C_n := \bigcap_{r \geq n} B_r$. Then $A_n \subset C_n$, $C_n \in \mathcal{L}$, $\mu^*(C_n) < \mu^*(A_n) + \varepsilon$, and $C_1 \subset C_2 \subset \cdots$. Let $C := \bigcup_n C_n$. Then by Theorem 3.1.1, $\mu(C_n) \uparrow \mu(C) \leq c + \varepsilon$. So since $A \subset C$, $\mu^*(A) \leq \mu(C) \leq c + \varepsilon$. Letting $\varepsilon \downarrow 0$ gives $\mu^*(A) \leq c$, so $\mu^*(A) = c$. $\qquad\qquad\qquad\qquad\qquad\qquad\qquad\qquad\qquad\qquad\qquad\qquad\qquad\quad\square$

Problems

1. Let (X, \mathcal{S}, μ) be a σ-finite measure space with $\mu(X) = +\infty$. Show that for any $M < \infty$ there is some $A \in \mathcal{S}$ with $M < \mu(A) < \infty$.

2. Let X be an infinite set. Let $m(A) = 0$ for any finite set A and $m(A) = +\infty$ for any infinite A. Show that m is finitely additive but not countably additive.

3. Let X be an infinite set. Let \mathcal{A} be the collection of subsets A of X such that either A is finite, and then set $m(A) := 0$, or the complement of A is finite, and then set $m(A) := 1$.
 (a) Show that \mathcal{A} is an algebra but not a σ-algebra.
 (b) Show that m is finitely additive on \mathcal{A}.
 (c) Under what condition on X can m be extended to a countably additive measure on a σ-algebra?

4. Let (X, \mathcal{T}) be a topological space and $p \in X$. A "punctured neighborhood" of p is a set B which includes $A \setminus \{p\}$ for some neighborhood A of p. In other words, B is a neighborhood of p except that B may not contain p itself. Let $m(B) = 1$ if B is a punctured neighborhood of p and $m(B) = 0$ if B is disjoint from some punctured neighborhood of p.
 (a) Show that m is defined on an algebra.
 (b) If (S, d) is a metric space and $\{p\}$ is not open, show that m is finitely additive.
 (c) If (S, d) is a metric space, show that m cannot be extended to a measure.

5. In $I := [0, 1]$, let \mathcal{F} be the collection of all sets of first category (countable unions of nowhere dense sets; see §2.5). For each $A \in \mathcal{F}$ let $m(A) = 0$ and $m(I \setminus A) = 1$.
 (a) Show that m is a measure defined on a σ-algebra.
 (b) Let $B = [0, 1/2]$. Is B measurable for m^*?

6. (a) Let \mathcal{S} be a σ-algebra of subsets of X. Let μ_1, \ldots, μ_n be measures on \mathcal{S}. Let c_j be nonnegative constants for $j = 1, \ldots, n$. Show that $c_1 \mu_1 + \cdots + c_n \mu_n$ is a measure on \mathcal{S}.
 (b) For any $x \in X$ and subset A of X let $\delta_x(A) := 1_A(x)$ ($= 1$ if $x \in A$ and 0 otherwise). Show that δ_x is a measure on 2^X.

7. Given a measure space (X, \mathcal{S}, μ), let A_j and B_j be subsets of X such that $\mu^*(A_j \triangle B_j) = 0$ for all $j = 1, 2, \ldots$. Show that $\mu^*(\bigcup_j A_j) = \mu^*(\bigcup_j B_j)$.

8. Let \mathcal{A} be the algebra of subsets of \mathbb{R} generated by the collection of all sets $A_{a,b} := [-b, -a) \cup (a, b]$ for $0 < a < b$. Let $m(A_{a,b}) := b - a$. Show that m extends to a countably additive measure on a σ-algebra. Is $[1, 2]$ measurable for m^*?

9. For any set $A \subset \mathbb{N}$ let $\#(A, n)$ be the number of elements in A which are less than n and $d(A) := \lim_{n\to\infty} \#(n, A)/n$ for any $A \in \mathcal{C}$, the collection of sets for which the limit exists. (Here $d(A)$ is called the *density* of A.)
 (a) Show that d is not countably additive on \mathcal{C}.
 (b) Show that \mathcal{C} is not a ring. *Hint:* Let A be the set of even numbers. Let B be the set of even numbers between 2^{2n} and 2^{2n+1} and odd numbers between 2^{2n-1} and 2^{2n} for any $n = 1, 2, \ldots$. Which of A, B, and $A \cap B$ are in \mathcal{C}? (Nevertheless, by the way, d can be extended to be finitely additive on $2^{\mathbb{N}}$.)

10. (a) Let \mathcal{U} be an ultrafilter of subsets of a set X. Let $m(A) = 1$ for all $A \in \mathcal{U}$ and $m(A) = 0$ otherwise. Show that m is finitely additive on 2^X.
 (b) Let X be an infinite set, and let m be finitely additive from 2^X into $\{0, 1\}$ with $m(X) = 1$. Show that $\{A: m(A) = 1\}$ is an ultrafilter.
 (c) Show that for any infinite set, there is a finitely additive function m defined on 2^X with $m(A) = 0$ for every finite set A and $m(X) = 1$. *Hint:* Use part (a).
 (d) If X is countable, show that m in (c) is not countably additive.
 (e) In a game, two players, Sam and Joe, each pick a nonnegative integer at random. For each, the probability that the number is in any set A is $m(A)$ for m from part (c) with $X = \mathbb{N}$. The one who gets the larger number wins. A coin is tossed to determine whose number you find out first. It's heads, so you find out Sam's and still don't know Joe's. Now, who do you think will win? (Strange things happen without countable additivity.)

3.2. Semirings and Rings

In §3.1, countably additive, nonnegative functions on rings were extended in 3.1.4 to such functions on σ-algebras (measures). Length was shown in 3.1.3 to be countably additive on the collection \mathcal{C} of all left open, right closed intervals $(a, b] := \{x \in \mathbb{R}: a < x \leq b\}$. There is still a missing link between

these facts, since \mathcal{C} is not a ring. To make the connection, \mathcal{C} will be shown to have the following property, which will be useful more generally in extending measures:

Definition. For any set X, a collection $\mathcal{D} \subset 2^X$ is called a *semiring* if $\emptyset \in \mathcal{D}$ and for any A and B in \mathcal{D}, we have $A \cap B \in \mathcal{D}$ and $A \backslash B = \bigcup_{1 \le j \le n} C_j$ for some finite n and disjoint $C_j \in \mathcal{D}$.

3.2.1. Proposition \mathcal{C} *is a semiring.*

Proof. We have $\emptyset = (a, a] \in \mathcal{C}$. The intersection of two intervals in \mathcal{C} is in \mathcal{C} (possibly empty). Now $(a, b] \backslash (c, d]$ is either in \mathcal{C} or, if $a < c < d < b$, it is a union of two disjoint intervals in \mathcal{C}. $\qquad\square$

Another kind of semiring will be useful in constructing measures on Cartesian products: if we think of \mathcal{A} and \mathcal{B} as semirings of intervals, then \mathcal{D} in the next fact will be a semiring of rectangles.

3.2.2. Proposition *Let* $X = Y \times Z$ *and let* \mathcal{A} *and* \mathcal{B} *be semirings of subsets of the sets* Y *and* Z, *respectively. Let* $\mathcal{D} := \{A \times B : A \in \mathcal{A}, B \in \mathcal{B}\}$. *Then* \mathcal{D} *is a semiring.*

Proof. First, $\emptyset = \emptyset \times \emptyset \in \mathcal{D}$. For intersections, we have

$$(A \times B) \cap (E \times F) = (A \cap E) \times (B \cap F)$$

for any $A, E \in \mathcal{A}$ and $B, F \in \mathcal{B}$. For differences,

$$D := (A \times B) \backslash (E \times F) = (A \times B) \backslash ((A \cap E) \times (B \cap F)),$$

so in evaluating D we may assume $E \subset A$ and $F \subset B$. Then

$$D = ((A \backslash E) \times B) \cup (E \times (B \backslash F))$$

$$= \left(\left(\bigcup_{1 \le j \le m} G_j \right) \times B \right) \cup \left(E \times \bigcup_{1 \le k \le n} H_k \right)$$

$$= \bigcup_{1 \le j \le m} (G_j \times B) \cup \bigcup_{1 \le k \le n} (E \times H_k)$$

for some finite m, n, disjoint $G_j \in \mathcal{A}$ and disjoint $H_k \in \mathcal{B}$. Thus D is a union of sets in \mathcal{D}, which are disjoint (since $G_j \cap E = \emptyset$ for all j). $\qquad\square$

Note that a semiring \mathcal{R} is a ring if for any $A, B \in \mathcal{R}$ we have $A \cup B \in \mathcal{R}$. Any semiring gives a ring as follows:

3.2.3. Proposition *For any semiring \mathcal{D}, let \mathcal{R} be the set of all finite disjoint unions of members of \mathcal{D}. Then \mathcal{R} is a ring.*

Proof. Clearly $\emptyset \in \mathcal{R}$. Let $A = \bigcup_{1 \le j \le m} A_j$, $B = \bigcup_{1 \le k \le n} B_k$, where the A_j are disjoint and in \mathcal{D}, as are the B_k. Then

$$A \cap B = \bigcup \{A_j \cap B_k : j = 1, \dots, m, k = 1, \dots, n\},$$

where the $A_j \cap B_k$ are disjoint and in \mathcal{D}, so $A \cap B \in \mathcal{R}$. A disjoint union of finitely many sets in \mathcal{R} is clearly in \mathcal{R}. Since $A \cup B = A \cup (B \backslash A)$, it remains only to prove $B \backslash A \in \mathcal{R}$. Now

$$B \backslash A = \bigcup_{1 \le k \le n} \left(B_k \ \backslash \bigcup_{1 \le j \le m} A_j \right),$$

a disjoint union over k, so it will be enough to prove that for each k, $B_k \backslash \bigcup_{1 \le j \le m} A_j = \bigcap_{1 \le j \le m} B_k \backslash A_j \in \mathcal{R}$. Each $B_k \backslash A_j \in \mathcal{R}$ by definition of semiring. Since the intersection of any two sets in \mathcal{R} is in \mathcal{R}, it follows by induction that so is the intersection of finitely many sets in \mathcal{R}. \square

For example, any union of two intervals $(a, b]$ and $(c, d]$ either is another left-open, right-closed interval $(\min(a, c), \max(b, d)]$ if the two intervals overlap, or it's a disjoint union, or both if $b = c$ or $a = d$.

For any collection $\mathcal{A} \subset 2^X$, the intersection of all rings including \mathcal{A} is a ring including \mathcal{A}, called the ring generated by \mathcal{A}. If \mathcal{A} is a semiring, then the ring generated by \mathcal{A} is clearly the one given by Proposition 3.2.3. An additive function extends from a semiring to the ring it generates:

3.2.4. Proposition *Let \mathcal{D} be any semiring and α an additive function from \mathcal{D} into $[0, \infty]$. For disjoint $C_j \in \mathcal{D}$ let*

$$\mu \left(\bigcup_{1 \le j \le m} C_j \right) := \sum_{1 \le j \le m} \alpha(C_j).$$

Then μ is well-defined and additive on the ring \mathcal{R} generated by \mathcal{D}. If α is countably additive on \mathcal{D}, so is μ on \mathcal{R}, and then μ extends to a measure on the σ-algebra generated by \mathcal{D} or \mathcal{R}.

Proof. Suppose some C_j are disjoint in \mathcal{D}, as are some D_k, and $\bigcup_{1 \le j \le m} C_j = \bigcup_{1 \le k \le n} D_k$. Then for each j, $C_j = \bigcup_{1 \le k \le n} C_j \cap D_k$, a disjoint union of sets

in \mathcal{D}, so by additivity

$$\sum_{j=1}^{m} \alpha(C_j) = \sum_{j=1}^{m}\sum_{k=1}^{n} \alpha(C_j \cap D_k) = \sum_{k=1}^{n} \alpha(D_k).$$

So μ is well-defined. For any disjoint B_1, \ldots, B_n in \mathcal{R}, let $B_i = \bigcup_{1 \le j \le m(i)} B_{ij}, i = 1, \ldots, n$, where for each i, the B_{ij} are disjoint, in \mathcal{D}, and $m(i) < \infty$. Thus all the B_{ij} are disjoint. Let B be their union over all i and j. Then since μ is well-defined,

$$\mu(B) = \sum_{i=1}^{n}\sum_{j=1}^{m(i)} \alpha(B_{ij}) = \sum_{i=1}^{n} \mu(B_i),$$

so μ is additive. Suppose α is countably additive, $B \in \mathcal{R}$, so $B = \bigcup_{1 \le i \le n} C_i$ for some disjoint $C_i \in \mathcal{D}$, and $B = \bigcup_{1 \le r < \infty} A_r$ for some disjoint $A_r \in \mathcal{R}$. Let $A_r = \bigcup_{1 \le j \le k(r)} A_{rj}$, for some disjoint $A_{rj} \in \mathcal{D}$. Then

$$B = \bigcup_{i=1}^{n}\bigcup_{r=1}^{\infty}\bigcup_{j=1}^{k(r)} C_i \cap A_{rj},$$

a union of disjoint sets in \mathcal{D}. Each A_r is a finite disjoint union of the sets $C_i \cap A_{rj}$. By countable additivity of α on \mathcal{D} and rearrangement of sums of nonnegative terms (3.1.2),

$$\mu(B) = \sum_{i=1}^{n} \alpha(C_i) = \sum_{i=1}^{n}\sum_{r=1}^{\infty}\sum_{j=1}^{k(r)} \alpha(C_i \cap A_{rj})$$

$$= \sum_{r=1}^{\infty}\sum_{i=1}^{n}\sum_{j=1}^{k(r)} \alpha(C_i \cap A_{rj}) = \sum_{r=1}^{\infty} \mu(A_r),$$

so μ is countably additive on \mathcal{R}. Then it extends to a measure on a σ-algebra by Theorem 3.1.4. $\qquad\square$

An algebra is a ring \mathcal{A} such that the complement of any set in \mathcal{A} is in \mathcal{A}. To get from a ring \mathcal{R} to the smallest algebra including it, we clearly have to put in the complements of sets in \mathcal{R}. That turns out to be enough since, for example, unions of complements are complements of intersections:

3.2.5. Proposition *Let \mathcal{R} be any ring of subsets of a set X. Let $\mathcal{A} := \mathcal{R} \cup \{X \backslash B : B \in \mathcal{R}\}$. Then \mathcal{A} is an algebra.*

Proof. Clearly $\emptyset \in \mathcal{A}$, and $A \in \mathcal{A}$ if and only if $X \backslash A \in \mathcal{A}$. Let $C \in \mathcal{A}$ and $D \in \mathcal{A}$. Then $C \cap D \in \mathcal{A}$ if

(a) $C, D \in \mathcal{R}$, so $C \cap D \in \mathcal{R}$;

(b) $C \in \mathcal{R}$ and $D = X \backslash B$, $B \in \mathcal{R}$, with $C \cap D = C \backslash B \in \mathcal{R}$;

(c) $X \backslash C$ and $X \backslash D \in \mathcal{R}$, for then $X \backslash (C \cap D) = (X \backslash C) \cup (X \backslash D) \in \mathcal{R}$.

\square

Now, as in §3.1, let G be any nondecreasing function from \mathbb{R} into itself, continuous from the right. The most important special case will be when G is the identity function, $G(x) \equiv x$. For simplicity, it may be easier to think just in terms of that case. On the semiring \mathcal{C} of all intervals $(a, b]$, we have $\mu := \mu_G$ defined by $\mu((a, b]) = G(b) - G(a)$ for $a \le b$ in \mathbb{R}. If G is the identity, then μ will be called λ, for Lebesgue (length).

The ring generated by \mathcal{C}, from Proposition 3.2.3, consists of all finite unions

$$A = \bigcup_{j=1}^{n} (a_j, b_j], \quad \text{with } -\infty < a_1 \le b_1 < \cdots \le b_n < +\infty.$$

The complement of A is

$$(-\infty, a_1] \cup (b_1, a_2] \cup \cdots \cup (b_n, +\infty).$$

Let \mathcal{B} be the σ-algebra of subsets of \mathbb{R} generated by \mathcal{C}. Note that any open interval (a, b) with $b < \infty$, as the union of the intervals $(a, b - 1/n]$ for $n = 1, 2, \ldots$, is in \mathcal{B}. The open intervals (q, r) with q and r rational form a countable base for the topology of \mathbb{R}. Thus all open sets are in \mathcal{B}. Conversely, any interval $(a, b]$ is the intersection of the open intervals $(a, b + 1/n)$ for $n = 1, 2, \ldots$. Thus \mathcal{B} is also the σ-algebra generated by the topology of \mathbb{R}. In any topological space, the σ-algebra generated by the topology is called the *Borel* σ-algebra; sets in it are called *Borel* sets.

The next fact yields a measure for each function G:

3.2.6. Theorem *For any nondecreasing function G continuous from the right, $\mu_G((a, b]) := G(b) - G(a)$ for $a \le b$, on the collection \mathcal{C} of intervals $(a, b]$, $a \le b$, extends uniquely to a measure on the Borel σ-algebra.*

Specifically, Lebesgue measure λ exists, namely, a measure on the Borel σ-algebra in \mathbb{R} with $\lambda((a, b]) = b - a$ for any real $a \le b$.

Proof. First, μ_G is countably additive on \mathcal{C} by Theorem 3.1.3. By Proposition 3.2.1, \mathcal{C} is a semiring. So μ_G extends to a measure on the Borel σ-algebra by Proposition 3.2.4. As $\mu_G((n, n + 1]) < \infty$ for all $n = 0, \pm 1, \pm 2, \ldots, \mu_G$ is σ-finite, and the uniqueness follows from Theorem 3.1.10 on each $(n, n + 1]$ and countable additivity. \square

More generally, when is a measure μ on a σ-algebra \mathcal{B} determined uniquely by its values on the sets in a class $\mathcal{C} \subset \mathcal{B}$? If this is true for every measure μ which has finite values on all sets in \mathcal{C}, then \mathcal{C} will be called a *determining class*. For example, if X is a finite set and $\mathcal{B} = 2^X$ the collection of all its subsets, then the set \mathcal{C} of all singletons $\{x\}$ for $x \in X$ is a determining class. Here are two sufficient conditions for a class \mathcal{C} to be a determining class:

3.2.7. Theorem *If X is a set and $\mathcal{C} \subset \mathcal{B}$ where \mathcal{B} is the σ-algebra of subsets of X generated by \mathcal{C} (that is, the smallest σ-algebra including \mathcal{C}), then \mathcal{C} is a determining class if \mathcal{C} is an algebra, or if X is the union of countably many sets in \mathcal{C} and \mathcal{C} is a semiring.*

Proof. If μ is a measure on \mathcal{B}, finite on all sets in \mathcal{C}, then by the assumptions μ is σ-finite. If \mathcal{C} is a semiring, then by Proposition 3.2.3, the values of μ on \mathcal{C} uniquely determine μ on the smallest ring including \mathcal{C}, so we can assume \mathcal{C} is a ring. By assumption, there exist sets $A_n \in \mathcal{C}$ with $A_n \uparrow X$. For any $B \in \mathcal{B}$, $\mu(B \cap A_n) \uparrow \mu(B)$, so it's enough to show that the values $\mu(B \cap A_n)$ are unique, and we can assume $\mu(X) < \infty$. Then, by Proposition 3.2.5, μ is unique and has finite values on the smallest algebra including \mathcal{C}, so we can assume \mathcal{C} is an algebra. Then the theorem follows from Theorem 3.1.10. $\qquad \square$

One reason for talking about semirings, rings, and algebras is that some such assumption is needed for the unique extension or determining class property; for \mathcal{C} just to generate \mathcal{B} is not enough, even if \mathcal{C} only contains two sets:

Example. Let $X = \{1, 2, 3, 4\}$ and $\mathcal{B} = 2^X$. Let $\mathcal{C} = \{\{1, 2\}, \{1, 3\}\}$. Then \mathcal{C} generates \mathcal{B}, but \mathcal{C} is not a determining class: if $\mu\{j\} = 1/4$ for $j = 1, 2, 3, 4$, and $\alpha\{1\} = \alpha\{4\} = 1/6$, while $\alpha\{2\} = \alpha\{3\} = 1/3$, then $\alpha = \mu$ on \mathcal{C}.

Although not needed for the construction of Lebesgue measure, another kind of collection of sets is sometimes useful. A collection \mathcal{L} of sets is called a *lattice* iff for every $A \in \mathcal{L}$ and $B \in \mathcal{L}$, we have $A \cap B \in \mathcal{L}$ and $A \cup B \in \mathcal{L}$. For example, in any topological space, the topology (collection of all open sets) is a lattice; so is the collection of all closed sets. The ring generated by a lattice is characterized as follows:

***3.2.8. Proposition** *Let \mathcal{L} be a lattice of sets with $\emptyset \in \mathcal{L}$. Then the smallest ring \mathcal{R} including \mathcal{L} is the collection \mathcal{C} of all finite unions $\bigcup_{1 \le i \le n} A_i \backslash B_i$ for $A_i \in \mathcal{L}$ and $B_i \in \mathcal{L}$, where the sets $A_i \backslash B_i$ are disjoint.*

Proof. Clearly $\mathcal{L} \subset \mathcal{C} \subset \mathcal{R}$. What has to be proved is that \mathcal{C} is a ring. We have $\varnothing \in \mathcal{C}$. Let $S_m := \bigcup_{1 \le i \le m}(A_i \backslash B_i)$ and $T_n := \bigcup_{i \le j \le n} C_j \backslash D_j$ where the sets A_i, B_i, C_j, and D_j are all in \mathcal{L}, the $A_i \backslash B_i$ are disjoint, and the $C_j \backslash D_j$ are disjoint. To show $S_m \backslash T_n \in \mathcal{C}$, by induction on n, note that for any sets U, V, and $W, U \backslash (V \cup W) = (U \backslash V) \backslash W$, so the proof for $n = 1$ will also give the induction step. Since the sets $(A_i \backslash B_i) \backslash (C_1 \backslash D_1)$ are disjoint, we can treat them separately. Now for any sets A, B, C, and D,

$$(A \backslash B) \backslash (C \backslash D) = (A \backslash (B \cup C)) \cup ((A \cap C \cap D) \backslash B),$$

a disjoint union, so $S_m \backslash T_n \in \mathcal{C}$.

Now $S_m \cup T_n = S_m \cup (T_n \backslash S_m)$, a disjoint union of sets in \mathcal{C}, so $S_m \cup T_n \in \mathcal{C}$ and \mathcal{C} is a ring. \square

Problems

1. (a) Show that the set of all open intervals (a, b) in \mathbb{R} is not a semiring or a lattice.
 (b) Show that the set of all intervals which may be open or closed on the right or left, $[a, b), [a, b], (a, b]$, or (a, b), is a semiring.

2. If \mathcal{D} is the collection of all rectangles of the form $(a, c] \times (b, d]$ in \mathbb{R}^2, show that \mathcal{D} is a semiring and find the smallest possible value of n in the definition of semiring.

3. Do Problem 2 for the collection of all sets $(a, b] \times (c, d] \times (e, f]$ in \mathbb{R}^3.

4. Call a collection \mathcal{D} of sets a *demiring* if for any A and $B \in \mathcal{D}$, both $A \backslash B$ (as for a semiring) and $A \cap B$ are finite disjoint unions of sets in \mathcal{D}. Let \mathcal{D} be the collection of all sets $h(a, b) := \{\theta \in \mathbb{R}: \text{for some integer } n \in \mathbb{Z}, a < \theta + 2n\pi \le b\}$ for any $a \le b$ in \mathbb{R}. Show that \mathcal{D} is a demiring, but not a semiring.

5. Let \mathcal{D} be a demiring (as defined in Problem 4) and \mathcal{R} the collection of finite disjoint unions of members of \mathcal{D}. Show that \mathcal{R} is a ring.

6. Extend Proposition 3.2.4 to demirings in place of semirings.

7. Let $G(x) = 0$ for $x < 1, G(x) = 1$ for $1 \le x < 2$, and $G(x) = 3$ for $x \ge 2$. Let μ be the measure obtained from G by Theorem 3.2.6. Evaluate μ as a finite linear combination of point masses δ_x, as in Problem 6 in §3.1.

8. Is there a collection of open sets in \mathbb{R}^2 which is a base for the topology of \mathbb{R}^2, and which is a semiring? Why, or why not?

9. Show that the collection of all half-lines $(-\infty, x]$ for $x \in \mathbb{R}$ is a determining class (for the σ-algebra it generates).

3.3. Completion of Measures

Let μ be a measure on a σ-algebra \mathcal{S} of subsets of a set X, so that (X, \mathcal{S}, μ) is a measure space. If \mathcal{S} is not all of 2^X, it can be useful to extend μ to a larger σ-algebra, especially if there are sets whose measure is somehow determined by the measures of sets already in \mathcal{S}. For example, if $A \subset E \subset C$, A and C are in \mathcal{S}, but E is not, and $\mu(A) = \mu(C)$, it seems that $\mu(E)$ should also equal $\mu(A)$.

This section will make such an extension of μ. The method has to do with the outer measure μ^*, as defined in §3.1. Starting with an arbitrary set E, the first step is to find a set $C \in \mathcal{S}$ with $E \subset C$ and $\mu(C)$ as small as possible:

3.3.1. Theorem *For any measure space (X, \mathcal{S}, μ) and any $E \subset X$, there is a $C \in \mathcal{S}$ with $E \subset C$ and $\mu^*(E) = \mu(C)$.*

Proof. If $E \subset \bigcup_{n \geq 1} A_n$ and $A_n \in \mathcal{S}$, let $A := \bigcup_{n \geq 1} A_n$. Then $E \subset A$, $A \in \mathcal{S}$, and as shown in the proof of Lemma 3.1.6, $\mu(A) \leq \sum_{n \geq 1} \mu(A_n)$. So $\mu^*(E) = \inf\{\mu(A): E \subset A, A \in \mathcal{S}\}$. For each $m = 1, 2, \ldots$, choose $C_m \in \mathcal{S}$ with $E \subset C_m$ and $\mu(C_m) \leq \mu^*(E) + 1/m$. Let $C := \bigcap_m C_m$. Then $C \in \mathcal{S}$, since $C = X \backslash (\bigcup_m X \backslash C_m)$, $E \subset C$, and $\mu^*(E) = \mu(C)$. $\qquad\square$

A set C as in Theorem 3.3.1 is called a *measurable cover* of E. If $E \in \mathcal{S}$, then E is a measurable cover of itself. On the other hand, if $C \in \mathcal{S}$, and the only subsets of C in \mathcal{S} are itself and \emptyset (but C may contain more than one point), then C is a measurable cover of any non-empty subset $E \subset C$. If there is going to be a set $A \in \mathcal{S}$ with $A \subset E \subset C$ and $\mu(C \backslash A) = 0$, then $\mu^*(E \backslash A) = \mu^*(C \backslash E) = 0$. The collection of "null" sets for μ is defined by $\mathcal{N}(\mu) := \{F \subset X : \mu^*(F) = 0\}$. Then any subset of a countable union of sets in $\mathcal{N}(\mu)$ is in $\mathcal{N}(\mu)$.

Let $\mathcal{S} \vee \mathcal{N}(\mu)$ be the σ-algebra generated by $\mathcal{S} \cup \mathcal{N}(\mu)$. Let $\mathcal{S}_\mu^* := \{E \subset X : E \, \triangle \, B \in \mathcal{N}(\mu) \text{ for some } B \in \mathcal{S}\}$, recalling $E \, \triangle \, B := (E \backslash B) \cup (B \backslash E)$. Then E will be called almost equal to B. \mathcal{S}_μ^* is the collection of sets which almost equal sets in \mathcal{S}. For example, if $A \subset E \subset C$ and $\mu(C \backslash A) = 0$, then $E \in \mathcal{S}_\mu^*$ and we can take $B = A$ or C in the definition. The two ways just defined of extending a σ-algebra by sets of measure 0 give the same result:

3.3.2. Proposition *For any measure space (X, \mathcal{S}, μ), and the collection \mathcal{S}_μ^* of sets almost equal to sets in \mathcal{S}, we have $\mathcal{S}_\mu^* = \mathcal{S} \vee \mathcal{N}(\mu)$, the smallest σ-algebra including \mathcal{S} and $\mathcal{N}(\mu)$.*

Proof. If $E \bigtriangleup B \in \mathcal{N}(\mu)$ and $B \in \mathcal{S}$, then $E \backslash B \in \mathcal{N}(\mu)$ and $B \backslash E \in \mathcal{N}(\mu)$, so $E = (B \cup (E \backslash B)) \backslash (B \backslash E) \in \mathcal{S} \vee \mathcal{N}(\mu)$. Thus $\mathcal{S}_\mu^* \subset \mathcal{S} \vee \mathcal{N}(\mu)$.

Conversely, $\mathcal{S} \subset \mathcal{S}_\mu^*$ and $\mathcal{N}(\mu) \subset \mathcal{S}_\mu^*$. If $A_n \bigtriangleup B_n \in \mathcal{N}(\mu)$ for all n, then

$$\left(\bigcup_n A_n \right) \bigtriangleup \left(\bigcup_n B_n \right) \subset \bigcup_n (A_n \bigtriangleup B_n) \in \mathcal{N}(\mu),$$

and $(X \backslash A_1) \bigtriangleup (X \backslash B_1) = A_1 \bigtriangleup B_1 \in \mathcal{N}(\mu)$. Thus \mathcal{S}_μ^* is a σ-algebra, so $\mathcal{S} \vee \mathcal{N}(\mu) \subset \mathcal{S}_\mu^*$. $\qquad\square$

For example, let μ be a measure on the set \mathbb{Z} of all integers with $\mu\{k\} > 0$ for each $k > 0$ and $\mu\{k\} = 0$ for all $k \leq 0$. Then $\mathcal{N}(\mu)$ is the ring of all subsets of the set $-\mathbb{N}$ of nonpositive integers. Let \mathcal{S} be the σ-algebra of all sets $A \subset \mathbb{Z}$ such that for some set B of positive integers, either $A = B$ or $A = -\mathbb{N} \cup B$. Then the two equal σ-algebras in Proposition 3.3.2 both equal the σ-algebra $2^{\mathbb{Z}}$ of all subsets of \mathbb{Z}.

Given a measure μ, the next fact will say that μ can be extended to a measure $\bar{\mu}$ so that whenever two sets are almost equal for μ, and μ is defined for at least one of them, then $\bar{\mu}$ will be defined and equal for both, as follows. If $A \bigtriangleup B \in \mathcal{N}(\mu)$ and $B \in \mathcal{S}$, let $\bar{\mu}(A) := \mu(B)$. Here A can be any set in the σ-algebra $\mathcal{S} \vee \mathcal{N}(\mu)$ which was just characterized in Proposition 3.3.2, and $\bar{\mu}$ is a measure, as follows:

3.3.3. Proposition *For any measure space (X, \mathcal{S}, μ), $\bar{\mu}$ is a well-defined measure on the σ-algebra $\mathcal{S} \vee \mathcal{N}(\mu)$ which equals μ on \mathcal{S}.*

Proof. To show that μ is well-defined, let $A \bigtriangleup B \in \mathcal{N}(\mu)$ and $A \bigtriangleup C \in \mathcal{N}(\mu)$, where B and $C \in \mathcal{S}$. Then $B \bigtriangleup C = (A \bigtriangleup B) \bigtriangleup (A \bigtriangleup C)$. (To see this, let D and E be any sets. Note that $1_{D \bigtriangleup E} = 1_D + 1_E$ where addition is modulo 2, that is, 2 is replaced by zero. This addition is commutative and associative. When sets are replaced by their indicator functions and \bigtriangleup by $+$, the desired equation results.) Thus $B \bigtriangleup C \in \mathcal{N}(\mu)$, so $\mu(B) = \mu(C)$, and $\bar{\mu}$ is well-defined.

Suppose $A_n \in \mathcal{S} \vee \mathcal{N}(\mu)$ and the A_n are disjoint. Take $B_n \in \mathcal{S}$ such that $C_n := A_n \bigtriangleup B_n \in \mathcal{N}(\mu)$. Then for $i \neq j$, $B_i \cap B_j \subset C_i \cup C_j$, so $\mu(B_i \cap B_j) = 0$. Let $D_n := B_n \backslash \bigcup_{i < n} B_i$ and $B := \bigcup_{i \geq 1} B_i$. Then for all n, $\mu(D_n) = \mu(B_n)$ and $\mu(B) = \sum_{n \geq 1} \mu(D_n) = \sum_n \mu(B_n)$. As shown in the proof of

Proposition 3.3.2, if $A := \bigcup_{n \geq 1} A_n$, then $A \bigtriangleup B \in \mathcal{N}(\mu)$, so $\bar{\mu}(A) = \mu(B) = \sum_n \bar{\mu}(A_n)$ and $\bar{\mu}$ is countably additive. Clearly, it equals μ on \mathcal{S}. □

The measure $\bar{\mu}$ is called the *completion* of μ.

In §3.1 we began with a countably additive nonnegative function μ on an algebra. Let $\tilde{\mu}$ be the restriction of μ^* to $\mathcal{M}(\mu^*)$, which is a measure and extends μ by Lemmas 3.1.6 and 3.1.8. This extension is usually also called μ. For the notation to make sense, we need to check that the outer measure μ^* is not changed by the extension:

3.3.4. Proposition *For any countably additive nonnegative function μ on an algebra \mathcal{A} in a set X, $\tilde{\mu}^* = \mu^*$ on 2^X.*

Proof. Since $\tilde{\mu} = \mu$ on \mathcal{A} (by Lemma 3.1.6), clearly $\tilde{\mu}^* \leq \mu^*$. Conversely, for any $A \subset X$, by Theorem 3.3.1 take $C \in \mathcal{M}(\mu^*)$ with $A \subset C$ and $\tilde{\mu}^*(A) = \tilde{\mu}(C)$. Then $\mu^*(A) \leq \mu^*(C) = \tilde{\mu}(C) = \tilde{\mu}^*(A)$. □

A measure space (X, \mathcal{S}, μ) is called *finite* iff $\mu(X) < +\infty$, and *σ-finite* iff there is a sequence of sets $E_n \in \mathcal{S}$ with $\mu(E_n) < \infty$ and $X = \bigcup_n E_n$. For such spaces, the σ-algebra treated in Proposition 3.3.2 equals the σ-algebra $\mathcal{M}(\mu^*)$ defined in §3.1 (recall that sets in $\mathcal{M}(\mu^*)$ are those that "split" all sets additively for μ^*):

3.3.5. Theorem *For any σ-finite measure space (X, \mathcal{S}, μ), $\mathcal{S} \vee \mathcal{N}(\mu) = \mathcal{M}(\mu^*)$.*

Proof. Let $X = \bigcup_n E_n$, $\mu(E_n) < \infty$, and $E \in \mathcal{M}(\mu^*)$. To show that $E \in \mathcal{S} \vee \mathcal{N}(\mu)$ it is enough to show that $E \cap E_n \in \mathcal{S} \vee \mathcal{N}(\mu)$ for all n. Thus we can assume μ is finite. Let A be a measurable cover of E, by Theorem 3.3.1. Then $\mu^*(E) = \mu(A) = \mu^*(A) = \mu^*(E) + \mu^*(A \backslash E)$, so $\mu^*(A \backslash E) = 0$ and $E = A \backslash (A \backslash E) \in \mathcal{S} \vee \mathcal{N}(\mu)$, so $\mathcal{M}(\mu^*) \subset \mathcal{S} \vee \mathcal{N}(\mu)$. The converse inclusion follows from Lemmas 3.1.7 and 3.1.8 and Proposition 3.1.9. □

Example. Let $\mathcal{S} := \{\emptyset, X\}$, $\mu(\emptyset) = 0$, and $\mu(X) = +\infty$. Then $\mu^*(E) = +\infty$ whenever $\emptyset \neq E \subset X$. Thus all subsets of X are in $\mathcal{M}(\mu^*)$ while $\mathcal{S} \vee \mathcal{N}(\mu) = \{\emptyset, X\}$, a smaller σ-algebra if X has more than one point. This indicates why σ-finiteness is needed in Theorem 3.3.5.

A topology induces a "relative" topology on an arbitrary subset (§2.1) and, similarly, a measure induces one on a subset, which may not be measurable:

***3.3.6. Theorem** *Let (X, \mathcal{S}, μ) be a measure space and Y any subset of X with finite outer measure: $\mu^*(Y) < \infty$. Let \mathcal{S}_Y be the collection of all sets $A \cap Y$ for $A \in \mathcal{S}$ and let $v(B) := \mu^*(B)$ for each $B \in \mathcal{S}_Y$. Then (Y, \mathcal{S}_Y, v) is a measure space.*

Proof. First, it is easily seen that \mathcal{S}_Y is a σ-algebra of subsets of Y. Let C be a measurable cover of Y (by Theorem 3.3.1). Let $A_n \in \mathcal{S}$ be such that $A_n \cap Y$ are disjoint sets for $n = 1, 2, \ldots$. To show that $v(\bigcup_n A_n \cap Y) = \sum_n v(A_n \cap Y)$, we can replace each A_n by $A_n \cap C$ without changing $A_n \cap Y$. Let $D_n := A_n \setminus \bigcup_{j<n} A_j$ for each n. Then $D_n \in \mathcal{S}$, and since the sets $A_n \cap Y$ are disjoint, we have $D_n \cap Y = A_n \cap Y$ for all n. So we can replace A_n by D_n and assume the A_n are disjoint as well as $A_n \subset C$. To show $v(A_n \cap Y) = \mu(A_n)$, we clearly have $v(A_n \cap Y) \leq \mu(A_n)$. Let F_n be a measurable cover of $A_n \cap Y$, so $v(A_n \cap Y) = \mu(F_n)$. We can assume $F_n \subset A_n$. If $\mu(F_n) < \mu(A_n)$, then $A_n \setminus F_n \in \mathcal{S}$, is disjoint from Y, and has positive measure, so that C could be replaced by the set $C \setminus (A_n \setminus F_n)$, contradicting the fact that C is a measurable cover of Y. So $v(A_n \cap Y) = \mu(A_n)$ for all n. Let $A := \bigcup_n A_n$. Then $\sum_n \mu(A_n) = \mu(A)$. We have $\mu(A \cap Y) = \mu(A)$ by the same argument as was used for each A_n, so v is countably additive. □

Problems

In all of the following problems, (X, \mathcal{S}, μ) is a measure space.

1. Show that the set of all $A \in \mathcal{S}$ with $\mu(A) < \infty$ is a ring.

2. (a) Show that symmetric difference of sets is associative: $A \triangle (B \triangle C) = (A \triangle B) \triangle C$ for any sets A, B, and C.

 (b) If A_1, A_2, \ldots, A_n are any sets, and $A := A_1 \triangle (A_2 \triangle (A_3 \triangle (\cdots (A_{n-1} \triangle A_n) \cdots))$, show that A is the set of points belonging to an odd number of the A_j.

3. If $\bar{\mu}$ is the completion of μ, show that a set E is in the σ-algebra on which $\bar{\mu}$ is defined if and only if there exist some A and C in the domain of μ with $A \subset E \subset C$ and $\mu(C \setminus A) = 0$.

4. Prove or disprove: If $A \subset E \subset C$ and $\mu(A) = \mu(C)$, then E is in the domain of the completion $\bar{\mu}$. *Hint:* What if $\mu(A) = \infty$?

5. Let μ be the measure on $[0, 1]$ which is 0 on all sets of first category (countable unions of nowhere dense sets) and 1 on all complements of such sets. What is the completion of μ?

6. For the inner measure μ_* defined by $\mu_*(A) := \sup\{\mu(B) : B \subset A\}$, show that for any set $A \subset X$, there is a measurable set $B \subset A$ with $\mu(B) = \mu_*(A)$.

7. If (X, \mathcal{S}, μ) is a measure space and A_1, A_2, \ldots, are any subsets of X, let C_j be a measurable cover of A_j for $j = 1, 2, \ldots$.
 (a) Show that $\bigcup_j C_j$ is a measurable cover of $\bigcup_j A_j$.
 (b) Give an example to show that $C_1 \cap C_2$ need not be a measurable cover of $A_1 \cap A_2$.

8. A collection \mathcal{R} of sets is called a σ-*ring* iff it is a ring and any countable union of sets in \mathcal{R} is in \mathcal{R}. Show that if \mathcal{R} is a σ-ring of subsets of a set X, then $\mathcal{R} \cup \{X \backslash A : A \in \mathcal{R}\}$ is a σ-algebra. If $X \notin \mathcal{R}$, so \mathcal{R} is not a σ-algebra, and μ is a countably additive function from \mathcal{R} into $[0, \infty]$, show that setting $\mu(X \backslash A) := +\infty$ for each $A \in \mathcal{R}$ makes μ a measure.

9. A collection \mathcal{N} of sets is called *hereditary* if $A \subset B \in \mathcal{N}$ always implies $A \in \mathcal{N}$. If, in addition, $\bigcup_n A_n \in \mathcal{N}$ for any sequence of sets $A_n \in \mathcal{N}$, then \mathcal{N} is called a *hereditary* σ-*ring*. Let \mathcal{N} be a hereditary σ-ring of subsets of X.
 (a) Show that \mathcal{N} is, in fact, a σ-ring.
 (b) Show that $\mathcal{N} \vee \mathcal{S}$ is the collection of all sets $A \subset X$ such that $A \triangle B \in \mathcal{N}$ for some $B \in \mathcal{S}$.
 (c) If $\mu(A) = 0$ for all $A \in \mathcal{N} \cap \mathcal{S}$, show that μ can be extended to a measure on $\mathcal{S} \vee \mathcal{N}$ so that $\mu(C) = 0$ for all $C \in \mathcal{N}$.

3.4. Lebesgue Measure and Nonmeasurable Sets

From the previous sections, it's not clear whether or not all subsets of \mathbb{R} are measurable for Lebesgue measure λ (as defined in Theorem 3.2.6). The proof that there are nonmeasurable sets for λ will need the axiom of choice. So let's look at some easy examples of nonmeasurable sets for other measures. First, let $X = \{0, 1\}$, and let \mathcal{S} be the trivial σ-algebra $\{\emptyset, X\}$. Let $\mu(X) = 1$. Then $\mu^*(\{0\}) = \mu^*(\{1\}) = 1$, so $\{0\}$ and $\{1\}$ are not measurable for μ^*. For another, slightly less trivial example, let X be any uncountable set and $\mathcal{S} := \{A \subset X : A \text{ or } X \backslash A \text{ is countable}\}$. Let $m(A) = 0$ for A countable and $m(A) = 1$ if A has countable complement. Then \mathcal{S} is a σ-algebra and m is a measure on it. Any uncountable set C with uncountable complement in X is nonmeasurable since

$$m^*(C) + m^*(X \backslash C) = 1 + 1 = 2 \neq m^*(X) = m(X) = 1.$$

A set $E \subset \mathbb{R}$ is called *Lebesgue measurable* iff $E \in \mathcal{M}(\lambda^*)$. We know from Lemmas 3.1.6 and 3.1.8 that λ^*, restricted to the σ-algebra $\mathcal{M}(\lambda^*)$ of Lebesgue measurable sets, is a measure which extends λ. This measure is also called λ, or Lebesgue measure. $\mathcal{M}(\lambda^*)$ is the largest domain on which λ is generally defined. (It can be extended to larger σ-algebras, but only in a rather arbitrary way: see Problems 3–5). From Proposition 3.3.2 and Theorem 3.3.5, we also know that for every Lebesgue measurable set E there is a Borel set B with $\lambda(E \, \Delta \, B) = 0$.

It is going to be proved that not all sets are Lebesgue measurable. One conceivable method would be to show that there are more sets (in terms of cardinality) than there are Lebesgue measurable sets. Now c is the cardinality of $[0, 1]$ or \mathbb{R}; in other words, to say that a set X has cardinality c is to say that there is a 1–1 function from $[0, 1]$ onto X. In the same sense, the cardinality of $2^{\mathbb{R}}$ is called 2^c. Now the method just suggested will not work because there are just as many Lebesgue measurable sets as there are sets in \mathbb{R}:

***3.4.1. Proposition** *There are 2^c Lebesgue measurable sets.*

Proof. Let C be the Cantor set,

$$C := \left\{ \sum_{n \geq 1} x_n / 3^n : x_n = 0 \text{ or } 2 \text{ for all } n \right\}.$$

For each $N = 1, 2, 3, \ldots$, let $C_N := \{\sum_{n \geq 1} x_n / 3^n : x_n = 0, 1, \text{ or } 2 \text{ for all } n$ and $x_n \neq 1$ for $n \leq N\}$. Thus C_1 is the unit interval $[0, 1]$ with the open "middle third" $(1/3, 2/3)$ removed. Then to get C_2, from the two remaining intervals, the open middle thirds $(1/9, 2/9)$ and $(7/9, 8/9)$ are removed. This process is iterated N times to give C_N. Thus $C_1 \supset C_2 \supset \cdots \supset C_N \supset \cdots$, and $\bigcap_{N \geq 1} C_N = C$. We have $\lambda(C_N) = (2/3)^N$ for all N. Thus $\lambda(C) = 0$. On the other hand, C has cardinal c. So C has 2^c subsets, all Lebesgue measurable with measure 0. Since \mathbb{R} has 2^c subsets, there are exactly 2^c Lebesgue measurable sets (by the equivalence theorem, 1.4.1). □

One might ask whether there is a Lebesgue measurable set A of positive measure whose complement also has positive measure and which is "spread evenly" over the line in the sense that for some constant r with $0 < r < 1$, for all intervals J with $0 < \lambda(J) < +\infty$, we have the ratio $\lambda(A \cap J)/\lambda(J) = r$. It turns out that there is no such set A. For any set of positive measure whose complement also has positive measure, some of the ratios must be close to 0 and others close to 1. Specifically:

3.4.2. Proposition *Let E be any Lebesgue measurable set in \mathbb{R} with $\lambda(E) >$
0. Then for any $\varepsilon > 0$ there is a finite, nontrivial interval J ($J = [a, b]$ for
some a, b with $-\infty < a < b < +\infty$) such that $\lambda(E \cap J) > (1 - \varepsilon)\lambda(J)$.*

Proof. We may assume $\lambda(E) < \infty$ and $\varepsilon < 1$. Take finite intervals $J_n :=$
$(a_n, b_n], a_n < b_n$, such that $E \subset \bigcup_n J_n$ and $\sum_{n \geq 1} \lambda(J_n) \leq \lambda(E)/(1 - \varepsilon)$.
Then $\lambda(E) \leq \sum_{n \geq 1} \lambda(J_n \cap E)$ and for some n, $(1 - \varepsilon)\lambda(J_n) \leq \lambda(J_n \cap E)$.
□

A set E can have positive Lebesgue measure, and in fact, its complement
can have measure 0, without E including any interval of positive length (take,
for example, the set of irrational numbers). But if we take all differences
between elements of a set of positive measure, the next fact shows we will
get a set including a nontrivial interval around 0. This fact about the structure
of measurable sets will then be used in showing next that nonmeasurable sets
exist.

3.4.3. Proposition *If E is any Lebesgue measurable set with $\lambda(E) > 0$, then
for some $\varepsilon > 0$,*

$$E - E := \{x - y : x, y \in E\} \supset [-\varepsilon, \varepsilon].$$

Proof. By Proposition 3.4.2 take an interval J with $\lambda(E \cap J) > 3\lambda(J)/4$. Let
$\varepsilon := \lambda(J)/2$. For any set $C \subset \mathbb{R}$ and $x \in \mathbb{R}$ let $C + x := \{y + x : y \in C\}$. Then
if $|x| \leq \varepsilon$,

$$(E \cap J) \cup ((E \cap J) + x) \subset J \cup (J + x), \quad \text{and}$$
$$\lambda(J \cup (J + x)) \leq 3\lambda(J)/2, \quad \text{while}$$
$$\lambda((E \cap J) + x) = \lambda(E \cap J), \quad \text{so}$$
$$((E \cap J) + x) \cap (E \cap J) \neq \varnothing, \quad \text{and}$$
$$x \in (E \cap J) - (E \cap J) \subset E - E. \qquad \qquad □$$

Next comes the main fact in this section, on existence of a nonmeasurable
set; specifically, a set E in $[0, 1]$ with outer measure 1, so E is "thick" in the
whole interval, but such that its complement is equally thick:

3.4.4. Theorem *Assuming the axiom of choice (as usual), there exists a
set $E \subset \mathbb{R}$ which is not Lebesgue measurable. In fact, there is a set $E \subset
I := [0, 1]$ with $\lambda^*(E) = \lambda^*(I \setminus E) = 1$.*

Proof. Recall that \mathbb{Z} is the set of all integers (positive, negative, or 0). Let α be a fixed irrational number, say $\alpha = 2^{1/2}$. Let G be the following additive subgroup of \mathbb{R}: $G := \mathbb{Z} + \mathbb{Z}\alpha := \{m + n\alpha : m, n \in \mathbb{Z}\}$. Let H be the subgroup $H := \{2m + n\alpha : m, n \in \mathbb{Z}\}$. To show that G is dense in \mathbb{R}, let $c := \inf\{g : g \in G, g > 0\}$. If $c = 0$, let $0 < g_n < 1/n$, $g_n \in G$. Then $G \supset \{mg_n : m \in \mathbb{Z}, n = 1, 2, \ldots\}$, a dense set. If $c > 0$ and $g_n \downarrow c$, $g_n \in G$, $g_n > c$, then $g_n - g_{n+1} > 0$, belong to G, and converge to 0, so $c = 0$, a contradiction. So $c \in G$ and $G = \{mc : m \in \mathbb{Z}\}$, a contradiction since α is irrational. Likewise, H and $H + 1$ are dense. The cosets $G + y$, $y \in \mathbb{R}$, are disjoint or identical. By the axiom of choice, let C be a set containing exactly one element of each coset. Let $X := C + H$. Then $\mathbb{R}\backslash X = C + H + 1$. Now $(X - X) \cap (H + 1) = \emptyset$. Since $H + 1$ is dense, by Proposition 3.4.3 X does not include any measurable set with positive Lebesgue measure. Let $E := X \cap I$. Then $\lambda^*(I\backslash E) = 1$. Likewise $(\mathbb{R}\backslash X) - (\mathbb{R}\backslash X) = (C + H + 1) - (C + H + 1) = (C + H) - (C + H)$ is disjoint from $H + 1$, so $\lambda^*(E) = 1$. \square

So Lebesgue measure is not defined on all subsets of I, but can it be extended, as a countably additive measure, to all subsets? The answer is no, at least if the continuum hypothesis is assumed (Appendix C).

Problems

1. Let E be a Lebesgue measurable set such that for all x in a dense subset of \mathbb{R}, $\lambda(E \triangle (E + x)) = 0$. Show that either $\lambda(E) = 0$ or $\lambda(\mathbb{R}\backslash E) = 0$.

2. Show that there exist sets $A_1 \supset A_2 \supset \cdots$ in $[0, 1]$ with $\lambda^*(A_k) = 1$ for all k and $\bigcap_k A_k = \emptyset$. *Hint:* With C and α as in the proof of Theorem 3.4.4, let $B_k := \{m + n\alpha : m, n \in \mathbb{Z}, |m| \geq 2k\}$ and $A_k := (C + \{m + n\alpha : m, n \in \mathbb{Z}$ and $|m| \geq k\}) \cap [0, 1]$. Show that B_k is dense in \mathbb{R} and then that $([0, 1]\backslash A_k) - ([0, 1]\backslash A_k)$ is disjoint from B_k.

3. If S is a σ-algebra of subsets of X and E any subset of X, show that the σ-algebra generated by $S \cup \{E\}$ is the collection of all sets of the form $(A \cap E) \cup (B\backslash E)$ for A and B in S.

4. Show that for any finite measure space (X, S, μ) and any set $E \subset X$, it is always possible to extend μ to a measure ρ on the σ-algebra $T := S \vee \{E\}$. *Hint:* In the form given in Problem 3, let $\rho(A \cap E) = \mu^*(A \cap E)$ and $\rho(B\backslash E) := \mu_*(B\backslash E) := \sup\{\mu(C) : C \in S, C \subset B\backslash E\}$. *Hint:* See Theorem 3.3.6.

5. Referring to Problem 4, show that one can extend μ to a measure α defined on E, where any value of $\alpha(E)$ in the interval $[\mu_*(E), \mu^*(E)]$ is possible.

6. If (X, \mathcal{S}, μ) is a measure space and A_n are sets in \mathcal{S} with $\mu(A_1) < \infty$ and $A_n \downarrow A$, that is, $A_1 \supset A_2 \supset \cdots$ with $\bigcap_n A_n = A$, show that $\lim_{n \to \infty} \mu(A_n) = \mu(A)$. *Hint:* The sets $A_n \backslash A_{n+1}$ are disjoint.

7. Let μ be a finite measure defined on the Borel σ-algebra of subsets of $[0, 1]$ with $\mu(\{p\}) = 0$ for each single point p. Let $\varepsilon > 0$.
 (a) Show that for any p there is an open interval J containing p with $\mu(J) < \varepsilon$.
 (b) Show that there is a dense open set U with $\mu(U) < \varepsilon$.

8. Let $\mu(A) = 0$ and $\mu(I \backslash A) = 1$ for every Borel set A of first category, as defined after Theorem 2.5.2, in $I := [0, 1]$. Show that μ cannot be extended to a measure on the Borel σ-algebra. *Hint:* Use Problem 7.

9. If (X, \mathcal{S}, μ) is a measure space and $\{E_n\}$ is a sequence of subsets of X, show that μ can always be extended to a measure on a σ-algebra containing E_1, \ldots, E_n but not necessarily all the E_n. *Hint:* Use Problem 4 and Problem 8, where $\{E_n\}$ are a base for the topology of $[0, 1]$.

10. If A_k are as in Problem 2, with $A_k \downarrow \emptyset$ and $\lambda^*(A_k) \equiv 1$, let $P_n(B \cap A_n) :=$ $\lambda^*(B \cap A_n) = \lambda(B)$ for every Borel set $B \subset [0, 1]$. Show that P_n is a countably additive measure on a σ-algebra of subsets of A_n such that for infinitely many n, A_{n+1} is not measurable for P_n in A_n. *Hint:* Use Theorem 3.3.6.

*3.5. Atomic and Nonatomic Measures

If (X, \mathcal{S}, μ) is a measure space, a set $A \in \mathcal{S}$ is called an *atom* of μ iff $0 < \mu(A) < \infty$ and for every $C \subset A$ with $C \in \mathcal{S}$, either $\mu(C) = 0$ or $\mu(C) = \mu(A)$. A measure without any atoms is called *nonatomic*.

The main examples of atoms are singletons $\{x\}$ that have positive finite measure. A set of positive finite measure is an atom if its only measurable subsets are itself and \emptyset. Here is a less trivial atom: let X be an uncountable set and let \mathcal{S} be the collection of sets A which either are countable, with $\mu(A) = 0$, or have countable complement, with $\mu(A) = 1$. Then μ is a measure and X is an atom. On the other hand, Lebesgue measure is nonatomic (the proof is left as a problem).

A measure space (X, \mathcal{S}, μ), or the measure μ, is called *purely atomic* iff there is a collection \mathcal{C} of atoms of μ such that for each $A \in \mathcal{S}$, $\mu(A)$ is the sum of the numbers $\mu(C)$ for all $C \in \mathcal{C}$ such that $\mu(A \cap C) = \mu(C)$. (The sum $\sum \{a_C : C \in \mathcal{C}\}$, for any nonnegative real numbers a_C, is defined as the supremum of the sums over all finite subsets of \mathcal{C}.) For the main examples of purely atomic measures, there is a function $f \geq 0$ such that

$\mu(A) = \sum\{f(x): x \in A\}$. Counting measures are purely atomic, with $f \equiv 1$. The most studied purely atomic measures on \mathbb{R} are concentrated in a countable set $\{x_n\}_{n \geq 1}$, with $\mu(A) = \sum_n c_n 1_A(x_n)$ for some $c_n \geq 0$.

Sets of infinite measure can be uninteresting, and/or cause some technical difficulties, unless they have subsets of arbitrarily large finite measure, which is true for σ-finite measures and those of the following more general kind. A measure space (X, \mathcal{S}, μ) is called *localizable* iff there is a collection \mathcal{A} of disjoint measurable sets of finite measure, whose union is all of X, such that for every set $B \subset X$, B is measurable if and only if $B \cap C \in \mathcal{S}$ for all $C \in \mathcal{A}$, and then $\mu(B) = \sum_{C \in \mathcal{A}} \mu(B \cap C)$. The most useful localizable measures are the σ-finite ones, with \mathcal{A} countable; counting measures on possibly uncountable sets provide other examples.

Most measures considered in practice are either purely atomic or non-atomic, but one can always add a purely atomic finite measure to a nonatomic one to get a measure for which the following decomposition is nontrivial:

3.5.1. Theorem *Let (X, \mathcal{S}, μ) be any localizable measure space. Then there exist measures ν and ρ on \mathcal{S} such that $\mu = \nu + \rho$, ν is purely atomic, and ρ is nonatomic.*

The proof of Theorem 3.5.1 will only be sketched, with the details left to Problems 1–7. First, one reduces to the case of finite measure spaces. Let \mathcal{C} be the collection of all atoms of μ. For two atoms A and B, define a relation $A \approx B$ iff $\mu(A \cap B) = \mu(A)$. This will be an equivalence relation. Let I be the set of all equivalence classes and choose one atom C_i in the equivalence class i for each $i \in I$. Let $\nu(A) = \sum_{i \in I} \mu(A \cap C_i)$, and $\rho = \mu - \nu$.

Problems

1. Show that if Theorem 3.5.1 holds for finite measure spaces, then it holds for all localizable measure spaces.

2. Show that in the definition of a localizable measure μ, either $\mu \equiv 0$ or \mathcal{A} can be chosen so that $\mu(C) > 0$ for all $C \in \mathcal{A}$.

3. For μ finite, show that \approx is an equivalence relation.

4. Still for μ finite, if A and B are two atoms not equivalent in this sense, show that $\mu(A \cap B) = 0$.

5. Show that ν, as defined above, is a purely atomic measure and $\nu \leq \mu$.

6. Show that for any measures $\nu \leq \mu$, there is a measure ρ with $\mu =$

$v + \rho$. *Hint:* This is easy for μ finite, but letting $\rho = \mu - v$ leaves ρ undefined for sets A with $\mu(A) = v(A) = \infty$. For such a set, let $\rho(A) := \sup\{(\mu - v)(B): v(B) < \infty \text{ and } B \subset A\}$.

7. With μ and v as in Problems 5–6, show that ρ is nonatomic.

8. Given a measure space (X, \mathcal{S}, μ), a measurable set A will be said to have *purely infinite* measure iff $\mu(A) = +\infty$ and for every measurable $B \subset A$, either $\mu(B) = 0$ or $\mu(B) = +\infty$. Say that two such sets, A and C, are equivalent iff $\mu(A \bigtriangleup C) = 0$. Give an example of a measure space and two purely infinite sets A and C which are not equivalent but for which $\mu(A \cap C) = +\infty$.

9. Show that Lebesgue measure is nonatomic.

10. Let X be a countable set. Show that any measure on X is purely atomic.

11. If (X, \mathcal{S}, μ) is a measure space, $\mu(X) = 1$, and μ is nonatomic, show that the range of μ is the whole interval $[0, 1]$. *Hints:* First show that $1/3 \leq \mu(C) \leq 2/3$ for some C; if not, show that a largest value $s < 1/3$ is attained, on a set B, and that the complement of B includes an atom. Then repeat the argument for μ restricted to C and to its complement to get sets of intermediate measure, and iterate to get a dense set of values of μ.

Notes

§3.1 Jordan (1892) defined a set to be "measurable" if its topological boundary has measure 0. So the set \mathbb{Q} of rational numbers is not measurable in Jordan's sense. Borel (1895, 1898) showed that length of intervals could be extended to a countably additive function on the σ-algebra generated by intervals, which contains all countable sets. Later, the σ-algebra was named for him. Fréchet (1965) wrote a biographical memoir on Borel. Borel wrote some 35 books and over 250 papers. His mathematical papers have been collected: Borel (1972). He also was elected to the French parliament (Chambre des Députés) from 1924 to 1936 and was Ministre de la Marine (Cabinet member for the Navy) in 1925. Some of Borel's less technical papers, many relating to the philosophy of mathematics and science, have been collected: Borel (1967).

Hawkins (1970, Chap. 4) reviews the historical development of measurable sets. Radon (1913) was, it seems, the first to define measures on general spaces (beyond \mathbb{R}^k). Caratheodory (1918) was apparently the first to define outer measures μ^* and the collection $\mathcal{M}(\mu^*)$ of measurable sets, to prove it is a σ-algebra, and to prove that μ^* restricted to it is a measure.

Why is *countable* additivity assumed? Length is not additive for arbitrary uncountable unions of closed intervals, since for example $[0, 1]$ is the union of c singletons $\{x\} = [x, x]$ of length 0. Thus additivity over such uncountable unions seems too strong an assumption. On the other hand, finite additivity is weak enough to allow some pathology, as in some of the problems at the end of §3.1. Probability is nowadays usually defined,

following Kolmogorov (1933), as a (countably additive) measure on a σ-algebra \mathcal{S} of subsets of a set X with $P(X) = 1$. Among the relatively few researchers in probability who work with finitely additive probability "measures," a notable work is that of Dubins and Savage (1965).

§3.2 The notion of "semiring," under the different name "type \mathcal{D}" collection of sets, was mentioned in some lecture notes of von Neumann (1940–1941, p. 79).

The (mesh) Riemann integral is defined, say for a continuous f on a finite interval $[a, b]$, as a limit of sums

$$\sum_{i=1}^{n} f(y_i)(x_i - x_{i-1}) \quad \text{where } a = x_0 \leq y_1 \leq x_1 \leq y_2 \leq \cdots \leq x_n = b,$$

as $\max_i(x_i - x_{i-1}) \to 0, n \to \infty$. Stieltjes (1894, pp. 68–76) defined an analogous integral $\int f \, dG$ for a function G, replacing $x_i - x_{i-1}$ by $G(x_i) - G(x_{i-1})$ in the sums. The resulting integrals have been called *Riemann-Stieltjes* integrals. The measures μ_G have been called Lebesgue-Stieltjes measures, although "measures" as such had not been defined in 1894.

§3.4 It seems that Vitali (1905) was the first to prove existence of a non-Lebesgue measurable set, according to Lebesgue (1907, p. 212). Van Vleck (1908) proved existence of a set E as in Theorem 3.4.4. Solovay (1970) has shown that the axiom of choice is indispensable here, and that countably many dependent choices are not enough, so that uncountably many choices are required to obtain a nonmeasurable set. (A precise statement of his results, however, involves conditions too technical to be given here.)

§3.5 Segal (1951) defined and studied localizable measure spaces.

References

An asterisk identifies works I have found discussed in secondary sources but have not seen in the original.

*Borel, Émile (1895). Sur quelques points de la théorie des fonctions. *Ann. Ecole Normale Sup.* (Ser. 3) *12*: 9–55, = *Œuvres*, I (CNRS, Paris, 1972), pp. 239–285.
———— (1898). *Leçons sur la théorie des fonctions*. Gauthier-Villars, Paris.
———— (1967). *Émile Borel: philosophe et homme d'action*. Selected, with a Préface, by Maurice Fréchet. Gauthier-Villars, Paris.
*———— (1972). *Œuvres*, 4 vols. Editions du Centre National de la Recherche Scientifique, Paris.
Caratheodory, Constantin (1918). *Vorlesungen über reelle Funktionen*. Teubner, Leipzig. 2d ed., 1927.
Dubins, Lester E., and Leonard Jimmie Savage (1965). *How to Gamble If You Must*. McGraw-Hill, New York.
Fréchet, Maurice (1965). *La vie et l'oeuvre d'Émile Borel*. L'Enseignement Math., Genève, also published in *L'Enseignement Math.* (Ser. 2) *11* (1965) 1–97.
Hawkins, Thomas (1970). *Lebesgue's Theory of Integration: Its Origins and Development*. Univ. Wisconsin Press.

Jordan, Camille (1892). Remarques sur les intégrales définies. *J. Math. pures appl.* (Ser. 4) *8*: 69–99.

Kolmogoroff, Andrei N. [Kolmogorov, A. N.] (1933). *Grundbegriffe der Wahrscheinlichkeitsrechnung.* Springer, Berlin. Published in English as *Foundations of the Theory of Probability*, 2d ed., ed. Nathan Morrison. Chelsea, New York, 1956.

Lebesgue, Henri (1907). Contribution a l'étude des correspondances de M. Zermelo. *Bull. Soc. Math. France 35*: 202–212.

von Neumann, John (1940–1941). Lectures on invariant measures. Unpublished lecture notes, Institute for Advanced Study, Princeton.

*Radon, Johann (1913). Theorie und Anwendungen der absolut additiven Mengenfunktionen. *Wien Akad. Sitzungsber. 122*: 1295–1438.

Segal, Irving Ezra (1951). Equivalences of measure spaces. *Amer. J. Math. 73*: 275–313.

Solovay, Robert M. (1970). A model of set-theory in which every set of reals is Lebesgue measurable. *Ann. Math.* (Ser. 2) *92*: 1–56.

Stieltjes, Thomas Jan (1894). Recherches sur les fractions continues. *Ann. Fac. Sci. Toulouse* (Ser. 1) *8*: 1–122 and *9* (1895): 1–47, = *Oeuvres complètes*, II (Nordhoff, Groningen, 1918), pp. 402–566.

van Vleck, Edward B. (1908). On non-measurable sets of points with an example. *Trans. Amer. Math. Soc. 9*: 237–244.

*Vitali, Giuseppe (1905). *Sul problema della misura dei gruppi di punti di una retta.* Gamberini e Parmeggiani, Bologna.

4

Integration

The classical, Riemann integral of the 19th century runs into difficulties with certain functions. For example:

(i) To integrate the function $x^{-1/2}$ from 0 to 1, the Riemann integral itself does not apply. One has to take a limit of Riemann integrals from ε to 1 as $\varepsilon \downarrow 0$.

(ii) Also, the Riemann integral lacks some completeness. For example, if functions f_n on $[0, 1]$ are continuous and $|f_n(x)| \leq 1$ for all n and x, while $f_n(x)$ converges for all x to some $f(x)$, then the Riemann integrals $\int_0^1 f_n(x)\,dx$ always converge, but the Riemann integral $\int_0^1 f(x)\,dx$ may not be defined.

The Lebesgue integral, to be defined and studied in this chapter, will make $\int_0^1 x^{-1/2}\,dx$ defined without any special, ad hoc limit process, and in (ii), $\int_0^1 f(x)\,dx$ will always be defined as a Lebesgue integral and will be the limit of the integrals of the f_n, while the Lebesgue integrals of Riemann integrable functions equal the Riemann integrals.

The Lebesgue integral also applies to functions on spaces much more general than \mathbb{R}, and with respect to general measures.

4.1. Simple Functions

A *measurable space* is a pair (X, \mathcal{S}) where X is a set and \mathcal{S} is a σ-algebra of subsets of X. Then a *simple function* on X is any finite sum

$$f = \sum_i a_i 1_{B(i)}, \quad \text{where } a_i \in \mathbb{R} \quad \text{and} \quad B(i) \in \mathcal{S}. \tag{4.1.1}$$

If μ is a measure on \mathcal{S}, we call f μ-*simple* iff it is simple and can be written in the form (4.1.1) with $\mu(B(i)) < \infty$ for all i. (If $\mu(X) = +\infty$, then $0 = a_1 1_{B(1)} + a_2 1_{B(2)}$ for $B(1) = B(2) = X$, $a_1 = 1$, and $a_2 = -1$, but 0 *is* a μ-simple function. So the definition of μ-simple requires only that there exist

114

$B(i)$ of finite measure and a_i for which (4.1.1) holds, not that all such $B(i)$ must have finite measure.) Some examples of simple functions on \mathbb{R} are the step functions, where each $B(i)$ is a finite interval.

Any finite collection of sets $B(1), \ldots, B(n)$ generates an algebra \mathcal{A}. A non-empty set A is called an *atom* of an algebra \mathcal{A} iff $A \in \mathcal{A}$ and for all $C \in \mathcal{A}$, either $A \subset C$ or $A \cap C = \emptyset$. For example, if $X = \{1, 2, 3, 4\}$, $B(1) = \{1, 2\}$, and $B(2) = \{1, 3\}$, then these two sets generate the algebra of all subsets of X, whose atoms are of course the singletons $\{1\}$, $\{2\}$, $\{3\}$, and $\{4\}$.

4.1.2. Proposition *Let X be any set and $B(1), \ldots, B(n)$ any subsets of X. Let \mathcal{A} be the smallest algebra of subsets of X containing the $B(i)$ for $i = 1, \ldots, n$. Let \mathcal{C} be the collection of all intersections $\bigcap_{1 \le i \le n} C(i)$ where for each i, either $C(i) = B(i)$ or $C(i) = X \backslash B(i)$. Then $\mathcal{C} \backslash \{\emptyset\}$ is the set of all atoms of \mathcal{A}, and every set in \mathcal{A} is the union of the atoms which it includes.*

Proof. Any two elements of \mathcal{C} are disjoint (for some i, one is included in $B(i)$, the other in $X \backslash B(i)$). The union of \mathcal{C} is X. Thus the set of all unions of members of \mathcal{C} is an algebra \mathcal{B}. Each $B(i)$ is the union of all the intersections in \mathcal{C} with $C(i) = B(i)$. Thus $B(i) \in \mathcal{B}$ and $\mathcal{A} \subset \mathcal{B}$. Clearly $\mathcal{B} \subset \mathcal{A}$, so $\mathcal{A} = \mathcal{B}$. Each non-empty set in \mathcal{C} is thus an atom of \mathcal{A}. A union of two or more distinct atoms is not an atom, so $\mathcal{C} \backslash \{\emptyset\}$ is the set of all atoms of \mathcal{A} and the rest follows. $\qquad\square$

Now, any simple function f can be written as $\sum_{1 \le j \le M} b_j 1_{A(j)}$ where the $A(j)$ are disjoint atoms of the algebra \mathcal{A} generated by the $B(i)$ in (4.1.1), and by Proposition 4.1.2, we have $M \le 2^n$. Thus in (4.1.1) we may assume that the $B(i)$ are disjoint. Then, if $f(x) \ge 0$ for all x, we will have $a_i \ge 0$ for all i. For example,

$$3 \cdot 1_{[1,3]} + 2 \cdot 1_{[2,4]} = 3 \cdot 1_{[1,2)} + 5 \cdot 1_{[2,3]} + 2 \cdot 1_{(3,4]}.$$

If (X, \mathcal{S}, μ) is any measure space and f any simple function on X, as in (4.1.1), with $a_i \ge 0$ for all i, the integral of f with respect to μ is defined by

$$\int f \, d\mu := \sum_i a_i \mu(B(i)) \in [0, \infty], \tag{4.1.3}$$

where $0 \cdot \infty$ is taken to be 0. We must first prove:

4.1.4. Proposition *For any nonnegative simple function f, $\int f \, d\mu$ is well-defined.*

Proof. Suppose $f = \sum_{i \in F} a_i 1_{E(i)} = \sum_{j \in G} b_j 1_{H(j)}$, $E_i := E(i)$, $H_j := H(j)$, where all a_i and b_j are nonnegative, F and G are finite, and $E(i)$ and $H(j)$ are in \mathcal{S}. Then we may assume that the $H(j)$ are disjoint atoms of the algebra generated by the $H(j)$ and $E(i)$. In that case, $b_j = \sum \{a_i : E_i \supset H_j\}$. Thus

$$\sum_j b_j \mu(H(j)) = \sum_j \mu(H(j)) \sum \{a_i : E_i \supset H_j\}$$

$$= \sum_i a_i \sum \{\mu(H_j) : H_j \subset E_i\} = \sum_i a_i \mu(E_i). \qquad \square$$

If f and g are simple, then clearly $f + g$, fg, $\max(f, g)$, and $\min(f, g)$ are all simple. It follows directly from (4.1.3) and Proposition 4.1.4 that if f and g are nonnegative simple functions and c is a constant, with $c > 0$, then $\int f + g \, d\mu = \int f \, d\mu + \int g \, d\mu$ and $\int cf \, d\mu = c \int f \, d\mu$. Also if $0 \le f \le g$, meaning that $0 \le f(x) \le g(x)$ for all x, then $\int f \, d\mu \le \int g \, d\mu$. For $E \in \mathcal{S}$ let $\int_E f \, d\mu := \int f 1_E \, d\mu$. Then, for example, $\int_E 1_A \, d\mu = \mu(A \cap E)$.

If (X, \mathcal{S}) and (Y, \mathcal{B}) are measurable spaces and f is a function from X into Y, then f is called *measurable* iff $f^{-1}(B) \in \mathcal{S}$ for all $B \in \mathcal{B}$. For example, if $X = Y$ and f is the identity function, measurability means that $\mathcal{B} \subset \mathcal{S}$. Similarly, in general, for measurability, the σ-algebra \mathcal{S} on the domain space needs to be large enough, and/or the σ-algebra \mathcal{B} on the range space not too large. If $Y = \mathbb{R}$ or $[-\infty, \infty]$, then the σ-algebra \mathcal{B} for measurability of functions into Y will (unless otherwise stated) be the σ-algebra of Borel sets generated by all (bounded or unbounded) intervals or open sets. Now given any measure space (X, \mathcal{S}, μ) and any measurable function f from X into $[0, +\infty]$, we define

$$\int f \, d\mu := \sup \left\{ \int g \, d\mu : 0 \le g \le f, g \text{ simple} \right\}.$$

For $a_n \in [-\infty, \infty]$, $a_n \uparrow$ means $a_n \le a_{n+1}$ for all n. Then $a_n \uparrow a$ means also $a_n \to a$. If $a = +\infty$, this means that for all $M < \infty$ there is a $K < \infty$ such that $a_n > M$ for all $n > K$. The following fact gives a handy approach to the integral of a nonnegative measurable function as the limit of a sequence, rather than a more general supremum:

4.1.5. Proposition *For any measurable $f \ge 0$, there exist simple f_n with $0 \le f_n \uparrow f$, meaning that $0 \le f_n(x) \uparrow f(x)$ for all x. For any such sequence f_n, $\int f_n \, d\mu \uparrow \int f \, d\mu$.*

g_2

g_1

Figure 4.1

Proof. For $n = 1, 2, \ldots$, and $j = 1, 2, \ldots, 2^n n - 1$, let $E_{nj} := f^{-1}((j/2^n, (j+1)/2^n])$, $E_n := f^{-1}((n, \infty])$. Let

$$f_n := n 1_{E_n} + \sum_{j=1}^{2^n n - 1} j 1_{E_{nj}} / 2^n.$$

In Figure 4.1, $g_n = f_n$ for the case $f(x) = x$ on $[0, \infty]$, so g_n does stairsteps of width and height $1/2^n$ for $0 \le x \le n$, with $g_n(x) \equiv n$ for $x > n$. Then for a general $f \ge 0$ we have $f_n = g_n \circ f$. Now $E_{nj} = E_{n+1,2j} \cup E_{n+1,2j+1}$, so on E_{nj} we have $f_n(x) = j/2^n = 2j/2^{n+1} < (2j+1)/2^{n+1}$, where $f_{n+1}(x)$ is one of the latter two terms. Thus $f_n(x) \le f_{n+1}(x)$ there. On E_n, $f_n(x) = n \le f_{n+1}(x)$. At points x not in E_n or in any E_{nj}, $f_n(x) = 0 \le f_{n+1}(x)$. Thus $f_n \le f_{n+1}$ everywhere. If $f(x) = +\infty$, then $f_n(x) = n$ for all n. Otherwise, for some $m \in \mathbb{N}$, $f(x) < m$. Then $f_n(x) \ge f(x) - 1/2^n$ for $n \ge m$. Thus $f_n(x) \to f(x)$ for all x.

Let g be any simple function with $0 \le g \le f$. If h_n are simple and $0 \le h_n \uparrow f$, then $\int h_n \, d\mu \uparrow c$ for some $c \in [0, \infty]$. To show that $c \ge \int g \, d\mu$, write $g = \sum_i a_i 1_{B(i)}$ where the $B(i)$ are disjoint and their union is all of X (so some a_i may be 0). For any simple function $h = \sum_j c_j 1_{A(j)}$, we have

$$h = \sum_i h 1_{B(i)} = \sum_{i,j} c_j 1_{B(i) \cap A(j)} \quad \text{and} \quad \int h \, d\mu = \sum_i \int_{B(i)} h \, d\mu$$

by Proposition 4.1.4. Thus it will be enough to show that for each i,

$$\lim_{n \to \infty} \int_{B(i)} h_n \, d\mu \ge a_i \mu(B(i)).$$

If $a_i = 0$, this is clear. Otherwise, dividing by a_i where $a_i > 0$, we may assume $g = 1_E$ for some $E \in \mathcal{S}$. Then, given $\varepsilon > 0$, let $F_n := \{x \in E : h_n(x) > 1 - \varepsilon\}$. Then $F_n \uparrow E$; in other words, $F_1 \subset F_2 \subset \cdots$ and $\bigcup_n F_n = E$. Thus by countable additivity, $\mu(E) = \mu(F_1) + \sum_{n \ge 1} \mu(F_{n+1} \setminus F_n)$, and $\mu(F_n) \uparrow \mu(E)$.

Hence $c \geq (1 - \varepsilon)\mu(E)$. Letting $\varepsilon \downarrow 0$ gives $c \geq \int g \, d\mu$. Thus $c \geq \int f \, d\mu$. Since $\int f_n \, d\mu \leq \int f \, d\mu$ for all n, we have $c = \int f \, d\mu$. □

A *σ-ring* is a collection \mathcal{R} of sets, with $\emptyset \in \mathcal{R}$, such that $A \backslash B \in \mathcal{R}$ for any $A \in \mathcal{R}$ and $B \in \mathcal{R}$, and such that $\bigcup_{j \geq 1} A_j \in \mathcal{R}$ whenever $A_j \in \mathcal{R}$ for $j = 1, 2, \ldots$ So any σ-algebra is a σ-ring. Conversely, a σ-ring \mathcal{R} of subsets of a set X is a σ-algebra in X if and only if $X \in \mathcal{R}$. For example, the set of all countable subsets of \mathbb{R} is a σ-ring which is not a σ-algebra. If f is a real-valued function on a set X, and \mathcal{R} is a σ-ring of subsets of X, then f is said to be *measurable* for \mathcal{R} iff $f^{-1}(B) \in \mathcal{R}$ for any Borel set $B \subset \mathbb{R}$ not containing 0. (If this is true for general Borel sets, then $f^{-1}(\mathbb{R}) = X \in \mathcal{R}$ implies \mathcal{R} is a σ-algebra.)

A σ-ring \mathcal{R} is said to be *generated* by \mathcal{C} iff \mathcal{R} is the smallest σ-ring including \mathcal{C}, just as for σ-algebras. The following fact makes it easier to check measurability of functions.

4.1.6. Theorem *Let (X, \mathcal{S}) and (Y, \mathcal{B}) be measurable spaces. Let \mathcal{B} be generated by \mathcal{C}. Then a function f from X into Y is measurable if and only if $f^{-1}(C) \in \mathcal{S}$ for all $C \in \mathcal{C}$. The same is true if X is a set, \mathcal{S} is a σ-ring of subsets of X, $Y = \mathbb{R}$, and \mathcal{B} is the σ-ring of Borel subsets of \mathbb{R} not containing 0.*

Proof. "Only if" is clear. To prove "if," let $\mathcal{D} := \{D \in \mathcal{B}: f^{-1}(D) \in \mathcal{S}\}$. We are assuming $\mathcal{C} \subset \mathcal{D}$. If $D_n \in \mathcal{D}$ for all n, then $f^{-1}(\bigcup_n D_n) = \bigcup_n f^{-1}(D_n)$, so $\bigcup_n D_n \in \mathcal{D}$. If $D \in \mathcal{D}$ and $E \in \mathcal{D}$, then $f^{-1}(E \backslash D) = f^{-1}(E) \backslash f^{-1}(D) \in \mathcal{S}$, so $E \backslash D \in \mathcal{D}$. Thus \mathcal{D} is a σ-ring and if \mathcal{S} is a σ-algebra, we have $f^{-1}(Y) = X \in \mathcal{S}$, so $Y \in \mathcal{D}$. In either case, $\mathcal{B} \subset \mathcal{D}$ and so $\mathcal{B} = \mathcal{D}$. □

A reasonably small collection \mathcal{C} of subsets of \mathbb{R}, which generates the whole Borel σ-algebra, is the set of all half-lines (t, ∞) for $t \in \mathbb{R}$. So to show that a real-valued function f is measurable, it's enough to show that $\{x: f(x) > t\}$ is measurable for each real t.

Let (X, \mathcal{A}), (Y, \mathcal{B}), and (Z, \mathcal{C}) be measurable spaces. If f is measurable from X into Y, and g from Y into Z, then for any $C \in \mathcal{C}$, $(g \circ f)^{-1}(C) = f^{-1}(g^{-1}(C)) \in \mathcal{A}$, since $g^{-1}(C) \in \mathcal{B}$. Thus $g \circ f$ is measurable from X into Z (the proof just given is essentially the same as the proof that the composition of continuous functions is continuous).

On the Cartesian product $Y \times Z$ let $\mathcal{B} \otimes \mathcal{C}$ be the σ-algebra generated by the set of all "rectangles" $B \times C$ with $B \in \mathcal{B}$ and $C \in \mathcal{C}$. Then $\mathcal{B} \otimes \mathcal{C}$ is called the *product σ-algebra* on $Y \times Z$. A function h from X into $Y \times Z$ is of the

form $h(x) = (f(x), g(x))$ for some function f from X into Y and function g from X into Z. By Theorem 4.1.6 we see that h is measurable if and only if both f and g are measurable, considering rectangles $B \times Z$ and $Y \times C$ for $B \in \mathcal{B}$ and $C \in \mathcal{C}$ (the set of such rectangles also generates $\mathcal{B} \otimes \mathcal{C}$).

Recall that a second-countable topology has a countable base (see Proposition 2.1.4) and that a Borel σ-algebra is generated by a topology. The next fact will be especially useful when $X = Y = \mathbb{R}$.

4.1.7. Proposition *Let (X, \mathcal{T}) and (Y, \mathcal{U}) be any two topological spaces. For any topological space (Z, \mathcal{V}) let its Borel σ-algebra be $\mathcal{B}(Z, \mathcal{V})$. Then the Borel σ-algebra \mathcal{C} of the product topology on $X \times Y$ includes the product σ-algebra $\mathcal{B}(X, \mathcal{T}) \otimes \mathcal{B}(Y, \mathcal{U})$. If both (X, \mathcal{T}) and (Y, \mathcal{U}) are second-countable, then the two σ-algebras on $X \times Y$ are equal.*

Proof. For any set $A \subset X$, let $\mathcal{U}(A)$ be the set of all $B \subset Y$ such that $A \times B \in \mathcal{C}$. If A is open, then $A \times Y \in \mathcal{C}$. Now $B \mapsto A \times B$ preserves set operations, specifically: for any $B \subset Y$, $A \times (Y \setminus B) = (A \times Y) \setminus (A \times B)$, and for any $B_n \subset Y$, $\bigcup_n (A \times B_n) = A \times \bigcup_n B_n$. It follows that $\mathcal{U}(A)$ is a σ-algebra of subsets of Y. It includes \mathcal{U} and hence $\mathcal{B}(Y, \mathcal{U})$. Then, for $B \in \mathcal{B}(Y, \mathcal{U})$, let $\mathcal{T}(B)$ be the set of all $A \subset X$ such that $A \times B \in \mathcal{C}$. Then $X \in \mathcal{T}(B)$, and $\mathcal{T}(B)$ is a σ-algebra. It includes \mathcal{T}, and hence $\mathcal{B}(X, \mathcal{T})$. Thus the product σ-algebra of the Borel σ-algebras is included in the Borel σ-algebra \mathcal{C} of the product.

In the other direction, suppose (X, \mathcal{T}) and (Y, \mathcal{U}) are second-countable. The product topology has a base \mathcal{W} consisting of all sets $A \times B$ where A belongs to a countable base of \mathcal{T} and B to a countable base of \mathcal{U}. Then the σ-algebra generated by \mathcal{W} is the Borel σ-algebra of the product topology. It is clearly included in the product σ-algebra. \square

The usual topology on \mathbb{R} is second-countable, by Proposition 2.1.4 (or since the intervals (a, b) for a and b rational form a base). Thus any continuous function from $\mathbb{R} \times \mathbb{R}$ into \mathbb{R} (or any topological space), being measurable for the Borel σ-algebras, is measurable for the product σ-algebra on $\mathbb{R} \times \mathbb{R}$ by Proposition 4.1.7. In particular, addition and multiplication are measurable from $\mathbb{R} \times \mathbb{R}$ into \mathbb{R}. Thus, for any measurable spaces (X, \mathcal{S}) and any two measurable real-valued functions f and g on X, $f + g$ and fg are measurable. Let $\mathcal{L}^0(X, \mathcal{S})$ denote the set of all measurable real-valued functions on X for \mathcal{S}. Then since constant functions are measurable, $\mathcal{L}^0(X, \mathcal{S})$ is a vector space over \mathbb{R} for the usual operations of addition and multiplication by constants, $(f + g)(x) := f(x) + g(x)$ and $(cf)(x) := cf(x)$ for any constant c. For nonnegative functions, integrals add:

4.1.8. Proposition *For any measure space* (X, S, μ), *and any two measurable functions* f *and* g *from* X *into* $[0, \infty]$,

$$\int f + g \, d\mu = \int f \, d\mu + \int g \, d\mu.$$

Proof. First, $(f + g)(x) = +\infty$ if and only if at least one of $f(x)$ or $g(x)$ is $+\infty$. The set where this happens is measurable, and $f + g$ is measurable on it. Restricted to the set where both f and g are finite, $f + g$ is measurable by the argument made just above. Thus $f + g$ is measurable. By Proposition 4.1.5, take simple functions f_n and g_n with $0 \le f_n \uparrow f$ and $0 \le g_n \uparrow g$. So for each n, $\int f_n + g_n \, d\mu = \int f_n \, d\mu + \int g_n \, d\mu$ by Proposition 4.1.4. Then by Proposition 4.1.5, $\int f + g \, d\mu = \lim_{n \to \infty} \int f_n + g_n \, d\mu = \lim_{n \to \infty} (\int f_n \, d\mu + \int g_n \, d\mu) = \int f \, d\mu + \int g \, d\mu$. \square

Proposition 4.1.8 extends, by induction, to any finite sum of nonnegative measurable functions.

Now given any measure space (X, S, μ) and measurable function f from X into $[-\infty, \infty]$, let $f^+ := \max(f, 0)$ and $f^- := -\min(f, 0)$. Then both f^+ and f^- are nonnegative and measurable (max and min, like plus and times, are continuous from $\mathbb{R} \times \mathbb{R}$ into \mathbb{R}). For all x, either $f^+(x) = 0$ or $f^-(x) = 0$, and $f(x) = f^+(x) - f^-(x)$, where this difference is always defined (not $\infty - \infty$). We say that the integral $\int f \, d\mu$ is defined if and only if $\int f^+ \, d\mu$ and $\int f^- \, d\mu$ are not both infinite. Then we define $\int f \, d\mu := \int f^+ \, d\mu - \int f^- \, d\mu$. Integrals are often written with variables, for example $\int f(x) \, d\mu(x) := \int f \, d\mu$. If, for example, $f(x) := x^2$, $\int x^2 \, d\mu(x) := \int f \, d\mu$. Also, if μ is Lebesgue measure λ, then $d\lambda(x)$ is often written as dx.

4.1.9. Lemma *For any measure space* (X, S, μ) *and two measurable functions* $f \le g$ *from* X *into* $[-\infty, \infty]$, *only the following cases are possible:*

(a) $\int f \, d\mu \le \int g \, d\mu$ *(both integrals defined).*
(b) $\int f \, d\mu$ *undefined,* $\int g \, d\mu = +\infty$.
(c) $\int f \, d\mu = -\infty$, $\int g \, d\mu$ *undefined.*
(d) *Both integrals undefined.*

Proof. If $f \ge 0$, it follows directly from the definitions that we must have case (a). In general, we have $f^+ \le g^+$ and $f^- \ge g^-$. Thus if both integrals are defined, (a) holds. If $\int f \, d\mu$ is undefined, then $+\infty = \int f^+ \, d\mu \le \int g^+ \, d\mu$, so $\int g^+ \, d\mu = +\infty$ and $\int g \, d\mu$ is undefined or $+\infty$. Likewise, if $\int g \, d\mu$ is undefined, $-\infty = \int -g^- \, d\mu \ge \int -f^- \, d\mu$, so $\int f \, d\mu$ is undefined or $-\infty$. \square

A measurable function f from X into \mathbb{R} such that $\int |f| \, d\mu < +\infty$ is called *integrable*. The set of all integrable functions for μ is called $\mathcal{L}^1(X, \mathcal{S}, \mu)$. This set may also be called $\mathcal{L}^1(\mu)$ or just \mathcal{L}^1.

4.1.10. Theorem *On* $\mathcal{L}^1(X, \mathcal{S}, \mu)$, $f \mapsto \int f \, d\mu$ *is linear, that is, for any* $f, g \in \mathcal{L}^1(X, \mathcal{S}, \mu)$ *and* $c \in \mathbb{R}$, $\int cf \, d\mu = c \int f \, d\mu$ *and* $\int f + g \, d\mu = \int f \, d\mu + \int g \, d\mu$. *The latter also holds if* $f \in \mathcal{L}^1(X, \mathcal{S}, \mu)$ *and* g *is any nonnegative, measurable function.*

Proof. Recall that $f + g$ and cf are measurable (see around Proposition 4.1.7). We have $\int cf \, d\mu = c \int f \, d\mu$ if $c = -1$ by the definitions. For $c \geq 0$ it follows from Proposition 4.1.5, so it holds for all $c \in \mathbb{R}$.

If f and $g \in \mathcal{L}^1$, then for $h := f + g$, we have $f^- + g^- + h^+ = f^+ + g^+ + h^-$. Thus $h^+ \leq f^+ + g^+$, since $h^- = 0$ where $h \geq 0$. So $\int h^+ \, d\mu < +\infty$ by Proposition 4.1.8 and Lemma 4.1.9. Likewise, $\int h^- \, d\mu < +\infty$, so $h \in \mathcal{L}^1(X, \mathcal{S}, \mu)$. Applying Proposition 4.1.8 and the definitions, we have $\int h \, d\mu = \int h^+ \, d\mu - \int h^- \, d\mu = \int f^+ \, d\mu + \int g^+ \, d\mu - \int f^- \, d\mu - \int g^- \, d\mu = \int f \, d\mu + \int g \, d\mu$.

If, instead, $g \geq 0$ and g is measurable, the remaining case is where $\int g \, d\mu = +\infty$. Then note that $g \leq (f+g)^+ + f^-$ (this is clear where $f \geq 0$; for $f < 0$, $g = (f + g) - f \leq (f + g)^+ + f^-$). Then by Proposition 4.1.8 and Lemma 4.1.9, $+\infty = \int g \, d\mu \leq \int (f+g)^+ d\mu + \int f^- \, d\mu$. Since $\int f^- \, d\mu$ is finite, this gives $\int (f + g)^+ \, d\mu = +\infty$. Next, $(f + g)^- \leq f^-$ implies $\int (f + g)^- \, d\mu$ is finite, so $\int f + g \, d\mu = +\infty = \int f \, d\mu + \int g \, d\mu$. $\qquad\square$

Functions, especially if they are not real-valued, may be called transformations, mappings, or maps. Let (X, \mathcal{S}, μ) be a measure space and (Y, \mathcal{B}) a measurable space. Let T be a measurable transformation from X into Y. Then let $(\mu \circ T^{-1})(A) := \mu(T^{-1}(A))$ for all $A \in \mathcal{B}$. Since $A \mapsto T^{-1}(A)$ preserves all set operations, such as countable unions, and preserves disjointness, $\mu \circ T^{-1}$ is a countably additive measure. It is finite if μ is, but not necessarily σ-finite if μ is (let T be a constant map). Here $\mu \circ T^{-1}$ is called the *image measure* of μ by T. For example, if μ is Lebesgue measure and $T(x) \equiv 2x$, then $\mu \circ T^{-1} = \mu/2$. Integrals for a measure and an image of it are related by a simple "change of variables" theorem:

4.1.11. Theorem *Let* f *be any measurable function from* Y *into* $[-\infty, \infty]$. *Then* $\int f \, d(\mu \circ T^{-1}) = \int f \circ T \, d\mu$ *if either integral is defined (possibly infinite).*

Proof. The result is clear if $f = c1_A$ for some A and $c \geq 0$. Thus by Proposition 4.1.8, it holds for any nonnegative simple function. It follows for any measurable $f \geq 0$ by Proposition 4.1.5. Then, taking f^+ and f^-, it holds for any measurable f from the definition of $\int f \, d\mu$, since $(f \circ T)^+ = f^+ \circ T$ and $(f \circ T)^- = f^- \circ T$. $\qquad\square$

Problems

1. For a measure space (X, \mathcal{S}, μ), let f be a simple function and g a μ-simple function. Show that fg is μ-simple.

2. In the construction of simple functions f_n for $f \equiv x$ on $[0, \infty)$ (Proposition 4.1.5), what is the largest value of f_4? How many different values does f_4 have in its range?

3. For f and g in $\mathcal{L}^1(X, \mathcal{S}, \mu)$ let $d(f, g) := \int |f - g| \, d\mu$. Show that d is a pseudometric on \mathcal{L}^1.

4. Show that the set of all μ-simple functions is dense in \mathcal{L}^1 for d.

5. On the set \mathbb{N} of nonnegative integers let c be counting measure: $c(E) = \text{card } E$ for E finite and $+\infty$ for E infinite. Show that for $f: \mathbb{N} \mapsto \mathbb{R}$, $f \in \mathcal{L}^1(\mathbb{N}, 2^{\mathbb{N}}, c)$ if and only if $\sum_n |f(n)| < +\infty$, and then $\int f \, dc = \sum_n f(n)$.

6. Let $f[A] := \{f(x) : x \in A\}$ for any set A. Given two sets B and C, let $D := f[B] \cup f[C]$, $E := f[B \cup C]$, $F := f[B] \cap f[C]$, and $G := f[B \cap C]$. Prove for all B and C, or disprove by counterexample, each of the following inclusions:
 (a) $D \subset E$; (b) $E \subset D$; (c) $F \subset G$; (d) $G \subset F$.

7. If (X, \mathcal{S}, μ) is a measure space and f a nonnegative measurable function on X, let $(f\mu)(A) := \int_A f \, d\mu$ for any set $A \in \mathcal{S}$.
 (a) Show that $f\mu$ is a measure.
 (b) If T is measurable and 1–1 from X onto Y for a measurable space (Y, \mathcal{A}), with a measurable inverse T^{-1}, show that $(f\mu) \circ T^{-1} = (f \circ T^{-1})(\mu \circ T^{-1})$.

8. Let f be a simple function on \mathbb{R}^2 defined by $f := \sum_{j=1}^n j1_{(j,j+2]\times(j,j+2]}$. Find the atoms of the algebra generated by the rectangles $(j, j+2] \times (j, j+2]$ for $j = 1, \ldots, n$ and express f as a sum of constants times indicator functions of such atoms.

9. Let \mathcal{R} be a σ-ring of subsets of a set X. Let \mathcal{S} be the σ-algebra generated by \mathcal{R}. Recall (§3.3, Problem 8) or prove that \mathcal{S} consists of all sets in \mathcal{R} and all complements of sets in \mathcal{R}.

(a) Let μ be countably additive from \mathcal{R} into $[0, \infty]$. For any set $C \subset X$ let $\mu_*(C) := \sup\{\mu(B): B \subset C, B \in \mathcal{R}\}$ (inner measure). Show that μ_* restricted to \mathcal{S} is a measure, which equals μ on \mathcal{R}.

(b) Show that the extension of μ to a measure on \mathcal{S} is unique if and only if either $\mathcal{S} = \mathcal{R}$ or $\mu_*(X \backslash A) = +\infty$ for all $A \in \mathcal{R}$.

10. Let (S, \mathcal{T}) be a second-countable topological space and (Y, d) any metric space. Show that the Borel σ-algebra in the product $S \times Y$ is the product σ-algebra of the Borel σ-algebras in S and in X. *Hint:* This improves on Proposition 4.1.7. Let V be any open set in $S \times Y$. Let $\{U_m\}_{m \geq 1}$ be a countable base for \mathcal{T}. For each m and $r > 0$ let

$$V_{mr} := \{y \in Y: \text{for some } \delta > 0, U_m \times B(y, r + \delta) \subset V\}$$

where $B(y, t) := \{v \in Y: d(y, v) < t\}$. Show that each V_{mr} is open in Y and $V = \bigcup_{m,n} U_m \times V_{m,1/n}$.

11. Show that for some topological spaces (X, \mathcal{S}) and (Y, \mathcal{T}), there is a closed set D in $X \times Y$ with product topology which is not in *any* product σ-algebra $\mathcal{A} \otimes \mathcal{B}$, for example if \mathcal{A} and \mathcal{B} are the Borel σ-algebras for the given topologies. *Hint:* Let $X = Y$ be a set with cardinality greater than c, for example, the set 2^I of all subsets of $I := [0, 1]$ (Theorem 1.4.2). Let D be the diagonal $\{(x, x): x \in X\}$. Show that for each $C \in \mathcal{A} \otimes \mathcal{B}$, there are sequences $\{A_n\} \subset \mathcal{A}$ and $\{B_n\} \subset \mathcal{B}$ such that C is in the σ-algebra generated by $\{A_n \times B_n\}_{n \geq 1}$. For each n, let $x =_n u$ mean that $x \in A_n$ if and only if $u \in A_n$. Define a relation $x \equiv u$ iff for all n, $x =_n u$. Show that this is an equivalence relation which has at most c different equivalence classes, and for any x, y, and u, if $x \equiv u$, then $(x, y) \in C$ if and only if $(u, y) \in C$. For $C = D$ and $y = x$, find a contradiction.

*4.2. Measurability

Let (Y, \mathcal{T}) be a topological space, with its σ-algebra of Borel sets $\mathcal{B} := \mathcal{B}(Y) := \mathcal{B}(Y, \mathcal{T})$ generated by \mathcal{T}. If (X, \mathcal{S}) is a measurable space, a function f from X into Y is called *measurable* iff $f^{-1}(B) \in \mathcal{S}$ for all $B \in \mathcal{B}$ (unless another σ-algebra in Y is specified). If X is the real line \mathbb{R}, with σ-algebras \mathcal{B} of Borel sets and \mathcal{L} of Lebesgue measurable sets, f is called *Borel* measurable iff it is measurable on $(\mathbb{R}, \mathcal{B})$, and *Lebesgue* measurable iff it is measurable on $(\mathbb{R}, \mathcal{L})$. Note that the Borel σ-algebra is used on the range space in both cases. In fact, the main themes of this section are that matters of measurability work out well if one takes \mathbb{R} or any complete separable metric space as range space and uses the Borel σ-algebra on it. The pathology – what specifically goes

wrong with other σ-algebras, or other range spaces – is less important at this stage. One example is Proposition 4.2.3 in this section. Further pathology is treated in Appendix E. It explains why there is much less about locally compact spaces in this book than in many past texts. The rest of the section could be skipped on first reading and used for reference later.

The following fact shows why the Lebesgue σ-algebra on \mathbb{R} as range may be too large:

4.2.1. Proposition *There exists a continuous, nondecreasing function f from $I := [0, 1]$ into itself and a Lebesgue measurable set L such that $f^{-1}(L)$ is not Lebesgue measurable (assuming the axiom of choice).*

Proof. Associated with the Cantor set C (as in the proof of Proposition 3.4.1) is the Cantor function g, defined as follows, from I into itself: g is nondecreasing and continuous, with $g = 1/2$ on $[1/3, 2/3]$, $1/4$ on $[1/9, 2/9]$, $3/4$ on $[7/9, 8/9]$, $1/8$ on $[1/27, 2/27]$, and so forth (see Figure 4.2).

Here g can be described as follows. Each $x \in [0, 1]$ has a ternary expansion $x = \sum_{n \geq 1} x_n/3^n$ where $x_n = 0, 1$, or 2 for all n. Numbers $m/3^n$, $m \in \mathbb{N}$, $0 < m < 3^n$, have two such expansions, while all other numbers in I have just one. Recall that C is the set of all x having an expansion with $x_n \neq 1$ for all n. For $x \notin C$, let $j(x)$ be the least j such that $x_j = 1$. If $x \in C$, let $j(x) = +\infty$. Then

$$g(x) = 1/2^{j(x)} + \sum_{i=1}^{j(x)-1} x_i/2^{i+1} \quad \text{for } 0 \leq x \leq 1.$$

One can show from this that g is nondecreasing and continuous (Halmos, 1950, p. 83, gives some hints), but these properties seem clear enough in the figure. Now g takes $I \setminus C$ onto the set of dyadic rationals $\{m/2^n$:

Figure 4.2

$m = 1, \ldots, 2^n - 1, n = 1, 2, \ldots$}. Since g takes I onto I, g must take C onto I (the value taken on each "middle third" in the complement of C is also taken at the endpoints, which are in C.) Let $h(x) := (g(x) + x)/2$ for $0 \le x \le 1$. Then h is continuous and strictly increasing, that is, $h(t) < h(u)$ for $t < u$, from I onto itself. It takes each open middle third interval in $I \setminus C$ onto an interval of half its length. Thus it takes $I \setminus C$ onto an open set U with $\lambda(U) = 1/2$ (recall from Proposition 3.4.1 that $\lambda(C) = 0$, so $\lambda(I \setminus C) = 1$). Let $f = h^{-1}$. Then f is continuous and strictly increasing from I onto itself, with $f^{-1}(C) = h[C] = I \setminus U := F$. Then $\lambda(F) = 1/2$ and every subset of F is of the form $f^{-1}(L)$, where $L \subset C$, so L is Lebesgue measurable, with $\lambda(L) = 0$. Let E be a nonmeasurable subset of I with $\lambda^*(E) = \lambda^*(I \setminus E) = 1$, by Theorem 3.4.4. Recall that (hence) neither E nor $I \setminus E$ includes any Lebesgue measurable set A with $\lambda(A) > 0$. Thus, neither $E \cap F$ nor $F \setminus E$ includes such a set. F is a measurable cover of $E \cap F$ (see §3.3) since if F is not and G is, $F \setminus E$ would include a measurable set $F \setminus G$ of positive measure. So $\lambda^*(E \cap F) = 1/2$. Likewise $\lambda^*(F \setminus E) = 1/2$, so $\lambda^*(E \cap F) + \lambda^*(F \setminus E) = 1 \ne \lambda(F) = 1/2$, and $E \cap F$ is not Lebesgue measurable. $\qquad\square$

The next two facts have to do with limits of sequences of measurable functions. To see that there is something not quite trivial involved here, let f_n be a sequence of real-valued functions on some set X such that for all $x \in X$, $f_n(x)$ converges to $f(x)$. Let U be an open interval (a, b). Note that possibly $x \in f_n^{-1}(U)$ for all n, but $x \notin f^{-1}(U)$ (if $f(x) = a$, say). Thus $f^{-1}(U)$ cannot be expressed in terms of the sets $f_n^{-1}(U)$.

4.2.2. Theorem *Let (X, S) be a measurable space and (Y, d) be a metric space. Let f_n be measurable functions from X into Y such that for all $x \in X$, $f_n(x) \to f(x)$ in Y. Then f is measurable.*

Proof. It will be enough to prove that $f^{-1}(U) \in S$ for any open U in Y (by Theorem 4.1.6). Let $F_m := \{y \in U : B(y, 1/m) \subset U\}$, where $B(y, r) := \{v : d(v, y) < r\}$. Then F_m is closed: if $y_j \in F_m$ for all j, $y_j \to y$, and $d(y, v) < 1/m$, then for j large enough, $d(y_j, v) < 1/m$, so $v \in U$. Now $f(x) \in U$ if and only if $f(x) \in F_m$ for some m, and then for n large enough, $d(f_n(x), f(x)) < 1/(2m)$ which implies $f_n(x) \in F_{2m}$ for n large enough. Conversely, if $f_n(x) \in F_m$ for n large enough, then $f(x) \in F_m \subset U$. Thus

$$f^{-1}(U) = \bigcup_m \bigcup_k \bigcap_{n \ge k} f_n^{-1}(F_m) \in S.$$

$\qquad\square$

Now let $I = [0, 1]$ with its usual topology. Then I^I with product topology is a compact Hausdorff space by Tychonoff's Theorem (2.2.8). Such spaces have many good properties, but Theorem 4.2.2 does not extend to them (as range spaces), according to the following fact. Its proof assumes the axiom of choice (as usual, especially when dealing with a space such as I^I).

4.2.3. Proposition *There exists a sequence of continuous (hence Borel measurable) functions f_n from I into I^I such that for all x in I, $f_n(x)$ converges in I^I to $f(x) \in I^I$, but f is not even Lebesgue measurable: there is an open set $W \subset I^I$ such that $f^{-1}(W)$ is not a Lebesgue measurable set in I.*

Proof. For x and y in I let $f_n(x)(y) := \max(0, 1 - n|x - y|)$. To check that f_n is continuous, it is enough to consider the usual subbase of the product topology. For any open $V \subset I$ and $y \in I$, $\{x \in I : f_n(x)(y) \in V\}$ is open in I as desired. Let

$$f(x)(y) := 1_{x=y} = 1 \text{ when } x = y, 0 \text{ otherwise.}$$

Then $f_n(x)(y) \to f(x)(y)$ as $n \to \infty$ for all x and y in I. Thus $f_n(x) \to f(x)$ in I^I for all $x \in I$. Now let E be any subset of I. Let $W := \{g \in I^I : g(y) > 1/2 \text{ for some } y \in E\}$. Then W is open in I^I and $f^{-1}(W) = E$, where E may not be Lebesgue measurable (Theorem 3.4.4). □

If (X, \mathcal{S}) is a measurable space and $A \subset X$, let $\mathcal{S}_A := \{B \cap A : B \in \mathcal{S}\}$. Then \mathcal{S}_A is a σ-algebra of subsets of A, and \mathcal{S}_A will be called the *relative σ-algebra* (of \mathcal{S} on A). The following straightforward fact is often used:

4.2.4. Lemma *Let (X, \mathcal{S}) and (Y, \mathcal{B}) be measurable spaces. Let E_n be disjoint sets in \mathcal{S} with $\bigcup_n E_n = X$. For each $n = 1, 2, \ldots$, let f_n be measurable from E_n, with relative σ-algebra, to Y. Define f by $f(x) = f_n(x)$ for each $x \in E_n$. Then f is measurable.*

Proof. Note that since each $E_n \in \mathcal{S}$, for any $B \in \mathcal{B}$, $f_n^{-1}(B) \in \mathcal{S}$, which is equivalent to $f_n^{-1}(B) = A_n \cap E_n$ for some $A_n \in \mathcal{S}$. Now

$$f^{-1}(B) = \bigcup_n f_n^{-1}(B) \in \mathcal{S}.$$ □

If f is any measurable function on X, then clearly the restriction of f to A is measurable for \mathcal{S}_A. Likewise, any continuous function, restricted to a subset, is continuous for the relative topology. But conversely, a continuous function for

the relative topology cannot always be extended to be continuous on a larger set. For example, $1/x$ on $(0, 1)$ cannot be extended to be continuous and real-valued on $[0, 1)$, nor can $\sin(1/x)$, which is bounded. (A continuous function into \mathbb{R} from a closed subset of a normal space X, such as a metric space, can always be extended to all of X, by the Tietze extension theorem (2.6.4).) A measurable function f, defined on a measurable set A, can be extended trivially to a function g measurable on X, letting g have, for example, some fixed value on $X \backslash A$. What is not so immediately obvious, but true, is that extension of real-valued measurable functions is always possible, even if A is not measurable:

4.2.5. Theorem *Let (X, \mathcal{S}) be any measurable space and A any subset of X (not necessarily in \mathcal{S}). Let f be a real-valued function on A measurable for \mathcal{S}_A. Then f can be extended to a real-valued function on all of X, measurable for \mathcal{S}.*

Proof. Let \mathcal{G} be the set of all \mathcal{S}_A-measurable real-valued functions on A which have \mathcal{S}-measurable extensions. Then clearly \mathcal{G} is a vector space, and $1_{A \cap S}$ has extension 1_S for each $S \in \mathcal{S}$, so \mathcal{G} contains all simple functions for \mathcal{S}_A. To prove $f \in \mathcal{G}$ we can assume $f \geq 0$, since if $f^+ \in \mathcal{G}$ and $f^- \in \mathcal{G}$, then $f \in \mathcal{G}$. Let f_n be simple functions (for \mathcal{S}_A) with $0 \leq f_n \uparrow f$, by Proposition 4.1.5. Let g_n extend f_n. Let $g(x) := \lim_{n \to \infty} g_n(x)$ whenever the limit exists (and is finite). Otherwise let $g(x) = 0$. Clearly, g extends f. The set of x for which $g_n(x)$ converges, or equivalently is a Cauchy sequence, is $G := \bigcap_{k \geq 1} \bigcup_{n \geq 1} \bigcap_{m \geq n} \{x : |g_m(x) - g_n(x)| < 1/k\}$. Hence $G \in \mathcal{S}$. Let $h_n := g_n$ on G, $h_n := 0$ on $X \backslash G$. Then by Lemma 4.2.4, each h_n is measurable, and $h_n(x) \to g(x)$ for all x. Thus by Theorem 4.2.2, g is \mathcal{S}-measurable. \square

The range space \mathbb{R} in Theorem 4.2.5 can be replaced by any complete separable metric space with its Borel σ-algebra, using Theorem 4.2.2 and the following fact:

4.2.6. Proposition *For any separable metric space (S, d), the identity function from S into itself is the pointwise limit of a sequence of Borel measurable functions f_n from S into itself where each f_n has finite range and $f_n(x) \to x$ as $n \to \infty$ for all x.*

Proof. Let $\{x_n\}$ be a countable dense set in S. For each $n = 1, 2, \ldots$, let $f_n(x)$ be the closest point to x among x_1, \ldots, x_n, or the point with lower index if two or more are equally close. Then the range of f_n is included in

$\{x_1, \dots, x_n\}$, and for each $j \le n$,

$$f_n^{-1}(\{x_j\}) = \bigcap_{i<j}\{x: d(x, x_i) > d(x, x_j)\} \cap \bigcap_{j \le i \le n}\{x: d(x, x_i) \ge d(x, x_j)\}.$$

The latter is an intersection of an open set and a closed set, hence a Borel set, so f_n is measurable. Clearly, the f_n converge pointwise to the identity. □

Given a measurable space (X, \mathcal{S}), a function g on X is called *simple* iff its range Y is finite and for each $y \in Y$, $g^{-1}(\{y\}) \in \mathcal{S}$.

4.2.7. Corollary *For any measurable space* (U, \mathcal{S}), $X \subset U$, *non-empty separable metric space* (S, d), *and* \mathcal{S}_X-*measurable function* g *from* X *into* S, *there are simple functions* g_n *from* X *into* S *with* $g_n(x) \to g(x)$ *for all* x. *If* S *is complete,* g *can be extended to all of* U *as an* \mathcal{S}-*measurable function.*

Proof. Let $g_n := f_n \circ g$ with f_n from Proposition 4.2.6. Then the g_n are simple and $g_n(x) \to g(x)$ for all x. Here each g_n can be defined on all of U. If S is complete, the rest of the proof is as for Theorem 4.2.5 (with 0 replaced by an arbitrary point of S). □

Now let (Y, \mathcal{B}) be a measurable space, X any set, and T a function from X into Y. Let $T^{-1}[\mathcal{B}] := \{T^{-1}(B): B \in \mathcal{B}\}$. Then $T^{-1}[\mathcal{B}]$ is a σ-algebra of subsets of X.

4.2.8. Theorem *Given a set* X, *a measurable space* (Y, \mathcal{B}), *and a function* T *from* X *into* Y, *a real-valued function* f *on* X *is* $T^{-1}[\mathcal{B}]$ *measurable on* X *if and only if* $f = g \circ T$ *for some* \mathcal{B}-*measurable function* g *on* Y.

Proof. "If" is clear. Conversely, if f is $T^{-1}[\mathcal{B}]$ measurable, then whenever $T(u) = T(v)$, we have $f(u) = f(v)$, for if not, let B be a Borel set in \mathbb{R} with $f(u) \in B$ and $f(v) \notin B$. Then $f^{-1}(B) = T^{-1}(C)$ for some $C \in \mathcal{B}$, with $T(u) \in C$ but $T(v) \notin C$, a contradiction. Thus, $f = g \circ T$ for some function g from $D :=$ range T into \mathbb{R}. For any Borel set $S \subset \mathbb{R}$, $T^{-1}(g^{-1}(S)) = f^{-1}(S) = T^{-1}(F)$ for some $F \in \mathcal{B}$, so $F \cap D = g^{-1}(S)$ and g is \mathcal{B}_D measurable. By Theorem 4.2.5, g has a \mathcal{B}-measurable extension to all of Y.
 □

Problems

1. Let (X, \mathcal{S}) be a measurable space and E_n measurable sets, not necessarily disjoint, whose union is X. Suppose that for each n, f_n is a measurable

real-valued function on E_n. Suppose that for any $x \in E_m \cap E_n$ for any m and n, $f_m(x) = f_n(x)$. Let $f(x) := f_n(x)$ for any $x \in E_n$ for any n. Show that f is measurable.

2. Let (X, \mathcal{S}) be a measurable space and f_n any sequence of measurable functions from X into $[-\infty, \infty]$. Show that
 (a) $f(x) := \sup_n f_n(x)$ defines a measurable function f.
 (b) $g(x) := \limsup_{n \to \infty} f_n(x) := \inf_m \sup_{m \geq n} f_m(x)$ defines a measurable function g, as does $\liminf_{n \to \infty} f_n := \sup_n \inf_{m \geq n} f_n$.

3. Prove or disprove: Let f be a continuous, strictly increasing function from $[0, 1]$ into itself such that the derivative $f'(x)$ exists for almost all x (Lebesgue measure). ("Strictly increasing" means $f(x) < f(y)$ for $0 \leq x < y \leq 1$.) Then $\int f'(t) \, dt = f(x) - f(0)$ for all x. *Hint:* See Proposition 4.2.1.

4. Let $f(x) := 1_{\{x\}}$, so that f defines a function from I into I^I, as in Proposition 4.2.3.
 (a) Show that the range of f is a Borel set in I^I.
 (b) Show that the graph of f is a Borel set in $I \times I^I$ (with product topology).

5. Prove or disprove: The function f in Problem 4 is the limit of a sequence of functions with finite range.

6. In Theorem 4.2.8, let $X = \mathbb{R}$, let Y be the unit circle in \mathbb{R}^2: $Y := \{(x, y): x^2 + y^2 = 1\}$, and \mathcal{B} the Borel σ-algebra on Y. Let $T(u) := (\cos u, \sin u)$ for all $u \in \mathbb{R}$. Find which of the following functions f on \mathbb{R} are $T^{-1}[\mathcal{B}]$ measurable, and for those that are, find a function g as in Theorem 4.2.8:
 (a) $f(t) = \cos(2t)$; (b) $f(t) = \sin(t/2)$; (c) $f(t) = \sin^2(t/2)$.

7. For the Cantor function g, as defined in the proof of Proposition 4.2.1, evaluate $g(k/8)$ for $k = 0, 1, 2, 3, 4, 5, 6, 7$, and 8.

8. Show that the collection of Borel sets in \mathbb{R} has the same cardinality c as \mathbb{R} does. *Hint:* Show easily that there are at least c Borel sets. Then take an uncountable well-ordered set $(J, <)$ such that for each $j \in J$, $\{i \in J: i < j\}$, is countable. Let α be the least element of J. Recursively, define families \mathcal{B}_j of Borel sets in \mathbb{R}: let \mathcal{B}_α be the collection of all open intervals with rational endpoints. For any $\beta \in J$, if β' is the next larger element, recursively let $\mathcal{B}_{\beta'}$ be the collection of all complements and countable unions of sets in \mathcal{B}_β. If $\gamma \in J$ does not have an immediate predecessor ($\gamma \neq \beta'$ for all β), $\gamma > \alpha$, let \mathcal{B}_γ be the union of \mathcal{B}_β for all $\beta < \gamma$. Show that the union of all the \mathcal{B}_β for $\beta \in J$ is the collection of all Borel sets, and so that its cardinality is c. (See Problem 5 in §1.4.)

9. Let f be a measurable function from X onto S where (X, \mathcal{A}) is a measurable space and (S, e) is a metric space with Borel σ-algebra. Let T be a subset of S with discrete relative topology (all subsets of T are open in T). Show that there is a measurable function g from X onto T. *Hint:* For $f(x)$ close enough to $t \in T$, let $g(x) = t$; otherwise, let $g(x) = t_o$ for a fixed $t_o \in T$.

10. Let f be a Borel measurable function from a separable metric space X onto a metric space S with metric e. Show that (S, e) is separable. *Hints:* As in Problem 8, X has at most c Borel sets. If S is not separable, then show that for some $\varepsilon > 0$ there is an uncountable subset T of S with $d(y, z) > \varepsilon$ for all $y \neq z$ in T. Use Problem 9 to get a measurable function g from X onto T. All $g^{-1}(A)$, $A \subset T$, are Borel sets in X.

4.3. Convergence Theorems for Integrals

Throughout this section let (X, \mathcal{S}, μ) be a measure space. A statement about $x \in X$ will be said to hold *almost everywhere*, or *a.e.*, iff it holds for all $x \notin A$ for some A with $\mu(A) = 0$. (The set of all x for which the statement holds will thus be measurable for the completion of μ, as in §3.3, but will not necessarily be in \mathcal{S}.) Such a statement will also be said to hold for almost all x. For example, $1_{[a,b]} = 1_{(a,b)}$ a.e. for Lebesgue measure.

4.3.1. Proposition *If f and g are two measurable functions from X into $[-\infty, \infty]$ such that $f(x) = g(x)$ a.e., then $\int f \, d\mu$ is defined if and only if $\int g \, d\mu$ is defined. When defined, the integrals are equal.*

Proof. Let $f = g$ on $X \backslash A$ where $\mu(A) = 0$. Let us show that $\int h \, d\mu = \int_{X \backslash A} h \, d\mu$, where h is any measurable function, and equality holds in the sense that the integrals are defined and equal if and only if either of them is defined. This is clearly true if h is an indicator function of a set in \mathcal{S}; then, if h is any nonnegative simple function; then, by Proposition 4.1.5, if h is any nonnegative measurable function; and thus for a general h, by definition of the integral. Letting $h = f$ and $h = g$ finishes the proof. \square

Let a function f be defined on a set $B \in \mathcal{S}$ with $\mu(X \backslash B) = 0$, where f has values in $[-\infty, \infty]$ and is measurable for \mathcal{S}_B. Then f can be extended to a measurable function for \mathcal{S} on X (let $f = 0$ on $X \backslash B$, for example). For any two extensions g and h of f to X, $g = h$ a.e. Thus we can define $\int f \, d\mu$ as $\int g \, d\mu$, if this is defined. Then $\int f \, d\mu$ is well-defined by Proposition 4.3.1. If $f_n = g_n$

a.e. for $n = 1, 2, \ldots$, then $\mu^*(\bigcup_n \{x : f_n(x) \neq g_n(x)\}) = 0$. Outside this set, $f_n = g_n$ for all n. Thus in theorems about integrals, even for sequences of functions as below, the hypotheses need only hold almost everywhere.

The three theorems in the rest of this section are among the most important and widely used in analysis.

4.3.2. Monotone Convergence Theorem *Let f_n be measurable functions from X into $[-\infty, \infty]$ such that $f_n \uparrow f$ and $\int f_1 \, d\mu > -\infty$. Then $\int f_n \, d\mu \uparrow \int f \, d\mu$.*

Proof. First, let us make sure f is measurable. For any $c \in \mathbb{R}$, $f^{-1}((c, \infty]) = \bigcup_{n \geq 1} f_n^{-1}((c, \infty]) \in \mathcal{S}$. It is easily seen that the set of all open half-lines $(c, \infty]$ generates the Borel σ-algebra of $[-\infty, \infty]$ (see the discussion just before Theorem 3.2.6). Then f is measurable by Theorem 4.1.6.

Next, suppose $f_1 \geq 0$. Then by Proposition 4.1.5, take simple $f_{nm} \uparrow f_n$ as $m \to \infty$ for each n. Let $g_n := \max(f_{1n}, f_{2n}, \ldots, f_{nn})$. Then each g_n is simple and $0 \leq g_n \uparrow f$. So by Proposition 4.1.5, $\int g_n \, d\mu \uparrow \int f \, d\mu$. Since $g_n \leq f_n \uparrow f$, we get $\int f_n \, d\mu \uparrow \int f \, d\mu$.

Next suppose $f \leq 0$. Let $g_n := -f_n \downarrow -f := g$. Then $0 \leq \int g \, d\mu \leq \int g_n \, d\mu < +\infty$ for all n (the middle inequality follows from Lemma 4.1.9). Now $0 \leq g_1 - g_n \uparrow g_1 - g$, so by the last paragraph, $\int g_1 - g_n \, d\mu \uparrow \int g_1 - g \, d\mu < +\infty$. These integrals all being finite, we can subtract them from $\int g_1 \, d\mu$, by Theorem 4.1.10 and get $\int g_n \, d\mu \downarrow \int g \, d\mu$, so $\int f_n \, d\mu \uparrow \int f \, d\mu$, as desired.

Now in the general case, we have $f_n^+ \uparrow f^+$ and $f_n^- \downarrow f^-$ with $\int f^- \, d\mu < +\infty$, so by the previous cases, $\int f_n^+ \, d\mu \uparrow \int f^+ \, d\mu$ and $+\infty > \int f_n^- \, d\mu \downarrow \int f^- \, d\mu \geq 0$. Thus $\int f_n \, d\mu \uparrow \int f \, d\mu$. \square

For example, let $f_n := -1_{[n,\infty)}$. Then $f_n \uparrow 0$ but $\int f_n \, d\mu \equiv -\infty$, not converging to 0. This shows why a hypothesis such as $\int f_1 \, d\mu > -\infty$ is needed in the monotone convergence theorem. There is a symmetric form of monotone convergence with $f_n \downarrow f$, $\int f_1 \, d\mu < +\infty$.

For any real a_n, $\liminf_{n \to \infty} a_n := \sup_m \inf_{n \geq m} a_n$ is defined (possibly infinite). So the next fact, though a one-sided inequality, is rather general:

4.3.3. Fatou's Lemma *Let f_n be any nonnegative measurable functions on X. Then $\int \liminf f_n \, d\mu \leq \liminf \int f_n \, d\mu$.*

Proof. Let $g_n(x) := \inf\{f_m(x): m \geq n\}$. Then $g_n \uparrow \liminf f_n$. Thus by monotone convergence, $\int g_n \, d\mu \uparrow \int \liminf f_n \, d\mu$. For all $m \geq n$, $g_n \leq f_m$, so $\int g_n \, d\mu \leq \int f_m \, d\mu$. Hence $\int g_n \, d\mu \leq \inf\{\int f_m \, d\mu: m \geq n\}$. Taking the limit of both sides as $n \to \infty$ finishes the proof. $\qquad\qquad\qquad\quad\square$

4.3.4. Corollary *Suppose that f_n are nonnegative and measurable, and that $f_n(x) \to f(x)$ for all x. Then $\int f \, d\mu \leq \sup_n \int f_n \, d\mu$.*

Example. Let $\mu = \lambda$, $f_n = 1_{[n,n+1]}$. Then $f_n(x) \to f(x) := 0$ for all x, while $\int f_n \, d\mu = 1$ for all n. This shows that the inequality in Fatou's lemma may be strict. Also, the example should help in remembering which way the inequality goes.

4.3.5. Dominated Convergence Theorem *Let f_n and g be in $\mathcal{L}^1(X, \mathcal{S}, \mu)$, $|f_n(x)| \leq g(x)$ and $f_n(x) \to f(x)$ for all x. Then $f \in \mathcal{L}^1$ and $\int f_n \, d\mu \to \int f \, d\mu$.*

Proof. Let $h_n(x) := \inf\{f_m(x): m \geq n\}$ and $j_n(x) := \sup\{f_m(x): m \geq n\}$. Then $h_n \leq f_n \leq j_n$. Since $h_n \uparrow f$, and $\int h_1 \, d\mu \geq -\int |g| \, d\mu > -\infty$, we have monotone convergence $\int h_n \, d\mu \uparrow \int f \, d\mu$ (by Theorem 4.3.2). Likewise considering $-j_n$, we have monotone convergence $\int j_n \, d\mu \downarrow \int f \, d\mu$. Since $\int h_n \, d\mu \leq \int f_n \, d\mu \leq \int j_n \, d\mu$, we get $\int f_n \, d\mu \to \int f \, d\mu$. $\qquad\quad\square$

Example. $1_{[n,2n]} \to 0$ but $\int 1_{[n,2n]} \, d\lambda = n \not\to 0$. This shows how the "domination" hypothesis $|f_n| \leq g \in \mathcal{L}^1$ is useful.

Remark. Sums can be considered as integrals for counting measures (which give measure 1 to each singleton), so the above convergence theorems can all be applied to sums.

Problems

1. Let $f_n \in \mathcal{L}^1(X, \mathcal{S}, \mu)$ satisfy $f_n \geq 0$, $f_n(x) \to f_0(x)$ as $n \to \infty$ for all x, and $\int f_n \, d\mu \to \int f_0 \, d\mu < \infty$. Show that $\int |f_n - f_0| \, d\mu \to 0$. *Hint:* $(f_n - f_0)^- \leq f_0$; use dominated convergence.

2. In the statement of Fatou's lemma, consider replacing "lim inf" by "lim sup," replacing "$f_n \geq 0$" by "$f_n \leq 0$," and replacing "\leq" by "\geq". Show that the statement remains true if all three changes are made, but give examples to show that the statement can fail if any one or two of the changes are made.

3. Suppose that $f_n(x) \geq 0$ and $f_n(x) \to f(x)$ as $n \to \infty$ for all x. Suppose also that $\int f_n \, d\mu$ converges to some $c > 0$. Show that $\int f \, d\mu$ is defined and in the interval $[0, c]$ but not necessarily equal to c. Show, by examples, that any value in $[0, c]$ is possible.

4. (a) Let $f_n := 1_{[0,n]}/n^2$. Is there an integrable function g (for Lebesgue measure on \mathbb{R}) which dominates these f_n?
 (b) Same question for $f_n := 1_{[0,n]}/(n \log n)$, $n \geq 2$.

5. Show that $\int_0^\infty \sin(e^x)/(1 + nx^2) \, dx \to 0$ as $n \to \infty$.

6. Show that $\int_0^1 (n \cos x)/(1 + n^2 x^{3/2}) \, dx \to 0$ as $n \to \infty$.

7. On a measure space (X, \mathcal{S}, μ) let f be a real, measurable function such that $\int f^2 \, d\mu < \infty$. Let g_n be measurable functions such that $|g_n(x)| \leq f(x)$ for all x and $g_n(x) \to g(x)$ for all x. Show that $\int (g_n + g)^2 \, d\mu \to 4 \int g^2 \, d\mu < \infty$ as $n \to \infty$.

8. Prove the dominated convergence theorem from Fatou's lemma by considering the sequences $g + f_n$ and $g - f_n$.

9. Let $g(x) := 1/(x \log x)$ for $x > 1$. Let $f_n := c_n 1_{A(n)}$ for some constants $c_n \geq 0$ and measurable subsets $A(n)$ of $[2, \infty)$. Prove or disprove: If $f_n(x) \to 0$ and $|f_n(x)| \leq g(x)$ for all x, then $\int_2^\infty f_n(x) \, dx \to 0$ as $n \to \infty$.

10. Let $f(x, y)$ be a measurable function of two real variables having a partial derivative $\partial f/\partial x$ which is bounded for $a < x < b$ and $c \leq y \leq d$, where c and d are finite and such that $\int_c^d |f(x, y)| \, dy < \infty$ for some $x \in (a, b)$. Prove that the integral is finite for all $x \in (a, b)$ and that we can "differentiate under the integral sign," that is, $(d/dx) \int_c^d f(x, y) \, dy = \int_c^d \partial f(x, y)/\partial x \, dy$, for $a < x < b$.

11. If $c = 0$ and $d = +\infty$, and if $\int_0^\infty |\partial f(x, y)/\partial x| \, dy < \infty$ for some $x = x_0$, show that the conclusion of Problem 10 need not hold for that x. *Hint:* Let $a = -1$, $b = 1$, $x_0 = 0$, and $f(x, y) = 0$ for $x \leq 1/(y + 1)$.

12. Let f_n and g_n be integrable functions for a measure μ with $|f_n| \leq g_n$. Suppose that as $n \to \infty$, $f_n(x) \to f(x)$ and $g_n(x) \to g(x)$ for almost all x. Show that if $\int g_n \, d\mu \to \int g \, d\mu < \infty$, then $\int f_n \, d\mu \to \int f \, d\mu$. *Hint:* See Problem 8.

13. Let (X, \mathcal{S}, μ) be a finite measure space, meaning that $\mu(X) < \infty$. A sequence $\{f_n\}$ of real-valued measurable functions on X is said to converge *in measure* to f if for every $\varepsilon > 0$, $\lim_{n \to \infty} \mu\{x : |f_n(x) - f(x)| > \varepsilon\} = 0$. Show that if $f_n \to f$ in measure and for some integrable function g, $|f_n| \leq g$ for all n, then $\int |f_n - f| \, d\mu \to 0$.

14. (a) If $f_n \to f$ in measure, $f_n \geq 0$ and $\int f_n \, d\mu \to \int f \, d\mu < \infty$, show that $\int |f_n - f| \, d\mu \to 0$. *Hint:* See Problem 1.

(b) If $f_n \to f$ in measure and $\int |f_n| \, d\mu \to \int |f| \, d\mu < \infty$, show that $\int |f_n - f| \, d\mu \to 0$.

15. For a finite measure space (X, \mathcal{S}, μ), a set \mathcal{F} of integrable functions is said to be *uniformly integrable* iff $\sup\{\int |f| \, d\mu \colon f \in \mathcal{F}\} < \infty$ and for every $\varepsilon > 0$, there is a $\delta > 0$ such that if $\mu(A) < \delta$, then $\int_A |f| \, d\mu < \varepsilon$ for every $f \in \mathcal{F}$. Prove that a sequence $\{f_n\}$ of integrable functions satisfies $\int |f_n - f| \, d\mu \to 0$ if and only if both $f_n \to f$ in measure and the f_n are uniformly integrable.

4.4. Product Measures

For any $a \leq b$ and $c \leq d$, the rectangle $[a, b] \times [c, d]$ in \mathbb{R}^2 has area $(d - c)(b - a)$, the product of the lengths of its sides. It's very familiar that area is defined for much more general sets. In this section, it will be defined even more generally, as a measure, and the Cartesian product will be defined for any two σ-finite measures in place of length on two real axes. Then the product will be extended to more than two factors, giving, for example, "volume" as a measure on \mathbb{R}^3.

Let (X, \mathcal{B}, μ) and (Y, \mathcal{C}, ν) be any two measure spaces. In $X \times Y$ let \mathcal{R} be the collection of all "rectangles" $B \times C$ with $B \in \mathcal{B}$ and $C \in \mathcal{C}$. For such sets let $\rho(B \times C) := \mu(B)\nu(C)$, where (in this case) we set $0 \cdot \infty := \infty \cdot 0 := 0$. \mathcal{R} is a semiring by Proposition 3.2.2.

4.4.1. Theorem ρ *is countably additive on* \mathcal{R}.

Proof. Suppose $B \times C = \bigcup_n B(n) \times C(n)$ in \mathcal{R} where the sets $B(n) \times C(n)$ are disjoint, $B(n) \in \mathcal{B}$, and $C(n) \in \mathcal{C}$ for all n. So for each $x \in X$ and $y \in Y$, $1_B(x)1_C(y) = \sum_n 1_{B(n)}(x)1_{C(n)}(y)$. Then integrating $d\nu(y)$ gives for each x, by countable additivity, $1_B(x)\nu(C) = \sum_n 1_{B(n)}(x)\nu(C(n))$. Now integrating $d\mu(x)$ gives, by additivity (Proposition 4.1.8) and monotone convergence (Theorem 4.3.2), $\mu(B)\nu(C) = \sum_n \mu(B(n))\nu(C(n))$. \square

Let \mathcal{A} be the ring generated by \mathcal{R}. Then \mathcal{A} consists of all unions of finitely many disjoint elements of \mathcal{R} (Proposition 3.2.3). Since $X \times Y \in \mathcal{R}$, \mathcal{A} is an algebra. For any disjoint $C_j \in \mathcal{R}$ and finite n let $\rho(\bigcup_{1 \leq j \leq n} C_j) := \sum_{1 \leq j \leq n} \rho(C_j)$. Here ρ is well-defined and countably additive on \mathcal{A} by Proposition 3.2.4 and Theorem 4.4.1. Then ρ can be extended to a countably additive measure on the product σ-algebra $\mathcal{B} \otimes \mathcal{C}$ (defined before Proposition 4.1.7) generated by \mathcal{R} or \mathcal{A} (Theorem 3.1.4) but, in general, such an extension

is not unique. The next steps will be to give conditions under which the extension is unique and can be written in terms of iterated integrals. For this, the following notion will be helpful. A collection \mathcal{M} of sets is called a *monotone class* iff whenever $M_n \in \mathcal{M}$ and $M_n \downarrow M$ or $M_n \uparrow M$, then $M \in \mathcal{M}$. For example, any σ-algebra is a monotone class, but in general a topology is not (an infinite intersection of open sets is usually not open). For any set X, 2^X is clearly a monotone class. The intersection of any set of monotone classes is a monotone class. Thus, for any collection \mathcal{D} of sets, there is a smallest monotone class including \mathcal{D}.

4.4.2. Theorem *If \mathcal{A} is an algebra of subsets of a set X, then the smallest monotone class \mathcal{M} including \mathcal{A} is a σ-algebra.*

Proof. Let $\mathcal{N} := \{E \in \mathcal{M}: X \backslash E \in \mathcal{M}\}$. Then $\mathcal{A} \subset \mathcal{N}$ and \mathcal{N} is a monotone class, so $\mathcal{N} = \mathcal{M}$. For each set $A \subset X$, let $\mathcal{M}_A := \{E: E \cap A \in \mathcal{M}\}$. Then for each A in \mathcal{A}, $\mathcal{A} \subset \mathcal{M}_A$ and \mathcal{M}_A is a monotone class, so $\mathcal{M} \subset \mathcal{M}_A$. Then for each $E \in \mathcal{M}$, \mathcal{M}_E is a monotone class including \mathcal{A}, so $\mathcal{M} \subset \mathcal{M}_E$. Thus \mathcal{M} is an algebra. Being a monotone class, it is a σ-algebra. $\quad\square$

The next fact says that the order of integration can be inverted for indicator functions of measurable sets (in the product σ-algebra). This will be the main step toward the construction of product measures in Theorem 4.4.4 and in interchange of integrals for more general functions (Theorem 4.4.5).

4.4.3. Theorem *Suppose $\mu(X) < +\infty$ and $\nu(Y) < +\infty$. Let*

$$\mathcal{F} := \left\{ E \subset X \times Y : \int \left[\int 1_E(x, y) \, d\mu(x) \right] d\nu(y) \right.$$
$$\left. = \int \left[\int 1_E(x, y) \, d\nu(y) \right] d\mu(x) \right\}.$$

Then $\mathcal{B} \otimes \mathcal{C} \subset \mathcal{F}$.

Proof. The definition of \mathcal{F} implies that all of the integrals appearing in it are defined, so that each function being integrated is measurable. It should be noted that this measurability holds at each step of the proof to follow. If $E = B \times C$ for some $B \in \mathcal{B}$ and $C \in \mathcal{C}$, then

$$\int\int 1_E \, d\mu \, d\nu = \mu(B) \int 1_C \, d\nu = \mu(B)\nu(C) = \int\int 1_E \, d\nu \, d\mu.$$

Thus $\mathcal{R} \subset \mathcal{F}$. If $E_n \in \mathcal{F}$, and $E_n \downarrow E$ or $E_n \uparrow E$, then $E \in \mathcal{F}$ by monotone convergence (Theorem 4.3.2), using finiteness. Thus \mathcal{F} is a monotone class. Also, any finite disjoint union of sets in \mathcal{F} is in \mathcal{F}. Thus $\mathcal{A} \subset \mathcal{F}$. Hence by Theorem 4.4.2, $\mathcal{B} \otimes \mathcal{C} \subset \mathcal{F}$. $\qquad\square$

4.4.4. Product Measure Existence Theorem *Let (X, \mathcal{B}, v) and (Y, \mathcal{C}, v) be two σ-finite measure spaces. Then ρ extends uniquely to a measure on $\mathcal{B} \otimes \mathcal{C}$ such that for all $E \in \mathcal{B} \otimes \mathcal{C}$,*

$$\rho(E) = \int\int 1_E(x, y)\, d\mu(x)\, dv(y) = \int\int 1_E(x, y)\, dv(y)\, d\mu(x).$$

Proof. First suppose μ and v are finite. Let $\alpha(E) := \int\int 1_E(x, y)\, d\mu(x)\, dv(y)$, $E \in \mathcal{B} \otimes \mathcal{C}$. Then by Theorem 4.4.3, α is defined and the order of integration can be reversed. Now α is finitely additive (for any finitely many disjoint sets in $\mathcal{B} \otimes \mathcal{C}$) by Proposition 4.1.8. Again, all functions being integrated will be measurable. Then, α is countably additive by monotone convergence (Theorem 4.3.2). For any other extension β of ρ to $\mathcal{B} \otimes \mathcal{C}$, the collection of sets on which $\alpha = \beta$ is a monotone class including \mathcal{A}, thus including $\mathcal{B} \otimes \mathcal{C}$. So the theorem holds for finite measures.

In general, let $X = \bigcup_m B_m$, $Y = \bigcup_n C_n$, where the B_m are disjoint in X and the C_n in Y, with $\mu(B_m) < \infty$ and $v(C_n) < \infty$ for all m and n. Let $E \in \mathcal{B} \otimes \mathcal{C}$ and $E(m, n) := E \cap (B_m \times C_n)$. Then for each m and n, by the finite case,

$$\int\int 1_{E(m,n)}\, d\mu\, dv = \int\int 1_{E(m,n)}\, dv\, d\mu.$$

This equation can be summed over all m and n (in any order, by Lemma 3.1.2). By countable additivity and monotone convergence, we get

$$\alpha(E) := \int\int 1_E\, d\mu\, dv = \int\int 1_E\, dv\, d\mu \text{ for any } E \in \mathcal{B} \otimes \mathcal{C}.$$

Then α is finitely additive, countably additive by monotone convergence, and thus a measure, which equals ρ on \mathcal{A}. If β is any other extension of ρ to a measure on $\mathcal{B} \otimes \mathcal{C}$, then for any $E \in \mathcal{B} \otimes \mathcal{C}$,

$$\beta(E) = \sum_{m,n} \beta(E(m, n)) = \sum_{m,n} \alpha(E(m, n)) = \alpha(E),$$

so the extension is unique. $\qquad\square$

Example. Let c be counting measure and λ Lebesgue measure on $I := [0, 1]$. In $I \times I$ let D be the diagonal $\{(x, x): x \in I\}$. Then D is measurable (it is closed and I is second-countable, so Proposition 4.1.7 applies), but $\int\int 1_D \, d\lambda \, dc = 0 \neq 1 = \int\int 1_D \, dc \, d\lambda$. This shows how σ-finiteness is useful in Theorem 4.4.4 (c is not σ-finite).

The measure ρ on $\mathcal{B} \otimes \mathcal{C}$ is called a *product measure* $\mu \times \nu$. Now here is the main theorem on integrals for product measures:

4.4.5. Theorem (Tonelli-Fubini) *Let (X, \mathcal{B}, μ) and (Y, \mathcal{C}, ν) be σ-finite, and let f be a function from $X \times Y$ into $[0, \infty]$ measurable for $\mathcal{B} \otimes \mathcal{C}$, or $f \in \mathcal{L}^1(X \times Y, \mathcal{B} \otimes \mathcal{C}, \mu \times \nu)$. Then*

$$\int f \, d(\mu \times \nu) = \int\int f(x, y) \, d\mu(x) \, d\nu(y) = \int\int f(x, y) \, d\nu(y) \, d\mu(x).$$

Here $\int f(x, y) \, d\mu(x)$ is defined for ν-almost all y and $\int f(x, y) \, d\nu(y)$ for μ-almost all x.

Proof. Recall that integrals are defined for functions only defined almost everywhere (as in and after Proposition 4.3.1). For nonnegative simple f the theorem follows from Theorem 4.4.4 and additivity of integrals (Proposition 4.1.8). Then for nonnegative measurable f it follows from monotone convergence (Proposition 4.1.5 and Theorem 4.3.2). Then, for $f \in \mathcal{L}^1(X \times Y, \mathcal{B} \otimes \mathcal{C}, \mu \times \nu)$, the theorem holds for f^+ and f^-. Thus, $\int f^+(x, y) \, d\mu(x) < \infty$ for almost all y (with respect to ν) and likewise for f^- and for μ and ν interchanged. For ν-almost all y, $\int |f(x, y)| \, d\mu(x) < \infty$, and then by Theorem 4.1.10,

$$\int f(x, y) \, d\mu(x) = \int f^+(x, y) \, d\mu(x) - \int f^-(x, y) \, d\mu(x),$$

with all three integrals being finite. Next, in integrating with respect to ν, the set of ν-measure 0 where an integral of f^+ or f^- is infinite doesn't matter, as in Proposition 4.3.1. So by Theorem 4.1.10 again, $\int\int f(x, y) \, d\mu(x) \, d\nu(y)$ is defined and finite and equals

$$\int\int f^+(x, y) \, d\mu(x) \, d\nu(y) - \int\int f^-(x, y) \, d\mu(x) \, d\nu(y).$$

Then the theorem for f^+ and f^- implies it for f. $\qquad\square$

0	0	0	0	0	1	−1	0	0
0	0	0	0	1	−1	0	0	0
0	0	0	1	−1	0	0	0	0
0	0	1	−1	0	0	0	0	0
0	1	−1	0	0	0	0	0	0
sums	1	0	0	0	0	0	0	0

Figure 4.4

Remark. To prove that $f \in \mathcal{L}^1(X \times Y, \mathcal{B} \otimes \mathcal{C}, \mu \times \nu)$ one can prove that f is $\mathcal{B} \otimes \mathcal{C}$-measurable and then that $\int\int |f| \, d\mu \, d\nu < +\infty$ or $\int\int |f| \, d\nu \, d\mu < +\infty$.

Example (a). Let $X = Y = \mathbb{N}$ and $\mu = \nu =$ counting measure. So for $f: \mathbb{N} \mapsto \mathbb{R}$ we have $\int f \, d\mu = \int f(n) \, d\mu(n) = \sum_n f(n)$, where $f \in \mathcal{L}^1(\mu)$ if and only if $\sum_n |f(n)| < +\infty$. (For counting measure, the σ-algebra is $2^{\mathbb{N}}$, so all functions are measurable.) On $\mathbb{N} \times \mathbb{N}$ let $g(n, n) := 1$, $g(n + 1, n) := -1$ for all $n \in \mathbb{N}$, and $g(m, n) := 0$ for $m \neq n$ and $m \neq n + 1$ (see Figure 4.4). Then g is bounded and measurable on $X \times Y$, $\int\int g(m, n) \, d\mu(m) \, d\nu(n) = (1 - 1) + (1 - 1) + \cdots = 0$, but $\int\int g(m, n) \, d\nu(n) \, d\mu(m) = 1 + (1 - 1) + (1 - 1) + \cdots = 1$. Thus the integrals cannot be interchanged (both g^+ and g^- have infinite integrals).

Example (b). For $x \in \mathbb{R}$ and $t > 0$ let

$$f(x, t) := (2\pi t)^{-1/2} \exp(-x^2/(2t)).$$

Let $g(x, t) := \partial f / \partial t$. Then $\partial f / \partial x = -xf/t$ and $\partial^2 f / \partial x^2 = (x^2 t^{-2} - t^{-1}) f = 2g$. Thus f satisfies the partial differential equation $2\partial f / \partial t = \partial^2 f / \partial x^2$, called a heat equation. (In fact, f is called a "fundamental solution" of the heat equation: Schwartz, 1966, p. 145.) For every $t > 0$ we have $\int_{-\infty}^{\infty} f(x, t) \, dx = 1$ (where $dx := d\lambda(x)$) since by polar coordinates (developed in Problem 6 below)

$$\left(\int_0^\infty \exp(-x^2/2) \, dx \right)^2 = (\pi/2) \int_0^\infty r \cdot \exp(-r^2/2) \, dr = \pi/2.$$

(Here $f(\cdot, t)$ is called a "normal" or "Gaussian" probability density. Such functions will have a major role in Chapter 9.) Now for any $s > 0$,

$$\int_{-\infty}^{\infty} \int_s^\infty g(x, t) \, dt \, dx = \int_{-\infty}^{\infty} -f(x, s) \, dx = -1, \quad \text{but}$$

$$\int_s^\infty \int_{-\infty}^{\infty} g(x, t) \, dx \, dt = \int_s^\infty \partial f / \partial x \big|_0^\infty \, dt = 0,$$

since $\partial f/\partial x \to 0$ as $|x| \to \infty$ (by L'Hospital's rule). Thus here again, the order of integration cannot be interchanged, and neither g^+ nor g^- is integrable. Some insight into this paradox comes from Schwartz's theory of distributions where although the functions $2\partial f/\partial t$ and $\partial^2 f/\partial x^2$ are smooth and equal for $t > 0$, their difference is not 0, but rather a measure concentrated at $(0, 0)$; see Hörmander (1983, pp. 80–81).

Let S_j be a σ-algebra of subsets of X_j for each $j = 1, \ldots, n$. Then the *product σ-algebra $S_1 \otimes \cdots \otimes S_n$* is defined as the smallest σ-algebra for which each coordinate function x_j is measurable. It is easily seen that this agrees with the previous definition for $n = 2$ and that for each $n \geq 2$, by induction on n, $S_1 \otimes \cdots \otimes S_n$ is the smallest σ-algebra of subsets of X containing all sets $A_1 \times \cdots \times A_n$ with $A_j \in S_j$ for each $j = 1, \ldots, n$. Just as a function continuous for a product topology is called jointly continuous, a function measurable for a product σ-algebra will be called *jointly measurable*.

4.4.6. Theorem *Let (X_j, S_j, μ_j) be σ-finite measure spaces for $j = 1, \ldots, n$. Then there is a unique measure μ on the product σ-algebra S in $X = X_1 \times \cdots \times X_n$ such that for any $A_j \in S_j$ for $j = 1, \ldots, n$, $\mu(A_1 \times \cdots \times A_n) = \mu_1(A_1)\mu_2(A_2) \cdots \mu_n(A_n)$, or 0 if any $\mu_j(A_j) = 0$, even if another is $+\infty$. If f is nonnegative and jointly measurable on X, or if $f \in \mathcal{L}^1(X, S, \mu)$, then*

$$\int f \, d\mu = \int \cdots \int f(x_1, \ldots, x_n) \, d\mu_1(x_1) \cdots d\mu_n(x_n),$$

where for $f \in \mathcal{L}^1(X, S, \mu)$, the iterated integral is defined recursively "from the outside" in the sense that for μ_n-almost all x_n, the iterated integral with respect to the other variables is defined and finite, so that except on a set of μ_{n-1} measure 0 (possibly depending on x_n) the iterated integral for the first $n - 2$ variables is defined and finite, and so on. The same holds if the integrations are done in any order.

Proof. The statement follows from Theorems 4.4.4 and 4.4.5 and induction on n. □

The best-known example of Theorem 4.4.6 is Lebesgue measure λ^n on \mathbb{R}^n, which is a product with $\mu_j = $ Lebesgue measure λ on \mathbb{R} for each j. Then λ is length, λ^2 is area, λ^3 is volume, and so forth.

Problems

1. Let (X, \mathcal{B}, μ) and (Y, \mathcal{C}, ν) be σ-finite measure spaces. Let $f \in \mathcal{L}^1(X, \mathcal{B}, \mu)$ and $g \in \mathcal{L}^1(Y, \mathcal{C}, \nu)$. Let $h(x, y) := f(x)g(y)$. Prove that h belongs to $\mathcal{L}^1(X \times Y, \mathcal{B} \otimes \mathcal{C}, \mu \times \nu)$ and $\int h \, d(\mu \times \nu) = \int f \, d\mu \int g \, d\nu$.

2. Let (X, \mathcal{B}, μ) be a σ-finite measure space and f a nonnegative measurable function on X. Prove that for $\lambda :=$ Lebesgue measure, $\int f \, d\mu = (\mu \times \lambda) \{(x, y) : 0 < y < f(x)\}$ ("the integral is the area under the curve").

3. Let (X, \mathcal{B}, μ) be σ-finite and f any measurable real-valued function on X. Prove that $(\mu \times \lambda)\{(x, y) : y = f(x)\} = 0$ (the graph of a real measurable function has measure 0).

4. Let (X, \le) be an uncountable well-ordered set such that for any $y \in X$, $\{x \in X : x < y\}$ is countable. For any $A \subset X$, let if $\mu(A) = 0$ if A is countable and $\mu(A) = 1$ if $X \backslash A$ is countable. Show that μ is a measure defined on a σ-algebra. Define T, the "ordinal triangle," by $T := \{(x, y) \in X \times X : y < x\}$. Evaluate the iterated integrals $\int\!\int 1_T(x, y) \, d\mu(y) \, d\mu(x)$ and $\int\!\int 1_T(x, y) \, d\mu(x) \, d\mu(y)$. How are the results consistent with the product measure theorems (4.4.4 and 4.4.5)?

5. For the product measure $\lambda \times \lambda$ on \mathbb{R}^2, the usual Lebesgue measure on the plane, it's very easy to show that rectangles parallel to the axes have $\lambda \times \lambda$ measure equal to their usual areas, the product of their sides. Prove this for rectangles that are not necessarily parallel to the axes.

6. *Polar coordinates.* Let T be the function from $X := [0, \infty) \times [0, 2\pi)$ onto \mathbb{R}^2 defined by $T(r, \theta) := (r \cdot \cos\theta, r \cdot \sin\theta)$. Show that T is 1–1 on $(0, \infty) \times [0, 2\pi)$. Let σ be the measure on $(0, \infty)$ defined by $\sigma(A) := \int_A r \, d\lambda(r) := \int 1_A(r) r \, d\lambda(r)$. Let $\mu := \sigma \times \lambda$ on X. Show that the image measure $\mu \circ T^{-1}$ is Lebesgue measure $\lambda^2 := \lambda \times \lambda$ on \mathbb{R}^2. *Hint:* Prove $\lambda^2(B) = \mu(T^{-1}(B))$ when $T^{-1}(B)$ is a rectangle where $s < r \le t$ and $\alpha < \theta \le \beta$ (B is a sector of an annulus). You might do this by calculus, or show that when $(t - s)/s$ and $\beta - \alpha$ are small, B can be well approximated inside and out by rectangles from the last problem; then assemble small sets B to make larger ones. Show that the set of finite disjoint unions of such sets B is a ring, which generates the σ-algebra of measurable sets in \mathbb{R}^2 (use Proposition 4.1.7). (Applying a Jacobian theorem isn't allowed.)

7. For a measurable real-valued function f on $(0, \infty)$ let $P(f) := \{p \in (0, \infty) : |f|^p \in \mathcal{L}^1(\mathbb{R}, \mathcal{B}, \lambda)\}$, where \mathcal{B} is the Borel σ-algebra. Show that for every subinterval J of $(0, \infty)$, which may be open or closed on the left and on the right, there is some f such that $P(f) = J$. *Hint:* Consider

functions such as $x^a |\log x|^b$ on $(0, 1]$ and on $[1, \infty)$, where $b = 0$ unless $a = -1$.

8. Let $B_k(0, r) := \{x \in \mathbb{R}^k : x_1^2 + \cdots + x_k^2 \le r^2\}$, a ball of radius r in \mathbb{R}^k. Let $v_k := \lambda^k(B_k(0, 1))$ (k-dimensional volume of the unit ball).
 (a) Show that for any $r \ge 0$, $\lambda^k(B_k(0, r)) = v_k r^k$.
 (b) Evaluate v_k for all k. *Hint:* Starting from the known values of v_1 and v_2, do induction from k to $k + 2$ for each $k = 1, 2, \ldots$. Use polar coordinates in place of x_{k+1} and x_{k+2} in $\mathbb{R}^{k+2} = \mathbb{R}^k \times \mathbb{R}^2$.

9. Let S^{k-1} denote the unit sphere (boundary of $B_k(0, 1)$) in \mathbb{R}^k. Continuing Problem 8, $x \mapsto (|x|, x/|x|)$ gives a 1–1 mapping T of $\mathbb{R}^k \backslash \{0\}$ onto $(0, \infty) \times S^{k-1}$. Show that the image measure $\lambda^k \circ T^{-1}$ is a product measure, with the measure on $(0, \infty)$ given by $\rho_k(A) := \int_A r^{k-1}\, dr$ for any Borel set $A \subset (0, \infty)$ and some measure ω_k on S^{k-1}. Find the total mass $\omega_k(S^{k-1})$. *Hint:* Let $\gamma_k(A, B) := \lambda^k(\{x : |x| \in A \text{ and } x/|x| \in B\})$ for any Borel sets $A \subset (0, \infty)$ and $B \subset S^{k-1}$. Define ω_k by $\omega_k(B) := k\gamma_k((0, 1), B)$. Show that $\gamma_k(A, B) = \rho_k(A)\omega_k(B)$ starting with simple sets A and progressing to general Borel sets.

10. Show that α, as defined in the proof of Theorem 4.4.4, is a countably additive measure even if μ and ν are not σ-finite.

11. If $(X_i, \mathcal{S}_i, \mu_i)$ are measure spaces for all i in some index set I, where the sets X_i are disjoint, the *direct sum* of these measure spaces is defined by taking $X = \bigcup_i X_i$, letting $\mathcal{S} := \{A \subset X : A \cap X_i \in \mathcal{S}_i \text{ for all } i\}$, and $\mu(A) := \sum_i \mu_i(A \cap X_i)$ for each $A \in \mathcal{S}$. Show that (X, \mathcal{S}, μ) is a measure space.

12. A measure space is called *localizable* iff it can be written as a direct sum of finite measure spaces (see also §3.5). Show that:
 (a) Any σ-finite measure space is localizable.
 (b) Any direct sum of σ-finite measure spaces is localizable.

13. Consider the unit square I^2 with the Borel σ-algebra. For each $x \in I := [0, 1]$ let I_x be the vertical interval $\{(x, y) : 0 \le y \le 1\}$. Let μ be the measure on I^2 given by the direct sum of the one-dimensional Lebesgue measures on each I_x. Likewise, let $J_y := \{(x, y) : 0 \le x \le 1\}$ and let ν be the direct sum measure for the one-dimensional Lebesgue measures on each J_y. Let \mathcal{B} be the collection of sets measurable for both direct sums μ and ν in I^2. Prove or disprove: $(I^2, \mathcal{B}, \mu + \nu)$ is localizable. *Suggestions:* Look at problem 4. Assume the continuum hypothesis (Appendix A.3).

14. (Bledsoe-Morse product measure.) Given two σ-finite measure spaces (X, \mathcal{S}, μ) and (Y, \mathcal{T}, ν) let $\mathcal{N}(\mu, \nu)$ be the collection of all sets $A \subset$

$X \times Y$ such that $\int 1_A(x, y) \, d\mu(x) = 0$ for ν-almost all y, and $\int 1_A(x, y) \, d\nu(y) = 0$ for μ-almost all x (for other values of y or x, respectively, the integrals may be undefined). Show that the product measure $\mu \times \nu$ on the product σ-algebra $\mathcal{S} \otimes \mathcal{T}$ can be extended to a measure ρ on the σ-algebra $\mathcal{U} = \mathcal{U}(\mu, \nu)$ generated by $\mathcal{S} \otimes \mathcal{T}$ and $\mathcal{N}(\mu, \nu)$, with $\rho = 0$ on $\mathcal{N}(\mu, \nu)$, and so that the Tonelli-Fubini theorem (4.4.5) holds with $\mathcal{S} \otimes \mathcal{T}$ replaced by $\mathcal{U}(\mu, \nu)$. *Hint:* $\mathcal{N}(\mu, \nu)$ is a hereditary σ-ring, as in problem 9 at the end of §3.3.

15. For $I := [0, 1]$ with Borel σ-algebra and Lebesgue measure λ, take the cube I^3 with product measure (volume) $\lambda^3 = \lambda \times \lambda \times \lambda$. Let $f(x, y, z) := 1/\sqrt{|y - z|}$ for $y \neq z$, $f(x, y, z) := +\infty$ for $y = z$. Show that f is integrable for λ^3, but that for each $z \in I$, the set of y such that $\int f(x, y, z) \, d\lambda(x) = +\infty$ is non-empty and depends on z.

*4.5. Daniell-Stone Integrals

Now that integrals have been defined with respect to measures, the process will be reversed, in a sense: given an "integral" operation with suitable properties, it will be shown that it can be represented as the integral with respect to some measure.

Let \mathcal{L} be a non-empty collection of real-valued functions on a set X. Then \mathcal{L} is a *real vector space* iff for all $f, g \in \mathcal{L}$ and $c \in \mathbb{R}$, $cf + g \in \mathcal{L}$. Let $f \vee g := \max(f, g)$, $f \wedge g := \min(f, g)$. A vector space \mathcal{L} of functions is called a *vector lattice* iff for all f and g in \mathcal{L}, $f \vee g \in \mathcal{L}$. Then also $f \wedge g \equiv -(-f \vee -g) \in \mathcal{L}$.

Examples. For any measure space (X, \mathcal{S}, μ), the set $\mathcal{L}^1(X, \mathcal{S}, \mu)$ of all μ-integrable real functions is a vector lattice, as is the set of all μ-simple functions. For any topological space (X, \mathcal{T}), the collection $C_b(X, \mathcal{T})$, of all bounded continuous real-valued functions on X is a vector lattice. On the other hand, let C^1 denote the set of all functions $f : \mathbb{R} \mapsto \mathbb{R}$ such that the derivative f' exists everywhere and is continuous. Then C^1 is a vector space but not a lattice.

Definition. Given a set X and a vector lattice \mathcal{L} of real functions on X, a *pre-integral* is a function I from \mathcal{L} into \mathbb{R} such that:

(a) I is linear: $I(cf + g) = cI(f) + I(g)$ for all $c \in \mathbb{R}$ and $f, g \in \mathcal{L}$.
(b) I is nonnegative, in the sense that whenever $f \in \mathcal{L}$ and $f \geq 0$ (everywhere on X), then $I(f) \geq 0$.
(c) $I(f_n) \downarrow 0$ whenever $f_n \in \mathcal{L}$ and $f_n(x) \downarrow 0$ for all x.

Remark. For any vector space \mathcal{L} of functions, the constant function 0 belongs to \mathcal{L}. Thus if \mathcal{L} is a vector lattice, then for any $f \in \mathcal{L}$, the functions $f^+ := \max(f, 0)$ and $f^- := -\min(f, 0)$ belong to \mathcal{L}, and are nonnegative. So there are enough nonnegative functions in \mathcal{L} so that (b) is not vacuous.

Example. Let $\mathcal{L} = C[0, 1]$, the space of all continuous real-valued functions on $[0, 1]$. Then \mathcal{L} is a vector lattice. Let $I(f)$ be the classical Riemann integral, $I(f) := \int_0^1 f(x)\,dx$ as in calculus. Then clearly I is linear and nonnegative. If $f_n \in C[0, 1]$ and $f_n \downarrow 0$, then $f_n \downarrow 0$ uniformly on $[0, 1]$ by Dini's theorem (2.4.10). Thus $I(f_n) \downarrow 0$ and (c) holds, so I is a pre-integral. Likewise, for any compact topological space K, and $C(K)$ the space of continuous real functions on K, if I is a linear, nonnegative function on $C(K)$, then I is a pre-integral (as will be treated in §7.4).

For the rest of this section, assume given a set X, a vector lattice \mathcal{L} of real functions on X, and a pre-integral I on \mathcal{L}. For any two functions f and g in \mathcal{L} with $f \leq g$ (that is, $f(x) \leq g(x)$ for all x), let

$$[f, g) := \{\langle x, t \rangle \in X \times \mathbb{R} : f(x) \leq t < g(x)\}.$$

Let \mathcal{S} be the collection of all $[f, g)$ for $f \leq g$ in \mathcal{L}. Define ν on \mathcal{S} by $\nu([f, g)) := I(g - f)$. So if $g \geq 0$, then $I(g) = \nu([0, g))$, and for any $f \in \mathcal{L}$, $I(f) = \nu([0, f^+)) - \nu([0, f^-))$.

The next fact will provide an efficient approach to the theorem of Daniell and Stone (4.5.2).

4.5.1. Theorem (A. C. Zaanen) *ν extends to a countably additive measure on the σ-algebra \mathcal{T} generated by \mathcal{S}.*

Proof. It will be proved that \mathcal{S} is a semiring, and ν is well-defined and countably additive on it. This will then imply the theorem, using Proposition 3.2.4.

\mathcal{S} is a semiring, just as in Proposition 3.2.1: $\emptyset \in \mathcal{S}$ for $f = g$, and for any $f \leq g$ and $h \leq j$ in \mathcal{S}, $[f, g) \cap [h, j) = [f \vee h, f \vee h \vee (g \wedge j))$, just as for intervals in \mathbb{R}. Also,

$$[f, g) \setminus [h, j) = [f, f \vee (g \wedge h)) \cup [g \wedge (j \vee f), g),$$

a union of two disjoint sets in \mathcal{S}.

Suppose $[f, g) = [h, j)$. Then for each x, if the interval $[f(x), g(x))$ is non-empty, it equals $[h(x), j(x))$, so $f(x) = h(x)$ and $g(x) = j(x)$, and $(g - f)(x) = (j - h)(x)$. On the other hand, if $[f(x), g(x))$ is empty, then so is

$[h(x), j(x))$, and $f(x) = g(x)$, $h(x) = j(x)$, so $(g - f)(x) = 0 = (j - h)(x)$. So $g - f \equiv j - h \in \mathcal{L}$ and $I(g - f) = I(j - h)$, so ν is well-defined on \mathcal{S}.

For countable additivity, if $[f, g) = \bigcup_n [f_n, g_n)$, where all the functions f, f_n, g, and g_n are in \mathcal{L}, and the sets $[f_n, g_n)$ are disjoint, then for each x, $[f(x), g(x)) = \bigcup_n [f_n(x), g_n(x))$, where the intervals $[f_n(x), g_n(x))$ are disjoint. The length of intervals is countably additive (Theorem 3.1.3 for $G(x) \equiv x$), for intervals $[a, b)$ just as for intervals $(a, b]$ by symmetry. Thus,

$$(g - f)(x) = \sum_n (g_n - f_n)(x) \qquad \text{for all } x.$$

Let $h_n := g - f - \sum_{1 \le j \le n}(g_j - f_j)$. Then $h_n \in \mathcal{L}$ and $h_n \downarrow 0$, so $I(h_n) \downarrow 0$. It follows that $I(g - f) = \sum_{j \ge 1} I(g_j - f_j)$, so that ν is countably additive on \mathcal{S}. \square

The extension of ν to a σ-algebra will also be called ν. A main point of M. H. Stone's contribution to the theory is to represent $I(f)$ as $\int f \, d\mu$ for a measure μ on X. To define such a μ, Stone found that an additional assumption was useful. The vector lattice \mathcal{L} will be called a *Stone* vector lattice iff for all $f \in \mathcal{L}$, $f \wedge 1 \in \mathcal{L}$. (Note that in general, constant functions need not belong to \mathcal{L}; any vector lattice containing the constant functions will be a Stone vector lattice.)

Example. Let $X := [0, 1]$, $f(x) \equiv x$, and $\mathcal{L} = \{cf : c \in \mathbb{R}\}$. Let $I(cf) = c$. Then \mathcal{L} is a vector lattice, but not a Stone vector lattice, and I is a pre-integral on \mathcal{L}.

In practice, though, vector lattices with integrals that are actually applied, such as spaces of continuous, integrable, or \mathcal{L}^p functions, are also Stone vector lattices, so Stone's condition is satisfied.

As defined before Theorem 4.1.6, a real-valued function f on X is called measurable for a σ-ring \mathcal{B} of subsets of X iff $f^{-1}(A) \in \mathcal{B}$ for every Borel set A in \mathbb{R} not containing 0. If \mathcal{B} is a σ-algebra, this is equivalent to the usual definition since $f^{-1}\{0\} = X \backslash f^{-1}(\mathbb{R} \backslash \{0\})$.

Here is the main theorem of this section:

4.5.2. Theorem (Stone-Daniell) *Let I be a pre-integral on a Stone vector lattice \mathcal{L}. Then there is a measure μ on X such that $I(f) = \int f \, d\mu$ for all $f \in \mathcal{L}$. The measure μ is uniquely determined on the smallest σ-ring \mathcal{B} for which all functions in \mathcal{L} are measurable.*

Proof. Let \mathcal{T} be as in Theorem 4.5.1. The proof of Theorem 3.1.4 gives a particular extension of ν to a countably additive measure on \mathcal{T} via the outer measure ν^*. This is what will be meant by ν on \mathcal{T}, although ν may not be σ-finite, so ν on \mathcal{S} could have other extensions to a measure on \mathcal{T}.

Let \mathcal{M} be the collection of all sets $f^{-1}((1, \infty))$ for $f \in \mathcal{L}$. Then \mathcal{M} contains, for any $f \in \mathcal{L}$ and $r > 0$, the sets $f^{-1}((r, \infty)) = (f/r)^{-1}((1, \infty))$ and $f^{-1}((-\infty, -r)) = (-f)^{-1}((r, \infty))$. Since the intervals $(-\infty, -r)$ and (r, ∞) for $r > 0$ generate the σ-ring of Borel subsets of \mathbb{R} not containing 0, Theorem 4.1.6 implies that \mathcal{M} generates the σ-ring \mathcal{B} defined in the statement of the theorem.

Let $f \geq 0$, $f \in \mathcal{L}$, and let $g_n := (n(f - f \wedge 1)) \wedge 1$. Then for any $c > 0$, $[0, cg_n) \uparrow f^{-1}((1, \infty)) \times [0, c)$. Thus for each $A \in \mathcal{M}$, $A \times [0, c) \in \mathcal{T}$. Let $\mu(A) := \nu(A \times [0, 1))$ for each $A \in \mathcal{B}$. Then μ is well-defined because $\{A: A \times [0, 1) \in \mathcal{T}\}$ is a σ-ring.

It will be shown next that for any $A \in \mathcal{B}$ and $c > 0$, $\nu(A \times [0, c)) = c\nu(A \times [0, 1))$. Let $M_c((x, t)) := (x, ct)$ for $x \in X$, $t \in \mathbb{R}$ and $0 < c < \infty$. Then M_c is one-to-one from $X \times \mathbb{R}$ onto itself. Let $M_c[E] := \{M_c((x, t)): (x, t) \in E\}$ for any $E \subset X \times \mathbb{R}$. Note that $M_c \equiv M_{1/c}^{-1}$ preserves all set operations. For any $f \leq g$ in \mathcal{L}, so that $D := [f, g) \in \mathcal{S}$, clearly $M_c[D] = [cf, cg) \in \mathcal{S}$ and $\nu(M_c[D]) = c\nu(D)$. It follows that $E \in \mathcal{T}$ if and only if $M_c[E] \in \mathcal{T}$. Since \mathcal{S} is a semiring, by Propositions 3.2.3 and 3.2.4, for the ring \mathcal{R} generated by \mathcal{S}, we have $\nu(M_c[E]) = c\nu(E)$ for all $E \in \mathcal{R}$. It follows by definition of outer measure that $\nu^*(M_c[E]) = c\nu^*(E)$ for all $E \subset X \times \mathbb{R}$. Thus ν satisfies $\nu(M_c[E]) \equiv c\nu(E)$ for all $E \in \mathcal{T}$.

If $E = [0, 1_A)$ for a set A, then for any $c > 0$, $M_c[E] = A \times [0, c)$. It follows that $\nu(A \times [0, c)) = c\nu(A \times [0, 1)) = c\mu(A)$ for all $A \in \mathcal{B}$.

Let f_k be simple functions for \mathcal{B} with $0 \leq f_k \uparrow f$ as given by Proposition 4.1.5. Then $[0, f_k) \uparrow [0, f)$. For any function $h \geq 0$ which is simple for \mathcal{B}, we have $h = \sum_i c_i 1_{A(i)}$ for some $c_i > 0$ and some disjoint sets $A(i) \in \mathcal{B}$. Thus $[0, h)$ is the disjoint union of the sets $[0, c_i 1_{A(i)})$, and $\nu([0, h)) = \sum_i \nu([0, c_i 1_{A(i)})) = \sum_i c_i \mu(A(i)) = \int h \, d\mu$. Applying this to $h = f_k$ for each k, letting $k \to \infty$ and applying monotone convergence, gives

$$I(f) = \nu([0, f)) = \lim_{k \to \infty} \nu([0, f_k)) = \lim_{k \to \infty} \int f_k \, d\mu = \int f \, d\mu.$$

Then for a general $f \in \mathcal{L}$, let $f = f^+ - f^-$ and apply $I(g) = \int g \, d\mu$ for $g = f^+$ and f^- to prove it for $g = f$.

Now for the uniqueness part, let \mathcal{E} be the collection of all sets A in \mathcal{B} such that for any two measures μ and γ for which $I(f) = \int f d\mu = \int f d\gamma$ for all $f \in \mathcal{L}$, we have $\mu(A) = \gamma(A)$. Earlier in the proof, for $A \in \mathcal{M}$ we

found $0 \leq g_n \in \mathcal{L}$ with $g_n \uparrow 1_A$. Thus $\mathcal{M} \subset \mathcal{E}$. Clearly $\mu(A) < \infty$ for each $A \in \mathcal{M}$. Now $f^{-1}((1, \infty)) \cup g^{-1}((1, \infty)) = (f \vee g)^{-1}((1, \infty))$, and $f^{-1}((1, \infty)) \cap g^{-1}((1, \infty)) = (f \wedge g)^{-1}((1, \infty))$, so if $C \in \mathcal{M}$ and $D \in \mathcal{M}$ then $C \cup D \in \mathcal{M}$ and $C \cap D \in \mathcal{M}$. Since $1_{C \setminus D} = 1_C - 1_{C \cap D}$, it follows that $C \setminus D \in \mathcal{E}$. Then according to Proposition 3.2.8, every set in the ring generated by \mathcal{M} is a finite, disjoint union of sets $C_i \setminus D_i$ for C_i and D_i in \mathcal{M}, so this ring is included in \mathcal{E}.

Now, every set in \mathcal{B} is included in a countable union of sets in \mathcal{M} (since the collection of all sets satisfying this condition is a σ-ring), and since sets in \mathcal{M} have finite measure (for μ or γ), it follows as in Theorem 3.1.10 that $\mu = \gamma$ on \mathcal{B}. □

Problems

1. Let f be a function on a set X and $\mathcal{L} := \{cf : c \in \mathbb{R}\}$.
 (a) Show that \mathcal{L} is a vector lattice if $f \geq 0$ or if $f \leq 0$ but not otherwise.
 (b) Under what conditions on f is \mathcal{L} a Stone vector lattice?
 (c) If I is a pre-integral on \mathcal{L}, is it always true that $I(f) = \int f \, d\mu$ for some measure μ?
 (d) If μ exists in part (c), under what conditions on f is it unique?

2. For $k = 1, \ldots, n$, let X_k be disjoint sets and let \mathcal{F}_k be a vector lattice of functions on X_k. Let $X = \bigcup_{1 \leq k \leq n} X_k$. Let \mathcal{F} be the set of all functions $f = (f_1, \ldots, f_n)$ on X such that for each $k = 1, \ldots, n$, $f_k \in \mathcal{F}_k$ and $f = f_k$ on X_k.
 (a) Show that \mathcal{F} is a vector lattice.
 (b) If for each k, I_k is a pre-integral on \mathcal{F}_k, for $f \in \mathcal{F}$ let $I(f) := \sum_{1 \leq k \leq n} I_k(f_k)$. Show that I is a pre-integral on \mathcal{F}. (Then (X, \mathcal{F}, I) is called the *direct sum* of the $(X_k, \mathcal{F}_k, I_k)$.)

3. Let I be a pre-integral on a vector lattice \mathcal{L}. Let \mathcal{U} be the set of all functions f such that for some $f_n \in \mathcal{L}$ for all n, $f_n(x) \uparrow f(x)$ for all x. Set $\overline{I}(f) := \lim_{n \to \infty} I(f_n) \leq \infty$ and show that \overline{I} is well-defined on \mathcal{U}.

4. Let $I^{\mathbb{N}}$ be the set of all sequences $\{x_n\}$ where $x_n \in I := [0, 1]$ for all n. So $I^{\mathbb{N}}$ is an infinite-dimensional unit cube. The problem is to construct an infinite-dimensional Lebesgue measure on $I^{\mathbb{N}}$. Let \mathcal{L} be the set of all real-valued functions f on $I^{\mathbb{N}}$ such that for some finite n and continuous g on I^n, $f(\{x_j\}_{j \geq 1}) \equiv g(\{x_j\}_{1 \leq j \leq n})$.
 (a) Prove that \mathcal{L} is a Stone vector lattice.
 (b) For f and g as above let $I(f) := \int_0^1 \cdots \int_0^1 g(x_1, \ldots, x_n) \, dx_1 \cdots dx_n$. Show that I is a pre-integral.

5. Let \mathcal{L} be the set of all sequences $\{x_n\}_{n\geq 1}$ of real numbers such that x_n converges as $n \to \infty$.
 (a) Show that \mathcal{L} is a Stone vector lattice.
 (b) Let $I(\{x_n\}_{n\geq 1}) = \lim_{n\to\infty} x_n$ on \mathcal{L}. Show that I is *not* a pre-integral.
 (c) Let \mathcal{M} be the set of all sequences $\{x_n\}_{n\geq 0}$ such that $x_n \to x_0$ as $n \to \infty$. Show that \mathcal{M} is a Stone vector lattice and I, as defined in part (b), is a pre-integral.

6. Let \mathcal{P}_2 be the set of all polynomials Q on \mathbb{R} of degree at most 2, so $Q(x) \equiv ax^2 + bx + c$ for some real $a, b,$ and c. Let $I(Q) := a$ for any such Q. Show that:
 (a) I is linear on the vector space \mathcal{P}_2, which is not a lattice.
 (b) For any $f \geq 0$ in \mathcal{P}_2, $I(f) \geq 0$.
 (c) For any $f_n \downarrow 0$ in \mathcal{P}_2, $I(f_n) \downarrow 0$.
 (d) Prove that there is no vector lattice \mathcal{L} including \mathcal{P}_2 such that I can be extended to a pre-integral on \mathcal{L}. *Hint:* Show that $I(g) = 0$ for g bounded, $g \in \mathcal{L}$.

7. If μ in Theorem 4.5.2 is σ-finite, prove that $\nu = \mu \times \lambda$ on \mathcal{T}.

8. Give an example where the measure μ in the Stone-Daniell theorem has more than one extension to the smallest σ-algebra \mathcal{A} for which all functions in \mathcal{L} are measurable, and where μ is bounded on the smallest σ-ring \mathcal{R} for which the functions are measurable. *Hint:* See Problem 9 of §4.1.

9. Give a similar example, but where μ is unbounded on \mathcal{R}.

10. Show that in the situation of the last two problems there is always a smallest extension ν of μ; that is, for any extension ρ of μ to a measure on \mathcal{A}, $\nu(B) \leq \rho(B)$ for all $B \in \mathcal{A}$.

11. Let g_1, \ldots, g_k be linearly independent real-valued functions on a set X, that is, if for real constants c_j, $\sum_{j=1}^{k} c_j g_j \equiv 0$, then $c_1 = c_2 = \cdots = 0$. Show that for some x_1, \ldots, x_k in X, g_1, \ldots, g_j are linearly independent on $\{x_1, \ldots, x_j\}$ for each $j = 1, \ldots, k$.

12. Let \mathcal{F} be a finite-dimensional vector space of real-valued functions on a set X. Suppose that $f_n \in \mathcal{F}$ and $f_n(x) \to f(x)$ for all $x \in X$. Show that $f \in \mathcal{F}$. *Hint:* Use Problem 11.

13. Let \mathcal{F} be a vector lattice of functions on a set $\{p, q, r\}$ of three points such that for some b and c, $f(p) \equiv bf(q) + cf(r)$ for all f in \mathcal{F}. Show that either \mathcal{F} is one-dimensional, as in Problem 1, or $b \geq 0$, $c \geq 0$, and $bc = 0$.

14. Let \mathcal{F} be a vector lattice of functions on a set X such that \mathcal{F} is a finite-dimensional vector space. Show that \mathcal{F} is a direct sum (as in Problem 2)

of one-dimensional vector lattices (as in Problem 1(a)). *Hint:* Use the results of Problems 11 and 13.

15. If \mathcal{F} is a finite-dimensional Stone vector lattice, show that in the direct sum representation in Problem 14, $\mathcal{F}_i = \{c1_{X(i)}: c \in \mathbb{R}\}$ for some set $X(i)$ for each i.

16. If I is a nonnegative linear function on a finite-dimensional vector lattice \mathcal{F}, show that I is a pre-integral and for some finite measure μ, $I(f) = \int f\,d\mu$ for all $f \in \mathcal{F}$. Show that μ is uniquely determined on the smallest σ-ring for which all functions in \mathcal{F} are measurable if and only if \mathcal{F} is a Stone vector lattice. *Hints:* Use the results of Problems 14 and 15.

17. If V is a vector space of real-valued functions such that for some $n < \infty$ and all $f \in V$, the cardinality of the range of f is at most n, then show that the dimension of V is at most n.

Notes

§4.1 The modern notion of integral is due to Henri Lebesgue (1902). Hawkins (1970) treats its history, and Medvedev (1975) Lebesgue's own work. Lebesgue's discovery of his integral was a breakthrough in analysis. Lebesgue's collected works have been published in five volumes (Lebesgue, 1972–73), of which the first two include his work on "intégration et dérivation." The first volume also includes three essays on Lebesgue's life and work by one or more of Arnaud Denjoy, Lucienne Félix, and Paul Montel, and a description of his own work and its reception through 1922 in some 80 pages by Lebesgue. May (1966) also gives a brief biography of Lebesgue, who lived from 1875 to 1941.

The image measure theorem, 4.1.11, can take a different form in Euclidean spaces (see the notes to §4.4).

§4.2 Proposition 4.2.1 appears as a sequence of exercises in Halmos (1950, p. 83). He does not give earlier references. Hausdorff (1914, pp. 390–392) proved Theorem 4.2.2. Proposition 4.2.3 appeared in Dudley (1971, Proposition 3). Theorem 4.2.5 is due to von Alexits (1930) and Sierpiński (1930). The proof given here is essentially as in Lehmann (1959, pp. 37–38). The extension to complete separable range spaces (done here via Proposition 4.2.6) is due to Kuratowski (1933; 1966, p. 434). Shortt (1983) considers more general range spaces. Theorem 4.2.8 appears in Lehmann (1959, p. 37, Lemma 1), who gives some earlier references, but I do not know when it first appeared. Many thanks to Deborah Allinger and Rae Shortt, who supplied information for these notes.

§4.3 All the limit theorems were first proved with respect to Lebesgue measure on bounded intervals or measurable subsets of \mathbb{R}. In that case: Beppo Levi (1906) proved the monotone convergence theorem (4.3.2). Fatou (1906) stated his lemma (4.3.3). According to Nathan (1971), although Fatou (1878–1929) was educated, and is best known, as a mathematician, he worked throughout his career as an astronomer at the Paris Observatory. Vitali (1907) proved a convergence theorem for uniformly integrable sequences

(see theorem 10.3.6 below) which includes dominated convergence (4.3.5). Lebesgue (1902, §25) had proved dominated convergence for a uniformly bounded sequence of functions, still on a bounded interval.

Lebesgue (1910, p. 375) explicitly formulated the dominated convergence theorem, now with respect to multidimensional Lebesgue measure λ^k.

§4.4 Lebesgue (1902, §§37–39), repr. in Lebesgue (1972–73, Vol. 2), showed that integrals $\int_c^b dx$ and $\int_c^d dy$, for finite a, b, c, and d, could be interchanged for any bounded, measurable function $f(x, y)$. According to Saks (1937, p. 77), for f continuous this had been "known since Cauchy." For f the indicator of a measurable set in \mathbb{R}^2 it gives the product measure existence theorem (4.4.4) for Lebesgue measure on bounded intervals. Lebesgue (1902, Introduction) stated that the extension to more than two variables (as in Theorem 4.4.6?) is immediate. Lebesgue (1902, §40) went on, unfortunately, to claim that when f is not necessarily bounded nor measurable, the integrals can still be interchanged provided all the integrals appearing exist. This can fail for measurable, unbounded functions in examples, also known since Cauchy (Hawkins, 1970, p. 91), similar to (a) and (b) in the text, and also for bounded, nonmeasurable functions (Problem 4).

Fubini (1907) stated a theorem on interchanging integrals (like 4.4.5), for Lebesgue measure, and Theorem 4.4.5 is generally known as "Fubini's theorem." But Fubini's proof was "defective" (Hawkins, 1970, p. 161). Apparently Tonelli (1909) gave the first correct proof, incorporating part of Fubini's.

Assuming the continuum hypothesis, Problem 4 applies to Lebesgue measure. H. Friedman (1980) showed that it is consistent with the usual (Zermelo-Fraenkel, see Appendix A) set theory, including the axiom of choice, that whenever a bounded (possibly nonmeasurable) function f on $[0, 1] \times [0, 1]$ is such that both iterated integrals $\int\int f \, dx \, dy$ and $\int\int f \, dy \, dx$ are defined, they are equal. So the continuum hypothesis assumption cannot be dispensed with. With it, there exist quite strange sets: Sierpiński (1920), by transfinite recursion, defines a set A in $[0, 1] \times [0, 1]$ which intersects every line (not only those parallel to the axes) in at most two points, but which has outer measure 1 for $\lambda \times \lambda$. Bledsoe and Morse (1955) defined their extended product measure (Problem 14), which does give Sierpiński's set measure 0. Assuming the continuum hypothesis, the Bledsoe-Morse product measure does not avoid violation of Friedman's statement, by the example in Problem 4.

For Lebesgue measure λ^k on \mathbb{R}^k, the image measure theorem, 4.1.11, has a more concrete form where T is a 1–1 function from \mathbb{R}^k into itself whose inverse T^{-1} has continuous first partial derivatives: then $\lambda^k \circ T^{-1}$ can be replaced, under some conditions, by the Jacobian of T^{-1} times λ^k; see, for example, Rudin (1976, p. 252).

§4.5 Daniell (1917–1918) developed the theory of his integral, for spaces of bounded functions, beginning with a pre-integral I and extending it to a large class of functions $\mathcal{L}^1(I)$, parallel to and in many cases agreeing with the theory of integrals with respect to measures. Later Stone (1948) showed that under his condition that $f \wedge 1 \in \mathcal{L}$ for each $f \in \mathcal{L}$, the pre-integral I can be represented as the integral with respect to a measure. Daniell is otherwise known for some contributions to mathematical statistics (see Stigler, 1973). A. C. Zaanen (1958, Section 13) proved Theorem 4.5.1 as part of his new, short proof of the Stone-Daniell theorem (4.5.2).

References

An asterisk identifies works I have found discussed in secondary sources but have not seen in the original.

Alexits, G. von (1930). Über die Erweiterung einer Baireschen Funktion. *Fund. Math.* *15*: 51–56.

Bledsoe, W. W., and Anthony P. Morse (1955). Product measures. *Trans. Amer. Math. Soc. 79*: 173–215.

Daniell, Percy J. (1917–1918). A general form of integral. *Ann. Math. 19*: 279–294.

Dudley, R. M. (1971). On measurability over product spaces. *Bull. Amer. Math. Soc. 77*: 271–274.

Fatou, Pierre Joseph Louis (1906). Séries trigonométriques et séries de Taylor. *Acta Math. 30*: 335–400.

Friedman, Harvey (1980). A consistent Fubini-Tonelli theorem for nonmeasurable functions. *Illinois J. Math. 24*: 390–395.

*Fubini, Guido (1907). Sugli integrali multipli. *Rendiconti Accad. Nazionale dei Lincei* (Rome) (Ser. 5) *16*: 608–614.

Halmos, Paul R. (1950). *Measure Theory*. Van Nostrand, Princeton. Repr. Springer, New York (1974).

Hausdorff, Felix (1914) (see references to Chapter 1).

Hawkins, T. (1970). *Lebesgue's Theory of Integration, Its Origins and Development*. University of Wisconsin Press.

Hörmander, Lars (1983). *The Analysis of Linear Partial Differential Operators*, I. Springer, Berlin.

Kuratowski, C. [Kazimierz] (1933). Sur les théorèmes topologiques de la théorie des fonctions de variables réelles. *Comptes Rendus Acad. Sci. Paris 197*: 19–20.

———— (1966). *Topology*, vol. 1. Academic Press, New York.

Lebesgue, Henri (1902). Intégrale, longueur, aire (thèse, Univ. Paris). *Annali Mat. pura e appl.* (Ser. 3) 7: 231–359. Also in Lebesgue (1972–1973) 2, pp. 11–154.

———— (1904). *Leçons sur l'intégration et la recherche des fonctions primitives*. Paris.

———— (1910). Sur l'intégration des fonctions discontinues. *Ann. Scient. Ecole Normale Sup.* (Ser. 3) *27*: 361–450. Also in Lebesgue (1972–1973), 2, pp. 185–274.

———— (1972–1973). *Oeuvres scientifiques*. 5 vols. L'Enseignement Math., Inst. Math., Univ. Genève.

Lehmann, Erich (1959). *Testing Statistical Hypotheses*. 2d ed. Wiley, New York (1986).

Levi, Beppo (1906). Sopra l'integrazione delle serie. *Rend. Istituto Lombardo di Sci. e Lett.* (Ser. 2) *39*: 775–780.

May, Kenneth O. (1966). Biographical sketch of Henri Lebesgue. In Lebesgue, H., *Measure and the Integral*, Ed. K. O. May, Holden-Day, San Francisco, transl. from *La Mesure des grandeurs, Enseignement Math 31–34* (1933–1936), repub. as a *Monographie* (1956).

Medvedev, F. A. (1975). The work of Henri Lebesgue on the theory of functions (on the occasion of his centenary). *Russian Math. Surveys 30*, no. 4: 179–191. Transl. from *Uspekhi Mat. Nauk 30*, no. 4: 227–238.

Nathan, Henry (1971). Fatou, Pierre Joseph Louis. In *Dictionary of Scientific Biography, 4*, pp. 547–548. Scribner's, New York.

Rudin, Walter (1976). *Principles of Mathematical Analysis*. 3d ed. McGraw-Hill, New York.

Saks, Stanisław (1937). *Theory of the Integral*. 2d ed. English transl. by L. C. Young. Hafner, New York. Repr. corrected Dover, New York (1964).

Schwartz, Laurent (1966). *Théorie des Distributions*. 2d ed. Hermann, Paris.

Shortt, R. M. (1983). The extension of measurable functions. *Proc. Amer. Math. Soc.* *87*: 444–446.

Sierpiński, Wacław (1920). Sur un problème concernant les ensembles measurables superficiellement. *Fundamenta Math. 1*: 112–115.

————— (1930). Sur l'extension des fonctions de Baire définies sur les ensembles linéaires quelconques. *Fund. Math. 16*: 81–89.

Stigler, Stephen M. (1973). Simon Newcomb, Percy Daniell, and the history of robust estimation 1885–1920. *J. Amer. Statist. Assoc. 68*: 872–879.

Stone, Marshall Harvey (1948). Notes on integration, II. *Proc. Nat. Acad. Sci. USA 34*: 447–455.

Tonelli, Leonida (1909). Sull'integrazione per parti. *Rendiconti Accad. Nazionale dei Lincei* (Ser. 5) *18*: 246–253. Reprinted in Tonelli, L., *Opere Scelte* (1960), *1*, pp. 156–165. Edizioni Cremonese, Rome.

*Vitali, Giuseppe (1904–05). Sulle funzioni integrali. *Atti Accad. Sci. Torino 40*: 1021–1034.

————— (1907). Sull'integrazione per serie. *Rend. Circolo Mat. Palermo 23*: 137–155.

Zaanen, Adriaan C. (1958). *Introduction to the Theory of Integration*. North-Holland, Amsterdam.

5

L^p Spaces; Introduction to Functional Analysis

The key idea of functional analysis is to consider functions as "points" in a space of functions. To begin with, consider bounded, measurable functions on a finite measure space (X, S, μ) such as the unit interval with Lebesgue measure. For any two such functions, f and g, we have a finite integral $\int fg\, d\mu = \int f(x)g(x)\, d\mu(x)$. If we consider functions as vectors, then this integral has the properties of an inner product or dot product (f, g): it is non-negative when $f = g$, symmetric in the sense that $(f, g) \equiv (g, f)$, and linear in f for fixed g. Using this inner product, one can develop an analogue of Euclidean geometry in a space of functions, with a distance $d(f, g) = (f - g, f - g)^{1/2}$, just as in a finite-dimensional vector space. In fact, if μ is counting measure on a finite sct with k elements, (f, g) becomes the usual inner product of vectors in \mathbb{R}^k. But if μ is Lebesgue measure on $[0, 1]$, for example, then for the metric space of functions with distance d to be complete, we will need to include some unbounded functions f such that $\int f^2\, d\mu < \infty$. Along the same lines, for each $p > 0$ and μ there is the collection of functions f which are measurable and for which $\int |f|^p\, d\mu < \infty$. This collection is called \mathcal{L}^p or $\mathcal{L}^p(\mu)$. It is not immediately clear that if f and g are in \mathcal{L}^p, then so is $f + g$, this will follow from an inequality for $\int |f + g|^p\, d\mu$ in terms of the corresponding integrals for f and g separately, which will be treated in §5.1. In §§5.3 and 5.4, the inner product idea, corresponding to $p = 2$, is further developed.

We have not finished with measure theory: a very important fact, the "Radon-Nikodym" theorem, about the relationship of two measures, will be proved in §5.5 using some functional analysis.

5.1. Inequalities for Integrals

On $(0, 1]$ with Lebesgue measure, the function $f(x) = x^p$ is integrable if and only if $p > -1$. Thus the product of two integrable functions, such as $x^{-1/2}$ times $x^{-1/2}$, need not be integrable. Conditions can be given for products to be integrable, however, as follows.

152

Definition. For any measure space (X, \mathcal{S}, μ) and $0 < p < \infty$, $\mathcal{L}^p(X, \mathcal{S}, \mu) :=$ $\mathcal{L}^p(X, \mathcal{S}, \mu, \mathbb{R})$ denotes the set of all measurable functions f on X such that $\int |f|^p \, d\mu < \infty$ and the values of f are real numbers except possibly on a set of measure 0, where f may be undefined or infinite. For $1 \leq p < \infty$, let $\|f\|_p := (\int |f|^p \, d\mu)^{1/p}$, called the "$L^p$ norm" or "p-norm" of f.

Let \mathbb{C} be the complex plane, as defined in Appendix B. A function f into \mathbb{C} can be written $f = g + ih$ where g and h are real-valued functions, called the real and imaginary parts of f, respectively. \mathbb{C} will have the usual topology of \mathbb{R}^2. Thus, f will be measurable if and only if both g and h are measurable (see the discussion before Proposition 4.1.7). Now, the space $\mathcal{L}^p(X, \mathcal{L}, \mu, \mathbb{C})$, called complex \mathcal{L}^p, as opposed to "real \mathcal{L}^p" of the last paragraph, is defined as the set of all measurable functions f on X, with values in \mathbb{C} except possibly on a set of measure 0 where f may be undefined, such that $\int |f|^p \, d\mu < \infty$. The p-norm is defined as before.

Here is a first, basic inequality:

5.1.1. Theorem *For any integrable function f with values in \mathbb{R} or \mathbb{C}, $|\int f \, d\mu| \leq \int |f| \, d\mu$.*

Proof. First, if f has real values,

$$\left| \int f \, d\mu \right| = \left| \int f^+ \, d\mu - \int f^- \, d\mu \right| \leq \left| \int f^+ \, d\mu + \int f^- \, d\mu \right|$$

$$= \int f^+ + f^- \, d\mu = \int |f| \, d\mu.$$

Now if f has complex values, $f = g + ih$ where g and h are real-valued, and we need to prove

$$\left\{ \left(\int g \, d\mu \right)^2 + \left(\int h \, d\mu \right)^2 \right\}^{1/2} \leq \int (g^2 + h^2)^{1/2} \, d\mu.$$

A rotation of coordinates in \mathbb{R}^2 by an angle θ replaces (g, h) by $(g \cos \theta - h \sin \theta, g \sin \theta + h \cos \theta)$. This preserves both sides of the inequality. So, rotating the vector $(\int g \, d\mu, \int h \, d\mu)$ to point along the positive x axis, we can assume that $\int h \, d\mu = 0$. Then from the real case, $|\int g \, d\mu| \leq \int |g| \, d\mu$, and $|g| \leq (g^2 + h^2)^{1/2}$, so the theorem follows (by Lemma 4.1.9). □

The inequality 5.1.1 becomes an equality if f is nonnegative or more generally if $f \equiv cg$ where $g \geq 0$ and c is a fixed complex number.

The next inequality may be somewhat unexpected, except possibly for $p = 2$ (Corollary 5.1.4). It will be used in the proof of the basic inequality 5.1.5 (subadditivity of $\|\cdot\|_p$).

5.1.2. Theorem (Rogers-Hölder Inequality) *If $1 < p < \infty$, $p^{-1} + q^{-1} = 1$, $f \in \mathcal{L}^p(X, \mathcal{S}, \mu)$ and $g \in \mathcal{L}^q(X, \mathcal{S}, \mu)$, then $fg \in \mathcal{L}^1(X, \mathcal{S}, \mu)$ and $|\int fg\,d\mu| \le \int |fg|\,d\mu \le \|f\|_p \|g\|_q$.*

Proof. If $\|f\|_p = 0$, then $f = 0$ a.e., so $fg = 0$ a.e., $\int |fg|\,d\mu = 0$, and the inequality holds, and likewise if $\|g\|_q = 0$. So we can assume these norms are not 0. Now, for any constant $c > 0$, $\|cf\|_p = c\|f\|_p$, so dividing out by the norms, we can assume $\|f\|_p = \|g\|_q = 1$. Note that fg is measurable (as shown just before Proposition 4.1.8). Now we will use:

5.1.3. Lemma *For any positive real numbers u and v and $0 < \alpha < 1$, $u^\alpha v^{1-\alpha} \le \alpha u + (1 - \alpha)v$.*

Proof. Dividing by v, we can assume $v = 1$. To prove $u^\alpha \le \alpha u + 1 - \alpha$, note that it holds for $u = 1$. Taking derivatives, we have $\alpha u^{\alpha-1} > \alpha$ for $u < 1$ and $\alpha u^{\alpha-1} < \alpha$ for $u > 1$. These facts imply the lemma via the mean value theorem of calculus. □

Now to continue the proof of Theorem 5.1.2, let $\alpha := 1/p$ and for each x let $u(x) := |f(x)|^p$ and $v(x) := |g(x)|^q$. Then $1 - \alpha = 1/q$ and the lemma gives $|f(x)g(x)| \le \alpha|f(x)|^p + (1 - \alpha)|g(x)|^q$ for all x. Integrating gives $\int |fg|\,d\mu \le \alpha + (1 - \alpha) = 1$, proving the main (second) inequality in Theorem 5.1.2 and showing that $fg \in \mathcal{L}^1(X, \mathcal{S}, \mu)$. Now the first inequality follows from Theorem 5.1.1, proving Theorem 5.1.2. □

The best-known and most often used case is for $p = 2$:

5.1.4. Corollary (Cauchy-Bunyakovsky-Schwarz Inequality) *For any f and g in $\mathcal{L}^2(X, \mathcal{S}, \mu)$, we have $fg \in \mathcal{L}^1(X, \mathcal{S}, \mu)$ and $|\int fg\,d\mu| \le \|f\|_2 \|g\|_2$.*

For some examples of the Rogers-Hölder inequality, let $X = [0, 1]$ with μ = Lebesgue measure. Let $f(x) = x^{-r}$ and $g(x) = x^{-s}$ for some $r > 0$ and $s > 0$. Then $fg \in \mathcal{L}^1$ if and only if $r + s < 1$. In the borderline case $r + s = 1$, if we set $p = 1/r$, and $p^{-1} + q^{-1} = 1$ as in Theorem 5.1.2, so $q = 1/s$, then $f \notin \mathcal{L}^p$, but $f \in \mathcal{L}^t$ for any $t < p$, and $g \notin \mathcal{L}^q$, but $g \in \mathcal{L}^u$ for any $u < q$. So the hypotheses of Theorem 5.1.2 both fail, but both are on the borderline

of holding, as the conclusion is. This suggests that for hypotheses of its type, Theorem 5.1.2 can hardly be improved. Theorem 6.4.1 will provide a more precise statement.

\mathcal{L}^p spaces will also be defined for $p = +\infty$. A measurable function f is called *essentially bounded* iff for some $M < \infty$, $|f| \leq M$ almost everywhere. The set of all essentially bounded real functions is called $\mathcal{L}^\infty(X, \mathcal{S}, \mu)$ or just \mathcal{L}^∞ if the measure space intended is clear. Likewise, $\mathcal{L}^\infty(X, \mathcal{S}, \mu, \mathbb{C})$ denotes the space of measurable, essentially bounded functions on X which are defined and have complex values almost everywhere. For any f let $\|f\|_\infty := \inf\{M : |f| \leq M \text{ a.e.}\}$. So for some numbers $M_n \downarrow \|f\|_\infty$, $|f| \leq M_n$ except on A_n with $\mu(A_n) = 0$. Thus except on the union of the A_n, and hence a.e., we have $|f| \leq M_n$ for all n and so $|f| \leq \|f\|_\infty$ a.e.

If f is continuous on $[0, 1]$ and μ is Lebesgue measure, then $\|f\|_\infty = \sup|f|$, since for any value of f, it takes nearby values on sets of positive measure.

If $f \in \mathcal{L}^1$ and $g \in \mathcal{L}^\infty$, then a.e. $|fg| \leq |f| \|g\|_\infty$, so $fg \in \mathcal{L}^1$ and $\int |fg| \, d\mu \leq \|f\|_1 \|g\|_\infty$. Thus the Rogers-Hölder inequality extends to the case $p = 1$, $q = \infty$. A pseudometric on \mathcal{L}^p for $p \geq 1$ will be defined by $d(f, g) := \|f - g\|_p$. To show that d satisfies the triangle inequality we need the next fact.

5.1.5. Theorem (Minkowski-Riesz Inequality) *For $1 \leq p \leq \infty$, if f and g are in $\mathcal{L}^p(X, \mathcal{S}, \mu)$, then $f + g \in \mathcal{L}^p(X, \mathcal{S}, \mu)$ and $\|f + g\|_p \leq \|f\|_p + \|g\|_p$.*

Proof. First, $f + g$ is measurable as in Proposition 4.1.8. Since $|f + g| \leq |f| + |g|$, we can replace f and g by their absolute values and so assume $f \geq 0$ and $g \geq 0$. If $f = 0$ a.e., or $g = 0$ a.e., the inequality is clear. For $p = 1$ or ∞ the inequality is straightforward.

For $1 < p < \infty$ we have $(f + g)^p \leq 2^p \max(f^p, g^p) \leq 2^p(f^p + g^p)$, so $f + g \in \mathcal{L}^p$. Then applying the Rogers-Hölder inequality (Theorem 5.1.2) gives $\|f + g\|_p^p = \int (f + g)^p \, d\mu = \int f(f + g)^{p-1} \, d\mu + \int g(f + g)^{p-1} \, d\mu \leq \|f\|_p \|(f + g)^{p-1}\|_q + \|g\|_p \|(f + g)^{p-1}\|_q$. Now $(p - 1)q = p$, so $\|f + g\|_p^p \leq (\|f\|_p + \|g\|_p)\|f + g\|_p^{p/q}$. Since $p - p/q = 1$, dividing by the last factor gives $\|f + g\|_p \leq \|f\|_p + \|g\|_p$. $\qquad \square$

If X is a finite set with counting measure and $p = 2$, then Theorem 5.1.5 reduces to the triangle inequality for the usual Euclidean distance.

Let X be a real vector space (as defined in linear algebra or, specifically, in Appendix B). A *seminorm* on X is a function $\|\cdot\|$ from X into $[0, \infty)$ such that

(i) $\|cx\| = |c|\|x\|$ for all $c \in \mathbb{R}$ and $x \in X$, and
(ii) $\|x + y\| \le \|x\| + \|y\|$ for all $x, y \in X$.

A seminorm $\|\cdot\|$ is called a *norm* iff $\|x\| = 0$ only for $x = 0$. (In any case $\|0\| = \|0 \cdot 0\| = 0 \cdot \|0\| = 0$.)

Examples. The Minkowski-Riesz inequality (Theorem 5.1.5) implies that for any measure space (X, \mathcal{S}, μ) and $1 \le p \le \infty$, $\mathcal{L}^p(X, \mathcal{S}, \mu)$ is a vector space and $\|\cdot\|_p$ is a seminorm on it. If there is a non-empty set A with $\mu(A) = 0$, then for each p, $\|1_A\|_p = 0$, and $\|\cdot\|_p$ is not a norm on \mathcal{L}^p. For any seminorm $\|\cdot\|$, let $d(x, y) := \|x - y\|$ for any $x, y \in X$. Then it is easily seen that d is a pseudometric, that is, it satisfies all conditions for a metric except that possibly $d(x, y) = 0$ for some $x \ne y$. So d is a metric if and only if $\|\cdot\|$ is a norm.

The next inequality is occasionally useful:

5.1.6. Arithmetic-Geometric Mean Inequality For any nonnegative numbers x_1, \ldots, x_n, $(x_1 x_2 \cdots x_n)^{1/n} \le (x_1 + \cdots + x_n)/n$.

Remark. Here $(x_1 x_2 \cdots x_n)^{1/n}$ is called the "geometric mean" of x_1, \ldots, x_n and $(x_1 + \cdots + x_n)/n$ is called the "arithmetic mean."

Proof. The proof will be by induction on n. The inequality is trivial for $n = 1$. It always holds if any of the x_i is 0, so assume $x_i > 0$, for all i. For the induction step, apply Lemma 5.1.3 with $u = (x_1 x_2 \cdots x_{n-1})^{1/(n-1)}$, $v = x_n$, and $\alpha = (n-1)/n$, so $1 - \alpha = 1/n$, and the inequality holds for n. □

Problems

1. Prove or disprove, for any complex numbers z_1, \ldots, z_n:
 (a) $|z_1 z_2 \cdots z_n|^{1/n} \le (|z_1| + \cdots + |z_n|)/n$;
 (b) $|z_1 z_2 \cdots z_n|^{1/n} \le |z_1 + \cdots + z_n|/n$.

2. Give another proof of Theorem 5.1.1, without proving it first for real-valued functions, using the complex polar decomposition of $\int f \, d\mu = re^{i\theta}$ where $r \ge 0$ and θ is real, and $f = \rho e^{i\varphi}$, where $\rho \ge 0$ and φ are real-valued functions.

3. Let Y be a function with values in \mathbb{R}^3, $Y = (f, g, h)$ where f, g, and h are integrable functions for a measure μ. Let $\int Y \, d\mu := (\int f \, d\mu,$

$\int g \, d\mu, h \, d\mu$). Show that $\|\int Y \, d\mu\| \le \int \|Y\| \, d\mu$ if $\|(x, y, z)\| =$
(a) $(x^2 + y^2 + z^2)^{1/2}$; (b) $|x| + |y| + |z|$;
(c) $\max(|x|, |y|, |z|)$; (d) $(|x|^p + |y|^p + |z|^p)^{1/p}$, where $1 < p < \infty$.

4. Let $(Y, \|\cdot\|)$ be a linear space with a seminorm $\|\cdot\|$. Assume that Y is separable for $d(y, z) := \|y - z\|$. Let (X, \mathcal{S}, μ) be a measure space and $1 \le p < \infty$. Let $\mathcal{L}^p(X, \mathcal{S}, \mu, Y)$ be the set of all functions f from X into Y which are measurable for the Borel σ-algebra on Y, generated by the open sets for d, and such that $\int \|f(x)\|^p \, d\mu(X) < \infty$. Let $\|f\|_p$ be the pth root of the integral. If $f \in \mathcal{L}^p(X, \mathcal{S}, \mu, Y)$ and $g \in \mathcal{L}^q(X, \mathcal{S}, \mu, \mathbb{R})$, where $1 < p < \infty$ and $p^{-1} + q^{-1} = 1$, show that $gf \in \mathcal{L}^1(X, \mathcal{S}, \mu, Y)$ and $\|gf\|_1 \le \|g\|_q \|f\|_p$. *Hint:* For measurability, show that $(c, y) \mapsto cy$ is jointly continuous: $\mathbb{R} \times Y \mapsto Y$ and thus jointly measurable by Proposition 4.1.7.

5. Show that for $1 \le p < \infty$, $\|\cdot\|_p$ is a seminorm on $\mathcal{L}^p(X, \mathcal{S}, \mu, Y)$.

6. For $1 \le p < \infty$ and two σ-finite measure spaces (X, \mathcal{S}, μ) and (Y, \mathcal{T}, ν), assume that $\mathcal{L}^p(Y, \mathcal{T}, \nu)$ is separable. Show that $\mathcal{L}^p(X, \mathcal{S}, \mu, \mathcal{L}^p(Y, \mathcal{T}, \nu, \mathbb{R}))$ is isometric to $\mathcal{L}^p(X \times Y, \mathcal{S} \otimes \mathcal{T}, \mu \times \nu, \mathbb{R})$, that is, there is a 1–1 linear function from one space onto the other, preserving the seminorms $\|\cdot\|_p$. *Hint:* Improve Corollary 4.2.7 with $U = X$ to obtain $\|g_n(x)\| \uparrow \|g(x)\|$ for all x.

7. For $0 < p < 1$, let $\mathcal{L}^p(X, \mathcal{S}, \mu)$ be the set of all real-valued measurable functions on X with $\int |f(x)|^p \, d\mu(x) < \infty$.
 (a) Show by an example that the pth root of the integral is *not* necessarily a seminorm on \mathcal{L}^p.
 (b) Show that for $0 < p \le 1$, $d_p(f, g) := \int |f - g|^p \, d\mu$ defines a pseudo-metric on $\mathcal{L}^p(X, \mathcal{S}, \mu)$.

8. Show by examples that for $1 < p < \infty$, d_p as defined in the last problem may not define a pseudometric on $\mathcal{L}^p(X, \mathcal{S}, \mu)$.

9. Let f and g be positive, measurable functions on X for a measure space (X, \mathcal{S}, μ). Let $0 < t < r < m < \infty$.
 (a) Show that if the integrals on the right are finite, the following holds (Rogers' inequality): $(\int fg^r d\mu)^{m-t} \le (\int fg^t d\mu)^{m-r}(\int fg^m d\mu)^{r-t}$. *Hint:* Use the Rogers-Hölder inequality (Theorem 5.1.2).
 (b) Show how, conversely, the Rogers-Hölder inequality follows from Rogers' inequality. *Hint:* Let $t = 1$ and $m = 2$.
 (c) Prove that for any $m > 1$, $(\int fg \, d\mu)^m \le (\int f \, d\mu)^{m-1}(\int fg^m \, d\mu)$ ("historical Hölder's inequality").

(d) Show how the Rogers-Hölder inequality follows from the inequality in (c).

5.2. Norms and Completeness of *Lp*

For any set X with a pseudometric d, let $x \sim y$ iff $d(x, y) = 0$. Then it is easy to check that \sim is an equivalence relation. For each $x \in X$ let $x^{\sim} := \{y : y \sim x\}$. Let X^{\sim} be the set of all equivalence classes x^{\sim} for $x \in X$. Let $d(x^{\sim}, y^{\sim}) := d(x, y)$ for any x and y in X. It is easy to see that d is well-defined and a metric on X^{\sim}. If X is a vector space with a seminorm $\|\cdot\|$, then $\{x \in X : \|x\| = 0\}$ is a vector subspace Z of X. For each $y \in X$ let $y + Z := \{y + z : z \in Z\}$. Then, clearly, $y^{\sim} = y + Z$. The factor space $X^{\sim} = \{x + Z : x \in X\}$, often called the quotient space or factor space X/Z, is then in a natural way also a vector space, on which $\|\cdot\|$ defines a norm. For each space \mathcal{L}^p, the factor space of equivalence classes defined in this way is called L^p, that is, $L^p(X, \mathcal{S}, \mu) := \{f^{\sim} : f \in \mathcal{L}^p(X, \mathcal{S}, \mu)\}$. Note that if $g \geq 0$ and g is measurable, then $\int g \, d\mu = 0$ if and only if $g = 0$ almost everywhere for μ. Thus $h \sim f$ if and only if $h = f$ a.e. Sometimes a function may be said to be in \mathcal{L}^p if it is undefined and/or infinite on a set of measure 0, as long as elsewhere it equals a real-valued function in \mathcal{L}^p a.e. In any case, the class L^p of equivalence classes remains the same, as each equivalence class contains a real-valued function. The equivalence classes f^{\sim} are sometimes called functionoids. Many authors treat functions equal a.e. as identical and do not distinguish too carefully between \mathcal{L}^p and L^p.

Definitions. If X is a vector space and $\|\cdot\|$ is a norm on it, then $(X, \|\cdot\|)$ is called a *normed linear space*. A *Banach space* is a normed linear space which is complete, for the metric defined by the norm. For a vector space over the field \mathbb{C} of complex numbers, the definition of seminorm is the same except that complex constants c are allowed in $\|cx\| = |c|\|x\|$, and the other definitions are unchanged.

Examples. The finite-dimensional space \mathbb{R}^k is a Banach space, with the usual norm $(x_1^2 + \cdots + x_k^2)^{1/2}$ or with the norm $|x_1| + \cdots + |x_k|$. Let $C_b(X)$ be the space of all bounded, continuous real-valued functions on a topological space X, with the supremum norm $\|f\|_{\infty} := \sup\{|f(x)| : x \in X\}$. Then $C_b(X)$ is a Banach space: its completeness was proved in Theorem 2.4.9.

5.2.1. Theorem (*Completeness of Lp*) *For any measure space* (X, \mathcal{S}, μ) *and* $1 \leq p \leq \infty$, $(L^p(X, \mathcal{S}, \mu), \|\cdot\|_p)$ *is a Banach space.*

Proof. As noted after the Minkowski-Riesz inequality (Theorem 5.1.5), $\|\cdot\|_p$ is a seminorm on \mathcal{L}^p; it follows easily that it defines a norm on L^p. To prove it is complete, let $\{f_n\}$ be a Cauchy sequence in \mathcal{L}^p. If $p = \infty$, then for almost all x, $|f_m(x) - f_n(x)| \le \|f_m - f_n\|_\infty$ for all m and n (taking a union of sets of measure 0 for countably many pairs (m, n)). For such x, $f_n(x)$ converges to some number, say $f(x)$. Let $f(x):=0$ for other values of x. Then f is measurable by Theorem 4.2.2. For almost all x and all m, $|f(x) - f_m(x)| \le \sup_{n\ge m}\|f_n - f_m\|_\infty \le 1$ for m large enough. Then $\|f\|_\infty \le \|f_m\|_\infty + 1$, so $f \in \mathcal{L}^\infty$ and $\|f - f_n\|_\infty \to 0$ as $n \to \infty$, as desired.

Now let $1 \le p < \infty$. In any metric space, a Cauchy sequence with a convergent subsequence is convergent to the same limit, so it is enough to prove convergence of a subsequence. Thus it can be assumed that $\|f_m - f_n\|_p < 1/2^n$ for all n and $m > n$. Let $A(n):=\{x: |f_n(x) - f_{n+1}(x)| \ge 1/n^2\}$. Then $1_{A(n)}/n^2 \le |f_n - f_{n+1}|$, so for all n, $\mu(A(n))/n^{2p} \le \int|f_n - f_{n+1}|^p\,d\mu < 2^{-np}$, and $\sum_n \mu(A(n)) \le \sum_n n^{2p}/2^{np} < \infty$. Thus for $B(n):= \bigcup_{m\ge n} A(m)$, $B(n) \downarrow$ and $\mu(B(n)) \to 0$ as $n \to \infty$. For any $x \notin \bigcap_{n=1}^\infty B(n)$, and so for almost all x, $|f_n(x) - f_{n+1}(x)| \le 1/n^2$ for all large enough n. Then for any $m > n$, $|f_m(x) - f_n(x)| \le \sum_{j=n}^\infty 1/j^2$. Since $\sum_{j=1}^\infty 1/j^2$ converges, $\sum_{j=n}^\infty 1/j^2 \to 0$ as $n \to \infty$. Thus for such x, $\{f_n(x)\}$ is a Cauchy sequence, with a limit $f(x)$. For other x, forming a set of measure 0, let $f(x) = 0$. Then, as before, f is measurable. By Fatou's Lemma (4.3.3), $\int|f|^p\,d\mu \le \liminf_{n\to\infty}\int|f_n|^p\,d\mu < \infty$ (recall that any Cauchy sequence is bounded). Likewise, $\int|f - f_n|^p\,d\mu \le \liminf_{m\to\infty}\int|f_m - f_n|^p\,d\mu \to 0$ as $n \to \infty$, so $\|f_n - f\|_p \to 0$. \square

Example. There is a sequence $\{f_n\} \subset \mathcal{L}^\infty([0, 1], \mathcal{B}, \lambda)$ with $f_n \to 0$ in \mathcal{L}^p for $1 \le p < \infty$ but $f_n(x)$ not converging to 0 for any x. Let $f_1 := 1_{[0,1]}$, $f_2 := 1_{[0,1/2]}$, $f_3 := 1_{[1/2,1]}$, $f_4 := 1_{[0,1/3]}$, \ldots, $f_k := 1_{[(j-1)/n, j/n]}$ for $k = j + n(n - 1)/2, n = 1, 2, \ldots, j = 1, \ldots, n$. Then $f_k \ge 0$, $\int f_k(x)^p\,d\lambda(x) = n^{-1} \to 0$ as $k \to \infty$ for $1 \le p < \infty$, while for all x,

$$\liminf_{k\to\infty} f_k(x) = 0 < 1 = \limsup_{k\to\infty} f_k(x).$$

This shows why a subsequence was needed to get pointwise convergence in the proof of completeness of L^p.

Problems

1. For $0 < p < 1$, $\mathcal{L}^p(X, \mathcal{S}, \mu)$ is the set of all measurable real functions on X with $\int|f|^p\,d\mu < \infty$, with a pseudometric $d_p(f, g):=\int|f - g|^p\,d\mu$. Show that \mathcal{L}^p is complete for d_p.

2. Let (X, S, μ) be a measure space and $(Y, \|\cdot\|)$ a Banach space. Show that $\mathcal{L}^p(X, S, \mu, Y)$ (as defined in Problem 4, §5.1) is complete.

3. For an alternate proof of completeness of \mathcal{L}^p, after picking a subsequence as in the proof given for Theorem 5.2.1, let f_n be a sequence of functions in $\mathcal{L}^p(X, S, \mu)$, where $1 \leq p < \infty$, such that $\sum_{n \geq 1} \|f_n - f_{n+1}\|_p < \infty$. Let $G_n := \sum_{1 \leq k < n} |f_{k+1} - f_k|$. Show that
 (a) the sequence G_n converges, almost everywhere, to a function G in \mathcal{L}^p.
 (b) G_n converges to G in \mathcal{L}^p.
 (c) Show that wherever $\{G_n\}$ converges, $\{f_n\}$ also converges.
 (d) Letting $f_n \to f$ a.e., show that $f_n \to f$ in \mathcal{L}^p, using dominated convergence.

4. For $1 \leq p < r < +\infty$ and $\lambda =$ Lebesgue measure on \mathbb{R}, let $S := \mathcal{L}^p(\mathbb{R}, \lambda) \cap \mathcal{L}^r(\mathbb{R}, \lambda)$ and $\|\cdot\| := \|\cdot\|_p + \|\cdot\|_r$. Prove or disprove: $(S, \|\cdot\|)$ is complete.

5. A function p from a real vector space V into $[0, \infty)$ is called *positive-homogeneous* if $p(cv) = cp(v)$ for every $v \in V$ and $c \geq 0$, and *subadditive* iff $p(u + v) \leq p(u) + p(v)$ for any u and v in V.
 (a) Give an example of a subadditive, positive-homogeneous function on \mathbb{R}, which is 0 only at 0, but which is not a seminorm, so not a norm.
 (b) If p is subadditive and positive-homogeneous on a real vector space X, show that $\|x\| := p(x) + p(-x)$ defines a seminorm on X.

6. Let $(X_i, \|\cdot\|_i)$ be Banach spaces for $i = 1, 2, \ldots$. Define the *direct sum* X as the set of all sequences $x := \{x_i\}_{i \geq 1}$ such that $x_i \in X_i$ for all i and $\|x\| := \sum_i \|x_i\|_i < \infty$. Show that $(X, \|\cdot\|)$ is a Banach space.

7. (Continuation.) For each $j = 1, 2, \ldots$, let X_j be a copy of \mathbb{R}^2 with its usual norm $\|(y_j, z_j)\| := (y_j^2 + z_j^2)^{1/2}$. Let U_j and V_j be one-dimensional subspaces of X_j where $U_j = \{(c, 0); c \in \mathbb{R}\}$ and $V_j = \{(jc, c): c \in \mathcal{R}\}$. Let U be the direct sum of the U_j, V of the V_j, and X of the X_j. Show that U and V are complete but $U + V := \{u + v: u \in U, v \in V\}$ in X is not. *Hint:* Consider $\{(0, j^{-2})\}_{j=1}^\infty$.

5.3. Hilbert Spaces

Let H be a vector space over a field K where $K = \mathbb{R}$ or \mathbb{C}, the field of complex numbers. For $z = x + iy \in \mathbb{C}$ (where x and y are real) let $\bar{z} := z^- := x - iy$, called the *complex conjugate* of z (see also Appendix B). A *semi-inner product* on H is a function (\cdot, \cdot) from $H \times H$ into K such that

(i) $(cf + g, h) = c(f, h) + (g, h)$ for all $c \in K$ and $f, g \in H$.

(ii) $(f, g) = (g, f)^-$ for all $f, g \in H$. (Thus if $K = \mathbb{R}, (f, g) = (g, f)$.)

(iii) (\cdot,\cdot) is nonnegative definite: that is, $(f, f) \geq 0$ for all $f \in H$.

In the real case, a semi-inner product is a (i) bilinear, (ii) symmetric, (iii) nonnegative definite form on $H \times H$. In the complex case, condition (ii) is called "conjugate-symmetric." Then by (i) and (ii), $(h, cf + g) = \bar{c}(h, f) + (h, g)$. Thus (h, f) is linear in h and "conjugate-linear" in f. (These conditions are sometimes called "sesquilinear"; "sesqui-" is a Latin prefix meaning "one and a half." For example, a sesquicentennial is a 150th anniversary.)

A semi-inner product is called an *inner product* iff $(f, f) = 0$ implies $f = 0$. A (semi-)inner product space is a pair $(V, (\cdot,\cdot))$ where V is a vector space over $K = \mathbb{R}$ or \mathbb{C} and (\cdot,\cdot) is a (semi-)inner product on V. A classical example of an inner product is the dot product of vectors: for $z, w \in K^n$ let $(z, w) := \sum_{1 \leq j \leq n} z_j \bar{w}_j$. Some of the main examples of inner products, mentioned in the introduction to this chapter, are as follows:

5.3.1. Theorem *For any measure space* (X, \mathcal{S}, μ),

$$(f, g) := \int f \bar{g} \, d\mu$$

defines a semi-inner product on real or complex $\mathcal{L}^2(X, \mathcal{L}, \mu)$.

Proof. First consider the real case. Then (f, g) is always defined and finite by Corollary 5.1.4. Properties (i), (ii), and (iii) are easily checked. For the complex case, suppose $f = g + ih$ and $\gamma = \alpha + i\beta$ are in complex \mathcal{L}^2, so that g, h, α, and β are all in real \mathcal{L}^2. Then (f, γ) is defined and finite, by applying the real case four times. Again, the defining properties of a semi-inner product are easy to check. \square

Given a semi-inner product (\cdot,\cdot), let $\|f\| := (f, f)^{1/2}$. As the notation indicates, $\|\cdot\|$ will be shown to be a seminorm. First:

5.3.2. Proposition $\|cf\| = |c|\|f\|$ *for all* $c \in K$ *and* $f \in H$.

Proof. $\|cf\| = (cf, cf)^{1/2} = (c\bar{c}(f, f))^{1/2} = (|c|^2(f, f))^{1/2} = |c|\|f\|.$ \square

Corollary 5.1.4 was a Cauchy inequality for inner products of the form $\int fg \, d\mu$, as in Theorem 5.3.1. The inequality is useful and true for any semi-inner product:

5.3.3. Cauchy-Bunyakovsky-Schwarz Inequality for Inner Products. For any semi-inner product (\cdot,\cdot) and $f, g \in H$,

$$|(f, g)|^2 \le (f, f)(g, g).$$

Proof. By Proposition 5.3.2, f can be multiplied by any complex c with $|c| = 1$ without changing either side of the inequality, so it can be assumed that (f, g) is real. Then for all real $x, 0 \le \|f + xg\|^2 = (f, f) + 2x(f, g) + x^2(g, g)$. A nonnegative quadratic $ax^2 + bx + c$ has $b^2 - 4ac \le 0$, so $4(f, g)^2 - 4(f, f)(g, g) \le 0$. □

Now the seminorm property can be proved:

5.3.4. Theorem *For any (semi-)inner product (\cdot,\cdot), and $\|x\| := (x, x)^{1/2}$, $\|\cdot\|$ is a (semi)norm.*

Proof. Using 5.3.3, we get for any $f, g \in H$, $\|f + g\|^2 = \|f\|^2 + (f, g) + (g, f) + \|g\|^2 \le (\|f\| + \|g\|)^2$. This and Proposition 5.3.2 give the conclusion. □

Definition. $(H, (\cdot,\cdot))$ is an *inner product space* iff H is a vector space over $K = \mathbb{R}$ or \mathbb{C} and (\cdot,\cdot) is an inner product on H. Then H is called a *Hilbert space* if it is complete (for the metric defined by the norm, $d(x, y) := \|x - y\| = (x - y, x - y)^{1/2}$).

Examples. Let (X, \mathcal{S}, μ) be a measure space. Then $L^2(X, \mathcal{S}, \mu)$ is a Hilbert space with the inner product $(f^\sim, g^\sim) := (f, g) := \int f(x)\bar{g}(x) d\mu(x)$ by Theorem 5.3.1. Completeness follows from Theorem 5.2.1. If X is any set with counting measure, then \mathcal{L}^p can be identified with L^p and is called $\ell^p(X)$, or "little ell p of X." Specifically, if X is the set of positive integers, $\ell^p(X)$ is called ℓ^p. Then ℓ^2 is a Hilbert space, the set of all "square-summable" sequences $\{z_k\}$ with $\sum_k |z_k|^2 < \infty$, and $(\{z_k\}, \{w_k\}) := \sum_k z_k \bar{w}_k$.

It would follow quite easily from 5.3.3 that the inner product (x, y) is continuous in x for fixed y, or in y for fixed x. The joint continuity in (x, y) takes a bit more proof:

5.3.5. Theorem *For any inner product space $(H, (\cdot,\cdot))$, the inner product is jointly continuous, that is, continuous for the product topology from $H \times H$ into K.*

Proof. Given u and $v \in H$, we have as $x \to u$ and $y \to v$, by 5.3.3 and 5.3.4,
$|(x, y) - (u, v)| \le |(x, y - v)| + |(x - u, v)| \le \|x\|\|y - v\| + \|x - u\|\|v\| \le (\|u\| + \|x - u\|)\|y - v\| + \|x - u\|\|v\| \to 0$, since $\|x - u\| \to 0$ and $\|y - v\| \to 0$. □

We say two elements f and g of H are *orthogonal* or *perpendicular*, written $f \perp g$, iff $(f, g) = 0$. For example, in a space $L^2(X, \mathcal{S}, \mu)$, the indicator functions of disjoint sets are perpendicular. The following two facts, which extend classical theorems of plane geometry, are straightforward (with the current definitions):

5.3.6. The Babylonian-Pythagorean Theorem *If* $f \perp g$, *then* $\|f + g\|^2 = \|f\|^2 + \|g\|^2$.

5.3.7. The Parallelogram Law For any seminorm $\|\cdot\|$ defined by a semi-inner product and any $f, g \in H$,

$$\|f + g\|^2 + \|f - g\|^2 = 2\|f\|^2 + 2\|g\|^2.$$

For any $A \subset H$ let $A^{\perp} := \{y : (x, y) = 0 \text{ for all } x \in A\}$. So in the plane \mathbb{R}^2, if A is a line through the origin, A^{\perp} is the perpendicular line. From any point p in the plane, we can drop a perpendicular to the line A, which will meet A in a point y. The vector $z = p - y \in A^{\perp}$, so $p = y + z$ decomposes p as a sum of two orthogonal vectors. This decomposition will now be extended to inner product spaces, with A being a linear subspace F. Note that completeness of F is needed for this geometrical construction to succeed.

5.3.8. Theorem (Orthogonal Decomposition) *For any inner product space* $(H, (\cdot, \cdot))$, *Hilbert (that is, complete linear) subspace* $F \subset H$, *and* $x \in H$ *there is a unique representation* $x = y + z$ *with* $y \in F$ *and* $z \in F^{\perp}$.

Proof. Let $c := \inf\{\|x - f\| : f \in F\}$. Take $f_n \in F$ with $\|x - f_n\| \downarrow c$. Then by the parallelogram law applied to $f_m - x$ and $f_n - x$, for any m and n,

$$\|f_m - f_n\|^2 = 2\|f_m - x\|^2 + 2\|f_n - x\|^2 - 4\|(f_m + f_n)/2 - x\|^2.$$

Since $(f_m + f_n)/2 \in F$, $\|(f_m + f_n)/2 - x\| \ge c$. Thus as m and $n \to \infty$, $\|f_m - f_n\| \to 0$. By assumption F is complete, so $f_n \to y$ for some $y \in F$ as $n \to \infty$. Let $z := x - y$. Then $\|z\| = c$ by continuity (Theorem 5.3.5). If $(z, f) \ne 0$ for some $f \in F$, let $u := (z, f)v$ with $v \downarrow 0$. Then $\|x - (y + uf)\|^2 = \|z - uf\|^2 = c^2 - 2v|(z, f)|^2 + v^2|(z, f)|^2\|f\|^2 < c^2$

for v small enough, while $y + uf \in F$, contradicting the definition of c. Thus $z \in F^\perp$. If also $x = g + h$, $g \in F$, $h \in F^\perp$, then $y - g + z - h = 0$ and by Theorem 5.3.6, $0 = \|y - g + z - h\|^2 = \|y - g\|^2 + \|z - h\|^2$, so $y = g$ and $z = h$. ☐

Remark. For each x in a Hilbert space H, and closed linear subspace F of H, let $\pi_F(x) := y \in F$ where $x = y + z$ is the orthogonal decomposition of x. Then π_F is linear and is called the orthogonal projection from H to F.

Many functions in \mathcal{L}^2 of Lebesgue measure, being unbounded, cannot be integrated with the classical Riemann integral. So spaces of Riemann integrable functions would not be complete in the \mathcal{L}^2 norm, and the orthogonal decomposition would not apply to them. This shows one of the advantages of Lebesgue integration.

Problems

1. In \mathbb{R}^2 with the usual inner product $((x, y), (u, v)) := xu + yv$, let $F := \{(x, y): x + 2y = 0\}$. What is F^\perp?

2. In \mathbb{R}^3 with usual inner product, let $F := \{(x, y, z): x + 2y + 3z = 0\}$. Let $u := (1, 0, 0)$. Find the orthogonal decomposition $u = v + w$, $v \in F$, $w \in F^\perp$.

3. A set C in a vector space is called *convex* iff for all x and y in C and $0 \le t \le 1$, we have $tx + (1 - t)y \in C$. Let C be a closed, convex set in a Hilbert space H. Show that for each $u \in H$ there is a unique nearest point $v \in C$, in other words $\|u - w\| > \|u - v\|$ for any w in C with $w \ne v$. *Hint:* See the proof of the orthogonal decomposition (Theorem 5.3.8).

4. In \mathbb{R}^2 with usual metric let C be the set $C := \{(x, y): x^2 + 2x + y^2 - 2y \le 3\}$. Let $u := (1, 2)$. Find the nearest point to u in C and prove it is the nearest.

5. An $n \times n$ matrix $A := \{A_{jk}\}_{1 \le j \le n, 1 \le k \le n}$ with entries $A_{jk} \in K$ defines a linear transformation from K^n into itself by $A(\{z_j\}_{1 \le j \le n}) := \{\sum_{1 \le k \le n} A_{jk} z_k\}_{1 \le j \le n}$. Then A is nonnegative definite if $(Az, z) \ge 0$ for all $z \in K^n$. Show that for $K = \mathbb{C}$, if A is nonnegative definite, $A_{jk} = \bar{A}_{kj}$ for all j and k, but if $K = \mathbb{R}$, this property may not hold (A need not be symmetric).

6. Show that if $(X, \|\cdot\|)$ is a real normed linear space where the norm satisfies the parallelogram law (as in 5.3.7), $\|x + y\|^2 + \|x - y\|^2 = 2\|x\|^2 + 2\|y\|^2$

for all x and y in X, then there is an inner product (\cdot,\cdot) with $\|x\| = (x, x)^{1/2}$ for all $x \in X$. *Hint:* Consider $\|x + y\|^2 - \|x - y\|^2$.

7. On $[0, 1]$ with Lebesgue measure λ, show that for $1 \le p \le \infty$, on $L^p([0, 1], \lambda)$, the norm $\|f\|_p \equiv (f, f)^{1/2}$ for some inner product (\cdot,\cdot) if and only if $p = 2$. *Hint:* Use the parallelogram law 5.3.7.

8. In $L^2([-1, 1], \lambda)$, let F be the set of f with $f(x) = f(-x)$ for almost all x. Show that F is a closed linear subspace, and that F^\perp consists of all g with $g(-x) = -g(x)$ for almost all x. Let $h(x) = (1 + x)^4$. Find $f \in F$ and $g \in F^\perp$ such that $h = g + f$.

9. Let X be a non-zero complex vector space. Under what conditions on complex numbers z and w is it true that $A(u, v) := z(u, v) + w((u, v))$ is always an inner product for any two inner products (\cdot,\cdot) and $((\cdot,\cdot))$ on X?

10. Let X be a non-zero complex vector space. Under what conditions on complex numbers u, v, and w is it true that $B(f, g) := u(f, g) + v(g, f) + w((f, g))$ is always an inner product for any two inner products (\cdot,\cdot) and $((\cdot,\cdot))$ on X?

11. Let F be the set of all sequences $\{x_n\}$ such that $\sup_n n|x_n| < \infty$. Prove or disprove:
 (a) F is a linear subspace of ℓ^2.
 (b) The intersection of F with ℓ^2 is closed in ℓ^2.

5.4. Orthonormal Sets and Bases

A set $\{e_i\}_{i \in I}$ in a semi-inner product space H is called *orthonormal* iff $(e_i, e_j) = 1$ whenever $i = j$, and 0 otherwise. For example, in \mathbb{R}^n the unit vectors e_i, where e_i has ith coordinate 1 and other coordinates 0, form an orthonormal basis. In \mathbb{R}^2, another orthonormal basis is $e_1 = 2^{-1/2}(1, 1)$, $e_2 = 2^{-1/2}(-1, 1)$. The Babylonian-Pythagorean theorem (5.3.6) extends to any finite number of terms by induction, giving the next fact:

5.4.1. Theorem *For any finite set $\{u_1, \ldots, u_n\} \subset H$ with $u_i \perp u_j$ whenever $i \ne j$, $\|\sum_{1 \le j \le n} u_j\|^2 = \sum_{1 \le j \le n} \|u_j\|^2$. Thus, for any finite orthonormal set $\{e_j\}_{i \le j \le n}$ and $z_j \in K = \mathbb{R}$ or \mathbb{C},*

$$\left\| \sum_{i \le j \le n} z_j e_j \right\|^2 = \sum_{1 \le j \le n} |z_j|^2.$$

Recall that a topological space (X, \mathcal{T}) is called *Hausdorff* iff for any $x \neq y$ in X there are disjoint open sets U and V with $x \in U$ and $y \in V$. In such a space, convergent nets as defined in §2.1 have unique limits. Given an index set I, and a function $i \mapsto x_i$ from I into a vector space X with a Hausdorff topology \mathcal{T}, sums $\sum_i x_i$ are defined as follows. Let \mathcal{F} be the collection of all finite subsets of I, directed by inclusion. Then $\sum_{i \in I} x_i = x$ iff the net $\{\sum_{i \in F} x_i\}_{F \in \mathcal{F}}$ converges to x for \mathcal{T}. This is called "unconditional" convergence, since it does not depend on any ordering of I. If $I = \mathbb{N}$, then $\sum_n x_n = x$ is defined to mean that $\sum_{0 \leq j \leq m} x_j \to x$ with respect to \mathcal{T} as $m \to \infty$. Then, series such as $\sum_n (-1)^n / n$ are convergent but not unconditionally convergent.

In \mathbb{R}, $\sum_{\alpha \in I} x_\alpha = x \in \mathbb{R}$ means that for any $\varepsilon > 0$, there is a finite set $F \subset I$ such that for any finite set G with $F \subset G \subset I$, $|x - \sum_{\alpha \in G} x_\alpha| < \varepsilon$. $\sum_{\alpha \in I} x_\alpha = +\infty$ means that for any finite M, there is a finite $F \subset I$ such that for any finite G with $F \subset G \subset I$, $\sum_{\alpha \in G} x_\alpha > M$.

Series of nonnegative terms can be added in any order, according to Lemma 3.1.2 (Appendix D). The next fact is another way of saying the same thing, but now for possibly uncountable index sets.

5.4.2. Lemma *In \mathbb{R}, if $x_\alpha \geq 0$ for all $\alpha \in I$, then*

$$\sum_{\alpha \in I} x_\alpha = S(\{x_\alpha\}) := \sup \left\{ \sum_{\alpha \in F} x_\alpha : F \text{ finite}, F \subset I \right\}.$$

Proof. First suppose $S(\{x_\alpha\}) < \infty$. Let $\varepsilon > 0$ and take a finite $F \subset I$ with $\sum_{\alpha \in F} x_\alpha > S(\{x_\alpha\}) - \varepsilon$. Then for any finite $G \subset I$ with $F \subset G$, $S(\{x_\alpha\}) - \varepsilon < \sum_{\alpha \in F} x_\alpha \leq \sum_{\alpha \in G} x_\alpha \leq S(\{x_\alpha\})$. This implies $\sum_{\alpha \in I} x_\alpha = S(\{x_\alpha\})$.

If $S(\{x_\alpha\}) = +\infty$, then for any $M < \infty$, there is a finite $F \subset I$ with $\sum_{\alpha \in F} x_\alpha > M$, and the same holds for any finite $G \supset F$ in place of F, so $\sum_{\alpha \in I} x_\alpha = +\infty$ by definition. \square

Suppose that the index set I is uncountable and that uncountably many of the real numbers x_α are different from 0. Then, for some n, either there are infinitely many α with $x_\alpha > 1/n$, or infinitely many with $x_\alpha < -1/n$, since a countable union of finite sets would be countable. In either case, $\sum_{\alpha \in I} x_\alpha$ could not converge to any finite limit. In practice, sums of real numbers are usually taken over countable index sets such as finite sets, \mathbb{N}, or the set of positive integers. If possibly uncountable index sets are treated, then, it is because it can be done at very little extra cost and/or all but countably many terms in the sums are 0. The next two facts will relate inner products to coordinates and sums over orthonormal bases.

5.4.3. Bessel's Inequality For any orthonormal set $\{e_\alpha\}_{\alpha \in I}$ and $x \in H$, $\|x\|^2 \geq \sum_{\alpha \in I} |(x, e_\alpha)|^2$.

Proof. By Lemma 5.4.2 it can be assumed that I is finite. Then let $y := \sum_{\alpha \in I} (x, e_\alpha) e_\alpha$. Now by linearity of the inner product and Theorem 5.4.1, $(x, y) = \sum_{\alpha \in I} |(x, e_\alpha)|^2 = (y, y)$. Thus $x - y \perp y$, so by Theorem 5.3.6, $\|x\|^2 = \|x - y\|^2 + \|y\|^2 \geq \|y\|^2 = (y, y)$. $\qquad\square$

In \mathbb{R}^n, the inner product of two vectors $x = (x_1, \ldots, x_n)$ and $y = (y_1, \ldots, y_n)$ is usually defined by $(x, y) = x_1 y_1 + \cdots + x_n y_n$. This can also be written as $\sum_{1 \leq j \leq n} (x, e_j)(y, e_j)$ for the usual orthonormal basis e_1, \ldots, e_n of \mathbb{R}^n. The latter sum turns out not to depend on the choice of orthonormal basis for \mathbb{R}^n, and it extends to orthonormal sets in general inner product spaces as follows:

5.4.4. Theorem (Parseval-Bessel Equality) *If* $\{e_\alpha\}$ *is an orthonormal set in* H, $x \in H$, $y \in H$, $x = \sum_{\alpha \in I} x_\alpha e_\alpha$, *and* $y = \sum_{\alpha \in I} y_\alpha e_\alpha$, *then* $x_\alpha = (x, e_\alpha)$ *and* $y_\alpha = (y, e_\alpha)$ *for all* α, *and* $(x, y) = \sum_{\alpha \in I} x_\alpha \bar{y}_\alpha$.

Proof. By continuity of the inner product (Theorem 5.3.5), we have for each $\beta \in I$, $(x, e_\beta) = (\sum_{\alpha \in I} x_\alpha e_\alpha, e_\beta) = x_\beta$, and likewise for y.

The last statement in the theorem holds if only finitely many of the x_α and y_α are not 0, by the linearity properties of the inner product. For any finite $F \subset I$ let $x_F := \sum_{\alpha \in F} x_\alpha e_\alpha$, $y_F := \sum_{\alpha \in F} y_\alpha e_\alpha$. Then the nets $x_F \to x$ and $y_F \to y$. By continuity again (Theorem 5.3.5), $(x_F, y_F) \to (x, y)$, implying the desired result. $\qquad\square$

5.4.5. Riesz-Fischer Theorem *For any Hilbert space* H, *any orthonormal set* $\{e_\alpha\}$ *and any* $x_\alpha \in K$, $\sum_{\alpha \in I} |x_\alpha|^2 < \infty$ *if and only if the sum* $\sum_{\alpha \in I} x_\alpha e_\alpha$ *converges to some* x *in* H.

Proof. "If" follows from 5.4.3 and 5.4.4. "Only if": for each $n = 1, 2, \ldots$, choose a finite set $F(n) \subset I$ such that $F(n)$ increases with n and $\sum_{\alpha \notin F(n)} |x_\alpha|^2 < 1/n^2$. Then by Theorem 5.4.1, $\{\sum_{\alpha \in F(n)} x_\alpha e_\alpha\}$ is a Cauchy sequence, convergent to some x since H is complete. Then, the net of all partial sums converges to the same limit. $\qquad\square$

In the space ℓ^2 of all sequences $x = \{x_n\}_{n \geq 1}$ such that $\sum_n |x_n|^2 < \infty$, let $\{e_n\}_{n \geq 1}$ be the standard basis, that is, $(e_n)_j = 1$ for $j = n$ and 0 otherwise. Then the Riesz-Fischer theorem implies that $\sum_n x_n e_n$ converges to x in ℓ^2. In

other words, if $(y^{(n)})_j = x_j$ for $j \leq n$ and 0 for $j > n$, then $y^{(n)} \to x$ in ℓ^2. (By contrast, consider the space ℓ^∞ of all bounded sequences $\{x_j\}_{j \geq 1}$ with supremum norm $\|x\|_\infty = \sup_j |x_j|$. Then $y^{(n)}$ do not converge to x for $\|\cdot\|_\infty$, say if $x_j = 1$ for all j.)

Given a subset S of a vector space H over K, the linear span of S is defined as the set of all sums $\sum_{x \in F} z_x x$ where F is any finite subset of S and $z_x \in K$ for each $x \in F$. Then recall that S is linearly independent if and only if no element y of S belongs to the linear span of $S \setminus \{y\}$. Given a linearly independent sequence, one can get an orthonormal set with the same linear span:

5.4.6. Gram-Schmidt Orthonormalization Theorem *Let H be any inner product space and $\{f_n\}$ any linearly independent sequence in H. Then there is an orthonormal sequence $\{e_n\}$ in H such that for each n, $\{f_n, \ldots, f_n\}$ and $\{e_1, \ldots, e_n\}$ have the same linear span.*

Proof. Let $e_1 := f_1 / \|f_1\|$. This works for $n = 1$. Then recursively, given e_1, \ldots, e_{n-1}, let

$$g_n := f_n - \sum_{1 \leq j \leq n-1} (f_n, e_j) e_j.$$

Then $g_n \perp e_j$ for all $j < n$, but g_n is not 0 since f_n is not in the linear span of $\{f_1, \ldots, f_{n-1}\}$, which is the same as the linear span of $\{e_i, \ldots, e_{n-1}\}$ by induction hypothesis. Let $e_n := g_n / \|g_n\|$. Then the desired properties hold. □

For example, in \mathbb{R}^3 let $f_1 = (1, 1, 1)$, $f_2 = (1, 1, 2)$, and $f_3 = (1, 2, 3)$. Then $e_1 = 3^{-1/2}(1, 1, 1)$, $e_2 = 6^{-1/2}(-1, -1, 2)$, and $e_3 = 2^{-1/2}(-1, 1, 0)$. In this case, there is only a one-dimensional subspace orthogonal to f_1 and f_2, so e_3 can be found without necessarily computing as in the proof.

In a finite-dimensional vector space S, a basis is a set $\{e_j\}_{1 \leq j \leq n}$ such that each element s of S can be written in one and only one way as a sum $\sum_{1 \leq j \leq n} s_j e_j$ for some numbers s_j. Then a linear transformation A from S into itself corresponds to a matrix $\{A_{jk}\}$ with $A(e_j) = \sum_{1 \leq k \leq n} A_{jk} e_k$ for each j. In general, bases in finite-dimensional spaces are quite useful. So it's natural to try to extend the idea of basis to infinite-dimensional spaces. But it turns out to be hard to keep all the good properties of bases of finite-dimensional spaces. In any vector space S, a *Hamel* basis is a set $\{e_\alpha\}_{\alpha \in I}$ such that every $s \in S$ can be written in one and only one way as $\sum_{\alpha \in I} s_\alpha e_\alpha$, where only finitely many of the numbers s_α are not 0. So "Hamel basis" is an algebraic notion, which does not relate to any topology on S. For a finite-dimensional

space, a Hamel basis is just a basis, but in an infinite-dimensional Banach space, such as a Hilbert space, there are no obvious Hamel bases, although they can be shown to exist with the axiom of choice. In analysis, one usually wants to deal with converging infinite sums, such as Taylor series or Fourier series, so the notion of Hamel basis is not usually appropriate. In a Banach space $(S, \|\cdot\|)$, an *unconditional basis* is a collection $\{e_\alpha\}_{\alpha \in I}$ such that for each $s \in S$, there is a unique set of numbers $\{s_\alpha\}_{\alpha \in I}$ such that $\sum_{\alpha \in I} s_\alpha e_\alpha = s$ (converging for $\|\cdot\|$). In a separable Banach spaces S, a *Schauder basis* is a sequence $\{f_n\}_{n \geq 1}$ such that for each $s \in S$, there is a unique sequence of numbers s_n such that $\|s - \sum_{1 \leq j \leq n} s_j f_j\| \to 0$ as $n \to \infty$. It is possible to find Schauder bases in the separable Banach spaces that are most useful in analysis, but Schauder bases may not be unconditional bases, and in general it may be hard to find unconditional bases. One of the several advantages of Hilbert spaces over general Banach spaces is that Hilbert spaces have bases with good properties, defined as follows.

Definition. Given an inner product space $(H, (\cdot, \cdot))$, an *orthonormal basis* for H is an orthonormal set $\{e_\alpha\}_{\alpha \in I}$ such that for each $x \in H, x = \sum_{\alpha \in I} (x, e_\alpha) e_\alpha$.

5.4.7. Theorem *Every Hilbert space has an orthonormal basis.*

Proof. If a collection of orthonormal sets is linearly ordered by inclusion (a chain), then their union is clearly an orthonormal set. Thus by Zorn's lemma (Theorem 1.5.1), let $\{e_\alpha\}_{\alpha \in I}$ be a maximal orthonormal set. Take any $x \in H$. Let $y := \sum_{\alpha \in I} (x, e_\alpha) e_\alpha$, where the sum converges by Bessel's inequality (5.4.3) and the Riesz-Fischer theorem (5.4.5). If $y = x$, we are done. If not, then $x - y \perp e_\alpha$ for all α, so we can adjoin a new element $(x - y)/\|x - y\|$, contradicting the maximality of the orthonormal set. \square

For example, in the Hilbert space ℓ^2, the "standard basis" $\{e_n\}_{n \geq 1}$ is orthonormal and is easily seen to be, in fact, a basis.

Given two metric spaces (S, d) and (T, e), recall that an *isometry* is a function f from S into T such that $e(f(x), f(y)) = d(x, y)$ for all $x, y \in S$. If f is onto T, then S and T are said to be *isometric*. For example, in \mathbb{R}^2 with usual metric, translations $x \mapsto x + v$, rotations around any center, and reflections in any line are all isometries.

5.4.8. Theorem *Every Hilbert space H is isometric to a space $\ell^2(X)$ for some set X.*

Proof. Let $\{e_\alpha\}_{\alpha \in X}$ be an orthonormal basis of H. Then $x \mapsto \{(x, e_\alpha)\}_{\alpha \in X}$ takes H into $\ell^2(X)$ by Bessel's inequality (5.4.3). This function preserves inner products by the Parseval-Bessel equality (5.4.4). It is onto $\ell^2(X)$ by the Riesz-Fischer theorem (5.4.5). □

Theorem 5.4.8, saying that every Hilbert space can be represented as L^2 of a counting measure, gives a kind of converse to the fact that every L^2 space is a Hilbert space. The next fact is a characterization of bases among orthonormal sets.

5.4.9. Theorem *For any inner product space $(H, (\cdot, \cdot))$, an orthonormal set $\{e_\alpha\}_{\alpha \in I}$ is an orthonormal basis of H if and only if its linear span S is dense in H.*

Proof. If the orthonormal set is a basis, then clearly S is dense. Conversely, suppose S is dense. For each $x \in H$ and finite set $J \subset I$, let $x_J := \sum_{\alpha \in J} (x, e_\alpha) e_\alpha$. Then $x - x_J \perp e_\alpha$ for each $\alpha \in J$. For any $c_\alpha \in K$ and $z := \sum_{\alpha \in J} c_\alpha e_\alpha$, we have $x - x_J \perp x_J - z$. Thus by the Babylonian-Pythagorean theorem (5.3.6),

$$\|x - z\|^2 = \|x - x_J\|^2 + \|x_J - z\|^2 \geq \|x - x_J\|^2.$$

For any $\varepsilon > 0$, there is a J and such a z with $\|x - z\| < \varepsilon$, and so $\|x - x_J\| < \varepsilon$. If M is another finite subset of I with $J \subset M$, then

$$x = x_J + (x_M - x_J) + (x - x_M)$$

where $x_M - x_J \perp x - x_M$. Then by Theorem 5.3.6 again,

$$\|x - x_J\|^2 = \|x_M - x_J\|^2 + \|x - x_M\|^2 \geq \|x - x_M\|^2.$$

So the net of real numbers $\|x - x_J\|$ converges to 0 as the finite set J increases. In other words, the net $\{x_J\}$ converges to x, and $\{e_\alpha\}_{\alpha \in I}$ is an orthonormal basis. □

5.4.10. Corollary *Let $(H, (\cdot, \cdot))$ be any inner product space with a countable set having a dense linear span. Then H has an orthonormal basis.*

Proof. By Gram-Schmidt orthonormalization (5.4.6), there is a countable orthonormal set having a dense linear span, which is a basis by Theorem 5.4.9. □

Remark. Having a countable set with dense linear span is actually equivalent to being separable (having a countable dense set), as one can take rational coefficients in the linear span and still have a dense set.

For the Lebesgue integral, L^2 spaces are already complete, but if an inner product were defined for continuous functions by $(f, g) = \int_0^1 f(x)g(x)\,dx$ (Riemann integral), then a completion would be needed to get a Hilbert space.

When a metric space is completed, the metric d extends to the completion. For a normed linear space, d is defined by the norm $\|\cdot\|$, which extends to the completion with $\|x\| = d(0, x)$. If the norm is defined by an inner product, the inner product could be extended via $(x, y) \equiv (\|x + y\|^2 - \|x - y\|^2)/4$ in the real case, with a more complicated formula in the complex case. Here is a different way to extend the inner product to the completion, giving a Hilbert space.

5.4.11. Theorem *Let $(J, (\cdot,\cdot))$ be any inner product space. Let S be the completion of J (for the usual metric from the norm defined by the inner product). Then the vector space structure and inner product both can be extended to S and it is a Hilbert space.*

Proof. Let H be the set of all continuous conjugate-linear functions from J into K. For $f \in H$, let $\|f\| := \sup\{|f(x)|: x \in J, \|x\| \le 1\}$. Then H is naturally a vector space over K and $\|\cdot\|$ is a norm on H. For $j \in J$ let $g(j) := (j, \cdot)$. Then by the Cauchy-Bunyakovsky-Schwarz inequality (5.3.3), g maps J into H and $\|g(j)\| \le \|j\|$ for all $j \in J$. On the other hand, for each $j \neq 0$ in J, letting $x = j/\|j\|$ shows that $\|g(j)\| \ge \|j\|$, so $\|g(j)\| = \|j\|$ and g is an isometry from J into H. It is easily seen that H is complete. Thus the closure of the range of g is a completion S of J (this closure is actually all of H, as will be proved soon, but that fact is not needed here). The joint continuity of inner products, and the stronger estimates given in its proof (Theorem 5.3.5), show that the inner product extends by continuity from J to H (the details are left to Problem 8). $\qquad\square$

Example (Fourier Series). Some sets of trigonometric functions may well be historically, and even now, the most important examples of infinite orthonormal sets. Let $X = [-\pi, \pi)$ with measure $\mu := \lambda/(2\pi)$, where λ is Lebesgue measure. Then X can be identified with the unit circle $\{z: |z| = 1\}$ in \mathbb{C}, via $z = e^{i\theta}$ for $-\pi \le \theta < \pi$. The functions $f_n(\theta) := e^{in\theta}$ on X, for all integers $n \in \mathbb{Z}$, are orthonormal. (It will be shown in Proposition 7.4.2 below, or by another method in Problem 6, that they form a basis of $L^2(X, \mu)$.) Then, the constant function 1 and the functions $\theta \mapsto 2^{1/2} \sin(n\theta)$ and $\theta \mapsto 2^{1/2} \cos(n\theta)$

for $n = 1, 2, \ldots$, form an orthonormal basis of real $L^2([-\pi, \pi), \mu)$ ("real Fourier series").

Problems

1. Let $H = L^2([0, 1], \lambda)$ where λ is Lebesgue measure on $[0, 1]$. Let $f_n(x) := x^{4+n}$ for $n = 1, 2, 3$. Find the corresponding orthonormal e_1, e_2, e_3 given by the Gram-Schmidt process (5.4.6).

2. Let $(H, (\cdot, \cdot))$ be an inner product space and F a complete linear subspace. Let $\{e_\alpha\}$ be an orthonormal basis of F. Show that $f(x) := \sum_\alpha (x, e_\alpha)e_\alpha$ gives the orthogonal projection of x into F.

3. If $(H, (\cdot, \cdot))$ is an inner product space and $\{x_n\}$ is a sequence with $x_m \perp x_n$ for all $m \neq n$, prove that for $\sum_{n \geq 1} x_n$, unconditional convergence is equivalent to ordinary convergence (convergence of the sequence $\sum_{1 \leq j \leq n} x_j$ as $n \to \infty$).

4. Show that the space $C([0, 1])$ of continuous real functions on $[0, 1]$ is dense in real $L^2([0, 1], \lambda)$. *Hint:* If $(f, g) = 0$ for all $g \in C([0, 1])$, show that $(f, 1_A) = 0$ for every measurable set A, starting with intervals A.

5. Assuming Problem 4, show that real $L^2([0, 1], \lambda)$ has an othonormal basis $\{P_n\}_{n \in \mathbb{N}}$ where each P_n is a polynomial of degree n ("Legendre polynomials"). *Hint:* Use the Weierstrass approximation theorem 2.4.12 and the Gram-Schmidt process.

6. Assuming Problem 4, prove that, as stated in the example following Theorem 5.4.11, $\{\theta \mapsto e^{in\theta}\}_{n \in \mathbb{Z}}$ is an orthonormal basis of complex $L^2([0, 2\pi), \lambda/(2\pi))$. *Hint:* Use the complex Stone-Weierstrass theorem 2.4.13.

7. Assuming Problem 6, prove that the functions $\theta \mapsto \sqrt{2}\sin(n\theta), n = 1, 2, \ldots$, form an orthonormal basis of real $L^2([0, \pi), \lambda/\pi)$. *Hint:* Consider even and odd functions on $[-\pi, \pi)$.

8. Show in detail how an inner product on an inner product space extends to its completion (proof of Theorem 5.4.11).

9. Let f_n be orthonormal in a Hilbert space H. For what values of α does the series $\sum_n f_n/(1 + n^2)^\alpha$ converge in H?

10. (Rademacher functions.) For $x \in [0, 1)$, let $f_n(x) := 1$ for $2k \leq 2^n x < 2k + 1$ and $f_n(x) := -1$ for $2k + 1 \leq 2^n x < 2k + 2$ for any integer k.
 (a) Show that the f_n are orthonormal in $H := L^2([0, 1], \lambda)$.
 (b) Show that the f_n are not a basis of H.

11. Let $f_{nk}(x) := a_{nk}$ for $2k \le 2^n x < 2k + 1$ and $f_{nk}(x) = b_{nk}$ for $2k + 1 \le x < 2k + 2$, for each $n = 0, 1, \dots$ and $k = 0, 1, \dots, K(n)$. Find values of $K(n)$, a_{nk}, and b_{nk} such that the set of all the functions f_{nk} is an orthonormal basis of $L^2([0, 1], \lambda)$. How uniquely are a_{nk} and b_{nk} determined?

12. Let (X, \mathcal{S}, μ) and (Y, \mathcal{T}, ν) be two measure spaces. Let $\{f_n\}$ be an orthonormal basis of $L^2(X, \mathcal{S}, \mu)$ and $\{g_n\}$ an orthonormal basis of $L^2(Y, \mathcal{T}, \nu)$. Show that the set of all functions $h_{mn}(x, y) := f_m(x) g_n(y)$ is an orthonormal basis of $L^2(X \times Y, \mathcal{S} \otimes \mathcal{T}, \mu \times \nu)$.

13. Let H be a Hilbert space with an orthonormal basis $\{e_n\}$. Let $h \in H$ with $h = \sum_n h_n e_n$. Show that for $0 \le r \le 1$, $\sum_n h_n r^n e_n$ converges in H to an element $h^{(r)}$ which converges to h as $r \uparrow 1$.

14. On the unit circle in \mathbb{C}, $\{e^{i\theta} : -\pi < \theta \le \pi\}$, with the measure $d\mu(\theta) = d\theta/(2\pi)$, the functions $\theta \mapsto e^{in\theta}$ for all $n \in \mathbb{Z}$ form an orthonormal basis, by Problem 6 above. In $H := L^2(\mu)$, let H^2 be the closed linear subspace spanned by the functions $e^{in\theta}$ for $n \ge 0$. Let $f(\theta) := \theta$ for $-\pi < \theta \le \pi$. Let h be the orthogonal projection of f into H^2. Show that h is unbounded. *Hints:* Using integration by parts, show that f has Fourier series $i \sum_{n \ge 1} (-1)^n (e^{in\theta} - e^{-in\theta})/n$. (The series for h diverges at $\theta = \pi$, by the way.) To prove $h \notin \mathcal{L}^\infty(\mu)$, let $1 + e^{i\theta} = \rho e^{i\varphi}$ where $|\varphi| \le \pi/2$, φ is real, and $\rho \ge 0$. Show that $h(\theta) = -i \cdot \log(1 + e^{i\theta}) := -i \cdot \log \rho + \varphi$ for almost all θ, by the method of Problem 13, as follows. The Taylor series of $\log(1 + z)$ for $|z| < 1$ (Appendix B) yields the Fourier series of $\log(1 + re^{i\theta})$ for $0 \le r < 1$. As $r \uparrow 1$ prove convergence to $\log(1 + e^{i\theta})$ uniformly on intervals $|\theta| \le \pi - \varepsilon$, $\varepsilon > 0$. (Since H^2 is a very natural subspace, the fact that projection onto it does not preserve boundedness shows a need for the Lebesgue integral.)

5.5. Linear Forms on Hilbert Spaces, Inclusions of L^p Spaces, and Relations between Two Measures

Let H be a Hilbert space over the field $K = \mathbb{R}$ or \mathbb{C}.

A linear function f from the finite-dimensional Euclidean space \mathbb{R}^k into \mathbb{R} is always continuous and can be written as $f(x) = f(\{x_j\}_{i \le j \le k}) = \sum_{1 \le j \le k} x_j h_j = (x, h)$, an inner product of x with h. Here $h_j = f(e_j)$ where e_j is the unit vector which is 1 in the jth place and 0 elsewhere. The representation of continuous linear forms as inner products extends to general Hilbert spaces:

5.5.1. Theorem (Riesz-Fréchet) *A function f from a Hilbert space H into K is linear and continuous if and only if for some $h \in H$, $f(x) = (x, h)$ for all $x \in H$. If so, then h is unique.*

Proof. "If" follows from Theorem 5.3.5 (continuity) and the definition of inner product (linearity). To prove "only if," let f be linear and continuous from H to K. If $f = 0$, let $h = 0$. Otherwise, let $F := \{x \in H : f(x) = 0\}$. Since f is continuous, F is closed and hence complete. Choose any $v \notin F$. By orthogonal decomposition (Theorem 5.3.8), $v = y + z$ where $y \in F$ and $z \in F^{\perp}$, $z \neq 0$. Then $f(z) \neq 0$. Let $u = z/f(z)$. Then $f(u) = 1$ and $u \in F^{\perp}$. For any $x \in H$, $x - f(x)u \in F$, so $(x, u) - f(x)(u, u) = 0$. Let $h := u/(u, u)$. Then $f(x) = (x, h)$ for all $x \in H$.

If $(x, h) = (x, g)$ for all $x \in H$, then $(x, h - g) = 0$. Setting $x = h - g$ gives $h = g$, so uniqueness follows. □

On a finite measure space, integrability of $|f|^p$ for a function f will be shown to be more restrictive for larger values of p. This fact is included here partly for use in a proof later in this section.

5.5.2. Theorem *Let (X, \mathcal{S}, μ) be any finite measure space ($\mu(X)$ is finite). Then for $1 \leq r < s \leq \infty$, $\mathcal{L}^s(X, \mathcal{S}, \mu) \subset \mathcal{L}^r(X, \mathcal{S}, \mu)$, and the identity function from L^s into L^r is continuous.*

Proof. Let $f \in \mathcal{L}^s$. If $s = \infty$, then $|f| \leq \|f\|_{\infty}$ a.e., so $\int |f|^r d\mu \leq \|f\|_{\infty}^r \mu(X)$ and $\|f\|_r \leq \|f\|_{\infty} \mu(X)^{1/r}$. Then for f and g in \mathcal{L}^{∞}, $\|f - g\|_r \leq \|f - g\|_{\infty} \mu(X)^{1/r}$, giving the continuity. If $s < \infty$, then by the Rogers-Hölder inequality (5.1.2) with $p := s/r$, and $q = p/(p - 1)$, we have $\int |f|^r d\mu = \int |f|^r \cdot 1 \, d\mu \leq (\int |f|^{r(s/r)} d\mu)^{r/s} \mu(X)^{1/q} < \infty$. Thus $\|f\|_r \leq \|f\|_s \mu(X)^{1/qr}$ for all $f \in \mathcal{L}^s$. This implies the continuity as stated. □

Remark. If $\mu(X) = 1$, so μ is a probability measure, then for any measurable f, $\|f\|_p$ is nondecreasing as p increases.

Now let ν and μ be two measures on the same measurable space (X, \mathcal{S}). Say ν is *absolutely continuous with respect to* μ, in symbols $\nu \prec \mu$, iff $\nu(A) = 0$ whenever $\mu(A) = 0$. Say that μ and ν are *singular*, or $\nu \perp \mu$, iff there is a measurable set A with $\mu(A) = \nu(X \backslash A) = 0$. Recall that $\int_A f \, d\mu := \int f 1_A \, d\mu$ for any measure μ, measurable set A, and f any integrable or nonnegative measurable function. If f is nonnegative and measurable, then $\nu(A) := \int_A f \, d\mu$ defines a measure absolutely continuous with respect to μ. If μ is σ-finite

and v is finite, the Radon-Nikodym theorem (5.5.4 below) will show that existence of such an f is necessary for absolute continuity; this is among the most important facts in measure theory. The next two theorems will be proved together.

5.5.3. Theorem (Lebesgue Decomposition) *Let (X, S) be a measurable space and μ and v two σ-finite measures on it. Then there are unique measures v_{ac} and v_s such that $v = v_{ac} + v_s$, $v_{ac} \prec \mu$ and $v_s \perp \mu$.*

For example, let μ be Lebesgue measure on $[0, 2]$ and v Lebesgue measure on $[1, 3]$. Then v_{ac} is Lebesgue measure on $[1, 2]$ and v_s is Lebesgue measure on $[2, 3]$.

5.5.4. Theorem (Radon-Nikodym) *On the measurable space (X, S) let μ be a σ-finite measure. Let v be a finite measure, absolutely continuous with respect to μ. Then for some $h \in \mathcal{L}^1(X, S, \mu)$,*

$$v(E) = \int_E h \, d\mu \qquad \text{for all } E \text{ in } S.$$

Any two such h are equal a.e. (μ).

Proof of Theorems 5.5.3 and 5.5.4 (J. von Neumann). Take a sequence of disjoint measurable sets A_k such that $X = \bigcup_k A_k$ and $\mu(A_k) < \infty$ for all k. Likewise, take a sequence of disjoint measurable sets B_m of finite measure for v, whose union is X. Then by taking all the intersections $A_k \cap B_m$, there is a sequence E_n of disjoint, measurable sets, whose union is X, such that $\mu(E_n) < \infty$ and $v(E_n) < \infty$ for all n.

For any measure ρ on S let $\rho_n(C) := \rho(C \cap E_n)$ for each $C \in S$. Then $\rho(C) = \sum_n \rho_n(C)$, and $\rho(C) = 0$ if and only if for all n, $\rho_n(C) = 0$. Hence $\rho \prec \mu$ if and only if (for all n, $\rho_n \prec \mu$) if and only if (for all n, $\rho_n \prec \mu_n$). Also, $\rho \perp \mu$ if and only if (for all n, $\rho_n \perp \mu$) if and only if (for all n, $\rho_n \perp \mu_n$).

For any measures α_n on S with $\alpha_n(X \setminus E_n) = 0$ for all n, $\alpha(C) := \sum_n \alpha_n(C)$ for each $C \in S$ defines a measure α with $\alpha \prec \mu$ if and only if $\alpha_n \prec \mu_n$ for all n, and $\alpha \perp \mu$ if and only if $\alpha_n \perp \mu_n$ for all n. Thus in proving the Lebesgue decomposition theorem (5.5.3), both μ and v can be assumed finite. For the Radon-Nikodym theorem (5.5.4), if it is proved on each E_n, with some function $h_n \geq 0$, let $h(x) := h_n(x)$ for each $x \in E_n$. Then h is measurable (Lemma 4.2.4) and

$$\int h \, d\mu = \sum_{n \geq 1} \int h_n \, d\mu = \sum_{n \geq 1} v(E_n) = v(X) < \infty$$

by assumption in Theorem 5.5.4, so $h \in \mathcal{L}^1(\mu)$. Thus we can assume μ (as well as ν) is finite in proving Theorem 5.5.4 also.

Now form the Hilbert space $H = L^2(X, \mathcal{S}, \mu + \nu)$. Then by Theorem 5.5.2, $L^2 \subset L^1$, and the identity from L^2 into L^1 is continuous. Also, $\nu \le \mu + \nu$. Thus the linear function $f \mapsto \int f \, d\nu$ is continuous from H to \mathbb{R}. Then by Theorem 5.5.1, there is a $g \in \mathcal{L}^2(X, \mathcal{S}, \mu + \nu)$ such that $\int f \, d\nu = \int fg \, d(\mu + \nu)$ for all $f \in \mathcal{L}^2(\mu + \nu)$. Thus

$$\int f(1 - g) \, d(\mu + \nu) = \int f \, d\mu \qquad \text{for all } f \in \mathcal{L}^2(\mu + \nu). \qquad (*)$$

Now $g \ge 0$ a.e. for $\mu + \nu$, since otherwise we can take f as the indicator of the set $\{x: g(x) < 0\}$ for a contradiction. Likewise, $(*)$ implies $g \le 1$ a.e. $(\mu + \nu)$. Thus we can assume $0 \le g(x) \le 1$ for all x. Then $(*)$ holds for all measurable $f \ge 0$ by monotone convergence.

Let $A := \{x: g(x) = 1\}$, $B := X \backslash A$. Then letting $f = 1_A$ in $(*)$ gives $\mu(A) = 0$. For all $E \in \mathcal{S}$, let

$$\nu_s(E) := \nu(E \cap A), \quad \text{and} \quad \nu_{ac}(E) := \nu(E \cap B).$$

Then ν_s and ν_{ac} are measures, $\nu = \nu_s + \nu_{ac}$, and $\nu_s \perp \mu$. If $\mu(E) = 0$ and $E \subset B$, then $\int_E (1 - g) \, d(\mu + \nu) = 0$ by $(*)$ with $1 - g > 0$ on E, so $(\mu + \nu)(E) = 0$ and $\nu(E) = \nu_{ac}(E) = 0$. Hence $\nu_{ac} \prec \mu$. So the existence of a Lebesgue decomposition is proved.

To prove uniqueness, we can still assume both measures are finite. Suppose $\nu = \rho + \sigma$ with $\rho \prec \mu$ and $\sigma \perp \mu$. Then $\rho(A) = 0$, since $\mu(A) = 0$. Thus for all $E \in \mathcal{S}$,

$$\nu_s(E) = \nu(E \cap A) = \sigma(E \cap A) \le \sigma(E).$$

Thus $\nu_s \le \sigma$ and $\rho \le \nu_{ac}$. Then $\sigma - \nu_s = \nu_{ac} - \rho$ is a measure both absolutely continuous and singular with respect to μ, so it is 0, $\rho = \nu_{ac}$, and $\sigma = \nu_s$. This proves uniqueness of the Lebesgue decomposition, finishing the proof of Theorem 5.5.3.

Now to prove Theorem 5.5.4, we have $\nu = \nu_{ac}$, so $\nu_s = 0$. Let $h := g/(1 - g)$ on B and $h := 0$ on A. For any $E \in \mathcal{S}$, let $f = h1_E$ in $(*)$. Then

$$\int_E h \, d\mu = \int_{B \cap E} g \, d(\mu + \nu) = \nu(B \cap E) = \nu(E).$$

Thus the existence of h is proved in the Radon-Nikodym theorem.

Suppose j is another function such that for all $E \in \mathcal{S}$, $\int_E j \, d\mu = \nu(E)$, so $\int_E j - h \, d\mu = 0$. Letting $E_1 := \{x: j(x) > h(x)\}$ and $E_2 := \{x: j(x) < h(x)\}$, integrating $j - h$ over these sets gives $\mu(E_1) = \mu(E_2) = 0$. In more detail,

one can consider the sets where $j - h > 1/n$, show that each has measure 0, hence so does their union, and likewise for the sets where $j - h < -1/n$. So $j = h$ a.e. (μ), completing the proof of Theorem 5.5.4. □

The function h in the Radon-Nikodym theorem is called the Radon-Nikodym *derivative* or *density* of ν with respect to μ and is written $h = d\nu/d\mu$. This (Leibnizian) notation leads to correct conclusions, as shown in some of the problems.

Problems

1. Show that if a finite measure ν is absolutely continuous with respect to a σ-finite measure μ, and f is any function integrable with respect to ν, then

$$\int f \frac{d\nu}{d\mu} d\mu = \int f \, d\nu.$$

2. If β, ν, and μ are finite measures such that $\beta \prec \nu$ and $\nu \prec \mu$, show that

$$\frac{d\beta}{d\mu} = \frac{d\beta}{d\nu} \cdot \frac{d\nu}{d\mu} \quad \text{almost everywhere for } \mu. \text{ } \textit{Hint:} \text{ Apply Problem 1.}$$

3. Let $g(x, y) = 2$ for $0 \le y \le x \le 1$ and $g = 0$ elsewhere. Let μ be the measure on \mathbb{R}^2 which is absolutely continuous with respect to Lebesgue measure λ^2 and has $d\mu/d\lambda^2 = g$. Let $T(x, y) := x$ from \mathbb{R}^2 onto \mathbb{R}^1 and let τ be the image measure $\mu \circ T^{-1}$. Find the Lebesgue decomposition of Lebesgue measure λ on \mathbb{R} with respect to τ and the Radon-Nikodym derivative $d\lambda_{ac}/d\tau$.

4. If μ and ν are finite measures on a σ-algebra \mathcal{S}, show that ν is absolutely continuous with respect to μ if and only if for every $\varepsilon > 0$ there is a $\delta > 0$ such that for any $A \in \mathcal{S}$, $\mu(A) < \delta$ implies $\nu(A) < \varepsilon$.

5. Lebesgue measure λ is absolutely continuous with respect to counting measure c on $[0, 1]$, which is not σ-finite. Show that the conclusion of the Radon-Nikodym theorem does not hold in this case (there is no "derivative" $d\lambda/dc$ having the properties of h in the Radon-Nikodym theorem).

6. For any measure space (X, \mathcal{S}, μ), show that if $p < r < s$, and $f \in \mathcal{L}^p(X, \mathcal{S}, \mu) \cap \mathcal{L}^s(X, \mathcal{S}, \mu)$, then $f \in \mathcal{L}^r(X, \mathcal{S}, \mu)$. *Hint:* Use Rogers-Hölder inequalities.

7. Recall the spaces ℓ^p which are the spaces $\mathcal{L}^p(\mathbb{N}, 2^{\mathbb{N}}, c)$ where c is counting measure on \mathbb{N}; in other words, $\ell^p := \{\{x_n\}: \sum_{n \ge 0} |x_n|^p < \infty\}$, with norm $(\sum_n |x_n|^p)^{1/p}$ for $1 \le p < \infty$. Show that $\ell^p \subset \ell^r$ for $0 < p \le r$ (note

that the inclusion goes in the opposite direction from the case of L^p of a finite measure).

8. Take the Cantor function g defined in the proof of Proposition 4.2.1. As a nondecreasing function, it defines a measure ν on $[0, 1]$ with $\nu((a, b]) = g(b) - g(a)$ for $0 \le a < b \le 1$ by Theorem 3.2.6. Show that $\nu \perp \lambda$ where λ is Lebesgue measure.

9. In $[0, 1]$ with Borel σ-algebra \mathcal{B}, let $\mu(A) = 0$ for all $A \in \mathcal{B}$ of first category and $\mu(A) = +\infty$ for all other $A \in \mathcal{B}$. Show that μ is a measure, and that if ν is a finite measure on \mathcal{B} absolutely continuous with respect to μ, then $\nu \equiv 0$. *Hint:* See §3.4, Problems 7–8. Thus, the conclusion of the Radon-Nikodym theorem holds with $h \equiv 0$, although μ is not σ-finite.

5.6. Signed Measures

Measures have been defined to be nonnegative, so one can think of them as distributions of mass. "Signed measures" will satisfy the definition of measure except that they have real, possibly negative values, such as distributions of electric charge.

Definition. Given a measurable space (X, \mathcal{S}), a *signed measure* is a function μ from \mathcal{S} into $[-\infty, \infty]$ which is countably additive, that is, $\mu(\emptyset) = 0$ and for any disjoint $A_n \in \mathcal{S}$, $\mu(\bigcup_n A_n) = \sum_n \mu(A_n)$.

Remarks. The sums in the definition are all supposed to be defined, so that we cannot have sets A and B with $\mu(A) = +\infty$ and $\mu(B) = -\infty$. This is clear if A and B are disjoint. Otherwise, $\mu(A \backslash B) + \mu(A \cap B) = +\infty$ and $\mu(B \backslash A) + \mu(A \cap B) = -\infty$. When a sum of two terms is infinite, at least one is infinite of the same sign, and the other is also or is finite. Thus $\mu(A \cap B)$ is finite, $\mu(A \backslash B) = +\infty$, and $\mu(B \backslash A) = -\infty$. But then $\mu(A \triangle B) = \mu(A \backslash B) + \mu(B \backslash A)$ is undefined, a contradiction. Thus each signed measure takes on at most one of the two values $-\infty$ and $+\infty$. In practice, spaces of signed measures usually consist of finite valued ones.

If ν and ρ are two measures, at least one of which is finite, then $\nu - \rho$ is a signed measure. The following fact shows that all signed measures are of this form, where ν and ρ can be taken to be singular:

5.6.1. Theorem (Hahn-Jordan Decomposition) *For any signed measure μ on a measurable space (X, \mathcal{S}) there is a set $D \in \mathcal{S}$ such that for all $E \in \mathcal{S}$,*

$\mu^+(E) := \mu(E \cap D) \geq 0$ and $\mu^-(E) := -\mu(E \backslash D) \geq 0$. Then μ^+ and μ^- are measures, at least one of which is finite, $\mu = \mu^+ - \mu^-$, and $\mu^+ \perp \mu^-$ (μ^+ and μ^- are singular). These properties uniquely determine μ^+ and μ^-. If F is any other set in \mathcal{S} with the properties of D, then $(\mu^+ + \mu^-)(F \triangle D) = 0$. For any $E \in \mathcal{S}$, $\mu^+(E) = \sup\{\mu(G): G \subset E\}$ and $\mu^-(E) = -\inf\{\mu(E): G \subset E\}$.

Proof. Replacing μ by $-\mu$ if necessary, we can assume that $\mu(E) > -\infty$ for all $E \in \mathcal{S}$. Let $c := \inf\{\mu(E): E \in \mathcal{S}\}$. Take $c_n \downarrow c$ and $E_n \in \mathcal{S}$ with $\mu(E_n) < c_n$ for all n. Let $A_1 := E_1$. Recursively define $A_{n+1} := A_n$ iff $\mu(E_{n+1} \backslash A_n) \geq 0$; otherwise, let $A_{n+1} := A_n \cup E_{n+1}$. Then for all n, $\mu(E_n \backslash A_n) \geq 0$, so $\mu(E_n \cap A_n) \leq \mu(E_n) < c_n$. Now $A_1 \subset A_2 \subset \cdots$ and $\mu(A_1) \geq \mu(A_2) \geq \cdots$. Let $B_1 := \bigcup_n A_n$. Then $\mu(B_1) < c_1$ and $\inf\{\mu(E): E \subset B_1\} = c$. Then by another recursion, there exists a decreasing sequence of sets B_n with $\mu(B_n) < c_n$ for all n. Let $C := \bigcap_n B_n$. Then $\mu(C) = c$. Thus $c > -\infty$. For any $E \in \mathcal{S}$, $\mu(C \cap E) \leq 0$ (otherwise $\mu(C \backslash E) < c$) and $\mu(E \backslash C) \geq 0$ (otherwise $\mu(C \cup E) < c$). Let $D := X \backslash C$. Then D, μ^+, and μ^- have the stated properties.

For any $E \in \mathcal{S}$, clearly $\mu^+(E) = \mu(E \cap D) \leq \sup\{\mu(G): G \subset E\}$. For any $G \subset E$, $\mu(G) = \mu^+(G) - \mu^-(G) \leq \mu^+(G) \leq \mu^+(E)$. Thus $\mu^+(E) = \sup\{\mu(G): G \subset E\}$. Likewise, $\mu^-(E) = -\inf\{\mu(G): G \subset E\}$. To prove uniqueness of μ^+ and μ^-, suppose $\mu = \rho - \sigma$ for nonnegative, singular measures ρ and σ. Then for any measurable set $G \subset E$, $\rho(E) \geq \rho(G) \geq \mu(G)$. Thus, $\rho \geq \mu^+$. Likewise, $\sigma \geq \mu^-$. Take $H \in \mathcal{S}$ with $\rho(X \backslash H) = \sigma(H) = 0$. Then for any $E \in \mathcal{S}$, $\rho(E) = \rho(E \cap H) = \mu(E \cap H) \leq \mu^+(E)$, so $\rho = \mu^+$ and $\sigma = \mu^-$, proving uniqueness.

If F is another set in \mathcal{S} with the properties of D, then for any $E \in \mathcal{S}$, $\mu^+(E) = \mu^+(E \cap D) = \mu(E \cap D) = \mu^+(E \cap F) = \mu(E \cap F)$, so $\mu((D \triangle F) \cap E) = 0$. Thus $\mu^+(D \triangle F) = 0 = \mu(D \triangle F) = \mu^-(D \triangle F)$, finishing the proof. \square

The decomposition of X into sets D and C where μ is nonnegative and nonpositive, respectively, is called a *Hahn decomposition* of X with respect to μ. The equation $\mu = \mu^+ - \mu^-$ is called the *Jordan decomposition* of μ. The measure $|\mu| := \mu^+ + \mu^-$ is called the *total variation* measure for μ. Two signed measures μ and ν are said to be *singular* iff $|\mu|$ and $|\nu|$ are singular.

5.6.2. Corollary *The Lebesgue decomposition and Radon-Nikodym theorems (5.5.3 and 5.5.4) also hold where ν is a finite signed measure.*

Proof. The Jordan decomposition (5.6.1) gives $\nu = \nu^+ - \nu^-$ where ν^+ and ν^- are finite measures. One can apply theorem 5.5.3 to ν and 5.5.4 to ν^+ and ν^-, then subtract the results. □

For example, if $\nu(A) = \int_A f\, d\mu$ for all A in a σ-algebra, where μ is a σ-finite measure and f is an integrable function, then ν is a finite signed measure, absolutely continuous with respect to μ. A Hahn decomposition for ν can be defined by the set $D = \{x: f(x) \geq 0\}$, with $\nu^+(A) = \int_A f^+\, d\mu$ and $\nu^-(A) = \int_A f^-\, d\mu$ for any measurable set A. Other sets D' defining a Hahn decomposition will satisfy $\int_{D \,\triangle\, D'} |f|\, d\mu = 0$.

Now let \mathcal{A} be any algebra of subsets of a set X. An extended real-valued function μ defined on \mathcal{A} is called *countably additive* if $\mu(\emptyset) = 0$ and for any disjoint sets A_1, A_2, \ldots, in \mathcal{A} such that $A := \bigcup_{n \geq 1} A_n$ happens to be in \mathcal{A}, we have $\mu(A) = \sum_n \mu(A_n)$. Then, as with signed measures, μ cannot take both values $+\infty$ and $-\infty$. We say μ is *bounded above* if for some $M < +\infty$, $\mu(A) \leq M$ for all $A \in \mathcal{A}$. Also, μ is called *bounded below* iff $-\mu$ is bounded above. Recall that every countably additive, nonnegative function defined on an algebra has a countably additive extension to the σ-algebra generated by \mathcal{A} (Theorem 3.1.4), giving a measure. To extend this fact to signed measures, boundedness above or below is needed.

5.6.3. Theorem *If μ is countably additive from an algebra \mathcal{A} into $[-\infty, \infty]$, then μ extends to a signed measure on the σ-algebra \mathcal{S} generated by \mathcal{A} if and only if μ is either bounded above or bounded below.*

Proof. "Only if": by the Hahn-Jordan decomposition theorem (5.6.1), for any signed measure μ on a σ-algebra, either μ^+ or μ^- is finite. Thus μ is bounded above or below. Conversely, suppose μ is bounded above or below. For any $A \in \mathcal{A}$ let $\mu^+(A) := \sup\{\mu(B): B \subset A\}$. Then $\mu^+ \geq 0$ (let $B = \emptyset$) and $\mu^+(\emptyset) = 0$. Suppose A_n are disjoint, in \mathcal{A}, and $A = \bigcup_n A_n \in \mathcal{A}$. Then clearly $\mu^+(A) \geq \sum_n \mu^+(A_n)$. Conversely, for any $B \subset A$ with $B \in \mathcal{A}$, $\mu(B) = \sum_{n \geq 1} \mu(B \cap A_n) \leq \sum_{n \geq 1} \mu^+(A_n)$, so $\mu^+(A) \leq \sum_{n \geq 1} \mu^+(A_n)$. Thus μ^+ is countably additive on \mathcal{A}. Letting $\mu^- := (-\mu)^+$, we see that μ^- is also countably additive and nonnegative. For each $A \in \mathcal{A}$, $\mu(A) = \mu^+(A) - \mu^-(A)$, since by assumption $\mu^+(A)$ and $\mu^-(A)$ cannot both be infinite. Then by Theorem 3.1.4, each of μ^+ and μ^- extends to a measure on \mathcal{S}, with at least one finite, so the difference of the two measures is a well-defined signed measure extending μ to \mathcal{S}. □

Next, an example shows why "bounded above or below," which follows from countable additivity on a σ-algebra, does not follow from it on an algebra:

5.6.4. Proposition *There exists a countably additive, real-valued function on an algebra \mathcal{A} which cannot be extended to a signed measure on the σ-algebra \mathcal{S} generated by \mathcal{A}.*

Proof. Let X be an uncountable set, say \mathbb{R}. Write $X = A \cup B$ where A and B are disjoint and uncountable, say $A = (-\infty, 0]$, $B = (0, \infty)$. Let \mathcal{A} be the algebra consisting of all finite sets and their complements. Let c be counting measure. For F finite let $\mu(F) := c(A \cap F) - c(B \cap F)$. For G with finite complement F let $\mu(G) := -\mu(F)$.

If $C_n \in \mathcal{A}$, C_n are disjoint, $C = \bigcup_n C_n$, and C is finite, then all C_n are finite and all but finitely many of them are empty, so clearly countable additivity holds. If C has finite complement, then so does exactly one of the C_n, say C_1. Then for $n > 1$, the C_n are finite, and all but finitely many are empty. We have

$$X \backslash C_1 = (X \backslash C) \cup \bigcup_{n \geq 2} C_n,$$

a disjoint union, with $X \backslash C_1$ finite, so $-\mu(C_1) = \mu(X \backslash C_1) = -\mu(C) + \sum_{n \geq 2} \mu(C_n)$, and $\mu(C) = \sum_{n \geq 1} \mu(C_n)$. Thus μ is countably additive on the algebra \mathcal{A}. Since it is unbounded above and below, it has no countably additive extension to \mathcal{S}, by Theorem 5.6.3. $\qquad \square$

For a finitely additive real-valued function μ on an algebra \mathcal{A}, countable additivity is equivalent to the condition that $\mu(A_n) \to 0$ whenever $A_n \in \mathcal{A}$ and $A_n \downarrow \emptyset$, as was proved in Theorem 3.1.1.

Problems

1. Show that for the algebra \mathcal{A} in the proof of Proposition 5.6.4, every finitely additive function on \mathcal{A} is countably additive.

2. Let $f(x) := x^2 - 6x + 5$ for all real x. Let $\nu(A) := \int_A f(x)\,dx$ for each Borel set A in \mathbb{R}. Find a Hahn decomposition of \mathbb{R} for ν and the Jordan decomposition of ν.

3. For what polynomials P can f in Problem 2 be replaced by P and give a well-defined signed measure on the Borel sets of \mathbb{R}?

4. A *Radon* measure μ on a topological space S is defined as a real-valued function defined on every Borel set with compact closure, and such that for every fixed compact set K, the restriction of μ to the Borel subsets of K is countably additive. Show that for every continuous function f on \mathbb{R}, $\mu(A) := \int_A f(x)\, d\lambda(x)$ defines a Radon measure. *Note:* A Radon measure is not a measure, at least as it stands, unless S is compact. Some authors may also impose further conditions in defining Radon measures, such as regularity; see §7.1.

5. Prove that any Radon measure μ on \mathbb{R} has a Jordan decomposition $\mu = \mu^+ - \mu^-$ where μ^+ and μ^- are nonnegative, σ-finite Radon measures on \mathbb{R} and can be extended to ordinary nonnegative measures on the σ-algebra of all Borel sets, in a unique way.

6. Carry out the decomposition in Problem 5 for the Radon measure $\mu(A) = \int_A x^3 - x\, dx$ on \mathbb{R}.

7. Let $\mu(A) := \int_A x\, d\lambda(x)$. Show that μ is a Radon measure on \mathbb{R} which cannot be extended to a signed measure on \mathbb{R}.

8. Let μ and ν be finite measures such that ν is absolutely continuous with respect to μ, and $f = d\nu/d\mu$. Show that for each $r > 0$, the Hahn decomposition of $\nu - r\mu$ gives sets A and A^c such that $f \geq r$ a.e. on A and $f \leq r$ a.e. on A^c.

9. Prove the Radon-Nikodym theorem, for finite measures, from the Hahn decomposition theorem. *Hint:* See Problem 8 on how to define f.

10. The *support* of a real-valued function f on a topological space is the closure of $\{x: f(x) \neq 0\}$. Let \mathcal{L} be the set of continuous real-valued functions on \mathbb{R} with compact support.
 (a) Show that $\int f\, d\mu$ is defined for any $f \in \mathcal{L}$ and Radon measure μ.
 (b) Show that \mathcal{L} is a Stone vector lattice as defined in §4.5.
 (c) Show that if I is linear from \mathcal{L} into \mathbb{R} and $I(f) \geq 0$ whenever $f \geq 0$, then L is a pre-integral.

Notes

§5.1 Cauchy (1821, p. 373) proved inequality 5.1.4 for finite sums, in other words where μ is counting measure on a finite set. Bunyakovsky (1859) proved it for Riemann integrals. H. A. Schwarz (1885, p. 351) rediscovered it much later. The inequality was commonly known as the "Schwarz inequality" or, more recently, as the "Cauchy-Schwarz" inequality. Inequality 5.1.2 is generally known as "Hölder's inequality," but Hölder (1889) made clear that he was indebted to a paper of L. J. Rogers (1888). Problem 9 relates the results I found in their papers to 5.1.2. Rogers (1888, p. 149, §3, (1) and (4)) proved "Rogers' inequality," of Problem 9, for finite sums and integrals $\int_a^b dx$.

Hölder (1889, p. 44) proved the "historical Hölder inequality" of Problem 9 for finite sums. F. Riesz (1910, p. 456) proved the current form for integrals. Rogers did other work of lasting interest; see Stanley (1971). Hölder is also well known for the "Hölder condition" on a function f, $|f(x) - f(y)| \le K|x - y|^{\alpha}$ for $0 < \alpha \le 1$, and for other work, including some in finite group theory: see van der Waerden (1939). Inequality 5.1.5 has generally been called Minkowski's inequality. H. Minkowski (1907, p. 95) proved it for finite sums. On Minkowski's life there is a memoir by his eldest daughter: Rüdenberg (1973). F. Riesz (1910, p. 456) extended the inequality to integrals. The above notes are partly based on Hardy, Littlewood, and Polya (1952), who modestly write that they "have never undertaken systematic bibliographic research." Hardy and Littlewood, two of the leading British mathematicians of their time, had one of the most fruitful mathematical collaborations, writing some 100 papers together.

§5.2 F. Riesz (1909, 1910) defined the L^p spaces (for $1 < p < 2$ and $2 < p < \infty$) and proved their completeness. Riesz (1906) defined the L^2 distance. E. Fischer (1907) proved completeness of L^2. On the related "Riesz-Fischer" theorems, see §5.4 and its notes. Banach (1922) defined normed linear spaces and Banach spaces (which he called "espaces de type B"). Wiener (1922), independently, also defined normed linear spaces. Banach (1932) led the development of the theory of such spaces, earning their being named for him. Steinhaus (1961) writes of Banach that "shortly after his birth he was, to be brought up, given to a washerwoman whose name was Banachowa, who lived in an attic . . . by the time he was fifteen Banach had to make his own living by giving lessons."

§5.3 Hilbert, in papers published from 1904 to 1910, developed "Hilbert space" methods, for use in solving integral equations. The definition of (complete) linear inner product space, or Hilbert space, which was only implicit in Hilbert's work, was made explicit and expanded on by E. Schmidt (1907, 1908), who emphasized the geometrical view of Hilbert spaces as "Euclidean" spaces. The proof of the orthogonal decomposition theorem (5.3.8) is from F. Riesz (1934), although the theorem itself is much older. Riesz gives credit for ideas to Levi (1906, §7).

To show that the distance in \mathbb{R}^2, for example, defined by $d((x, y), (u, v)) := ((x - u)^2 + (y - v)^2)^{1/2}$, is invariant under rotations, one would use the trigonometric identity $\cos^2 \theta + \sin^2 \theta = 1$, which rests, in turn, on the classical Pythagorean theorem of plane geometry. In this sense, it would be circular reasoning to claim that Theorem 5.3.6 actually proves Pythagoras' theorem. The classical Pythagorean theorem has long been known through Euclid's Elements of Geometry: Euclid flourished around 300 B.C., and only fragments of earlier Greek books of geometry have been preserved. Pythagoras, of Samos, lived from about 560 to 480 B.C. He founded a kind of sect, or secret society, with interests in mathematics, among other things. The word "mathematics" derives from the Greek $\mu\alpha\theta\eta\mu\alpha\tau\iota\kappa o\iota$, people who had elevated status among the Pythagoreans. O. Neugebauer (1935, I, p. 180; 1957, p. 36) interpreted cuneiform texts of Babylonia as showing that "Pythagoras' theorem," or at least many examples of it, had been known from the time of Hammurabi—that is, before 1600 B.C. (!) See also Buck (1980). It is not known who first actually proved the theorem.

§5.4 There exists an (incomplete, uncountable-dimensional) inner product space without an orthonormal basis (Dixmier, 1953). The first infinite orthonormal sets to be studied were, apparently, the trigonometric functions appearing in Fourier series. The functions

$\cos(kx), k = 0, 1, \ldots$, are orthogonal in $\mathcal{L}^2([0, \pi])$ for the usual length (Lebesgue) measure. Parseval (1799, 1801) discovered the identity (Theorem 5.4.4) in this special case, except that a different numerical factor is required to make these functions orthonormal for $k = 0$ than for $k = 1, 2, \ldots$. Parseval cites Euler (1755) for some ideas. F. W. Bessel (1784–1846) is best known in mathematics in connection with "Bessel functions," which include functions $J_n(r)$ of the radius r in polar coordinates that are eigenfunctions of the Laplace operator,

$$((\partial^2/\partial x^2) + (\partial^2/\partial y^2))J_n(r) = c_n J_n(r), \qquad c_n \in \mathbb{R},$$

and related functions. On Bessel functions, beside many books of tables, there are at least seven monographs, for example Watson (1944). Bessel worked mainly in astronomy (his work was collected in Bessel, 1875). Among other achievements, he was the first to calculate the distance to a star other than the sun (Fricke, 1970, p. 100). Riesz (1907a) says Bessel had stated his inequality (5.4.3) for continuous functions.

"Gram-Schmidt" orthonormalization (Theorem 5.4.6) was first discovered, apparently, by the Danish statistician J. P. Gram (1879; 1883, p. 48), on whom Schweder (1980, p. 118) gives a biographical note. Orthonormalization became much better known from an exposition by E. Schmidt (1907, p. 442). The theory of orthonormal bases developed quickly after 1900, first for trigonometric series in, among others, papers of Hurwitz (1903) and Fatou (1906). Hilbert and Schmidt (see the notes to the previous section) both used orthonormal bases. The Riesz-Fischer theorem (5.4.5) is named on the basis of papers of Fischer (1907) (on completeness of L^2) and, specifically, Riesz (1907a). On its history see Siegmund-Schultze (1982, Kap. 5). The orthogonal decomposition theorem (5.3.8) can also be proved as follows: given $x \in H$, and an orthonormal basis $\{e_\alpha\}$ of the subspace F, let $y := \sum_\alpha (x, e_\alpha) e_\alpha$. The theorem was proved first when F is separable, where the Gram-Schmidt process (5.4.6) gives an orthonormal basis of F.

§5.5 The Riesz-Fréchet theorem (5.5.1), generally known as a "Riesz representation theorem," was stated by Riesz (1907b) and Fréchet (1907) in separate notes in the same issue of the *Comptes Rendus*. The Radon-Nikodym theorem is due to Radon (1913, pp. 1342–1351), Daniell (1920), and Nikodym (1930). The combined, rather short proof of the Lebesgue decomposition and Radon-Nikodym theorem was given by von Neumann (1940).

Problems 5 and 9 show that the Radon-Nikodym theorem can fail, but does not always fail, when μ is not σ-finite. See Bell and Hagood (1981).

§5.6 For a real-valued function f on an interval $[a, b]$, the *total variation* of f is defined as the supremum of all sums

$$\sum_{1 \le j \le n} |f(x_j) - f(x_{j-1})|,$$

where $a \le x_0 \le x_1 \le \cdots \le x_n \le b$. A function is said to be of *bounded variation* on $[a, b]$ iff its total variation there is finite. Jordan (1881) defined the notion of function of bounded variation and proved that every such function is a difference of two nondecreasing functions, say $f = g - h$. The "Jordan decomposition," in more general situations, is named in honor of this result and may be considered an extension of it. If μ is a finite signed measure on $[a, b]$ and $f(x) = \mu([a, x])$, then we can take $g(x) = \mu^+([a, x])$ and $h(x) = \mu^-([a, x])$ for $a \le x \le b$.

Jordan made substantial contributions to many different branches of mathematics, including topology and finite group theory. (See his *Oeuvres*, referenced below.) Lebesgue (1910, pp. 381–382) defined signed measures and studied such measures of the form

$$\mu(A) = \int_A f(x)\, dm(x),$$

where $f \in \mathcal{L}^1(m)$ for a measure m. Then μ^+ and μ^- are given by such "indefinite integrals" of f^+ and f^-, respectively. The "Hahn decomposition" and the Jordan decomposition in the present generality were proved by Hahn (1921, pp. 393–406). Although Hahn called his 1921 book a first volume, the second volume was not published before he died in 1934. Most of it appeared eventually in the form of Hahn and Rosenthal (1948). On Arthur Rosenthal, who lived from 1887 to 1959, see Haupt (1960).

References

An asterisk identifies works I have found discussed in secondary sources but have not seen in the original.

Banach, Stefan (1922). Sur les opérations dans les ensembles abstraits et leurs applications aux équations intégrales. *Fundamenta Math. 3*: 133–181.

———— (1932). *Théorie des operations linéaires*. Monografje Matematyczne, Warsaw. 2d ed. (posth.), Chelsea, New York, 1963.

Bell, W. C., and Hagood, J. W. (1981). The necessity of sigma-finiteness in the Radon-Nikodym theorem. *Mathematika 28*: 99–101.

*Bessel, Friedrich Wilhelm (1875). *Abhandlungen*. Ed. Rudolf Engelmann. 3 vols. Leipzig.

Buck, R. Creighton (1980). Sherlock Holmes in Babylon. *Amer. Math. Monthly 87*: 335–345.

*Bunyakovsky, Viktor Yakovlevich (1859). Sur quelques inégalités concernant les intégrales ordinaires et les intégrales aux différences finies. *Mémoires de l'Acad. de St.-Petersbourg* (Ser. 7) *1*, no. 9.

Cauchy, Augustin Louis (1821). *Cours d'analyse de l'École Royale Polytechnique* (Paris). Also in *Oeuvres complètes d'Augustin Cauchy* (Ser. 2) *3*. Gauthier-Villars, Paris (1897).

Daniell, Percy J. (1920). Stieltjes derivatives. *Bull. Amer. Math. Soc. 26*: 444–448.

Dixmier, Jacques (1953). Sur les bases orthonormales dans les espaces préhilbertiens. *Acta Sci. Math. Szeged 15*: 29–30.

Euler, Leonhard (1755). *Institutiones calculi differentialis*. Acad. Imp. Sci. Petropolitanae, St. Petersburg; also in *Opera Omnia* (Ser. 1) *10*.

Fatou, Pierre (1906). Séries trigonométriques et séries de Taylor. *Acta Math. 30*: 335–400.

Fischer, Ernst (1907). Sur la convergence en moyenne. *Comptes Rendus Acad. Sci. Paris 144*: 1022–1024.

Fréchet, Maurice (1907). Sur les ensembles de fonctions et les opérations linéaires. *Comptes Rendus Acad. Sci. Paris 144*: 1414–1416.

Fricke, Walter (1970). Bessel, Friedrich Wilhelm. *Dictionary of Scientific Biography, 2*, pp. 97–102.

*Gram, Jørgen Pedersen (1879). *Om Räkkeudviklinger, bestemte ved Hjälp of de mindste Kvadraters Methode* (Doctordissertation). Höst, Copenhagen.

——— (1883). Uber die Entwickelung reeller Functionen in Reihen mittelst der Methode der Kleinsten Quadrate. *Journal f. reine u. angew. Math. 94*: 41–73.

Hahn, Hans (1921). *Theorie der reellen Funktionen, "I. Band."* Julius Springer, Berlin.

——— (posth.) and Arthur Rosenthal (1948). *Set Functions*. Univ. New Mexico Press.

Hardy, Godfrey Harold, John Edensor Littlewood, and George Polya (1952). *Inequalities*. 2d ed. Cambridge Univ. Press. Repr. 1967.

Haupt, O. (1960). Arthur Rosenthal (in German). *Jahresb. deutsche Math.-Vereinig. 63*: 89–96.

Hilbert, David (1904–1910). Grundzüge einer allgemeinen Theorie der linearen Integralgleichungen. *Nachr. Ges. Wiss. Göttingen Math.-Phys. Kl. 1904*: 49–91, 213–259; *1905*: 307–338; *1906*: 157–227, 439–480; *1910*: 355–417. Also published as a book by Teubner, Leipzig, 1912, repr. Chelsea, New York. 1952.

Hölder, Otto (1889). Über einen Mittelwerthssatz. *Nachr. Akad. Wiss. Göttingen Math.-Phys. Kl. 1889*: 38–47.

Hurwitz, A. (1903). Über die Fourierschen Konstanten integrierbarer Funktionen. *Math. Annalen 57*: 425–446.

Jordan, Camille (1881). Sur la série de Fourier. *Comptes Rendus Acad. Sci. Paris 92*: 228–230. Also in Jordan (1961–1964), *4*, pp. 393–395.

——— (1961–1964). *Oeuvres de Camille Jordan*. J. Dieudonné and R. Garnier, eds. 4 vols. Gauthier-Villars, Paris.

Lebesgue, Henri (1910). Sur l'intégration des fonctions discontinues. *Ann. Scient. Ecole Normale Sup.* (Ser. 3) *27*: 361–450. Also in Lebesgue (1972–1973) *2*, pp. 185–274.

——— (1972–1973). *Oeuvres scientifiques*. 5 vols. L'Enseignement Mathématique. Institut de Mathématique, Univ. Genève.

Levi, Beppo (1906). Sul principio de Dirichlet. *Rendiconti Circ. Mat. Palermo 22*: 293–360.

Minkowski, Hermann (1907). *Diophantische Approximationen*. Teubner, Leipzig.

——— (1973, posth.). *Briefe an David Hilbert*. Ed. Lily Rüdenberg and Hans Zassenhaus. Springer, Berlin.

Neugebauer, Otto (1935). *Mathematische Keilschrift-Texte*. 2 vols. Springer, Berlin.

——— (1957). *The Exact Sciences in Antiquity*. 2d ed. Brown Univ. Press. Repr. Dover, New York (1969).

von Neumann, John [Johann] (1940). On rings of operators, III. *Ann. Math. 41*: 94–161.

Nikodym, Otton Martin (1930). Sur une généralisation des mesures de M. J. Radon. *Fundamenta Math. 15*: 131–179.

*Parseval des Chênes, Marc-Antoine (1799). Mémoire sur les séries et sur l'intégration complète d'une équation aux différences partielles linéaires du second ordre, à coefficiens constans. *Mémoires présentés à l'Institut des Sciences, Lettres et Arts, par divers savans, et lus dans ses assemblées. Sciences math. et phys. (savans étrangers) 1* (1806): 638–648.

*——— (1801). Intégration générale et complète des équations de la propagation du son, l'air étant considéré avec ses trois dimensions. Ibid., pp. 379–398.

*Radon, Johann (1913). Theorie und Anwendungen der absolut additiven Mengenfunktionen. *Sitzungsber. Akad. Wiss. Wien Abt. IIa*: 1295–1438.

Riesz, Frédéric [Frigyes] (1906). Sur les ensembles de fonctions. *Comptes Rendus Acad. Sci. Paris 143*: 738–741.

——— (1907a). Sur les systèmes orthogonaux de fonctions. *Comptes Rendus Acad. Sci. Paris 144*: 615–619.

——— (1907b). Sur une espèce de géométrie analytique des fonctions sommables. *Comptes Rendus Acad. Sci. Paris 144*: 1409–1411.

——— (1909). Sur les suites de fonctions mesurables. *Comptes Rendus Acad. Sci. Paris 148*: 1303–1305.

——— (1910). Untersuchungen über Systeme integrierbarer Funktionen. *Math. Annalen 69*: 449–497.

——— (1934). Zur Theorie des Hilbertschen Raumes. *Acta Sci. Math. Szeged 7*: 34–38.

Rogers, Leonard James (1888). An extension of a certain theorem in inequalities. *Messenger of Math. 17*: 145–150.

Rüdenberg, Lily (1973). Erinnerungen an H. Minkowski. In Minkowski (1973), pp. 9–16.

Schmidt, Erhard (1907). Zur Theorie der linearen und nichtlinearen Integralgleichungen. I. Teil: Entwicklung willkürlicher Funktionen nach Systemen vorgeschriebener. *Math. Annalen 63*: 433–476.

——— (1907). Auflösung der allgemeinen linearen Integralgleichung. *Math. Annalen 64*: 161–174.

——— (1908). Über die Auflösung linearer Gleichungen mit unendlich vielen Unbekannten. *Rend. Circ. Mat. Palermo 25*: 53–77.

*Schwarz, Hermann Amandus (1885). Über ein die Flächen kleinsten Flächeninhalts betreffendes Problem der Variationsrechnung. *Acta Soc. Scient. Fenn. 15*: 315–362.

Schweder, Tore (1980). Scandinavian statistics, some early lines of development. *Scand. J. Statist 7*: 113–129.

Siegmund-Schultze, Reinhard (1982). Die Anfänge der Funktionalanalysis und ihr Platz im Umwälzungsprozess der Mathematik um 1900. *Arch. Hist. Exact Sci. 26*: 13–71.

Stanley, Richard P. (1971). Theory and application of plane partitions. *Studies in Applied Math. 50*: 167–188.

Steinhaus, Hugo (1961). Stefan Banach, 1892–1945. *Scripta Math. 26*: 93–100.

van der Waerden, Bartel Leendert (1939). Nachruf auf Otto Hölder. *Math. Ann. 116*: 157–165.

Watson, George Neville (1944). *A Treatise on the Theory of Bessel Functions*, 2d ed. Cambridge Univ. Press, repr. 1966; 1st ed., 1922.

Wiener, Norbert (1922). Limit in terms of continuous transformation. *Bull. Soc. Math. France 50*: 119–134.

6

Convex Sets and Duality of Normed Spaces

Functional analysis is concerned with infinite-dimensional linear spaces, such as Banach spaces and Hilbert spaces, which most often consist of functions or equivalence classes of functions. Each Banach space X has a dual space X' defined as the set of all continuous linear functions from X into the field \mathbb{R} or \mathbb{C}.

One of the main examples of duality is for L^p spaces. Let (X, S, μ) be a measure space. Let $1 < p < \infty$ and $1/p + 1/q = 1$. Then it turns out that L^p and L^q are dual to each other via the linear functional $f \mapsto \int fg \, d\mu$ for f in \mathcal{L}^p and g in \mathcal{L}^q. For $p = q = 2$, L^2 is a Hilbert space, where it was shown previously that any continuous linear form on a Hilbert space H is given by inner product with a fixed element of H (Theorem 5.5.1).

Other than linear subspaces, some of the most natural and frequently applied subsets of a vector space S are the convex subsets C, such that for any x and y in C, and $0 < t < 1$, we have $tx + (1-t)y \in C$. These sets are treated in §§6.2 and 6.6. A function for which the region above its graph is convex is called a convex function. §6.3 deals with convex functions. Convex sets and functions are among the main subjects of modern real analysis.

6.1. Lipschitz, Continuous, and Bounded Functionals

Let (S, d) and (T, e) be metric spaces. A function f from S into T is called *Lipschitzian*, or *Lipschitz*, iff for some $K < \infty$,

$$e(f(x), f(y)) \le K d(x, y) \qquad \text{for all } x, y \in S.$$

Then let $\|f\|_L$ denote the smallest such K (which exists). Most often $T = \mathbb{R}$ with usual metric.

Recall that a continuous real function on a closed subset of a metric (or normal) space can be extended to the whole space by the Tietze-Urysohn extension theorem (2.6.4). A continuous function on a non-closed subset cannot necessarily be extended: for example, the function $x \mapsto \sin(1/x)$

on the open interval $(0, 1)$ cannot be extended to be continuous at 0. For a Lipschitz function, extension from an arbitrary subset is possible:

6.1.1. Theorem *Let (S, d) be any metric space, E any subset of S, and f any real-valued Lipschitzian function on E. Then f can be extended to S without increasing $\|f\|_L$.*

Proof. Let $M := \|f\|_L$. Suppose we have an inclusion-chain of functions f_α from subsets E_α of S into \mathbb{R} with $\|f_\alpha\|_L \leq M$ for all α. Let g be the union of the f_α. Then g is a function with $\|g\|_L \leq M$. Thus by Zorn's Lemma it suffices to extend f to one additional point $x \in S \backslash E$.

A real number y is a possible value of $f(x)$ if and only if $|y - f(u)| \leq M d(u, x)$ for all $u \in E$, or equivalently the following two conditions both hold:

 (i) $-M d(u, x) \leq y - f(u)$ for all $u \in E$, and
 (ii) $y - f(v) \leq M d(x, v)$ for all $v \in E$.

Such a y exists if and only if

$$(*) \quad \sup_{u \in E}(f(u) - M d(u, x)) \leq \inf_{v \in E}(f(v) + M d(v, x)).$$

Now by assumption, for all $u, v \in E$,

$$f(u) - f(v) \leq M d(u, v) \leq M d(u, x) + M d(x, v), \quad \text{and}$$

$$f(u) - M d(u, x) \leq f(v) + M d(x, v).$$

This implies $(*)$, so the extension to x is possible. $\qquad \square$

For example, if E is not a closed subset of S, let $\{x_n\}$ be any sequence in E converging to a point x of $S \backslash E$. Then $\{f(x_n)\}$ is a Cauchy sequence, converging to some real number which can be defined as $f(x)$ since it does not depend on the particular sequence $\{x_n\}$, and the choice of $f(x)$ is unique. If x is not in the closure of E, then the choice of $f(x)$ may not be unique (see Problem 2 below).

Let $(X, |\cdot|)$ and $(Y, \|\cdot\|)$ be two normed linear spaces (as defined in §5.2). A linear function T from X into Y, often called an *operator*, is called *bounded* iff it is Lipschitzian. (*Note:* Usually a nonlinear function is called bounded if its range is bounded, but the range of a linear function T is a bounded set if and only if $T \equiv 0$.)

6.1.2. Theorem *Given a linear operator T from X into Y, where $(X, |\cdot|)$ and $(Y, \|\cdot\|)$ are normed linear spaces, the following are equivalent:*

(I) T is bounded.
(II) T is continuous.
(III) T is continuous at 0.
(IV) The range of T restricted to $\{x \in X: |x| \leq 1\}$ is bounded.

Proof. Every Lipschitzian function (linear or not) is continuous, so (I) implies (II), which implies (III). If (III) holds, take $\delta > 0$ such that $|x| \leq \delta$ implies $\|Tx\| \leq 1$. Then for any x in X with $0 < |x| \leq 1$, $\|Tx\| = \|(|x|/\delta)T(\delta x/|x|)\| \leq |x|/\delta \leq 1/\delta$, and $\|T0\| = 0$, so (IV) holds.

Now if (IV) holds, say $\|Tx\| \leq M$ whenever $|x| \leq 1$, then for all $x \neq u \in X$, $\|Tx - Tu\| = \||x - u|T((x-u)/|x-u|)\| \leq M|x-u|$, so (I) holds, completing the proof. $\qquad\square$

For example, let H be a Hilbert space $L^2(X, \mathcal{S}, \mu)$. Let M be a function in $\mathcal{L}^2(X \times X, \mu \times \mu)$. For each $f \in H$ let $T(f)(x) := \int M(x, y)f(y)\,d\mu(y)$. By the Tonelli-Fubini theorem and Cauchy-Bunyakovsky-Schwarz inequality, $T(f)(x)$ is well-defined for almost all x and

$$\int T(f)(x)^2\,d\mu(x) \leq \int \left(\int M(x,y)^2\,d\mu(y)\right)\int f(t)^2\,d\mu(t)\,d\mu(x)$$

$$\leq \int f^2\,d\mu \int M^2\,d(\mu \times \mu) < \infty,$$

so T is a bounded linear operator.

Let K be either \mathbb{R} or \mathbb{C}. A function f defined on a linear space X will be called *linear over K* if $f(cx + y) = cf(x) + f(y)$ for all c in K and all x and y in X. A function linear over \mathbb{R} will be called *real linear* and a function linear over \mathbb{C} will be called *complex linear*. For any normed linear space $(X, \|\cdot\|)$ over K, let X' denote the set of all continuous linear functions (often called *functionals* or *forms*) from X into K. For each $f \in X'$, let

$$\|f\|' := \sup\{|f(x)|/\|x\|: 0 \neq x \in X\}.$$

Then $\|f\|' < \infty$ by Theorem 6.1.2, and $(X', \|\cdot\|')$ will be called the *dual* or *dual space* of the normed linear space $(X, \|\cdot\|)$.

Now if (X, \mathcal{S}, μ) is a measure space, $1 < p < \infty$ and $q = p/(p-1)$, then the space $L^p(X, \mathcal{S}, \mu)$ of equivalence classes of measurable functions f with

$\|f\|_p^p := \int |f|^p \, d\mu < \infty$ and with the norm $\|\cdot\|_p$ is a Banach space (by Theorems 5.1.5 and 5.2.1) and for each $g \in \mathcal{L}^q$, $f \mapsto \int fg \, d\mu$ is a bounded linear form on L^p by the Rogers-Hölder inequality (5.1.2). In other words, this linear form belongs to $(L^p)'$. (It is shown in §6.4 that all elements of $(L^p)'$ arise in this way.)

6.1.3. Theorem *For any normed linear space* $(X, \|\cdot\|)$ *over* $K = \mathbb{R}$ *or* \mathbb{C}, $(X', \|\cdot\|')$ *is a Banach space.*

Proof. Clearly, X' is a linear space over K. For any $f \in X'$ and $c \in K$, $\|cf\|' = |c| \|f\|'$. If also $g \in X'$, then for all $x \in X$, $|(f+g)(x)| \le |f(x)| + |g(x)|$, so $\|f + g\|' \le \|f\|' + \|g\|'$. Thus $\|\cdot\|'$ is a seminorm. If $f \ne 0$ in X', then $f(x) \ne 0$ for some x in X, so $\|f\|' > 0$ and $\|\cdot\|'$ is a norm. Let $\{f_n\}$ be a Cauchy sequence for it. Then for each $x \in X$,

$$|f_n(x) - f_m(x)| \le \|f_n - f_m\|' \|x\|,$$

so $\{f_n(x)\}$ is a Cauchy sequence in K and converges to some $f(x)$. As a pointwise limit of linear functions, f is linear. Now $\{f_n\}$, being a Cauchy sequence, is bounded, so for some $M < \infty$, $\|f_n\|' \le M$ for all n. Then for all $x \in X$ with $x \ne 0$,

$$|f(x)| / \|x\| = \lim_{n \to \infty} |f_n(x)| / \|x\| \le M,$$

so $f \in X'$ and $\|f\|' \le M$. Likewise, for any m,

$$\|f_m - f\|' \le \limsup_{n \to \infty} \|f_m - f_n\|' \to 0 \quad \text{as} \quad m \to \infty,$$

so the Cauchy sequence converges to f, and X' is complete. □

The following extension theorem for bounded linear forms on a normed space is crucial to duality theory and is one of the three or four preeminent theorems in functional analysis:

6.1.4. Hahn-Banach Theorem Let $(X, \|\cdot\|)$ be any normed linear space over $K = \mathbb{R}$ or \mathbb{C}, E any linear subspace (with the same norm), and $f \in E'$. Then f can be extended to an element of X' with the same norm $\|f\|'$.

Proof. Let $M := \|f\|'$ on E. Then f is Lipschitzian on E with $\|f\|_L \le M$ by the proof of Theorem 6.1.2. First suppose $K = \mathbb{R}$. Then, for any $x \in X \setminus E$, f can be extended to a Lipschitzian function on $E \cup \{x\}$ without increasing $\|f\|_L$, by Theorem 6.1.1. Each element of the smallest linear subspace F

including $E \cup \{x\}$ can be written as $u + cx$ for some unique $u \in E$ and $c \in \mathbb{R}$. Set $f(u + cx) := f(u) + cf(x)$. Then f is linear, and $|f(u + cx)| \leq M\|u + cx\|$ holds if $c = 0$, so suppose $c \neq 0$. Then

$$|f(u + cx)| = \left| -cf\left(-\frac{u}{c} - x \right) \right| \leq |c|M \left\| -\frac{u}{c} - x \right\| = M\|u + cx\|.$$

So $\|f\|' \leq M$ on F. For any inclusion-chain of extensions G of f to linear functions on linear subspaces with $\|G\|' \leq M$, the union is an extension with the same properties. Thus by Zorn's lemma, as in the proof of Theorem 6.1.1, the theorem is proved for $K = \mathbb{R}$.

If $K = \mathbb{C}$, let g be the real part of f. Then g is a real linear form, and $\|g\|' \leq \|f\|'$ on E. Let h be the imaginary part of f, so that h is a real-valued, real linear form on E with $f(u) = g(u) + ih(u)$ for all $u \in E$. Then $f(iu) = g(iu) + ih(iu) = if(u) = ig(u) - h(u)$, so $h(u) = -g(iu)$ and $f(u) = g(u) - ig(iu)$ for all $u \in E$. By the real case, extend g to a real linear form on all of X with $\|g\|' \leq M$, and define $f(x) := g(x) - ig(ix)$ for all $x \in X$. Then f is linear over \mathbb{R} since g is, and for any x,

$$f(ix) = g(ix) - ig(-x) = ig(x) + g(ix) = if(x),$$

so f is complex linear. For any $x \in X$ with $f(x) \neq 0$, let $\gamma := (g(x) + ig(ix))/|f(x)|$. Then $|\gamma| = 1$ and $f(\gamma x) \geq 0$, so $|f(x)| = |f(\gamma x)/\gamma| = |f(\gamma x)| = f(\gamma x) = g(\gamma x) \leq M\|\gamma x\| = M\|x\|$, so $\|f\|' \leq M$, finishing the proof. \square

For any normed linear space $(X, \|\cdot\|)$ there is a natural map I'' of X into $X'' := (X')'$, defined by $I''(x)(f) := f(x)$ for each $x \in X$ and $f \in X'$. Clearly I'' is linear. Let $\|\cdot\|''$ be the norm on X'' as dual of X'. It follows from the definition of $\|\cdot\|'$ that for all $x \in X$, $\|I''x\|'' \leq \|x\|$.

6.1.5. Corollary *For any normed space $(X, \|\cdot\|)$ and any $x \in X$ with $x \neq 0$, there is an $f \in X'$ with $\|f\|' = 1$ and $f(x) = \|x\|$. Thus $\|I''x\|'' = \|x\|$ for all $x \in X$.*

Proof. Given $x \neq 0$, let $f(cx) := c\|x\|$ for all $c \in K$ to define f on the one-dimensional subspace J spanned by x. Then f has the desired properties on J. By the Hahn-Banach theorem it can be extended to all of X, keeping $\|f\|' = 1$. Using this f in the definition of $\|\cdot\|''$ gives $\|I''x\|'' \geq \|x\|$, so $\|I''x\|'' = \|x\|$.
\square

Definition. A normed linear space $(X, \|\cdot\|)$ is called *reflexive* iff I'' takes X onto X''.

In any case, I'' is a linear isometry (preserving the metrics defined by norms) and is thus one-to-one. It follows that every finite-dimensional normed space is reflexive. Since dual spaces such as X'' are complete (by Theorem 6.1.3), a reflexive normed space must be a Banach space. Hilbert spaces are all reflexive, as follows from Theorem 5.5.1.

Here is an example of a nonreflexive space. Recall that ℓ^1 is the set of all sequences $\{x_n\}$ of real numbers such that the norm

$$\|\{x_n\}\|_1 := \sum_n |x_n| < \infty.$$

Thus ℓ^1 is L^1 of counting measure on the set of positive integers. Let y be any element of the dual space $(\ell^1)'$. Let e_n be the sequence with 1 in the nth place and 0 elsewhere. Set $y_n := y(e_n)$. As finite linear combinations of the e_n are dense in ℓ^1, the map from y to $\{y_n\}$ is one-to-one. Let ℓ^∞ be the set of all bounded sequences $\{y_n\}$ of real numbers, with supremum norm $\|\{y_n\}\|_\infty := \sup_n |y_n|$. Then the dual of ℓ^1 can be identified with ℓ^∞, and $\|y\|_1' = \|\{y_n\}\|_\infty$.

Let c denote the subspace of ℓ^∞ consisting of all convergent sequences. On c, a linear form f is defined by $f(y) := \lim_{n\to\infty} y_n$. Then $\|f\|_\infty' \le 1$ and by the Hahn-Banach theorem, f can be extended to an element of $(\ell^\infty)'$. Now, f is not in the range of I'' on ℓ^1, so ℓ^1 is not reflexive.

Problems

1. Which of the following functions are Lipschitzian on the indicated subsets of \mathbb{R}?
 (a) $f(x) = x^2$ on $[0, 1]$.
 (b) $f(x) = x^2$ on $[1, \infty)$.
 (c) $f(x) = x^{1/2}$ on $[0, 1]$.
 (d) $f(x) = x^{1/2}$ on $[1, \infty)$.

2. Let f be a Lipschitzian function on X and E a subset of X where $\|f\|_L$ on X is the same as for its restriction to E. Let $x \in X \backslash E$. Does f on E uniquely determine f on X?
 (a) Prove that the answer is always yes if x is in the closure \overline{E} of E.
 (b) Give an example where the answer is yes for some f even though x is not in \overline{E}.
 (c) Give an example of X, E, f, and x where the answer is no.

3. Prove that a Banach space X is reflexive if and only if its dual X' is reflexive. *Note:* "Only if" is easier.

4. Let $(X, \|\cdot\|)$ be a normed linear space, E a closed linear subspace, and $x \in X \backslash E$. Prove that for some $f \in X'$, $f \equiv 0$ on E and $f(x) = 1$.

5. Show that the extension theorem for Lipschitz functions (6.1.1) does not hold for complex-valued functions. *Hint:* Let $S = \mathbb{R}^2$ with the norm $\|(x, y)\| := \max(|x|, |y|)$. Let $E := \{(0, 0), (0, 2), (2, 0)\}$, with $f(0, 0) = 0$, $f(2, 0) = 2$, and $f(0, 2) = 1 + 3^{1/2}i$. How to extend f to $(1, 1)$?

6. Let (S, d) be a metric space, $E \subset S$, and f a complex-valued Lipschitz function defined on E, with $|f(x) - f(y)| \le Kd(x, y)$ for all x and y in E. Show that f can be extended to all of S with $|f(u) - f(v)| \le 2^{1/2}Kd(u, v)$ for all u and v in S.

7. Let c_0 denote the space of all sequences $\{x_n\}$ of real numbers which converge to 0 as $n \to \infty$, with the norm $\|\{x_n\}\|_s := \sup_n |x_n|$. Show that $(c_0, \|\cdot\|_s)$ is a Banach space and that its dual space is isometric to the space ℓ^1 of all summable sequences (L^1 of the integers with counting measure). Show that c_0 is not reflexive.

8. Let X be a Banach space with dual space X' and T a subset of X. Let $T^\perp := \{f \in X': f(t) = 0 \text{ for all } t \in T\}$. Likewise for $S \subset X'$, let $S^\perp := \{x \in X: f(x) = 0 \text{ for all } f \in S\}$. Show that for any $T \subset X$,
 (a) T^\perp is a closed linear subspace of X'.
 (b) $T \subset (T^\perp)^\perp$.
 (c) $T = (T^\perp)^\perp$ if T is a closed linear subspace.
 Hint: See Problem 4.

9. Let $(X, \|\cdot\|)$ be a Banach space with dual space X'. Then the *weak* topology* on X' is the smallest topology such that for each $x \in X$, the function $I''(x)$ defined by $I''(x)(f) := f(x)$ for $f \in X'$ is continuous on X'. Show that $E_1' := \{y \in X': \|y\|' \le 1\}$ is compact for the weak* topology (Alaoglu's theorem). *Hint:* Use Tychonoff's theorem and show that E_1' is closed for the product topology in the set of all real functions on X.

10. Let (X, \mathcal{S}, μ) be a measure space and $(S, \|\cdot\|)$ a separable Banach space. A function f from X into S will be called *simple* if it takes only finitely many values, each on a set in \mathcal{S}. Let g be any measurable function from X into S such that $\int \|g\| \, d\mu < \infty$. Show that there exist simple functions f_n with $\int \|f_n - g\| \, d\mu \to 0$ as $n \to \infty$. If f is simple with $f(x) = \sum_{1 \le i \le n} 1_{A(i)}(x)s_i$ for $s_i \in S$ and $A(i) \in \mathcal{S}$, let $\int f \, d\mu = \sum_{1 \le i \le n} \mu(A(i))s_i \in S$. Show that $\int f \, d\mu$ is well-defined and if $\int \|f_n - g\| \, d\mu \to 0$, then $\int f_n \, d\mu$ converge to an element of S depending only on g, which will be called $\int g \, d\mu$ (Bochner integral).

11. (Continuation.) If g is a function from X into S, then t is called the *Pettis integral* of g if for all $u \in S'$, $\int u(g) \, d\mu = u(t)$. Show that the Bochner

integral of g, when it exists, as defined in Problem 10, is also the Pettis integral.

12. Prove that every Hilbert space H over the complex field \mathbb{C} is reflexive. *Hints:* Let $C(h)(f) := (f, h)$ for $f, h \in H$. Then C takes H onto H' by Theorem 5.5.1. Show using (5.3.3) that $\|C(h)\|' = \|h\|$ for all $h \in H$. Let $(C(f), C(h))' := (h, f)$ for all $f, h \in H$. Show that this defines an inner product $(\cdot, \cdot)'$ on $H' \times H'$ such that $[(\psi, \psi)']^{1/2} = \|\psi\|'$ for all $\psi \in H'$; and H' is then a Hilbert space. Apply Theorem 5.5.1 to H' and $(H')'$ to conclude that I'' is onto $(H')'$.

6.2. Convex Sets and Their Separation

Let V be a real vector space. A set C in X is called *convex* iff for any $x, y \in C$ and $\lambda \in [0, 1]$, we have $\lambda x + (1 - \lambda)y \in C$. Then $\lambda x + (1 - \lambda)y$ is called a *convex combination* of x and y. Also, C is called a *cone* if for all $x \in C$, we have $tx \in C$ for all $t \geq 0$. (See Figure 6.2.) For any seminorm $\|\cdot\|$ on a vector space X, $r \geq 0$ and $y \in X$, the sets $\{x : \|x\| \leq r\}$ and $\{x : \|x - y\| \leq r\}$ are convex.

A set A in V is called *radial at* $x \in A$ iff for every $y \in V$, there is a $\delta > 0$ such that $x + ty \in A$ whenever $|t| < \delta$. Thus, on every line L through x, $A \cap L$ includes an open interval in L, for the usual topology of a line, containing x. The set A will be called *radial* iff it is radial at each of its points. Most often, facts about radial sets will be applied to open sets for some topology, but it will also be useful to develop some of the facts without requiring a topology.

For example, in polar coordinates (r, θ) in \mathbb{R}^2, let A be the union of all the lines $\{(r, \theta) : r \geq 0\}$ for θ/π irrational, and of all the line segments L_θ

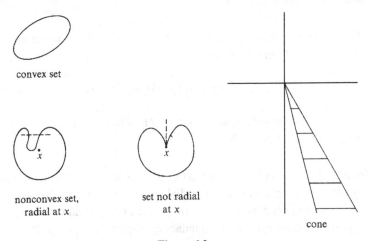

convex set

nonconvex set,
radial at x

set not radial
at x

cone

Figure 6.2

where if $0 \leq \theta < 2\pi$ and $\theta/\pi = m/n$ is rational and in lowest terms, then $L_\theta = \{(r, \theta) : 0 \leq r < 1/n\}$. This set A is radial at 0 and nowhere else.

The *radial kernel* of a set A is defined as the largest radial set A_o included in A. Clearly the union of radial sets is radial, so A_o is well-defined, although it may be empty. Note that A_o is not necessarily the set A_r of points at which A is radial; in the example just given, $A_r = \{0\}$ but $\{0\}$ is not a radial set. For convex sets the situation is better:

6.2.1. Lemma　　*If A is convex, then $A_r = A_o$.*

Proof. Clearly $A_o \subset A_r$. Conversely, let $x \in A_r$, and take any w and z in V. Then for some $\delta > 0$, $x + sw \in A$ whenever $|s| < \delta$, and $x + tz \in A$ whenever $|t| < \delta$. Then by convexity $(x + sw + x + tz)/2 \in A$, so $x + aw + bz \in A$ whenever $|a| < \delta/2$ and $|b| < \delta/2$. Thus since z is arbitrary, A is radial at $x + aw$ whenever $|a| < \delta/2$, so $x + aw \in A_r$. Then since w is arbitrary, A_r is radial at x, so A_r is a radial set, $A_r \subset A_o$, and $A_r = A_o$.　　□

In \mathbb{R}^k, clearly the interior of any set is included in its radial kernel. Often the radial kernel will equal the interior. An exception is the set A in \mathbb{R}^2 whose complement consists of all points (x, y) with $x > 0$ and $x^2 \leq y \leq 2x^2$. Then $A = A_o$ but A is not open, as it includes no neighborhood of $(0, 0)$. If A is convex, then the next lemma shows A_o is convex and more, taking any y in A.

6.2.2. Lemma　　*If A is convex and radial at x, then $(1 - t)x + ty \in A_o$ whenever $y \in A$ and $0 \leq t < 1$.*

Proof. For each $z \in V$, take $\varepsilon > 0$ such that $x + uz \in A$ whenever $|u| < \varepsilon$. Let $v := (1 - t)u$. Then by convexity,

$$(1 - t)(x + uz) + ty = ((1 - t)x + ty) + vz \in A.$$

Here, v may be any number with $|v| < (1 - t)\varepsilon$. Thus $(1 - t)x + ty$ is in A_r, which equals A_o by Lemma 6.2.1.　　□

The next theorem is the main fact in this section. It implies that two disjoint convex sets, at least one of which is radial somewhere, can be separated into two half-spaces defined by a linear form f, $\{x: f(x) \geq c\}$ and $\{x: f(x) \leq c\}$, which are disjoint except for the boundary hyperplane $\{x: f(x) = c\}$.

6.2.3. Separation Theorem *Let A and B be non-empty convex subsets of a real vector space V such that A is radial at some point x and $A_o \cap B = \emptyset$. Then there is a real linear functional f on V, not identically 0, such that*

$$\inf_{b \in B} f(b) \geq \sup_{a \in A} f(a).$$

Remarks. By Lemma 6.2.2, A_o is non-empty. Since $f \neq 0$, there is some z with $f(z) \neq 0$. Then $f(x + tz) = f(x) + tf(z)$ as t varies, so f is not constant on A. Here f is said to *separate* A and B. For example, let $A = \{(x, y) \in \mathbb{R}^2: x^2 + y^2 \leq 1\}$ and $B = \{(1, y): y \in \mathbb{R}.$ Then $A_o = \{(x, y): x^2 + y^2 < 1\}$, which is disjoint from B, and a separating linear functional f is given by $f(x, y) = cx$ for any $c > 0$.

Proof. Let $C := \{t(b - a): t \geq 0, b \in B, a \in A\}$. Then C is a convex cone. By Lemma 6.2.1, $x \in A_o$. Take $y \in B$. If $x - y \in C$, take $u \in A$, $v \in B$, and $t \geq 0$ such that $x - y = t(v - u)$, so $x + tu = y + tv$, and $(x + tu)/(1 + t) = (y + tv)/(1 + t)$. By Lemma 6.2.2, $(x + tu)/(1 + t) \in A_o$, but $(y + tv)/(1 + t) \in B$, a contradiction. So $x - y \notin C$. Thus $C \neq V$. Let $z := y - x$, so $-z \notin C$. Two lemmas, 6.2.4 and 6.2.5, will form part of the proof of Theorem 6.2.3.

6.2.4. Lemma *For any vector space V, convex cone $C \subset V$ and $p \in V \backslash C$, there is a maximal convex cone M including C and not containing p. For each $v \in V$, either $v \in M$ or $-v \in M$ (or both).*

Example. Let $V = \mathbb{R}^2$, $C = \{(\xi, \eta): |\eta| \leq \xi\}$ and $p = (-1, 0)$. Then M will be a half-plane $\{(\xi, \eta): \xi \geq c\eta\}$ where $|c| \leq 1$.

Proof. If C_α are convex cones, linearly ordered by inclusion, which do not contain p, then their union is also a convex cone not containing p. So, by Zorn's Lemma, there is a maximal convex cone M, including C and not containing p. For any convex cone M and $v \in V$, $\{tv + u: u \in M, t \geq 0\}$ is the smallest convex cone including M and containing v. Thus if the latter conclusion in Lemma 6.2.4 fails, $p = u + tv$ for some $u \in M$ and $t > 0$. Likewise, $p = w - sv$ for some $w \in M$ and $s > 0$. Then

$$(s + t)p = s(u + tv) + t(w - sv) = su + tw = (s + t)\left(\frac{su}{s + t} + \frac{tw}{s + t}\right) \in M,$$

since M is a convex cone. But then $p = (s + t)p/(s + t) \in M$ because $1/(s + t) > 0$, a contradiction, proving Lemma 6.2.4. \square

Now for C and z as defined at the beginning of the proof of Theorem 6.2.3 ($z = y - x$), the set $\{y - u: u \in A\}$ is radial at z, so $z \in C_o$ by Lemma 6.2.1. Let $p := -z$. By Lemma 6.2.4, take a maximal convex cone $M \supset C$ with $p \notin M$. Then $z \in M_o$.

A linear subspace E of the vector space V is said to have *codimension k* iff there is a k-dimensional linear subspace F such that $V = E + F := \{\xi + \eta: \xi \in E, \eta \in F\}$, and k is the smallest dimension for which this is true.

Now in the example for Lemma 6.2.4, $M \backslash M_o$ is the line $\xi = c\eta$, which is a linear subspace of codimension 1. This holds more generally:

6.2.5. Lemma *If $z \in M_o$ where M is a maximal convex cone not containing $-z$, then $M \backslash M_o$ is a linear subspace of V of codimension 1.*

Proof. For any $u \notin M$, $au + m = -z$ for some $a > 0$ and $m \in M$. Then $-au/2 = (z + m)/2$, so M is radial at $-au/2$ by Lemma 6.2.2 with $t = 1/2$. Thus M is radial at $-u$ since M is a cone. So $-u \in M_o$ by Lemma 6.2.1. If $0 \in M_o$, then $M = V$, contradicting $-z \notin M$. So $0 \notin M_o$. For any $v \in M_o$, we have $-v \notin M$, since otherwise $0 = (v + (-v))/2 \in M_o$ by Lemma 6.2.2. Thus $M_o = \{-u: u \notin M\}$.

Now if $v \in M \backslash M_o$, then also $-v \in M \backslash M_o$. So $M \cap - M = M \backslash M_o$. Since M is a convex cone, $M \cap -M$ is a linear subspace. To show that it has codimension 1, for the 1-dimensional subspace $\{cz: c \in \mathbb{R}\}$, take any $g \in V$. If $g \in -M_o$, then since M_o and $-M_o$ are radial, $S := \{t: tg + (1 - t)z \in M_o\}$ and $T := \{t: tg + (1 - t)z \in -M_o\}$ are non-empty, open sets of real numbers, with $0 \in S$ and $1 \in T$. Since S and T are disjoint, their union cannot include all of $[0, 1]$, a connected set (the supremum of $S \cap [0, 1]$ cannot be in S or T). Thus $tg + (1 - t)z \in M \cap -M$ for some t, $0 < t < 1$. Then $g = -(1 - t)z/t + w$ for some $w \in M \cap -M$, as desired. If, on the other hand, $g \notin -M_o$, then $g \in M$, and either $g \in M \cap -M = M \backslash M_o$ or $g \in M_o$, so $-g \in -M_o$ and $-g$, thus g, is in the linear span of $M \cap -M$ and $\{z\}$. So this holds for all g, proving that $M \cap -M$ has codimension 1 and thus Lemma 6.2.5. □

Now define a linear form f on V with $f(z) = 1$ and $f = 0$ on $M \cap -M$. Then $f = 0$ exactly on $M \cap -M$, and $f > 0$ on M_o while $f < 0$ on $-M_o$. Then $f(b) \geq f(a)$ for all $b \in B$ and $a \in A$, proving Theorem 6.2.3. □

Remark. Under the other hypotheses of the theorem, the condition that A_o and B be disjoint is also necessary for the existence of a separating f, since if $x \in A_o \cap B$ and $f(v) > 0$, then $x + cv \in A$ for c small enough, and $f(x + cv) > f(x)$.

Given a set S in a topological space X with closure \overline{S} and interior int S, recall that the *boundary* of S is defined as $\partial S := \overline{S} \backslash \text{int } S$. In a normed linear space X, a closed *hyperplane* is defined as a set $f^{-1}\{c\}$ for some continuous linear form f, not identically 0, and constant c. For a set $S \subset X$, a *support hyperplane* at a point $x \in \partial S$ is a closed hyperplane $f^{-1}\{c\}$ containing x such that either $f \geq c$ on S or $f \leq c$ on S. For example, in \mathbb{R}^2 the set S where $y \leq x^2$ does not have a support hyperplane, which in this case would be a line, since every line through a point on the boundary has points of S on both sides of it. On the other hand, tangent lines to the boundary are support hyperplanes for the complement of S.

Note that on \mathbb{R}^k, all linear forms are continuous, so any hyperplane is closed. Now some facts about convex sets in \mathbb{R}^k will be developed.

6.2.6. Theorem *For any convex set C in \mathbb{R}^k, either C has non-empty interior or C is included in some hyperplane.*

Proof. Take any $x \in C$ and let W be the linear span of (smallest linear subspace containing) all $y - x$ for $y \in C$. If W is not all of \mathbb{R}^k, then there is a linear f on \mathbb{R}^k which is not identically 0 but which is 0 on W. Then C is included in the hyperplane $f^{-1}\{f(x)\}$.

Conversely, if $W = \mathbb{R}^k$, let $y_j - x$ be linearly independent for $j = 1, \ldots, k$, with $y_j \in C$. The set of all convex combinations $p_0 x + \sum_{1 \leq j \leq k} p_j y_j$, where $p_j \geq 0$ and $\sum_{0 \leq j \leq k} p_j = 1$, is called a *simplex*. (For example, if $k = 2$, it is a triangle.) Then C includes this simplex, which has non-empty interior, consisting of those points with $p_j > 0$ for all j. $\qquad\square$

If a convex set is an island, then from each point on the coast, there is at least a $180°$ unobstructed view of the ocean. This fact extends to general convex sets as follows.

6.2.7. Theorem *For any convex set C in \mathbb{R}^k and any $x \in \partial C$, C has at least one support hyperplane $f^{-1}\{c\}$ at x. If y is in the interior of C, and $f \leq c$ on C, then $f(y) < c$.*

Remark. By definition of support hyperplane, either $f \leq c$ on C or $f \geq c$ on C. If necessary, replacing f by $-f$ and c by $-c$, we can assume that $f \leq c$ on C.

Proof. Let $x \in \partial C$. If C has empty interior, then by Theorem 6.2.6, it is included in some hyperplane, which is a support hyperplane. So suppose C has a non-empty interior containing a point y, so C is radial at y. Let

$g(t) := x + t(x - y)$. If $g(t) \in C$ for some $t > 0$, note that

$$x = \frac{ty}{1+t} + \frac{g(t)}{1+t}, \qquad \text{a convex combination of } y \text{ and } g(t).$$

If y is replaced by each point in a neighborhood of y in C, then since $t > 0$, the same convex combinations give all points in a neighborhood of x, included in C, contradicting $x \in \partial C$. Thus if we set $B := \{g(t): t > 0\}$, then $B \cap C = \emptyset$ and *a fortiori* $B \cap C_o = \emptyset$.

So apply the separation theorem (6.2.3) to the set C (with $C_o \neq \emptyset$) and B, obtaining a nonzero linear form f with $\inf_B f \geq \sup_C f$. Now as $x \in \overline{C}$, and f is uniformly continuous, $f(x) \leq \sup_C f$, and $x \in \overline{B}$ implies $f(x) \geq \inf_B f$. Thus $\inf_B f = f(x) = \sup_C f$. Let $c := f(x)$ and let H be the hyperplane $\{u \in \mathbb{R}^k: f(u) = f(x)\}$. Then H is a support hyperplane to C at x.

If $f(y) \geq c$, then since f is not constant, it would have values larger than c in every neighborhood of y, and thus on C, a contradiction. So $f(y) < c$. $\quad\square$

6.2.8. Proposition *For f and g as in the above proof, $f(g(t)) > f(x)$ for all $t > 0$.*

Proof. Since $f(y) < f(x)$, it follows that $f(g(t))$ is a strictly increasing function of t, which implies the proposition. $\quad\square$

Now a closed *half-space* in \mathbb{R}^k is defined as a set $\{x: f(x) \leq c\}$ where f is a non-zero linear form and $c \in \mathbb{R}$. Note that $\{x: f(x) \geq c\} = \{x: -f(x) \leq -c\}$, which is also a closed half-space as defined. It's easy to see that any half-space is convex, so any intersection of half-spaces is convex. Conversely, here is a characterization of closed, convex sets in \mathbb{R}^k.

6.2.9. Theorem *A set in \mathbb{R}^k is closed and convex if and only if it is an intersection of closed half-spaces.*

Examples. A convex polygon in \mathbb{R}^2 with k sides is an intersection of k half-spaces. The disk $\{(x, y): x^2 + y^2 \leq 1\}$ is the intersection of all the half-spaces $\{(x, y): sx + ty \leq 1\}$ where $s^2 + t^2 = 1$.

Proof. Clearly, an intersection of closed half-spaces is closed and convex. Conversely, let C be closed and convex. If $k = 1$, then C is a closed interval (an intersection of two half-spaces) or half-line or the whole line. The whole line is the intersection of the empty set of half-spaces (by definition), so the

conclusion holds for $k = 1$. Let us proceed by induction on k. If C has no interior, then by Theorem 6.2.6 it is included in a hyperplane $H = f^{-1}\{c\}$ for some non-zero linear form f and $c \in \mathbb{R}$. Then

$$H = \{x: f(x) \geq c\} \cap \{x: f(x) \leq c\},$$

an intersection of two half-spaces. Also,

$$H = u + V := \{u + v: v \in V\}$$

for some $(k - 1)$-dimensional linear subspace V and $u \in \mathbb{R}^k$. By induction assumption, $C - u$ is an intersection of half-spaces $\{v \in V: f_\alpha(v) \leq c_\alpha\}$. The linear forms f_α on V can be assumed to be defined on \mathbb{R}^k. Then

$$C = \{x \in H: f_\alpha(x) \leq c_\alpha + f_\alpha(u) \text{ for all } \alpha\},$$

so C is an intersection of closed half-spaces.

Thus, suppose C has an interior, so C_o is non-empty, containing a point y, say. For each point z not in C, the line segment L joining y to z must intersect ∂C at some point x, since the interior and complement of C are open sets each having non-empty intersection with L (as in the proof of Lemma 6.2.5). Let $f^{-1}\{c\}$ be a support hyperplane to C at x, by Theorem 6.2.7, where we can take C to be included in $\{u: f(u) \leq c\}$. Then by Proposition 6.2.8, since $z = g(t)$ for some $t > 0$, $f(z) > f(x) = c$, and $z \notin \{u: f(u) \leq c\}$. So the intersection of all such half-spaces $\{u: f(u) \leq c\}$ is exactly C. □

A union of two adjoining but disjoint open intervals in \mathbb{R}, for example $(0, 1) \cup (1, 2)$, is not convex and is smaller than the interior of its closure. For convex sets the latter cannot happen:

***6.2.10. Proposition** *Any convex open set C in \mathbb{R}^k is the interior of its closure \overline{C}.*

Proof. Every open set is included in the interior of its closure. Suppose $x \in (\text{int } \overline{C}) \backslash C$. Apply the separation theorem (6.2.3) to C and $\{x\}$, so that there is a non-zero linear form f such that $f(x) \geq \sup_{y \in C} f(y)$. Then since int \overline{C} is open, there is some u in int \overline{C} with $f(u) > f(x)$. Since f is continuous, there are some $v_n \in C$ with $f(v_n) \to f(u)$, so $f(v_n) > f(x)$, a contradiction. □

Next, here is a fact intermediate between the separation theorem (6.2.3) and the Hahn-Banach theorem (6.1.4), as is shown in Problem 6 below.

6.2.11. Theorem *Let X be a linear space and E a linear subspace. Let U be a convex subset of X which is radial at some point in E. Let h be a non-zero real linear form on E which is bounded above on $U \cap E$. Then h can be extended to a real linear form g on X with $\sup_{x \in U} g(x) = \sup_{y \in U \cap E} h(y)$.*

Proof. Let $t := \sup_{y \in E \cap U} h(y)$ and $B := \{y \in E : h(y) > t\}$. Then U and B are convex and disjoint. The separation theorem (6.2.3) applies with $A = U$ to give a non-zero linear form f on X with $\sup_{x \in U} f(x) \le \inf_{v \in B} f(v)$.

Let U be radial at $x_o \in U \cap E$. Then $f(x_o) < \sup_{x \in U} f(x)$, so $f(v) > f(x_o)$ for any $v \in B$ and f is not constant (zero) on E. Let $F := \{x \in E: f(x) = f(x_o)\}$. Suppose that h takes two different values at points of F. Then h takes all real values on the line joining these points, so the line intersects B. But $f \equiv f(x_o)$ on this line, giving a contradiction. Thus h is constant on F, say $h \equiv c$ on F. The smallest linear subspace including F and any point w of $E \setminus F$ is E. If $c = 0$, then since h is not constant on E, we have $0 \in F$ and $f(x_o) = 0$. Taking $w \in E \setminus F$, $h(w) \ne 0 \ne f(w)$ and $f(x) = f(w)h(x)/h(w)$ for $x = w$ and all $x \in F$, thus for all $x \in E$. On the other hand, if $c \ne 0$, then 0 is not in F, so $f(x_o) \ne 0$, and $f(x) = f(x_o)h(x)/c$ for all $x \in F$ and for $x = 0$, hence for all $x \in E$. So in either case, $f \equiv \alpha h$ on E for some $\alpha \ne 0$. Since both f and h are smaller at x_o than on B, $\alpha > 0$. Let $g := f/\alpha$. Then g is a linear form extending h to X, with

$$\sup_{x \in U} g(x) \le \inf_{v \in B} f(v)/\alpha = \inf_{v \in B} h(v) = t. \qquad \square$$

Problems

1. Give an example of a set A in \mathbb{R}^2 which is radial at every point of the interval $\{(x, 0): |x| < 1\}$ but such that this interval is not included in the interior of A.

2. Show that the ellipsoid $x^2/a^2 + y^2/b^2 + z^2/c^2 \le 1$ is convex. Find a support plane at each point of its boundary and show that it is unique.

3. In \mathbb{R}^3, which planes through the origin are support planes of the unit cube $[0, 1]^3$? Find all the support planes of the cube.

4. The Banach space c_0 of all sequences of real numbers converging to 0 has the norm $\|\{x_n\}\| := \sup_n |x_n|$. Show that each support hyperplane H at a point x of the boundary ∂B of the unit ball $B := \{y: \|y\| \le 1\}$ also contains other points of ∂B. *Hint:* See Problem 7 of §6.1.

5. Give an example of a Banach space X, a closed convex set C in X,

and a point $u \in X$ which does not have a unique nearest point in C. *Hint:* Let $X = \mathbb{R}^2$ with the norm $\|(x, y)\| := \max(|x|, |y|)$.

6. Let $(X, \|\cdot\|)$ be a real normed space, E a linear subspace, and $h \in E'$. Give a proof that h can be extended to be a member of X' (the Hahn-Banach theorem, 6.1.4) based on Theorem 6.2.11. *Hint:* Let $U := \{x \in X: \|x\| < 1\}$. (Because of such relationships, separation theorems for convex sets are sometimes called "geometric forms" of the Hahn-Banach theorem.)

7. Show that in any finite dimensional Banach space (\mathbb{R}^k with any norm), for any closed, convex set C and any point x not in C, there is at least one nearest point y in C; in other words, $\|y - x\| = \inf_{z \in C} \|z - x\|$.

8. Give an example of a closed, convex set C in a Banach space S and an $x \in S$ which has no nearest point in C. *Hint:* Let S be the space ℓ^1 of absolutely summable sequences with norm $\|\{x_n\}\|_1 := \sum_n |x_n|$. Let $C := \{\{t_j(1 + j)/j\}_{j \geq 1}: t_j \geq 0, \sum_j t_j = 1\}$ and $x = 0$.

9. ("Geometry of numbers"). A set C in a vector space V is called *symmetric* iff $-x \in C$ whenever $x \in C$. Suppose C is a convex, symmetric set in \mathbb{R}^k with Lebesgue volume $\lambda(C) > 2^k$. Show that C contains at least one point $z = (z_1, \ldots, z_k) \in \mathbb{Z}^k$, that is, z has integer coordinates $z_i \in \mathbb{Z}$ for all i, with $z \neq 0$. *Hint:* Write each $x \in \mathbb{R}^k$ as $x = y + z$ where $z/2 \in \mathbb{Z}^k$ and $-1 < y_i \leq 1$ for all i. Show that there must be x and $x' \neq x$ in C with the same y, and consider $(x - x')/2$.

10. An open half-space is a set of the form $\{x: f(x) > c\}$ where f is a continuous linear function. Show that in \mathbb{R}^k, any open convex set is an intersection of open half-spaces.

6.3. Convex Functions

Let V be a real vector space and C a convex set in V. A real-valued function f on C is called *convex* iff

$$f(\lambda x + (1 - \lambda)y) \leq \lambda f(x) + (1 - \lambda) f(y) \tag{6.3.1}$$

for every x and y in C and $0 \leq \lambda \geq 1$. Here the line segment of all ordered pairs $\langle \lambda x + (1 - \lambda)y, \lambda f(x) + (1 - \lambda) f(y) \rangle$, $0 \leq \lambda \leq 1$, is a *chord* joining the two points $\langle x, f(x) \rangle$ and $\langle y, f(y) \rangle$ on the graph of f. Thus a convex function f is one for which the chords are all on or above the graph of f. For example, $f(x) = x^2$ is convex on \mathbb{R} and $g(x) = -x^2$ is not convex.

In \mathbb{R}, a convex set is an interval (which may be closed or open, bounded or unbounded at either end). Convex functions have "increasing difference-quotients," as follows:

$$t \qquad\qquad u \qquad\qquad v$$

Figure 6.3A

6.3.2. Proposition *Let f be a convex function on an interval J in \mathbb{R}, and $t < u < v$ in J. Then*

$$\frac{f(u) - f(t)}{u - t} \leq \frac{f(v) - f(t)}{v - t} \leq \frac{f(v) - f(u)}{v - u}.$$

Proof. (See Figure 6.3A.) Let $x := t$ and $y := v$. Then we find that $u = \lambda x + (1 - \lambda)y$ for $\lambda = (v - u)/(v - t)$, where $0 < \lambda < 1$. Then applying (6.3.1) gives

$$f(u) \leq \frac{v - u}{v - t} f(t) + \frac{u - t}{v - t} f(v).$$

The relations between the slopes as stated are clear in the figure. More analytically, the latter inequality can be written

$$(v - t)f(u) \leq (v - u)f(t) + (u - t)f(v), \qquad \text{which implies}$$

$$(v - t)(f(u) - f(t)) \leq (u - t)(f(v) - f(t)),$$

giving the left inequality in Proposition 6.3.2. The right inequality says

$$(v - u)(f(v) - f(t)) \leq (v - t)(f(v) - f(u)),$$

which on canceling $vf(v)$ terms also follows. □

Convex functions are not necessarily differentiable everywhere. For example, $f(x) := |x|$ is not differentiable at 0, although it has left and right derivatives there. Except possibly at endpoints of their domains of definition, convex functions always have finite one-sided derivatives:

6.3.3. Corollary *For any convex function f defined on an interval including $[a, b]$ with $a < b$, the right-hand difference quotients $(f(a + h) - f(a))/h$ are nonincreasing as $h \downarrow 0$, having a limit*

$$f'(a^+) := \lim_{h \downarrow 0} (f(a + h) - f(a))/h \geq -\infty.$$

If f is defined at any $t < a$, then all the above difference quotients, and hence their limit, are at least $(f(a) - f(t))/(a - t)$, so that $f'(a^+)$ is finite. Likewise, the difference quotients $(f(b) - f(b - h))/h$ are nondecreasing as $h \downarrow 0$, having a limit $f'(b^-) \le +\infty$, and which is $\le (f(c) - f(b))/(c - b)$ if f is defined at some $c > b$.

Further, $f'(x^+)$ is a nondecreasing function of x on $[a, b)$, and $f'(x^-)$ is nondecreasing on $(a, b]$, with $f'(x^-) \le f'(x^+)$ for all x in $[a, b]$.

On (a, b), since both left and right derivatives exist and are finite, f is continuous.

Proof. These properties follow straightforwardly from Proposition 6.3.2.

\square

Examples. Let $f(0) := f(1) := 1$ and $f(x) := 0$ for $0 < x < 1$. Then f is convex on $[0, 1]$ but not continuous at the endpoints. Also, let $f(t) := t^2$ for all t and $x = 0$. Then the right-hand difference quotients are $h^2/h = h$, which decrease to 0 as $h \downarrow 0$. The left-hand difference quotients are $-h^2/h = -h$, which increase to 0 as $h \downarrow 0$.

A convex function f on \mathbb{R}^k, restricted to a line, is convex, so it must have one-sided directional derivatives, as in Corollary 6.3.3. These derivatives will be shown to be bounded in absolute value, uniformly on compact subsets of an open set where f is defined:

6.3.4. Theorem *Let f be a convex real-valued function on a convex open set U in \mathbb{R}^k. Then at each point x of U and each $v \in \mathbb{R}^k$, f has a finite directional derivative in the direction v,*

$$D_v f(x) := \lim_{h \downarrow 0} (f(x + hv) - f(x))/h.$$

On any compact convex set K included in U, and for v bounded, say $|v| \le 1$, these directional derivatives are bounded, and f is Lipschitzian on K. Thus f is continuous on U.

Proof. Existence of directional derivatives follows directly from Corollary 6.3.3. The rest of the proof will use induction on k. For $k = 1$, let $a < b < c < d$, with f convex on (a, d). Let $a < t < b$ and $c < v < d$. Then all difference-quotients of f on $[b, c]$ are bounded below by $(f(b) - f(t))/(b - t)$ and above by $(f(v) - f(c))/(v - c)$, by Proposition 6.3.2 iterated, so that f is Lipschitzian on $[b, c]$ and the left and right derivatives on $[b, c]$ are uniformly bounded by Corollary 6.3.3.

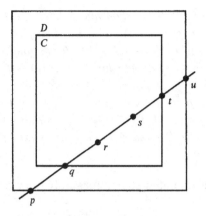

Figure 6.3B

Now in higher dimensions, suppose C is a closed cube $\prod_{j=1}^{k}[c_j, c_j + s]$ included in U. Then for some $\delta > 0$, C is included in the interior of another closed cube $D := \prod_{j=1}^{k}[c_j - \delta, c_j + s + \delta]$ which is also included in U. On the faces of these cubes, which are cubes of lower dimension in the interior of open sets where f is convex, f is Lipschitzian, hence continuous and bounded, by induction hypothesis. Thus for some $M < \infty$, $\sup_{\partial D} f - \inf_{\partial C} f \leq M$ and $\sup_{\partial C} f - \inf_{\partial D} f \leq M$. For any two points r and s in int C, let the line L through r and s first meet ∂D at p, then ∂C at q, then r, then s, then ∂C at t, then ∂D at u (see Figure 6.3B). Then

$$-\frac{M}{\delta} \leq \frac{f(q) - f(p)}{|q - p|} \leq \frac{f(s) - f(r)}{|s - r|} \leq \frac{f(u) - f(t)}{|u - t|} \leq \frac{M}{\delta}.$$

(To see the second inequality, for example, one can insert an intermediate term $(f(r) - f(q))/|r - q|$ and apply Proposition 6.3.2.) Thus $|f(s) - f(r)| \leq M|s - r|/\delta$ and f is Lipschitzian on C. So $|D_v f| \leq M/\delta$ on int C for $|v| \leq 1$.

Now, any compact convex set K included in U has an open cover by the interiors of such cubes C, and there is a finite subcover by interiors of cubes $C_i, i = 1, \ldots, n$. Taking the maximum of finitely many bounds, there is an $N < \infty$ such that the directional derivatives $D_v f$ for $|v| = 1$ are bounded in length by N on each C_i and thus on K. It follows that f is Lipschitzian on K with $|f(x) - f(y)| \leq N|x - y|$ for all $x, y \in K$. For each $x \in U$ and some $t > 0$, $|y - x| < t$ implies $y \in U$. Then taking $K := \{y: |y - x| \leq t/2\}$, which is compact and included in U, f is continuous at x, so f is continuous on U. \square

Example. Let $f(x) := 1/x$ on $(0, \infty)$. Then f is convex and continuous, and Lipschitzian on closed subsets $[c, \infty)$, $c > 0$, but f is not Lipschitzian on all of $(0, \infty)$ and cannot be defined so as to be continuous at 0.

There are further facts about convex functions in §10.2 below.

Problems

1. Suppose that a real-valued function f on an open interval J in \mathbb{R} has a second derivative f'' on J. Show that f is convex if and only if $f'' \geq 0$ everywhere on J.

2. Let $f(x, y) := x^2 y^2$ for all x and y. Show that although f is convex in x for each y, and in y for each x, it is not convex on \mathbb{R}^2.

3. If f and g are two convex functions on the same domain, show that $f + g$ and $\max(f, g)$ are convex. Give an example to show that $\min(f, g)$ need not be.

4. For any set F and point x in a metric space recall that $d(x, F) :=$ $\inf\{d(x, y): y \in F\}$. Let F be a closed set in a normed linear space S with the usual distance $d(x, y) := \|x - y\|$. Show that $d(\cdot, F)$ is a convex function if and only if F is a convex set.

5. Let U be a convex open set in \mathbb{R}^k. For any set $A \subset \mathbb{R}^k$ let $-A := \{-x: x \in A\}$. Assume $U = -U$, so that $0 \in U$. Let μ be a measure defined on the Borel subsets of U with $0 < \mu(U) < \infty$ and $\mu(B) \equiv \mu(-B)$. Let f be a convex function on U. Show that $f(0) \leq \int f \, d\mu / \mu(U)$. *Hint:* Use the image measure theorem 4.2.8 with $T(x) \equiv -x$.

6. Let f be a real function on a convex open set U in \mathbb{R}^2 for which the second partial derivatives $D_{ij} f := \partial^2 f / \partial x_i \partial x_j$ exist and are continuous everywhere in U for i and $j = 1$ and 2. Show that f is convex if the matrix $\{D_{ij} f\}_{i,j=1,2}$ is nonnegative definite at all points of U. *Hint:* Consider the restrictions of f to lines and use the result of Problem 1.

7. Show that for any convex function f defined on an open interval including a closed interval $[a, b]$, $f(b) - f(a) = \int_a^b f'(x^+) \, dx = \int_a^b f'(x^-) \, dx$.

8. Suppose in the definition of convex function the value $-\infty$ is allowed. Let f be a convex function defined on a convex set $A \subset \mathbb{R}^k$ with $f(x) = -\infty$ for some $x \in A$. Show that $f(y) = -\infty$ for all $y \in A$.

9. Let $f(x) := |x|^p$ for all real x. For what values of p is it true that (a) f is convex; (b) the derivative $f'(x)$ exists for all x; (c) the second derivative f'' exists for all x?

10. Let f be a convex function defined on a convex set $A \subset \mathbb{R}^k$. For some fixed x and y in \mathbb{R}^k let $g(t) := f(x + ty) + f(x - ty)$ whenever this is defined. Show that g is a nondecreasing function defined on an interval (possibly empty) in \mathbb{R}.

11. Recall that a *Radon measure* on \mathbb{R} is a function μ into \mathbb{R}, defined on all

bounded Borel sets and countably additive on the Borel subsets of any fixed bounded interval. Show that for any convex function f on an open interval $U \subset \mathbb{R}$, there is a unique nonnegative Radon measure μ such that $f'(y^+) - f'(x^+) = \mu((x, y])$ and $f'(y^-) - f'(x^-) = \mu([x, y))$ for all $x < y$ in U.

12. (Continuation.) Conversely, show that for any nonnegative Radon measure μ on an open interval $U \subset \mathbb{R}$, there exists a convex function f on U satisfying the relations in Problem 11. If g is another such convex function (for the same μ), what condition must the difference $f - g$ satisfy?

*6.4. Duality of L^p Spaces

Recall that any continuous linear function F from a Hilbert space H to its field of scalars (\mathbb{R} or \mathbb{C}) can be written as an inner product: for some g in H, $F(h) = (h, g)$ for all h in H (Theorem 5.5.1). Specifically, if H is a space $L^2(X, \mathcal{B}, \mu)$ for some measure μ, then each continuous linear form F on L^2 can be written, for some g in \mathcal{L}^2, as $F(h) = \int h \bar{g} \, d\mu$ for all h in L^2. Conversely, every such integral defines a linear form which is continuous, since by the Cauchy-Bunyakovsky-Schwarz inequality (5.3.3), $|\int(h_1 - h_2)\bar{g} \, d\mu| \le \|h_1 - h_2\| \, \|g\|$. These facts will be extended to L^p spaces for $p \ne 2$ in a way indicated by the Rogers-Hölder inequality (5.1.2). It says that if $1/p + 1/q = 1$, where $p \ge 1$, and if $f \in \mathcal{L}^p(\mu)$ and $g \in \mathcal{L}^q(\mu)$, then $|\int fg \, d\mu| \le \|f\|_p \|g\|_q$. So setting $F(f) = \int fg \, d\mu$, for $g \in \mathcal{L}^q(\mu)$, defines a continuous linear form F on L^p, since $|\int(f_1 - f_2)g \, d\mu| \le \|f_1 - f_2\|_p \|g\|_q$. For $p = q = 2$, we just noticed that every continuous linear form can be represented this way. This representation will now be extended to $1 \le p < \infty$ (it is not true for $p = \infty$ in general). The next theorem, then, provides a kind of converse to the Rogers-Hölder inequality. (Recall the notions of dual Banach space and dual norm $\|\cdot\|'$ defined in §6.1.) The theorem will give an isometry of the dual space $(L^p)'$ and L^q. In this sense, the dual of L^p is L^q for $1 \le p < \infty$.

6.4.1. Riesz Representation Theorem *For any σ-finite measure space (X, \mathcal{S}, μ), $1 \le p < \infty$ and $p^{-1} + q^{-1} = 1$, the map T defined by*

$$T(g)(f) := \int fg \, d\mu$$

gives a linear isometry of $(L^q, \|\cdot\|_q)$ onto $((L^p)', \|\cdot\|'_p)$.

Proof. First suppose $1 < p < \infty$. Then T takes \mathcal{L}^q into $(L^p)'$ by the Rogers-Hölder inequality (5.1.2), with $\|T(g)\|'_p \le \|g\|_q$. Given g in \mathcal{L}^q, let $f(x) :=$ $|g(x)|^q / g(x)$ if $g(x) \ne 0$ or 0 if $g(x) = 0$. Then $\int f g \, d\mu = \int |g|^q \, d\mu$ and $\int |f|^p \, d\mu = \int |g|^{p(q-1)} \, d\mu = \int |g|^q \, d\mu$, so $\|T(g)\|'_p \ge (\int |g|^q \, d\mu)^{1-1/p} =$ $\|g\|_q$. So $\|T(g)\|'_p = \|g\|_q$ for all g in \mathcal{L}^q and T is an isometry into, that is, $\|T(g) - T(\gamma)\|'_p = \|g - \gamma\|_q$ for all g and γ in \mathcal{L}^q.

Now suppose $L \in (L^p)'$. If $K = \mathbb{C}$, then for some M and N in real $(L_p)'$, $L(f) = M(f) + iN(f)$ for all f in real \mathcal{L}^p. So to prove T is onto $(L^p)'$, we can assume $K = \mathbb{R}$.

For $f \ge 0$, $f \in \mathcal{L}^p$, let $L^+(f) := \sup\{L(h) : 0 \le h \le f\}$. Since $0 \le h \le f$ implies $\|h\|_p \le \|f\|_p$, $L^+(f)$ is finite. Clearly $L^+(cf) = cL^+(f)$ for any $c \ge 0$.

6.4.2. Lemma *If $g \ge 0$, $f_1 \ge 0$, and $f_2 \ge 0$, with all three functions in \mathcal{L}^p, then $0 \le g \le f_1 + f_2$ if and only if $g = g_1 + g_2$ for some measurable g_i with $0 \le g_i \le f_i$ for $i = 1, 2$.*

Proof. "If" is clear. Conversely, suppose $0 \le g \le f_1 + f_2$. Let $g_1 :=$ $\min(f_1, g)$ and $g_2 := g - g_1$. Then clearly $g_i \ge 0$ for $i = 1$ and 2 and $g_1 \le f_1$. Now $g_2 \le f_2$ since otherwise, at some x, $g_2 > f_2 \ge 0$ implies $g_1 = f_1$ so $f_1 + f_2 < g_1 + g_2 = g$, a contradiction, proving the lemma. \square

Clearly, if $f_i \ge 0$ for $i = 1, 2$, then $L^+(f_1 + f_2) \ge L^+(f_1) + L^+(f_2)$. Conversely, the lemma implies $L^+(f_1 + f_2) \le L^+(f_1) + L^+(f_2)$, so $L^+(f_1 + f_2) =$ $L^+(f_1) + L^+(f_2)$. Let us define $L^+(f_1 - f_2) = L^+(f_1) - L^+(f_2)$. This is well-defined since if $f_1 - f_2 = g - h$, where g and h are also nonnegative functions in \mathcal{L}^p, then $f_1 + h = f_2 + g$, so $L^+(f_1) + L^+(h) = L^+(f_1 + h) =$ $L^+(f_2 + g) = L^+(f_2) + L^+(g)$, and $L^+(f_1) - L^+(f_2) = L^+(g) - L^+(h)$, as desired.

Thus L^+ is defined and linear on \mathcal{L}^p. Let $L^- := L^+ - L$. Then L^- is linear, $L^-(f) \ge 0$ for all $f \ge 0$ in \mathcal{L}^p, and $L = L^+ - L^-$.

\mathcal{L}^p is a Stone vector lattice (as defined in §4.5). If $f_n(x) \downarrow 0$ for all x, with $f_1 \in \mathcal{L}^p$, then $\|f_n\|_p \to 0$ by dominated or monotone covergence (§4.3), using $p < \infty$. So by the Stone-Daniell theorem (4.5.2), there are measures β^+ and β^- with $L^+(f) = \int f \, d\beta^+$ and $L^-(f) = \int f \, d\beta^-$ for all $f \in \mathcal{L}^p$. Clearly β^+ and β^- are absolutely continuous with respect to μ. First suppose μ is finite. Then β^+ and β^- are finite (letting $f = 1$), so by the Radon-Nikodym theorem (5.5.4), there are nonnegative measurable functions g^+ and g^- with $\beta^+(A) = \int_A g^+ \, d\mu$ and likewise for β^- and g^- for

any measurable set A. Considering simple functions and their monotone limits as usual (Proposition 4.1.5), we have $L^+(f) = \int fg^+ \, d\mu$ for all $f \in \mathcal{L}^p$ (first considering $f \geq 0$, then letting $f = f^+ - f^-$ as usual). Likewise $L^-(f) = \int fg^- \, d\mu$ for all $f \in \mathcal{L}^p$. Let $g := g^+ - g^-$. The function g can be defined consistently on a sequence of sets of finite μ-measure whose union is (almost) all of X, although the integrability properties of g so far are only given on (some) sets of finite measure. We have $L(f) = \int fg \, d\mu$ whenever $f = 0$ outside some set of finite measure.

To show that $g \in \mathcal{L}^q$, let $g_n := g(n) := g$ for $|g| \leq n$ and $g_n := 0$ elsewhere. For a set $E_n := E(n)$ with $\mu(E_n) < \infty$, let $f_n := 1_{E(n)} 1_{g(n) \neq 0} |g_n|^q / g_n$. Then $f_n \in \mathcal{L}^p$ and

$$\|L\|'_p \geq |L(f_n)| / \|f_n\|_p = \left(\int_{E(n)} |g_n|^q \, d\mu \right)^{1-1/p} = \left\| 1_{E(n)} g_n \right\|_q.$$

Letting $n \to \infty$ we can get $E(n) \uparrow X$, since μ is σ-finite. Thus by Fatou's Lemma (4.3.3), $g \in \mathcal{L}^q$. Then $L(f) = \int fg \, d\mu$ for all $f \in \mathcal{L}^p$ by dominated convergence (4.3.5) and the Rogers-Hölder inequality (5.1.2). This finishes the proof for $1 < p < \infty$.

The remaining case is $p = 1$, $q = \infty$. If $g \in \mathcal{L}^\infty$, then $T(g)$ is in $(L^1)'$ and $\|T(g)\|'_1 \leq \|g\|_\infty$. For any $\varepsilon > 0$, there is a set A with $0 < \mu(A) < \infty$ and $|g(x)| > \|g\|_\infty - \varepsilon$ for all $x \in A$. We can assume $\varepsilon < \|g\|_\infty$ (if $\|g\|_\infty = 0$, then $g = 0$ a.e.). Let $f := 1_A |g| / g$. Then $f \in \mathcal{L}^1$ and $|\int fg \, d\mu| \geq (\|g\|_\infty - \varepsilon) \|f\|_1$. Letting $\varepsilon \downarrow 0$ gives $\|T(g)\|'_1 \geq \|g\|_\infty$, so as for $p > 1$, T is an isometry from \mathcal{L}^∞ into $(L^1)'$.

To prove T is onto, let $L \in (L^1)'$. As before, we can define the decomposition $L = L^+ - L^-$ and find a measurable function g such that $L(f) = \int fg \, d\mu$ for all those f in \mathcal{L}^1 which are 0 outside some set of finite measure. If g is not in \mathcal{L}^∞, then for all $n = 1, 2, \ldots$, there is a set $A_n := A(n)$ with $0 < \mu(A_n) < \infty$ and $|g(x)| > n$ for all x in A_n. Let $f = 1_{A(n)} |g| / g$. Then

$$\|L\|'_1 \geq |L(f)| / \|f\|_1 = \left| \int fg \, d\mu \right| \Big/ \mu(A_n) \geq n$$

for all n, a contradiction. So $g \in \mathcal{L}^\infty$ and T is onto. $\qquad\square$

Problems

1. (a) Show that $L^1([0, 1]), \lambda)$ is not reflexive. *Hint:* $C[0, 1] \subset L^\infty \subset (L^1)'$; if $T(f) := f(0)$, then $T \in C[0, 1]'$. Use the Hahn-Banach theorem (6.1.4).

 (b) Show that $\ell^1 := L^1(\mathbb{N}, 2^{\mathbb{N}}, c)$ is not reflexive, where c is counting measure on \mathbb{N}.

2. Recall the spaces $L^p(X, \mathcal{S}, \mu)$ defined in Problem 1 of §5.2 for $0 < p < 1$.
 (a) Show that for $X = [0, 1]$ and $\mu = \lambda$, the only continuous linear function from L^p to \mathbb{R} is identically 0.
 (b) For X an infinite set and $\mu =$ counting measure, show that there exist non-zero continuous linear real functions on $\ell^p := L^p(X, 2^X, \mu)$ for $0 < p < 1$.

3. Let S be a vector lattice of real-valued functions on a set X, as defined in §4.5. Let L be a linear function from S to \mathbb{R}. Let $L^+(f) := \sup\{L(g): 0 \leq g \leq f\}$ for $f \geq 0$ in S.
 (a) Show that $L^+(f)$ may be infinite for some $f \in S$. *Hint:* See Proposition 5.6.4.
 (b) If $L^+(f)$ is finite for all $f \geq 0$, show that L^+ and $L^- := L^+ - L$ can be extended to linear functions from S to \mathbb{R}.

4. For $1 \leq r < p < \infty$ let V be the identity from L^p into L^r where $L^s := L^s([0, 1]), \lambda)$ for $s = p$ or r. Let U be the transpose of V from $(L^r)'$ into $(L^p)'$, $U(h) := h \circ V$ for each $h \in (L^r)'$. Show that U is not onto $(L^p)'$. *Hint:* Consider functions $h(x) := 1/x^a, 0 < x \leq 1$, for $0 < a < 1$.

6.5. Uniform Boundedness and Closed Graphs

This section will prove three of perhaps the four main theorems of classical functional analysis (the fourth is the Hahn-Banach theorem, 6.1.4). Let $(X, \|\cdot\|)$ and $(Y, |\cdot|)$ be two normed linear spaces. A linear function T from X into Y is often called a linear *transformation* or *operator*. If X and Y are finite-dimensional, with X having a basis of linearly independent elements e_1, \ldots, e_m and Y having a basis f_1, \ldots, f_n, then T determines, and is determined by, the matrix $\{T_{ij}\}_{1 \leq i \leq n, 1 \leq j \leq m}$ where $T(e_j) = \sum_{1 \leq i \leq n} T_{ij} f_i$. The facts to be developed in this section are interesting for infinite-dimensional Banach spaces, where even though some sort of basis might exist (for example, an orthonormal basis of a Hilbert space, §5.3), operators are less often studied in terms of bases and matrices. Recall (from Theorem 6.1.2) that a linear operator T is continuous if and only if it is *bounded* in the sense that

$$\|T\| := \sup\{|T(x)|: \|x\| \leq 1\} < \infty.$$

Here $\|T\|$ is called the *operator norm* of T. The first main theorem will say that a pointwise bounded set of continuous linear operators on a Banach space is bounded in operator norm. This would not be surprising for a finite-dimensional space where one could take a maximum over a finite basis. For general, infinite-dimensional spaces it is much more remarkable:

6.5.1. Theorem (Uniform Boundedness Principle) *Let* $(X, |\cdot|)$ *be a Banach space. For each* α *in an index set* I, *let* T_α *be a bounded linear operator from* X *into some normed linear space* $(S_\alpha, |\cdot|_\alpha)$. *Suppose that for each* x *in* X, $\sup_{\alpha \in I} |T_\alpha(x)|_\alpha < \infty$. *Then* $\sup_{\alpha \in I} \|T_\alpha\| < \infty$.

Proof. Let $A := \{x \in X : |T_\alpha(x)|_\alpha \le 1 \text{ for all } \alpha \in I\}$. Then by the hypothesis, $\bigcup_n nA = X$ where $nA := \{nx : x \in A\}$. By the category theorem (2.5.2), some nA is dense somewhere. Since multiplication by n is a homeomorphism, A is dense somewhere. Now A is closed, so for some $x \in A$ and $\delta > 0$, the ball

$$B(x, \delta) := \{y : |y - x| < \delta\}$$

is included in A. Since A is symmetric, $B(-x, \delta)$ is also included in A. Also, A is convex. So for any $u \in B(0, \delta)$,

$$u = \tfrac{1}{2}(x + u + (-x + u)) \in A.$$

It follows that $\|T_\alpha\| \le 1/\delta$ for all α. □

Example. Let H be a real Hilbert space with an orthonormal basis $\{e_n\}_{n=1}^\infty$. Let X be the set of all finite linear combinations of the e_n, an incomplete inner product space. Let $T_n(x) = (x, ne_n)$ for all x. Then $T_n(x) \to 0$ as $n \to \infty$ for all $x \in X$, and $\sup_n |T_n(x)| < \infty$. Let $S_n = \mathbb{R}$ for all n. Then $\|T_n\| = n \to \infty$ as $n \to \infty$. This shows how the completeness assumption was useful in Theorem 6.5.1.

In most applications of Theorem 6.5.1, the spaces S_α are all the same. Often they are all equal to the field K, so that $T_\alpha \in X'$. In that case, the theorem says that for any Banach space X, if $\sup_\alpha |T_\alpha(x)| < \infty$ for all x in X, then $\sup_\alpha \|T_\alpha\|' < \infty$.

An *F-space* is a vector space V over the field K ($= \mathbb{R}$ or \mathbb{C}) together with a complete metric d which is invariant: $d(x, y) = d(x + z, y + z)$ for all x, y, and z in V, and for which addition in V and multiplication by scalars in K are jointly continuous into V for d. Clearly, a Banach space with its usual metric is an F-space.

A function T from one topological space into another is called *open* iff for every open set U in its domain, $T(U) := \{T(x) : x \in U\}$ is open in the range space. In the following theorem, the hypotheses that T be onto Y (not just into Y) and that Y be complete are crucial. For example, let H be a Hilbert space with an orthonormal basis $\{e_n\}_{n \ge 1}$. Let T be the linear operator with

$T(e_n) = e_n/n$ for all n. Then T is continuous, with $\|T\| = 1$, but not onto H (though of course onto its range, which is incomplete) and not open.

6.5.2. Open Mapping Theorem *Let (X, d) and (Y, e) be F-spaces. Let T be a continuous linear operator from X onto Y. Then T is open.*

Proof. For $r > 0$ let $X_r := \{x \in X : d(0, x) < r\}$, and likewise $Y_r := \{y \in Y : e(0, y) < r\}$. For a given, fixed value of r let $V := X_{r/2}$. Then by continuity of scalar multiplication, $\bigcup_n nV = X$. Thus by the category theorem (2.5.2), some set $T(nV) = nT(V)$ is dense in some non-empty open set in Y, and $T(V)$ is dense in some ball $y + Y_\varepsilon$, $\varepsilon > 0$. Now V is symmetric since $d(0, x) = d(-x, 0) = d(0, -x)$ for any x, and $V - V := \{u - v : u \in V \text{ and } v \in V\}$ is included in X_r. Thus $T(X_r)$ is dense in Y_ε.

Now let s be any number larger than r. Let $r_1 := r$ and $r_n := (s - r)/2^{n-1}$ for $n \geq 2$. Then $\sum_n r_n = s$. Let $r(j) := r_j$ for each j. Then for each $j = 1, 2, \ldots, T(X_{r(j)})$ is dense in $Y_{\varepsilon(j)}$ for some $\varepsilon_j := \varepsilon(j) > 0$. It will be proved that $T(X_s)$ includes Y_ε. Given any $y \in Y_\varepsilon$, take $x_1 \in X_{r(1)}$ such that $e(T(x_1), y) < \varepsilon_2$. Then choose $x_2 \in X_{r(2)}$ such that $e(T(x_2), y - T(x_1)) < \varepsilon_3$, so that $e(T(x_1 + x_2), y) < \varepsilon_3$. Thus, recursively, there are $x_j \in X_{r(j)}$ for $j = 1, 2, \ldots$, such that for $n = 1, 2, \ldots, e(T(x_1 + \cdots + x_n), y) < \varepsilon_{n+1}$.

Since X is complete, $\sum_{j=1}^{\infty} x_j$ converges in X to some x with $T(x) = y$ and $x \in X_s$, so $T(X_s) \supset Y_\varepsilon$, as desired.

So for any neighborhood U of 0 in X, and $x \in X$, $T(U)$ includes a neighborhood of 0 in Y, and $T(x + U)$ includes a neighborhood of $T(x)$. So T is open. \square

The next fact is more often applied than the open mapping theorem and follows rather easily from it:

6.5.3. Closed Graph Theorem *Let X and Y be F-spaces and T a linear operator from X into Y. Then T is continuous if and only if it (that is, its graph $\{\langle x, Tx \rangle : x \in X\}$) is closed, for the product topology on $X \times Y$.*

Proof. "Only if" is clear, since if $\langle x_n, T(x_n) \rangle \to \langle x, y \rangle$, then by continuity $T(x_n) \to T(x) = y$. Conversely, suppose T is closed. Then it, as a closed linear subspace of the F-space $X \times Y$, is itself an F-space. The projection $P(x, y) := x$ takes T onto X and is continuous. So by the open mapping theorem (6.5.2), P is open. Since it is 1–1, it is a homeomorphism. Thus its inverse $x \mapsto \langle x, T(x) \rangle$ is continuous, so T is continuous. \square

6.5.4. Corollary *For any continuous, linear, 1–1 operator T from an F-space X onto an F-space Y, the inverse T^{-1} is also continuous.*

Proof. This is a corollary of the open mapping theorem (6.5.2) or of the closed graph theorem (the graph of T^{-1} in $Y \times X$ is closed, being the graph of T in $X \times Y$). □

Again, the assumptions that T is onto the range space Y and that Y is complete are both important: see Problem 2 below.

Problems

1. Show that for any two normed linear spaces $(X, \|\cdot\|)$ and $(Y, |\cdot|)$, and $L(X, Y)$ the set of all bounded linear operators from X into Y, the operator norm $T \mapsto \|T\|$ is in fact a norm on $L(X, Y)$.

2. Prove in detail that the operator T with $T(e_n) = e_n/n$, where $\{e_n\}$ is an orthonormal basis of a Hilbert space, is continuous but not open and not onto. Show that its range is dense.

3. Prove that in the open mapping theorem (6.5.2) it is enough to assume, rather than T onto, that the range of T is of second category in Y, that is, it is not a countable union of nowhere dense sets.

4. Let $(X, \|\cdot\|)$ and $(Y, |\cdot|)$ be Banach spaces. Let T_n be a sequence of bounded linear operators from X into Y such that for all x in X, $T_n(x)$ converges to some $T(x)$. Show that T is a bounded linear operator.

5. Show that T in Problem 4 need not be bounded if the sequence T_n is replaced by a net $\{T_\alpha\}_{\alpha \in I}$. In fact, show that every linear transformation from X into Y (continuous or not) is the limit of some net of bounded linear operators. *Hint:* Show, using Zorn's lemma, that X has a *Hamel basis B*, that is, each point in X is a unique finite linear combination of elements of B. Use Hamel bases to show that there are unbounded linear functions from any infinite-dimensional Banach space into \mathbb{R}.

6. A real vector space (linear space) S with a topology \mathcal{T} is called a *topological vector space* iff addition is continuous from $S \times S$ into S and multiplication by scalars is continuous from $\mathbb{R} \times S$ into S. Show that the only Hausdorff topology on \mathbb{R} making it a topological vector space is the usual topology \mathcal{U}. *Hint:* On compact subsets of \mathbb{R} for \mathcal{U}, use Theorem 2.2.11. Then, if $\mathcal{T} \neq \mathcal{U}$, there is a net $x_\alpha \to 0$ for \mathcal{T} with $|x_\alpha| \to \infty$, so $1 = (1/x_\alpha)x_\alpha \to 0$.

7. Let T^2 be the set of all ordered pairs of complex numbers (z, w) with $|z| = |w| = 1$. Let f be the function from \mathbb{R} into T^2 defined by $f(x) = (e^{ix}, e^{i\gamma x})$, where γ is irrational, say $\gamma = 2^{1/2}$. Show that f is 1–1 and continuous but not open. Let \mathcal{T} be the set of all $f^{-1}(U)$ where U is open in T^2. Show that \mathcal{T} is a topology and that for \mathcal{T}, addition is continuous, but scalar multiplication $x \mapsto cx$ is not continuous from $(\mathbb{R}, \mathcal{T})$ to $(\mathbb{R}, \mathcal{T})$ even for a fixed c such as $c = \sqrt{3}$. *Hint:* $\{(e^{ix}, e^{i\gamma x}, e^{icx}) : x \in \mathbb{R}\}$ is dense in T^3.

8. Show that the only Hausdorff topology on \mathbb{R}^k making it a topological vector space (for k finite) is the usual topology. *Hint:* The case $k = 1$ is Problem 6.

9. Let F be a finite-dimensional linear subspace of an F-space X, so that there is some finite set x_1, \ldots, x_n such that each $x \in F$ is of the form $x = a_1 x_1 + \cdots + a_n x_n$ for some numbers a_j. Show that F is closed. *Hint:* Use Problem 8 and the notion of uniformity (§2.7).

10. In any infinite-dimensional F-space, show that any compact set has empty interior.

11. A continuous linear operator T from one Banach space X into another one Y is called *compact* iff $T\{x \in X : \|x\| \leq 1\}$ has compact closure. Show that if X and Y are infinite-dimensional, then T cannot be onto Y. *Hint:* Apply Problem 10.

12. Show that there is a separable Banach space X and a compact linear operator T from X into itself such that $T\{x : \|x\| \leq 1\}$ is not compact. *Hint:* Let $X = c_0$, the space of all sequences converging to 0, with norm $\|\{x_n\}\| := \sup_n |x_n|$.

13. Let $(F_n, d_n)_{n \geq 1}$ be any sequence of F-spaces. Show that the Cartesian product $\Pi_n F_n$ with product topology (and linear structure defined by $c\{x_n\} + \{y_n\} = \{cx_n + y_n\}$) is an F-space. *Hint:* See Theorem 2.5.7.

14. In Problem 13, if each F_n is \mathbb{R} with its usual metric, show that there is no norm $\|\cdot\|$ on the product space for which the metric $d(x, y) := \|x - y\|$ metrizes the product topology.

*6.6. The Brunn-Minkowski Inequality

For any two sets A and B in a vector space let $A + B := \{x + y : x \in A, y \in B\}$. For any constant c let $cA := \{cx : x \in A\}$. In \mathbb{R}^k, if A and B are compact, then $A + B$ is compact, being the range of the continuous function "$+$" on the compact set $A \times B$. Thus if A and B are countable unions of compact sets,

$A + B$ is also a countable union of compact sets, so it is Borel measurable. Let V denote Lebesgue measure (volume) in \mathbb{R}^k, given by Theorem 4.4.6. Here is one of the main theorems about convexity:

6.6.1. Theorem (Brunn-Minkowski Inequality) *Let A and B be any two non-empty compact sets in \mathbb{R}^k. Then*

(a) $V(A + B)^{1/k} \geq V(A)^{1/k} + V(B)^{1/k}$.

(b) *For $0 < \lambda < 1$, $V(\lambda A + (1 - \lambda)B)^{1/k} \geq \lambda V(A)^{1/k} + (1 - \lambda)V(B)^{1/k}$.*

Proof. Clearly $V(cA) \equiv c^k V(A)$ for $c > 0$. Thus (a) is equivalent to the special case of (b) where $\lambda = 1/2$. Also, (b) follows from (a), replacing A by λA and B by $(1 - \lambda)B$. So it will be enough to prove (a). It clearly holds when $V(A) = 0$ or $V(B) = 0$.

So assume that both A and B have positive volume. Two sets C and D in \mathbb{R}^k will be called *quasi-disjoint* if their intersection is included in some $(k - 1)$-dimensional hyperplane H. (For example, if $C \subset \{x: x_1 \leq 3\}$ and $D \subset \{x: x_1 \geq 3\}$, then C and D are quasi-disjoint.) Then $V(C \cup D) = V(C) + V(D)$. Such a hyperplane H splits A into a union of two quasi-disjoint closed sets A' and A'', which are the intersections of A with the closed half-spaces on each side of H. Likewise let B be split into sets B' and B'' by a hyperplane J parallel to H, chosen so that $r := V(B')/V(B) = V(A')/V(A)$, where B' is the intersection of B with the closed half-space on the same side of J as A' is of H. In the usual coordinates $x = \langle x_1, \ldots, x_k \rangle$ of \mathbb{R}^k, only hyperplanes $\{x: x_i = c\}$ for $c \in \mathbb{R}$ and $i = 1, \ldots, k$ will be needed below. If $H = \{x: x_i = c\}$, then $J = \{x: x_i = d\}$ for some d, and $H + J = \{x: x_i = c + d\}$. Letting $A' := \{x \in A: x_i \leq c\}$, we then have $B' = \{x \in B: x_i \leq d\}$, etc.

Now $A' + B'$ and $A'' + B''$ are quasi-disjoint, as $A' + B'$ is included in $\{x: x_i \leq c + d\}$ and $A'' + B''$ in $\{x: x_i \geq c + d\}$, so

$$V(A) = V(A') + V(A''), \quad V(B) = V(B') + V(B'') \quad \text{and}$$

$$V(A + B) \geq V(A' + B') + V(A'' + B'').$$

Let $p(A, B) := V(A + B) - (V(A)^{1/k} + V(B)^{1/k})^k$. So we need to prove $p(A, B) \geq 0$. We have $V(A'')/V(A) = 1 - r = V(B'')/V(B)$. Then

$$s := V(B)/V(A) = V(B')/V(A') = V(B'')/V(A'') \quad \text{and}$$

$$\left(V(A)^{1/k} + V(B)^{1/k}\right)^k = V(A)\left(1 + s^{1/k}\right)^k.$$

Corresponding equations hold for A', B' and A'', B''. It then follows that

$$p(A, B) \geq p(A', B') + p(A'', B'').$$

A *block* will be a Cartesian product of closed intervals. (In \mathbb{R}^2, a block is a closed rectangle parallel to the axes.) Let us first show that $p(A, B) \geq 0$ when A and B are blocks. Let the lengths of sides of A be a_1, \ldots, a_k and those of B, b_1, \ldots, b_k. Let $\alpha_i := a_i/(a_i + b_i)$, $i = 1, \ldots, k$, and $C := A + B$. Then C is also a block, with sides of lengths $a_i + b_i$, and we have

$$p(A, B) = V(C)\left\{1 - (\alpha_1 \cdots \alpha_k)^{1/k} - ((1 - \alpha_1) \cdots (1 - \alpha_k))^{1/k}\right\}.$$

Since geometric means are smaller than arithmetic means (Theorem 5.1.6), we get $p(A, B) \geq 0$.

Next suppose A is a union of m quasi-disjoint blocks and B is a union of n such blocks. The proof for this case will be by induction on m and n. It is done for $m = n = 1$. For an induction step, suppose the theorem holds for unions of at most $m - 1$ quasi-disjoint blocks plus unions of at most n, with $m \geq 2$. For any two of the quasi-disjoint blocks, say C and D, in the representation of A, there must be some i such that the intervals forming the ith factors of C and D are quasi-disjoint, so that for some c, C or D is included in $\{x : x_i \leq c\}$ and the other in $\{x : x_i \geq c\}$. Intersecting all the blocks in the representation of A with these half-spaces, we get a splitting $A = A' \cup A''$ as above where each of A' and A'' is a union of at most $m - 1$ quasi-disjoint blocks. Take the corresponding splitting of B by a hyperplane J as described above. Each of B' and B'' then still consists of a union of at most n blocks. So the induction hypothesis applies and

$$p(A, B) \geq P(A', B') + p(A'', B'') \geq 0,$$

completing the proof for finite unions of quasi-disjoint blocks.

Now any compact set A is a decreasing intersection of finite unions U_n of quasi-disjoint blocks—for example, the cubes which intersect A whose sides are intervals $[k_i/2^n, (k_i + 1)/2^n]$, $k_i \in \mathbb{Z}$. Let V_n be the corresponding unions for B. Then $U_n + V_n$ decreases to $A + B$. Since these sets have finite volume, monotone convergence applies, and the volumes of the converging sets converge, giving the inequality in the general case of any compact A and B. $\qquad\square$

The Brunn-Minkowski inequality has another form, applicable to sections of one convex set. Let C be a set in \mathbb{R}^{k+1}. For $x \in \mathbb{R}^k$ and $u \in \mathbb{R}$ we have $\langle x, u \rangle \in \mathbb{R}^{k+1}$. Let $C_u := \{x \in \mathbb{R}^k : \langle x, u \rangle \in C\}$. Let V_k denote k-dimensional Lebesgue measure. Then:

6.6.2. Theorem (Brunn-Minkowski Inequality for sections) *Let C be a convex, compact set in \mathbb{R}^{k+1}. Let $f(t) := V_k(C_t)^{1/k}$. Then the set on which*

$f > 0$ *is an interval J, and on this interval f is concave:* $f(\lambda u + (1-\lambda)v) \geq \lambda f(u) + (1-\lambda)f(v)$ *for all* $u, v \in J$, *and* $0 \leq \lambda \leq 1$.

Proof. If $\langle x, u \rangle \in C$ and $\langle y, v \rangle \in C$, then since C is convex, $\langle \lambda x + (1-\lambda)y, \lambda u + (1-\lambda)v \rangle \in C$. Thus

$$C_{\lambda u+(1-\lambda)v} \supset \lambda C_u + (1-\lambda)C_v.$$

The conclusion then follows from Theorem 6.6.1(b). □

Problems

1. Show that for some closed sets A and B, $A + B$ is not closed. *Hint:* Let $A = \{\langle x, y \rangle : xy = 1\}$, $B = \{\langle x, y \rangle : xy = -1\}$.

2. Let C be the part of a cone in \mathbb{R}^{k+1} given by

$$C := \{\langle x, u \rangle : x \in \mathbb{R}^k, u \in \mathbb{R}, |x| \leq u \leq 1\}.$$

Evaluate the function $f(t)$ in Theorem 6.6.2 for this C and show that $V_k(C_t)^p$ is concave if and only if $p \leq 1/k$, so that the exponent $1/k$ in the definition of f is the best possible.

3. Let A be the triangle in \mathbb{R}^3 with vertices $(0, 0, 0)$, $(1, 0, 0)$, and $(1/2, 1, 0)$. Let B be the triangle with vertices $(1/2, 0, 1)$, $(0, 1, 1)$, and $(1, 1, 1)$. Let C be the convex hull of A and B. Evaluate $V(C_t)$ for $0 \leq t \leq 1$.

Notes

§6.1 The Hahn-Banach theorem (6.1.4) resulted from work of Hahn (1927) and Banach (1929, 1932). Banach (1932) made extensive contributions to functional analysis.

Lipschitz (1864) defined the class of functions named for him. The proof of the extension theorem for Lipschitz functions (6.1.1) is essentially a subset of the original proof of the Hahn-Banach theorem, but the case of nonlinear Lipschitz functions seems to have been first treated explicitly by Kirszbraun (1934) and independently by McShane (1934). Minty (1970) gives and reviews some extensions of the Kirszbraun-McShane theorem.

§6.2 It seems that convex sets (in two and three dimensions) were first studied systematically by H. Brunn (1887, 1889) and then by H. Minkowski (1897). Brunn (1910) and Minkowski (1910) proved the existence of support (hyper)planes. Bonnesen and Fenchel (1934), in Copenhagen, gave a further exposition with many references. Minkowski (1910) found uses for convex sets in number theory, creating the "Geometry of Numbers."

The separation theorem for convex sets (6.2.3) is due to Dieudonné (1941). Earlier, Eidelheit (1936) proved a separation theorem for convex sets "without common inner points" in a normed linear space. The above proof is based on Kelley and Namioka (1963, pp. 19–23). Eidelheit, working in Lwow, Poland, published several papers in the

Polish journal *Studia Mathematica* beginning in 1936. His career was cut short by the 1939 invasion (in 1940, papers of his appeared in *Annals of Math.* in the United States and in *Revista Ci. Lima*, Peru). When *Studia* was able to resume publication in 1948, it included a posthumous paper of Eidelheit, with a footnote saying he was killed in March 1943. Dunford and Schwartz (1958, p. 460) give further history of separation theorems. In accepting the Steele Prize for mathematical exposition for Dunford and Schwartz (1958), Dunford (1981) said that Robert G. Bartle wrote most of the historical Notes and Remarks.

§6.3 According to Hardy, Littlewood, and Polya (1934, 1952), "the foundations of the theory of convex functions are due to Jensen." Jensen (1906, p. 191) wrote: "Il me semble que la notion 'fonction convexe' est à peu près aussi fondamentale que celles-ci: 'fonction positive', 'fonction croissante'. Si je ne me trompe pas en ceci la notion devra trouver sa place dans les expositions élémentaires de la théorie des fonctions réelles"—and I agree. For more recent developments, see, for example, Roberts and Varberg (1973) and Rockafellar (1970).

§6.4 The fact that the dual of L^p is L^q for $1 \le p < \infty$ and $p^{-1} + q^{-1} = 1$ (Theorem 6.4.1) was first proved when μ is Lebesgue measure on [0, 1] by F. Riesz (1910).

§6.5 The following notes are based on Dunford and Schwartz (1958, §II.5). Hahn (1922) proved the uniform boundedness principle for sequences of linear forms on a Banach space. Then Hildebrandt (1923) proved it more generally. Banach and Steinhaus (1927) showed that it was sufficient for the $T_\alpha x$ to be bounded for x in a set of second category. Thus the theorem has been called the "Banach-Steinhaus" theorem.

Schauder (1930) proved the open mapping theorem (6.5.2) in Banach spaces, and Banach (1932) proved it for F-spaces. Banach (1931, 1932) also proved the closed graph theorem (6.5.3) and extended both theorems to certain topological groups.

The open mapping and Hahn-Banach theorems have found uses in the theory of partial differential equations (Hörmander, 1964, pp. 65, 87, 98, 101).

§6.6 The relatively elementary proof given for the Brunn-Minkowski inequality (6.6.1) is due to Hadwiger and Ohmann (1956). There are several other proofs (cf. Bonnesen and Fenchel, 1934). The inequality is due to Brunn (1887, Kap. III; 1889, Kap. III) at least for $k = 2, 3$. Minkowski (1910) proved that equality holds in Theorem 6.6.1, if $V(A) > 0$ and $V(B) > 0$, if and only if A and B are homothetic ($B = cA + v$ for some $c \in \mathbb{R}$ and $v \in \mathbb{R}^k$). These early references are according to Bonnesen and Fenchel (1934, pp. 90–91).

References

An asterisk identifies works I have found discussed in secondary sources but have not seen in the original.

Banach, Stefan (1929). Sur les fonctionelles linéaires I, II. *Studia Math. 1*: 211–216, 223–239.

——— (1931). Über metrische Gruppen. *Studia Math. 3*: 101–113.

——— (1932). *Théorie des Opérations Linéaires*. Monografje Matematyczne I, Warsaw.

———— and Hugo Steinhaus (1927). Sur le principe de la condensation de singularités. *Fund. Math. 9*: 50–61.

Bonnesen, Tommy, and Werner Fenchel (1934). *Theorie der konvexen Körper.* Springer, Berlin. Repub. Chelsea, New York, 1948.

*Brunn, Hermann (1887). Über Ovale and Eiflächen. Inauguraldissertation, Univ. München.

*———— (1889). Über Kurven ohne Wendepunkte. Habilitationsschrift, Univ. München.

*———— (1910). Zur Theorie der Eigebiete. *Arch. Math. Phys.* (Ser. 3) *17*: 289–300.

Dieudonné, Jean (1941). Sur le théorème de Hahn-Banach. *Rev. Sci. 79*: 642–643.

———— (1943). Sur la séparation des ensembles convexes dans un espace de Banach. *Rev. Sci. 81*: 277–278.

Dunford, Nelson (1981). [Response]. *Amer. Math. Soc. Notices 28*: 509–510.

———— and Jacob T. Schwartz, with the assistance of William G. Badé and Robert G. Bartle (1958). *Linear Operators: Part I, General Theory.* Interscience, New York.

Eidelheit, Maks (1936). Zur Theorie der konvexen Mengen in linearen normierten Räumen. *Studia Math. 6*: 104–111.

Hadwiger, H., and D. Ohmann (1956). Brunn-Minkowskischer Satz und Isoperimetrie. *Math. Zeitschr. 66*: 1–8.

Hahn, Hans (1922). Über Folgen linearer Operationen. *Monatsh. für Math. und Physik 32*: 3–88.

———— (1927). Über lineare Gleichungssyteme in linearen Räumen. *J. für die reine und angew. Math. 157*: 214–229.

Hardy, Godfrey Harold, John Edensor Littlewood, and George Polya (1934). *Inequalities.* Cambridge University Press 2d ed., 1952, repr. 1967.

Hildebrandt, T. H. (1923). On uniform limitedness of sets of functional operations. *Bull. Amer. Math. Soc. 29*: 309–315.

Hörmander, Lars (1964). *Linear partial differential operators.* Springer, Berlin.

Jensen, Johan Ludvig William Valdemar (1906). Sur les fonctions convexes et les inégalites entre les valeurs moyennes. *Acta Math. 30*: 175–193.

Kelley, John L., Isaac Namioka, and eight co-authors (1963). *Linear Topological Spaces.* Van Nostrand, Princeton. Repr. Springer, New York (1976).

Kirszbraun, M. D. (1934). Über die zusammenziehenden und Lipschitzschen Transformationen. *Fund. Math. 22*: 77–108.

Lipschitz, R. O. S. (1864). Recherches sur le développement en séries trigonométriques des fonctions arbitraires et principalement de celles qui, dans un intervalle fini, admettent une infinité de maxima et de minima (in Latin). *J. reine angew. Math. 63*: 296–308; French transl. in *Acta Math. 36* (1913): 281–295.

McShane, Edward J. (1934). Extension of range of functions. *Bull. Amer. Math. Soc. 40*: 837–842.

Minkowski, Hermann (1897). Allgemeine Lehrsätze über die konvexen Polyeder. *Nachr. Ges. Wiss. Göttingen Math. Phys. Kl. 1897*: 198–219; *Gesammelte Abhandlungen, 2* (Leipzig and Berlin, 1911, repr. Chelsea, New York, 1967), pp. 103–121.

———— (1901). Sur les surfaces convexes fermées. *C. R. Acad. Sci. Paris 132*: 21–24; *Gesammelte Abhandlungen, 2*, pp. 128–130.

———— (1910). *Geometrie der Zahlen.* Teubner; Leipzig; repr. Chelsea, New York, 1953.

——— (1911, posth.). Theorie der konvexen Körper, insbesondere Begründung des Oberflächenbegriffs. *Gesammelte Abhandlungen, 2*, pp. 131–229.

Minty, George J. (1970). On the extension of Lipschitz, Lipschitz-Hölder continuous, and monotone functions. *Bull. Amer. Math. Soc. 76*: 334–339.

Riesz, F. (1910). Untersuchungen über Systeme integrierbarer Funktionen. *Math. Annalen 69*: 449–497.

Roberts, Arthur Wayne, and Dale E. Varberg (1973). *Convex Functions*. Academic Press, New York.

Rockafellar, R. Tyrrell (1970). *Convex Analysis*. Princeton University Press.

Schauder, Juliusz (1930). Über die Umkehrung linearer, stetiger Funktionaloperationen. *Studia Math. 2*: 1–6.

7

Measure, Topology, and Differentiation

Nearly every measure used in mathematics is defined on a space where there is also a topology such that the domain of the measure is either the Borel σ-algebra generated by the topology, its completion for the measure, or perhaps an intermediate σ-algebra. Defining the integrals of real-valued functions on a measure space did not involve any topology as such on the domain space, although structures on the range space \mathbb{R} (order as well as topology) were used. Section 7.1 will explore relations between measures and topologies.

The derivative of one measure with respect to another, $d\gamma/d\mu = f$, is defined by the Radon-Nikodym theorem (§5.5) in case γ is absolutely continuous with respect to μ. A natural question is whether differentiation is valid in the sense that then $\gamma(A)/\mu(A)$ converges to $f(x)$ as the set A shrinks down to $\{x\}$. For this, it is clearly not enough that the sets A contain x and their measures approach 0, as most of the sets might be far from x. One would expect that the sets should be included in neighborhoods of x forming a filter base converging to x. In \mathbb{R}, for the usual differentiation, the sets A are intervals, usually with an endpoint at x. It turns out that it is not enough for the sets A to converge to x. For example, in \mathbb{R}^2, if the sets A are rectangles containing x, there are still counterexamples to differentiability of measures (absolutely continuous with respect to Lebesgue measure) if the ratios of the sides of the rectangles are unbounded. Section 7.2 treats differentiation.

The other sections will deal with further relations of measure and topology.

7.1. Baire and Borel σ-Algebras and Regularity of Measures

For any topological space (X, \mathcal{T}), the *Borel* σ-algebra $\mathcal{B}(X, \mathcal{T})$ is defined as the σ-algebra generated by \mathcal{T}. Sets in this σ-algebra are called *Borel* sets. The σ-algebra $Ba(X, \mathcal{T})$ of *Baire* sets is defined as the smallest σ-algebra for

which all continuous real functions are measurable, with, as usual, the Borel σ-algebra on the range space \mathbb{R}.

Let $C_b(X, \mathcal{T})$ be the space of all bounded continuous real functions on X. For any real function f, the bounded function arc tan f is measurable if and only if f is measurable (for any σ-algebra on the domain of f), and continuous iff f is continuous. Thus $Ba(X, \mathcal{T})$ is also the smallest σ-algebra for which all f in $C_b(X, \mathcal{T})$ are measurable.

Clearly every Baire set is a Borel set: $Ba(X, \mathcal{T}) \subset \mathcal{B}(X, \mathcal{T})$. The two σ-algebras will be shown to be equal in metric spaces, but not in general.

7.1.1. Theorem *In any metric space (S, d), every Borel set is a Baire set, so $Ba(S, \mathcal{T}) = \mathcal{B}(S, \mathcal{T})$ for the metric topology \mathcal{T}.*

Proof. Let F be any closed set in S. Let

$$f(x) := d(x, F) := \inf\{d(x, y): y \in F\}, \quad x \in S.$$

Then for any x and u in S, $|d(x, F) - d(u, F)| \leq d(x, u)$, as shown in (2.5.3). Thus f is continuous. Since F is closed, $F = f^{-1}\{0\}$, so F is a Baire set, hence so is its complement, and the conclusion follows. □

Example. In a Cartesian product K of uncountably many compact Hausdorff spaces each having more than one point, a singleton is closed and is hence a Borel set, but it is not a Baire set, as follows. By the Stone-Weierstrass theorem (2.4.11), the set $C_F(K)$ of continuous real-valued functions depending on only finitely many coordinates is dense in the set $C(K)$ of all continuous real functions. Thus for any $f \in C(K)$, there are f_n in $C_F(K)$ with $\sup_{x \in K} |(f - f_n)(x)| < 1/n$. It follows that f depends at most on countably many coordinates, taking the union of the sets of coordinates that the f_n depend on. The collection of Baire sets depending on only countably many coordinates thus includes a collection generating the σ-algebra and is closed under taking complements and countable unions (again taking a countable union of sets of indices), so it is the entire Baire σ-algebra, which thus does not contain singletons.

It is often useful to approximate general measurable sets by more tractable sets such as closed or compact sets. In doing this it will be good to recall some relations between being closed and compact: any closed subset of a compact set is compact (Theorem 2.2.2), and in a Hausdorff space, any compact set is closed (Proposition 2.2.9).

Definitions. Let (X, \mathcal{T}) be a topological space and μ a measure on some σ-algebra \mathcal{S}. Then a set A in \mathcal{S} will be called *regular* if

$$\mu(A) = \sup\{\mu(K)\colon K \text{ compact}, \, K \subset A, \, K \in \mathcal{S}\}.$$

If μ is finite, it is called *tight* iff X is regular. Likewise, A is called *closed regular* iff

$$\mu(A) = \sup\{\mu(F)\colon F \text{ closed}, \, F \subset A, \, F \in \mathcal{S}\}.$$

Then μ is called (closed) *regular* iff every set in \mathcal{S} is (closed) regular.

Examples. Any finite measure on the Borel σ-algebra of \mathbb{R}^k is tight, since \mathbb{R}^k is a countable union of compact sets $K_n := \{x\colon |x| \leq n\}$.

Let A be the set $\mathbb{R} \backslash \mathbb{Q}$ of irrational numbers, and let $\mu(B) := \lambda(B \cap [0, 1])$ for any Borel set $B \subset \mathbb{R}$ where λ is Lebesgue measure. To see that A is regular, given $\varepsilon > 0$, let $\{q_n\}$ be an enumeration of the rational numbers and take an open interval U_n of length $\varepsilon/2^n$ around q_n for each n. Let U be the union of all the U_n and $K := [0, 1] \backslash U$. Then $K \subset A$ and $\mu(A \backslash K) < \varepsilon$. Letting $\varepsilon \downarrow 0$ shows that A is regular. (It will be shown in Theorem 7.1.3 that all Borel sets in \mathbb{R} are regular.)

The notion "regular" will be applied usually, though not always, to σ-algebras \mathcal{S} including the Baire σ-algebra $Ba(X, \mathcal{T})$. The next fact will help to show that all sets in some σ-algebras are regular:

7.1.2. Proposition *Let (X, \mathcal{T}) be a Hausdorff topological space, \mathcal{S} a σ-algebra of subsets of X, and μ a finite, tight measure on \mathcal{S}. Let*

$$\mathcal{R} := \{A \in \mathcal{S}\colon A \text{ and } X \backslash A \text{ are regular for } \mu\}.$$

Then \mathcal{R} is a σ-algebra. The same is true if "regular" is replaced by "closed regular."

Proof. By definition, $A \in \mathcal{R}$ if and only if $X \backslash A \in \mathcal{R}$. Let A_1, A_2, \ldots, be in \mathcal{R}. Let $A := \bigcup_{n \geq 1} A_n$. Given $\varepsilon > 0$, take compact sets K_n included in A_n and L_n in $X \backslash A_n$ with $\mu(A_n \backslash K_n) < \varepsilon/3^n$ and $\mu((X \backslash A_n) \backslash L_n) < \varepsilon/2^n$ for all n. Then for some $M < \infty$, $\mu(\bigcup_{1 \leq n \leq M} A_n) > \mu(A) - \varepsilon/2$. Let $K := \bigcup_{1 \leq n < M} K_n$. Then K is compact, $K \subset A$, and $\mu(K) \geq \mu(A) - \varepsilon$. Let $L := \bigcap_{1 \leq n < \infty} L_n$. Then L is compact and $\mu((X \backslash A) \backslash L) \leq \sum_n \varepsilon/2^n = \varepsilon$. Thus A and $X \backslash A$ are regular. $X \in \mathcal{R}$ since μ is tight, and \mathcal{R} is a σ-algebra. The same holds for "closed regular" since finite unions and countable intersections of closed sets are

closed. (The analogue of "tight" with "closed" instead of "compact" always holds since X itself is closed.) □

A measure on the Baire σ-algebra will be called a *Baire* measure, and a measure on the Borel σ-algebra will be called a *Borel* measure. In a metric space S, if S is regular, then so are all Borel sets:

7.1.3. Theorem *On any metric space (S, d), any finite Borel measure μ is closed regular. If μ is tight, then it is regular.*

Proof. Let U be any open set and F its complement. Let
$$F_n := \{x : d(x, F) \geq 1/n\}, \qquad n = 1, 2, \ldots$$
(as in the proofs of Theorems 7.1.1 and 4.2.2), so that the F_n are closed and their union is U. Thus the σ-algebra \mathcal{R} in Proposition 7.1.2 for closed regularity contains all open sets and hence all Borel sets, so μ is closed regular.

If μ is tight, then given $\varepsilon > 0$, take a compact set K with $\mu(S \setminus K) < \varepsilon/2$. For any Borel set B, take a closed $F \subset B$ with $\mu(B \setminus F) < \varepsilon/2$. Let $L := K \cap F$. Then L is compact (Theorem 2.2.2), $L \subset B$, and $\mu(B \setminus L) < \varepsilon$, so B is regular and μ is regular. □

Next, the regularity of S, and so of all Borel sets, always holds if the metric space is complete and separable:

7.1.4. Ulam's Theorem *On any complete separable metric space (S, d), any finite Borel measure is regular.*

Proof. To show μ is tight, let $\{x_n\}_{n \geq 1}$ be a sequence dense in S. For any $\delta > 0$ and $x \in S$ let $\overline{B}(x, \delta) := \{y : d(x, y) \leq \delta\}$. Given $\varepsilon > 0$, for each $m = 1, 2, \ldots$, take $n(m) < \infty$ such that
$$\mu\left(S \setminus \bigcup_{n=1}^{n(m)} \overline{B}\left(x_n, \frac{1}{m}\right)\right) < \frac{\varepsilon}{2^m}. \quad \text{Let}$$
$$K := \bigcap_{m \geq 1} \bigcup_{n=1}^{n(m)} \overline{B}\left(x_n, \frac{1}{m}\right).$$
Then K is totally bounded and closed in a complete space and is hence compact by Proposition 2.4.1 and Theorem 2.3.1. Now
$$\mu(S \setminus K) \leq \sum_{m=1}^{\infty} \varepsilon/2^m = \varepsilon.$$
So μ is tight, and the theorem follows from Theorem 7.1.3. □

For all finite Borel measures on a metric space S to be regular, it is not necessary for S to be complete: first, recall that some noncomplete metric spaces, such as $(0, 1)$ with usual metric, are homeomorphic to complete ones ("topologically complete": Theorem 2.5.4). On the other hand, a topological space is called σ-compact iff it is a union of countably many compact sets. It follows directly from Theorem 7.1.3 that a finite Borel measure on a σ-compact metric space is always regular, although some σ-compact spaces, such as the space \mathbb{Q} of rational numbers, are not even topologically complete (Corollary 2.5.6). Ulam's theorem is more interesting for metric spaces, such as separable, infinite-dimensional Banach spaces, which are not σ-compact. On a separable metric space which is "bad" enough, to the point of being a non-Lebesgue measurable subset of $[0, 1]$, for example, not all finite Borel measures are regular (Problem 9 below).

Now among spaces which may not be metrizable, some of the most popular are (locally) compact spaces. After a first fact here, regularity in compact Hausdorff spaces will be developed in §7.3.

7.1.5. Theorem *For any compact Hausdorff space K, any finite Baire measure μ on K is regular.*

Proof. Let f be any continuous real function on K and F any closed set in \mathbb{R}. Then $f^{-1}(F)$ is a compact Baire set and hence regular. Now $\mathbb{R}\backslash F$ is a countable union of closed sets F_n, as in the proof of Theorem 7.1.3. Thus

$$K\backslash f^{-1}(F) = \bigcup_n f^{-1}(F_n),$$

a countable union of compact sets. So $K\backslash f^{-1}(F)$ is regular. Since such sets $f^{-1}(F)$ generate the Baire σ-algebra, by Proposition 7.1.2 all Baire sets are regular. $\qquad\square$

Problems

1. For a finite measure space (X, \mathcal{S}, μ), suppose there are points x_i and numbers $t_i > 0$ such that for any $A \in \mathcal{S}, \mu(A) = \sum\{t_i : x_i \in A\}$. (Then μ is purely atomic, as defined in §3.5.) Suppose that the singleton $\{x\} \in \mathcal{S}$ for each $x \in X$. Show that every set in \mathcal{S} is regular.

2. Let μ be a finite Borel measure on a separable metric space S. Assume $\mu(\{x\}) = 0$ for all $x \in S$. Prove that there is a set A of first category (a countable union of nowhere dense sets; see Theorem 2.5.2) with $\mu(S\backslash A) = 0$.

3. If μ is a Borel measure on a topological space S, then a closed set F is called the *support* of μ if it is the smallest closed set H such that $\mu(S\backslash H) = 0$. Prove that if S is metrizable and separable, the support of μ always exists. *Hint:* Use Proposition 2.1.4.

4. For any closed set F in a separable metric space, prove that there exists a finite Borel measure μ with support F. *Hints:* Apply §2.1, Problem 5. Define a measure as in Problem 1.

5. Let (S, d) be a complete separable metric space with S non-empty. Suppose S has no isolated points, that is, for each $x \in S$, x is in the closure of $S\backslash\{x\}$. Prove that there exists a Borel measure μ on S with $\mu(S) = 1$ and $\mu(\{x\}) = 0$ for all $x \in S$. *Hint:* Define a 1–1 Borel measurable function f from $[0, 1]$ into S and let $\mu = \lambda \circ f^{-1}$.

6. Show that there is a (non-Hausdorff) topology on a set X of two points for which the Baire and Borel σ-algebras are different.

7. A collection \mathcal{L} of subsets of a set X will be called a *lattice* iff $\emptyset \in \mathcal{L}$, $X \in \mathcal{L}$, and for any A and B in \mathcal{L}, $A \cup B \in \mathcal{L}$ and $A \cap B \in \mathcal{L}$. Show that then the collection \mathcal{D} of all sets $A\backslash B$ for A and B in \mathcal{L} is a semiring (§3.2). Then show that the algebra \mathcal{A} generated by (smallest algebra including) a lattice \mathcal{L} is the collection of all finite unions $\bigcup_{1\le i\le n} A_i\backslash B_i$ for A_i and B_i in \mathcal{L}, where the sets $A_i\backslash B_i$ can be taken to be disjoint for different i. (In any topological space, the collection of all open sets forms a lattice, as does the collection of all closed sets.)

8. A lattice \mathcal{L} of sets is called a *σ-lattice* iff for any sequence $\{A_n\} \subset \mathcal{L}$, we have $\bigcup_{n\ge 1} A_n \in \mathcal{L}$ and $\bigcap_{n\ge 1} A_n \in \mathcal{L}$. For any Hausdorff topological space (X, \mathcal{T}) and σ-algebra \mathcal{S} of subsets of S, and for any finite measure μ on \mathcal{S}, show that the collection of all regular sets in \mathcal{S} for μ is a σ-lattice, and so is the collection of all closed regular sets. *Hint:* See the proof of Proposition 7.1.2.

9. Let A be a nonmeasurable set for Lebesgue measure λ on $[0, 1]$ with outer measure $\lambda^*(A) = 1$ by Theorem 3.4.4. Define a measure μ on sets B which are Borel subsets of A by $\mu(B) = \lambda^*(B)$, using Theorem 3.3.6. Show that the collection of regular measurable sets for μ is not an algebra. *Hint:* Is A regular?

10. Let X be a countable set. Show that every σ-algebra \mathcal{A} of subsets of X is the Borel σ-algebra for some topology \mathcal{T} on X. *Hint:* Try $\mathcal{T} = \mathcal{A}$.

11. Let $I := [0, 1]$ Show that the set $C[0, 1]$ of all continuous real functions on I is a Borel set in \mathbb{R}^I, the set of all real functions on I, with product topology. *Hint:* Show that $C[0, 1]$ is a countable intersection of countable

unions of closed sets F_{mn}, where $F_{mn} := \{f : |x - y| \le 1/m$ implies $|f(x) - f(y)| \le 1/n\}$.

12. Let μ be a finite Borel measure on a separable metric space. Show that for every atom A of μ, as defined in §3.5, there is an $x \in A$ with $\mu(A) = \mu(\{x\})$. Thus, show that μ is purely atomic if and only if it has the form given in Problem 1, and μ is nonatomic if and only if $\mu(\{x\}) = 0$ for all x.

*7.2. Lebesgue's Differentiation Theorems

Let λ denote Lebesgue measure, $dx := d\lambda(x)$. The first theorem is a form of the fundamental theorem of calculus for the Lebesgue integral. In the classical form of the theorem, the integrand g is a continuous function and the derivative of the indefinite integral equals g everywhere. For a Lebesgue integrable function, such as the indicator function of the set of irrational numbers, we can't expect the equation to hold everywhere, but it will hold almost everywhere:

7.2.1. Theorem *If a function g is integrable on an interval $[a, b]$ for Lebesgue measure, then for λ-almost all $x \in (a, b)$,*

$$\frac{d}{dx} \int_a^x g(t)\, dt = g(x).$$

Proof. Two lemmas will be useful.

7.2.2. Lemma *Let \mathcal{U} be a collection of open intervals in \mathbb{R} with bounded union W. Then for any $t < \lambda(W)$ there is a finite, disjoint subcollection $\{V_1, \ldots, V_q\} \subset \mathcal{U}$ such that*

$$\sum_{i=1}^{q} \lambda(V_i) > \frac{t}{3}.$$

Proof. By regularity (Theorem 7.1.4) take a compact $K \subset W$ with $\lambda(K) > t$. Then take finitely many $U_1, \ldots, U_n \in \mathcal{U}$ such that $K \subset U_1 \cup \cdots \cup U_n$, numbered so that $\lambda(U_1) \ge \lambda(U_2) \ge \cdots \ge \lambda(U_n)$. Let $V_1 := U_1$. For $j = 2, 3, \ldots$, define V_j recursively by $V_j := U_m := U_{m(j)}$ for the least m such that U_m does not intersect any V_i for $i < j$. Let W_i be the interval with the same center as V_i, but three times as long. Then for each $r \ge 2$, either U_r is one of the V_j or U_r intersects $V_i = U_k$ for some $k < r$, so $\lambda(U_r) \le \lambda(U_k)$ and U_r is included in W_i (see Figure 7.2).

Figure 7.2

Hence if q is the largest i such that V_i is defined,

$$t < \lambda(K) < \lambda\left(\bigcup_{j=1}^{n} U_j\right) \le \sum_{i=1}^{q} \lambda(W_i) = 3\sum_{i=1}^{q} \lambda(V_i). \qquad \square$$

7.2.3. Lemma *If μ is a finite Borel measure on an interval $[a, b]$ and A is a Borel subset of $[a, b]$ with $\mu(A) = 0$, then for λ-almost all $x \in A$, $(d/dx)\mu([a, x]) = 0$.*

Proof. It will be enough to show that for λ-almost all $x \in A$, $\lim_{h\downarrow 0} \mu((x - h, x + h))/h = 0$. For $j = 1, 2, \ldots$, let

$$P_j := \left\{x \in A: \limsup_{h\downarrow 0} \mu((x - h, x + h))/h > 1/j\right\}.$$

Here the lim sup can be restricted to rational $h > 0$ since the function of h in question is continuous from the left in h. Also, the function $1_{(x-h, x+h)}(y)$ is jointly Borel measurable in x, h, and y, so by the Tonelli-Fubini theorem (and its proof) $\mu((x - h, x + h))$ is jointly Borel measurable in x and h. Thus the lim sup is measurable and P_j is measurable.

Given $\varepsilon > 0$, take an open $V \supset A$ with $\mu(V) < \varepsilon$ (such a V exists by closed regularity). For each $x \in P_j$ there is an $h > 0$ such that $(x - h, x + h) \subset V$ and $\mu((x - h, x + h)) > h/j$. Such intervals $(x - h, x + h)$ cover P_j, so by Lemma 7.2.2, for any $t < \lambda(P_j)$ there are finitely many disjoint intervals $J_1, \ldots, J_q \subset V$ with

$$t \le 3\sum_{i=1}^{q} \lambda(J_i) \le 6j\sum_{i=1}^{q} \mu(J_i) \le 6j\mu(V) \le 6j\varepsilon.$$

Thus $\lambda(P_j) \le 6j\varepsilon$. Letting $\varepsilon \downarrow 0$ gives $\lambda(P_j) = 0$, $j = 1, 2, \ldots$, and letting $j \to \infty$ proves the lemma. $\qquad \square$

Now to prove Theorem 7.2.1, for each rational r let

$$g_r := \max(g - r, 0), \qquad f_r(x) := \int_a^x g_r(t)\, dt.$$

Then by Lemma 7.2.3, applied to μ where $\mu(E) := \int_E g_r(t)\,d\lambda(t)$ for any Borel set E, $df_r(x)/dx = 0$ for λ–almost all x such that $g(x) \le r$. Let B be the union over all rational r of the sets of measure 0 where $g(x) \le r$ but $df_r(x)/dx$ is not 0. Then $\lambda(B) = 0$. Whenever $a \le x < x + h \le b$,

$$\int_x^{x+h} g(t)\,dt - rh = \int_x^{x+h} g(t) - r\,dt \le \int_x^{x+h} g_r(t)\,dt, \quad \text{so}$$

$$\frac{1}{h}\int_x^{x+h} g(t)\,dt \le r + \frac{1}{h}\int_x^{x+h} g_r(t)\,dt. \quad \text{Likewise if } a \le x - h < x \le b,$$

$$\frac{1}{h}\int_{x-h}^x g(t)\,dt \le r + \frac{1}{h}\int_{x-h}^x g_r(t).$$

For a function f from an open interval containing a point x into \mathbb{R}, the upper and lower, left and right derived numbers at x will be defined as follows:

$$UR(f, x) := \limsup_{h\downarrow 0}(f(x+h) - f(x))/h \ge$$

$$LR(f, x) := \liminf_{h\downarrow 0}(f(x+h) - f(x))/h, \quad \text{and}$$

$$UL(f, x) := \limsup_{h\downarrow 0}(f(x) - f(x-h))/h \ge$$

$$LL(f, x) := \liminf_{h\downarrow 0}(f(x) - f(x-h))/h.$$

These quantities are always defined but possibly $+\infty$ or $-\infty$. Now f has a (finite) derivative at x if and only if the four derived numbers are all equal (and finite).

If x is not in B and $r > g(x)$, then we have $df_r(x)/dx = 0$. Next, for $f(x) := \int_a^x g(t)\,dt$, letting $h \downarrow 0$, we have $UR(f, x) \le r$ and $UL(f, x) \le r$. Letting $r \downarrow g(x)$ then gives $UR(f, x) \le g(x)$ and $UL(f, x) \le g(x)$. Considering $-g$ likewise then gives $UR(-f, x)$ and $UL(-f, x)$ both $\le -g(x)$, so $LL(f, x) \ge g(x)$ and $LR(f, x) \ge g(x)$. For such x, all four derived numbers thus equal $g(x)$, so we have

$$\frac{d}{dx}\int_a^x g(t)\,dt = g(x), \qquad \text{proving Theorem 7.2.1.} \qquad \square$$

Other functions which will be shown (Theorem 7.2.7) to have derivatives almost everywhere (although they may not be the indefinite integrals of them) are the functions of bounded variation, defined as follows. For any real-valued function f on a set $J \subset \mathbb{R}$, its *total variation* on J is defined by

$$\text{var}_f J := \sup\left\{\sum_{i=1}^n |f(x_i) - f(x_{i-1})| : x_0 < x_1 < \cdots < x_n, x_i \in J\right\},$$

where $x_i \in J$ for $i = 0, 1, \ldots, n$. If $\text{var}_f J < +\infty$, f is said to be of *bounded variation* on J. Often J is a closed interval $[a, b]$.

Functions of bounded variation are characterized as differences of nondecreasing functions:

7.2.4. Theorem *A real function f on $[a, b]$ is of bounded variation if and only if $f = F - G$ where F and G are nondecreasing real functions on $[a, b]$.*

Proof. If F is nondecreasing (written "$F\!\uparrow$"), meaning that $F(x) \leq F(y)$ for $a \leq x \leq y \leq b$, then clearly $\text{var}_F[a, b] = F(b) - F(a) < +\infty$. Then if also $G\!\uparrow$, $\text{var}_{F-G}[a, b] \leq \text{var}_F[a, b] + \text{var}_G[a, b] < +\infty$.

Conversely, if f is any real function of bounded variation on $[a, b]$, let $F(x) := \text{var}_f[a, x]$. Then clearly $F\!\uparrow$. Let $G := F - f$. Then for $a \leq x \leq y \leq b$,

$$G(y) - G(x) = F(y) - F(x) - (f(y) - f(x)) \geq 0$$

because $\text{var}_f[a, x] + f(y) - f(x) \leq \text{var}_f[a, y]$. So $G\!\uparrow$ and $f = F - G$. $\quad\square$

For any countable set $\{q_n\} \subset \mathbb{R}$, one can define a nondecreasing function F, with $0 \leq F \leq 1$, by $F(t) := \sum\{2^{-n} : q_n \leq t\}$. Then F is discontinuous, with a jump of height 2^{-n} at each q_n. Here $\{q_n\}$ might be the dense set of rational numbers. This shows that any countable set can be a set of discontinuities, as in the next fact:

7.2.5. Theorem *If f is a real function of bounded variation on $[a, b]$, then f is continuous except at most on a countable set.*

Proof. By Theorem 7.2.4 we can assume $f\!\uparrow$. Then for $a < x \leq b$ we have limits $f(x^-) := \lim_{u\uparrow x} f(u)$ and for $a \leq x < b$, $f(x^+) := \lim_{v\downarrow x} f(v)$. If $a \leq u < x < v \leq b$, then $\text{var}_f[a, b] \geq |f(x) - f(u)| + |f(v) - f(x)|$. Letting $u \uparrow x$ and $v \downarrow x$ gives

$$\text{var}_f[a, b] \geq |f(x) - f(x^-)| + |f(x^+) - f(x)|.$$

Similarly, for any finite set $E \subset (a, b)$, we have

$$\text{var}_f[a, b] \geq \sum_{x \in E} |f(x^+) - f(x)| + |f(x) - f(x^-)|.$$

Thus for $n = 1, 2, \ldots$, there are at most finitely many $x \in [a, b]$ with $|f(x^+) - f(x)| > 1/n$ or $|f(x) - f(x^-)| > 1/n$. Thus, except at most on a countable set, we have $f(x^+) = f(x) = f(x^-)$, so f is continuous at x. $\quad\square$

Recall that f is *continuous from the right* at x iff $f(x^+) = f(x)$ (as in §§3.1 and 3.2). Thus $1_{[0,1)}$ is continuous from the right, but $1_{[0,1]}$ is not, for example. For any finite, nonnegative measure μ, and $a \in \mathbb{R}$, $f(x) := \mu((a, x])$ gives a nondecreasing function f, continuous from the right. Conversely, given such an f, there is such a μ by Theorem 3.2.6. More precisely, this relationship between measures μ and functions f can be extended to signed measures and functions of bounded variation, as follows:

7.2.6. Theorem *Given a real function f on $[a, b]$, the following are equivalent:*

(i) There is a countably additive, finite signed measure μ on the Borel subsets of $[a, b]$ such that $\mu((a, x]) = f(x) - f(a)$ for $a \le x \le b$;

(ii) f is of bounded variation, and continuous from the right on $[a, b]$.

Proof. If (i) holds, then f is continuous from the right on $[a, b)$ by countable additivity. By the Hahn-Jordan decomposition (Theorem 5.6.1) it can be assumed that μ is a nonnegative, finite measure. But then f is nondecreasing, so its total variation is $f(b) - f(a) < +\infty$, as desired.

Assuming (ii), then by Theorem 7.2.4, we have $f = g - h$ for some nondecreasing g and h, and the conclusion for g and h will imply it for f, so it can be assumed that f is nondecreasing. Then the result holds by Theorem 3.2.6. \square

Before continuing, let's recall the example of the Cantor function g defined in the proof of Proposition 4.2.1. This is a nondecreasing, continuous function from $[0, 1]$ onto itself, which is constant on each of the "middle third" intervals: $g = 1/2$ on $[1/3, 2/3]$, $g = 1/4$ on $[1/9, 2/9]$, and so forth. Then $g'(x) = 0$ for $1/3 < x < 2/3$, for $1/9 < x < 2/9$, for $7/9 < x < 8/9$, and so forth. So $g'(x) = 0$ almost everywhere for Lebesgue measure on $[0, 1]$, but g is not an indefinite integral of its derivative as in Theorem 7.2.1, as it is not constant, although it is continuous. (Physicists generally ignored such functions and the measures they define until the advent of "strange attractors," on which see Grebogi et al., 1987.)

Here is another main fact in Lebesgue's theory. Recall that "almost everywhere" ("a.e.") means except on a set of Lebesgue measure 0.

7.2.7. Theorem *For any real function f of bounded variation on an interval (a, b), the derivative $f'(x)$ exists a.e. and is in \mathcal{L}^1 of Lebesgue measure on (a, b).*

Proof. Let $g(x) := f(x^+)$ and $h(x) := g(x) - f(x)$ for $a < x < b$. Then $h(x) = 0$ except for x in a countable set C by Theorem 7.2.5. Let $V := \mathrm{var}_f(a, b)$. For any finite $F \subset C$ and $x \in F$, take $y_x > x$ close enough to x so that the intervals $[x, y_x]$ are disjoint, so $\sum_{x \in F} |f(y_x) - f(x)| \leq V$. Letting $y_x \downarrow x$ for each $x \in F$ gives $\sum_{x \in F} |h(x)| \leq V$. Then letting $F \uparrow C$ gives $\sum_{x \in C} |h(x)| \leq V$, so h and $g \equiv f + h$ are of bounded variation and it will be enough to prove the theorem for g and for h. First, for h, let $v(A) := \sum_{x \in A \cap C} |h(x)|$, a finite measure defined on all Borel subsets A of (a, b). Since v is concentrated on the countable set C, we have for any $s \in (a, b)$ that $(d/dx)v([s, x]) = 0$ for $x \in B$ where $\lambda((s, b) \setminus B) = 0$. If $s < t < u < v$ in (a, b) then $|h(v) - h(u)|/(v - u) \leq v((u, v])/(v - u)$ if $h(u) = 0$ or $\leq v((t, v])/(v - u)$ if $h(v) = 0$. Letting $t \uparrow v$ if $v \in B$ or $v \downarrow u$ if $u \in B$ we get that $h' = 0$ on B. Letting $s \downarrow a$ we then have $h' = 0$ λ-almost everywhere on (a, b).

So, it will be enough to prove the theorem for g, in other words, when f is right-continuous. Again by Theorem 7.2.4, it can be assumed that $f \uparrow$. We can extend f to $[a, b]$ without changing its total variation by setting $f(a) := f(a^+)$, $f(b) := f(b^-)$. Let μ be the measure on the Borel sets of $[a, b]$ with $\mu((a, x]) = f(x) - f(a)$ for $a \leq x \leq b$ (and $\mu(\{a\}) = 0$), given by Theorem 7.2.6. By the Lebesgue decomposition (Theorem 5.5.3), we have $\mu = \mu_{ac} + \mu_{sing}$ where μ_{ac} is absolutely continuous and μ_{sing} is singular with respect to Lebesgue measure. Let $g(x) := \mu_{ac}((a, x])$, $h(x) := \mu_{sing}((a, x])$, for $a < x \leq b$. Then by Lemma 7.2.3, $h'(x) = 0$ a.e. By the Radon-Nikodym theorem (5.5.4), there is a function $j \in \mathcal{L}^1([a, b], \lambda)$ with

$$g(x) = \int_a^x j(t)\, dt, \qquad a < x \leq b.$$

Thus a.e., by Theorem 7.2.1, $g'(x)$ exists and equals $j(x)$, and $f'(x) = j(x)$ a.e. $\qquad \square$

A function f on an interval $[a, b]$ is called *absolutely continuous* iff for every $\varepsilon > 0$ there is a $\delta > 0$ such that for any disjoint subintervals $[a_i, b_i)$ of $[a, b]$ for $i = 1, 2, \ldots$, with $a_i < b_i$ and $\sum_i (b_i - a_i) < \delta$, we have $\sum_i |f(b_i) - f(a_i)| < \varepsilon$.

The first four problems have to do with absolute continuity of functions. Together with Theorem 7.2.1 and the proof of Theorem 7.2.7, they show that a function f on $[a, b]$ is absolutely continuous iff $f(x) - f(a) \equiv \mu((a, x])$ for some signed measure μ absolutely continuous with respect to Lebesgue measure.

Problems

1. Show that any absolutely continuous function f is of bounded variation. *Hint:* If f has unbounded variation, it does so on arbitrarily short intervals.

2. Assume Problem 1. Let f be a function of bounded variation, continuous from the right on $[a, b]$. Show that f is absolutely continuous if and only if the signed measure μ defined by Theorem 7.2.6 is absolutely continuous with respect to Lebesgue measure. *Hint:* Compare Problem 4 in §5.5.

3. Assume Problems 1 and 2. Show that f is absolutely continuous on $[a, b]$ if and only if for some $j \in \mathcal{L}^1([a, b], \lambda)$, $f(x) = f(a) + \int_a^x j(t)\, dt$ for all $x \in [a, b]$. *Hint:* This was a historical predecessor of the Radon-Nikodym theorem 5.5.4.

4. Show that if f and g are absolutely continuous on $[a, b]$, so is their product fg.

5. Let $\{J_i\}_{1 \le i \le n}$ be a finite collection of bounded open intervals in \mathbb{R}. Show that there is a subcollection, consisting of disjoint intervals, whose total length is at least $\lambda(\bigcup_i J_i)/2$. Use this to give another proof of Lemma 7.2.2. *Hints:* We can assume one of the J_i is empty. Let each non-empty $J_i = (a_i, b_i)$. Choose V_1 with the smallest value of a_i and among such, with the largest value of b_i. Let $V_j = (c_j, d_j)$, so far for $j = 1$. Recursively, given V_j, choose V_{j+1} as a J_i, if one exists, such that $d_j \in J_i$, and among such, with the largest b_i. If no such i exists, let $V_{j+1} := \emptyset$ and let V_{j+2} be a J_i, if one exists, with $a_i \ge d_j$, and among such J_i, one with smallest a_i and then with largest b_i. If no such J_i exists either, the construction ends. Take the collection of disjoint intervals as either $V_1, V_3, \ldots,$ or $V_2, V_4, \ldots,$ where any empty interval can be deleted.

6. Let \mathcal{S} be a collection of open subintervals of \mathbb{R} with union U. Suppose that for each $\varepsilon > 0$, every point of U is contained in an interval in \mathcal{S} with length less than ε. Show that there is a sequence $\{V_n\} \subset \mathcal{S}$, consisting of disjoint intervals, such that the union $V := \bigcup_n V_n$ is almost all of U, in other words, $\lambda(U \setminus V) = 0$. *Hint:* Use Problem 5 and an iteration.

7. Let $\{D_i\}_{1 \le i \le n}$ be any finite collection of open disks in \mathbb{R}^2. Thus $D_i := B(p_i, r_i) := \{z \in \mathbb{R}^2 : |z - p_i| < r_i\}$ where $r_i > 0$, $p_i \in \mathbb{R}^2$, and $|\cdot|$ is the usual Euclidean length of a vector in \mathbb{R}^2. Show that there is a collection \mathcal{D} of disjoint D_i whose union V has area $\lambda^2(V) \ge \lambda^2(\bigcup_i D_i)/4$. *Hint:* Use induction. Take a D_i of largest radius, put it in \mathcal{D}, exclude all D_j which intersect it, and apply the induction assumption to the rest.

8. Let \mathcal{E} be a collection of open disks in \mathbb{R}^2 with union U such that for every point $x \in U$ and $\varepsilon > 0$, there is a disk $D \in \mathcal{E}$ containing x of radius

less than ε. Show that there is a disjoint subcollection \mathcal{D} of \mathcal{E} with union V such that $\lambda^2(U \backslash V) = 0$. *Hint:* See Problems 6 and 7.

9. Extend Problems 5–8 to \mathbb{R}^d for $d \geq 3$, with constant $1/2^d$.

10. Let μ be a measure on \mathbb{R}^k, absolutely continuous with respect to Lebesgue measure λ^k with Radon-Nikodym derivative $f = d\mu/d\lambda^k$. Show that for λ^k-almost all x, and any $\varepsilon > 0$, there is a $\delta > 0$ such that for any ball $B(p, r)$ containing x $(|x - p| < r)$ with $r < \delta$, we have $|f(x) - \mu(B(p, r))/\lambda^k(B(p, r))| < \varepsilon$.

*7.3. The Regularity Extension

Recall that a topological space (S, \mathcal{T}) is called *locally compact* if and only if each point x of S has a compact neighborhood (that is, for some open V and compact L, $x \in V \subset L$). Some of the main examples of compact Hausdorff nonmetrizable spaces are products of uncountably many compact Hausdorff spaces (Tychonoff's theorem, 2.2.8). For measure theory on locally compact but nonmetrizable spaces, the following theorem is basic. Its proof will occupy the rest of the section. The theorem is not useful for metric spaces, where all Borel sets are Baire sets (Theorem 7.1.1).

7.3.1. Theorem *Let K be a compact Hausdorff space and μ any finite Baire measure on K. Then μ has an extension to a Borel measure on K, and a unique regular Borel extension.*

Proof. For any compact set L let $\mu^*(L) := \inf\{\mu(A): A \supset L\}$ where A runs over Baire sets.

7.3.2. Lemma *For any disjoint compact L and M,*

$$\mu^*(L \cup M) = \mu^*(L) + \mu^*(M).$$

Proof. We know that a compact Hausdorff space is normal (Theorem 2.6.2) and that any two disjoint compact sets can be separated by a continuous real function (Lemma 2.6.3). Thus there exist disjoint Baire sets A and B with $A \supset L$ and $B \supset M$. Then for any Baire set $C \supset L \cup M$,

$$\mu(C) \geq \mu(C \cap A) + \mu(C \cap B) \geq \mu^*(L) + \mu^*(M),$$

so $\mu^*(L \cup M) \geq \mu^*(L) + \mu^*(M)$. The converse inequality always holds (see Lemma 3.1.5), so we have equality. $\qquad\square$

Now for any Borel set B let

$$\nu(B) := \sup\{\mu^*(L): L \text{ compact}, L \subset B\}.$$

If B is compact, clearly $\nu(B) = \mu^*(B)$. If $B \subset C$, then $\nu(B) \leq \nu(C)$. Let $\mathcal{M}(\nu)$ be the collection of all Borel sets B such that $\nu(A) = \nu(A \cap B) + \nu(A \backslash B)$ for all Borel sets A. By Lemma 7.3.2, we always have $\nu(A) \geq \nu(A \cap B) + \nu(A \backslash B)$, so

(7.3.3) $B \in \mathcal{M}(\nu)$ if and only if $\nu(A) \leq \nu(A \cap B) + \nu(A \backslash B)$ for all Borel sets A.

Also in (7.3.3), we can replace "Borel" by "compact," since in one direction, all compact sets are Borel; in the other, if $\nu(L) \leq \nu(L \cap B) + \nu(L \backslash B)$ for all compact $L \subset A$, then since $\mu^*(L) = \nu(L)$, the definition of ν gives the inequality in (7.3.3). Now $\mathcal{M}(\nu)$ is an algebra and ν is finitely additive on it, by a subset of the proof of Lemma 3.1.8.

7.3.4. Lemma *For any closed $L \subset K$, $L \in \mathcal{M}(\nu)$.*

Proof. We need to show that for any compact $M \subset K$,

$$\mu^*(M) \leq \mu^*(M \cap L) + \sup\{\mu^*(N): N \subset M \backslash L, N \text{ compact}\}.$$

Given $\varepsilon > 0$, take a Baire set $V \supset M \cap L$ with $\mu(V) < \mu^*(M \cap L) + \varepsilon$. By Theorem 7.1.5, we can take V to be open. Then $M \backslash V$ is compact and included in $M \backslash L$, so

$$\mu^*(M \cap L) + \nu(M \backslash L) \geq \mu(V) - \varepsilon + \mu^*(M \backslash V).$$

For any Baire set $W \supset M \backslash V$, $M \subset V \cup W$ so $\mu^*(M) \leq \mu(V) + \mu(W)$. Letting $\mu(W) \downarrow \mu^*(M \backslash V)$ and $\varepsilon \downarrow 0$ gives the desired result. $\qquad \square$

7.3.5. Lemma *$\mathcal{M}(\nu)$ is the σ-algebra of all Borel sets and ν is a measure on it.*

Proof. We know that $\mathcal{M}(\nu)$ is an algebra and from Lemma 7.3.4 that it contains all open, closed, and compact sets. It remains to check monotone convergence properties.

First let B_n be Borel sets, $B_n \downarrow \emptyset$. Suppose $\nu(B_n) \geq \varepsilon > 0$. Take compact $K_n \subset B_n$ with $\mu^*(K_n) \geq \nu(B_n) - \varepsilon/3^n$. Let $L_n := \cap_{1 \leq j \leq n} K_j$. Since L_n and

K_j are in $\mathcal{M}(\nu)$, being compact, we have $\nu(B_n \backslash K_n) \leq \varepsilon/3^n$ for all n. Then since

$$B_n \backslash L_n \subset \bigcup_{i \leq j \leq n} B_j \backslash K_j,$$

it follows that $\nu(B_n \backslash L_n) \leq \varepsilon/2$ for all n, and $\nu(L_n) \geq \varepsilon/2 > 0$. Thus $L_n \neq \emptyset$. But $L_n \subset B_n \downarrow \emptyset$ implies $L_n \downarrow \emptyset$, contradicting compactness: $K \subset \bigcup_n K \backslash L_n$ with no finite subcover. Thus $\nu(B_n) \downarrow 0$.

Now let $E_n \in \mathcal{M}(\nu)$ and $E_n \uparrow E$. We want to prove $E \in \mathcal{M}(\nu)$ and $\nu(E_n) \uparrow \nu(E)$. Since E is a Borel set, for each n, $\nu(E) = \nu(E_n) + \nu(E \backslash E_n)$, and $E \backslash E_n \downarrow \emptyset$, so $\nu(E \backslash E_n) \downarrow 0$, as was just proved. Thus $\nu(E_n) \uparrow \nu(E)$.

For any Borel set A, and all n,

$$\nu(A) = \nu(A \cap E_n) + \nu(A \backslash E_n) \leq \nu(A \cap E) + \nu(A \backslash E_n).$$

Let $D_n := A \backslash E_n \downarrow D := A \backslash E$. We want to show $\nu(D_n) \downarrow \nu(D)$. Clearly $\nu(D_n) \downarrow \alpha$ for some $\alpha \geq \nu(D)$. If $\alpha > \nu(D)$, take $0 < \varepsilon < \alpha - \nu(D)$. For each $n = 1, 2, \dots$, take a compact $C_n \subset D_n$ with $\mu^*(C_n) > \nu(D_n) - \varepsilon/3^n$. Let $F_n := \bigcap_{1 \leq j \leq n} C_j$. Then for each n,

$$C_n \subset F_n \cup \bigcup_{1 \leq i \leq n} C_n \backslash C_i.$$

Since all compact sets are in the algebra $\mathcal{M}(\nu)$,

$$\nu(C_n) \leq \nu(F_n) + \sum_{1 \leq i \leq n} \nu(C_n \backslash C_i).$$

Since $C_n \subset D_n \subset D_i$ for $i \leq n$,

$$\nu(C_n) \leq \nu(F_n) + \sum_{1 \leq i \leq n} \varepsilon/3^i.$$

Hence $\nu(F_n) \geq \nu(C_n) - \varepsilon/2 \geq \alpha - \varepsilon$. Let $F := \bigcap_{1 \leq n < \infty} F_n = \bigcap_{1 \leq n < \infty} C_n$. Then $F \subset D$ and F is compact, so $\nu(D) \geq \mu^*(F)$. Since F_n and F are in $\mathcal{M}(\nu)$, and $F_n \downarrow F$, we have $F_n \backslash F \downarrow \emptyset$ and $\nu(F_n) \downarrow \nu(F)$ by the first part of the proof. But then $\mu^*(F) = \nu(F) \geq \alpha - \varepsilon > \nu(D)$, a contradiction. So $\nu(D_n) \downarrow \nu(D)$, and $\nu(A) \leq \nu(A \cap E) + \nu(D)$, so $E \in \mathcal{M}(\nu)$. A Borel set $E \in \mathcal{M}(\nu)$ iff $K \backslash E \in \mathcal{M}(\nu)$. Thus if $E_n \in \mathcal{M}(\nu)$ and $E_n \downarrow E$ then $E \in \mathcal{M}(\nu)$. So $\mathcal{M}(\nu)$ is a monotone class and by Theorem 4.4.2 $\mathcal{M}(\nu)$ is a σ-algebra and ν is a measure on it, proving Lemma 7.3.5. $\qquad\square$

Now returning to the proof of Theorem 7.3.1, for any Baire set B and compact $L \subset B$, we have $\mu(B) \geq \mu^*(L)$, so $\mu(B) \geq \nu(B)$. Likewise $\mu(K \backslash B) \geq$

$v(K \setminus B)$, so $\mu(B) \leq v(B)$. Then $\mu(B) = v(B)$. So v extends μ. Since $v = \mu^*$ on compact sets, v is regular by definition.

Now to prove uniqueness, let ρ be another regular extension of μ to the Borel sets. Then for any compact L, $\rho(L) \leq \mu^*(L)$, so for any Borel set B, $\rho(B) \leq v(B)$ by regularity. Likewise $\rho(K \setminus B) \leq v(K \setminus B)$, and $\rho(K) = v(K) = \mu(K)$, so $\rho(B) = v(B)$, proving Theorem 7.3.1. □

Problems

1. Suppose μ is finitely additive and nonnegative on an algebra \mathcal{A} of subsets of X with $\mu(X) < \infty$. Let (X, \mathcal{F}) be a Hausdorff topological space. Suppose that μ is regular in the sense that for each $A \in \mathcal{A}$, $\mu(A) = \sup \{\mu(K): K \subset A, K \in \mathcal{A}, K \text{ compact}\}$. Show that μ is countably additive on \mathcal{A}. *Hint:* Use Theorem 3.1.1.

2. Let (K, \leq) be an uncountable well-ordered set with a largest element Ω such that for any $x < \Omega$, $\{y: y \leq x\}$ is countable. On K, take the interval topology, a subbase for which is given by $\{\{x: x < \beta\}: \beta \in K\} \cup \{\{x: x > \alpha\}: \alpha \in K\}$. Then K is a compact Hausdorff space. For any Borel set A, let $\mu(A) = 1$ if A includes an uncountable set relatively closed in $\{x: x < \Omega\}$, $\mu(A) = 0$ otherwise. Show that μ is a nonregular measure and does not have a support (F is the *support* of μ iff F is the smallest closed set whose complement has measure 0). *Hints:* If C and D are uncountable closed sets, show that $C \cap D$ is uncountable: for any $a \in K$, take $a < c_1 < d_1 < c_2 < d_2 < \cdots, c_i \in C$ and $d_i \in D$. Then $\sup c_i = \sup d_i \in C \cap D$. Let $\mathcal{C} := \{A: A \text{ or } K \setminus A \text{ includes an uncountable closed set}\}$. Use monotone classes (Theorem 4.4.2) to show that \mathcal{C} contains all Borel sets. Use Theorem 3.1.1 to show that μ is countably additive.

3. Show that there exists a compact Hausdorff space X which is not the support of any finite regular Borel measure. *Hint:* Let X be uncountable with $\{x\}$ open for all x except one x_o. Let the neighborhoods of x_o be the sets containing x_o and all but finitely many other points. Then X is compact. Show that a subset A of X is the support of a finite regular Borel measure on X if and only if A is countable and, if infinite, contains x_o.

4. Let $I := [0, 1]$ and $S := 2^I$ with product topology. Show that the Baire σ-algebra \mathcal{A} in S is not the Borel σ-algebra of any topology \mathcal{F}. *Hint:* Suppose that it is. Recall the example just after the proof of Theorem 7.1.1. Here is a sketch:
 (a) For each $x \in S$, let D_x be the closure of $\{x\}$ for \mathcal{T}. Show that for some countable $C(x)$, $y \in D_x$ if $y(t) = x(t)$ for all $t \in C(x)$.

(b) Let $x \leq y$ if and only if $y \in D_x$. Show that this is a partial ordering, and $x \leq y$ if and only if $D_y \subset D_x$.

(c) By recursion, define a set $W \subset S$ well-ordered for \leq, where for each $w \in W$, $\{v \in W : v \leq w\}$ is countable. Show that W can be taken to be uncountable, and then the intersection of the D_x for $x \in W$ is closed for \mathcal{T} but not in \mathcal{A}, a contradiction.

5. During the proof of Lemma 7.3.5, it is found that for any Borel sets $B_n \downarrow \emptyset$, we have $\nu(B_n) \downarrow 0$. Can the proof be finished directly, at that point, by applying Theorem 3.1.1? Explain.

*7.4. The Dual of $C(K)$ and Fourier Series

For any compact Hausdorff space K, let $C(K)$ denote the linear space of all continuous real-valued functions on K, with the supremum norm $\|f\|_\infty := \sup\{|f(x)| : x \in K\}$. Then $(C(K), \|\cdot\|_\infty)$ is a Banach space by Theorem 2.4.9, since all continuous functions on K are bounded.

Let (X, \mathcal{S}) be a measurable space, f a real measurable function on X, and μ a signed measure on \mathcal{S}. We take the Jordan decomposition $\mu = \mu^+ - \mu^-$ (Theorem 5.6.1). For any real measurable function f, let $\int f \, d\mu := \int f \, d\mu^+ - \int f \, d\mu^-$ whenever this is defined, so that if $\int f^+ \, d\mu^+ = +\infty$ or $\int f^- \, d\mu^- = +\infty$, then $\int f^- \, d\mu^+$ and $\int f^+ \, d\mu^-$ are finite.

In particular if (X, \mathcal{T}) is a topological space, with $C_b(X, \mathcal{T})$ the space of bounded continuous real functions on X, $f \in C_b(X, \mathcal{T})$, and μ is a finite signed Baire measure on X, then $\int f \, d\mu$ is always defined and finite. Let $I_\mu(f) := \int f \, d\mu$. Then I_μ belongs to the dual Banach space $C_b(X, \mathcal{T})'$. For example, if μ is a point mass δ_x, with $\delta_x(A) := 1_A(x)$ for any (Baire) set A, then $I_\mu(f) = f(x)$ for any function f.

Representations of elements of the dual spaces of the L^p spaces (Theorem 6.4.1) are often called "Riesz representation" theorems, but so is the next theorem:

7.4.1. Theorem *For any compact Hausdorff space X, $\mu \mapsto I_\mu$ is a linear isometry of the space $M(X, \mathcal{T})$ of all finite signed Baire measures on X, with norm $\|\mu\| := |\mu|(X)$, onto $C(X)'$ with dual norm $\|\cdot\|_\infty'$, where $\|T\|_\infty' := \sup\{|T(f)| : \|f\|_\infty \leq 1\}$.*

Proof. Clearly $\mu \mapsto I_\mu$ is a linear map of $M(X, \mathcal{T})$ into $C(X)'$. Given μ, take a Hahn decomposition $X = A \cup (X \backslash A)$ with $\mu^+(X \backslash A) = \mu^-(A) = 0$, by Theorem 5.6.1, where A is a Baire set. By regularity of Baire measures

(Theorem 7.1.5), given $\varepsilon > 0$ take compact $K \subset A$ and $L \subset X \backslash A$ with $\mu^+(A \backslash K) < \varepsilon/4$ and $\mu^-((X \backslash A) \backslash L) < \varepsilon/4$. By Urysohn's lemma (2.6.3) take $f \in C(X)$ with $-1 \le f(x) \le 1$ for all x, $f = 1$ on K, and $f = -1$ on L. Then $\|f\|_\infty \le 1$ and $|\int f \, d\mu| \ge |\mu|(X) - \varepsilon$. Letting $\varepsilon \downarrow 0$ gives that $\|I_\mu\|' = |\mu|(X)$, so $\mu \mapsto I_\mu$ is an isometry.

To prove it is onto, let $L \in C(X)'$. As in the proof of Theorem 6.4.1, around Lemma 6.4.2, we have $L = L^+ - L^-$ where L^+ and L^- are both in $C(X)'$, and for all $f \ge 0$ in $C(X)$, $L^+(f) \ge 0$ and $L^-(f) \ge 0$. Clearly $C(X)$ is a Stone vector lattice, as defined in §4.5. Then L^+ and L^- are pre-integrals by Dini's theorem (2.4.10). Thus by the Stone-Daniell theorem (4.5.2), there are measures ρ and σ with $L^+(f) = \int f \, d\rho$ and $L^-(f) = \int f \, d\sigma$ for all $f \in C(X)$. Here ρ and σ are finite measures since the constant $1 \in C(K)$. Then $\mu := \rho - \sigma \in M(X, \mathcal{T})$ and $L(f) = \int f \, d\mu$ for all $f \in C(X)$, proving Theorem 7.4.1. $\qquad\square$

It then follows from the regularity extension (Theorem 7.3.1) that μ could be taken to be a regular Borel measure. The theorem is often stated in that form.

The rest of this section will be devoted to Fourier series in relation to the uniform boundedness principle. Let S^1 be the unit circle $\{z : |z| = 1\} \subset \mathbb{C}$, where the complex plane \mathbb{C} is defined in Appendix B. Let $C(S^1)$ be the Banach space of all continuous complex-valued functions on S^1 with supremum norm $\|\cdot\|_\infty$. By the complex Stone-Weierstrass theorem (2.4.13), finite linear combinations of powers $z \mapsto z^m$, $m \in \mathbb{Z}$, are dense in $C(S^1)$ for $\|\cdot\|_\infty$ (uniform convergence). Let μ be the natural rotationally invariant probability measure on S^1, namely, $\mu = \lambda \circ e^{-1}$ where $e(x) := e^{2\pi i x}$ and λ is Lebesgue measure on $[0, 1]$. We can also write $d\mu := d\theta/(2\pi)$ where θ is the usual angular coordinate on the circle, $0 \le \theta < 2\pi$ ($\theta = 2\pi x$). For any measure space (X, μ) and $1 \le p < \infty$, recall that the L^p spaces $L^p(X, \mu, K)$ are defined where the field $K = \mathbb{R}$ or \mathbb{C}.

7.4.2. Proposition $\{z^m\}_{m \in \mathbb{Z}}$ *form an orthonormal basis of* $L^2(S^1, \mu, \mathbb{C})$.

Proof. L^2 is complete (a Hilbert space) by Theorem 5.2.1. It is easily checked that the z^m for $m \in \mathbb{Z}$ are orthonormal. If they are not a basis, then by Theorems 5.3.8 and 5.4.9, there is some $f \in \mathcal{L}^2(S^1, \mu)$ with $f \perp z^m$ for all $m \in \mathbb{Z}$ and $\|f\|_2 > 0$. Now $L^2 \subset L^1$ by Theorem 5.5.2. Let \mathcal{A} be the set of all finite linear combinations of powers z^m. Then for any $g \in C(S^1)$, there are $P_n \in \mathcal{A}$ with $\|P_n - g\|_\infty \to 0$. Thus $\int f \cdot (P_n - g) \, d\mu \to 0$. Since $\int f P_n \, d\mu = 0$ for all n (note that the complex conjugates $\overline{P}_n \in \mathcal{A}$ also) we have $\int f g \, d\mu = 0$.

Let $\nu(A) := \int_A f\,d\mu$ for any Borel set $A \subset S^1$. Then by Theorem 7.4.1, $\|I_\nu\|'$ is the total variation of ν, which is clearly $\int |f|\,d\mu > 0$, but $I_\nu = 0$, a contradiction, proving Proposition 7.4.2. □

The *Fourier series* of a function $f \in \mathcal{L}^1(S^1, \mu)$ is defined as the series $\sum_{n\in\mathbb{Z}} a_n z^n$ where $a_n := a_n(f) := \int f(z)z^{-n}\,d\mu(z)$. The series will converge, in different senses, under different conditions. For example, if $f \in \mathcal{L}^2$, its Fourier series converges to f in the \mathcal{L}^2 norm by Proposition 7.4.2 and the definition of orthonormal basis (just above Theorem 5.4.7). For uniform convergence, however, the situation is more complicated:

7.4.3. Proposition *There is an $f \in C(S^1)$ such that the Fourier series of f does not converge to f at $z = 1$ and so does not converge uniformly to f.*

Proof. Let $S_n(f)$ be the value of the nth partial sum of the Fourier series of f at 1, namely,

$$S_n(f) := \sum_{m=-n}^{n} \int f(z)z^{-m}\,d\mu(z) = \int f(z)D_n(z)\,d\mu(z),$$

where $D_n(z) := \sum_{m=-n}^{n} z^m = (z^{n+1}-z^{-n})/(z-1)$, $z \neq 1$, which is real-valued since $z^m + z^{-m}$ is for each m. Then S_n is a continuous linear form on $C(S^1)$, and by Theorem 7.4.1, $\|S_n\|'$ equals the total variation of the signed measure ν defined by $\nu(A) = \int_A D_n(z)\,d\mu(z)$. This total variation is $\int |D_n(z)|\,d\mu(z)$. Then, in view of the uniform boundedness principle (6.5.1), it is enough to prove $\|D_n\|_1 \to \infty$ as $n \to \infty$. By the change of variables $z = e^{ix}$, we want to prove

$$\int_{-\pi}^{\pi} \left|\left(e^{i(n+1)x} - e^{-inx}\right)/(e^{ix} - 1)\right| dx \to \infty$$

or equivalently, dividing numerator and denominator by $e^{ix/2}$,

$$\int_0^{\pi} \left|\sin\left(\left(n+\frac{1}{2}\right)x\right)\right| \Big/ \sin\left(\frac{x}{2}\right) dx \to \infty.$$

Now $\sin(nx + x/2) = \sin(nx)\cos(x/2) + \cos(nx)\sin(x/2)$. For $\theta := x/2$, we have $0 \leq \theta \leq \pi/2$, so $\cos^2\theta \leq \cos\theta$, $1 - 2\cos\theta + \cos^2\theta \leq \sin^2\theta$, and $|1 - \cos\theta| \leq \sin\theta$. So it will be enough to show that

$$\int_0^{\pi} |\sin(nx)| \Big/ \sin\left(\frac{x}{2}\right) dx \to \infty.$$

For this we can use:

7.4.4. Lemma *For any continuous real function g on an interval* $[a, b]$,

$$\lim_{n\to\infty} \int_a^b |\sin(nx)| g(x)\, dx = \frac{2}{\pi} \int_a^b g(x)\, dx.$$

Proof. We can assume that $|g(x)| \le 1$ for $a \le x \le b$. Given $\varepsilon > 0$, take $\delta > 0$ small enough so that $|g(x) - g(y)| < \varepsilon$ whenever $|x - y| < \delta$. Take n large enough so that $2\pi/n < \delta$. For that n, we can decompose $[a, b]$ into disjoint intervals $I(j) := [a_j, b_j)$ of length $2\pi/n$ each, with a leftover interval I_o of length $\lambda(I_o) < 2\pi/n$. The contribution of I_o to the integrals approaches 0 as $n \to \infty$. For each j,

$$\left| \int_{I(j)} |\sin(nx)| (g(x) - g(a_j))\, dx \right| \le 2\pi\varepsilon/n.$$

The sum of these differences over all j is at most $(b - a)\varepsilon$. For each j, the average of $|\sin(nx)|$ over $I(j)$ equals the average of $|\sin t|$ over an interval of length 2π, a complete cycle of periodicity. This average is $2/\pi$. The lemma then follows. □

Next, to continue proving Proposition 7.4.3, it follows that for any $c > 0$, as $n \to \infty$

$$\int_c^\pi \frac{|\sin(nx)|}{\sin(x/2)}\, dx \to \frac{2}{\pi} \int_c^\pi \frac{1}{\sin(x/2)}\, dx.$$

Since $(\sin x)/x \to 1$ as $x \to 0$ and $\int_0^1 1/x\, dx = +\infty$, we can let $c \downarrow 0$ to infer $\|D_n\|_1 \to \infty$, which implies Proposition 7.4.3. □

Problems

1. Let $\mu = \delta_1 - \delta_i + i\delta_{-1}$ on S^1. Under what condition on a complex-valued function $f \in C_b(S^1)$ with $\|f\|_\infty \le 1$ will it be true that $\int f\, d\mu = \|I_\mu\|'$?

2. Let K be any compact Hausdorff space. Prove that the space $C(K)$ of continuous functions on K is dense in the space $L^p(K, \mu)$ for each p with $1 \le p < \infty$ and any finite Baire measure μ on K. *Hint:* If not, then apply Problem 4 of §6.1 and Theorems 6.4.1 and 7.4.1.

3. Show that there is an orthonormal basis of $L^2([0, 2\pi), \lambda, \mathbb{R})$ consisting of functions $x \mapsto a_n \sin(nx)$, $n \ge 1$, and $x \mapsto b_n \cos(nx)$, $n \ge 0$, and evaluate a_n and b_n for all n.

4. Let S be a locally compact Hausdorff space. Let $C_0(S)$ be the space of all continuous real functions on S with compact support. Let L be a linear

functional on $C_0(S)$ with $L(f) \geq 0$ whenever $f \geq 0$. Show that there is a Radon measure μ, as defined in §5.6, Problems 4–5, with $L(f) = \int f \, d\mu$ for all $f \in C_0(S)$.

5. Let $\alpha = \nu + i\rho$ be a finite, complex-valued measure on S^1, where ν and ρ are finite, real-valued signed measures. Define the Fourier series of α as the (formal) series $\sum_{m \in \mathbb{Z}} a_m z^m$ where $a_m := \int z^{-m} d\alpha(z)$. A sequence of finite complex measures α_n is said to converge *weakly* to such a measure α if $\int f \, d\alpha_n \to \int f \, d\alpha$ as $n \to \infty$ for all $f \in C(S^1)$. Recall that $d\mu(\theta) = d\theta/2\pi$. Any function $g \in \mathcal{L}^1(S^1, \mu)$ defines a complex measure $[g]$ by $[g](A) := \int_A g \, d\mu$ for every Borel set $A \subset S^1$. For the finite complex measure α, let $\alpha_n := [\sum_{|m| \leq n} a_m z^m]$. Show that α_n converges weakly to α if $\alpha = [h]$ for some $h \in \mathcal{L}^2(S^1, \mu)$, but not if $\alpha = \delta_w$ for any $w \in S^1$. *Hint:* See the proof of Proposition 7.4.3.

6. Show that as $n \to \infty$, the complex measures $[z^n]$ converge weakly, as defined in Problem 5, and find their limit. *Hint:* See the proof of Lemma 7.4.4.

7. Let (X, \mathcal{T}) be a topological space and μ and ν two finite measures on the Baire σ-algebra \mathcal{S} for \mathcal{T}. Let $1 \leq p < \infty$. Under what conditions is the linear form $I_\nu: f \mapsto \int f \, d\nu$ on $C_b(X, \mathcal{T})$ continuous for the topology of the $\mathcal{L}^p(\mu)$ seminorm $(\int |f|^p \, d\mu)^{1/p}$? (Give the conditions in terms of the Lebesgue decomposition and Radon-Nikodym theorem for ν and μ.)

*7.5. Almost Uniform Convergence and Lusin's Theorem

A measurable function is not necessarily continuous anywhere—take, for example, the indicator function of the set of rational numbers. This function, however, restricted to the complement of a set of measure 0, becomes continuous. For another example, given $\varepsilon > 0$, take an open interval of length $\varepsilon/2^n$ around the nth rational number in an enumeration of the rational numbers. The union of the intervals is a dense open set of measure $< \varepsilon$. Its indicator is not continuous even when restricted to the complement of any set of measure 0. Still, it is continuous when restricted to the complement of a set of *small* measure. This will be proved for rather general measurable functions (Theorem 7.5.2 below). First, pointwise convergence is shown to be uniform except on small sets:

7.5.1. Egoroff's Theorem *Let (X, \mathcal{S}, μ) be finite measure space. Let f_n and f be measurable functions from X into a separable metric space S with metric d. Suppose $f_n(x) \to f(x)$ for μ–almost all x. Then for any $\varepsilon > 0$ there is a*

set A with $\mu(X \backslash A) < \varepsilon$ such that $f_n \to f$ uniformly on A, that is,

$$\lim_{n \to \infty} \sup\{d(f_n(x), f(x)): x \in A\} = 0.$$

Proof. For $m, n = 1, 2, \ldots$, let

$$A_{mn} := \{x: d(f_k(x), f(x)) \leq 1/m \text{ for all } k \geq n\}.$$

Each A_{mn} is measurable by Proposition 4.1.7. Then for each m, $\mu(X \backslash A_{mn}) \downarrow 0$ as $n \to \infty$. Choose $n(m)$ such that $\mu(X \backslash A_{mn(m)}) < \varepsilon/2^m$. Let $A := \bigcap_{m \geq 1} A_{mn(m)}$. Then $\mu(X \backslash A) < \varepsilon$ and $f_n \to f$ uniformly on A. $\qquad \square$

For example, the functions $x^n \to 0$ everywhere on $[0, 1)$, and uniformly on any interval $[0, 1 - \varepsilon)$, $\varepsilon > 0$.

Recall that a finite measure μ on the Borel σ-algebra of a topological space is called closed regular if for each Borel set B, $\mu(B) = \sup\{\mu(F): F \text{ closed}, F \subset B\}$. Any finite Borel measure on a metric space is closed regular by Theorem 7.1.3.

The following is an extension to more general domain and range spaces of a theorem of Lusin for real-valued functions of a real variable.

7.5.2. Lusin's Theorem *Let (X, \mathcal{T}) be any topological space and μ a finite, closed regular Borel measure on X. Let (S, d) be a separable metric space and let f be a Borel measurable function from X into S. Then for any $\varepsilon > 0$ there is a closed set $F \subset X$ such that $\mu(X \backslash F) < \varepsilon$ and the restriction of f to F is continuous.*

Proof. Let $\{s_n\}_{n \geq 1}$ be a countable dense set in S. For $m = 1, 2, \ldots$, and any $x \in X$, let $f_m(x) = s_n$ for the least n such that $d(f(x), s_n) < 1/m$. Then f_m is measurable and defined on all of X. For each m, let $n(m)$ be large enough so that

$$\mu\{x: d(f(x), s_n) \geq 1/m \text{ for all } n \leq n(m)\} \leq 1/2^m.$$

By closed regularity, for $n = 1, \ldots, n(m)$, take a closed set $F_{mn} \subset f_m^{-1}\{s_n\}$ with

$$\mu\left(f_m^{-1}\{s_n\} \backslash F_{mn}\right) < \frac{1}{2^m n(m)}.$$

For each fixed m, the sets F_{mn} are disjoint for different values of n. Let $F_m := \bigcup_{n=1}^{n(m)} F_{mn}$. Then f_m is continuous on F_m. By choice of $n(m)$ and F_{mn}, $\mu(X \backslash F_m) < 2/2^m$.

Since $d(f_m, f) < 1/m$ everywhere, clearly $f_m \to f$ uniformly (so Egoroff's theorem 7.5.1 is not needed). For $r = 1, 2, \ldots$, let $H_r := \bigcap_{m=r}^{\infty} F_m$.

Then H_r is a closed set and $\mu(X \backslash H_r) \leq 4/2^r$. Take r large enough so that $4/2^r < \varepsilon$. Then f restricted to H_r is continuous as the uniform limit of continuous functions f_m on $H_r \subset F_m$ for $m \geq r$, so we can let $F = H_r$ and the proof is complete. $\qquad\qquad\square$

Problems

1. Let (X, \mathcal{T}) be a normal topological space and μ a finite, closed regular measure on the Baire σ-algebra in X. Let $C_b(X, \mathcal{T})$ be the space of bounded continuous complex functions on X for \mathcal{T}. For each $g \in \mathcal{L}^1(X, \mu)$ and $f \in C_b(X, \mathcal{T})$, let $T_g(f) := \int gf\, d\mu$. Then show that $T_g \in C_b(X, \mathcal{T})'$ and $\|T_g\|' = \int |g|\, d\mu$.

2. A set A in a topological space is said to have the *Baire* property if there are an open set U and sets B and C of first category (countable unions of nowhere dense sets) such that $A = (U \backslash B) \cup C$. Show that the collection of all sets with the Baire property is a σ-algebra, which includes the Borel σ-algebra. *Hint:* Don't treat countable intersections directly.

3. Show that for any Borel measurable function f from a separable metric space S into \mathbb{R}, there is a set D of first category such that f restricted to $S \backslash D$ is continuous. *Hint:* Use the result of Problem 2. (This is a "category analogue" of Lusin's theorem, in a sense stronger in the category case than the measure case.)

4. Let (X, \mathcal{S}, μ) be a finite measure space. Suppose for some $M < \infty$, $\{f_n\}$ is a sequence of measurable functions with $|f_n(x)| \leq M$ for all n and x. Show that $\int f_n\, d\mu \to \int f\, d\mu$ (the bounded covergence theorem, a special case of Lebesgue's dominated convergence theorem) using Egoroff's theorem.

5. Let f be the indicator function of the Cantor set. Define a specific closed set $F \subset [0, 1]$, with $\lambda(F) > 7/8$, such that the restriction of f to F is continuous.

6. Do the same where f is the indicator function of the set of irrational numbers.

Notes

§7.1 A Borel set, now generally defined as a set in the σ-algebra generated by a topology, was previously defined, in a locally compact space, as a set in the σ-ring generated by compact sets (Halmos, 1950, p. 219), with a corresponding definition of Baire set.

Ulam's theorem was stated by Oxtoby and Ulam (1939), who attribute it specifically to Ulam, saying that a separate publication was planned (in *Comptes Rendus Acad. Sci. Paris*), but apparently Ulam did not complete the separate paper. Ulam (1976) wrote an autobiography.

J. von Neumann (1932) had proved the weaker statement where in the proof of Theorem 7.1.4 one has the union from 1 to $n(m)$ but not the intersection over m, so that K is closed and is a finite union of sets of diameter $\leq 2/m$ (the diameter of a set A is $\sup_{x,y \in A} d(x, y)$).

In unpublished lecture notes (of which I obtained a copy from the Institute for Advanced Study), von Neumann (1940–1941, pp. 85–91) developed the notion of regular measure. Halmos (1950, pp. 292, 295) cites the 1940–1941 notes as a primary source on regular measures.

§7.2 Lebesgue (1904, pp. 120–125) proved Theorem 7.2.1, the fundamental theorem of calculus for the Lebesgue integral. The proof given for it, including Lemmas 7.2.2 and 7.2.3, is based on the proof in Rudin (1974, pp. 162–165). Vitali (1904–05) proved that a function is absolutely continuous iff it is an indefinite integral of an \mathcal{L}^1 function. The decomposition of a function of bounded variation as a difference of two nondecreasing functions (Theorem 7.2.4) is due to Camille Jordan (see the notes to §5.6, above). Lebesgue (1904, p. 128) proved his theorem (7.2.7) on differentiability almost everywhere of functions of bounded variation. On proofs of the theorem which do not use measure theory (except for the notion of "set of measure 0," essential to the statement), see Riesz (1930–32) and Riesz and Nagy (1953, pp. 3–10).

Differentiation of absolutely continuous measures on \mathbb{R}^k, as in some of the problems, is due to Vitali (1908) and Lebesgue (1910).

§7.3 Halmos (1950, §§51–54) treats the regularity extension, giving as references Ambrose (1946), Kakutani and Kodaira (1944), Kodaira (1941), and von Neumann (1940–1941) for various aspects of the development. I thank Dorothy Maharam Stone for telling me of some examples in the problems.

§7.4 According to Riesz and Nagy (1953, p. 110), the representation theorem 7.4.1 for the case of $C[0, 1]$ is due to F. Riesz (1909, 1914). On various recent proofs of the general theorem see, for example, Garling (1986).

The Fourier series of an \mathcal{L}^2 function on S^1 converges almost everywhere (for μ) by a difficult theorem of L. Carleson (1966). R. A. Hunt (1968) proved the a.e. convergence of Fourier series if $f \in \mathcal{L}^p(S^1, \mu)$ for some $p > 1$ or more generally if

$$\int |f(z)|(\max(0, \log|f(z)|))^2 \, d\mu(z) < \infty.$$

C. Fefferman (1971) gave an extension of Carleson's result to functions of two variables (on $S^1 \times S^1$, also called the torus T^2).

Kolmogorov (1923, 1926) found functions in $\mathcal{L}^1(\mu)$ whose Fourier series diverge almost everywhere (μ), then everywhere. Zygmund (1959, pp. 306–310) gives an exposition.

Proposition 7.4.3 was first proved by du Bois-Reymond (1876), of course by a different method.

The original work on Fourier series was that of Fourier in 1807, part of a long memoir on the theory of heat, submitted to the Institut de France and ultimately published and annotated in Grattan-Guinness (1972). Fourier claimed "la résolution d'une fonction arbitraire en sinus ou en cosinus d'arcs multiples" (Grattan-Guinness, 1972, p. 193). The series certainly can represent discontinuous functions; Fourier showed

that

$$\cos(u) - \frac{1}{3}\cos(3u) + \frac{1}{5}\cos(5u) - \cdots = \begin{cases} \pi/4 & \text{for } |u| < \pi/2 \\ 0 & \text{for } u = \pm\pi/2 \\ -\pi/4 & \text{for } \pi/2 < |u| < 3\pi/2, \text{etc.} \end{cases}$$

On the other hand, Fourier's calculations used Taylor series for some range of the argument u. Fourier's memoir was refereed by Laplace, Lagrange, Monge, and S. F. Lacroix and not accepted as it stood. Apparently, Lagrange in particular objected to a lack of rigor in the claim to represent an "arbitrary" function by trigonometric series (Grattan-Guinness, 1972, p. 24). The Institut proposed the propagation of heat in solids as a grand prize topic in mathematics for 1811. The committee of judges was Lagrange, Laplace, Legendre, R. J. Haüy, and E. Malus (Herivel, 1975, p. 156). Fourier won the prize. The prize essay and his further work (Fourier, 1822) were eventually published (Fourier, 1824, 1826). On Fourier see also Ravetz and Grattan-Guinness (1972).

§7.5 Theorems 7.5.1 and 7.5.2 for real functions of a real variable were published respectively by Egoroff (1911) and Lusin (1912). These references are from Saks (1937). On Egoroff, see Paplauskas (1971), and on Lusin, see the notes to §13.2.

Lusin's theorem for measurable functions with values in any separable metric space, on a space with a closed regular finite measure, was first proved, as far as I know, by Schaerf (1947), who proved it for f with values in any second-countable topological space (i.e., a space having a countable base for its topology) and for more general domain spaces ("neighborhood spaces"). See also Schaerf (1948) and Zakon (1965) for more extensions.

References

An asterisk identifies works I have found discussed in secondary sources but have not seen in the original.

*Ambrose, Warren (1946). *Lectures on topological groups* (unpublished). Ann Arbor.
*du Bois-Reymond, P. (1876). Untersuchungen über die Convergenz und Divergenz der Fourierschen Darstellungsformeln. *Abh. Akad. München 12*: 1–103.
Carleson, Lennart (1966). On convergence and growth of partial sums of Fourier series. *Acta Math. 116*: 135–157.
Egoroff, Dmitri (1911). Sur les suites de fonctions mesurables. *C. R. Acad. Sci. Paris 152*: 244–246.
Fefferman, Charles (1971). On the convergence of multiple Fourier series. *Bull. Amer. Math. Soc. 77*: 744–745.
Fourier, Jean Baptiste Joseph (1822). *Théorie analytique de la chaleur.* F. Didot, Paris.
*———— (1824). Théorie du mouvement de la chaleur dans les corps solides. *Mémoires de l'Acad. Royale des Sciences 4* (1819–1820; publ. 1824): 185–555.
*———— (1826). Suite du mémoire intitulé: "Théorie du mouvement de la chaleur dans les corps solides." *Mémoires de l'Acad. Royale des Sciences 5* (1821–1822; publ. 1826): 153–246; *Oeuvres de Fourier, 2*, pp. 1–94.
———— (1888–1890, posth.) *Oeuvres.* Ed. G. Darboux. Gauthier-Villars, Paris.
Garling, David J. H. (1986). Another 'short' proof of the Riesz representation theorem. *Math. Proc. Camb. Phil. Soc. 99*: 261–262.

Grattan-Guinness, Ivor (1972). *Joseph Fourier, 1768–1830.* MIT Press, Cambridge, Mass.

Grebogi, Celso, Edward Ott, and James A. Yorke (1987). Chaos, strange attractors, and fractal basin boundaries in nonlinear dynamics. *Science 238*: 632–638.

Halmos, Paul (1950). *Measure Theory.* Van Nostrand, Princeton.

Herivel, John (1975). *Joseph Fourier: The Man and the Physicist.* Clarendon Press, Oxford.

Hunt, R. A. (1968). On the convergence of Fourier series. In *Orthogonal Expansions and Their Continuous Analogues*, pp. 235–255. Southern Illinois Univ. Press.

*Kakutani, Shizuo, and Kunihiko Kodaira (1944). Über das Haarsche Mass in der lokal bikompakten Gruppe. *Proc. Imp. Acad. Tokyo 20*: 444–450.

*Kodaira, Kunihiko (1941). Über die Beziehung zwischen den Massen und Topologien in einer Gruppe. *Proc. Math.-Phys. Soc. Japan 23*: 67–119.

Kolmogoroff, Andrei N. [Kolmogorov, A. N.] (1923). Une série de Fourier-Lebesgue divergente presque partout. *Fund. Math. 4*: 324–328.

———— (1926). Une série de Fourier-Lebesgue divergente partout. *C. R. Acad. Sci. Paris 183*: 1327–1328.

Lebesgue, Henri Léon (1904). Leçons sur l'intégration et la recherche des fonctions primitives, Paris. 2d ed., 1928. Repr. in *Oeuvres Scientifiques 2*, pp. 111–154.

———— (1910). Sur l'intégration des fonctions discontinues. *Ann. Ecole Norm. Sup.* (Ser. 3) *27*: 361–450. Repr. in *Oeuvres Scientifiques 2*, pp. 185–274.

Lusin, Nikolai (1912). Sur les propriétés des fonctions mesurables. *C. R. Acad. Sci. Paris 154*: 1688–1690.

von Neumann, Johann (1932). Einige Sätze über messbare Abbildungen. *Ann. Math. 33*: 574–586, and *Collected Works* [1961, below], 2, no. 16, p. 297.

———— (1940–1941). *Lectures on invariant measures.* Notes by Paul R. Halmos. Unpublished. Institute for Advanced Study, Princeton.

———— (1961–1963). *Collected Works.* Ed. A. H. Taub. Pergamon Press, London.

Oxtoby, John C., and Stanislaw M. Ulam (1939). On the existence of a measure invariant under a transformation. *Ann. Math.* (Ser. 2) *40*: 560–566.

Paplauskas, A. B. (1971). Egorov, Dimitry Fedorovich. *Dictionary of Scientific Biography, 4*, pp. 287–288.

Ravetz, Jerome R., and Ivor Grattan-Guinness (1972). Fourier, Jean Baptiste Joseph. *Dictionary of Scientific Biography 5*, pp. 93–99.

Riesz, Frigyes [Frédéric] (1909). Sur les opérations fonctionnelles linéaires. *Comptes Rendus Acad. Sci. Paris 149*: 974–977.

*———— (1914). Démonstration nouvelle d'un théorème concernant les opérations. *Ann. Ecole Normale Sup.* (Ser. 3) *31*: 9–14.

———— (1930–1932). Sur l'existence de la dérivée des fonctions monotones et sur quelques problèmes qui s'y rattachent. *Acta Sci. Math. Szeged 5*: 208–221.

———— and Béla Szőkefalvi-Nagy (1953). *Functional Analysis.* Ungar, New York (1955). Transl. L. F. Boron from 2d French ed., *Leçons d'analyse fonctionelle.* 5th French ed., Gauthier-Villars, Paris (1968).

Rudin, Walter (1966, 1974, 1987). *Real and Complex Analysis.* 1st, 2d and 3d eds. McGraw-Hill, New York.

———— (1976). *Principles of Mathematical Analysis.* 3d ed. McGraw-Hill, New York.

Saks, Stanisław (1937). *Theory of the Integral.* 2d ed. Monografie Matematyczne, Warsaw; English transl. L. C. Young. Hafner, New York. Repr. Dover, New York (1964).

Schaerf, H. M. (1947). On the continuity of measurable functions in neighborhood spaces. *Portugal. Math. 6*: 33–44.

——— (1948). On the continuity of measurable functions in neighborhood spaces II. *Ibid. 7*: 91–92.

Ulam, Stanislaw Marcin (1976). *Adventures of a Mathematician.* Scribner's, New York.

*Vitali, Giuseppe (1904–05). Sulle funzioni integrali. *Atti Accad. Sci. Torino 40*: 1021–1034.

*——— (1908). Sui gruppi di punti e sulle funzioni di variabili reali. *Ibid. 43*: 75–92.

Zakon, Elias (1965). On "essentially metrizable" spaces and on measurable functions with values in such spaces. *Trans. Amer. Math. Soc. 119*: 443–453.

Zygmund, Antoni (1959). *Trigonometric Series.* 2 vols. 2d ed. Cambridge University Press.

8

Introduction to Probability Theory

Probabilities are easiest to define on finite sets. For example, consider a toss of a fair coin. Here "fair" means that heads and tails are equally likely. The situation may be represented by a set with two points H and T where $H = $ "heads" and $T = $ "tails." The total probability of all possible outcomes is set equal to 1. Let "$P(\ldots)$" denote "the probability of \ldots." If two possible outcomes cannot both happen, then one assumes that their probabilities add. Thus $P(H) + P(T) = 1$. By assumption $P(H) = P(T)$, so $P(H) = P(T) = 1/2$.

Now suppose the coin is tossed twice. There are then four possible outcomes of the two tosses: HH, HT, TH, and TT. Considering these four as equally likely, they must each have probability $1/4$. Likewise, if the coin is tossed n times, we have 2^n possible strings of n letters H and T, where each string has probability $1/2^n$.

Next let n go to infinity. Then we have all possible infinite sequences of H's and T's. Each individual sequence has probability 0, but this does not determine the probabilities of other interesting sets of possible outcomes, as it did when n was finite. To consider such sets, first let us replace H by 1 and T by 0, precede the sequence by a "binary point" (as in decimal point), and regard the sequence as a binary expansion. For example,

$$0.0101010101 \ldots \text{ is an expansion of } 1/4 + 1/16 + \cdots = 1/3,$$

where in general, if $d_n = 0$ or 1 for all n, the sequence or binary expansion $0.d_1 d_2 d_3 \ldots = \sum_{1 \leq n < \infty} d_n / 2^n$. Every number x in the interval $[0, 1]$ has such an expansion. The dyadic rationals $k/2^n$, if k is an integer with $0 < k < 2^n$, have two such expansions, one being the usual expansion of $k/2^n$ followed by all 0's, the other being the usual expansion of $(k - 1)/2^n$ followed by an infinite string of 1's. There are just countably many such dyadic rationals. All other numbers in $[0, 1]$ have unique binary expansions.

On $[0, 1]$, there is the usual measure λ, called Lebesgue or uniform measure, which assigns to each subinterval its length. Then λ is a countably additive

measure defined on the σ-algebra \mathcal{B} of all Borel sets, or on the larger σ-algebra \mathcal{L} of all Lebesgue measurable sets. The countable set of dyadic rationals has measure 0 for λ.

A possible sequence of outcomes on the first n tosses then corresponds to the set of all infinite sequences beginning with a given sequence of n 0's or 1's, and in turn to a subinterval of $[0, 1]$ of length $1/2^n$. Any set of m different sequences of outcomes for the first n tosses corresponds to a set in $[0, 1]$ with Lebesgue measure $m/2^n$. To this extent, at least, Lebesgue measure corresponds to the probabilities previously defined. This is an example of the general relation of measures and probabilities to be defined next.

8.1. Basic Definitions

A *measurable space* (Ω, \mathcal{S}) is a set Ω with a σ-algebra \mathcal{S} of subsets of Ω. A *probability measure* P is a measure on \mathcal{S} with $P(\Omega) = 1$. (Recall that a measure, by definition, is nonnegative and countably additive.) Then (Ω, \mathcal{S}, P) is called a *probability space*.

If Ω is finite, as in the case of finitely many coin tosses, or even countable, usually the σ-algebra \mathcal{S} is 2^Ω, the set of all subsets of Ω. If Ω is uncountable, for example $\Omega = [0, 1]$, usually \mathcal{S} is a proper subset of 2^Ω. Members of \mathcal{S}, called *measurable sets* in measure theory, are also called *events* in a probability space. For an event A, sometimes $P(A)$ may be written PA.

On a finite set F with m elements, there is a natural probability measure P, called the (discrete) *uniform* measure on 2^F, which assigns probability $1/m$ to the singleton $\{x\}$ for each x in F. Coin tossing gave us examples with $m = 2^n$, $n = 1, 2, \ldots$. Another classical example is a fair *die*, a perfect cube which is thrown at random so that each of the six faces, marked with the integers from 1 to 6, has equal probability $1/6$ of coming up.

If (Ω, \mathcal{A}, P) is a probability space and (S, \mathcal{B}) any measurable space, a measurable function X from Ω into S is called a *random variable*. Then the image measure $P \circ X^{-1}$ defined on \mathcal{B} is a probability measure which may be called the *law* of X, or $\mathcal{L}(X)$.

Often S is the real line \mathbb{R} and \mathcal{B} is the σ-algebra of Borel sets. The *expectation* or *mean* EX of a real-valued random variable X is defined as $\int X \, dP$ iff (meaning if and only if) the integral exists. Like any integral, the expectation is linear:

8.1.1. Theorem *For any two random variables X and Y such that the expectations EX and EY are both defined and finite, and any constant c,*
$$E(cX + Y) = cEX + EY.$$

Proof. This is part of Theorem 4.1.10. □

The *variance* $\sigma^2(X)$ of a random variable X is defined by

$$\operatorname{var}(X):=\sigma^2(X):=\begin{cases} E(X-EX)^2 = EX^2 - (EX)^2, & \text{if } EX^2 < \infty; \\ +\infty & \text{if } EX^2 = +\infty. \end{cases}$$

(Here, as throughout, ":=" means "equals by definition.") If $E(X^2) < \infty$, then $\sigma(X):=\sqrt{\sigma^2(X)}$ is called the *standard deviation* of X. By the Cauchy-Bunyakovsky-Schwarz inequality (5.1.4), $E|X| = E(|X| \cdot 1) \leq (EX^2)^{1/2}$, so that if X is a random variable with $E(X^2) < \infty$, then EX is defined and finite.

Events are often given in the form $\{\omega: \ldots \omega \ldots\}$, the set of all ω such that a given condition holds. In probability, an event is often written just as the condition, without the argument ω. For example, if X and Y are two random variables, instead of the longer notation $P(\{\omega: X(\omega) > Y(\omega)\})$, one usually just writes $P(X > Y)$.

In coin tossing, supposing that the first toss gave H, then we assumed that the second toss was still equally likely to give H or T, as HH and HT both were given probability $1/4$. In other words, the outcome of the second toss was assumed to be independent of the outcome of the first. This is an example of a rather crucial notion in probability, which will be defined more generally.

Two events A and B are called *independent* (for a probability measure P) iff $P(A \cap B) = P(A)P(B)$. Let X and Y be two random variables on the same probability space, with values in S and T respectively, where (S, \mathcal{U}) and (T, \mathcal{V}) are the measurable spaces for which X and Y are measurable. Then we can form a "vector" random variable $\langle X, Y \rangle$ where $\langle X, Y \rangle(\omega) := \langle X(\omega), Y(\omega) \rangle$ for each ω in Ω. Then X and Y are called *independent* iff the law $\mathcal{L}(\langle X, Y \rangle)$ equals the product measure $\mathcal{L}(X) \times \mathcal{L}(Y)$. In other words, for any measurable sets $U \in \mathcal{U}$ and $V \in \mathcal{V}$, $P(X \in U \text{ and } Y \in V) = P(X \in U)P(Y \in V)$. If $S = T = \mathbb{R}$, X and Y are independent, $E|X| < \infty$, and $E|Y| < \infty$, then $E(XY) = EXEY$ by the Tonelli-Fubini theorem (4.4.5) (and Theorem 4.1.11).

Given any probability spaces $(\Omega_j, \mathcal{S}_j, P_j)$, $j = 1, \ldots, n$, we can form the Cartesian product $\Omega_1 \times \Omega_2 \times \cdots \times \Omega_n$ with the product σ-algebra of the \mathcal{S}_j and the product probability measure $P_1 \times \cdots \times P_n$ given by Theorem 4.4.6. Random variables X_1, \ldots, X_n on one probability space (Ω, \mathcal{S}, P) are called *independent*, or more specifically *jointly independent*, if the law $\mathcal{L}(\langle X_1, \ldots, X_n \rangle)$ equals the product measure $\mathcal{L}(X_1) \times \cdots \times \mathcal{L}(X_n)$. So, for example, if the X_i are to be real-valued, given any probability measures μ_1, \ldots, μ_n on the Borel σ-algebra \mathcal{B} of \mathbb{R}, we can take the product of the probability spaces $(\mathbb{R}, \mathcal{B}, \mu_i)$ by Theorem 4.4.6 to get a product measure μ

on \mathbb{R}^n for which the coordinates X_1, \ldots, X_n are independent with the given laws.

Any set of random variables is called *independent* iff every finite subset of it is independent. Random variables X_j are called *pairwise independent* iff for all $i \neq j$, X_i and X_j are independent. (Note that all definitions of independence are with respect to some probability measure P; random variables may be independent for some P but not for another.)

The *covariance* of two random variables X and Y having finite variances, on a probability space (Ω, P), is defined by

$$\mathrm{cov}(X, Y) := E((X - EX)(Y - EY)) = E(XY) - EX\,EY.$$

(This is the inner product of $X - EX$ and $Y - EY$ in the Hilbert space $L^2(\Omega, P)$.) So if X and Y are independent, their covariance is 0. Some of the benefits of independence have to do with properties of the covariance and variance:

8.1.2. Theorem *For any random variables* X_1, \ldots, X_n *with finite variances on one probability space, let* $S_n := X_1 + \cdots + X_n$. *Then*

$$\mathrm{var}(S_n) = \sum_{1 \leq i \leq n} \mathrm{var}(X_i) + 2 \sum_{1 \leq i < j \leq n} \mathrm{cov}(X_i, X_j).$$

If the covariances are all 0, *and thus if the* X_i *are independent or just pairwise independent,*

$$\mathrm{var}(X_1 + \cdots + X_n) = \mathrm{var}(X_1) + \cdots + \mathrm{var}(X_n).$$

Proof. If we replace each X_i by $X_i - EX_i$, then none of the variances or covariances is changed, nor is $\mathrm{var}(S_n)$, in view of the linearity of the expectation (Theorem 8.1.1). So we can assume that $EX_i = 0$ for all i. Then

$$\mathrm{var}(S_n) = E(X_1 + \cdots + X_n)(X_1 + \cdots + X_n)$$

and the first equation in the theorem follows, and then the second. \square

Recall that for any event A the *indicator function* 1_A is defined by

$$1_A(x) = \delta_x(A) = \begin{cases} 1 & \text{if } x \in A \\ 0 & \text{otherwise.} \end{cases}$$

(Probabilists generally use "characteristic function" to refer to a Fourier transform of a probability measure P; for example, on \mathbb{R}, $f(t) = \int e^{ixt} dP(x)$ is the characteristic function of P.) A set of events is called *independent* iff their indicator functions are independent.

The situation of coin tossing can be generalized. Suppose A_1, \ldots, A_n are independent events, all having the same probability $p = P(A_j)$, $j = 1, \ldots, n$. Let $q := 1 - p$. For example, one may throw a "biased" coin n times, where the probability of heads is p. Then let A_j be the event that the coin comes up heads on the jth toss. Any particular sequence of n outcomes, in which k of the A_j occur and the other $n - k$ do not, has probability $p^k q^{n-k}$ by independence. Thus the probability that exactly k of the A_j occur is the sum of $\binom{n}{k}$ such probabilities, that is, $b(k, n, p) := \binom{n}{k} p^k q^{n-k}$, where

$$\binom{n}{k} = \frac{n!}{k!(n-k)!} \qquad \text{for any integers } 0 \leq k \leq n.$$

Then $b(k, n, p)$ is called a *binomial* probability. It is also described as "the probability of k successes in n independent trials with probability p of success on each trial." For a more extensive treatment of such probabilities, and in general for more concrete, combinatorial aspects of probability theory, one reference is the classic book of W. Feller (1968).

Problems

1. The *uniform probability* on $\{1, \ldots, n\}$ is defined by $P\{j\} = 1/n$ for $j = 1, \ldots, n$. For the identity random variable $X(j) \equiv j$, find the mean EX and the variance $\sigma^2(X)$.

2. Let X be a random variable whose distribution P is uniform on the interval $[1, 4]$, that is, $P(A) = \lambda(A \cap [1, 4])/3$ for any Borel set A. Find the mean and variance of X.

3. Let X and Y be independent real random variables with finite variances. Let $Z = aX + bY + c$. Find the mean and variance of Z in terms of those of X and Y and a, b, and c.

4. Find a probability space with three random variables X, Y, and Z which are pairwise independent but not independent. *Hint:* Take the space as in Problem 1 with $n = 4$ and X, Y, and Z indicator functions.

5. Find a measurable space (Ω, \mathcal{S}) with two probability measures P and Q on \mathcal{S} and two sets A and B in \mathcal{S} which are independent for P but not for Q.

6. A plane is ruled by a set of parallel lines at distance d apart. A needle of length s is thrown at random onto the plane. In detail, let X be the distance from the center of the needle to the nearest line. Let θ be the smallest nonnegative angle between the line along the needle and the lines ruling the plane. Then for some c and γ, $P(a \leq X \leq b) = (b - a)/c$ for

$0 \leq a \leq b \leq c$, $P(\alpha \leq \theta \leq \beta) = (\beta - \alpha)/\gamma$ for $0 \leq \alpha \leq \beta \leq \gamma$, and θ is independent of X.

(a) For a "random" throw, what should c and γ be?

(b) Find the probability that the needle intersects one or more lines. *Hint:* Distinguish $s \geq d$ from $s < d$.

7. Suppose instead of a needle, a circular coin of radius r is thrown. What is the probability that it meets at least one line?

8. Find the probability of exactly 3 successes in 10 independent trials with probability 0.3 of success on each trial.

9. Let $E(k, n, p)$ be the probability of k or more successes in n independent trials with probability p of success on each trial, so $E(k, n, p) = \sum_{k \leq j \leq n} b(j, n, p)$. Show that
 (a) for $k = 1, \ldots, n$, $E(k, n, p) = 1 - E(n - k + 1, n, 1 - p)$;
 (b) for $k = 1, 2, \ldots$, $E(k, 2k - 1, 1/2) = 1/2$.

10. For random variables X and Y on the same space (Ω, P), let $r(X, Y) = \text{cov}(X, Y)/(\text{var}(X)\text{var}(Y))^{1/2}$ ("correlation coefficient") if $\text{var}(X) > 0$ and $\text{var}(Y) > 0$; $r(X, Y)$ is undefined if $\text{var}(X)$ or $\text{var}(Y)$ is 0 or ∞.
 (a) Show that if $r(X, Y)$ is defined, then $r(X, Y) = \cos \theta$, where θ is the angle between the vectors $X - EX$ and $Y - EY$ in the subspace of $L^2(P)$ spanned by these two functions, so $-1 \leq r(X, Y) \leq 1$.
 (b) If $r(X, Y) = .7$ and $r(Y, Z) = .8$, find the smallest and largest possible values of $r(X, Z)$.

8.2. Infinite Products of Probability Spaces

For a sequence of n repeated, independent trials of an experiment, some probability distributions and variables converge as n tends to ∞. In proving such limit theorems, it is useful to be able to construct a probability space on which a sequence of independent random variables is defined in a natural way; specifically, as coordinates for a countable Cartesian product.

The Cartesian product of finitely many σ-finite measure spaces gives a σ-finite measure space (Theorem 4.4.6). For example, Cartesian products of Lebesgue measure on the line give Lebesgue measure on finite-dimensional Euclidean spaces. But suppose we take a measure space $\{0, 1\}$ with two points each having measure 1, $\mu(\{0\}) = \mu(\{1\}) = 1$, and form a countable Cartesian product of copies of this space, so that the measure of any countable product of sets equals the product of their measures. Then we would get an uncountable space in which all singletons have measure 1, giving the measure usually called "counting measure." An uncountable set with counting measure is not

a σ-finite measure space, although in this example it was a countable product of finite measure spaces. By contrast, the countable product of probability spaces will again be a probability space. Here are some definitions.

For each $n = 1, 2, \ldots$, let $(\Omega_n, \mathcal{S}_n, P_n)$ be a probability space. Let Ω be the Cartesian product $\Pi_{1 \leq n < \infty} \Omega_n$, that is, the set of all sequences $\{\omega_n\}_{1 \leq n < \infty}$ with $\omega_n \in \Omega_n$ for all n. Let π_n be the natural projection of Ω onto Ω_n for each n: $\pi_n(\{\omega_m\}_{m \geq 1}) = \omega_n$ for all n. Let \mathcal{S} be the smallest σ-algebra of subsets of Ω such that for all m, π_m is measurable from (Ω, \mathcal{S}) to $(\Omega_m, \mathcal{S}_m)$. In other words, \mathcal{S} is the smallest σ-algebra containing all sets $\pi_n^{-1}(A)$ for all n and all $A \in \mathcal{S}_n$.

For example, the interval $[0, 1]$ is related to a product through decimal or binary expansions: for each $n = 1, 2, \ldots$, let Ω_n be the set of ten digits $\{0, 1, \ldots, 9\}$, with $P_n(\{j\}) = 1/10$ for $j = 0, 1, \ldots, 9$. Then points $\omega = \{j_n\}_{n \geq 1}$ of Ω yield numbers $x(\omega) = \sum_{n \geq 1} j_n/10^n$, where the j_n are the digits in a decimal expansion of x. The set of all positive integers can be written as a union of infinitely many countably infinite sets A_k (for example, let A_k be the set of integers divisible by 2^{k-1} but not by 2^k for $k = 1, 2, \ldots$). Let $A_k := \{n(k, i)\}_{i \geq 1}$. Then each number x in $[0, 1]$, with decimal expansion $\sum_n j_n(x)/10^n$, can encode an infinite sequence of such numbers $y_k := \sum_i j_{n(k,i)}/10^i$. Let $T(x) := \{y_k\}_{k \geq 1}$ and $T_k(x) := y_k$. Let λ be Lebesgue measure on $[0, 1]$. Then $\lambda \circ T^{-1}$ gives a probability measure on the product of a sequence of copies of $[0, 1]$ for which the distribution of any (y_1, \ldots, y_k) is Lebesgue measure λ^k on the unit cube I^k in \mathbb{R}^k. If probability measures P_n on spaces Ω_n can be represented as $P_n = \lambda \circ V_n^{-1}$ for some functions V_n from $[0, 1]$ to Ω_n, as is true for a great many P_n, then the product of the measures P_n can be written as $\lambda \circ W^{-1}$ where $W(x) := \{V_n(T_n(x))\}_{n \geq 1}$.

On the other hand, there are probability measures which cannot be written as images of one-dimensional Lebesgue measure. For measures on product spaces other than product measures, a consistent definition of measures on any finite Cartesian product of factor spaces Ω_n does not guarantee existence of a consistent measure on the infinite product (§12.1, problem 2, below). So it's worth noting that this section will give a Cartesian product measure for a completely arbitrary sequence of probability spaces. This product can't be proved to exist (at least not directly) by classical methods such as decimal expansions. Having noted its generality, let's return to the construction.

Let \mathcal{R} be the collection of all sets $\Pi_n A_n \subset \Omega$ where $A_n \in \mathcal{S}_n$ for all n and $A_n = \Omega_n$ except for at most finitely many values of n. Elements of \mathcal{R} will be called *rectangles* or *finite-dimensional rectangles*. Now recall the notion of semiring from §3.2. \mathcal{R} has this property:

8.2.1. Proposition *The collection \mathcal{R} of finite-dimensional rectangles in the infinite product Ω is a semiring. The algebra \mathcal{A} generated by \mathcal{R} is the collection of finite disjoint unions of elements of \mathcal{R}.*

Proof. If C and D are any two rectangles, then clearly $C \cap D$ is a rectangle. In a product of two spaces, the collection of rectangles is a semiring by Proposition 3.2.2. Specifically, a difference of two rectangles is a union of two disjoint rectangles:

$$(A \times B) \backslash (E \times F) = ((A \backslash E) \times B) \cup ((A \cap E) \times (B \backslash F)).$$

It follows by induction that in any finite Cartesian product, any difference $C \backslash D$ of two rectangles is a finite disjoint union of rectangles. Thus \mathcal{R} is a semiring. We have $\Omega \in \mathcal{R}$, so the ring \mathcal{A} generated by \mathcal{R}, as in Proposition 3.2.3, is an algebra. Since every algebra is a ring, \mathcal{A} is the algebra generated by \mathcal{R}. By Proposition 3.2.3, \mathcal{A} consists of all finite disjoint unions of elements of \mathcal{R}. $\qquad\square$

Now for $A = \Pi_n A_n \in \mathcal{R}$, let $P(A) := \Pi_n P_n(A_n)$. The product converges since all but finitely many factors are 1. Here is the main theorem to be proved in the rest of this section:

8.2.2. Theorem (Existence Theorem for Infinite Product Probabilities) *P on \mathcal{R} extends uniquely to a (countably additive) probability measure on \mathcal{S}.*

Proof. For each $A \in \mathcal{A}$, write A as a finite disjoint union of sets in \mathcal{R} (Proposition 8.2.1), say $A = \bigcup_{1 \le r \le N} B_r$, and define $P(A) := \sum_{1 \le r \le N} P(B_r)$. Let us first show that P is well-defined and finitely additive on such finite disjoint unions. Each B_r is a product of sets $A_{rn} \in \mathcal{S}_n$ with $A_{rn} = \Omega_n$ for all $n \ge n(r)$ for some $n(r) < \infty$. Let m be the maximum of the $n(r)$ for $r = 1, \ldots, N$. Then since all the A_{rn} equal Ω_n for $n > m$, properties of P on such sets are equivalent to properties of the finite product measure on $\Omega_1 \times \cdots \times \Omega_m$. To show that P is well-defined, if a set is equal to two different finite disjoint unions of sets in \mathcal{R}, we can take the maximum of the values of m for the two unions and still have a finite product. So P is well-defined and finitely additive on \mathcal{A} by the finite product measure theorem (4.4.6).

If P is countably additive on \mathcal{A}, then it has a unique countably additive extension to \mathcal{S} by Theorem 3.1.10. So it's enough to prove countable additivity on \mathcal{A}. Equivalently, by Theorem 3.1.1, if $A_j \in \mathcal{A}$, $A_1 \supset A_2 \supset \cdots$, and $\bigcap_j A_j = \emptyset$, we want to prove $P(A_j) \downarrow 0$. In other words, if A_j is a

decreasing sequence of sets in \mathcal{A} and for some $\varepsilon > 0$, $P(A_j) \geq \varepsilon$ for all j, we must show $\bigcap_j A_j \neq \emptyset$.

Let $P^{(0)} := P$ on \mathcal{A}. For each $n \geq 1$, let $\Omega^{(n)} := \Pi_{m>n}\Omega_m$. Let $\mathcal{A}^{(n)}$ and $P^{(n)}$ be defined on $\Omega^{(n)}$ just as \mathcal{A} and P were on Ω. For each $E \subset \Omega$ and $x_i \in \Omega_i$, $i = 1, \ldots, n$, let $E^{(n)}(x_1, \ldots, x_n) := \{\{x_m\}_{m>n} \in \Omega^{(n)} : x = \{x_i\}_{i\geq 1} \in E\}$.

For a set A in a product space $X \times Y$ and $x \in X$, let $A_x := \{y \in Y : \langle x, y \rangle \in A\}$ (see Figure 8.2A). If A is in a product σ-algebra $\mathcal{S} \otimes \mathcal{T}$ then $A_x \in \mathcal{T}$ by the proof of Theorem 4.4.3. For any $E \in \mathcal{A}$ there is an N large enough so that $E = F \times \Pi_{n>N}\Omega_n$ for some $F \subset \Pi_{n\leq N}\Omega_n$. (Since E is a finite union of rectangles with this property, take the maximum of the values of N for the rectangles.) Then $F = \bigcup_{k=1}^{m} F_k$, where $F_k = \Pi_{i=1}^{N}F_{ki}$ for some $F_{ki} \in \mathcal{S}_i$, $i = 1, \ldots, N$, $k = 1, \ldots, m$. Now for any $n < N$ and $x_i \in \Omega_i, i = 1, \ldots, n$, $E^{(n)}(x_1, \ldots, x_n) = G \times \Omega^{(N)}$ where G is the union of those sets $\Pi_{n<i\leq N}F_{ki}$ such that $x_i \in F_{ki}$ for all $i = 1, \ldots, n$. Thus $E^{(n)}(x_1, \ldots, x_n) \in \mathcal{A}^{(n)}$, so $P^{(n)}$ of it is defined. Then by the Tonelli-Fubini theorem in $\Omega_1 \times \cdots \times \Omega_n \times \Pi_{n<i\leq N}\Omega_i$ (Theorem 4.4.6), we have

$$P(E) = \int P^{(n)}\big(E^{(n)}(x_1, \ldots, x_n)\big)\Pi_{1\leq j\leq n}\, dP_j(x_j). \qquad (8.2.3)$$

For the ε with $P(A_j) \geq \varepsilon$ for all j, let

$$F_j := \big\{x_1 \in \Omega_1 : P^{(1)}\big(A_j^{(1)}(x_1)\big) > \varepsilon/2\big\}.$$

For each j, apply (8.2.3) to $E = A_j, n = 1$. Then

$$\varepsilon \leq P(A_j) = \int P^{(1)}\big(A_j^{(1)}(x_1)\big)\, dP_1(x_1)$$

$$= \left(\int_{F_j} + \int_{\Omega_1\setminus F_j}\right) P^{(1)}\big(A_j^{(1)}(x_1)\big)\, dP_1(x_1)$$

$$\leq P_1(F_j) + \varepsilon/2$$

(see Figure 8.2B). Thus $P_1(F_j) \geq \varepsilon/2$ for all j. As j increases, the sets A_j decrease; thus, so do the $A_j^{(1)}$ and the F_j.

Figure 8.2A

Figure 8.2B

Since P_1 is countably additive, $P_1(\bigcap_j F_j) \geq \varepsilon/2 > 0$ (by monotone convergence of indicator functions), so $\bigcap_j F_j \neq \emptyset$. Take any $y_1 \in \bigcap_j F_j$. Let $f_j(y, x) := P^{(2)}(A_j^{(2)}(y, x))$ and $G_j := \{x_2 \in \Omega_2 : f_j(y_1, x_2) > \varepsilon/4\}$. Then G_j decrease as j increases,

$$\varepsilon/2 < P^{(1)}\big(A_j^{(1)}(y_1)\big) = \int f_j(y_1, x)\, dP_2(x) \qquad \text{for all } j,$$

and $P_2(G_j) > \varepsilon/4$ for all j, so the intersection of all the G_j is non-empty in Ω_2 and we can choose y_2 in it.

Inductively, by the same argument there are $y_n \in \Omega_n$ for all n such that $P^{(n)}(A_j^{(n)}(y_1, \ldots, y_n)) \geq \varepsilon/2^n$ for all j and n. Let $y := \{y_n\}_{n\geq 1} \in \Omega$. To prove that $y \in A_j$ for each j, choose n large enough (depending on j) so that for all x_1, \ldots, x_n, $A_j^{(n)}(x_1, \ldots, x_n) = \emptyset$ or $\Omega^{(n)}$. This is possible since $A_j \in \mathcal{A}$. Then $A_j^{(n)}(y_1, \ldots, y_n) = \Omega^{(n)}$, so $y \in A_j$. Hence $\bigcap_j A_j \neq \emptyset$. $\qquad \square$

Actually, Theorem 8.2.2 holds for arbitrary (not necessarily countable) products of probability spaces. The proof needs no major change, since each set in the σ-algebra \mathcal{S} depends on only countably many coordinates. In other words, given a product $\Pi_{i\in I}\Omega_i$, where I is a possibly uncountable index set, for each set A in \mathcal{S} there is a countable subset J of I and a set $B \subset \Pi_{i\in J}\Omega_i$ such that $A = B \times \Pi_{i\notin J}\Omega_i$.

Problems

1. In \mathbb{R}^2, for ordinary rectangles which are Cartesian products of intervals (which may be open or closed at either end), show that for any two rectangles C and D, $C\backslash D$ is a union of at most k rectangles for some finite k, and find the smallest possible value of k.

2. Similarly as in Theorem 3.2.7, for any σ-algebra \mathcal{B}, a collection $\mathcal{C} \subset \mathcal{B}$ will be called a *probability determining class* if any two probability measures on \mathcal{B}, equal on \mathcal{C}, are also equal on \mathcal{B}. Recall that for \mathcal{C} to generate \mathcal{B} is not sufficient for \mathcal{C} to be a (probability) determining class.
 (a) Show that in a countable product (Ω, \mathcal{S}) of spaces $(\Omega_n, \mathcal{S}_n)$, the set of all rectangles is a probability determining class for \mathcal{S}.
 (b) Let \mathcal{R}_m be the set of all rectangles $\bigcap_{j\in F} \pi_j^{-1}(A_j)$ for $A_j \in \mathcal{S}_j$ and where F contains m indices. Show that \mathcal{R}_m is not a probability determining class for any finite m, for example if each Ω_n is a two-point set $\{0, 1\}$ and \mathcal{S}_n is the σ-algebra of all its subsets. *Hint:* Let $\omega_2, \omega_3, \ldots,$ be independent and each equal to 0 or 1 with probability $1/2$ each. Let $\omega_1 = 0$ or 1 where $\omega_1 \equiv S_m := \omega_2 + \cdots + \omega_{m+1} \mod 2$ (that is,

$\omega_1 - S_m$ is divisible by 2). Show that this gives the same probabilities to each set in \mathcal{R}_m as a law with all ω_j independent.

3. Show that Theorem 8.2.2 is true for arbitrary (uncountable) products of probability spaces, as suggested at the end of the section. *Hint:* Show that the collection of all sets A as described is a σ-algebra, using the fact that a countable union of sets J_m is countable.

4. Let (X, \mathcal{S}, P) be a probability space and (Y, \mathcal{T}) a measurable space. Suppose that for each $x \in X$, Q_x is a probability measure on (Y, \mathcal{T}). Suppose that for each $C \in \mathcal{T}$, the function $x \mapsto Q_x(C)$ is measurable for \mathcal{S}. Show that there is a probability measure μ on $(X \times Y, \mathcal{S} \otimes \mathcal{T})$ such that for any set $B \in \mathcal{S} \otimes \mathcal{T}$, $\mu(B) = \int \int 1_B(x, y) \, dQ_x(y) \, dP(x)$.

5. Suppose that for $n = 1, 2, \ldots$ and $0 \le t \le 1$, $P_{n,t}$ is a probability measure on the Borel σ-algebra in \mathbb{R} such that for each interval $[a, b] \subset \mathbb{R}$ and each n, if $f(t) := P_{n,t}([a, b])$, then f is Borel measurable. Let P_t be the product measure $\Pi_{1 \le n < \infty} P_{n,t}$ on \mathbb{R}^∞. Show that there exists a probability measure Q on \mathbb{R}^∞ such that for every measurable set $C \subset \mathbb{R}^\infty$, $Q(C) = \int_0^1 P_t(C) \, dt$.

6. Let $X := \{0, 1\}$, $\mathcal{S} = 2^X$ and $P(\{0\}) = P(\{1\}) = 1/2$. Let (Ω, \mathcal{B}, Q) be a countable product of copies of (X, \mathcal{S}, P). For each $\{x_n\}_{n \ge 1} \in \Omega$, where $x_n = 0$ or 1 for all n, let $f(\{x_n\}) = \sum_n x_n/2^n$. Show that the image measure $Q \circ f^{-1}$ is Lebesgue measure λ on $[0, 1]$.

7. Let (X_j, \mathcal{S}_j) be measurable spaces for $j = 1, 2, \ldots$. Let P_1 be a probability measure on \mathcal{S}_1. Suppose that for each n and each $x_j \in X_j$ for $j = 1, \ldots, n$, $P(x_1, \ldots, x_n)(\cdot)$ is a probability measure on $(X_{n+1}, \mathcal{S}_{n+1})$, with the property that for each $A \in \mathcal{S}_{n+1}$, $(x_1, \ldots, x_n) \mapsto P(x_1, \ldots, x_n)(A)$ is measurable for the product σ-algebra $\mathcal{S}_1 \otimes \cdots \otimes \mathcal{S}_n$. Show that there exists a probability measure P on the product σ-algebra in $\Pi_{n \ge 1} X_n$ such that for each n, and each set B in the product σ-algebra in $X_1 \times \cdots \times X_n$, and $C := B \times X_{n+1} \times X_{n+2} \times \cdots$, $P(C) = \int \cdots \int 1_B(x_1, \ldots, x_n) \, dP(x_1, \ldots, x_{n-1})(x_n) \ldots dP(x_1)(x_2) \, dP_1(x_1)$. *Hints:* First prove this for finite products (n fixed). The case $n = 2$ is Problem 4. Prove it for general finite n by induction on n, writing $X_1 \times X_2 \times X_3$ as $X_1 \times (X_2 \times X_3)$, and so on. Go from finite to infinite products as in the proof of Theorem 8.2.2.

8.3. Laws of Large Numbers

Let (Ω, \mathcal{S}, P) be a probability space. Let X_1, X_2, \ldots, be real random variables on Ω. Let $S_n := X_1 + \cdots + X_n$. Any event with probability 1 is said to

happen *almost surely* (*a.s.*). Thus a sequence Y_n of (real) random variables is said to converge *a.s.* to a random variable Y iff $P(Y_n \to Y) = 1$. (The set on which a sequence of random variables converges is measurable, as shown in the proof of Theorem 4.2.5.) The sequence Y_n is said to converge *in probability* to Y iff for every $\varepsilon > 0$, $\lim_{n \to \infty} P(|Y_n - Y| > \varepsilon) = 0$. Clearly, almost sure convergence implies convergence in probability. The sequence X_1, X_2, \ldots, is said to satisfy the *strong law of large numbers* iff for some constant c, S_n/n converges to c almost surely. The sequence is said to satisfy the *weak law of large numbers* iff for some constant c, S_n/n converges to c in probability.

A basic example of a law of large numbers is convergence of relative frequencies to probabilities. Suppose that X_j are independent variables with $P(X_j = 1) = p = 1 - P(X_j = 0)$ for all j. If $X_j = 1$, say we have a success on the jth trial, otherwise a failure. Then S_n is the number of successes and S_n/n is the relative frequency of success in the first n trials. Laws of large numbers say that the relative frequency of success converges to the probability p of success.

As an introduction, a weak law of large numbers will be proved quickly under useful although not weakest possible hypotheses. First we note a classical and very often used fact:

8.3.1. The Bienaymé-Chebyshev Inequality *For any real random variable X and $t > 0$, $P(|X| \geq t) \leq EX^2/t^2$.*

Proof. $EX^2 \geq E\left(X^2 1_{\{|X| \geq t\}}\right) \geq t^2 P(|X| \geq t)$. □

8.3.2. Theorem *If X_1, X_2, \ldots, are random variables with mean 0, $EX_j^2 = 1$ and $EX_i X_j = 0$ for all $i \neq j$, then the weak law of large numbers holds for them.*

Remark. The hypothesis says X_i are orthonormal in the Hilbert space $L^2(\Omega, P)$.

Proof. We have $ES_n^2 = n$, using Theorem 8.1.2. Thus $E((S_n/n)^2) = 1/n$, so for any $\varepsilon > 0$, by Chebyshev's inequality,

$$P(|S_n/n| \geq \varepsilon) \leq 1/(n\varepsilon^2) \to 0 \qquad \text{as } n \to \infty. \qquad \square$$

Random variables X_j are called *identically distributed* iff $\mathcal{L}(X_n) = \mathcal{L}(X_1)$ for all n. "Independent and identically distributed" is abbreviated "i.i.d." For example, if X_1, X_2, \ldots, are i.i.d. variables with mean μ and variance σ^2,

$\sigma > 0$, then Theorem 8.3.2 applies to the variables $(X_j - \mu)/\sigma$, implying that S_n/n converges to μ in probability. In laws of large numbers

$$(X_1 + \cdots + X_n)/n \to c$$

for i.i.d. variables, usually $c = EX_1$, so that the average of X_1, \ldots, X_n converges to the "true" average EX_1.

In dealing with independence and almost sure convergence the next two facts will be useful.

8.3.3. Lemma *If* $0 \le p_n < 1$ *for all* n, *then* $\Pi_n(1 - p_n) = 0$ *if and only if* $\sum_n p_n = +\infty$.

Proof. If $\limsup_{n\to\infty} p_n > 0$, then clearly $\sum_n p_n = +\infty$ and $\Pi_n(1 - p_n) = 0$. So assume $p_n \to 0$ as $n \to \infty$. Since $1 - p \le e^{-p}$ for $0 \le p \le 1$ (with equality at 0, and the derivatives $-1 < -e^{-p}$), $\sum p_n = +\infty$ implies $\Pi_n(1 - p_n) = 0$. For the converse, $1 - p \ge e^{-2p}$ for $0 \le p \le 1/2$ (the inequality holds for $p = 0$ and $1/2$, and the function $f(p) := 1 - p - e^{-2p}$ has a derivative $f'(p)$ which is 0 at just one point $p = (\ln 2)/2$, a relative maximum of f). Taking M large enough so that $p_n < 1/2$ for $n \ge M$, if $\Pi_n(1 - p_n) = 0$, then $0 = \Pi_{n>M}(1 - p_n) \ge \Pi_{n>M} \exp(-2p_n) = \exp(-2\sum_{n>M} p_n) \ge 0$, so $\sum_n p_n = +\infty$. \square

Definition. Given a probability space (Ω, \mathcal{S}, P) and a sequence of events A_n, let $\limsup A_n$ be the event $\bigcap_{m\ge 1} \bigcup_{n\ge m} A_n$.

The event $\limsup A_n$ is sometimes called "A_n i.o.," meaning that A_n occur "infinitely often" as $n \to \infty$. The event $\limsup A_n$ occurs if and only if infinitely many of the A_n occur. For example, let Y_n be random variables for $n = 0, 1, \ldots$, and for each $\varepsilon > 0$, let $A_n(\varepsilon)$ be the event $\{|Y_n - Y_0| > \varepsilon\}$. Then $Y_n \to Y_0$ a.s. if and only if $A_n(\varepsilon)$ do not occur i.o. for any $\varepsilon > 0$ or equivalently for any $\varepsilon = 1/m$, $m = 1, 2, \ldots$.

Next is one of the most often used facts in probability theory:

8.3.4. Theorem (Borel-Cantelli Lemma) *If* A_n *are any events with* $\sum_n P(A_n) < \infty$, *then* $P(\limsup A_n) = 0$. *If the* A_n *are independent and* $\sum_n P(A_n) = +\infty$, *then* $P(\limsup A_n) = 1$.

Proof. The first part holds since for each m,

$$P(\limsup A_n) \le P\left(\bigcup_{n\ge m} A_n\right) \le \sum_{n\ge m} P(A_n) \to 0 \qquad \text{as } m \to \infty,$$

where $P(\bigcup_n A_n) \leq \sum_n P(A_n)$ ("Boole's inequality") by Lemma 3.1.5 and Theorem 3.1.10.

If the A_n are independent and $\sum_n P(A_n) = +\infty$, then for each m,

$$P\left(\Omega \setminus \bigcup_{n \geq m} A_n\right) = \Pi_{n \geq m}(1 - P(A_n)) = 0 \qquad \text{using Lemma 8.3.3.}$$

Thus $P(\bigcup_{n \geq m} A_n) = 1$ for all m. Let $m \to \infty$ to finish the proof. $\qquad \square$

In n tosses of a fair coin, the probability that tails comes up every time is $1/2^n$. Since the sum of these probabilities converges, almost surely heads will eventually come up.

The next theorem is the strong law of large numbers for i.i.d. variables, the main theorem of this section. It shows that $E|X_1| < \infty$ is necessary, as well as sufficient, for the strong law to hold. (It is not quite necessary for the weak law; the notes to this section go into this fine point.)

8.3.5. Theorem *For independent, identically distributed real X_j, if $E|X_1| < \infty$, then the strong law of large numbers holds, that is, $S_n/n \to EX_1$ a.s. If $E|X_1| = +\infty$, then a.s. S_n/n does not converge to any finite limit.*

Proof. First, a lemma will be of use.

8.3.6. Lemma *For any nonnegative random variable Y,*

$$EY \leq \sum_{n \geq 0} P(Y > n) \leq EY + 1.$$

Thus $EY < \infty$ if and only if $\sum_{n \geq 0} P(Y > n) < \infty$.

Proof. Let $A(k) := A_k := \{k < Y \leq k + 1\}, k = 0, 1, \ldots$. Then

$$\sum_{n \geq 0} P(Y > n) = \sum_{n \geq 0} \sum_{k \geq n} P(A_k) = \sum_{k \geq 0}(k + 1)P(A_k),$$

rearranging sums of nonnegative terms by Lemma 3.1.2. Let $U := \sum_{k \geq 0} k 1_{A(k)}$. Then $U \leq Y \leq U + 1$, so $EU \leq EY \leq EU + 1 \leq EY + 1$, and the lemma follows. $\qquad \square$

Now continuing with the proof of Theorem 8.3.5, first suppose $E|X_1| < \infty$. Note that for any independent random variables X_j and Borel measurable functions f_j, the variables $f_j(X_j)$ are also independent. Specifically, the positive parts $X_j^+ := \max(X_j, 0)$ are independent. Likewise, the

$X_j^- := -\min(X_j, 0)$ are independent of each other. Thus it will be enough to prove convergence separately for the positive and negative parts. So we can assume that $X_n \geq 0$ for all n.

Let $Y_j := X_j 1\{X_j \leq j\}$ where $1\{\ldots\}$ is the indicator function $1_{\{\ldots\}}$. Let $T_n := \sum_{1 \leq j \leq n} Y_j$. Given any number $\alpha > 1$, let $k(n) := [\alpha^n]$ where $[x]$ denotes the greatest integer $\leq x$. Then

$$1 \leq k(n) \leq \alpha^n < k(n) + 1 \leq 2k(n), \qquad \text{so } k(n)^{-2} \leq 4\alpha^{-2n}.$$

For $x \geq 1$, $[x] \geq x/2$. Take any $\varepsilon > 0$. Recall that $\text{var}(X) := E((X - EX)^2)$ denotes the variance of a random variable X. By Chebyshev's inequality (8.3.1) and Theorem 8.1.2, there exist finite constants c_1, c_2, \ldots, depending only on ε and α, such that

$$\sum := \sum_{n \geq 1} P\{|T_{k(n)} - ET_{k(n)}| > \varepsilon k(n)\} \leq c_1 \sum_{n \geq 1} \text{var}(T_{k(n)})/k(n)^2$$

$$= c_1 \sum_{n \geq 1} k(n)^{-2} \sum_{1 \leq i \leq k(n)} \text{var}(Y_i) = c_1 \sum_{i \geq 1} \text{var}(Y_i) \sum_{k(n) \geq i} k(n)^{-2},$$

and $\sum_n k(n)^{-2} 1_{k(n) \geq i} \leq 4 \sum_n \alpha^{-2n} 1\{\alpha^n \geq i\} \leq 4i^{-2}/(1 - \alpha^{-2}) \leq c_2 i^{-2}$, so if F is the law of X_1, $F(x) \equiv P(X_1 \leq x)$,

$$\sum \leq c_3 \sum_{i \geq 1} EY_i^2/i^2 = c_3 \sum_{i \geq 1} \int_0^i x^2 \, dF(x)/i^2$$

$$= c_3 \sum_{i \geq 1} i^{-2} \left(\sum_{0 \leq k < i} \int_k^{k+1} x^2 \, dF(x) \right) \leq c_4 \sum_{k \geq 0} \int_k^{k+1} x^2 \, dF(x)/(k + 1),$$

since $Q_k := \sum_{i > k} i^{-2} < \int_k^\infty x^{-2} \, dx = 1/k$, which yields, if $k \geq 1$, that $Q_k \leq 2/(k + 1)$, while $Q_0 = 1 + Q_1 < 2 = 2/(k + 1)$ for $k = 0$. Thus

$$\sum \leq c_4 \sum_{k \geq 0} \int_k^{k+1} x \, dF(x) = c_4 EX_1 < \infty.$$

So by the Borel-Cantelli lemma (8.3.4), $|T_{k(n)} - ET_{k(n)}|/k(n) \to 0$ a.s. As $n \to \infty$, $EY_n \uparrow EX_1$. It follows that $ET_{k(n)}/k(n) \uparrow EX_1$. Thus $T_{k(n)}/k(n) \to EX_1$ a.s. Now,

$$\sum_j P(X_j \neq Y_j) = \sum_j P(X_j > j) < \infty \qquad \text{by Lemma 8.3.6,}$$

so a.s. $X_j = Y_j$ for all large enough j, say $j > m(\omega)$. As $n \to \infty$,

$S_{m(\omega)}/k(n) \to 0$ and $T_{m(\omega)}/k(n) \to 0$, so $S_{k(n)}/k(n) \to EX_1$ a.s. As $n \to \infty$, $k(n+1)/k(n) \to \alpha$, so for n large enough, $1 \le k(n+1)/k(n) < \alpha^2$. Then for $k(n) < j \le k(n+1)$,

$$S_{k(n)}/k(n) \le \alpha^2 S_j/j \le \alpha^4 S_{k(n+1)}/k(n+1).$$

Thus a.s.,

$$\alpha^{-2} EX_1 \le \liminf_{j \to \infty} S_j/j \le \limsup_{j \to \infty} S_j/j \le \alpha^2 EX_1.$$

Letting $\alpha \downarrow 1$ finishes the proof in case $E|X_1| < \infty$.

Conversely, if S_n/n converges to a finite limit, then clearly $S_n/(n+1)$ converges to the same limit, as does S_{n-1}/n, so $X_n/n = (S_n - S_{n-1})/n \to 0$. But if $E|X_1| = +\infty$, then by Lemma 8.3.6 and the Borel-Cantelli lemma (8.3.4), a.s. $|X_n| > n$ for infinitely many n. Thus the probability that $S_n(\omega)/n$ is a Cauchy sequence is 0. $\qquad\square$

The proof of the half of the Borel-Cantelli lemma (8.3.4) without independence, $\sum P(A_n) < \infty$ implies $P(A_n \text{ i.o.}) = 0$, used subadditivity $P(\bigcup_n A_n) \le \sum_n P(A_n)$. Here is a lower bound for $P(\bigcup_{1 \le i \le n} A_i)$ that also does not require independence:

8.3.7. Theorem (Bonferroni Inequality) *For any events $A_i := A(i)$,*

$$P\left(\bigcup_{i=1}^n A_i\right) \ge \sum_{i=1}^n P(A_i) - \sum_{1 \le i < j \le n} P(A_i \cap A_j).$$

Proof. Let's prove for indicator functions

$$\sum_{i=1}^n 1_{A_i} \le 1_{\bigcup_{1 \le i \le n} A_i} + \sum_{1 \le i < j \le n} 1_{A_i \cap A_j}.$$

A given point ω belongs to A_i for k values of i for some $k = 0, 1, \ldots, n$. The inequality for indicators holds if $k = 0$ $(0 \le 0)$, if $k = 1$ $(1 \le 1)$, and if $k \ge 2$: $k - 1 \le k(k-1)/2$. Then integrating both sides with respect to P finishes the proof. $\qquad\square$

The Bonferroni inequality is most useful when the latter sum in it is small, so that the lower bound given for the probability of the union is close to the upper bound $\sum_i P(A_i)$. For example, suppose that $p_i := P(A_i) < \varepsilon$ for all i and that the A_i are pairwise independent, that is, A_i is independent of A_j whenever $i \ne j$. Then $2 \sum_{1 \le i < j \le n} P(A_i \cap A_j) \le n(n-1)\varepsilon^2$, where if

266 *Introduction to Probability Theory*

$n\varepsilon$ is small, $n^2\varepsilon^2$ is even smaller. If the A_i are actually independent, then the probability of their union is $1 - \Pi_i(1 - p_i)$. If the product is expanded, the linear and quadratic terms in the p_i correspond to the two sums in the Bonferroni inequality.

Beside the following problems, there will be others on laws of large numbers at the end of §9.7 after further techniques have been developed.

Problems

1. Let $A(n)$ be a sequence of independent events and $p_n := P(A(n))$. Under what conditions on p_n does $1_{A(n)} \to 0$
 (a) in probability?
 (b) almost surely? *Hint:* Use the Borel-Cantelli lemma.

2. Show that for any real random variable X, $p > 0$, and $t > 0$, $P(|X| \geq t) \leq E|X|^p/t^p$.

3. If $X_1, X_2, \ldots,$ are random variables with mean 0, $EX_j^2 = 1$ and for all $i \neq j$ $EX_iX_j = 0$, show that for any $\alpha > 1$, $S_n/n^\alpha \to 0$ almost surely.

4. If $X_1, X_2, \ldots,$ are random variables with $EX_iX_j = 0$ for all $i \neq j$, and $\sup_j EX_j^2 < \infty$, show that for any $\alpha > 1/2$, $S_n/n^\alpha \to 0$ in probability.

5. Let Y have a standard exponential distribution, meaning that $P(Y > t) = e^{-t}$ for all $t \geq 0$. Evaluate EY and $\sum_{n \geq 0} P(Y > n)$ and verify the inequalities in Lemma 8.3.6 in this case.

6. Show that for any $c_n > 0$ with $c_n \to 0$ (no matter how fast), there exist random variables $V_n \to 0$ in probability such that $c_n V_n$ does not approach 0 a.s. *Hint:* Let A_n be independent events with $P(A_n) = 1/n$. Let $V_n = 1/c_n$ on A_n and 0 elsewhere.

7. Why not prove the strong law of large numbers by using the Borel-Cantelli lemma and showing that for every $\varepsilon > 0$, $\sum_n P(|S_n/n| > \varepsilon) < \infty$? Because the latter series diverges in general: let $X_1, X_2, \ldots,$ be i.i.d. with a density f given by $f(x) := |x|^{-3}$ for $|x| \geq 1$ and $f(x) = 0$ otherwise, so $P(X_j \in A) = \int_A f\,dx$ for each Borel set A and $EX_1 = 0$. For $n = 1, 2, \ldots,$ and $j = 1, \ldots, n$, let A_{nj} be the event $\{X_j > n\}$ and B_{nj} the event $\{\sum_{1 \leq i \leq n} X_i \geq X_j\}$. Let $C_{nj} := A_{nj} \cap B_{nj}$.
 (a) Show that $P(C_{nj}) = n^{-2}/4$ for each $n \geq 2$ and j.
 (b) Show that $\{S_n/n > 1\} \supset \bigcup_{1 \leq j \leq n} C_{nj}$ for each n and $\sum_n P(S_n > n)$ diverges. *Hint:* Use Bonferroni's inequality and $P(C_{ni} \cap C_{nj}) \leq P(A_{ni} \cap A_{nj})$.

*8.4. Ergodic Theorems

Laws of large numbers, or similar facts, can also be proved without any independence or orthogonality conditions on the variables, and the measure space need not be finite. Let (X, \mathcal{A}, μ) be a σ-finite measure space. Let T be a measurable transformation (= function) from X into itself, that is, $T^{-1}(B) \in \mathcal{A}$ for each $B \in \mathcal{A}$. T is called *measure-preserving* iff $\mu(T^{-1}(B)) = \mu(B)$ for all $B \in \mathcal{A}$. A set Y is called *T-invariant* iff $T^{-1}(Y) = Y$. Since T^{-1} preserves all set operations such as unions and complements, the collection of all T-invariant sets is a σ-algebra. Since the intersection of any two σ-algebras is a σ-algebra, the collection $\mathcal{A}_{\text{inv}(T)}$ of all measurable T-invariant sets is a σ-algebra. A measure-preserving transformation T is called *ergodic* iff for every $Y \in \mathcal{A}_{\text{inv}(T)}$, either $\mu(Y) = 0$ or $\mu(X \backslash Y) = 0$.

Examples

1. For Lebesgue measure on \mathbb{R}, each translation $x \mapsto x + y$ is measure-preserving (but not ergodic: see Problem 1).
2. Let X be the unit circle in the plane, $X = \{(\cos \theta, \sin \theta) : 0 \leq \theta < 2\pi\}$, with the measure μ given by $d\mu(\theta) = d\theta$. Then any rotation T of X by an angle α is measure-preserving (ergodic for some α and not others: see Problem 4).
3. For Lebesgue measure λ^k on \mathbb{R}^k, translations, rotations around any axis, and reflections in any plane are measure-preserving. These transformations are not ergodic; for example, for $k = 2$, a rotation around the origin is not ergodic: annuli $a < r < b$ for the polar coordinate r are invariant sets which violate ergodicity.
4. If we take an infinite product of copies of one probability space (X, \mathcal{S}, P), as in §8.2, then the "shift" transformation $\{x_n\}_{n \geq 1} \mapsto \{x_{n+1}\}_{n \geq 1}$ is measure-preserving. (It will be shown in Theorem 8.4.5 that it is ergodic.)
5. An example of a measure-preserving transformation which is not one-to-one: on the unit interval $[0, 1)$ with Lebesgue measure, let $f(x) = 2x$ for $0 \leq x < 1/2$ and $f(x) = 2x - 1$ for $1/2 \leq x < 1$.

Let $f \in \mathcal{L}^1(X, \mathcal{A}, \mu)$; in other words, f is a measurable real-valued function on X with $\int |f| \, d\mu < \infty$. For a measure-preserving transformation T, let $f_0 := f$ and $f_j := f \circ T^j$, $j = 1, 2, \ldots$ (where the exponent j denotes composition, $T^j = T \circ T \circ \cdots \circ T$, to j terms). Let $S_n := f_0 + f_1 + \cdots + f_{n-1}$. Then S_n/n will be shown to converge; if T is ergodic, the limit will be a constant (as in laws of large numbers):

8.4.1. Ergodic Theorem *Let T be any measure-preserving transformation of a σ-finite measure space (X, \mathcal{A}, μ). Then for any $f \in \mathcal{L}^1(X, \mathcal{A}, \mu)$ there is a function $\varphi \in \mathcal{L}^1(X, \mathcal{A}_{\text{inv}(T)}, \mu)$ such that $\lim_{n\to\infty} S_n(x)/n = \varphi(x)$ for μ-almost all x, with $\int |\varphi| \, d\mu \leq \int |f| \, d\mu$. If T is ergodic, then φ equals (almost everywhere) some constant c. If $\mu(X) < \infty$, then S_n/n converges to φ in \mathcal{L}^1; that is, $\lim_{n\to\infty} \int |\varphi - S_n/n| \, d\mu = 0$, so $\int \varphi \, d\mu = \int f \, d\mu$.*

The proof of the ergodic theorem will depend on another fact. A function U from $\mathcal{L}^1(X, \mathcal{A}, \mu)$ into itself is called *linear* iff $U(cf+g) = cU(f)+U(g)$ for any real c and functions f and g in \mathcal{L}^1. It is called *positive* iff whenever $f \geq 0$ (meaning that $f(x) \geq 0$ for all x), and $f \in \mathcal{L}^1$, then $Uf \geq 0$. U is called a *contraction* iff $\int |Uf| \, d\mu \leq \int |f| \, d\mu$ for all $f \in \mathcal{L}^1$. For any measure-preserving transformation T from X into itself, $Uf := f \circ T$ defines a positive linear contraction U (by the image measure theorem, 4.1.11).

Given such a U and an $f \in \mathcal{L}^1$, let $S_0(f) := 0$ and $S_n(f) := f + Uf + \cdots + U^{n-1}(f)$, $n \geq 1$. Let $S_n^+(f) := \max_{0 \leq j \leq n} S_j(f)$.

8.4.2. Maximal Ergodic Lemma *For any positive linear contraction U of $\mathcal{L}^1(X, \mathcal{A}, \mu)$, any $f \in \mathcal{L}^1(X, \mathcal{A}, \mu)$ and $n = 0, 1, \ldots$, let $A := \{x \colon S_n^+(f) > 0\}$. Then $\int_A f \, d\mu \geq 0$.*

Proof. For $r = 0, 1, \ldots$, $f + US_r(f) = S_{r+1}(f)$. Note that since U is positive and linear, $g \geq h$ implies $Ug \geq Uh$. Thus for $j = 1, \ldots, n$, $S_j(f) = f + US_{j-1}(f) \leq f + US_n^+(f)$.

If $x \in A$, then $S_n^+(f)(x) = \max_{1 \leq j \leq n} S_j(f)(x)$. Combining gives for all $x \in A$, $f \geq S_n^+(f) - US_n^+(f)$. Since $S_n^+(f) \geq 0$ on X and $S_n^+ = 0$ outside A, we have

$$\int_A f \, d\mu \geq \int_A (S_n^+(f) - US_n^+(f)) \, d\mu = \int_X S_n^+(f) \, d\mu - \int_A US_n^+(f) \, d\mu$$

$$\geq \int_X S_n^+(f) \, d\mu - \int_X US_n^+(f) \, d\mu \geq 0$$

since U is a contraction. \square

Now to prove the ergodic theorem (8.4.1), for any real $a < b$ let $Y := Y(a, b) := \{x \colon \liminf_{n\to\infty} S_n/n < a < b < \limsup_{n\to\infty} S_n/n\}$. Then Y is measurable. If in this definition S_n is replaced by $S_n \circ T$, we get $T^{-1}(Y)$ instead of Y. Now $S_n \circ T = S_{n+1} - f$, and $f/n \to 0$ as $n \to \infty$, so we can put S_{n+1} in place of $S_n \circ T$. But in the original definition of Y, S_n/n can be replaced equivalently by $S_{n+1}/(n+1)$ and then by S_{n+1}/n. So Y is an invariant set.

To prove that $\mu(Y) = 0$, we can assume $b > 0$ since otherwise $a < 0$ and we can consider $-f$ and $-a$ in place of f and b. Suppose $C \subset Y$, $C \in \mathcal{A}$, and $\mu(C) < \infty$. Let $g := f - b1_C$ and $A(n) := \{x: S_n^+(g)(x) > 0\}$. Then by the maximal ergodic lemma (8.4.2),

$$\int_{A(n)} f - b1_C \, d\mu \geq 0$$

for $n = 1, 2, \ldots$. Since $g \geq f - b$, it follows that for all j and x, $S_j(g) \geq S_j(f) - jb$. On Y, $\sup_n(S_n(f) - nb) > 0$. Thus $Y \subset A := \bigcup_n A(n)$. Since $A(n) \uparrow A$ as $n \to \infty$, the dominated convergence theorem gives $\int_A f - b1_C \, d\mu \geq 0$, so that $\int_A f \, d\mu \geq b\mu(A \cap C) = b\mu(C)$. Since μ is σ-finite, we can let C increase up to Y and get $\int f^+ d\mu \geq b\mu(Y)$. Thus since $f \in \mathcal{L}^1$, we see that $\mu(Y) < \infty$.

Now since Y is invariant, we can restrict everything to Y and assume $X = Y = C = A$. Then since we have a finite measure space, $f - b \in \mathcal{L}^1$ and $\int_Y f - b \, d\mu \geq 0$. For $a - f$, the maximal ergodic lemma and dominated convergence as above give $\int_Y a - f \, d\mu \geq 0$. Summing gives $\int_Y a - b \, d\mu \geq 0$. Since $a < b$, we conclude that $\mu(Y) = 0$.

Taking all pairs of rational numbers $a < b$ gives that $S_n(f)/n$ converges almost everywhere (a.e.) for μ to some function φ with values in $[-\infty, \infty]$. T^n is measure-preserving for each n, and

$$\int |f \circ T^n| \, d\mu = \int |f| \circ T^n \, d\mu = \int |f| \, d(\mu \circ (T^n)^{-1}) = \int |f| \, d\mu,$$

where the middle equation is given by the image measure theorem (4.1.11). It follows that $\int |S_n/n| \, d\mu \leq \int |f| \, d\mu$. Thus by Fatou's lemma (4.3.3), $\int |\varphi| \, d\mu \leq \int |f| \, d\mu < \infty$. So φ is finite a.e. Specifically, let $\varphi := \limsup_{n \to \infty} S_n(f)/n$ if this is finite; otherwise, set $\varphi = 0$. Then φ is measurable. Also, φ is T-invariant, $\varphi = \varphi \circ T$, as in the proof that Y is T-invariant. Thus for any Borel set B of real numbers, $\varphi^{-1}(B) = (\varphi \circ T)^{-1}(B) = T^{-1}(\varphi^{-1}(B))$, so $\varphi^{-1}(B) \in \mathcal{A}_{\mathrm{inv}(T)}$ and φ is measurable for $\mathcal{A}_{\mathrm{inv}(T)}$, proving the first conclusion in Theorem 8.4.1.

If T is ergodic, let D be the set of those y such that $\mu(\{x: \varphi(x) > y\}) > 0$. Let $c := \sup D$. Then for $n = 1, 2, \ldots$, $\mu(\{x: \varphi(x) > c + 1/n\}) = 0$, so $\varphi \leq c$ a.e. On the other hand, take $y_n \in D$ with $y_n < c$ and $y_n \uparrow c$. For each n, since T is ergodic, $\mu(\{x: \varphi(x) \leq y_n\}) = 0$, so $\varphi \geq c$ a.e. and $\varphi = c$ a.e. Thus c is finite, and if $\mu(X) = +\infty$, then $c = 0$.

If $\mu(X) < \infty$ (with T not necessarily ergodic), it remains to prove $\int |\varphi - S_n/n| \, d\mu \to 0$. If f is bounded, say $|f| \leq K$ a.e., then for all n, $|S_n/n| \leq K$ a.e., and the conclusion follows from dominated convergence (Theorem

4.3.5). For a general f in \mathcal{L}^1, let $f_K := \max(-K, \min(f, K))$. Then each f_K is bounded and $\int |f_K - f|\, d\mu \to 0$ as $K \to \infty$. Given $\varepsilon > 0$, let $g = f_K$ for K large enough so that $\int |f - g|\, d\mu < \varepsilon/3$. Then $S_n(g)/n$ converges a.e. and in \mathcal{L}^1 to some function j. For all n, $\int |S_n(f) - S_n(g)|/n\, d\mu \le \varepsilon/3$. Then by Fatou's Lemma (4.3.3), $\int |\varphi - j|\, d\mu \le \varepsilon/3$. For n large enough, $\int |S_n(g)/n - j|\, d\mu < \varepsilon/3$, so $\limsup_{n\to\infty} \int |S_n(f)/n - \varphi|\, d\mu \le \varepsilon$. Letting $\varepsilon \downarrow 0$ gives that the latter lim sup (and hence limit) is 0. Then since $\int S_n(f)/n\, d\mu = \int f\, d\mu$ for all n, we have $\int \varphi\, d\mu = \int f\, d\mu$, finishing the proof of the ergodic theorem (8.4.1). $\qquad\square$

The following can be proved in the same way:

8.4.3. Corollary *If in Theorem 8.4.1 $\mu(X) < \infty$ (with T not necessarily ergodic) and $f \in \mathcal{L}^p$ for some p, $1 < p < \infty$, that is $\int |f|^p\, d\mu < \infty$, then $S_n(f)/n$ converges to φ in \mathcal{L}^p; in other words,*

$$\int |S_n(f)/n - \varphi|^p\, d\mu \to 0 \qquad \text{as } n \to \infty.$$

Next, a fact about product probabilities will be useful. Let I be an index set which will be either the set \mathbb{N} of all nonnegative integers or the set \mathbb{Z} of all integers. For each $i \in I$ let (Ω_i, S_i, P_i) be a probability space. Let $\Omega := \Pi_{i \in I}\Omega_i = \{\{x_i\}_{i \in I}: x_i \in \Omega_i \text{ for each } i\}$. Then $\{x_i\}_{i \in I} = \{x_i\}_{i \ge 0}$ for $i = \mathbb{N}$ or a two-sided sequence $\{\ldots, x_{-2}, x_{-1}, x_0, x_1, x_2, \ldots\}$ for $I = \mathbb{Z}$. On Ω we have the product σ-algebra \mathcal{S}, the smallest σ-algebra making each coordinate x_i measurable, and the product probability $Q := Q^{(I)} := \Pi_{i \in I} P_i$ given by Theorem 8.2.2.

For $n = 1, 2, \ldots$, let \mathcal{B}_n be the smallest σ-algebra of subsets of Ω such that x_j are measurable for all j with $|j| \le n$. Let $\mathcal{B}^{(n)}$ be the smallest σ-algebra such that x_j are measurable for all $j > n$. Let $\mathcal{B}^{(\infty)} := \bigcap_{n=1}^{\infty} \mathcal{B}^{(n)}$. Then $\mathcal{B}^{(\infty)}$ is called the σ-algebra of *tail events*. For example, if $\Omega_i = \mathbb{R}$ for all i with Borel σ-algebra and $S_n := x_1 + \cdots + x_n$, then events such as $\{\limsup_{n\to\infty} S_n/n > 0\}$ are in $\mathcal{B}^{(\infty)}$.

8.4.4. Kolmogorov's 0–1 Law *For any product probability Q and $A \in \mathcal{B}^{(\infty)}$, $Q(A) = 0$ or 1.*

Proof. Let \mathcal{C} be the collection of all sets D in \mathcal{S} such that for each $\delta > 0$ there is some n and $B \in \mathcal{B}_n$ with $Q(B \bigtriangleup D) < \delta$, where $B \bigtriangleup D$ is the symmetric difference $(B \setminus D) \cup (D \setminus B)$. Then $\mathcal{C} \supset \mathcal{B}_n$ for all n. If $D \in \mathcal{C}$, then for the complement D^c and any B we have $B^c \bigtriangleup D^c = B \bigtriangleup D$, so $D^c \in \mathcal{C}$. For any $D_j \in \mathcal{C}$, $j = 1, 2, \ldots$, let $D := \bigcup_j D_j$. Given $\delta > 0$, there is an m such

that $Q(D\setminus\bigcup_{j=1}^{m} D_j) < \delta/2$. For each $j = 1, \ldots, m$ there is an $n(j)$ and a $B_j \in \mathcal{B}_{n(j)}$ such that $Q(D_j \triangle B_j) < \delta/(2m)$. Let $n := \max_{j \le m} n(j)$ and $B := \bigcup_{j \le m} B_j$. Then $B \in \mathcal{B}_n$ and $Q(D \triangle B) < \delta$, so $D \in \mathcal{C}$. It follows that \mathcal{C} is a σ-algebra, so $\mathcal{C} = \mathcal{S}$.

Given $A \in \mathcal{B}^{(\infty)}$, take $B_n \in \mathcal{B}_n$ with $Q(B_n \triangle A) \to 0$ as $n \to \infty$. Since Q is a product probability and $A \in \mathcal{B}^{(n)}$, $Q(B_n \cap A) = Q(B_n)Q(A)$ for all n. Letting $n \to \infty$ gives $Q(A) = Q(A)^2$, so $Q(A) = 0$ or 1. \square

Now consider the special case where all $(\Omega_i, \mathcal{S}_i, P_i)$ are copies of one probability space (X, \mathcal{A}, P). Then let $X^I := \Omega$ and $P^I := Q$, so that the coordinates x_i are i.i.d. (P). The *shift* transformation T is defined by $T(\{x_i\}_{i \in I}) := \{x_{i+1}\}_{i \in I}$. Then T is a well-defined measure-preserving transformation of X^I onto itself for the measure P^I, with $I = \mathbb{N}$ or \mathbb{Z}. T is called a *unilateral shift* for $I = \mathbb{N}$ and a *bilateral shift* for $I = \mathbb{Z}$.

8.4.5. Theorem *For $I = \mathbb{N}$ or \mathbb{Z} and any probability space (X, \mathcal{A}, P), the shift T is always ergodic on X^I for P^I.*

Proof. If $I = \mathbb{N}$, then for any $Y \in \mathcal{B}$, $T^{-1}(Y) \in \mathcal{B}^{(0)}$, $T^{-1}(T^{-1}(Y)) \in \mathcal{B}^{(1)}$, and so forth, so if Y is an invariant set, then $Y \in \mathcal{B}^{(\infty)}$. Then the 0–1 law (8.4.4) implies that T is ergodic.

If $I = \mathbb{Z}$, then given $A \in \mathcal{B}$, as in the proof of 8.4.4, take $B_n \in \mathcal{B}_n$ with $P^I(B_n \triangle A) \to 0$ as $n \to \infty$. If A is invariant, $A = T^{-2n-1}A$ so $P^I(T^{-2n-1}(B_n \triangle A)) = P^I(T^{-2n-1}B_n \triangle A) \to 0$. On the other hand, $T^{-2n-1}B_n \in \mathcal{B}^{(n)}$, so $T^{-2n-1}B_n$ and B_n are independent. Letting $n \to \infty$ then again gives $P^I(A)^2 = P^I(A)$ and $P^I(A) = 0$ or 1. \square

The ergodic theorem gives another proof of the strong law of large numbers for i.i.d. variables X^n with finite mean (Theorem 8.3.5) as follows. Let I be the set of positive integers. Consider the function Y defined by $Y(\omega) := \{X_n(\omega)\}_{n \ge 1}$ from Ω into \mathbb{R}^I. Then the image measure $P \circ Y^{-1}$ equals Q^I where $Q := \mathcal{L}(X_1)$, by independence and the uniqueness in Theorem 8.2.2. So we can assume $\Omega = \mathbb{R}^I$, X_n are the coordinates, and $P = Q^I$. By Theorem 8.4.5, T is ergodic, so that the i.i.d. strong law (8.3.5) follows from the ergodic theorem (8.4.1).

A set A in a product X^I is called *symmetric* if $T(A) = A$ for any transformation T with $T(\{x_n\}_{n \in I}) = \{x_{\pi(n)}\}_{n \in I}$ for some 1–1 function π from I onto itself, where $\pi(j) = j$ for all but finitely many j. Then π is called a *finite permutation*.

8.4.6. Theorem (The Hewitt-Savage 0–1 Law). *Let I be any countably infinite index set, A any measurable symmetric set in a product space X^I, and P any probability measure on X. Then $P^I(A) = 0$ or 1.*

Proof. We can assume $I = \mathbb{Z}$. For each $n = 1, 2, \ldots$, and $j \in \mathbb{Z}$, let $\pi_n(j) = j$ if $|j| > n$, or $j + 1$ if $-n \leq j < n$, and $\pi_n(n) := -n$. Then π_n is a finite permutation. Let \mathcal{F}_n be the smallest σ-algebra for which X_i are measurable for $|i| \leq n$. Given $\varepsilon > 0$, as in the proof of 8.4.4, there are n and $B_n \in \mathcal{F}_n$ with $P^I(A \triangle B_n) < \varepsilon/2$. Let $\zeta_n(x) := \{x_{\pi_n(j)}\}_{j \in \mathbb{Z}}$. Let $\pi_\infty(j) := j + 1$ for all j, so ζ_∞ is the shift. Each ζ_n preserves P^I and A, so that $\varepsilon/2 > P^I(A \triangle \zeta_{n+2}(B_n)) = P^I(A \triangle \zeta_\infty(B_n)) = P^I(\zeta_\infty^{-1} A \triangle B_n)$, and $P^I(\zeta_\infty^{-1}(A) \triangle A) < \varepsilon$, so $P^I(\zeta_\infty^{-1} A \triangle A) = 0$. By problem 7 below, $P^I(A \triangle B) = 0$ for an invariant set B, and $P(B) = 0$ or 1 by Theorem 8.4.5, and so likewise for A. $\qquad\square$

Problems

1. On \mathbb{R} with Lebesgue measure, show that the measure-preserving transformation $x \mapsto x + y$ is not ergodic.

2. Prove Corollary 8.4.3.

3. In $X = \mathbb{R}^{\mathbb{N}}$ or $\mathbb{R}^{\mathbb{Z}}$, let A be the set of all sequences $\{x_n\}$ such that $(x_1 + \cdots + x_n)/n \to 1$ as $n \to +\infty$. Show why A is or is not
 (a) invariant for the unilateral shift $\{x_j\} \mapsto \{x_{j+1}\}$ in $\mathbb{R}^{\mathbb{N}}$ or $\mathbb{R}^{\mathbb{Z}}$;
 (b) symmetric (as in the Hewitt-Savage 0–1 law) in $\mathbb{R}^{\mathbb{N}}$ or $\mathbb{R}^{\mathbb{Z}}$.

4. Let S^1 be the unit circle in \mathbb{R}^2, $S^1 := \{(x, y): x^2 + y^2 = 1\}$, with the arc length measure μ: $d\mu(\theta) = d\theta$ for the usual polar coordinate θ. Let T be the rotation of S^1 by an angle α. Show that T is ergodic if and only if α/π is irrational. *Hints:* If $\varepsilon > 0$ and $\mu(A) > 0$, there is an interval $J: a < \theta < b$ with $\mu(A \cap J) > (1 - \varepsilon)\mu(J)$, by Proposition 3.4.2. If α/π is irrational, show that for any $z \in S^1$, $\{T^n z\}_{n \geq 0}$ is dense in S^1.

5. (a) On \mathbb{Z} with counting measure, show that the shift $n \mapsto n+1$ is ergodic.
 (b) Let π be a 1–1 function from \mathbb{N} onto itself which is an ergodic transformation of counting measure on \mathbb{N}. Let (X, \mathcal{B}, P) be a probability space. Define T_π from $X^{\mathbb{N}}$ onto itself by $T_\pi(\{x_i\}_{i \in \mathbb{N}}) := \{x_{\pi(i)}\}_{i \in \mathbb{N}}$. Show that T_π is ergodic.

6. On the unit interval $[0, 1)$, take Lebesgue measure. Represent numbers by their decimal expansions $x = 0.x_1 x_2 x_3 \ldots$, where each x_j is an integer from 0 to 9, and for those numbers with two expansions, ending with an infinite string of 0's or 9's, choose the one with 0's.

(a) Show that the shift transformation of the digits, $\{x_n\} \mapsto \{x_{n+1}\}$, is measure-preserving.

(b) Show that this transformation is ergodic.

7. Given a σ-finite measure space (X, \mathcal{S}, μ) and a measure-preserving transformation T of X into itself, a set $A \in \mathcal{S}$ is called *almost invariant* iff $\mu(A \bigtriangleup T^{-1}(A)) = 0$. Show that for any almost invariant set A, there is an invariant measurable set B with $\mu(A \bigtriangleup B) = 0$. *Hint:* Let $T^{-n-1}(A) := T^{-1}(T^{-n}(A))$ for $n \geq 1$ and $B := \bigcap_{m=1}^{\infty} \bigcup_{n=m}^{\infty} T^{-n}(A)$.

8. If T is an ergodic measure-preserving transformation of X for a σ-finite measure space (X, \mathcal{S}, μ), show that T has the same properties for the completion of μ, where the σ-algebra \mathcal{S} is extended to contain all subsets of sets of measure 0 for μ.

9. Let $(X_j, \mathcal{S}_j, \mu_j)$ be σ-finite measure spaces for $j = 1, 2$ and let T_j be a measure-preserving transformation of X_j. Let $T(x_1, x_2) := (T_1(x_1), T_2(x_2))$.

(a) Show that T is a measure-preserving transformation of $(X_1 \times X_2, \mathcal{S}_1 \otimes \mathcal{S}_2, \mu_1 \times \mu_2)$.

(b) Show that even if T_1 and T_2 are both ergodic, T may not be. *Hint:* Let each X_j be the circle and let each T_j be a rotation by the same angle α (as in Problem 4) with α/π irrational.

Notes

§8.1 For a fair die, with probability of each face exactly $1/6$, we would have $P(\{5, 6\}) = 1/3$. In the dice actually used, the numbers are marked by hollowed-out pips on each face. The 6 face is lightened the most and is opposite the 1 face, which is lightened least. Similarly, the 5 face is lighter than the opposite 2 face. Some actual experiments gave the result $P(\{5, 6\}) = 0.3377 \pm 0.0008$ ("Weldon's dice data"; see Feller, 1968, pp. 148–149).

A set function P is called *finitely additive* iff for any finite n and disjoint sets A_1, \ldots, A_n in the domain of P,

$$P(A_1 \cup \cdots \cup A_n) = \sum_{j=1}^{n} P(A_j).$$

Some problems at the end of §3.1 indicated the pathology that can occur with only finite additivity rather than countable additivity. The definition of probability as a (countably additive, nonnegative) measure of total mass 1 on a σ-algebra in a general space is adopted by the overwhelming majority of researchers on probability. The book of Kolmogorov (1933) first made the definition widely known (see, e.g., Bingham (2000)). The paper by Barone and Novikoff (1978), whose projected Part II apparently has not (yet) appeared, is primarily about Borel (1909) and secondarily about some probability examples in Hausdorff (1914).

Kolmogorov (1929b) gave an axiomatization of probability on general spaces, beginning with a finitely additive function on a collection of sets that need not be an algebra,

such as the density of sets of integers (Problem 9 of §3.1). In Axiom I, probabilities are required to be ">0" (presumably meaning ≥ 0). A "normal" probability was defined as countably additive on an algebra and thus extends uniquely to a σ-algebra, as in Theorem 3.1.10. But Kolmogorov (1929b) does not yet restrict probabilities to be "normal." The paper was little noticed; it was not listed in the mathematics review journals of the time, *Jahrbuch über die Fortschritte der Mathematik* or *Revue Semestrielle des Publications Mathématiques*. These reviews seem to have covered few Russian-language papers, even in journals more likely to attract mathematicians' attention. Kolmogorov (1929b) was included in Kolmogorov (1986) and then mentioned by Shiryayev (1989, p. 884).

Thus Ulam (1932) may have been the first to give the "Kolmogorov (1933)" definition of probability to an international audience, apparently independently of Kolmogorov. Ulam required in addition that singletons be measurable. Ulam's note was an announcement of the joint paper Lomnicki and Ulam (1934), in which Kolmogorov (1933) is cited. In the two previous decades, probability measures were defined on particular spaces such as Euclidean spaces and under further restrictions. Often [0, 1] with Lebesgue measure was taken as a basic probability space.

Kolmogorov (1933) also included, of importance, a definition of conditional expectation and an existence theorem for stochastic processes (see the notes to §§10.1 and 12.1). Other works of Kolmogorov's are mentioned in this and later chapters. As Gnedenko and Smirnov (1963) wrote, "In the contemporary theory of probability, A. N. Kolmogorov is duly considered as the accepted leader." Kolmogorov shared the Wolf Prize (with Henri Cartan) in mathematics for 1980, the third year the prize was awarded (*Notices*, 1981). Other articles in his honor were Gnedenko (1973) and *The Times'* obituary (1987).

Dubins and Savage (1965) treat probabilities which are only finitely additive. Other developments of finitely additive "probabilities" lead in different directions and are often not even called probabilities in the literature.

One main example is "invariant means." Let G be an Abelian group: that is, a function $+$ is defined from $G \times G$ into G such that for any x, y, and z in G, $x + y = y + x$ and $x + (y + z) = (x + y) + z$, and there is a $0 \in G$ such that for each $x \in G$, $x + 0 = x$ and there is a $y \in G$ with $x + y = 0$. Then an *invariant mean* is a function d defined on all subsets of G, nonnegative and finitely additive, with $d(G) = 1$, and with $d(A + m) = d(A)$ for all $A \subset G$ and $m \in G$, where $A + m := \{a + m : a \in A\}$. Invariant means can also be defined on semigroups, such as the set \mathbb{Z}^+ of all positive integers. On \mathbb{N} or \mathbb{Z}^+, invariant means are extensions of the "density" treated in Problem 9 at the end of §3.1. On the theory of invariant means, some references are Banach (1923), von Neumann (1929), Day (1942), and Greenleaf (1969).

The problem of throwing a needle onto a ruled plane is associated with Buffon (1733, pp. 43–45; 1778, pp. 147–153). I owe these references to Jordan (1972, p. 23). Jordan, incidentally, published a number of papers in French, over the name Charles Jordan; some in German, with the name Karl Jordan; and many in Hungarian, with the name Jordan Károly.

§8.2 P. J. Daniell (1919) first proved existence of a countably additive product probability on an infinite product in case each factor is Lebesgue measure λ on the unit interval [0, 1]. Daniell's proof used his approach to integration (§4.5 above), compactness, and

Dini's theorem (2.4.10), starting with continuous functions depending on only finitely many coordinates as the class \mathcal{L} in the Daniell integral theory.

Nearly all probability measures used in applications can be represented as images $\lambda \circ f^{-1}$ of Lebesgue measure for a measurable function f. From Daniell's theorem one rather easily gets products of such images. Kolmogorov (1933) constructed probabilities (not only product probabilities) on products of spaces such as the real line or locally compact spaces: see §12.1 below.

Apparently Łomnicki and Ulam (1934) first proved existence of a product of arbitrary probability spaces (Theorem 8.2.2) without any topology. Ulam (1932) had announced this as a joint result with Łomnicki. J. von Neumann (1935) proved the theorem, apparently independently, in lecture notes from the Institute for Advanced Study (Princeton) which were distributed but not actually published until 1950. Meanwhile, B. Jessen (1939) published the theorem in Danish after others, including E. S. Andersen, had given incorrect proofs. Kakutani (1943) also proved the result. Finally Andersen and Jessen (1946, p. 20) published Jessen's proof in English, and the theorem was often called the Andersen-Jessen theorem. The extension to measures on products where the measure on X_n depends on x_1, \ldots, x_{n-1} is usually attributed to Ionescu Tulcea (1949–1950). Andersen and Jessen (1948, p. 5) state that Doob was also aware of such a result and that a paper by Doob and Jessen on it was planned, but I found no joint paper by Doob and Jessen in the cumulative index for *Mathematical Reviews*, 1940–1959. I am indebted to Lucien Le Cam for some of the references.

§8.3 Révész (1968) wrote a book on laws of large numbers. Gnedenko and Kolmogorov (1949, 1968) is a classic on limit theorems, including laws of large numbers. Here are some of the known results. If $X_1, X_2, \ldots,$ are i.i.d., then a theorem of Kolmogorov (1929a) states that there exist constants a_n with $S_n/n - a_n \to 0$ in probability if and only if $M P(|X_1| > M) \to 0$ as $M \to +\infty$ (Révész, 1968, p. 51). Let φ be the *characteristic function* of X_1, $\varphi(t) := E \exp(i X_1 t)$. Then there is a constant c with $S_n/n \to c$ in probability if and only if φ has a derivative at 0, and $\varphi'(0) = ic$, a theorem of Ehrenfeucht and Fisz (1960) also presented in Révész (1968, p. 52). It can happen that this weak law holds but not the strong law. For example, let

$$P(X_1 \in B) = \beta \int_B 1_{\{|x| \geq 2\}} x^{-2} (\log |x|)^{-\gamma} \, dx$$

for any Borel set $B \subset \mathbb{R}$, where the constant β is chosen so that the integral over the whole line is 1. If $X_1, X_2, \ldots,$ are i.i.d., then the strong law of large numbers holds if and only if $\gamma > 1$. The weak law, but not the strong law, holds for $0 < \gamma \leq 1$, with $c = 0$. For $\gamma \leq 0$, the weak law does not hold, even with variable a_n, as Kolmogorov's condition is violated. If the logarithmic factor is removed and x^{-2} replaced by $|x|^{-p}$ for some other $p > 1$ (necessary to obtain a finite measure and then a probability measure), then the strong law holds for $p > 2$ and not even the weak law holds for $p \leq 2$. So the weak law without the strong law has a rather narrow range of applicability.

Bienaymé (1853, p. 321) apparently gave the earliest, somewhat imperfect, forms of the inequality (8.3.1) and the resulting weak law of large numbers. Chebyshev (1867, p. 183) proved more precise statements. He is credited with giving the first rigorous proof of a general limit theorem in probability, among other achievements in mathematics and technology (Youschkevitch, 1971). Bienaymé (1853, p. 315) had also noted that for

independent variables the variance of a sum equals the sum of the variances. On
Bienaymé see Heyde and Seneta (1977). Inequalities related to Bienaymé-Chebyshev's
have been surveyed by Godwin (1955, 1964) and Savage (1961), who says his bibliog-
raphy is "as complete as possible."

Jakob Bernoulli (1713) first proved a weak law of large numbers, for i.i.d. X_j each
taking just two values—the binomial or "Bernoulli" case. Sheynin (1968) surveys the
early history of weak laws.

The Borel-Cantelli lemma (8.3.4) was stated, with an insufficient proof, by Borel
(1909, p. 252) in case the events are independent. Cantelli (1917a, 1917b) noticed that
one half holds without independence, as had Hausdorff (1914, p. 421) in a special case.
Barone and Novikoff (1978) note the landmark quality of Borel's work on the founda-
tions of probability but make several criticisms of his proofs. They note that Borel (1903,
Thme. XI bis) came close (within ε, in the case of geometric probabilities) to giving
"Cantelli's" half of the lemma. Cohn (1972) gives some extensions of the Borel-Cantelli
lemma and references to others. Erdös and Rényi (1959) showed that the Borel-Cantelli
lemma holds for events which are only pairwise independent. For a proof see also Chung
(1974, Theorem 4.2.5, p. 76). On the history of the Borel-Cantelli lemma see Móri and
Székely (1983). Cantelli lived from 1875 to 1966. Benzi (1988) reviews some of his
work, especially on the foundations of probability.

The condition $EX_j^2 = 1$ in Theorem 8.3.2 can be replaced by $\sum_{j=1}^{n} EX_j^2/n^2 \to 0$
as $n \to \infty$, as A. A. Markov (1899) noted. On Markov, for whom Markov chains and
Markov processes are named, see Youschkevitch (1974).

The first strong law of large numbers, also in the Bernoulli case, was stated by Borel
(1909), and again without a correct proof. For independent variables X_n with $EX_n^4 < \infty$,
and under further restrictions, which include i.i.d. variables, Cantelli (1917a) proved the
strong law, according to Seneta (1992), who gives further history.

Kolmogorov (1930) showed that if X_n are independent random variables with
$EX_n = 0$ and $\sum_{n \geq 1} EX_n^2/n^2 < \infty$, then the strong law holds for them. Kolmogorov
(1933) gave Theorem 8.3.5, the strong law for i.i.d. variables if and only if $E|X_1| < \infty$.
On the relation between the 1930 and 1933 theorems, see the notes to Chapter 9 below.
The strong law 8.3.5 also follows from the ergodic theorem of G. D. Birkhoff (1932)
and Khinchin (1933), using another theorem of Kolmogorov, as shown in §8.4.

The rather short proof of the strong law (8.3.5) given above is due to Etemadi (1981).
He states and proves the result for identically distributed variables which need only be
pairwise independent. In applications, it seems that one rarely encounters sequences of
variables which are pairwise independent but not independent.

§8.4 L. Boltzmann (1887, p. 208) coined the word "ergodic" in connection with statis-
tical mechanics. For a gas of n molecules in a closed vessel, the set of possible positions
and momenta of all the molecules is a "phase space" of dimension $d = 6n$. If rotations,
etc., of individual molecules are considered, the dimension is still larger. Let $z(t) \in \mathbb{R}^d$
be the state of the gas at time t. Then $z(t)$ remains for all t on a compact surface S where
the energy is constant (as are, perhaps, certain other variables called "integrals of the
motion"). For each t, there is a transformation T_t taking $z(s)$ into $z(s+t)$ for all s and all
possible trajectories $z(\cdot)$, according to classical mechanics. There is a probability mea-
sure P on S for which all the T_t are measure-preserving transformations. Boltzmann's
original "ergodic hypothesis" was that $z(t)$ would run through all points of S, except

in special cases. Then, for any continuous function f on S, a long-term time average would equal an expectation,

$$\lim_{T \to \infty} T^{-1} \int_0^T f(z(t))\, dt = \int_S f(z)\, dP(z).$$

But M. Plancherel (1912, 1913) and A. Rosenthal (1913) independently showed that a smooth curve such as $z(\cdot)$ cannot fill up a manifold of dimension larger than 1 (although a continuous Peano curve can: see §2.4, Problem 9). One way to see this is that z, restricted to a finite time interval, has a range nowhere dense in S. Since S is complete, the range of z cannot be all of S by the category theorem (2.5.2 above). On the work of R. Baire in general, including this application of the category theorem generally known by his name, see Dugac (1976). Another method was to show that the range of z has measure 0 in S.

It was then asked whether the range of z might be dense in S, and the time and space averages still be equal. As Boltzmann himself pointed out, for the orbit to come close to all points in S may take an inordinately long time. Brush (1976, Book 2, pp. 363–385) reviews the history of the ergodic hypothesis. Brush (1976, Book 1, p. 80) remarks that ergodic theory "now seems to be of interest to mathematicians rather than physicists." See also D. ter Haar (1954, pp. 123–125 and 331–385, especially 356 ff).

The ergodic theorem (8.4.1) is a discrete-time version of the equality of "space" and long-term time averages. Birkhoff (1932) first proved it for indicator functions and a special class of measure spaces and transformations. Khinchin (1933) pointed out that the proof extends straightforwardly to \mathcal{L}^1 functions on any finite measure space. E. Hopf (1937) proved the maximal ergodic lemma (8.4.2), usually called the maximal ergodic theorem, for measure-preserving transformations. Yosida and Kakutani (1939) showed how the maximal ergodic lemma could be used to prove the ergodic theorem. The short proof of the maximal ergodic lemma, as given, is due to Garsia (1965).

Recall $\mathcal{L}^\infty(X, \mathcal{A}, \mu)$, the space of measurable real functions f on X for which $\|f\|_\infty < \infty$, where

$$\|f\|_\infty := \inf\{M \colon \mu\{x \colon |f(x)| > M\} = 0\}.$$

A contraction U of \mathcal{L}^1 is called *strong* iff it is also a contraction in the \mathcal{L}^∞ seminorm, so that $\|Uf\|_\infty \le \|f\|_\infty$ for all $f \in \mathcal{L}^1 \cap \mathcal{L}^\infty$. Dunford and Schwartz (1956) proved that if U is a strong contraction, then $n^{-1} \sum_{1 \le i \le n} U^i f$ converges almost everywhere as $n \to \infty$. Chacon (1961) gave a shorter proof and Jacobs (1963, pp. 371–376) another exposition.

If U is any positive contraction of \mathcal{L}^1 and g is a nonnegative function in \mathcal{L}^1, then the quotients $\sum_{0 \le i \le n} U^i f / \sum_{0 \le i \le n} U^i g$ converge almost everywhere on the set where the denominators are positive for some n (and thus almost everywhere if $g > 0$ everywhere). Hopf (1937) proved this for $Uf = f \circ T$, T a measure-preserving transformation. Chacon and Ornstein (1960) proved it in general. Their proof was rather long (it is also given in Jacobs, 1963, pp. 381–400). Brunel (1963) and Garsia (1967) shortened the proof. Note that if $\mu(X) < \infty$ we can let $g = 1$, so that the denominator becomes n as in the previous theorems.

Halmos (1953), Jacobs (1963), and Billingsley (1965) gave general expositions of ergodic theory. Later, substantial progress has been made (see Ornstein et al., 1982) on classifying measure-preserving transformations up to isomorphism.

Hewitt and Savage (1955, Theorem 11.3) proved their 0–1 law (8.4.6). They gave three proofs, saying that Halmos had told them the short proof given above.

References

An asterisk identifies works I have found discussed in secondary sources but have not seen in the original.

Andersen, Eric Sparre, and Borge Jessen (1946). Some limit theorems on integrals in an abstract set. *Danske Vid. Selsk. Mat.-Fys. Medd. 22*, no. 14. 29 pp.

———— and ———— (1948). On the introduction of measures in infinite product sets. *Danske Vid. Selsk. Mat.-Fys. Medd. 25*, no. 4. 8 pp.

Banach, Stefan (1923). Sur le problème de la mesure. *Fundamenta Mathematicae 4*: 7–33.

Barone, Jack, and Albert Novikoff (1978). A history of the axiomatic formulation of probability from Borel to Kolmogorov: Part I. *Arch. Hist. Exact Sci. 18*: 123–190.

Benzi, Margherita (1988). A "neoclassical probability theorist:" Francesco Paolo Cantelli (in Italian). *Historia Math. 15*: 53–72.

Bernoulli, Jakob (1713, posth.). *Ars Conjectandi*. Thurnisiorum, Basel. Repr. in *Die Werke von Jakob Bernoulli 3* (1975), pp. 107–286. Birkhäuser, Basel.

Bienaymé, Irenée-Jules (1853). Considérations a l'appui de la découverte de Laplace sur la loi de probabilité dans la méthode des moindres carrés. *C.R. Acad. Sci. Paris 37*: 309–324. Repr. in *J. math. pures appl.* (Ser. 2) *12* (1867): 158–276.

Billingsley, Patrick (1965). *Ergodic Theory and Information*. Wiley, New York.

Bingham, N. H. (2000). Studies in the history of probability and statistics XLVI. Measure into probability: From Lebesgue to Kolmogorov. *Biometrika 87*: 145–156.

Birkhoff, George D. (1932). Proof of the ergodic theorem. *Proc. Nat. Acad. Sci. USA 17*: 656–660.

Boltzmann, Ludwig (1887). Ueber die mechanischen Analogien des zweiten Hauptsatzes der Thermodynamik. *J. für die reine und angew. Math. 100*: 201–212.

Borel, Émile (1903). Contribution à l'analyse arithmétique du continu. *J. math. pures appl.* (Ser. 2) *9*: 329–375.

———— (1909). Les probabilités dénombrables et leurs applications arithmétiques. *Rendiconti Circolo Mat. Palermo 27*: 247–271.

Brunel, Antoine (1963). Sur un lemme ergodique voisin du lemme de E. Hopf, et sur une de ses applications. *C. R. Acad. Sci. Paris 256*: 5481–5484.

Brush, Stephen G. (1976). *The Kind of Motion We Call Heat*. 2 vols. North-Holland, Amsterdam.

*Buffon, Georges L. L. (1733). *Histoire de l' Académie*. Paris.

*———— (1778). X Essai d'Arithmétique Morale. In *Histoire Naturelle*, pp. 67–216. Paris.

*Cantelli, Francesco Paolo (1917a). Sulla probabilità come limite della frequenza. *Accad. Lincei, Roma, Cl. Sci. Fis., Mat., Nat., Rendiconti* (Ser. 5) *26*: 39–45.

*———— (1917b). Su due applicazione di un teorema di G. Boole alla statistica matematica. Ibid., pp. 295–302.

Chacon, Rafael V. (1961). On the ergodic theorem without assumption of positivity. *Bull. Amer. Math. Soc. 67*: 186–190.

——— and Donald S. Ornstein (1960). A general ergodic theorem. *Illinois J. Math. 4*: 153–160.

Chebyshev, Pafnuti Lvovich (1867). Des valeurs moyennes. *J. math. pures appl. 12* (1867): 177–184. Transl. from *Mat. Sbornik 2* (1867): 1–9. Repr. in *Oeuvres de P. L. Tchebychef 1*, pp. 687–694. Acad. Sci. St. Petersburg (1899–1907).

Chung, Kai Lai (1974). *A Course in Probability Theory*. 2d ed. Academic Press, New York.

Cohn, Harry (1972). On the Borel-Cantelli Lemma. *Israel J. Math. 12*: 11–16.

Daniell, Percy J. (1919). Integrals in an infinite number of dimensions. *Ann. Math. 20*: 281–288.

Day, Mahlon M. (1942). Ergodic theorems for Abelian semi-groups. *Trans. Amer. Math. Soc. 51*: 399–412.

Dubins, Lester E., and Leonard Jimmie Savage (1965). *How to Gamble If You Must; Inequalities for Stochastic Processes*. McGraw-Hill, New York.

Dugac, P. (1976). Notes et documents sur la vie et l'oeuvre de René Baire. *Arch. Hist. Exact Sci. 15*: 297–383.

Dunford, Nelson, and Jacob T. Schwartz (1956). Convergence almost everywhere of operator averages. *J. Rat. Mech. Anal.* (Indiana Univ.) *5*: 129–178.

Ehrenfeucht, Andrzej, and Marek Fisz (1960). A necessary and sufficient condition for the validity of the weak law of large numbers. *Bull. Acad. Polon. Sci. Ser. Math. Astron. Phys. 8*: 583–585.

*Erdös, Paul, and Alfréd Rényi (1959). On Cantor's series with divergent $\sum 1/q_n$. *Ann. Univ. Sci. Budapest. Eötvös Sect. Math. 2*: 93–109.

Etemadi, Nasrollah (1981). An elementary proof of the strong law of large numbers. *Z. Wahrsch. verw. Geb. 55*: 119–122.

Feller, William (1968). *An Introduction to Probability Theory and Its Applications*. Vol. 1, 3d ed. Wiley, New York.

Garsia, Adriano M. (1965). A simple proof of E. Hopf's maximal ergodic theorem. *J. Math. and Mech.* (Indiana Univ.) *14*: 381–382.

——— (1967). More about the maximal ergodic lemma of Brunel. *Proc. Nat. Acad. Sci. USA 57*: 21–24.

Gnedenko, Boris V. (1973). Andrei Nikolaevich Kolmogorov (On his 70th birthday). *Russian Math. Surveys 28*, no. *5*: 5–17. Transl. from *Uspekhi Mat. Nauk 28*, no. 5: 5–15.

——— and Andrei Nikolaevich Kolmogorov (1949). *Limit Distributions for Sums of Independent Random Variables*. Translated, annotated, and revised by Kai Lai Chung, with appendices by Joseph L. Doob and Pao Lo Hsu. Addison-Wesley, Reading, Mass. 1st ed. 1954, 2d ed. 1968.

——— and N. V. Smirnov (1963). On the work of A. N. Kolmogorov in the theory of probability. *Theory Probability Appls. 8*: 157–164.

Godwin, H. J. (1955). On generalizations of Tchebychef's inequality. *J. Amer. Statist. Assoc. 50*: 923–945.

——— (1964). *Inequalities on Distribution Functions*. Griffin, London.

Greenleaf, Frederick P. (1969). *Invariant Means on Topological Groups and their Applications*. Van Nostrand, New York.

ter Haar, D. (1954). *Elements of Statistical Mechanics.* Rinehart, New York.

Halmos, Paul R. (1953). *Lectures on Ergodic Theory.* Math. Soc. of Japan, Tokyo.

Hausdorff, Felix (1914). *Grundzüge der Mengenlehre,* 1st ed. Von Veit, Leipzig, repr. Chelsea, New York, 1949.

Hewitt, Edwin, and Leonard Jimmie Savage (1955). Symmetric measures on Cartesian products. *Trans. Amer. Math. Soc. 80*: 470–501.

Heyde, Christopher C., and Eugene Seneta (1977). *I. J. Bienaymé: Statistical Theory Anticipated.* Springer, New York.

Hopf, Eberhard (1937). *Ergodentheorie.* Springer, Berlin.

———— (1954). The general temporally discrete Markoff process. *J. Rat. Mech. Anal.* (Indiana Univ.) *3*: 13–45.

Ionescu Tulcea, Cassius (1949–1950). Mesures dans les espaces produits. *Atti Accad. Naz. Lincei Rend. Cl. Sci. Fis. Mat. Nat.* (Ser. 8) *7*: 208–211.

Jacobs, Konrad (1963). *Lecture Notes on Ergodic Theory.* Aarhus Universitet, Matematisk Institut.

Jessen, Borge (1939). Abstrakt Maal- og Integralteori 4. *Mat. Tidsskrift (B) 1939*, pp. 7–21.

Jordan, Károly (1972). *Chapters on the Classical Calculus of Probability.* Akadémiai Kiadó, Budapest.

Kakutani, Shizuo (1943). Notes on infinite product measures, *I. Proc. Imp. Acad. Tokyo* (became *Japan Academy Proceedings*) *19*: 148–151.

Khinchin, A. Ya. (1933). Zu Birkhoffs Lösung des Ergodenproblems. *Math. Ann. 107*: 485–488.

Kolmogorov, Andrei Nikolaevich (1929a). Bemerkungen zu meiner Arbeit "Über die Summen zufälliger Grössen." *Math. Ann. 102*: 484–488.

———— (1929b). The general theory of measure and the calculus of probability (in Russian). In *Coll. Works, Math. Sect.* (Communist Acad., Sect. Nat. Exact Sci.) *1*, 8–21. Izd. Komm. Akad., Moscow. Repr. in Kolmogorov (1986), pp. 48–58.

———— (1930). Sur la loi forte des grandes nombres. *Comptes Rendus Acad. Sci. Paris 191*: 910–912.

———— (1933). *Grundbegriffe der Wahrscheinlichkeitsrechnung,* Ergebnisse der Math., Springer, Berlin. English transl. *Foundations of the Theory of Probability,* Chelsea, New York, 1956.

———— (1986). *Probability Theory and Mathematical Statistics* [in Russian; selected works], ed. Yu. V. Prohorov. Nauka, Moscow.

Łomnicki, Z., and Stanisław Ulam (1934). Sur la théorie de la mesure dans les espaces combinatoires et son application au calcul des probabilités: I. Variables indépendantes. *Fund. Math. 23*: 237–278.

*Markov, Andrei Andreevich (1899). The law of large numbers and the method of least squares (in Russian). *Izv. Fiz.-Mat. Obshch. Kazan Univ.* (Ser. 2) *8*: 110–128.

Móri, T. F., and G. J. Székely (1983). On the Erdös-Rényi generalization of the Borel-Cantelli lemma. *Stud. Sci. Math. Hungar. 18*: 173–182.

von Neumann, Johann (1929). Zur allgemeinen Theorie des Masses. *Fund. Math. 13*: 73–116.

———— (1935). *Functional Operators.* Mimeographed lecture notes. Institute for Advanced Study, Princeton, N.J. Published in *Ann. Math. Studies* no. 21, *Functional Operators,* vol. I, *Measures and Integrals.* Princeton University Press, 1950.

Notices, Amer. Math. Soc. 28 (1981, p. 84; unsigned). 1980 Wolf Prize.

Ornstein, Donald S., D. J. Rudolph, and B. Weiss (1982). Equivalence of measure preserving transformations. *Amer. Math. Soc. Memoirs 262.*

Plancherel, Michel (1912). Sur l'incompatibilité de l'hypothèse ergodique et des équations d'Hamilton. *Archives sci. phys. nat.* (Ser. 4) *33*: 254–255.

———— (1913). Beweis der Unmöglichkeit ergodischer mechanischer Systeme. *Ann. Phys.* (Ser. 4) *42*: 1061–1063.

Révész, Pal (1968). *The Laws of Large Numbers.* Academic Press, New York.

Rosenthal, Arthur (1913). Beweis der Unmöglichkeit ergodischer Gassysteme. *Ann. Phys.* (Ser. 4) *42*: 796–806.

Savage, I. Richard (1961). Probability inequalities of the Tchebycheff type. *J. Research Nat. Bur. Standards 65B*, pp. 211–222.

Seneta, Eugene (1992). On the history of the strong law of large numbers and Boole's inequality. *Historia Mathematica 19*: 24–39.

Sheynin, O. B. (1968). On the early history of the law of large numbers. *Biometrika 55*, pp. 459–467.

Shiryayev, A. N. (1989). Kolmogorov: Life and creative activities. *Ann. Probab. 17*: 866–944.

The Times [London, unsigned] (26 October 1987). Andrei Nikolaevich Kolmogorov: 1903–1987. Repr. in *Inst. Math. Statist. Bulletin 16*: 324–325.

Ulam, Stanisław (1932). Zum Massbegriffe in Produkträumen. *Proc. International Cong. of Mathematicians* (Zürich), *2*, pp. 118–119.

Yosida, Kosaku, and Shizuo Kakutani (1939). Birkhoff's ergodic theorem and the maximal ergodic theorem. *Proc. Imp. Acad.* (Tokyo) *15*: 165–168.

Youschkevitch [Yushkevich], Alexander A. (1974). Markov, Andrei Andreevich. *Dictionary of Scientific Biography, 9*, pp. 124–130.

Youschkevitch, A. P. (1971). Chebyshev, Pafnuti Lvovich. *Dictionary of Scientific Biography, 3*, pp. 222–232.

9

Convergence of Laws and Central Limit Theorems

Let X_j be independent, identically distributed random variables with mean 0 and variance 1, and $S_n := X_1 + \cdots + X_n$ as always. Then one of the main theorems of probability theory, the central limit theorem, to be proved in this chapter, states that the laws of $S_n/n^{1/2}$ converge, in the sense that for every real x,

$$\lim_{n \to \infty} P\left(\frac{S_n}{\sqrt{n}} \le x\right) = \frac{1}{\sqrt{2\pi}} \int_{-\infty}^{x} e^{-t^2/2} \, dt.$$

The assumptions of mean 0 and variance 1 do not really restrict the generality of the theorem, since for any independent, identically distributed variables Y_i with mean m and positive, finite variance σ^2, we can form $X_i = (Y_i - m)/\sigma$, apply the theorem as just stated to the X_i, and conclude that $(Y_1 + \cdots + Y_n - nm)/(\sigma n^{1/2})$ has a distribution converging to the one given on the right above, called a "standard normal" distribution. So, to be able to apply the central limit theorem, the only real restriction on the distribution of the variables (given that they are nonconstant, independent, and identically distributed) is that $EY_j^2 < \infty$. Then, whatever the original form of the distribution of Y_j (discrete or continuous, symmetric or skewed, etc.), the limit distribution of their partial sums has the same, normal form—a remarkable fact. The requirement that the variables be identically distributed can be relaxed (Lindeberg's theorem, §9.6 below), and the central limit theorem, in a somewhat different form, will be stated and proved for multidimensional variables, in §9.5. (By the way, the variables $S_n/n^{1/2}$ do not converge in probability—in fact, if n is much larger than k, then S_n is nearly independent of S_k.)

9.1. Distribution Functions and Densities

Definitions. A *law* on \mathbb{R} (or any separable metric space) will be any probability measure defined on the Borel σ-algebra. Given a real-valued random variable X on some probability space (Ω, \mathcal{S}, P) with law $\mathcal{L}(X) := P \circ X^{-1} = Q$, the

(cumulative) *distribution function* of X or Q is the function F on \mathbb{R} given by

$$F(x) := Q((-\infty, x]) = P(X \leq x).$$

Such functions can be characterized:

9.1.1. Theorem *A function F on \mathbb{R} is a distribution function if and only if it is nondecreasing, is continuous from the right, and satisfies $\lim_{x \to -\infty} F(x) = 0$ and $\lim_{x \to +\infty} F(x) = 1$. There is a unique law on \mathbb{R} with a given distribution function F.*

Proof. If F is a distribution function, then the given properties follow from the properties of a probability measure (nonnegative, countably additive, and of total mass 1). For the converse, if F has the given properties, it follows from Theorem 3.2.6 that there is a unique law with distribution function F. \square

Examples. The function $F(x) = 0$ for $x < 0$ and $F(x) = 1$ for $x \geq 0$ is the distribution function of the law δ_0, and of any random variable X with $P(X = 0) = 1$. If the definition of F at 0 were changed to any value other than 1, such as 0, it would no longer be continuous from the right and so not a distribution function.

The function $F(x) = 0$ for $x \leq 0$, $F(x) = x$ for $0 < x \leq 1$, and $F(x) = 1$ for $x > 1$ is a distribution function, for the uniform distribution (law) on $[0, 1]$.

To find a random variable with a given law on any space, such as \mathbb{R}, one can always take that space as the probability space and the identity as the random variable. Sometimes, however, it is useful to define the random variable on a fixed probability space, as follows. Take (Ω, \mathcal{S}, P) to be the open interval $(0, 1)$ with Borel σ-algebra and Lebesgue measure. For any distribution function F on \mathbb{R} let

$$X(t) := X_F(t) := \inf\{x \colon F(x) \geq t\}, \qquad 0 < t < 1.$$

In the examples just given, note that for $F(x) = 1_{\{x \geq 0\}}$, $X_F(t) = 0$ for $0 < t < 1$, while if F is the distribution function of the uniform distribution on $[0, 1]$, $X_F(t) = t$ for $0 < t < 1$.

9.1.2. Proposition *For any distribution function F, X_F is a random variable with distribution function F.*

Proof. Clearly if $t \le F(x)$, then $X(t) \le x$. Conversely, if $X(t) \le x$, then since F is nondecreasing, $F(y) \ge t$ for all $y > x$, and hence $F(x) \ge t$ by right continuity. Since X is also nondecreasing, it is measurable. Thus

$$P\{t: X(t) \le x\} = P\{t: t \le F(x)\} = F(x). \qquad \square$$

Adding independent random variables results in the following operation on their laws, as will be shown:

Definition. Given two finite signed measures μ and ν on \mathbb{R}^k, their *convolution* is defined on each Borel set in \mathbb{R}^k by

$$(\mu * \nu)(A) := \int_{\mathbb{R}^k} \nu(A - x)\,d\mu(x), \quad \text{where } A - x := \{z - x : z \in A\}.$$

9.1.3. Theorem *If X and Y are independent random variables with values in \mathbb{R}^k, $\mathcal{L}(X) = \mu$ and $\mathcal{L}(Y) = \nu$, then $\mathcal{L}(X + Y) = \mu * \nu$. Thus $\mu * \nu$ is a probability measure.*

Proof. By independence, $\mathcal{L}(\langle X, Y \rangle)$ on \mathbb{R}^{2k} is the product measure $\mu \times \nu$. Now $X + Y \in A$ if and only if $Y \in A - X$, so $1_A(x + y) = 1_{A-x}(y)$, and the Tonelli-Fubini theorem gives the conclusion. $\qquad \square$

9.1.4. Theorem *Convolution of finite Borel measures is a commutative, associative operation:*

$$\mu * \nu \equiv \nu * \mu \quad \text{and} \quad (\mu * \nu) * \rho \equiv \mu * (\nu * \rho).$$

Proof. Given finite Borel measures μ, ν, and ρ, and any Borel set A, we have

$$((\mu * \nu) * \rho)(A) = (\mu \times \nu \times \rho)\{\langle x, y, z \rangle : x + y + z \in A\},$$

and this is preserved by changing the order of operations. $\qquad \square$

A law P on \mathbb{R} is said to have a *density* f iff P is absolutely continuous with respect to Lebesgue measure λ and has Radon-Nikodym derivative $dP/d\lambda = f$. In other words, $P(A) = \int_A f(x)\,dx$ for all Borel sets A. Then if F is the distribution function of P,

$$F(x) = \int_{-\infty}^{x} f(t)\,dt \qquad \text{for all } x \in \mathbb{R}$$

and for λ–almost all x, $F'(x)$ exists and equals $f(x)$ (by Theorem 7.2.1). A function f on \mathbb{R} is a probability density function if and only if it is measurable, nonnegative a.e. for Lebesgue measure, and

$$\int_{-\infty}^{\infty} f(x)\,dx = 1.$$

9.1.5. Proposition *If P has a density f, then for any measurable function g, $\int g\,dP = \int gf\,d\lambda$, where each side is defined (and finite) if and only if the other is.*

Proof. The proof is left to Problem 11. □

Convolution can be written in terms of densities when they exist:

9.1.6. Proposition *If P and Q are two laws on \mathbb{R} and P has a density f, then $P * Q$ has a density $h(x) := \int f(x - y)\,dQ(y)$. If Q also has a density g, then $h(x) \equiv \int f(x - y)g(y)\,dy$.*

Proof. For any Borel set A,

$$(P * Q)(A) = \int P(A - y)\,dQ(y) = \iint 1_A(x + y)\,dP(x)\,dQ(y)$$

$$= \iint 1_A(x + y)f(x)\,dx\,dQ(y) = \iint 1_A(u)f(u - y)\,du\,dQ(y)$$

$$= \int_A h(u)\,du$$

by the Tonelli-Fubini theorem and Proposition 9.1.5. The form of h if Q has density g also follows. □

Example. Let P be the standard exponential distribution, having density $f(x) := e^{-x}$ for $x \geq 0$ and 0 for $x < 0$. Then $P * P$ has density $(f * f)(x) = \int_0^{\infty} e^{-(x-y)} 1_{\{x-y \geq 0\}} e^{-y}\,dy = xe^{-x}$ for $x \geq 0$ and 0 for $x < 0$. This example will be extended in problem 12.

Problems

1. Let X be uniformly distributed on the interval $[2, 6]$, so that for any Borel set $A \subset R$, $P(X \in A) = \lambda(A \cap [2, 6])/4$ where λ is Lebesgue measure. Find the distribution function and density function of X.

2. Let k be the number of successes in 4 independent trials with probability 0.4 of success on each trial. Evaluate the distribution function of k for all real x.

3. Let X and Y be independent and both have the uniform distribution on $[0, 1]$, so that for any Borel set $B \subset \mathbb{R}$, $P(X \in B) = P(Y \in B) = \lambda(B \cap [0, 1])$. Find the distribution function and density function of $X + Y$.

4. Find the distribution function of $X + Y$ if X and Y are independent and $P(X = j) = P(Y = j) = 1/n$ for $j = 1, \ldots, n$.

5. If X is a random variable with law P having distribution function F, show that $F(X)$ has law λ on $[0, 1]$ if and only if P is nonatomic, that is, $P(\{x\}) = 0$ for any single point x.

6. Let X_n be independent real random variables with mean 0 and variance 1. Let Y be a real random variable with $EY^2 < \infty$. Show that $E(X_n Y) \to 0$ as $n \to \infty$. *Hint:* Use Bessel's inequality (5.4.3) and recall facts around Theorem 8.1.2.

7. Let X_1, X_2, \ldots be independent random variables with the same distribution function F.
 (a) Find the distribution function of $M_n := \max(X_1, \ldots, X_n)$.
 (b) If $F(x) = 1 - e^{-x}$ for $x > 0$ and 0 for $x \le 0$, show that the distribution function of $M_n - \log n$ converges as $n \to \infty$ and find its limit.

8. For any real random variable X let F_X be its distribution function. Express F_{-X} in terms of F_X.

9. What are the possible ranges $\{F(x): x \in \mathbb{R}\}$ of
 (a) continuous distribution functions F?
 (b) general distribution functions F?

10. Let F be the distribution function of a probability law P concentrated on the set \mathbb{Q} of rational numbers, so that $P(\mathbb{Q}) = 1$. Let H be the range of F. Prove or disprove:
 (a) H is always countable. *Hint:* Invert g in Proposition 4.2.1.
 (b) H always has Lebesgue measure 0.

11. Prove Proposition 9.1.5.

12. The *gamma function* is defined by $\Gamma(a) := \int_0^\infty x^{a-1} e^{-x}\, dx$ for any $a > 0$, so that $f_a(x) := e^{-x} x^{a-1} 1_{\{x>0\}} / \Gamma(a)$ is the density of a law Γ_a. The *beta function* is defined by $B(a, b) := \int_0^1 x^{a-1} (1 - x)^{b-1}\, dx$ for any $a > 0$ and $b > 0$. Show that then (i) $\Gamma_a * \Gamma_b = \Gamma_{a+b}$ and (ii) $B(a, b) \equiv \Gamma(a)\Gamma(b)/\Gamma(a + b)$. *Hints:* The last example in the section showed that $\Gamma_1 * \Gamma_1 = \Gamma_2$. In finding $\Gamma_a * \Gamma_b$, show that $\int_0^x (x - y)^{a-1} y^{b-1}\, dy =$

$x^{a+b-1}B(a, b)$ via the substitution $y = xu$. Note that in Γ_c, the normalizing constant $1/\Gamma(c)$ is unique, and use this for $c = a + b$ to prove (ii), and finish (i).

13. (a) Show by induction on k that $k! = \Gamma(k + 1)$ for $k = 0, 1, \ldots$
 (b) For the Poisson probability $Q(k, \lambda) := \sum_{j=k}^{\infty} e^{-\lambda}\lambda^j/j!$ and gamma density f_a from Problem 12, prove $Q(k, \lambda) = \int_0^{\lambda} f_k(x)\,dx$ for any $\lambda > 0$ and $k = 1, 2, \ldots$.
 (c) For the binomial probability $E(k, n, p) := \sum_{j=k}^{n} \binom{n}{j} p^j (1 - p)^{n-j}$ prove $E(k, n, p) = \int_0^p x^{k-1}(1 - x)^{n-k}\,dx/B(k, n - k + 1)$ for $0 < p < 1$ and $k = 1, \ldots, n$.

9.2. Convergence of Random Variables

In most of this chapter and the next, the random variables will have values in \mathbb{R} or finite-dimensional Euclidean spaces \mathbb{R}^k, but definitions and some facts will be given more generally, often in (separable) metric spaces.

Let (S, \mathcal{T}) be a topological space and (Ω, \mathcal{A}, P) a probability space. Let Y_0, Y_1, \ldots, be random variables on Ω with values in S. Recall that for Y to be measurable, for the σ-algebra of Borel sets in S generated by \mathcal{T}, it is equivalent that $Y^{-1}(U) \in \mathcal{A}$ for all open sets U (by Theorem 4.1.6). Then $Y_n \to Y_0$ *almost surely* (a.s.) means, as usual, that for almost all ω, $Y_n(\omega) \to Y_0(\omega)$. Recall also that if (S, d) is a metric space, Y_n are assumed to be (measurable) random variables only for $n \geq 1$, and $Y_n \to Y_0$ a.s., it follows that Y_0 is also a random variable (at least for the completion of P), by Theorem 4.2.2. This can fail in more general topological spaces (Proposition 4.2.3). Metric spaces also have the good property that the Baire and Borel σ-algebras are equal (Theorem 7.1.1).

If (S, d) is a metric space which is separable (has a countable dense set), then its topology is second-countable (Proposition 2.1.4). Thus in the Cartesian product of two such spaces, the Borel σ-algebra of the product topology equals the product σ-algebra of the Borel σ-algebras in the individual spaces (Proposition 4.1.7). The metric d is continuous, thus Borel measurable, on the product of S with itself, and so jointly measurable for the product σ-algebra. It follows that for any two random variables X and Y with values in S, $d(X, Y)$ is a random variable. So the following makes sense:

Definition. For any separable metric space (S, d), random variables Y_n on a probability space (Ω, \mathcal{S}, P) with values in S converge to Y_0 *in probability* iff for every $\varepsilon > 0$, $P\{d(Y_n, Y_0) > \varepsilon\} \to 0$ as $n \to \infty$.

For variables with values in any separable metric space, a.s. convergence clearly implies convergence in probability, but the converse does not always hold: for example, let $A(1), A(2), \ldots$, be any events such that $P(A(n)) \to 0$ as $n \to \infty$. Then $1_{A(n)} \to 0$ in probability. If the events are independent, then $1_{A(n)} \to 0$ a.s. if and only if $\sum_n P(A(n)) < \infty$, by the Borel-Cantelli lemma (8.3.4). So, if $P(A(n)) = 1/n$ for all n, the sequence converges to 0 in probability but not a.s. Nevertheless, there is a kind of converse in terms of subsequences, giving a close relationship between the two kinds of convergence:

9.2.1. Theorem *For any random variables X_n and X from a probability space (Ω, S, P) into a separable metric space (S, d), $X_n \to X$ in probability if and only if for every subsequence $X_{n(k)}$ there is a subsubsequence $X_{n(k(r))} \to X$ a.s.*

Proof. If $X_n \to X$ in probability, so does any subsequence. Given $X_{n(k)}$, if $k(r)$ is chosen so that

$$ P\big\{ d\big(X_{n(k(r))}, X \big) > 1/r \big\} < 1/r^2, \qquad r = 1, 2, \ldots, k(r) \to \infty, $$

then by the Borel-Cantelli lemma (8.3.4), $d(X_{n(k(r))}, X) \le 1/r$ for r large enough (depending on ω) a.s., so $X_{n(k(r))} \to X$ a.s.

Conversely, if X_n does not converge to X in probability, then there is an $\varepsilon > 0$ and a subsequence $X_{n(k)}$ such that

$$ P\big\{ d\big(X_{n(k)}, X \big) > \varepsilon \big\} \ge \varepsilon \qquad \text{for all } k. $$

This subsequence has no subsubsequence converging to X a.s. □

Example. Let $A(1), A(2), \ldots$, be any events with $P(A(n)) \to 0$ as $n \to \infty$, so that $1_{A(n)} \to 0$ in probability but not necessarily a.s. Then for some subsequence $n(k)$, $P(A(n(k))) < 1/k^2$ for all $k = 1, 2, \ldots$, and $1_{A(n(k))} \to 0$ a.s.

For any measurable spaces (Ω, \mathcal{A}) and (S, \mathcal{B}), let $\mathcal{L}^0(\Omega, \mathcal{A}; S, \mathcal{B})$ denote the set of all measurable functions from Ω into S for the given σ-algebras. If μ is a measure on \mathcal{A}, let $L^0(\Omega, \mathcal{A}, \mu; S, \mathcal{B})$ be the set of all equivalence classes of elements of $\mathcal{L}^0(\Omega, \mathcal{A}; S, \mathcal{B})$ for the relation of equality a.e. for μ. If S is a separable metric space, and the σ-algebra \mathcal{B} is omitted from the notation, it will be understood to be the Borel σ-algebra. Likewise, once \mathcal{A} and μ have been specified, they need not always be mentioned, so that one

can write $L^0(\Omega, S)$. If S is not specified, it will be understood to be the real line \mathbb{R}. In that case one may write $L^0(\mu)$ or $L^0(\Omega)$.

Convergence a.s. or in probability is unaffected if some of the variables are replaced by others equal to them a.s., so that these modes of convergence are defined on L^0 as well as on \mathcal{L}^0.

For any separable metric space (S, d), probability space (Ω, \mathcal{A}, P) and $X, Y \in \mathcal{L}^0(\Omega, S)$, let

$$\alpha(X, Y) := \inf\{\varepsilon \geq 0 \colon P(d(X, Y) > \varepsilon) \leq \varepsilon\}.$$

Thus, for some $\varepsilon_k \downarrow \alpha := \alpha(X, Y)$, $P(d(X, Y) > \varepsilon_k) \leq \varepsilon_k \leq \varepsilon_j$ for all $k \geq j$. As $k \to \infty$, the indicator function of $\{d(X, Y) > \varepsilon_k\}$ increases up to that of $\{d(X, Y) > \alpha\}$, so by monotone convergence, $P(d(X, Y) > \alpha) \leq \varepsilon_j$ for all j, and $P(d(X, Y) > \alpha) \leq \alpha$. In other words, the infimum in the definition of $\alpha(X, Y)$ is attained.

Examples. For two constant functions $X \equiv a$ and $Y \equiv b$, $\alpha(X, Y) = \min(1, |a - b|)$. For any event B, $\alpha(1_B, 0) = P(B)$.

9.2.2. Theorem *On $L^0(\Omega, S)$, α is a metric, which metrizes convergence in probability, so that $\alpha(X_n, X) \to 0$ if and only if $X_n \to X$ in probability.*

Proof. Clearly α is symmetric and nonnegative, and $\alpha(X, Y) = 0$ if and only if $X = Y$ a.s. For the triangle inequality, given random variables X, Y, and Z, we have, except on a set of probability at most $\alpha(X, Y)$, that $d(X, Y) \leq \alpha(X, Y)$, and likewise for Y and Z. Thus by the triangle inequality in S,

$$d(X, Z) \leq d(X, Y) + d(Y, Z) \leq \alpha(X, Y) + \alpha(Y, Z)$$

except on a set of probability at most $\alpha(X, Y) + \alpha(Y, Z)$. Thus $\alpha(X, Z) \leq \alpha(X, Y) + \alpha(Y, Z)$, and α is a metric on L^0. If $X_n \to X$ in probability, then for each $m = 1, 2, \ldots$, $P(d(X_n, X) > 1/m) \leq 1/m$ for n large enough, say $n \geq n(m)$. Then $\alpha(X_n, X) \to 0$ as $n \to \infty$. Conversely, if $X_n \to X$ for α, then for any $\delta > 0$ we have for n large enough $\alpha(X_n, X) < \delta$. Then $P(d(X_n, X) > \delta) < \delta$, so $X_n \to X$ in probability. $\quad\square$

The metric α is called the *Ky Fan* metric.

It follows from Theorems 9.2.1 and 9.2.2 that a.s. convergence is metrizable if and only if it coincides with convergence in probability. This is true if (Ω, P) is purely atomic, so that there are $\omega_k \in \Omega$ with $\sum_k P\{\omega_k\} = 1$, but usually false (see Problems 1–3).

9.2.3. Theorem *If (S, d) is a complete separable metric space and (Ω, \mathcal{S}, P) any probability space, then $L^0(\Omega, S)$ is complete for the Ky Fan metric α.*

Proof. Given a Cauchy sequence $\{X_n\}$, let $\{X_{n(r)}\}$ be a subsequence such that $\sup_{m \geq n(r)} \alpha(X_m, X_{n(r)}) \leq 1/r^2$. Then for all $s \geq r$, $P\{d(X_{n(r)}, X_{n(s)}) > 1/r^2\} \leq 1/r^2$. Since $\sum r^{-2}$ converges, the Borel-Cantelli lemma implies that almost surely $d(X_{n(r)}, X_{n(r+1)}) \leq 1/r^2$ for r large enough. Then for all $s \geq r$, $d(X_{n(r)}, X_{n(s)}) \leq \sum_{j \geq r} j^{-2} < 1/(r-1)$. Then $\{X_{n(r)}(\omega)\}_{r \geq 1}$ is a Cauchy sequence, convergent to some $Y(\omega)$, since S is complete. (In the event that $X_{n(r)}$ does not converge, define Y arbitrarily.) Then $\alpha(X_{n(r)}, Y) \to 0$. In any metric space, a Cauchy sequence with a convergent subsequence converges to the same limit, so $\alpha(X_m, Y) \to 0$. □

The following lemma provides a "Cauchy criterion" for almost sure convergence. Although it is not surprising, some may prefer to see it proved since a.s. convergence is not metrizable, for example, for Lebesgue measure on $[0, 1]$.

9.2.4. Lemma *Let (S, d) be a complete separable metric space and (Ω, \mathcal{A}, P) a probability space. Let Y_1, Y_2, \ldots, be random variables from Ω into S such that for every $\varepsilon > 0$,*

$$P\left\{\sup_{k \geq n} d(Y_n, Y_k) \geq \varepsilon\right\} \to 0 \qquad \text{as } n \to \infty.$$

Then for some random variable Y, $Y_n \to Y$ almost surely.

Proof. Let $A_{mn} := \{\sup\{d(Y_j, Y_k): j, k \geq n\} \leq 1/m\}$. For each m, A_{mn} increases as n increases, and A_{mn} is measurable. Applying the hypothesis for $\varepsilon = 1/(2m)$ and the triangle inequality $d(Y_j, Y_k) \leq d(Y_j, Y_n) + d(Y_n, Y_k)$ gives $P(A_{mn}) \uparrow 1$ as $n \to \infty$. Let $A_m := \bigcup_n A_{mn}$. Then $P(A_m) = 1$ for all m and $P(\bigcap_m A_m) = 1$. For any $\omega \in \bigcap_m A_m$, and thus almost surely, $Y_n(\omega)$ converges to some $Y(\omega)$. For ω not in $\bigcap_m A_m$, define $Y(\omega)$ as any fixed point of S. Then $Y(\cdot)$ is measurable by Theorem 4.2.2 (and Lemma 4.2.4). The proof is complete. □

Problems

1. For $n = 1, 2, \ldots$, and $2^{n-1} \leq j < 2^n$, let $f_j(x) = 1$ for $j/2^{n-1} - 1 \leq x \leq (j + 1)/2^{n-1} - 1$ and $f_j = 0$ elsewhere. Show that for the uniform

(Lebesgue) probability measure on $[0, 1]$, $f_j \to 0$ as $j \to \infty$ in probability but not a.s. Find a subsequence $f_{j(r)}$, $r = 1, 2, \ldots$, that converges to 0 a.s. as $r \to \infty$.

2. Referring to the definitions in §2.1, Problem 10, show that a.s. convergence is an L-convergence on L^0 and convergence in probability is an L^*-convergence. Assuming the results of §2.1, Problem 10, and Problem 1 above, show that there is no topology on L^0 such that a.s. convergence on $[0, 1]$ with Lebesgue measure is convergence for the topology.

3. Let $(X, 2^X, P)$ be a probability space where X is a countable set and 2^X is the power set (σ-algebra of all subsets of X). Show in detail that on X, convergence a.s. is equivalent to convergence in probability.

4. Let $f(x) := x/(1 + x)$ and $\tau(X, Y) := \int f(d(X, Y)) \, dP$. Show that τ is a metric on L^0 and that it metrizes convergence in probability. *Hint:* See Propositions 2.4.3 and 2.4.4.

5. For X and Y in $L^0(\Omega, S)$ for a metric space (S, d), let $\varphi(X, Y) := \inf\{\varepsilon + P\{\omega: d(X, Y) > \varepsilon\}: \varepsilon > 0\}$. Let α be the Ky Fan metric, with τ as in the previous problem. Find finite constants C_1 to C_4 such that for all (S, d) and random variables X, Y with values in S, and $\alpha := \alpha(X, Y)$, $\varphi := \varphi(X, Y)$, and $\tau := \tau(X, Y)$, we have $\varphi \leq C_1\alpha$, $\alpha \leq C_2\varphi$, $\tau \leq C_3\alpha$ and $\alpha \leq C_4\tau^{1/2}$.

6. Show that in $L^0(\lambda)$, where λ is Lebesgue measure on $[0, 1]$, there is no $C < \infty$ such that $\alpha \leq C\tau$ for all random variables X, Y as in Problem 5.

7. Let P and Q be two probability measures on the same σ-algebra \mathcal{A} which are equivalent, meaning that $P(A) = 0$ if and only if $Q(A) = 0$ for all $A \in \mathcal{A}$. Show that convergence in probability for P is equivalent to convergence in probability for Q.

8. For P and Q as in Problem 7, let α_P and α_Q be the corresponding Ky Fan metrics. Give an example of equivalent P and Q such that there is no $C < \infty$ for which $\alpha_P(X, Y) \leq C\alpha_Q(X, Y)$ for all real random variables measurable for \mathcal{A}.

9.3. Convergence of Laws

Let (S, \mathcal{T}) be a topological space. Let P_n be a sequence of laws, that is, probability measures on the Borel σ-algebra \mathcal{B} generated by \mathcal{T}. To define convergence of P_n to P_0, one way is as follows. For any signed measure μ we have the Jordan decomposition, $\mu = \mu^+ - \mu^-$ (Theorem 5.6.1), and the

total variation measure, $|\mu| := \mu^+ + \mu^- \geq 0$. We say that $P_n \to P_0$ in *total variation* iff as $n \to \infty$, $|P_n - P_0|(S) \to 0$, or equivalently

$$\sup_{A \in \mathcal{B}} |(P_n - P_0)(A)| \to 0.$$

Convergence in total variation, however, is too strong for most purposes. For example, let $x(n)$ be a sequence of points in S converging to x. Recall the point masses defined by $\delta_y(A) := 1_A(y)$ for each $y \in S$. Suppose that the singleton $\{x\}$ is a Borel set (as is true, for example, if S is metrizable or just Hausdorff). Then $\delta_{x(n)}$ do not converge to δ_x in total variation unless $x(n) = x$ for n large enough. In fact, it is not true that $\delta_{1/n}(A)$ converges to $\delta_0(A)$ for every Borel set $A \subset \mathbb{R}$, or for every closed set or open set (consider $A = (-\infty, 0]$ or $(0, \infty)$). $P_n \to \delta_0$ in total variation if and only if $P_n(\{0\}) \to 1$. The following definitions give a convergence better related to convergence in S, so that $\delta_{x(n)}$ will converge to δ_x whenever $x(n) \to x$.

Definitions. Let $C_b(S)$ be the set of all bounded, continuous, real-valued functions on S. We say that the laws P_n converge to a law P, written $P_n \to P$, or just $P_n \to P$, iff for every $f \in C_b(S)$, $\int f \, dP_n \to \int f \, dP$ as $n \to \infty$.\mathcal{L}

Note that any $f \in C_b(S)$, being bounded and measurable, is integrable for any law. If $x(n) \to x$, then $\delta_{x(n)} \to \delta_x$. Convergence of laws is convergence for a topology, specifically a product topology, the topology of pointwise convergence, restricted to a subset of the set of all real functions on $C_b(S)$. This implies the following, although a specific proof will be given.

9.3.1. Proposition *If P_n and P are laws such that for every subsequence $P_{n(k)}$ there is a subsubsequence $P_{n(k(r))} \underset{\mathcal{L}}{\to} P$, then $P_n \underset{\mathcal{L}}{\to} P$.*

Proof. If not, then for some $f \in C_b$, $\int f \, dP_n \not\to \int f \, dP$. Then for some $\varepsilon > 0$ and sequence $n(k)$, $|\int f \, dP_{n(k)} - \int f \, dP| > \varepsilon$ for all k. Then $P_{n(k(r))} \underset{\mathcal{L}}{\to} P$ gives a contradiction. □

Now several facts will be proved about (convergence of) laws on metric spaces.

9.3.2. Lemma *If (S, d) is a metric space, P and Q are two laws on S, and $\int f \, dP = \int f \, dQ$ for all $f \in C_b(S)$, then $P = Q$.*

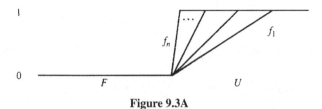

Figure 9.3A

Proof. Let U be any open subset of S with complement F and consider the distance $d(x, F)$ from F as in (2.5.3). For $n = 1, 2, \ldots$, let $f_n(x) :=$ $\min(1, nd(x, F))$ (see Figure 9.3A). Then $f_n \in C_b(S)$ and as $n \to \infty$, $f_n \uparrow 1_U$. So by monotone convergence, $P(U) = Q(U)$, so $P(F) = Q(F)$. Then by Theorem 7.1.3 (closed regularity), $P = Q$. □

It follows that a convergent sequence of laws has a unique limit.

A set \mathcal{P} of laws on a topological space S is called *uniformly tight* iff for every $\varepsilon > 0$ there is a compact $K \subset S$ such that $P(K) > 1 - \varepsilon$ for all $P \in \mathcal{P}$. Thus one law P is tight (as defined in §7.1) if and only if $\{P\}$ is uniformly tight.

Example. The sequence $\{\delta_n\}_{n \geq 1}$ of laws on \mathbb{R} is not uniformly tight.

The next fact will be proved for now only in \mathbb{R}^k. It actually holds in any metric space (Theorem 11.5.4 will prove it in complete separable metric spaces).

9.3.3. Theorem *Let $\{P_n\}$ be a uniformly tight sequence of laws on \mathbb{R}^k. Then there is a subsequence $P_{n(j)} \to P$ for some law P.*

Proof. For each compact set $K \subset \mathbb{R}^k$, all continuous real functions on K are bounded, so $C_b(K)$ equals the space $C(K)$ of all continuous real functions on K. Polynomials (in k variables) are dense in $C(K)$, for the supremum distance $s(f, g) := \sup_{x \in K} |f(x) - g(x)|$, by the (Stone-) Weierstrass approximation theorem (Corollary 2.4.12). Thus $C(K)$ is separable, taking polynomials with rational coefficients. Let D be a countable dense set in $C(K)$. Then $\{\{\int f \, dP_n\}_{f \in D}\}_{n \geq 1}$ is a sequence in the countable Cartesian product $L := \Pi_{f \in D}[\inf f, \sup f]$ of compact intervals. Now L with product topology is compact (Tychonoff's theorem, 2.2.8) and metrizable (Proposition 2.4.4), so (by Theorem 2.3.1) there is a subsequence $P_{m(j)}$ such that $\int f \, dP_{m(j)}$ converges for all $f \in D$. If $f \in C(K)$ and $\varepsilon > 0$, take $g \in D$ with $s(f, g) < \varepsilon$.

Then

$$\left| \int f \, dP_{m(j)} - \int g \, dP_{m(j)} \right| < \varepsilon \quad \text{for all } j, \text{ so}$$

$$\limsup_{j \to \infty} \int f \, dP_{m(j)} - \liminf_{j \to \infty} \int f \, dP_{m(j)} \le 2\varepsilon.$$

Letting $\varepsilon \downarrow 0$, we see that $\int f \, dP_{m(j)}$ converges for all $f \in C(K)$.

For each $r = 1, 2, \ldots$, take compact sets K such that $P_n(K_r) > 1 - 1/r$ for all n. Let $K(r) := K_r$. The previous part of the proof can be applied to $K = K_r$ for each r. The problem is to get a subsequence which converges for all r. This can be done as follows. For each r, let $D(r)$ be a countable dense set in $C(K_r)$. Then $\{\{\int f \, dP_n\}_{f \in D(r), r \ge 1}\}_{n \ge 1}$ is a sequence in the countable product $\Pi_{r \ge 1} \Pi_{f \in D(r)} [\inf f, \sup f]$. As before, this sequence has a convergent subsequence, say $P_{n(j)}$, and also as before $\int_{K(r)} f \, dP_{n(j)}$ converges for every $f \in C(K(r))$ and so for every $f \in C_b(\mathbb{R}^k)$.

Each such integral differs from $\int f \, dP_{n(j)}$ by at most $\sup |f|/r$, which approaches 0 as $r \to \infty$. Thus $\int f \, dP_{n(j)}$ converges as $j \to \infty$ for each $f \in C_b(\mathbb{R}^k)$. Its limit will be called $L(f)$.

To apply the Stone-Daniell theorem (4.5.2) to L, note that $C_b(\mathbb{R}^k)$ is a Stone vector lattice, L is linear, and $L(f) \ge 0$ whenever $f \ge 0$. Suppose $f_n \downarrow 0$ (pointwise). Let $M := \sup_x f_1(x)$. Then we have $0 \le f_n(x) \le M$ for all n and x. Given $\varepsilon > 0$, take r large enough so that $1/r < \varepsilon/(2M)$. By Dini's theorem (2.4.10), $f_n \downarrow 0$ uniformly on K_r, so for some J and all $n \ge J$, $f_n(x) \le \varepsilon/2$ for all $x \in K_r$. Then $\int f_n dP_m \le \varepsilon$ for all m, so $L(f_n) \le \varepsilon$. Letting $\varepsilon \downarrow 0$, we have $L(f_n) \downarrow 0$. Thus by the Stone-Daniell theorem (4.5.2), there is a nonnegative measure P on \mathbb{R}^k such that $L(f) = \int f \, dP$ for all $f \in C_b(\mathbb{R}^k)$. Taking $f = 1$ shows that P is a probability measure. Now $P_{n(j)} \to P$. $\qquad \square$

A fact in the converse direction to the last one will also be useful:

9.3.4. Proposition *Any converging sequence of laws $P_n \to P$ on \mathbb{R}^k is uniformly tight.*

Proof. Given $\varepsilon > 0$, take an $M < \infty$ such that $P(|x| > M) < \varepsilon$. Let f be a continuous real function such that $f(x) = 0$ for $|x| \le M$, $f(x) = 1$ for $|x| \ge 2M$, and $0 \le f \le 1$, for example,

$$f(x) := \max(0, \min(1, |x|/M - 1)).$$

Then $P_n(|x| > 2M) \leq \int f \, dP_n < \varepsilon$ for n large enough, say for $n > r$. For each $n \leq r$, take M_n such that $P_n(|x| > M_n) < \varepsilon$. Let $J := \max(2M, M_1, \ldots, M_r)$. Then for all n, $P_n(|x| > J) < \varepsilon$. ◻

Random variables X_n with values in a topological space S are said to converge *in law* or *in distribution* to a random variable X iff $\mathcal{L}(X_n) \to \mathcal{L}(X)$. For example, if X_n are i.i.d. (independent and identically distributed) non-constant random variables, then they converge in law but not in probability (and so not a.s.). For convergence in law, the X_n could be defined on different probability spaces, although usually they will be on the same space. Here, and on some other occasions, the \mathcal{L} under \to has been omitted, although $X_n \underset{\mathcal{L}}{\to} X$ would mean $\mathcal{L}(X_n) \underset{\mathcal{L}}{\to} \mathcal{L}(X)$.

Convergence in probability implies convergence in law:

9.3.5. Proposition *If (S, d) is a separable metric space and X_n are random variables with values in S such that $X_n \to X$ in probability, then $\mathcal{L}(X_n) \to \mathcal{L}(X)$.*

Proof. For any subsequence $X_{n(k)}$ take a subsubsequence $X_{n(k(r))} \to X$ a.s. by Theorem 9.2.1. Let $f \in C_b(S)$. Then by dominated convergence, $\int f(X_{n(k(r))}) \, dP \to \int f(X) \, dP$. Thus by Proposition 9.3.1, the conclusion follows. ◻

Although random variables converging in law very often do not converge in probability, there is a kind of converse to Proposition 9.3.5, saying that if on a separable metric space, some laws P_n converge to P_0, then on some probability space (such as $[0, 1]$ with Lebesgue measure) there exist random variables X_n with $\mathcal{L}(X_n) = P_n$ for all n and $X_n \to X_0$ a.s. That will be proved in §11.7.

In the real line, convergence of laws implies convergence of distribution functions in the following sense.

9.3.6. Theorem *If laws $P_n \to P_0$ on \mathbb{R}, and F_n is the distribution function of P_n for each n, then $F_n(t) \to F_0(t)$ for all t at which F_0 is continuous.*

Proof. For any $t, x \in \mathbb{R}$ and $\delta > 0$, let

$$f_{t,\delta}(x) := \min(1, \max(0, (t - x)/\delta)) \qquad \text{(see Figure 9.3B)}.$$

Then $f_{t,\delta}(x) = 0$ for $x \geq t$, $f_{t,\delta}(x) = 1$ for $x \leq t - \delta$, $0 \leq f_{t,\delta} \leq 1$, and

Figure 9.3B

$f_{t,\delta} \in C_b(\mathbb{R})$. So as $n \to \infty$,

$$F_n(t) \geq \int f_{t,\delta}\, dP_n \to \int f_{t,\delta}\, dP_0 \geq F_0(t - \delta) \quad \text{and}$$

$$F_n(t) \leq \int f_{t+\delta,\delta}\, dP_n \to \int f_{t+\delta,\delta}\, dP_0 \leq F_0(t + \delta).$$

Thus $\liminf_{n \to \infty} F_n(t) \geq F_0(t - \delta)$ and $\limsup_{n \to \infty} F_n(t) \leq F_0(t + \delta)$. Let $\delta \downarrow 0$. Then continuity of F_0 at t implies the conclusion. □

Example. $\delta_{1/n} \to \delta_0$ but the corresponding distribution functions do not converge at 0. This shows why continuity at t is needed in Theorem 9.3.6.

Recall that a *mapping* is a function; the word is usually used about continuous functions with values in general spaces. Continuous mappings always preserve convergence of laws:

9.3.7. Theorem (Continuous Mapping Theorem) *Let (X, \mathcal{T}) and (Y, \mathcal{U}) be any two topological spaces and G a continuous function from X into Y. Let P_n be laws on X with $P_n \to P_0$. Then on Y, the image laws converge: $P_n \circ G^{-1} \to P_0 \circ G^{-1}$.*

Proof. It is enough to note that for any bounded continuous real function f on Y, $f \circ G$ is such a function on X, and apply the image measure theorem (4.1.11). □

Problems

1. Let P_n be laws on the set \mathbb{Z} on integers with $P_n \underset{\mathcal{L}}{\to} P_0$. Show that $P_n \to P_0$ in total variation.

2. Let (S, d) be a metric space. Let T be the set of all point masses δ_x for $x \in S$, with the total variation distance. Show that the topology on T is discrete (and so does not depend on d).

3. For what values of x do the distribution functions of $\delta_{-1/n}$ converge to that of δ_0?

4. Let P be a law on \mathbb{R}^2 such that each line parallel to the coordinate axes has 0 probability, in symbols $P(x = c) = P(y = c) = 0$ for each constant c. Let P_n be laws converging to P. Show that $(P_n - P)((-\infty, a] \times (-\infty, b]) \to 0$ as $n \to \infty$ for all real a and b.

5. For any $a < b < c < d$ let $f_{a,b,c,d}$ be a function which equals 0 on $(-\infty, a]$ and $[d, \infty)$, which equals 1 on $[b, c]$, and whose graph on each of $[a, b]$ and $[c, d]$ is a line segment, making f a continuous, "piecewise linear" function. Let P and Q be two laws on \mathbb{R} such that $\int f_{a,b,c,d} d(P - Q) = 0$ for all $a < b < c < d$. Prove that $P = Q$.

6. Let X_n be random variables with values in a separable metric space S such that $X_n \to c$ in law where c is a point in S. Show that $X_n \to c$ in probability.

7. Let F be the indicator function of the half-line $[0, \infty)$. Given an example of a converging sequence of laws $P_n \to P_0$ on \mathbb{R} such that $P_n \circ F^{-1}$ does not converge to $P_0 \circ F^{-1}$.

8. Let $f_n(x) = 2$ for $(2k - 1)/2^n \le x < 2k/2^n$, $k = 1, \dots, 2^{n-1}$, and $f_n(x) = 0$ elsewhere. Let P_n be the law with density f_n (with respect to Lebesgue measure λ on $[0, 1]$).
 (a) Show that $P_n \to \lambda$.
 (b) Show that $\int f\, dP_n \to \int f\, d\lambda$ for every bounded measurable f on $[0, 1]$. *Hint:* The $f_n - 1$ are orthogonal for λ; use (5.4.3).
 (c) Find the total variation distance $\| P_n - \lambda \|$ for all n.

9. For the following sequences of laws P_n on \mathbb{R} having densities f_n with respect to Lebesgue measure, which are uniformly tight?
 (a) $f_n = 1_{[0,n]}/n$.
 (b) $f_n(x) = ne^{-nx} 1_{[0,\infty)}(x)$.
 (c) $f_n(x) = e^{-x/n} 1_{[0,\infty)}(x)/n$.

10. Express each positive integer n in prime factorization, as $n = 2^{n_2} 3^{n_3} 5^{n_5} 7^{n_7} 11^{n_{11}} \cdots$ for nonnegative integers n_2, n_3, \dots. Let $m(n) := n_2 + n_3 + \cdots$, a converging series. Let P_n be the law which puts mass $n_p/m(n)$ at p for each prime $p = 2, 3, 5, \dots$.
 (a) Show that $\{P_n\}_{n \ge 1}$ is not uniformly tight.
 (b) Show that the sequence $\{P_n\}$ has a convergent subsequence $\{P_{n(k)}\}$.
 (c) Find a convergent subsequence $P_{n(k)}$ such that for every prime p there is a k with $n(k)_p > 0$.

9.4. Characteristic Functions

Just as in \mathbb{R}^1, a function f on \mathbb{R}^k is called a *probability density* if $f \geq 0$, f is measurable for Lebesgue measure λ^k (the product measure of k copies of Lebesgue measure λ on \mathbb{R}, as in Theorem 4.4.6), and $\int f \, d\lambda^k = 1$. Then a law P on \mathbb{R}^k is given by $P(A) := \int_A f \, d\lambda^k$ for all Borel (or Lebesgue) measurable sets A. Thus f is the Radon-Nikodym derivative $dP/d\lambda^k$, and f is called the *density of P*.

Complex numbers, in case they may be unfamiliar, are defined in Appendix B. If f is a complex-valued function, $f \equiv g + ih$ where g and h are real-valued functions. Then for any measure μ, $\int f \, d\mu$ is defined as $\int g \, d\mu + i \int h \, d\mu$.

Let (x, t) be the usual inner product of two vectors x and t in \mathbb{R}^k. Given any random variable X with values in \mathbb{R}^k and law P, the *characteristic function* of X or P is defined by

$$f_P(t) := E e^{i(X,t)} = \int e^{i(x,t)} \, dP(x) \qquad \text{for all } t \in \mathbb{R}^k.$$

(Recall that for any real u, $e^{iu} = \cos(u) + i \cdot \sin(u)$.) Thus f_P is a Fourier transform of P, but without a constant multiplier such as $(2\pi)^{-k/2}$ which is used in much of Fourier analysis. Note that $f_P(0) = 1$ for any law P. For example, the law P giving measure $1/2$ each to $+1$ and -1 has characteristic function $f_P(t) \equiv \cos t$.

The usage of "characteristic function," as just defined, explains why workers in probability and statistics tend to refer to 1_A as the indicator function of a set A (rather than characteristic function as in much of the rest of mathematics).

The following densities are going to appear as limiting densities of suitably normalized partial sums S_n of i.i.d. variables X_j in \mathbb{R}^1 with $E X_1^2 < \infty$.

9.4.1. Proposition *For any $m \in \mathbb{R}$ and $\sigma > 0$, the function*

$$x \mapsto \varphi(m, \sigma^2, x) := \frac{1}{\sigma\sqrt{2\pi}} \exp(-(x - m)^2/(2\sigma^2))$$

is a probability density on \mathbb{R}.

Proof. By changes of variables it can be assumed that $m = 0$ and $\sigma = 1$ (if $u := (x - m)/\sigma$, then $du = dx/\sigma$). Now using the Tonelli-Fubini theorem

and polar coordinates (§4.4, Problem 6),

$$\left(\int_{-\infty}^{\infty} \exp(-x^2/2)\, dx \right)^2 = \int_{-\infty}^{\infty} \int_{-\infty}^{\infty} \exp((-x^2 - y^2)/2)\, dx\, dy$$

$$= \int_{0}^{2\pi} d\theta \int_{0}^{\infty} r \exp(-r^2/2)\, dr$$

$$= 2\pi(-\exp(-r^2/2))|_0^{\infty} = 2\pi. \qquad \square$$

Let $N(m, \sigma^2)$ be the law with density $\varphi(m, \sigma^2, \cdot)$ for any $m \in \mathbb{R}$ and $\sigma > 0$. In integrals, sometimes $P(dx)$ may be written instead of $dP(x)$, when P depends on other parameters as does $N(m, \sigma^2)$. Integrals for $N(m, \sigma^2)$ can be transformed by a linear change of variables to integrals for $N(0, 1)$. Then

$$\int_{-\infty}^{\infty} x N(0, 1)(dx) = (2\pi)^{-1/2} \int_{-\infty}^{\infty} - d(\exp(-x^2/2)) = 0$$

as in the integral with respect to r above. Next,

$$\int_{-\infty}^{\infty} x^2 N(0, 1)(dx) = (2\pi)^{-1/2} \int_{-\infty}^{\infty} - x d(\exp(-x^2/2)) = 1$$

using integration by parts and Proposition 9.4.1. It follows then that the mean and variance of $N(m, \sigma^2)$ (in other words, the mean and variance of any random variable with this law) are given by

$$\int_{-\infty}^{\infty} x N(m, \sigma^2)(dx) = m \quad \text{and} \quad \int_{-\infty}^{\infty} (x - m)^2 N(m, \sigma^2)(dx) = \sigma^2.$$

The law $N(m, \sigma^2)$ is called a *normal* or *Gaussian* law on \mathbb{R} with mean m and variance σ^2, and $N(0, 1)$ is called a *standard normal* distribution.

For any law P on \mathbb{R}, the integrals $\int_{-\infty}^{\infty} x^n \, dP(x)$, if defined, are called the *moments* of P. Thus the first moment is the mean, and if the mean is 0, the second moment is the variance. The function $g(u) = \int_{-\infty}^{\infty} e^{xu} \, dP(x)$, for whatever values of u it is defined and finite, is called the *moment generating function* of P. If it is defined and finite in a neighborhood of 0, a Taylor expansion of e^{xu} around $x = 0$ gives the nth moment of P as the nth derivative of g at 0, as will be shown in case $P = N(0, 1)$ in the next proof. In this sense, g generates the moments.

9.4.2. Proposition *(a) The characteristic functions of normal laws on \mathbb{R} are given by*

$$\int_{-\infty}^{\infty} e^{ixu} N(m, \sigma^2)(dx) = \exp(imu - \sigma^2 u^2/2).$$

(b) $\displaystyle \int_{-\infty}^{\infty} e^{xu} N(0, 1)(dx) = \exp(u^2/2)$ *for all real u;*

(c) $\displaystyle \int_{-\infty}^{\infty} x^{2n} N(0, 1)(dx) = (2n)!/(2^n n!)$ *for $n = 0, 1, \ldots,$*

$$= 1 \cdot 3 \cdot 5 \cdots (2n - 1), n \geq 1, \quad \text{and}$$

$$\int_{-\infty}^{\infty} x^m N(0, 1)(dx) = 0 \quad \text{for all odd } m = 1, 3, 5, \ldots.$$

Proof. First, for (b), $xu - x^2/2 \equiv u^2/2 - (x - u)^2/2$, so the integral on the left in (b) equals $\exp(u^2/2)N(u, 1)(\mathbb{R}) = \exp(u^2/2)$ as stated.

Next, for (c), since $\exp(|xu|) \leq \exp(xu) + \exp((-u)x)$, $\exp(|xu|)$ is integrable for $N(0, 1)$. The Taylor series of e^z, for $z = |xu|$, is a series of positive terms, so it can be integrated term by term with respect to $N(0, 1)(dx)$ by dominated convergence. It follows that the Taylor series of e^{xu} (as a function of x) can also be integrated termwise. The coefficient of u^n in a power series expansion of a function $g(u)$ is unique, being $g^{(n)}(0)/n!$. Expand both sides of (b), putting $z = u^2/2$ in the Taylor series of e^z, and equate the coefficients of each power of u. The coefficients of odd powers, and so the odd moments of $N(0, 1)$, are 0. For even powers, we have on the right $u^{2n}/(2^n n!)$ and on the left $u^{2n}/(2n)!$ times the $2n$th moment of $N(0, 1)$, which implies (c).

Now for (a), let $v = (x - m)/\sigma$, so $e^{ixu} = e^{imu} e^{iv\sigma u}$. This substitution reduces the general case to the case $m = 0$ and $\sigma = 1$. We just want to prove that in (b), the real number u can be replaced by an imaginary number iu. To justify this, expanding e^{ixu} by the Taylor series of e^z, the series (or its real and imaginary parts) can be integrated term by term using dominated convergence: $|e^{ixu}| = 1 \leq e^{|xu|}$, and the absolute values of corresponding terms in the Taylor series of e^z for $z = ixu$ and $z = |xu|$ are equal. The coefficients of $(iu)^n$ for each n are then given by (c). Summing the series gives (a). □

Example. Let X be a binomial random variable, the number of successes in n independent trials with probability p of success, so

$$P(X = k) = \binom{n}{k} p^k q^{n-k}, \quad k = 0, 1, \ldots, n, \quad \text{where } q := 1 - p.$$

The moment generating function of this distribution is

$$\sum_{0 \le k \le n} e^{kt} P(X = k) = (pe^t + q)^n$$

by the binomial theorem.

The sum of independent real random variables corresponds to the convolution of their laws (Theorem 9.1.3); now it will be shown to correspond to the product of their characteristic functions, a simpler operation:

9.4.3. Theorem *If X_1, X_2, ..., are independent random variables in \mathbb{R}^k, and $S_n := X_1 + \cdots + X_n$, then for each $t \in \mathbb{R}^k$,*

$$E \exp(i(S_n, t)) = \Pi_{j=1}^n E \exp(i(X_j, t)).$$

So if the X_j are identically distributed, with characteristic function f, then the characteristic function of S_n is $f(t)^n$.

Proof. As $\exp(i(S_n, t)) = \Pi_{1 \le j \le n} \exp(i(X_j, t))$, independence and the Tonelli-Fubini theorem give the conclusion. $\quad\square$

Next, a global property, namely, a finite absolute moment $E|X|^r$, implies a local, differentiability property of the characteristic function:

9.4.4. Theorem *If X is a real random variables with $E|X|^r < \infty$ for some integer $r \ge 0$, then the characteristic function $f(t) := E \exp(iXt)$ is C^r, that is, f has continuous derivatives through order r on \mathbb{R}, where for $Q := \mathcal{L}(X) := P \circ X^{-1}$,*

$$f^{(j)}(t) = \int_{-\infty}^{\infty} (ix)^j e^{ixt} dQ(x), \quad j = 0, 1, \ldots, r,$$

$$f^{(j)}(0) = \int_{-\infty}^{\infty} (ix)^j dQ(x).$$

Proof. For $r = 0$, f is continuous by dominated convergence since e^{ixt} is continuous in t and uniformly bounded. For $r = 1$, we have for any real x, t, and h, $|e^{ix(t+h)} - e^{ixt}| \le |xh|$, since for any a and b, with $g(u) := e^{iu}$, $|g(b) - g(a)| = |\int_a^b g'(t) dt| \le |b - a| \sup_{a \le x \le b} |g'(x)| = |b - a|$. Then we can differentiate the definition of f under the integral sign using dominated convergence. This can be iterated through the rth derivative. $\quad\square$

9.4.5. Theorem *Let X_1, X_2, ..., be i.i.d real random variables with $E X_1 = 0$ and $E X_1^2 := \sigma^2 < \infty$. Let $S_n := X_1 + \cdots + X_n$. Then for all real t,*

$$\lim_{n \to \infty} E \exp \left(i S_n t / n^{1/2} \right) = \exp(-\sigma^2 t^2 / 2).$$

Remark. This last theorem says that the characteristic functions of the laws of $S_n/n^{1/2}$ converge pointwise on \mathbb{R} to that of the law $N(0, \sigma^2)$. This will be a step in proving convergence of the laws.

Proof. Let $f(t) := E \exp(i X_1 t)$. Then by Theorem 9.4.4, $f(0) = 1$, $f'(0) = 0$, and $f''(0) = -\sigma^2$. Thus by Taylor's theorem with remainder, $f(t) = 1 - \sigma^2 t^2/2 + o(t^2)$ as $t \to 0$, where the o notation means, as usual, $o(t^2)/t^2 \to 0$ as $t \to 0$. Now by Theorem 9.4.3, for any fixed t and $n \to \infty$,

$$E \exp\left(i S_n t / n^{1/2}\right) = f\left(t/n^{1/2}\right)^n = (1 - \sigma^2 t^2/(2n) + o(t^2/n))^n$$
$$\to \exp(-\sigma^2 t^2/2) \qquad \text{as } n \to \infty,$$

as can be seen by taking logarithms and their Taylor series. □

Problems

1. Evaluate: (a) $\int_{-\infty}^{\infty} x^2 N(3, 4)(dx)$; (b) $\int_{-\infty}^{\infty} x^4 N(1, 4)(dx)$.

2. Find the moment generating function (where defined) and moments of the law P having density e^{-x} for $x > 0$ and 0 for $x < 0$ (exponential distribution).

3. Find the characteristic function of the exponential distribution in Problem 2, and also of the law having density $e^{-|x|}/2$ for all x. In the latter case, evaluate the characteristic function of $S_n/n^{1/2}$ as in Theorem 9.4.5 and show directly that it converges as $n \to \infty$.

4. (a) If X_j are independent real random variables, each having a moment generating function $f_j(t)$ defined and finite for $|t| \le h$ for some $h > 0$, show that the moment generating function of S_n is the product of those for X_1, \ldots, X_n if $|t| \le h$.
 (b) If X is a nonnegative random variable, show that for any $t \ge 0$, $P(X \ge t) \le \inf_{u \ge 0} e^{-tu} g(u)$, where g is the moment generating function of X.
 (c) Find the moment generating function of $X(n, p)$, the number of successes in n independent trials with probability p of success on each trial, as in the example just before 9.4.3, using part (a). *Hint:* X is a sum of n independent, simpler variables.
 (d) Prove that $E(k, n, p) := P(X(n, p) \ge k) \le (np/k)^k (nq/(n-k))^{n-k}$ for $k \ge np$. *Hint:* Apply (b) with $t = k$ and $e^u = kq/((n-k)p)$.

5. Evaluate the moment generating functions of the following laws:
 (a) P is a Poisson distribution: for some $\lambda > 0$, $P(\{k\}) = e^{-\lambda}\lambda^k/k!$ for $k = 0, 1, \ldots$.
 (b) P is the uniform distribution (Lebesgue measure) on $[0, 1]$.

6. Find the characteristic function of the law on \mathbb{R} having the density $f = 1_{[-1,1]}/2$ with respect to Lebesgue measure.

7. Let X have a binomial distribution, as in the example just before Theorem 9.4.3. Evaluate EX, EX^2, and EX^3 from the moment generating function.

8. Let f be a complex-valued function on \mathbb{R} such that $f'(t)$ is finite for t in a neighborhood of 0 and $f''(0)$ is finite.
 (a) Show that $f''(0) = \lim_{t \to 0}(f(t) - 2f(0) + f(-t))/t^2$. *Hint:* Use a Taylor series with remainder (Appendix B, Theorem B.5).
 (b) If f is the characteristic function of a law P, show that $\int x^2 \, dP(x) < \infty$. *Hint:* Apply (a) and Fatou's lemma.

9. Give an example of a law P on \mathbb{R} with a characteristic function f such that $f'(0)$ is finite but $\int |x| \, dP(x) = +\infty$. *Hint:* Let P have a density $c 1_{[3,\infty)}(|x|)/(x^2 \log |x|)$ for the suitable constant c. Show that f is real-valued and $|1 - f(t)| = o(|t|)$ as $|t| \to 0$, using $1 - \cos(xt) \leq x^2 t^2/2$ for $|x| \leq 2/|t|$ and $|1 - \cos(xt)| \leq 2$ for $|x| > 2/|t|$.

9.5. Uniqueness of Characteristic Functions and a Central Limit Theorem

In this section C_b will denote the set of bounded, continuous *complex*-valued functions. One step in showing that convergence of characteristic functions (as in Theorem 9.4.5) implies convergence of laws will be that the correspondence of laws and characteristic functions is 1–1:

9.5.1. Uniqueness Theorem *If P and Q are laws on \mathbb{R}^k with the same characteristic function g on \mathbb{R}^k, then $P = Q$.*

In problems 7–9, it will be shown that characteristic functions may be equal in a neighborhood of 0 but not everywhere.

In proving Theorem 9.5.1, it will be helpful to convolve laws with normal distributions. Let $N(0, \sigma^2 I)$ be the law on \mathbb{R}^k which is the product of copies of $N(0, \sigma^2)$, so that the coordinates x_1, \ldots, x_k are i.i.d. with law $N(0, \sigma^2)$. As $\sigma \downarrow 0$, a convolution $P^{(\sigma)} := P * N(0, \sigma^2 I)$ will converge to the law P; such convolutions will always have densities, and the Fourier analysis (evaluation of characteristic functions) can be done explicitly for normal laws. The details will be given in the next two lemmas.

9.5.2. Lemma *Let P be a probability law on \mathbb{R}^k with characteristic function $g(t) = \int e^{i(x,t)} \, dP(x)$. Then $P^{(\sigma)}$ has a density $f^{(\sigma)}$ which satisfies*

$$f^{(\sigma)}(x) = (2\pi)^{-k} \int g(t) \exp(-i(x, t) - \sigma^2 |t|^2/2) \, dt.$$

Proof. Let φ_σ be the density of $(N, \sigma^2 I)$. First suppose $k = 1$. For any $x \in \mathbb{R}$, we can write $f^{(\sigma)}(x)$ as

$$\int \varphi_\sigma(x - y) \, dP(y) \qquad\qquad \text{by (9.1.6)}$$

$$= \int \varphi_\sigma(y - x) \, dP(y) \qquad\qquad (\varphi_\sigma \text{ is even})$$

$$= \int (2\pi)^{-1} \int \exp(i(y - x)t - \sigma^2 t^2 / 2) \, dt \, dP(y) \qquad\qquad \text{by (9.4.2a)},$$

$$\text{for } m = 0 \text{ and with } \sigma \text{ and } 1/\sigma \text{ interchanged}$$

$$= (2\pi)^{-1} \int \int e^{iyt} \, dP(y) \exp(-ixt - \sigma^2 t^2 / 2) \, dt \qquad\qquad \text{by Tonelli-Fubini}$$

$$= (2\pi)^{-1} \int g(t) \exp(-ixt - \sigma^2 t^2 / 2) \, dt,$$

as desired. For $k > 1$ the proof is essentially the same, with, for example, xt replaced by (x, t), t^2 by $|t|^2$, and $(2\pi)^{-1}$ by $(2\pi)^{-k}$. □

9.5.3. Lemma *For any law P on \mathbb{R}^k, the laws $P^{(\sigma)}$ converge to P as $\sigma \downarrow 0$.*

Proof. Let X have the law P and let Y be independent of X and have law $N(0, I)$ on \mathbb{R}^k. Then for each $\sigma > 0$, σY has law $N(0, \sigma^2 I)$, so $X + \sigma Y$ has law $P^{(\sigma)}$ by Theorem 9.1.3, and $X + \sigma Y$ converges to X a.s. and so in probability as $\sigma \downarrow 0$. So the laws $P^{(\sigma)}$ converge to P by Proposition 9.3.5.

 □

Proof of the Uniqueness Theorem (9.5.1). The laws $P^{(\sigma)}$, by Lemma 9.5.2, have densities determined by g, so $P^{(\sigma)} = Q^{(\sigma)}$ for all $\sigma > 0$. These laws converge to P and Q respectively as $\sigma \downarrow 0$, by Lemma 9.5.3. Limits of converging laws are unique by Lemma 9.3.2, so $P = Q$. □

The assignment of characteristic functions to laws has an inverse by Theorem 9.5.1. For some laws the inverse is given by an integral formula:

9.5.4. Fourier Inversion Theorem *Let f be a probability density on \mathbb{R}^k. Let g be its characteristic function,*

$$g(t) := \int f(x) e^{i(x,t)} \, dx \qquad \text{for each } t \in \mathbb{R}^k.$$

If g is integrable for Lebesgue measure dt on \mathbb{R}^k, then for Lebesgue almost all $x \in \mathbb{R}^k$,

$$f(x) = (2\pi)^{-k} \int g(t)e^{-i(x,t)} \, dt.$$

For example, normal densities f, as shown in Proposition 9.4.2(a) for $k = 1$, have characteristic functions which are integrable, so that Theorem 9.5.4 will apply to them. On the other hand, any characteristic function is continuous (by dominated convergence), and likewise the last integral in Theorem 9.5.4 represents a continuous function of x for g integrable. So Theorem 9.5.4 can apply only to densities equal almost everywhere to continuous functions, which is not true for many densities such as the uniform density $1_{[0,1]}$ or the exponential density $e^{-x}1_{[0,\infty)}(x)$.

Proof. In Lemma 9.5.2, as $\sigma \downarrow 0$, the right side converges uniformly in x to $h(x) := (2\pi)^{-k} \int g(t)e^{-i(x,t)} \, dt$, since g is integrable, and

$$\int |g(t)| \big| 1 - \exp(-\sigma^2|t|^2/2) \big| \, dt \to 0$$

by dominated convergence. For any continuous function v on \mathbb{R}^k with compact support, $\int v \, dP^{(\sigma)}$ converges as $\sigma \downarrow 0$ to $\int vh \, dx$, and also to $\int vf \, dx$ by Lemma 9.5.3, so these two integrals are equal. It follows, as in Lemma 9.3.2, that $\int_U f \, dx = \int_U h \, dx$ for every bounded open set U, then any bounded closed set, or any bounded Borel set. Restricted to $|x| < m$ for each m, f and h are thus both densities of the same measure with respect to Lebesgue measure, so $f = h$ a.e. for $|x| < m$ by uniqueness in the Radon-Nikodym theorem (5.5.4). It follows that $f = h$ a.e. on \mathbb{R}^k. $\qquad\square$

It will be shown in §9.8 that convergence of characteristic functions to a continuous limit function implies uniform tightness. That will not be needed as yet; the following will suffice for the present.

9.5.5. Lemma *Let P_n be a uniformly tight sequence of laws on \mathbb{R}^k with characteristic functions $f_n(t)$ converging for all t to a function $f(t)$. Then $P_n \to P$ for a law P having characteristic function f.*

Proof. By Theorem 9.3.3, any subsequence of P_n has a subsubsequence converging to some law. All the limit laws have characteristic function f,

so all are equal to some P by the uniqueness theorem. Then $P_n \to P$ by Proposition 9.3.1. \square

Example. The functions $f_n := \exp(-nt^2/2)$ are the characteristic functions of the laws $N(0, n)$, which are not uniformly tight, and do not converge, while $f_n(t)$ converges for all t to $1_{\{0\}}(t)$. This shows why uniform tightness is helpful in Lemma 9.5.5.

For a random variable $Y = \langle Y_1, \ldots, Y_k \rangle$ with values in \mathbb{R}^k, and law P, the *mean* or *expectation* of Y, or of P, is defined by $EY := \langle EY_1, \ldots, EY_k \rangle$ iff all these means exist and are finite. Also, we have $|Y|^2 := Y_1^2 + \cdots + Y_k^2$. As usual, $S_n := X_1 + \cdots + X_n$. Recall (from §8.1) that for any two real random variables X and Y with $EX^2 < \infty$ and $EY^2 < \infty$, the *covariance* of X and Y is defined by

$$\mathrm{cov}(X, Y) := E((X - EX)(Y - EY)) = E(XY) - EX\,EY.$$

Thus the covariance of X with itself is its variance. If X is a random variable with values in \mathbb{R}^k and $E|X|^2 < \infty$, then the *covariance matrix* of X, or its law, is the $k \times k$ matrix $\mathrm{cov}(X_i, X_j)$, $i, j = 1, \ldots, k$. For any constant vector V, clearly X and $X + V$ have the same covariance.

Example. Let P be the law on \mathbb{R}^2 with $P((0, 0)) = P((0, 1)) = P((1, 1)) = 1/3$. Then the mean of P is $(1/3, 2/3)$, and its covariance matrix is

$$\begin{pmatrix} 2/9 & 1/9 \\ 1/9 & 2/9 \end{pmatrix}.$$

Now, here is one of the main theorems in all of probability theory. The more classical one-dimensional case will be stated as the second part of the theorem.

9.5.6. Central Limit Theorem (a) *Let* $X_1, X_2, \ldots,$ *be independent, identically distributed random variables with values in* \mathbb{R}^k, $EX_1 = 0$, *and* $0 < E|X_1|^2 < \infty$. *Then as* $n \to \infty$, $\mathcal{L}(S_n/n^{1/2}) \to P$ *where* P *has the characteristic function*

$$f_P(t) = \exp\left(-\frac{1}{2} \sum_{r,s=1}^{k} C_{rs} t_r t_s \right) \quad \text{and} \quad C_{rs} := EX_{1r} X_{1s}$$

(*C is the covariance matrix of the random vector* X_1).

(b) If the hypotheses of part (a) hold, $k = 1$, and $EX_1^2 = 1$, then for any real x, as $n \to \infty$

$$P\left(S_n/n^{1/2} \leq x\right) \to \Phi(x) := (2\pi)^{-1/2} \int_{-\infty}^{x} \exp(-t^2/2)\, dt.$$

Proof. The vectors X_i are independent and have mean 0, so $E(X_i, X_j) = 0$ for $i \neq j$. It follows that $E|S_n/n^{1/2}|^2 = E|X_1|^2$ for all n, using Theorem 8.1.2 for $k = 1$. Given $\varepsilon > 0$, for M large enough, $E|X_1|^2/M^2 < \varepsilon$. Then the laws of $S_n/n^{1/2}$ are uniformly tight since by Chebyshev's inequality (8.3.1), $P(|S_n/n^{1/2}| > M) < \varepsilon$ for all n.

Now for each $t \in \mathbb{R}^k$, the random variables (t, X_i) are i.i.d. real-valued with mean 0 and $E(t, X_1)^2 < \infty$. Thus by Theorem 9.4.5,

$$E \exp\left(i\left(t, S_n/n^{1/2}\right)\right) \to \exp(-E(t, X_1)^2/2) \qquad \text{as } n \to \infty$$

for all $t \in \mathbb{R}^k$ where the function on the right equals f_P as given. So by Lemma 9.5.5, $\mathcal{L}(S_n/n^{1/2}) \to P$ for a law P with characteristic function f_P, proving part (a). Part (b) then follows by way of Theorem 9.3.6, since Φ is continuous for all x, and by the uniqueness theorem, 9.5.1, and the fact that the characteristic function of $N(0, 1)$ is $\exp(-t^2/2)$ as given by Proposition 9.4.2(a). $\qquad\square$

The covariance matrix C is symmetric: $C_{rs} = C_{sr}$ for any r and s. Also, C is nonnegative definite: for any $t \in \mathbb{R}^k$,

$$\sum_{r,s=1}^{k} C_{rs} t_r t_s = E \left(\sum_{r=1}^{k} t_r X_{1r} \right)^2 \geq 0.$$

Examples. Let

$$A = \begin{pmatrix} 1 & 2 \\ 2 & 1 \end{pmatrix}, \quad B = \begin{pmatrix} 3 & 1 \\ 2 & 4 \end{pmatrix}, \quad \text{and} \quad C = \begin{pmatrix} 5 & -2 \\ -2 & 4 \end{pmatrix}.$$

Then A and B are not covariance matrices because B is not symmetric and A is not nonnegative definite: $(At, t) < 0$ for $t = (-1, 1)$. C is a covariance matrix.

A probability law on \mathbb{R}^k with characteristic function as given in the central limit theorem will be called $N(0, C)$, for "normal law with mean 0 and covariance C." The next fact shows that such laws exist and do have covariance C.

9.5.7. Theorem *For any* $k \times k$ *nonnegative definite, symmetric matrix* C, *a probability law* $N(0, C)$ *on* \mathbb{R}^k *exists, having mean* 0 *and covariance*

$$\int x_r x_s \, dN(0, C)(x) = C_{rs}, \qquad r, s = 1, \dots, k. \text{ Let}$$

$$H := \text{range } C = \left\{ \left\{ \sum_{s=1}^{k} C_{rs} y_s \right\}_{r=1}^{k} : y \in \mathbb{R}^k \right\}.$$

Then $N(0, C)(H) = 1$. *Let* j *be the dimension of* H, *so* $j = \text{rank}(C)$. *The linear transformation* D *defined by* C, *restricted to* H, *gives a linear transformation* D_H *of* H *onto itself.* $N(0, C)$ *has a density (Radon-Nikodym derivative) with respect to Lebesgue measure in* H *for a basis diagonalizing* C *given by*

$$(2\pi)^{-j/2} (\det D_H)^{-1/2} \exp\left(-\left(D_H^{-1}x, x\right)/2\right), \qquad x \in H. \qquad (9.5.8)$$

Examples. For $k = 1$, recall (Propositions 9.4.1, 9.4.2) that the law $N(0, \sigma^2)$ with density $(2\pi)^{-1/2}\sigma^{-1} \exp(-x^2/(2\sigma^2))$ has characteristic function $\exp(-\sigma^2 t^2/2)$. Products of such laws give the cases of (9.5.8) where D_H is a diagonal matrix.

Proof. Since C is symmetric and real, D is self-adjoint, which means that $(Dx, y) = (x, Dy)$ for all $x, y \in \mathbb{R}^k$. If $z \perp H$, that is, $(z, h) = 0$ for all $h \in H$, then for all x, $0 = (Dx, z) = (x, Dz)$, so $Dz = 0$ (take $x = Dz$). Conversely, if $Dz = 0$, then $z \perp H$, so $H^\perp := \{z : z \perp H\} = \{z : Dz = 0\}$.

Now D, being self-adjoint, has an orthonormal basis of eigenvectors $e_i : D(e_i) = d_i e_i$, $i = 1, \dots, k$ (see, for example, Hoffman and Kunze, 1971, p. 314, Thm. 18). The eigenvectors e_i with $d_i \neq 0$, and thus $d_i > 0$, are exactly those in H and form an orthonormal basis of H. Let them be numbered as e_1, \dots, e_j. Restricting D to H, we find that the resulting D_H takes H into itself, and onto itself since $d_i > 0$ for $i = 1, \dots, j$. For the given basis, D_H is represented by a diagonal matrix with d_i on the diagonal.

In one dimension, by Proposition 9.4.1, for any $\sigma > 0$ there is a law $N(0, \sigma^2)$ on \mathbb{R} having density $(2\pi\sigma^2)^{-1/2} \exp(-x^2/2\sigma^2)$ with respect to Lebesgue measure. By Proposition 9.4.2, its characteristic function is $\exp(-\sigma^2 t^2/2)$. For $\sigma = 0$, $N(0, 0)$ is defined as the point mass δ_0 at 0; its characteristic function is the constant 1. Now with respect to the basis $\{e_r\}_{1 \leq r \leq j}$, H can be represented as \mathbb{R}^j. Let P be the Cartesian product for $r = 1$ to k of the laws $N(0, d_r)$. Then since $d_r = 0$ for $r > j$, $P(H) = 1$. By the Tonelli-Fubini

theorem, the characteristic function of P is

$$f_P(t) = \exp\left(-\sum_{r=1}^{j} d_r t_r^2 / 2\right) = \exp(-(Dt, t)/2).$$

Thus P has the characteristic function of a law $N(0, C)$ as defined, so a law $N(0, C)$ exists and by the uniqueness theorem $P = N(0, C)$.

Now on H, D_H^{-1} is represented for the given basis $\{e_r\}_{1 \le r \le j}$ by a diagonal matrix with $1/d_r$ as the rth diagonal entry, $r = 1, \ldots, j$. The determinant of D_H is $\Pi_{1 \le r \le j} d_j$. Thus P has the density given in (9.5.8), by the definitions and Proposition 9.4.1.

To find the mean and covariance of P we can again use a basis where D is diagonalized, and the one-dimensional case, just before Proposition 9.4.2. The mean is clearly 0, and we have

$$\int (x, y)(x, z) \, dN(0, C)(x) = (Dy, z)$$

in the diagonalizing coordinates and hence for any other basis, such as the original one. Thus Theorem 9.5.7 is proved. $\qquad\square$

For any law P on \mathbb{R}^k and $t \in \mathbb{R}^k$ we have a translation P_t of P by t, where for any measurable set A, $P_t(A) := P(A - t)$, setting $A - t := \{a - t : a \in A\}$. If X is a random variable with law P, then $X + t$ has law P_t. Thus if $EX = m$, it follows that $\int y \, dP_t(y) = E(X + t) = m + t$. Note that $P_t = P * \delta_t$. For characteristic functions we have

$$\int e^{i(x,y)} \, dP_t(x) = \int e^{i(x,y)} \, dP(x - t) = \int e^{i(u+t,y)} \, dP(u)$$

$$= e^{i(t,y)} \int e^{i(u,y)} \, dP(u).$$

So translation by t multiplies the characteristic function (written as a function of y) by $e^{i(t,y)}$, just as in one dimension for normal laws.

We can also write $P_t = P \circ \tau_t^{-1}$ where $\tau_t(x) := t + x$ for all x. For any $m \in \mathbb{R}^k$ and law $N(0, C)$, $N(m, C)$ will be the translate of $N(0, C)$ by m; in other words, $N(m, C) := N(0, C) \circ \tau_m^{-1}$. Thus if X is a random variable with law $N(0, C)$, $X + m$ will have law $N(m, C)$. Now $N(m, C)$ is a law with mean m and covariance C, called the *normal law with mean m and covariance C*.

Continuity of Cartesian product and convolution of laws can be proved without characteristic functions, but more easily with them:

9.5.9. Theorem *For any positive integers m and k, if P_n are laws on \mathbb{R}^m and Q_n on \mathbb{R}^k, $P_n \to P_0$ and $Q_n \to Q_0$, then the Cartesian product measures converge, $P_n \times Q_n \to P_0 \times Q_0$ on \mathbb{R}^{m+k}. If $k = m$, then $P_n * Q_n \to P_0 * Q_0$ on \mathbb{R}^k.*

Proof. First, by Proposition 9.3.4, the P_n are uniformly tight on \mathbb{R}^m, as are the Q_n on \mathbb{R}^k. Given $\varepsilon > 0$, take compact sets J and K such that $P_n(J) > 1 - \varepsilon/2$ and $Q_n(K) > 1 - \varepsilon/2$ for all n. Then $J \times K$ is compact and $(P_n \times Q_n)(J \times K) > 1 - \varepsilon$ for all n, so the laws $P_n \times Q_n$ are uniformly tight. Now, each function $e^{i(t,x)}$ on \mathbb{R}^{k+m} is a product of such functions on \mathbb{R}^k and \mathbb{R}^m. Thus by the Tonelli-Fubini theorem the characteristic functions of $P_n \times Q_n$ are products of those of P_n and Q_n and converge to that of $P_0 \times Q_0$. Then by Lemma 9.5.5, $P_n \times Q_n \to P_0 \times Q_0$.

Now for convolutions, each $P_n * Q_n$ is the image measure of $P_n \times Q_n$ on \mathbb{R}^{2k} by $\langle x, y \rangle \mapsto x + y$, so an application of the continuous mapping theorem (9.3.7) finishes the proof. □

The law δ_0 which puts all its probability at 0 acts as an "identity" for convolution with any other law P, in the sense that $\delta_0 * P = P * \delta_0 = P$, and no law except δ_0 acts as an identity for even one other law μ, as the following shows (although in general there is no "unique factorization" for convolution):

9.5.10. Proposition *If μ and Q are laws on \mathbb{R}, then $\mu * Q = \mu$ if and only if $Q = \delta_0$.*

Proof. "If" is clear. To prove "only if," we have characteristic functions $f_\mu(t) f_Q(t) = f_\mu(t)$ for all t. Since f_μ is continuous (by Theorem 9.4.4 with $r = 0$) and $f_\mu(0) = 1$, there is a neighborhood U of 0 in which $f_\mu(t) \neq 0$ and so $f_Q(t) = 1$.

Now $f_Q(t) = 1$ for $t \neq 0$ implies that $\int \cos(tx) \, dQ(x) = 1$ and, since $\cos(xt) \leq 1$ for all x, it follows that $\cos(xt) = 1$ and so $xt/(2\pi)$ is an integer for Q-almost all x, $Q(2\pi \mathbb{Z}/t) = 1$. For any other u in U we also have $Q(2\pi \mathbb{Z}/u) = 1$. Taking u with u/t irrational, it follows that $Q = \delta_0$. □

Next, it will be shown that Lebesgue measure λ^k on \mathbb{R}^k is invariant under suitable linear transformations. A linear transformation T from \mathbb{R}^k into itself is called *orthogonal* iff $(Tx, Ty) = (x, y)$ for all x and y in \mathbb{R}^k, where (\cdot, \cdot) is the usual inner product in \mathbb{R}^k. In \mathbb{R}^2, for example, orthogonal transformations may be rotations (around the origin) or reflections (in lines through the origin). It is "well known," say from school geometry, that the usual area measure λ^2 is invariant under such transformations, but an actual proof might be nontrivial. Here is a rather easy proof, in k dimensions, using probability:

9.5.11. Theorem *For* $k = 1, 2, \ldots,$ *and any orthogonal transformation* T *of* \mathbb{R}^k *into itself,* $\lambda^k \circ T^{-1} = \lambda^k$.

Proof. For $k = 2$, problem 6 of Sec. 4.4 (polar coordinates) indicates a proof and was already used for normal densities in the proof of Prop. 9.4.1.

The linear transformation T has a transpose (adjoint) T', satisfying $(Tx, y) = (x, T'y)$ for all x and y. Now T is orthogonal if and only if $T'T = I$ (the identity). This implies (since k is finite) that T must be onto and $T' = T^{-1}$. It follows that $T = (T')^{-1}$, so $TT' = I$, and T' is also orthogonal. From the image measure theorem (4.1.11), for the characteristic function of $N(0, I) \circ T^{-1}$ we have

$$\int e^{i(u,y)} \, dN(0, I) \circ T^{-1}(y) = \int e^{i(u,Tx)} \, dN(0, I)(x)$$

$$= \int e^{i(T'u,x)} \, dN(0, I)(x) = \exp(-|T'u|^2/2) = \exp(-|u|^2/2).$$

By uniqueness of characteristic functions (9.5.1), $N(0, I) \circ T^{-1} = N(0, I)$ ($N(0, I)$ is orthogonally invariant.) Now for any Borel set B, with $T^{-1}(B) = A$, using Theorem 4.1.11 in the next-to-last step,

$$\lambda^k \circ T^{-1}(B) = (2\pi)^{k/2} \int_A \exp(|x|^2/2) \, dN(0, I)(x)$$

$$= (2\pi)^{k/2} \int 1_B(Tx) \exp(|Tx|^2/2) \, dN(0, I)(x)$$

$$= (2\pi)^{k/2} \int 1_B(y) \exp(|y|^2/2) \, dN(0, I)(y)$$

$$= \lambda^k(B). \qquad \square$$

Next, the image of a normal law by an affine transformation is normal:

9.5.12. Proposition *Let* $N(m, C)$ *be a normal law on* \mathbb{R}^k *and let* A *be an affine transformation from* \mathbb{R}^k *into some* \mathbb{R}^j, *so that for some linear transformation* L *of* \mathbb{R}^k *into* \mathbb{R}^j *and* $w \in \mathbb{R}$, $Ax = Lx + w$ *for all* $x \in \mathbb{R}^k$. *Then* $N(m, C) \circ A^{-1}$ *is a normal law on* \mathbb{R}^j, *specifically* $N(u, LCL')$ *where* $u := Lm + w$.

Note. Here δ_m is considered as the normal law $N(m, 0)$, which can happen, for example, if $C = 0$ or $L = 0$. L is identified with its matrix, a $j \times k$ matrix, so that its transpose L' is a $k \times j$ matrix, with $(L')_{rs} = L_{sr}$ for $s = 1, \ldots, j$ and $r = 1, \ldots, k$.

Proof. Let y be a random variable in \mathbb{R}^k with law $N(0, C)$ and $x = y + m$. Consider the characteristic function, for $t \in \mathbb{R}^j$,

$$E \exp(i(t, Lx + w)) = \exp(i(t, u))E \exp(i(L't, y)).$$

From Theorems 9.5.6 and 9.5.7, y has the characteristic function

$$E \exp(i(v, y)) = \exp(-v'Cv/2).$$

For $v = L't$, $v'Cv = (L't)'CL't = t'LCL't$, so (by uniqueness of characteristic functions) $N(m, C) \circ A^{-1} = N(u, LCL')$, where LCL' is a symmetric, nonnegative definite $j \times j$ matrix. $\qquad\square$

If (X_1, \ldots, X_k) has a normal law $N(m, C)$ on \mathbb{R}^k, then X_1, \ldots, X_k are said to have a normal joint distribution or to be jointly normal.

9.5.13. Theorem *A random variable $X = (X_1, \ldots, X_k)$ in \mathbb{R}^k has a normal distribution if and only if for each $t \in \mathbb{R}^k$, $(t, X) = t_1 X_1 + \cdots + t_k X_k$ has a normal distribution in \mathbb{R}.*

Proof. "Only if" follows from Proposition 9.5.12 with $j = 1$. To prove "if," note first that $E|X|^2 = EX_1^2 + \cdots + EX_k^2 < \infty$, taking t's equal to the usual basis vectors of \mathbb{R}^k. Thus X has a well-defined, finite covariance matrix C. We can assume $EX = 0$. Then the characteristic function $E \exp(i(t, X)) = \exp(-(Ct, t)/2)$ since (X, t) has a normal distribution on \mathbb{R} with mean 0 and variance (Ct, t). By uniqueness of characteristic functions (Theorem 9.5.1) and the form of normal characteristic functions (Theorems 9.5.6 and 9.5.7), it follows that X has a normal distribution. $\qquad\square$

For jointly normal variables, independence is equivalent to having a zero covariance:

9.5.14. Theorem *Let $(X, Y) = (X_1, \ldots, X_k, Y_1, \ldots, Y_m)$ be jointly normal in \mathbb{R}^{k+m}. Then $X = (X_1, \ldots, X_k)$ is independent of (Y_1, \ldots, Y_m) if and only if $\mathrm{cov}(X_i, Y_j) = 0$ for all $i = 1, \ldots, k$ and $j = 1, \ldots, m$. So jointly normal real random variables X and Y are independent if and only if $\mathrm{cov}(X, Y) = 0$.*

Proof. Independent variables which are square-integrable always have zero covariance, as mentioned just before Theorem 8.1.2. Conversely, if (X, Y) have a joint normal distribution $N(m, C)$ with $\mathrm{cov}(X_i, Y_j) = 0$ for all i and j, let $m = (\mu, \nu)$ with $EX = \mu$ and $EY = \nu$. Consider the characteristic function $f(t, u) := E \exp(i((t, u) \cdot (X, Y)))$. As noted after the proof of Theorem 9.5.7,

we have $f(t, u) \equiv \exp(i(t\mu + uv))g(t, u)$ where $g(t, u)$ is the characteristic function of $N(0, C)$. Because of the 0 covariances C is of the form

$$C = \begin{pmatrix} D & 0 \\ 0 & F \end{pmatrix}$$

where D is the $k \times k$ covariance matrix of X and F the $m \times m$ covariance matrix of Y. By the form of normal characteristic functions (Theorems 9.5.6 and 9.5.7) we have $g(t, u) \equiv h(t)j(u)$ where h is the characteristic function of $N(0, D)$ and j of $N(0, F)$. It follows that $f(t, u) \equiv r(t)s(u)$ where r is the characteristic function of $N(\mu, D)$ and s of $N(v, F)$. By uniqueness of characteristic functions (Theorem 9.5.1), $N(m, C) = N(\mu, D) \times N(v, F)$, in other words X and Y are independent. $\qquad\square$

Problems

1. Let X and Y be independent real random variables with $\mathcal{L}(X) = N(m, \sigma^2)$ and $\mathcal{L}(Y) = N(\mu, \tau^2)$. Show that $\mathcal{L}(X + Y) = N(m + \mu, \sigma^2 + \tau^2)$.

2. (Random walks in \mathbb{Z}^2.) Let X_1, X_2, \ldots, be independent, identically distributed random variables with values in \mathbb{Z}^2, in other words $X_j = (k_j, m_j)$ where $k_j \in \mathbb{Z}$ and $m_j \in \mathbb{Z}$. Then S_n can be called the position after n steps in a random walk, starting at $(0, 0)$, where X_j is the jth step. Find the limit of the law of $S_n / n^{1/2}$ as $n \to \infty$ in the following cases, and give the densities of the limit laws.
 (a) X_1 has four possible values, each with probability $1/4$: $(1, 0)$, $(0, 1)$, $(-1, 0)$ and $(0, -1)$.
 (b) X_1 has 9 possible values, each with probability $1/9$, (k, m) where each of k and m is $-1, 0$, or 1.
 (c) X_1 has law P where $P((-1, 0)) = P((1, 0)) = 1/3$ and $P((-1, 1)) = P((1, -1)) = 1/6$. Find the eigenvectors of the covariance matrix in this case.

3. Let X_1, X_2, \ldots, be i.i.d. in \mathbb{R}^k, with $EX_1 = 0$ and $0 < E|X_1|^2 < \infty$. Let u_j be i.i.d. and independent of all the X_i with $P(u_1 = 1) = p = 1 - P(u_1 = 0)$ and $0 < p < 1$. Let $Y_j = u_j X_j$ for all j and $T_n := Y_1 + \cdots + Y_n$. What is the relation between the limits of the laws of $S_n / n^{1/2}$ and of $T_n / n^{1/2}$ as $n \to \infty$?

4. Let X_1, X_2, \ldots, be i.i.d. and uniformly distributed on the interval $[-1, 1]$, having density $1_{[-1,1]}/2$.
 (a) Let $S_n = X_1 + \cdots + X_n$. Find the densities of S_2 and S_3 (explicitly, not just as convolution integrals).
 (b) Find the limit of the laws of $S_n / n^{1/2}$ as $n \to \infty$.

5. Let an experiment have three possible disjoint outcomes A, B, and C with non-zero probabilities p, r, and s, respectively, $p + r + s = 1$. In n independent repetitions of the experiment, let n_A be the number of times A occurs, etc. Show that as $n \to \infty$, $\mathcal{L}((n_A - np, n_B - nr, n_C - ns)/n^{1/2})$ converges. Find the density of the limit law with respect to the measure μ in the plane $x + y + z = 0$ given by $\mu = \lambda^2 \circ T^{-1}$ where $T(x, y) = (x, y, -x - y)$.

6. Let P be a law with $P(F) = 1$ for some finite set F. Suppose that $P = \mu * \rho$ for some laws μ and ρ. Show that there are finite sets G and H with $\mu(G) = \rho(H) = 1$.

7. (a) Find the characteristic function of the law on \mathbb{R} with density $\max(1 - |x|, 0)$.

 (b) Show that $g(t) := \max(1 - |t|, 0)$ is the characteristic function of some law on \mathbb{R}. *Hint:* Use (a) and Theorem 9.5.4.

8. Let h be a continuous function on \mathbb{R} which is periodic of period 2π, so that $h(t) = h(t + 2\pi)$ for all t, with $h(0) = 1$. Show that h is a characteristic function of a law P on \mathbb{R} if and only if all the *Fourier coefficients* $a_n = \frac{1}{2\pi} \int_0^{2\pi} h(t)e^{int}\, dt$ are nonnegative, $n = \ldots -1, 0, 1, \ldots$. Then show that $P(\mathbb{Z}) = 1$. (Assume the result of Problem 6 of §5.4.)

9. Let g be as in Problem 7(b) ("tent function"). Let h be the function periodic of period 2 on \mathbb{R} with $h(t) = g(t)$ for $|t| \leq 1$ ("sawtooth function"). Show that h is a characteristic function. *Hint:* Apply Problem 8 with a change of scale from 2π to 2. *Note:* Then g and h are two different characteristic functions which are equal on $[-1, 1]$.

10. The laws δ_x will be called *point masses*. Call a probability law P on \mathbb{R} *indecomposable* if the only ways to represent P as a convolution of two laws, $P = \mu * \rho$, are to take μ or ρ as a point mass. Let $P = p\delta_u + r\delta_v + s\delta_w$ where $u < v < w$. Under what conditions on p, r, s, u, v, and w will P be indecomposable? *Hint:* If P is decomposable, show that $P = \mu * \rho$ where μ and ρ are each concentrated in two points, and $v = (u + w)/2$. Then the conditions on p, r, s are those for a quadratic polynomial to be a product of linear polynomials with nonnegative coefficients.

11. Let f be the characteristic function of a law P on \mathbb{R} and $|f(s)| = |f(t)| = 1$ where s/t is irrational and $t \neq 0$. Prove that $P = \delta_c$ for some c. *Hint:* Show that $e^{isx} = f(s)$ for P–almost all x. Consult the proof of Proposition 9.5.10.

12. Let H be a k-dimensional real Hilbert space with an orthonormal basis

e_1, \ldots, e_k, so for each $x \in H$, $x = x_1 e_1 + \cdots + x_k e_k$. Define a Lebesgue measure λ_H on H by $d\lambda_H = dx_1 \cdots dx_k$. Show that λ_H does not depend on the choice of basis (a corollary of Theorem 9.5.11). Let J be another k-dimensional real Hilbert space and T a linear transformation from H onto J. Define $|\det T|$ as the absolute value of the determinant of the matrix representing T for some choice of orthonormal bases of H and J. Show that $\lambda_H \circ T^{-1} = |\det T| \lambda_J$, so that $|\det T|$ does not depend on the choice of basis. *Hints:* Define T' from J into H such that $(Tx, y)_J = (x, T'y)_H$ for all $x \in H$ and $y \in J$. Obtain a basis of H for which $T'T$ is diagonalized. Recall from linear algebra that $\det(AB) = (\det A)(\det B)$ for any $k \times k$ matrices A and B.

13. For any measurable complex-valued function f on \mathbb{R} such that $\int_{-\infty}^{\infty} |f| + |f|^2 \, dx$ is finite, let $(Tf)(y) = (2\pi)^{-1/2} \int_{-\infty}^{\infty} f(x) e^{ixy} \, dx$. Show that the domain and range of T are dense subsets of complex $L^2(\mathbb{R}, \lambda)$, with $\int (Tf)(y)(Tg)(y)^* \, dy = \int f(x) g(x)^* \, dx$ for any f and g in the domain of T, where $*$ denotes complex conjugate. *Hint:* Prove the equation for linear combinations of normal densities, then use them to approximate other functions. (This is a "Plancherel theorem.")

14. Show that X_1 and X_2 may be normally distributed but not jointly normal. *Hint:* Let the joint distribution give measure $1/2$ each to the lines $x_1 = x_2$ and $x_1 = -x_2$.

15. If X_1, \ldots, X_k are independent with distribution $N(0, 1)$, so that $X = (X_1, \ldots, X_k)$ has distribution $N(0, I)$ on \mathbb{R}^k, then $\chi_k^2 := |X|^2 = X_1^2 + \cdots + X_k^2$ is said to have a "chi-squared distribution with k degrees of freedom." Show that $\chi_2^2/2$ has a standard exponential distribution (with density e^{-t} for $t > 0$ and 0 for $t < 0$). *Hint:* Use polar coordinates.

9.6. Triangular Arrays and Lindeberg's Theorem

It is often assumed that small measurement errors are normally distributed, at least approximately. It is not clear that such errors would be approximated by sums of i.i.d. variables. Much more plausibly, the errors might be sums of a number of terms which are independent but not necessarily identically distributed. This section will prove a central limit theorem for independent, nonidentically distributed variables satisfying some conditions.

A *triangular array* is an indexed collection of random variables X_{nj}, $n = 1, 2, \ldots, j = 1, \ldots, k(n)$. A *row* in the array consists of the X_{nj} for a given value of n. The *row sum* S_n is here defined by $S_n := \sum_{1 \le j \le k(n)} X_{nj}$. The aim is to prove that the laws $\mathcal{L}(S_n)$ converge as $n \to \infty$ under suitable

conditions. Here it will be assumed that the random variables are real-valued
and independent within each row. Random variables in different rows need
not be independent and even could be defined on different probability spaces,
since only convergence in law is at issue. The normalized partial sums of
§9.5 can be considered in terms of triangular arrays by setting $k(n) := n$ and
$X_{nj} := X_j/n^{1/2}, j = 1, \ldots, n, n = 1, 2, \ldots$.

If each row contained only one nonzero term (either $k(n) = 1$ or $X_{nj} = 0$
for $j \neq i$ for some i), then $\mathcal{L}(S_n)$ could be arbitrary. To show that $\mathcal{L}(S_n)$
converges, assumptions will be made which imply that for large n, any single
term X_{nj} will make only a relatively small contribution to the sum S_n of the
nth row. Thus the length $k(n)$ of the rows must go to ∞. This is what gives
the array a "triangular" shape, especially for $k = n$. Here is the extension of
the central limit theorem (9.5.6) to triangular arrays.

9.6.1. Lindeberg's Theorem *Assume that for each fixed* $n = 1, 2, \ldots, X_{nj}$
are independent real random variables for $j = 1, \ldots, k(n)$, *with* $EX_{nj} =$
$0, \sigma_{nj}^2 := EX_{nj}^2$ *and* $\sum_{1 \leq j \leq k(n)} \sigma_{nj}^2 = 1$. *Let* $\mu_{nj} := \mathcal{L}(X_{nj})$. *For any* $\varepsilon > 0$ *let*

$$E_{nj\varepsilon} := \int_{|x|>\varepsilon} x^2 \, d\mu_{nj}(x).$$

Assume that $\lim_{n\to\infty} \sum_{1 \leq j \leq k(n)} E_{nj\varepsilon} = 0$ *for each* $\varepsilon > 0$. *Then as* $n \to \infty$,
$\mathcal{L}(S_n) \to N(0, 1)$.

Proof. Since all the S_n have mean 0 and variance 1, the laws $\mathcal{L}(S_n)$ are uni-
formly tight by Chebyshev's inequality, as in the proof of Theorem 9.5.6.
Thus by Lemma 9.5.5, it will be enough to prove that the characteristic func-
tion of S_n converges pointwise to that of $N(0, 1)$. We have by independence
(Theorem 9.4.3)

$$E \exp(itS_n) = \prod_{1 \leq j \leq k(n)} E \exp(itX_{nj}).$$

Below, $\theta_r, r = 1, 2, \ldots$, denote complex numbers such that $|\theta_r| \leq 1$ and which
may depend on the other variables. We have the following partial Taylor
expansions (see Appendix B): for all real u,

$$e^{iu} = 1 + iu + \theta_1 u^2/2 = 1 + iu - u^2/2 + \theta_2 |u|^3/6$$

by Corollary B.4, since for $f(u) := e^{iu}, |f^{(n)}(u)| = |i^n e^{iu}| \equiv 1$. For all com-
plex z with $|z| \leq 1/2$, for the principal branch plog of the logarithm (see
Appendix B, Example (b) after Corollary B.4)

$$\text{plog}(1 + z) = z + \theta_3 z^2. \tag{9.6.2}$$

Take an $\varepsilon > 0$. Let $\theta_4 := \theta_2(tx)$ and $\theta_5 := \theta_1(tx)$. Then for each n and j,

$$E \exp(it X_{nj}) = \int 1 + itx \, d\mu_{nj}(x) + \int_{|x| \leq \varepsilon} -t^2 x^2/2 + \theta_4 |tx|^3/6 \, d\mu_{nj}(x)$$

$$+ \int_{|x| > \varepsilon} \theta_5 x^2 t^2/2 \, d\mu_{nj}(x).$$

The first integral equals 1 since $E X_{nj} = 0$. In the second integral, the integral of the first term equals $-t^2(\sigma_{nj}^2 - E_{nj\varepsilon})/2$. The second term, since $|x|^3 \leq \varepsilon x^2$ for $|x| \leq \varepsilon$, has absolute value at most $|t|^3 \sigma_{nj}^2 \varepsilon/6$. The last integral has absolute value at most $t^2 E_{nj\varepsilon}/2$. It follows that

$$E \exp(it X_{nj}) = 1 - t^2\left(\sigma_{nj}^2 - E_{nj\varepsilon}(1 - \theta_6)\right)/2 + \theta_7 |t|^3 \varepsilon \sigma_{nj}^2/6.$$

Since $0 \leq E_{nj\varepsilon} \leq \sigma_{nj}^2$, $|\sigma_{nj}^2 - E_{nj\varepsilon}(1 - \theta_6)| \leq \sigma_{nj}^2 - E_{nj\varepsilon} + E_{nj\varepsilon} = \sigma_{nj}^2$.

We have $\max_j \sigma_{nj}^2 \leq \varepsilon^2 + \sum_j E_{nj\varepsilon}$. For any real t, we can thus choose ε small enough and then n large enough so that for all j, and $z_{nj} := E \exp(it X_{nj}) - 1$, we have $|z_{nj}| \leq 1/2$. Then applying (9.6.2), for each fixed t, we have as $n \to \infty$, letting $o(1)$ denote any term that converges to 0,

$$E \exp\left(it \sum_{j=1}^{k(n)} X_{nj}\right) = \exp(-t^2(1 + o(1))/2 + \theta_8 |t|^3 \varepsilon/6$$

$$+ \theta_9 \sum_{j=1}^{k(n)} \left[\theta_{10} t^2 \sigma_{nj}^2(1 + |t|\varepsilon/6)\right]^2\big).$$

Now

$$\sum_{j=1}^{k(n)} \sigma_{nj}^4 \leq \left(\varepsilon^2 + \sum_j E_{nj\varepsilon}\right) \sum_j \sigma_{nj}^2 = \varepsilon^2 + o(1),$$

which substituted in the previous expression yields

$$\exp(-t^2/2 + o(1) + \theta_8 \varepsilon |t|^3/6 + \theta_{11} t^4 [(\varepsilon^2 + o(1))(1 + |t|\varepsilon/6)^2]).$$

Since $\varepsilon > 0$ was arbitrary, we can let $\varepsilon \downarrow 0$ and obtain $\exp(-t^2/2 + o(1))$. So $E \exp(it S_n) \to \exp(-t^2/2)$ as $n \to \infty$. $\qquad\square$

The central limit theorem for i.i.d. variables X_1, X_2, \ldots, in \mathbb{R} with $E X_1^2 < \infty$, previously proved in Theorem 9.5.6, will now be proved from Lindeberg's theorem as an example of its application. For simplicity suppose $E X_1^2 = 1$.

Let $k(n) = n$ and $X_{nj} = X_j/n^{1/2}$. Then $\sigma_{nj}^2 = 1/n$, $\sum_{1 \le j \le n} \sigma_{nj}^2 = 1$, and for $\varepsilon > 0$,

$$E_{nj\varepsilon} := \int_{|x| > \varepsilon} x^2 \, d\mathcal{L}(X_{nj})(x) = E X_{nj}^2 1_{\{|X_{nj}| > \varepsilon\}} = E X_1^2 1_{\{|X_1| > \varepsilon\sqrt{n}\}}/n,$$

so $\displaystyle\sum_{j=1}^n E_{nj\varepsilon} = E X_1^2 1_{\{|X_1| > \varepsilon\sqrt{n}\}} \to 0$ as $n \to \infty$

by dominated convergence.

The nth convolution power of a finite measure μ is defined by $\mu^{*n} = \mu * \mu \cdots * \mu$ (to n factors), with μ^{*0} defined as δ_0 where $\delta_0(A) := 1_A(0)$ for any set A. So for a probability measure μ, μ^{*n} is the law of $S_n = X_1 + \cdots + X_n$ where X_1, \ldots, X_n are i.i.d. with law μ. The *exponential* of a finite measure μ is defined by $\exp(\mu) := \sum_{n \ge 0} \mu^{*n}/n!$. These notions will be explored in the problems.

For $\lambda > 0$, the Poisson distribution P_λ on \mathbb{N} is defined by $P_\lambda(k) := e^{-\lambda}\lambda^k/k!$, $k = 0, 1, \ldots$. Poisson distributions arise as limits in some triangular arrays (Problem 2), especially for binomial distributions (Problem 3), and have been applied to data on, among other things, radioactive disintegrations, chromosome interchanges, bacteria counts, and events in a telephone network; see, for example, Feller (1968, pp. 159–164).

Problems

1. Let $X_{nk} = k s_k/n$ for $k = 1, \ldots, n$, where s_k are i.i.d. variables with $P(s_k = -1) = P(s_k = 1) = 1/2$. Find numbers σ_n such that Lindeberg's theorem applies to the variables X_{nk}/σ_n, $k = 1, \ldots, n$.

2. Let X_{nj} be a triangular array of random variables, independent for each n, $j = 1, \ldots, k(n)$, with $X_{nj} = 1_{A(n,j)}$ for some events $A(n, j)$ with $P(A(n, j)) = p_{nj}$. Let S_n be the sum of the nth row. If as $n \to \infty$, $\max_j p_{nj} \to 0$ and $\sum_j p_{nj} \to \lambda$, prove that $\mathcal{L}(S_n) \to P_\lambda$. *Hint:* Use characteristic functions.

3. Assuming the result of Problem 2, show that as $n \to \infty$ and $p = p_n \to 0$ so that $np_n \to \lambda$, the binomial distribution with $B_{n,p}(k) := \binom{n}{k} p^k (1 - p)^{n-k}$, $k = 0, 1, \ldots, n$, converges to P_λ.

4. Show that for any $\lambda > 0$, $e^{-\lambda} \exp(\lambda \delta_1)$ is the Poisson law P_λ. (So Poisson laws are multiples of exponentials of some of the simplest nonzero measures; on the other hand, the exponential gives the result of the limit operation in Problems 2 and 3.)

5. Show that for any finite (nonnegative) measure on \mathbb{R}, $e^{-\mu(\mathbb{R})}\exp(\mu)$ is a probability measure on \mathbb{R}. Find its characteristic function in terms of the function $f(t) = \int e^{ixt}d\mu(x)$.

6. (Continuation.) Show that for any two finite measures μ and ν on \mathbb{R}, $\exp(\mu + \nu) = \exp(\mu) * \exp(\nu)$.

7. A probability law P on \mathbb{R} is called *infinitely divisible* iff for each $n = 1, 2, \ldots$, there is a law P_n whose nth convolution power $P_n^{*n} = P$. Show that any normal law is infinitely divisible.

8. Show that for any finite measure μ on \mathbb{R}, the law $e^{-\mu(\mathbb{R})}\exp(\mu)$ of Problem 5 is infinitely divisible.

9. A law P on \mathbb{R} is called *stable* iff for every $n = 1, 2, \ldots$, $P^{*n} = P \circ A_n^{-1}$ for some affine function $A_n(x) = a_n x + b_n$. Show that all normal laws are stable.

10. Show that all stable laws are infinitely divisible. (There will be more about infinitely divisible and stable laws in §9.8.)

11. *Permutations.* For each $n = 1, 2, \ldots$, a *permutation* is a 1–1 function from $\{1, \ldots, n\}$ into itself. The set of all such permutations for a given n is called the *symmetric group* (on n "letters"). This group is usually called S_n and will here be called \mathcal{S}_n. It has $n!$ members. Let μ_n be the uniform law on it, with $\mu_n(\{s\}) = 1/n!$ for each single permutation s. For each $s \in \mathcal{S}_n$, consider the sequence $a_1 := 1$, and $a_j := s(a_{j-1})$ for each $j = 2, 3, \ldots$, as long as $a_j \neq a_i$ for all $i < j$. Let $J(1)$ be the largest j for which this is true.

 (a) Show that $s(a_{J(1)}) = 1$. The numbers $a_1, \ldots, a_{J(1)}$ are said to form a *cycle* of the permutation s. If $J(1) < n$, let i be the smallest number in $\{1, \ldots, n\}\setminus\{a_1, \ldots, a_{J(1)}\}$ and let $a_{J(1)+1} = i$, then apply s repeatedly to form a second cycle ending in $a_{J(2)}$ with $s(a_{J(2)}) = i$, and repeat the process until a_1, \ldots, a_n are defined and $J(k) = n$ for some k. Let $Y_{nr}(s) = 1$ if $r = J(m)$ for some m (a cycle is completed at the rth step), otherwise $Y_{nr}(s) = 0$.

 (b) Show that for each n, the random variables Y_{nr} on (\mathcal{S}_n, μ_n) are independent for $r = 1, \ldots, n$.

 (c) Find EY_{nr} and $\text{var}(Y_{nr})$ for each n and r. Let $T_n := \sum_{1 \leq r \leq n} Y_{nr}$. Then $T_n(s)$ is just the number of different cycles in a permutation s. Let $\sigma(T_n) := (\text{var}(T_n))^{1/2}$. Let $X_{nj} := (Y_{nj} - EY_{nj})/\sigma(T_n)$.

 (d) Show that the Lindeberg theorem applies to the X_{nj}, and so find constants a_n and b_n such that $\mathcal{L}((T_n - a_n)/b_n) \to N(0, 1)$ as $n \to \infty$, with $a_n = (\log n)^\alpha$ and $b_n = (\log n)^\beta$ for some α and β.

9.7. Sums of Independent Real Random Variables

When a series of real numbers is said to *converge*, without specifying the sum, it will mean that the sum is finite. A convergent sum of random variables will be finite a.s. Convergence of a sum $\sum_{n\geq 1} X_n$ of random variables, in a given sense (such as a.s., in probability or in law), will mean convergence of the sequence of partial sums $S_n := X_1 + \cdots + X_n$ in the given sense. For independent summands, we have:

9.7.1. Lévy's Equivalence Theorem For a series of independent real-valued random variables $X_1, X_2, \ldots,$ the following are equivalent:

(I) $\sum_{n=1}^{\infty} X_n$ converges a.s.
(II) $\sum_{n=1}^{\infty} X_n$ converges in probability.
(III) $\sum_{n=1}^{\infty} X_n$ converges in law, in other words, $\mathcal{L}(S_n)$ converges to some law μ on \mathbb{R} as $n \to \infty$.

If these conditions fail, then $\sum_{n=1}^{\infty} X_n$ diverges a.s.

Remark. If the X_j are always nonnegative, so that the S_n are nondecreasing in n, equivalence of the three kinds of convergence holds without the independence (the details are left as Problem 1). So the independence will be needed only when the X_j may have different signs for different j.

Proof. (I) implies (II) and (II) implies (III) for general sequences (Theorem 9.2.1, Proposition 9.3.5). Next, (III) implies (II): assuming (III), the laws $\mathcal{L}(S_n)$ are uniformly tight by Proposition 9.3.4. Thus for any $\varepsilon > 0$, there is an $M < \infty$ such that $P(|S_n| > M) < \varepsilon/2$ for all n. Then $P(|S_n - S_k| > 2M) < \varepsilon$ for all n and k, so the set of all $\mathcal{L}(S_n - S_k)$ is uniformly tight. To prove (II), it is enough by Theorems 9.2.2 and 9.2.3 to prove that $\{S_n\}$ is a Cauchy sequence for the Ky Fan metric α. If not, there is a $\delta > 0$ and a subsequence $n(i)$ such that $\alpha(S_{n(i)}, S_{n(i+1)}) > \delta$ for all i. Let $Y_i := S_{n(i+1)} - S_{n(i)}$. Then the laws $\mathcal{L}(Y_i)$ are uniformly tight, so by Theorem 9.3.3, they have a subsequence $Q_j := \mathcal{L}(Y_{i(j)})$ converging to some law Q. Note that $\alpha(Y_i, 0) > \delta$ for all i, so $P(|Y_i| > \delta) > \delta$. Take f to be continuous with $f(0) = 0, 0 \leq f \leq 1$, and $f(x) = 1$ for $|x| > \delta$. Then $Ef(Y_i) > \delta$, so $Ef(Y_{i(j)})$ does not converge to 0, and $Q \neq \delta_0$.

Let $P_j := \mathcal{L}(S_{n(i(j))})$. Then $P_j \to \mu$, $P_j * Q_j \to \mu$, and by continuity of convolution (Theorem 9.5.9), $P_j * Q_j \to \mu * Q$, so (by Lemma 9.3.2) $\mu = \mu * Q$, but this contradicts Proposition 9.5.10, so the proof that (III) implies (II) is complete.

Norms were defined in general after Theorem 5.1.5. The following inequality about norms of random variables will be used in the proof that (II) implies (I). The statement and proof extend even to infinite-dimensional normed spaces, but actually the result is needed here only for the usual Euclidean norm $\|x\| = (x_1^2 + \cdots + x_k^2)^{1/2}$.

9.7.2. Ottaviani's Inequality Let X_1, X_2, \ldots, X_n be independent random variables with values in \mathbb{R}^k and $S_j := X_1 + \cdots + X_j$, $j = 1, \ldots, n$. Let $\|\cdot\|$ be any norm on R^k. Suppose that for some $\alpha > 0$ and $c < 1$, $\max_{j \le n} P(\|S_n - S_j\| > \alpha) \le c$. Then

$$P\left(\max_{j \le n} \|S_j\| \ge 2\alpha\right) \le (1-c)^{-1} P(\|S_n\| \ge \alpha).$$

Proof. Let $m := m(\omega)$ be the least $j \le n$ such that $\|S_j\| \ge 2\alpha$, or $m := n+1$ if there is no such j. Then

$$P(\|S_n\| \ge \alpha) \ge P\left(\|S_n\| \ge \alpha, \max_{j \le n} \|S_j\| \ge 2\alpha\right) = \sum_{j=1}^{n} P(\|S_n\| \ge \alpha, m = j)$$

$$\ge \sum_{j=1}^{n} P(\|S_n - S_j\| \le \alpha, m = j)$$

since $m = j$ implies $\|S_j\| \ge 2\alpha$, which, with $\|S_n - S_j\| \le \alpha$, implies $\|S_n\| \ge \alpha$. As $\|S_n - S_j\|$ is a function of X_{j+1}, \ldots, X_n, it is independent of the event $\{m = j\}$, which is a function of X_1, \ldots, X_j. So the last sum equals

$$\sum_{j=1}^{n} P(\|S_n - S_j\| \le \alpha) P(m = j) \ge (1-c) \sum_{j=1}^{n} P(m = j)$$

$$= (1-c) P\left(\max_{j \le n} \|S_j\| \ge 2\alpha\right), \qquad \text{proving the inequality.} \qquad \square$$

Now assuming (II), let $c = 1/2$. Given $0 < \varepsilon < 1$, take K large enough so that for all $n \ge K$, $P(|S_n - S_K| \ge \varepsilon/2) < \varepsilon/2$. Apply Ottaviani's inequality with $\alpha = \varepsilon/2$ to the variables X_{K+1}, X_{K+2}, \ldots. Then for any $n \ge K$, $P(\max_{K \le j \le n} \|S_j - S_K\| \ge \varepsilon) \le \varepsilon$. Here we can let $n \to \infty$, then $\varepsilon \downarrow 0$ and $K \to \infty$. By Lemma 9.2.4, the original series converges a.s., proving (I), so (I) to (III) are equivalent.

Lastly, $\sum_{n=1}^{\infty} X_n$ either converges a.s. or diverges a.s. by the Kolmogorov 0–1 law (8.4.4), so Theorem 9.7.1 is proved. $\qquad \square$

Now, more concrete conditions for convergence will be brought in. For any real random variable X, the truncation of X at ± 1 will be defined by

$$X^1 := \begin{cases} X & \text{if } |X| \le 1 \\ 0 & \text{if } |X| > 1. \end{cases} \qquad \text{or}$$

Recall that for any real random variable X, EX is defined and finite if and only if $E|X| < \infty$, and the variance $\sigma^2(X)$ is defined, if $EX^2 < \infty$, by $\sigma^2(X) := E((X - EX)^2) = EX^2 - (EX)^2$. Here is a criterion:

9.7.3. Three-Series Theorem *For a series of independent real random variables X_j as in Theorem 9.7.1, convergence (almost surely, or equivalently in probability or in law) is equivalent to the following condition:*

(IV) All three of the following series (of real numbers) converge:

(a) $\displaystyle\sum_{n=1}^{\infty} P(|X_n| < 1).$

(b) $\displaystyle\sum_{n=1}^{\infty} EX_n^1.$

(c) $\displaystyle\sum_{n=1}^{\infty} \sigma^2\left(X_n^1\right).$

Note. In (b), absolute convergence—that is, convergence of $\sum_n |EX_n^1|$—is not necessary. For example, the series $\sum_n (-1)^n/n$ of constants converges and satisfies (a), (b), and (c).

Proof. Since (I) to (III) are equivalent in Theorem 9.7.1, it will be enough to prove that (IV) implies (II) and then that (I) implies (IV).

Assume (IV). Then by (a) and the Borel-Cantelli lemma (8.3.4), $P(X_n \neq X_n^1$ for some $n \ge m) \to 0$ as $m \to \infty$. So $\sum_n X_n$ converges if and only if $\sum_n X_n^1$ converges. Thus in proving convergence in probability, we can assume that $X_n = X_n^1$, in other words, $|X_n(\omega)| \le 1$ for all n and ω. Then, by series (b), the sum of the EX_n converges. Without affecting the convergence, divide all the X_n by 2, so that $|X_n| \le 1/2$ and $|EX_n| \le 1/2$. Now $\sum_n X_n$ converges if and only if $\sum_n X_n - EX_n$ converges. Taking $X_n - EX_n$, one can assume that $EX_n = 0$ for all n. Then by series (c), $\sum_n EX_n^2 < \infty$, and still $|X_n| \le 1$ for all n and ω.

For any $n \ge m$ we have, by independence (Theorem 8.1.2),

$$E((S_n - S_m)^2) = \sum_{m < j \le n} EX_j^2 \to 0 \qquad \text{as } m \to \infty.$$

Then by Chebyshev's inequality (8.3.1), for any $\varepsilon > 0$, $P(|S_n - S_m| > \varepsilon) \to 0$, so that the series converges in probability (using Theorem 9.2.3), so (II) holds.

Now to show that (I) implies (IV), assume (I). Then by the Borel-Cantelli lemma, here using independence, series (a) converges, and again one can take $|X_n| \le 1$. If (c) diverges, then for each $m \le n$, let $S_{mn} := \sum_{m \le j \le n} X_j$. Choose m large enough so that for all $n \ge m$, $P(|S_{mn}| > 1) < 0.1$. Let

$$\sigma_{mn} := (\text{var}(S_{mn}))^{1/2} = \left(\sum_{m \le j \le n} \sigma^2(X_j) \right)^{1/2} \to \infty \qquad \text{as } n \to \infty.$$

Let $T_{mn} := (S_{mn} - E S_{mn})/\sigma_{mn}$, or 0 if $\sigma_{mn} = 0$. For each m, $\mathcal{L}(T_{mn}) \to N(0, 1)$ as $n \to \infty$ by Lindeberg's theorem (9.6.1), where $E_{nj\varepsilon} = 0$ for n large since $|X_j - E X_j| \le 2$ and the sum of the variances diverges. The plan now is to show that since $\sum_{m \le j \le n} X_j$ is approximately normal with large variance, it cannot be small in probability. For converging laws, the distribution functions converge where the limit function is continuous (Theorem 9.3.6), so as $n \to \infty$,

$$P(S_{mn} \ge E S_{mn} + \sigma_{mn}) = P(T_{mn} \ge 1) \to N(0, 1)([1, \infty)) > .15 > .1.$$

Then by choice of m, $E S_{mn} \le -\sigma_{mn} + 1 \to -\infty$ as $n \to \infty$. Likewise,

$$P(S_{mn} \le E S_{mn} - \sigma_{mn}) = P(T_{mn} \le -1) \to N(0, 1)((-\infty, -1]) > .15 > .1$$

implies $E S_{mn} \to +\infty$, which is a contradiction. So series (c) converges. Now consider the series $\sum_n (X_n - E X_n)/2$. For it, all three series converge. Since, as previously shown, (IV) implies (II) implies (I), it follows that $\sum_n X_n - E X_n$ converges a.s. Then by subtraction, $\sum_n E X_n$ converges, so series (b) converges, (IV) follows, and the proof of the three-series theorem is complete. \square

Although it turned out not to be needed in the above proof, the following interesting improvement on Chebyshev's inequality was part of Kolmogorov's original proof of the three-series theorem.

9.7.4. Kolmogorov's Inequality If X_1, \ldots, X_n are independent real random variables with mean 0, $\varepsilon > 0$, and $S_k := X_1 + \cdots + X_k$, then

$$P\left(\max_{1 \le k \le n} |S_k| > \varepsilon \right) \le \varepsilon^{-2} \sum_{j=1}^{n} \sigma^2(X_j).$$

Proof. If $ES_n^2 = +\infty$, the inequality holds since the right side is infinite, so suppose $ES_n^2 < \infty$. Let A_n be the event $\{\max_{k \le n} |S_k| > \varepsilon\}$. For $k = 1, \ldots, n$, let $B(k)$ be the event $\{\max_{j < k} |S_j| \le \varepsilon < |S_k|\}$. Then A_n is the union of the disjoint events $B(k)$. By independence, $ES_k 1_{B(k)}(S_n - S_k) = 0$, and $1_{B(k)} = 1^2_{B(k)}$, so

$$\int_{B(k)} S_n^2 \, dP = E\big((S_k 1_{B(k)})^2\big) + E\big(\{(S_n - S_k)1_{B(k)}\}^2\big)$$

$$\ge ES_k^2 1_{B(k)} \ge \varepsilon^2 P(B_k). \qquad \text{Hence,}$$

$$P(A_n) = \sum_{k=1}^n P(B(k)) \le \varepsilon^{-2} \sum_{k=1}^n \int_{B(k)} S_n^2 \, dP \le \varepsilon^{-2} ES_n^2. \qquad \square$$

Problems

1. Prove the remark after Theorem 9.7.1, that for any nonnegative random variables X_j (not necessarily independent) the three kinds of convergence of the series $\sum_j X_j$ are equivalent.

2. If X_1, X_2, \ldots, are independent random variables, $\sum_n EX_n$ converges, and $\sum_n \sigma^2(X_n) < \infty$, show that $\sum_n X_n$ converges a.s. *Hint:* Don't truncate (i.e., don't consider X_n^1); apply part of the proof, rather than the statement, of the three-series theorem, and 9.7.4.

3. Let $A(n)$ be independent events with $P(A(n)) = 1/n$. Under what conditions, if any, on constants c_n does $\sum_n 1_{A(n)} - c_n$ converge a.s.?

4. Let X_1, X_2, \ldots, be i.i.d. real random variables with $EX_1 = 0$ and $0 < EX_1^2 < \infty$. Under what conditions on constants c_n does $\sum_n c_n X_n$ converge?

5. Let X_n be i.i.d. with the Cauchy distribution $P(X_n \in A) = \pi^{-1} \int_A dx / (1 + x^2)$. Under what conditions on constants a_n does $\sum_n a_n X_n$ converge a.s.? You can use the fact that the Cauchy distribution has characteristic function $e^{-|t|}$.

6. Let X_j be random variables with $EX_j^2 \le M < \infty$ for all j and $EX_j = EX_iX_j = 0$ for $i \ne j$. Show that X_j satisfy the strong law of large numbers. *Hints:*
 (a) Let $s(n) := n^2$. Prove $S_{s(n)}/n^2 \to 0$ a.s.
 (b) Let $D_n := \max\{|S_k - S_{s(n)}| : n^2 \le k < (n+1)^2\}$. Show that $D_n/n^2 \to 0$ a.s.
 (c) Combine (a) and (b) to finish the proof.

7. Let $\sum_j x_j$ be a convergent series of real numbers and b_n a sequence of positive numbers with $b_n \uparrow +\infty$ as $n \to \infty$. Show that $\lim_{n\to\infty}(\sum_{1\le k\le n} b_k x_k)/b_n = 0$, as follows.
 (a) If $x_k \ge 0$ for all k, show that this follows from dominated convergence.
 (b) Hints for the general case: $\sum_{1\le k\le n} b_k x_k = b_1(x_1 + \cdots + x_n) + (b_2 - b_1)(x_2 + \cdots + x_n) + \cdots + (b_n - b_{n-1})x_n$ (a kind of summation by parts). The partial sums $\sum_{1\le j\le n} x_j$ are bounded in absolute value, say by M. Given $\varepsilon > 0$, choose r large enough so that $|x_r + \cdots + x_n| < \varepsilon/2$ for all $n \ge r$. Then for n large enough so that $Mb_r/b_n < \varepsilon/2$ show that $|\sum_{1\le k\le n} b_k x_k|/b_n < \varepsilon$.

8. Let $X_1, X_2, \ldots,$ be independent real random variables with $EX_k = 0$ and $EX_k^2 < \infty$ for all k. Let $0 < b_n \uparrow +\infty$. If $\sum_k EX_k^2/b_k^2 < \infty$, show that $S_n/b_n \to 0$ a.s. *Hint:* First show, by the results of this section, that $\sum_k X_k/b_k$ converges a.s., then apply the result of Problem 7.

9. If $X_1, X_2, \ldots,$ are i.i.d. with $EX_1 = 0$ and $EX_1^2 < \infty$, give a proof that $S_n/(n^{1/2}(\log n)^{.5+\delta}) \to 0$ a.s. for any $\delta > 0$. (Use the result of Problem 8.) Note that this is in one way an improvement on the strong law of large numbers (8.3.5), although EX_k^2 rather than just $E|X_k|$ must be finite. (§12.5 gives still sharper b_n, with $\log n$ replaced by $\log \log n$.)

*9.8. The Lévy Continuity Theorem; Infinitely Divisible and Stable Laws

The Lévy continuity theorem says that if characteristic functions converge (to a continuous function), then the corresponding laws also converge. This was proved in Lemma 9.5.5 under the assumption that the sequence of laws was uniformly tight. But here it will be shown that the convergence of characteristic functions to a continuous limit implies the uniform tightness. The proof will be based on the following.

9.8.1. Truncation Inequality Let P be a probability measure on \mathbb{R} with characteristic function $f(t) := \int e^{ixt} dP(x)$. Then for any u with $0 < u < \infty$,

$$P\left(|x| \ge \frac{1}{u}\right) \le \frac{7}{u}\int_0^u 1 - \mathrm{Re}\, f(v)\, dv.$$

Proof. Let us first prove a

Claim. For any t with $|t| \ge 1$, we have $(\sin t)/t \le \sin 1$.

Proof. We may assume $t \geq 0$ since $(\sin t)/t$ is an even function. Now $\sin 1 \approx 0.84 > 0.8$, so the claim holds for $|t| \geq 1.3$, while for $1 \leq t \leq 1.3$, $(\sin t)/t$ is decreasing, proving the claim. □

Then by the Tonelli-Fubini theorem (4.4.5),

$$\frac{1}{u} \int_0^u \int_{-\infty}^{\infty} 1 - \cos(vx)\, dP(x)\, dv = \int_{-\infty}^{\infty} 1 - \frac{\sin(ux)}{ux}\, dP(x)$$
$$\geq (1 - \sin 1) P(|x| \geq 1/u).$$

Since $\sin 1 < 6/7$, this yields (9.8.1). □

9.8.2. Lévy Continuity Theorem *If P_n are laws on \mathbb{R}^k whose characteristic functions $f_n(t)$ converge for all t to some $f(t)$, where f is continuous at 0 along each coordinate axis, then $P_n \underset{\mathcal{L}}{\to} P$ for a law P with characteristic function f.*

Proof. Consider the coordinate axis lines $t = (0, \ldots, 0, t_j, 0, \ldots, 0)$ for $j = 1, \ldots, k$. Given $\varepsilon > 0$, take $\delta > 0$ small enough so that $|f(t) - 1| < \varepsilon/(7k)$ for $|t| \leq \delta$ and t on the axes. Then (9.8.1) and the dominated convergence theorem imply that $P_n(|x_j| \geq 1/\delta) < \varepsilon/k$ for n large enough, say $n \geq n_0$, and all j. For each $n < n_0$, by countable additivity there is some $M_n < \infty$ such that $P_n(|x_j| \geq M_n) < \varepsilon/k$ for $j = 1, \ldots, k$. Let $M := \max(1/\delta, \max\{M_j : 1 \leq j < n_0\})$. Then $P_n(|x_j| \geq M) < \varepsilon/k$ for all n and j. If C is the cube $\{x : |x_j| \leq M$ for $j = 1, \ldots, k\}$, then $P_n(C) > 1 - \varepsilon$ for all n. So the P_n are uniformly tight. By Lemma 9.5.5, the conclusion follows. □

Definition. A law P on \mathbb{R} is called *infinitely divisible* if for every $n = 1, 2, \ldots$, there is a law P_n such that the nth convolution power $P_n^{*n} = P$. In other words, there exist i.i.d. random variables X_{n1}, \ldots, X_{nn} such that $\mathcal{L}(X_{n1} + \cdots + X_{nn}) = P$. (Let $\mathcal{L}(X_{nj}) = P_n$ for each j.)

For example, normal laws are infinitely divisible since $N(m, \sigma^2) = N(m/n, \sigma^2/n)^{*n}$ for each n. From the properties of characteristic functions (multiplication and uniqueness), a law P with characteristic function f is infinitely divisible if and only if for each $n = 1, 2, \ldots$, there is a characteristic function f_n (of some law P_n) such that $f(t) = f_n(t)^n$ for all t. In this case the characteristic function f will be called infinitely divisible.

The following treatment of infinitely divisible and stable laws will state some of the main facts, leaving out most of the proofs. For more details see Breiman (1968, Chapter 9) and Loève (1977, §23).

The Poisson law P_λ on \mathbb{N} with parameter λ, where $P_\lambda(k) = e^{-\lambda}\lambda^k/k!$ for $k = 0, 1, \ldots$, is infinitely divisible with $(P_{\lambda/n})^{*n} = P_\lambda$ for all n. Its characteristic function is $\exp(\lambda(e^{it} - 1))$. If a random variable X has an infinitely divisible law, so does $bX + c$ for any real b and c. So $\exp(ict + \lambda(e^{ibt} - 1))$ is an infinitely divisible characteristic function. Also, the product of any two infinitely divisible characteristic functions is infinitely divisible. It follows that $g(t) := \exp(ict + \int e^{itx} - 1 \, d\mu(x))$ is an infinitely divisible characteristic function for any finite measure μ on \mathbb{R} (first suppose μ is concentrated in one point, and then in a finite set; then take a limit by the Lévy continuity theorem).

In the converse direction, an infinitely divisible characteristic function f is never 0 (see Problem 6), so that a unique continuous version of $\log f$ can be defined with $\log f(0) = 0$. Then $f = f_n^n$ for the characteristic function f_n of some law P_n. By continuity, $f_n \equiv \exp((\log f)/n)$, so $n(f_n - 1)$ converges to $\log f$ as $n \to \infty$, and

$$n(f_n - 1)(t) = n \int_{-\infty}^{\infty} e^{itx} - 1 \, dP_n(x).$$

So f must be a limit of characteristic functions of the form of g.

In fact, it is not necessary for μ to be finite near 0: as long as $\mu\{|x| > 1\}$ is finite, and $\int_{|x|<1} |x| \, d\mu(x) < \infty$, the integral in the definition of $g(t)$ will converge and give an infinitely divisible characteristic function. Also, if limits are taken in an appropriate way, the terms ict for variable c can yield a term $-itx$ in the integrand which then allows μ near 0 to satisfy the still weaker condition $\int_{|x|<1} |x|^2 \, d\mu(x) < \infty$. Then the two conditions on μ can be combined by saying that $d\mu(x) = (1 + x^2) dG(x)/x^2$ where G is a finite measure. This (via details given in the above-cited sources) yields the following form:

9.8.3. Lévy-Khinchin Formula A characteristic function f on \mathbb{R} is infinitely divisible if and only if it is of the form $e^{h(t)}$ where

$$h(t) = ict + \int_{-\infty}^{\infty} \left(e^{itx} - 1 - \frac{itx}{1 + x^2}\right) \frac{1 + x^2}{x^2} \, dG(x)$$

where $c \in \mathbb{R}$, G is a finite, nonnegative measure on \mathbb{R}, and the integrand is defined by continuity at $x = 0$ with value $-t^2/2$.

For the characteristic function of a normal measure $N(\mu, \sigma^2)$, we have $c = \mu$, and G is a point mass of size σ^2 at 0.

A law with characteristic function f, or f itself, is called *stable* iff for all $n = 1, 2, \ldots$, there are some a_n and b_n with $f(t)^n = f(a_n t) \exp(i b_n t)$ for all t (as in §9.6, Problem 9). Symmetric stable distributions take a rather simple form:

9.8.4. Theorem *A symmetric law $P(d P(-x) \equiv d P(x))$ with characteristic function f is stable if and only if for some $c \geq 0$ and p with $0 < p \leq 2$, $f(t) = \exp(-c|t|^p)$ for all t.*

Here p is called the index of the stable law, except in the degenerate case $c = 0$ where $f \equiv 1$ and the law is a point mass. Symmetric normal laws $N(0, \sigma^2)$ are stable with $p = 2$ and $c = \sigma^2/2$. A possibly nonsymmetric stable law is said to have index p if we can take $a_n = n^{1/p}$ in the definition of stability. It turns out that every stable law has an index p, with $0 < p \leq 2$. The stable laws of index 2 are just the normal laws. A random variable X with stable law of index $p < 2$ has $E|X|^r < \infty$ for $r > 0$ if and only if $r < p$. For example, the Cauchy distribution with density $1/(\pi(1 + x^2))$ is stable of index 1, with characteristic function $e^{-|t|}$. For most values of the index, symmetric stable densities are not known in closed form. For general (possibly nonsymmetric) stable characteristic functions we have the following:

9.8.5. Theorem *A characteristic function f is stable if and only if either it is normal, or it is stable of index p, $0 < p < 2$, and then $f(t) \equiv e^{h(t)}$ where $h(t) \equiv imt - c|t|^p(1 - i\beta \, \mathrm{sgn}(t)\omega(t, p))$ for some $m \in \mathbb{R}$, $c > 0$, $|\beta| \leq 1$, and $\omega(t, p) = \tan(\pi p/2)$ if $p \neq 1$, or $-(2/\pi)\log|t|$ if $p = 1$, and $\mathrm{sgn}(t) = 1$ if $t > 0$, 0 if $t = 0$, and -1 if $t < 0$.*

Along with the previous facts on infinitely divisible and stable laws, this theorem is proved in Loéve (1977, §23). In consulting other sources, *caveat lector* (Hall, 1981).

Problems

1. Show that the functions $\cos^{2n}(t)$ are characteristic functions for all $n = 1, 2, \ldots$, and converge as $n \to \infty$ for all t to a function $f(t)$ which is not a characteristic function, and the corresponding laws do not converge.

2. Prove that $(1 - \cos t)/t^2 \leq 1 - \cos 1$ whenever $|t| \geq 1$.

3. For an alternate proof of the uniqueness theorem for characteristic functions (9.5.1), let $|x|^2 := (x, x)$ for $x \in \mathbb{R}^k$. Let f_P be the characteristic function of P.

(a) Show that for any $t \in \mathbb{R}^k$ and $c > 0$, $\int f_P(t - u) \exp(-c|u|^2/2) \, du = A(c, k) \times \int \exp(-|x|^2/2c) + i(x, t) \, dP(x)$ for some constant $A(c, k)$.

(b) Let \mathcal{F} be the set of all complex linear combinations of constants and functions of the form $x \mapsto \exp(i(x, t) - b|x|^2)$ for $b > 0$ and $t \in \mathbb{R}^k$. If $f_P \equiv f_Q$ for two laws P and Q, show that $\int g \, dP = \int g \, dQ$ for all $g \in \mathcal{F}$.

(c) Take the one-point compactification K of \mathbb{R}^k (Theorem 2.8.1) and show that all functions in \mathcal{F} extend to continuous functions on K. Apply the Stone-Weierstrass theorem to conclude $\int g \, dP = \int g \, dQ$ for any continuous g on K, then dominated convergence to get the same for any $g \in C_b(\mathbb{R}^k)$, so $P = Q$ by Lemma 9.3.2.

4. For an alternate proof of the Lévy continuity theorem without the truncation inequalities, assuming that the characteristic functions of laws P_n converge to f_P for a law P (as is true for the central limit theorem), let $P(n) := P_n$. Then $\int f_{P(n)}(t - u) \exp(-c|u|^2) \, du \to \int f_P(t - u) \exp(-c|u|^2) \, du$ as $n \to \infty$ for all $t \in \mathbb{R}^k$, by dominated convergence. As in Problem 3, show that $\int g \, dP_n \to \int g \, dP$ for any continuous function g on \mathbb{R}^k such that $g = 0$ outside some compact set D. For each $\varepsilon > 0$, take such a $g = 1$ on a set C with $P(C) > 1 - \varepsilon$. Obtain that the P_n are uniformly tight and apply Lemma 9.5.5 and Theorem 9.5.1.

5. Show that for any law P on \mathbb{R} with characteristic function f, and $u > 0$,

$$\int_{|x|<1/u} x^2 \, dP(x) \le 3u^{-2}(1 - \operatorname{Re} f(u)).$$

Hints: Show that for all real t, $1 - \cos(t) \ge t^2/2 - t^4/24$, using derivatives, starting at $t = 0$. Then prove

$$\int_{-\infty}^{\infty} 1 - \cos(ux) \, dP(x) \ge \int_{|x|<1/u} \frac{1}{2} u^2 x^2 \left(1 - \frac{u^2 x^2}{12} \right) dP(x).$$

Bound the last part of the integrand on the right.

6. If f is an infinitely divisible characteristic function, show that $f(t) \ne 0$ for all t. *Hints:* Let f_n be characteristic functions with $f_n^n \equiv f$. Show that the complex conjugates of f and f_n are characteristic functions, and so are $|f|^2$ and $|f_n|^2$. Then $|f_n|^2(t) = |f|^{2/n}(t) \to H(t) = 0$ or 1 according as $f(t) = 0$ or not. Show that H is a characteristic function, so that $H \equiv 1$ by continuity (Theorem 9.4.4).

7. Let f be a continuous function on \mathbb{R} with $f(0) = 1$, $f(t) = f(-t)$ for all t, and such that f restricted to $[0, \infty)$ is convex, with $f(t) \to 0$ as $t \to \infty$. Show that f is a characteristic function. *Hints:* First suppose the graph of f consists of finitely many straight line segments; those above

[0, ∞) will have negative, increasing slopes and end at the t axis. Extend the rightmost segment back to $t = 0$ and take its mirror image for $t < 0$ to define a new function g. Consider $f - g$ and use induction on the number of line segments. Approximate a general f by such piecewise linear ones.

8. (a) Show that $\exp(-|t|^p)$ is a characteristic function for any p with $0 < p \le 1$, from the result of Problem 7. (It is true for different reasons if $1 < p \le 2$.)

 (b) Show that the functions in part (a) are characteristic functions of stable laws, directly from the definition.

9. Show that for $p > 2$, $\exp(-|t|^p)$ is not a characteristic function. *Hint:* See Problem 8(b) in §9.4.

Notes

§9.1 Who first invented the notion of distribution function? The idea seems to have developed in stages. Cumulative probabilities, such as the probability of at most k successes in n independent trials, appeared in the work of Jakob Bernoulli and others. Continuous distribution functions, as integrals of density functions, were treated by Laplace (1812) in his book and earlier work, if not still earlier by others.

§9.2 For any finite measure space $(\Omega, \mathcal{S}, \mu)$ and separable metric space (S, d), convergence *in measure* is defined on $L^0(\Omega, S)$ just as when μ is a probability measure. F. Riesz (1909) defined convergence in measure, for real-valued functions, and noted (as had Lebesgue) that a pointwise (a.e.) convergent sequence converges in measure. Riesz proved that any sequence converging in measure has a subsequence converging a.e.

Fréchet (1921, pp. 199–200) showed that convergence in measure is metrizable, by the metric $\varphi(X, Y) := \inf\{\varepsilon + \mu\{\omega : d(X, Y) > \varepsilon\} : \varepsilon > 0\}$. See also Fréchet (1928, pp. 90–91; 1937, p. 191). Ky Fan (1944) defined his metric. Born in Hangzhou. China, in 1914, he received degrees in Beijing (1936) and Paris (1941), where he lived until 1945, when he moved to the United States.

§9.3 Convergence of laws, in terms of convergence of one-dimensional distribution functions, is rather a classical idea. Convergence of laws in \mathbb{R}^2 was also first treated in terms of two-dimensional distribution functions $P_n((-\infty, x] \times (-\infty, y])$ (Khinchin, 1933, pp. 11–16). Apparently A. D. Alexandroff (1940, 1941, 1943) began the general theory of convergence of laws. His book-length series of papers proved several fundamental results. He is also known for work on the geometry of convex surfaces. For a review of Alexandroff's work, see Efimov et al. (1962).

§9.4 A. de Moivre (1733) discovered normal distributions in connection with a form of the central limit theorem, for binomial distributions where the independent X_i each have just two possible values, 1 and 0, with probabilities p and $1 - p$, respectively. De Moivre gave most attention to the case $p = 1/2$ but did treat general p in his original publication on the topic, de Moivre (1733), as is evident from its title (with "$(a + b)^n$" rather than, for example, "$(1 + 1)^n$"). In 1733, de Moivre was already 66 years of age; he never held any academic position but gave private lessons and consultations

(Schneider, 1968–1969). Daw and E. S. Pearson (1972) wrote: "Six copies of the 1733 Note have been located... five of which are or were bound up with the *Miscellanea Analytica*" (de Moivre, 1730), see also Archibald (1926). De Moivre also translated his 1733 work into English and included it in the second (1738) edition of his book *The Doctrine of Chances*; de Moivre (1756, repr. 1967) was the third and final edition and is the most accessible source; there see p. 250 for $p \neq 1/2$. Todhunter (1865, pp. 192–193) only noticed the treatment of $p = 1/2$ and so gave credit for the cases $p \neq 1/2$ to Laplace, an error which gained some currency. Pearson (1924) pointed out that de Moivre treated general p. Many probabilists still refer to normal distributions as "Gaussian." According to Pearson, this usage began because Gauss had published on these distributions a few years before the book of Laplace (1812), but normal distributions appeared in papers of Laplace in the 1770s and de Moivre's work another 40 years earlier, and for that matter, in the more extensively treated case $p = 1/2$. On the other hand, the notations "π" (3.14...) and "e" (2.71828...), due to Euler, did not exist yet for de Moivre, who used more roundabout expressions. Following Pearson, most statisticians use "normal" in preference to "Gaussian."

One library catalog lists at least 19 books by Todhunter, of which only the history, Todhunter (1865), is on probabilities or statistics. Baron (1976) says "Many boys went through school and university studying mathematics entirely from Todhunter's textbooks.... Todhunter was not an original mathematician." Todhunter (1865, p. 193) does write of de Moivre, "It will not be doubted that the Theory of Probability owes more to him than to any other mathematician, with the sole exception of Laplace," and lists as one of de Moivre's three principal contributions (if only for $p = 1/2$) the asymptotic normality of binomial distributions, which Todhunter calls de Moivre's "extension of the value of Bernoulli's theorem by the aid of Stirling's theorem," "[Jakob] Bernoulli's theorem" being the law of large numbers for binomial distributions. The formula $n! \sim (n/e)^n (2\pi n)^{1/2}$ as $n \to \infty$ had only just been discovered (Stirling, 1730) when de Moivre, who had nearly found the formula himself, put it to use. Stirling evaluated the constant $(2\pi)^{1/2}$ in the formula, for which de Moivre at first had only an approximate, and eventually divergent, expression; after learning the constant from his friend Stirling, de Moivre (1730) gave a different proof of the formula.

De Moivre wrote, in effect, in terms of the ratios of values of normal density functions at two points. Laplace (1774) gave a more explicit treatment of normal densities and proved the normalization 9.4.1, see Stigler (1986, p. 360).

The use of characteristic functions in probability can be traced back at least to Cauchy (1853). Their importance in the modern theory resulted largely from work of P. Lévy (1925). In Lévy's *Oeuvres* (1973–1980) the probability papers begin in vol. 3.

§9.5 After de Moivre (1733) first stated a form of the central limit theorem, for binomial variables, Laplace (1812) extended the theorem substantially. Lyapunoff (1901) proved the theorem for independent real variables X_i with mean 0 and $E|X_i|^{2+\delta} < \infty$ for some $\delta > 0$, which could be identically distributed or somewhat more general. These hypotheses were much weaker than those previously used. Lindeberg (1922a) first proved a theorem, given in §9.6, which implies the central limit theorem in the present form in \mathbb{R}, with hypotheses just $EX_1 = 0$ and $EX_1^2 < \infty$. The theorem and proof as given resulted from work of Lindeberg (1922a, 1922b) and Lévy (1922a, 1922b, 1922c). Lindeberg's proof did not use characteristic functions. Lévy developed the method of characteristic

functions which, after some errors were corrected, led to proofs much shorter than Lindeberg's. Lévy (1922c) first gave the uniqueness theorem (9.5.1) and a form of the continuity theorem (9.5.5). Both Lévy and Lindeberg used the method of adding a small, independent normal variable Y as in the proof of the uniqueness theorem.

The central limit theorem in \mathbb{R}^k was used, in a sense, by Pearson (1900). Apparently S. Bernstein (1927, pp. 44–45) gave the first general, rigorous proof for $k = 2$. Khinchin (1933, pp. 11–16) treats $k = 2$ and states that the general case can be treated just as well. Bernstein's and Khinchin's results cover some nonidentically distributed and/or dependent variables. An eight-page mathematical obituary of Bernstein (Alexandrov et al., 1969) mentions some work in genetics but otherwise spends only a short paragraph on Bernstein's work in probability. Bernstein, in his thesis, solved Hilbert's 19th problem, on analyticity of solutions of regular analytic variational problems. Bernstein's best-known work is on best approximation of continuous functions by polynomials.

§9.6 Lindeberg's theorem (9.6.1) has a converse: if as $n \to \infty$ $\max_j \sigma_{nj}^2 \to 0$, $\mathcal{L}(S_n)$ converges to $N(0, 1)$, $EX_{nj} \equiv 0$ and $\sum_j \sigma_{nj}^2 = 1$, then $\lim_{n\to\infty} \sum_j E_{nj\varepsilon} = 0$. Thus the latter condition, perhaps surprisingly, is sharp, as was shown independently by Feller (1935) and Lévy (1935, pp. 386–388).

Lindeberg (1922a, Satz III) proved his central limit theorem (9.6.1), in a slightly different form, and thereby for the first time proved the central limit theorem for independent, identically distributed real random variables X_1, X_2, \ldots, under the simple conditions $EX_1 = 0$ and $EX_1^2 < \infty$.

Both Lindeberg and Lévy, in their 1922 papers, actually treated triangular arrays, in effect if not formally. Lévy (1970, p. 76), in his autobiography, comments on the 1922 work. Schweder (1980, pp. 120, 127) gives a biographical sketch on Lindeberg. See also Cramér (1976, p. 514).

The distributions named for Poisson (1837) can be found in de Moivre (1711–1712); see Hald (1984). Poisson lived from 1781 to 1840. Sheynin (1977) reviews Poisson's work in probability.

Gnedenko and Kolmogorov (1949) is a classic monograph on limit theorems for sums of independent variables. Gikhman, Kolmogorov, and Korolyuk (1962) wrote in honor of Gnedenko's work.

§9.7 The three-series theorem resulted from work of Khinchin and Kolmogorov (1925) and then of Kolmogorov (1928, 1929), Lévy (1937, Chapter 6) proved his equivalence theorem. Ottaviani (1939) proved his inequality. The result of Problem 7 is due to Kronecker and of Problem 8 to Kolmogorov.

§9.8 The truncation inequality is as in the book of Loève (1977, p. 209), which refers in turn to Lévy (1937, 1954), Lévy (1922c) proved his continuity theorem for $k = 1$, and where the limit function is continuous everywhere, according to Bochner (1932, pp. 72, 223–224). Various extensions (to several dimensions, allowing convergence only a.e., letting the limit function be continuous only on the axes) are not too difficult; their priority is not addressed here.

The theory of infinitely divisible and stable laws was developed by Lévy (1924, 1925, 1937) and Khinchin and Lévy (1936). Hall (1981) gives a critical review of the history and several later expositions. Hall emphasizes "errors in the sign of β" in the proofs (which, in any case, do not affect the correctness of the result being proved).

References

An asterisk identifies works I have found discussed in secondary sources but have not seen in the original.

Alexandroff, A. D. (1940, 1941, 1943). Additive set-functions in abstract spaces. *Mat. Sbornik* (N.S.) *8*: 307–348; *9*: 563–628; *13*: 169–238.

Alexandroff, P. S., N. I. Akhiezer, B. V. Gnedenko, and A. N. Kolmogorov (1969). Sergei Natanovich Bernstein: Obituary. *Russian Math. Surveys 24*, no. 3, pp. 169–176. Transl. from *Uspekhi Mat. Nauk 24*, no. 3, pp. 211–218.

Archibald, R. C. (1926). A rare pamphlet of Moivre and some of his discoveries. *Isis 8*: 671–676.

Baron, Margaret E. (1976). Todhunter, Isaac. *Dictionary of Scientific Biography, 13*, 426–428.

Bernstein, Serge [Sergei Natanovich] (1927). Sur l'extension du théorème limite du calcul des probabilités aux sommes de quantités dépendantes. *Math. Ann. 97*: 1–59.

Bochner, Salomon (1932). *Vorlesungen über Fouriersche Integrale*. Akademische Verlagsges. M.B.H., Leipzig. Repr. Chelsea, New York, 1948.

Breiman, Leo (1968). *Probability*. Addison-Wesley, Reading, Mass.

Cauchy, Augustin-Louis (1853). Sur les resultats moyens d'observations de même nature, et sur les resultats les plus probables. *Comptes Rendus Acad. Sci. Paris 37*: 198–206. Sur la probabilité des erreurs qui affectent des resultats moyens d'observations de même nature. Ibid., pp. 264–272. Mémoire sur les resultats moyens d'un très-grand nombre d'observations [Summary]. Ibid., pp. 381–385.

Cramér, Harald (1976). Half a century with probability theory: Some personal recollections. *Ann. Probability 4*: 509–546.

Daw, R. H., and E. S. Pearson (1972). Studies in the history of probability and statistics, XXX. Abraham De Moivre's 1733 derivation of the normal curve: a bibliographical note. *Biometrika 59*: 677–680.

Efimov, N. V., V. A. Zalgaller, and A. V. Pogorelov (1962). Alexander Danilovich Alexandrov (On his 50th birthday). *Russian Math. Surveys 17*, no. 6: 127–133. Transl. from *Uspekhi Mat. Nauk 17*, no. 6: 171–184.

Fan, Ky (1944). Entfernung zweier zufälligen Grössen und die Konvergenz nach Wahrscheinlichkeit. *Math. Zeitschr. 49*: 681–683.

Feller, Willy (1935). Über den zentralen Grenzwertsatz der Wahrscheinlichkeitsrechnung. *Math. Zeitschr. 40*: 521–559.

——— (1968). *An Introduction to Probability Theory and Its Applications*. Vol. 1, 3d ed. Wiley, New York.

Fréchet, Maurice (1921). Sur divers modes de convergence d'une suite de fonctions d'une variable. *Bull. Calcutta Math. Soc. 11*: 187–206.

——— (1928). *Les espaces abstraits*... Gauthier-Villars, Paris.

——— (1937). *Recherches théoriques modernes sur la théorie des probabilités I*. Paris.

Gikhman, I. I., A. N. Kolmogorov, and V. S. Korolyuk (1962). Boris Vladimirovich Gnedenko (On his 50th birthday). *Russian Math. Surveys 17*, no. 4: 105–113. Transl. from *Uspekhi Mat. Nauk 17*, no. 4: 191–200.

Gnedenko, Boris V., and Andrei N. Kolmogorov (1949). *Limit Distributions for Sums of Independent Random Variables*. Translated, annotated, and revised by Kai Lai

Chung. Addison-Wesley, Reading, Mass., 1954, 2d ed. 1968, with appendices by J. L. Doob and P. L. Hsu.

Hald, Anders (1984). A. de Moivre: "De Mensura Sortis" or "On the measurement of chance." Transl. with commentary. *Internat. Statist. Rev. 52*: 229–262.

Hall, Peter (1981). A comedy of errors: The canonical form for a stable characteristic function. *Bull. London Math. Soc. 13*: 23–27.

Hoffman, Kenneth, and Ray Kunze (1971). *Linear Algebra.* 2d ed. Prentice-Hall, Englewood Cliffs, N.J.

Khinchin, A. Ya. [Aleksandr Yakovlevich] (1933). *Asymptotische Gesetze der Wahrscheinlichkeitsrechnung.* Springer, Berlin.

———— , and A. N. Kolmogoroff (1925). Über Konvergenz von Reihen deren Glieder durch den Zufall bestimmt werden. *Mat. Sbornik (Recueil Math.) 32*: 668–676. Russ. Summary, p. 677.

———— , and P. Lévy (1936). Sur les lois stables. *C. R. Acad. Sci. Paris 202*: 374–376.

Kolmogoroff, A. N. [Kolmogorov, Andrei Nikolaevich] (1928, 1929). Über die Summen durch den Zufall bestimmter unabhängiger Grössen. *Math. Annalen 99*: 309–319. Bemerkungen zu meiner Arbeit . . . Ibid. *102*: 484–488.

Laplace, Pierre Simon de (1774). Mémoire sur la probabilité des causes par les événemens. *Mémoires de mathématique et de physique présentés a l'Académie royale des sciences, par divers savans & lus dans ses assemblées 6*: 621–656. Repr. in *Oeuvres complètes de Laplace, 8*, 325–366. Transl. into English and included (except an "unrelated" §7, on differential equations) in Stigler (1986, pp. 364–378).

———— (1812). *Théorie analytique des probabilités.* 3d ed. (1820). Mme Ve Courcier, Paris. Repr. in *Oeuvres complètes de Laplace, 7.*

———— (1886–1912). *Oeuvres complètes de Laplace.* Gauthier-Villars, Paris.

Lévy, Paul (1922a). Sur le rôle de la loi de Gauss dans la théorie des erreurs. *C. R. Acad. Sci. Paris 174*: 855–857.

———— (1922b). Sur la loi de Gauss. *C. R. Acad. Sci. Paris 174*: 1682–1684.

———— (1922c). Sur la détermination des lois de probabilité par leurs fonctions caractéristiques. *C. R. Acad. Sci. Paris 175*: 854–856.

———— (1924). Théorie des erreurs. La loi de Gauss et les lois exceptionelles. *Bull. Soc. Math. France 52*: 49–85.

———— (1925). *Calcul des probabilités.* Gauthier-Villars, Paris.

———— (1935). Propriétés asymptotiques des sommes de variables indépendantes ou enchaînées. *J. math. pures appl.* (Ser. 9) *14*: 347–402.

———— (1937). *Théorie de l'addition des variables aléatoires.* Gauthier-Villars, Paris. 2d ed. 1954.

———— (1970). *Quelques aspects de la pensée d'un mathématicien.* Blanchard, Paris.

———— (1973–1980). *Oeuvres de Paul Lévy.* 6 vols. Gauthier-Villars, Paris.

Lindeberg, J. W. (1922a). Eine neue Herleitung des Exponentialgesetzes in der Wahrscheinlichkeitsrechnung. *Math. Zeitschr. 15*: 211–225.

———— (1922b). Sur la loi de Gauss. *C. R. Acad. Sci. Paris 174*: 1400–1402.

Loève, Michel (1977). *Probability Theory I.* 4th ed. Springer, New York.

*Lyapunoff, A. M. (1901). Nouvelle forme du théorème sur la limite des probabilités. *Mem. Acad. Sci. St. Petersburg 12*, no. 5: 1–24.

de Moivre, Abraham (1711–1712). De Mensura Sortis seu, de Probabilitate Eventuum

in Ludis a Casu Fortuito Pendentibus. *Phil. Trans. Roy. Soc. London 27*, whole no. 329. Transl. in Hald (1984).

——— (1730). *Miscellanea Analytica*. London.

——— (1733). *Approximatio ad Summam Terminorum Binomii* $(a + b)^n$ *in Seriem Expansi*. 7 pp. London. Repr. in Archibald (1926).

——— (1756, posth.). *The Doctrine of Chances*, 3d. ed. London. Repr. F. Case, London, and Chelsea, New York, 1967.

Ottaviani, G. (1939). Sulla teoria astratta del calcolo delle probabilità proposita dal Cantelli. *Giorn. Ist. Ital. Attuari 10*: 10–40.

Pearson, Karl (1900). On the criterion that a given system of deviations from the probable in the case of a correlated system of variables is such that it can be reasonably supposed to have arisen from random sampling. *Phil. Mag.* (Ser. 5) *50*: 157–175.

——— (1924). Historical note on the origin of the normal curve of errors. *Biometrika 16*: 402–404.

Poisson, Siméon D. (1837). *Recherches sur la probabilité des jugements en matière criminelle et en matière civile précédés des règles générales du calcul des probabilités*. Bachelier, Paris.

Riesz, F. (1909). Sur les suites de fonctions mesurables. *Comptes Rendus Acad. Sci. Paris 148*: 1303–1305.

Schneider, Ivo (1968–1969). Der Mathematiker Abraham de Moivre, 1667–1754. *Arch. Hist. Exact. Sci. 5*: 177–317.

Schweder, Tore (1980). Scandinavian statistics, some early lines of development. *Scand. J. Statist. 7*: 113–129.

Sheynin, O. B. (1977). S. D. Poisson's work in probability. *Arch. Hist. Exact Sci. 18*: 245–300.

Stigler, Stephen M. (1986). Laplace's 1774 memoir on inverse probability. *Statistical Science 1*: 359–378.

Stirling, James (1730). *Methodus Differentialis*. London.

Todhunter, Isaac (1865). *A History of the Mathematical Theory of Probability from the Time of Pascal to That of Laplace*. Macmillan, London.

10

Conditional Expectations and Martingales

Conditional expectations provide a useful tool for doing a number of calculations in probability. Also, martingales, which are sequences of random variables related by conditional expectations, satisfy some limit theorems which are among the most useful of all those that require conditions weaker than independence of random variables.

10.1. Conditional Expectations

Given a probability space (Ω, \mathcal{S}, P) and two events A and B in \mathcal{S} with $P(B) > 0$, the *conditional probability* of A given B is defined as $P(A \mid B) := P(A \cap B)/P(B)$. Here $P(A \mid B)$ is read as "the [conditional] probability of A given B," where "conditional" can be left out because "given" is enough to convey the idea of conditional probability. For example, in three tosses of a fair coin, the probability of at least two heads, given that the first toss is a head, is $3/4$.

The *conditional expectation* of a random variable X given the event B is defined (when it exists) as

$$ E(X \mid B) := \left(\int_B X \, dP \right) \Big/ P(B). $$

More important, however, is the *conditional expectation of X given* a sub-σ-algebra \mathcal{A} of \mathcal{S}, which is defined as a random variable Y, measurable for \mathcal{A}, such that for all $A \in \mathcal{A}$, $\int_A Y \, dP = \int_A X \, dP$, if such a Y exists. Then we write $Y := E(X \mid \mathcal{A})$.

If $\mathcal{A} = \mathcal{S}$, or more generally if X is measurable with respect to \mathcal{A}, then X itself satisfies the definition of $E(X \mid \mathcal{A})$. The more interesting, nontrivial case of conditional expectation is where X is not measurable for \mathcal{A}.

Examples

(i) In two tosses of a fair coin, there is a set Ω of four possible outcomes: HH, HT, TH, TT, each with probability $1/4$. Let X be the number of heads: 0, 1, or 2. Let \mathcal{A} be the σ-algebra of events determined by the first toss, namely $\mathcal{A} = \{\emptyset, \{HH, HT\}, \{TH, TT\}, \Omega\}$. Then $E(X \mid \mathcal{A}) = 3/2$ if the first toss is heads and $1/2$ if it's tails.

(ii) Let X and Y be independent random variables on a probability space (Ω, \mathcal{S}, P) and let \mathcal{A} be the smallest sub-σ-algebra for which Y is measurable. Then $E(X \mid \mathcal{A}) = EX$, since \mathcal{A} consists of all sets $A = Y^{-1}(B)$ for measurable sets B in the range of Y; for any such A, 1_A is independent of X and $E(1_A X) = E(1_A)EX$. So conditional expectation reduces to ordinary expectation in this case.

Let $\mathcal{L}^1 := \mathcal{L}^1(\Omega, \mathcal{S}, P)$. If all the integrals in the definition of conditional expectation are to be defined and finite, then taking $A = \Omega$, it is necessary that $X \in \mathcal{L}^1$. This condition is also sufficient for a conditional expectation to exist:

10.1.1. Theorem *For any $X \in \mathcal{L}^1$, and any sub-σ-algebra \mathcal{A} of \mathcal{S}, a conditional expectation $E(X \mid \mathcal{A})$ exists, and any two conditional expectations Y and Z given \mathcal{A} are equal a.s.*

Proof. Let $\mu(A) := \int_A X \, dP$ for all $A \in \mathcal{A}$. Then μ is a countably additive signed measure on \mathcal{A}, absolutely continuous with respect to P (restricted to \mathcal{A}), so by the Radon-Nikodym theorem for signed measures (Corollary 5.6.2), there is a $Y \in \mathcal{L}^1(\Omega, \mathcal{A}, P)$ such that $\mu(A) = \int_A Y \, dP$ for all $A \in \mathcal{A}$. Such a Y satisfies the definition of $E(X \mid \mathcal{A})$. If Z also satisfies it, then $\int_A Y - Z \, dP = 0$ for all $A \in \mathcal{A}$. Since Y and Z are \mathcal{A}-measurable, the events $\{Y > Z\} := \{\omega : Y(\omega) > Z(\omega)\}$ and $\{Y < Z\}$ are in \mathcal{A}. Integrating over these sets shows that they have probability 0 for P (if $P\{Y > Z\} > 0$, then for some n, $P\{Y > Z + 1/n\} > 0$), so $Y = Z$ a.s. $\qquad\square$

Many of the useful properties of expectations extend to conditional expectations. First, here are some precautions. Conditional expectations are defined only up to a.s. equality. Recall that a member of \mathcal{L}^1 is an integrable function, and a member of L^1 is an equivalence class of such functions. A conditional expectation is really an equivalence class in L^1 rather than a specific function in \mathcal{L}^1. As usual, such an equivalence class will be denoted by any member of it (a random variable X, etc.) in a slight abuse of language. Equations or inequalities in conditional expectations will be understood to hold only almost surely. The following is straightforward:

10.1.2. Theorem $E(cX + V \mid \mathcal{A}) = cE(X \mid \mathcal{A}) + E(V \mid \mathcal{A})$ *for any* $X, V \in L^1$ *and real* c.

10.1.3. Theorem *If* \mathcal{C} *is a sub-σ-algebra of* \mathcal{A}, *then (directly from the definitions), we have* $E(X \mid \mathcal{C}) = E(E(X \mid \mathcal{A}) \mid \mathcal{C})$ *for any* $X \in L^1$.

For the trivial σ-algebra $\{\emptyset, \Omega\}$ we have

10.1.4. Proposition $E(X \mid \{\emptyset, \Omega\}) = EX$ *for any* $X \in L^1$.

On the other hand, for $\mathcal{A} = \mathcal{S}$, clearly

10.1.5. Proposition $E(X \mid \mathcal{S}) = X$ *for any* $X \in L^1(\Omega, \mathcal{S}, P)$.

10.1.6. Proposition *If* $f \leq g$ *a.s., for functions* f *and* g *in* L^1, *then* $E(f \mid \mathcal{A}) \leq E(g \mid \mathcal{A})$.

Proof. Let B be the event $\{E(f \mid \mathcal{A}) > E(g \mid \mathcal{A})\}$. Since the conditional expectations are \mathcal{A}-measurable functions, B must be in \mathcal{A}. If B has positive probability, then integrating both conditional expectations and functions over B gives $E(1_B f) > E(1_B g)$ and a contradictory inequality. \square

The two main convergence theorems for integrals both hold for conditional expectations:

10.1.7. Monotone Convergence Theorem for Conditional Expectations
Suppose f_n *and* f *are in* \mathcal{L}^1 *and* $f_n \uparrow f$ *a.s. Then* $E(f_n \mid \mathcal{A}) \uparrow E(f \mid \mathcal{A})$ *a.s.*

Proof. By Proposition 10.1.6, $E(f_n \mid \mathcal{A}) \uparrow g$ a.s. for some $g \leq E(f \mid \mathcal{A})$. For each $A \in \mathcal{A}$, $\int_A g \, dP = \int_A f \, dP$ by ordinary monotone convergence (Theorem 4.3.2). So, by uniqueness in Theorem 10.1.1, $g = E(f \mid \mathcal{A})$. \square

10.1.8. Dominated Convergence Theorem for Conditional Expectations
If $|f_n| \leq h \in \mathcal{L}^1$, $f_n \in \mathcal{L}^1$, *and* $f_n \to f$ *a.s., then* $E(f_n \mid \mathcal{A}) \to E(f \mid \mathcal{A})$ *a.s.*

Proof. Just as in the proof of the dominated convergence theorem for integrals (Theorem 4.3.5), let

$$g_n := \inf_{m \geq n} f_m \leq f_n \leq \sup_{m \geq n} f_m := h_n.$$

Then $-h \leq g_n \uparrow f \leq h$, so by 10.1.7, $E(g_n \mid \mathcal{A}) \uparrow E(f \mid \mathcal{A})$ a.s. Likewise $h \geq h_n \downarrow f \geq -h$, so $E(h_n \mid \mathcal{A}) \downarrow E(f \mid \mathcal{A})$ a.s. Now $E(g_n \mid \mathcal{A}) \leq E(f_n \mid \mathcal{A}) \leq E(h_n \mid \mathcal{A})$ implies $E(f_n \mid \mathcal{A}) \to E(f \mid \mathcal{A})$ a.s. $\qquad \square$

Constant multiples can of course be interchanged with integral signs. More interestingly, multiplication by suitable nonconstant functions can be interchanged with conditional expectation:

10.1.9. Theorem *If g and $fg \in \mathcal{L}^1$ and f is measurable for \mathcal{A}, then* $E(fg \mid \mathcal{A}) = f E(g \mid \mathcal{A}).$

Proof. By Theorem 10.1.2 we can assume $g \geq 0$ and $f \geq 0$. Then by Proposition 10.1.6, both conditional expectations are nonnegative. By Proposition 4.1.5, there are nonnegative simple functions $f_n \uparrow f$. Then $f_n g \uparrow fg$ and by Theorem 10.1.7, $E(f_n \mid \mathcal{A}) \uparrow E(f \mid \mathcal{A})$, so $g E(f_n \mid \mathcal{A}) \uparrow g E(f \mid \mathcal{A})$, and $E(f_n g \mid \mathcal{A}) \uparrow E(fg \mid \mathcal{A})$. So it will be enough to treat the case that f is a simple function. Then by linearity (Theorem 10.1.2) we can assume that $f = 1_C$ for some $C \in \mathcal{A}$. Then for any $B \in \mathcal{A}$, since $C \cap B \in \mathcal{A}$,

$$\int_B 1_C E(g \mid \mathcal{A}) \, dP = \int_{C \cap B} E(g \mid \mathcal{A}) \, dP = \int_{C \cap B} g \, dP = \int_B 1_C g \, dP,$$

so by uniqueness of conditional expectation (Theorem 10.1.1),

$$1_C E(g \mid \mathcal{A}) = E(1_C g \mid \mathcal{A}) \qquad \text{a.s.} \qquad \square$$

One way to show that both g and fg are integrable, as needed for Theorem 10.1.9, is for f and g to be integrable and independent. Another way is for both to be square-integrable, applying the Cauchy-Bunyakovsky-Schwarz inequality (5.1.4) to fg to see that $fg \in \mathcal{L}^1$ and to $g \cdot 1$ to see that $g \in \mathcal{L}^1$.

For any random variable X in \mathcal{L}^1 and any sub-σ-algebra \mathcal{A}, $EX = EE(X \mid \mathcal{A})$, applying the definition of conditional expectation to the whole space. This relationship can be helpful in finding expectations, since the work of evaluating EX in complicated cases may be broken into parts, first finding $E(X \mid \mathcal{A})$, then its expectation. This is like finding a double integral as an iterated integral (Tonelli-Fubini theorem), but more general. Choosing the right σ-algebra \mathcal{A} to simplify the calculations is much like choosing the right coordinates (for example, polar or spherical coordinates) and doing the integration in the right order to make it easier.

One of the most important cases of conditional expectation, and the earliest historically, is conditional expectation given a random variable. Let (Ω, \mathcal{S}, P) be a probability space and X and Y two real-valued random

variables on Ω, with $E|Y| < \infty$. Let \mathcal{B} be the Borel σ-algebra in \mathbb{R}. Then $X^{-1}[\mathcal{B}] := \{X^{-1}(A): A \in \mathcal{B}\}$ is a sub-σ-algebra of \mathcal{S} and the smallest σ-algebra for which X is measurable. The conditional expectation of Y given X, written $E(Y \mid X)$, is defined as $E(Y \mid X^{-1}[\mathcal{B}])$. According to Theorem 4.2.8, for any two real random variables X and Y with $E|Y| < \infty$, there is a Borel measurable function g from \mathbb{R} into itself with $E(Y \mid X) = g(X)$, so that conditional expectations given X are functions of X.

Problems

1. If X is a random variable with $|X| \le c$ a.s. for a constant c, and \mathcal{A} is any σ-algebra of measurable sets, show that $|E(X \mid \mathcal{A})| \le c$ a.s.

2. Let T be the triangle in \mathbb{R}^2 where $0 \le y \le x \le 1$, so T has vertices $(0,0)$, $(1,0)$, and $(1,1)$. Let P be the uniform distribution on T, having density with respect to planar Lebesgue measure equal to 2 on T and 0 elsewhere. Let (X, Y) have distribution P. Let \mathcal{A} be the smallest σ-algebra for which X is measurable. Show that $E(Y \mid \mathcal{A}) = X/2$ a.s.

3. Give an example of measurable sets A and B and a σ-algebra \mathcal{A} of measurable sets for which it is not true that $E(E(1_A \mid \mathcal{A})E(1_B \mid \mathcal{A})) \le E(1_A 1_B) = P(A \cap B)$.

4. Let X_1, X_2, \ldots, be i.i.d. real random variables with $E|X_1| < \infty$ and $S_n := X_1 + \cdots + X_n$. Let \mathcal{C}_n be the smallest σ-algebra for which S_k are measurable for all $k \ge n$. Show that for $j = 1, \ldots, n$, $E(X_j \mid \mathcal{C}_n) = S_n/n$.

5. A random variable X is said to be *independent* of a σ-algebra \mathcal{A} iff for every A in \mathcal{A} and measurable B in the range of X, $P(X^{-1}(B) \cap A) = P(X^{-1}(B))P(A)$. If X is real-valued and independent of \mathcal{A}, with $E|X| < \infty$, show that $E(X \mid \mathcal{A}) = EX$ a.s.

6. (*Conditional Fatou's lemma.*) Let $f_n \ge 0$ be random variables on a probability space (Ω, \mathcal{S}, P) with $Ef_n < \infty$ for all n and $E \liminf_{n\to\infty} f_n < \infty$. Show that for any σ-algebra $\mathcal{A} \subset \mathcal{S}$, $E(\liminf f_n \mid \mathcal{A}) \le \liminf_{n\to\infty} E(f_n \mid \mathcal{A})$.

7. For any random variable $f \ge 0$ (not necessarily integrable) and σ-algebra \mathcal{A}, show that there is a random variable g, measurable for \mathcal{A}, with $0 \le g \le +\infty$, such that $Ef1_A = Eg1_A$ for all $A \in \mathcal{A}$. Show that it may happen that $f < \infty$ a.s. but $g = \infty$ a.s. *Hint:* Let Ω be the unit square $0 \le x \le 1, 0 \le y \le 1$, and let \mathcal{A} be the smallest σ-algebra for which x is measurable.

8. Let (X, Y) be a random variable with values in \mathbb{R}^2. Suppose that $\mathcal{L}(X, Y)$ has a density $f(\cdot, \cdot)$ with respect to Lebesgue measure λ^2 on \mathbb{R}^2. Let \mathcal{C}

be the smallest σ-algebra for which Y is measurable. Show that if $E|X| < \infty$,

$$E(X \mid C) = \int_{-\infty}^{\infty} xf(x, Y) dx \Big/ \int_{-\infty}^{\infty} f(x, Y) dx,$$

where the right side is defined almost surely.

9. Let (Ω, \mathcal{B}, P) be a probability space, (U, \mathcal{G}) and (V, \mathcal{H}) two measurable spaces, X and Y measurable functions from Ω to U and V respectively, and f a jointly measurable real-valued function on $U \times V$ with $\int |f(X, Y)| \, dP(\omega) < \infty$. Let \mathcal{A} be a sub-σ-algebra of \mathcal{B}. Suppose X is independent of \mathcal{A} and Y is measurable for \mathcal{A}. Let $\mathcal{L}(X) = \mu$ on (U, \mathcal{G}). Show that

$$E(f(X, Y) \mid \mathcal{A}) = \int_U f(x, Y) d\mu(x).$$

10.2. Regular Conditional Probabilities and Jensen's Inequality

For any probability space (Ω, \mathcal{A}, P) and sub-σ-algebra $C \subset \mathcal{A}$, a *conditional probability* given C is defined by $P(B \mid C)(\omega) := E(1_B \mid C)(\omega)$ for each $B \in \mathcal{A}$ and ω in Ω. For any $B \in \mathcal{A}$ with $P(B) > 0$, $P(A \mid B) := P(A \cap B)/P(B)$ for each $A \in \mathcal{A}$ defines a probability measure $P(\cdot \mid B)$ on \mathcal{A}. We might expect a conditional probability $P(\cdot \mid C)(\omega)$ to be a probability measure on \mathcal{A} as well, since for any specific sequence of disjoint events, countable additivity will hold for almost all ω by Theorems 10.1.2 and 10.1.7 or 10.1.8. But the set of zero probability where countable additivity fails might depend on the sequence, and the union of all such sets might cover Ω. A suitable choice of conditional probabilities to give conditional probability measures, when it exists, is defined as follows.

Let $P_{|C}$ be the restriction of P to C. A function $P(\cdot \mid C)(\cdot)$ defined on $\mathcal{A} \times \Omega$ is called a *regular conditional probability* iff both

(a) it is a conditional probability; more precisely, for each $B \in \mathcal{A}$, $P(B \mid C)(\cdot) = E(1_B \mid C)(\cdot) \ P_{|C}$–almost surely, where $P(B \mid C)(\cdot)$ is measurable for C, and

(b) for $P_{|C}$–almost every $\omega \in \Omega$, $P(\cdot \mid C)(\omega)$ is a probability measure on \mathcal{A}.

Notes. Regular conditional probabilities do not always exist (see Problem 6), but they exist in many cases, as will be shown.

In (b) of the last definition, "almost every" can be replaced if desired by "every": let (a) and (b) hold with $C \in C$, $P(C) = 0$, and $P(\cdot \mid C)(\omega)$ a probability measure for all $\omega \notin C$. Let ζ be a fixed point of Ω and redefine

$P(B \mid \mathcal{C})(\omega) := \delta_\zeta(B) := 1_B(\zeta)$ for all $\omega \in C$ and $B \in \mathcal{A}$. Then (a) still holds and now (b) holds for all $\omega \in \Omega$.

Suppose we have (X, Y) with a law on the Borel sets of \mathbb{R}^2 (where X and Y may not be independent) and want to define a conditional distribution for Y given X (or, given the smallest σ-algebra for which X is measurable). So far, conditional probabilities would be defined on sets in \mathbb{R}^2 (or even on some other probability space Ω where random variables X and Y are defined). Given $X = x$, we will have a vertical line in \mathbb{R}^2 as y varies. It is usually more convenient if the conditional distribution is defined as a distribution of Y in \mathbb{R} rather than of (x, Y) in lines in \mathbb{R}^2.

Suppose, for example, that X and Y take only integer values. Then the conditional distribution of Y given X can be written as $P(Y = j \mid X = i)$ whenever $P(X = i) > 0$. One notation for the conditional probability is $P_{Y \mid X}(j, i) := P(Y = j \mid X = i)$. This situation will be extended in several ways: Y will take values in some range space T in place of \mathbb{R}; Ω need not be a product of T with any space; the values of Y and X need not be discrete; and instead of conditioning on a value of X, we will condition on a σ-algebra \mathcal{C}. If \mathcal{C} is the smallest σ-algebra for which X is measurable, and X is real-valued, then a function is \mathcal{C}-measurable if and only if it is a Borel measurable function of X, by Theorem 4.2.8.

Here is a general definition of the conditional distribution of a random variable Y on its range space T, given a σ-algebra \mathcal{C}. Let (Ω, \mathcal{A}, P) be a probability space and (T, \mathcal{B}) a measurable space. Let Y be a measurable function from Ω into T. Let \mathcal{C} be a sub-σ-algebra of \mathcal{A}. Then a *conditional distribution for Y, given \mathcal{C},* is a function $P_{Y \mid \mathcal{C}}$ from $\mathcal{B} \times \Omega$ into $[0, 1]$ such that

(i) for $P_{\mid \mathcal{C}}$–almost all ω, $P_{Y \mid \mathcal{C}}(\cdot, \omega)$ is a probability measure on \mathcal{B}, and
(ii) for each $B \in \mathcal{B}$, $P_{Y \mid \mathcal{C}}(B, \cdot) = P(Y^{-1}(B) \mid \mathcal{C})(\cdot)$ almost surely for $P_{\mid \mathcal{C}}$, where $P_{Y \mid \mathcal{C}}(B, \cdot)$ is \mathcal{C}-measurable.

If $(T, \mathcal{B}) = (\Omega, \mathcal{A})$ and Y is the identity function $Y(\omega) \equiv \omega$, then a conditional distribution $P_{Y \mid \mathcal{C}}(\cdot, \cdot)$ clearly gives a regular conditional probability $P(\cdot \mid \mathcal{C})(\cdot)$.

The *product space case* of the definition of $P_{Y \mid \mathcal{C}}$ will mean the following. Let (S, \mathcal{D}) and (T, \mathcal{B}) be two measurable spaces. Let P be a probability measure on the product space $\Omega = S \times T$ with product σ-algebra $\mathcal{A} = \mathcal{D} \otimes \mathcal{B}$. Let X and Y be the usual projections of $S \times T$ onto S and T respectively, $X(x, y) \equiv x$, $Y(x, y) \equiv y$. Let $\mathcal{C} = X^{-1}[\mathcal{D}] := \{X^{-1}(D) : D \in \mathcal{D}\}$, the smallest σ-algebra for which X is measurable. Note that $X^{-1}(D) = D \times T$

for any $D \subset S$. Let $\mu := \mathcal{L}(X) := P \circ X^{-1}$ on \mathcal{D}, called the *marginal distribution* of P on S. It will be said that *conditional distributions P_x exist for P and for $x \in S$* if:

(a) For each $x \in S$, P_x is a probability measure on (T, \mathcal{B});
(b) For each $B \in \mathcal{B}$, $x \mapsto P_x(B)$ is \mathcal{D}-measurable from S into \mathbb{R};
(c) For each $D \in \mathcal{D}$ and $B \in \mathcal{B}$, $P(D \times B) = \int_D P_x(B) \, d\mu(x)$.

If for example P is a product measure $\mu \times \nu$, then we can just take $P_x \equiv \nu$.

Examples (Classical Conditional Distributions). (I) (Discrete Distributions). Let S and T be countable, $\mathcal{D} = 2^S, \mathcal{B} = 2^T$. For any $x \in S$ with $\mu(\{x\}) > 0$ and $D \subset T$ let $P_x(D) := P(\{\langle x, y \rangle : y \in D\})/\mu(\{x\})$. Fix any $t \in T$ and for $\mu(\{x\}) = 0$ let $P_x := \delta_t$, recalling that $\delta_t(B) := 1_B(t)$. These P_x are conditional distributions for P.

(II) (Continuous Distributions). Let $S = T = \mathbb{R}$ and let $\mathcal{D} = \mathcal{B}$ be the Borel σ-algebra. Suppose P has a density f with respect to Lebesgue measure on \mathbb{R}^2. Then μ has a density f_X, namely $f_X(x) := \int_{-\infty}^{\infty} f(x, y) \, dy$, with respect to Lebesgue measure λ on \mathbb{R}. For each $x \in \mathbb{R}$ with $0 < f_X(x) < \infty$ and $y \in \mathbb{R}$, let $f_{Y|X}(y \mid x) := f(x, y)/f_X(x)$. Let P_x be the law having density $f_{Y|X}(\cdot \mid x)$ with respect to λ. If $f_X(x) = 0$ or ∞ let P_x be a law having a fixed density with respect to λ, for example $1_{[0,1]}$. Then P_x are conditional distributions for P.

More specifically, let P be the uniform distribution on the triangle $V : 0 \le y \le x \le 1$ in \mathbb{R}^2, so that $f(x, y) \equiv 2 1_V(x, y)$. Then $f_X(x) = 2x$ for $0 \le x \le 1$ and 0 otherwise, and $f_{Y|X}(y \mid x) = 1/x$ for $0 \le y \le x$ if $0 < x \le 1$. P_0 can be chosen arbitrarily.

That a conditional probability is regular, or a conditional distribution exists, corresponds to the fact that for each x, P_x is a countably additive probability measure—in the last example, for $0 < x \le 1$, the uniform distribution on $[0, x]$, namely Lebesgue measure on $[0, x]$ divided by x.

Conditional distributions are useful in representing integrals (expectations) as iterated integrals, even without independence:

10.2.1. Theorem *(I) If P is a probability measure in the product space case, then a conditional distribution $P_{Y|C}$ exists if and only if conditional distributions P_x exist for P and for $x \in S$, and we can set*

$$P_{Y|C}(B, (x, y)) = P_x(B)$$

for all $B \in \mathcal{B}$, $x \in S$ and $y \in T$.

(II) If such conditional distributions P_x exist, then for any integrable function g for P,

$$\int g\,dP = \iint g(x, y)\,dP_x(y)\,d\mu(x).$$

Proof. To prove (I), we have that for a real-valued function on $S \times T$ to be \mathcal{C} measurable is equivalent to being a \mathcal{D}-measurable function of x, by Theorem 4.2.8. Thus if conditional distributions P_x exist, we get $P_{Y|\mathcal{C}}$ where actually (i) holds for all $\omega = (x, y)$. Conversely, if $P_{Y|\mathcal{C}}$ exists, then by (i), there is a C with $P(C) = 0$ and $C \in \mathcal{C}$, so $C = X^{-1}(D)$ for some $D \in \mathcal{D}$ with $\mu(D) = 0$, such that $P_{Y|\mathcal{C}}(\cdot, (x, y))$ is a probability measure on \mathcal{B} for all $x \notin D$. Fix a $t \in T$ and redefine $P_{Y|\mathcal{C}}(B, (x, y)) := \delta_t(B) := 1_B(t)$ for all $x \in D$. Then $P_{Y|\mathcal{C}}$ is still a conditional distribution for Y given \mathcal{C}, and now (i) holds for all $\omega = (x, y)$. By (ii), for each $B \in \mathcal{B}$ and all $(x, y) \in S \times T$, $P_{Y|\mathcal{C}}(B, (x, y)) = f_B(x)$ for some \mathcal{D}-measurable f_B. Let $P_x(B) := f_B(x)$. Then (a) and (b) hold for P_x. To prove (c), for any $x \in S$, $u \in T$, and $B \in \mathcal{B}$, we have

$$
\begin{aligned}
P_x(B) &= P_{Y|\mathcal{C}}(B, (x, u)) && \text{by definition of } P_x \\
&= P(Y^{-1}(B)\,|\,\mathcal{C})(x, u) && \text{a.s., by (ii) for } P_{Y|\mathcal{C}} \\
&= E(1_{S \times B}\,|\,\mathcal{C})(x, u).
\end{aligned}
$$

For any $D \in \mathcal{D}$, integrating the last function over $D \times T \in \mathcal{C}$ gives

$$P(D \times B) = \int_{D \times T} P_x(B)\,dP(x, u) = \int_D P_x(B)\,d\mu(x)$$

by definition of μ and the image measure theorem 4.1.11 for the mapping X, so (c) and (I) hold.

To prove part (II), we have the conclusion where $g = 1_{D \times B}$ for each $D \in \mathcal{D}$ and $B \in \mathcal{B}$. We can then pass from rectangles A to finite disjoint unions of rectangles and then, by monotone classes (Theorem 4.4.2), to general measurable sets $A \in \mathcal{A}$ as in §4.4. Then $g = 1_A$ can be replaced in turn by simple functions, nonnegative measurable functions, and general integrable functions. □

A topological space (S, \mathcal{T}) is called a *Polish space* iff \mathcal{T} is metrizable by some metric d such that (S, d) is a complete separable metric space. Here is a main theorem on existence and uniqueness of regular conditional probabilities:

10.2.2. Theorem *Let T be any Polish space, B its σ-algebra of Borel sets,* (Ω, \mathcal{A}, P) *any probability space, Y any measurable function from* Ω *into T, and C any sub-σ-algebra of* \mathcal{A}. *Then a conditional distribution* $P_{Y|C}(\cdot,\cdot)$ *on* $\mathcal{B} \times \Omega$ *exists. It is unique in the sense that if* $P'(\cdot,\cdot)$ *also satisfies the definition of* $P_{Y|C}$, *then for* $P_{|C}$*–almost all* ω, *the two laws* $P'(\cdot, \omega)$ *and* $P_{Y|C}(\cdot, \omega)$ *are identical.*

Proof. Here \mathcal{B} is generated by a countable subset \mathcal{U} such as the open balls with rational radii and centers in a countable dense subset (Proposition 2.1.4). The algebra generated by a finite set is finite. Thus the algebra \mathcal{V} generated by \mathcal{U} can be written as a countable increasing union of finite algebras \mathcal{V}_n. By Ulam's theorem (7.1.4), applied to the law $Q := P \circ Y^{-1}$ on (T, \mathcal{B}), for each $B \in \mathcal{V}$ there exist compact $B_j \subset B$ such that $Q(B_j) \uparrow Q(B)$ as $j \uparrow \infty$. Since finite unions of compact sets are compact, we can choose a specific sequence $\{B_j\}$ for B such that $B_1 \subset B_2 \subset \cdots$. Then by conditional monotone convergence (Theorem 10.1.7),

$$P(Y^{-1}(B_j)|\mathcal{C}) \uparrow P(Y^{-1}(B)|\mathcal{C}) \qquad \text{a.s.} \qquad (10.2.3)$$

The union of \mathcal{V} and the set of all B_j for all $B \in \mathcal{V}$ is countable, so it generates a countable algebra \mathcal{D}. For a given choice of $P(Y^{-1}(D)|\mathcal{C})(\omega)$ for all $D \in \mathcal{D}$ and $\omega \in \Omega$, which is \mathcal{C}-measurable in ω for each D, we have

(i) for each $D \in \mathcal{D}$, $P(Y^{-1}(D)|\mathcal{C})(\omega) \geq 0$ a.s.,

(ii) $P(Y^{-1}(T)|\mathcal{C})(\omega) = 1$ a.s., $P(\emptyset|\mathcal{C})(\omega) = 0$ a.s.,

(iii) for any $k = 1, 2, \ldots$, and disjoint D_1, \ldots, D_k in \mathcal{D},

$$P\left(Y^{-1}\left(\bigcup_{1 \leq j \leq k} D_j\right) \Big| \mathcal{C}\right) = \sum_{1 \leq j \leq k} P(Y^{-1}(D_j)|\mathcal{C}) \qquad \text{a.s., and}$$

(iv) For any $B \in \mathcal{V}$ and for the specific sequence B_j as chosen, (10.2.3) holds a.s.

Now (i) through (iv) consist of countably many equations or inequalities in terms of \mathcal{C}-measurable functions, including limits of sequences of \mathcal{C}-measurable functions, each holding a.s. Thus there is some W in \mathcal{C} with $P(W) = 0$ such that in (i) through (iv), "a.s." can be replaced by "for all $\omega \notin W$" in all cases.

Now given any two algebras $\mathcal{V} \subset \mathcal{D}$ of subsets of a set S, where (S, \mathcal{T}) is a Hausdorff topological space, and a finite, nonnegative, finitely additive function μ defined on \mathcal{D}, it will be said that μ is *regular on* \mathcal{V} *for* \mathcal{D} iff for

every $B \in \mathcal{V}$,

$$\mu(B) = \sup\{\mu(K): K \subset B, K \in \mathcal{D}, K \text{ compact}\}.$$

This holds for $P(Y^{-1}(\cdot) \mid C)(\omega)$ for the given \mathcal{V} and \mathcal{D} if $\omega \notin W$.

Example. In \mathbb{R}, let \mathcal{V} be the algebra generated by all left closed, right open intervals $[a, b)$, and let \mathcal{D} be the algebra generated by all intervals. Then any countably additive μ on \mathcal{D} will be regular on \mathcal{V} for \mathcal{D}, approximating intervals $[a, b)$ from inside by closed intervals $[a, b_n]$, $b_n \uparrow b$.

The next fact gives a way of proving countable additivity.

10.2.4. Lemma *If $\mathcal{V} \subset \mathcal{D}$ are two algebras, μ is finite and finitely additive on \mathcal{D}, and μ is regular on \mathcal{V} for \mathcal{D}, then μ is countably additive on \mathcal{V}.*

Proof. If not, then by Theorem 3.1.1, for some $\delta > 0$, there are sets $C_i \in \mathcal{V}$ with $C_i \downarrow \emptyset$ and $\mu(C_i) > \delta$ for all i. For each i, take a compact $K_i \in \mathcal{D}$ with $K_i \subset C_i$ and $\mu(C_i \setminus K_i) < \delta/3^i$. Then for each n, $\mu(K_1 \cap \cdots \cap K_n) \geq \mu(C_n) - \sum_{1 \leq i \leq n} \delta/3^i > \delta/2$. Thus each $J_n := K_1 \cap \cdots \cap K_n$ is non-empty. If the intersection of all the K_i is empty, then the complements of the sets J_n form an open cover of K_1, so they have a finite subcover, but since the complements increase with n, this means some J_n is disjoint from K_1, contradicting $J_n \neq \emptyset$. So the intersection of all the J_n is non-empty, but this contradicts $C_i \downarrow \emptyset$. \square

Problem 17 of Section 2.2 shows why the Hausdorff property is needed in the last proof (T_1 is not sufficient).

Now continuing with the proof of Theorem 10.2.2, for $\omega \notin W$, $P(Y^{-1}(\cdot) \mid C)(\omega)$ is countably additive on the algebra \mathcal{V} and hence extends to a countably additive probability measure μ_ω on the σ-algebra \mathcal{B} (Theorem 3.1.4). For $\omega \in W$, let μ_ω be any fixed probability measure, for example Q. Let \mathcal{E} be the collection of all sets $B \in \mathcal{B}$ such that $\omega \mapsto \mu_\omega(B)$ satisfies the definition of $P(Y^{-1}(B) \mid C)$. Then \mathcal{E} includes the algebra \mathcal{V}, and \mathcal{E} is a monotone class (by Theorem 10.1.7) and thus all of the σ-algebra \mathcal{B} by Theorem 4.4.2. Thus $P_{Y|C}(B, \omega) := \mu_\omega(B)$ for $B \in \mathcal{B}$ and $\omega \in \Omega$ defines a conditional distribution; note that $W \in C$ to see the C-measurability in ω. So the existence in Theorem 10.2.2 is proved.

For the uniqueness, $P_{|C}$–almost surely the two conditional distributions are equal for all sets in the countable algebra \mathcal{V} which generates \mathcal{B}. For all such

ω, the collection of sets on which there is agreement is a monotone class, so it is all of \mathcal{B}, proving Theorem 10.2.2. $\qquad\square$

Sometimes the integral of a function h with respect to a measure μ, usually written as $\int h\,d\mu$ or $\int h(x)\,d\mu(x)$, will be written as $\int h(x)\mu(dx)$. This can make clearer what is the variable of integration when, as with conditional distributions, the measure μ depends on other parameters. Next, it will be shown that when a conditional distribution exists, conditional expectations can be written as integrals for each ω:

10.2.5. Theorem *Let (Ω, \mathcal{A}, P) be a probability space, (T, \mathcal{B}) a measurable space, Y a measurable function from Ω into T, \mathcal{C} a sub-σ-algebra of \mathcal{A}, $P_{Y|\mathcal{C}}(\cdot, \cdot)$ a conditional distribution on $\mathcal{B} \times \Omega$, and g a measurable function from T into \mathbb{R}^k such that $E|g(Y)| < \infty$. Then for $P_{|\mathcal{C}}$–almost all ω, g is integrable with respect to $P_{Y|\mathcal{C}}(\cdot, \omega)$, and*

$$E(g \circ Y \mid \mathcal{C})(\omega) = \int g(y) P_{Y|\mathcal{C}}(dy, \omega).$$

Proof. If the theorem holds for each coordinate of g, it holds for g, so we can assume $k = 1$. First suppose $g = 1_B$ for some $B \in \mathcal{B}$. Then a.s.

$$E(1_B \circ Y \mid \mathcal{C})(\omega) = E\left(1_{Y^{-1}(B)} \mid \mathcal{C}\right)(\omega)$$

and $\int 1_B(y) P_{Y|\mathcal{C}}(dy, \omega) = P_{Y|\mathcal{C}}(B, \omega) = P(Y^{-1}(B) | \mathcal{C})(\omega)$ by the definitions.

Next, since both sides of the stated equation in Theorem 10.2.5 are linear in g, it holds for any simple function g. If $g \geq 0$, let $0 \leq g_n \uparrow g$ with g_n simple. Then $0 \leq g_n \circ Y \uparrow g \circ Y$. By assumption, $E|g(Y)|$ is finite, so by monotone convergence for conditional expectations (Theorem 10.1.7), $E(g_n \circ Y \mid \mathcal{C}) \uparrow E(g \circ Y \mid \mathcal{C})$ a.s., which has finite values. The monotone convergence theorem for the conditional distributions for each ω gives convergence of the integrals on the right in the theorem for g_n to those for g, so the theorem holds for measurable $g \geq 0$. It thus holds for any g with $E|g \circ Y| < \infty$ by linearity and $g = g^+ - g^-$. $\qquad\square$

For the next fact, recall (§§6.2–6.3) that a set $C \subset \mathbb{R}^k$ is called *convex* iff for every x and y in C and $0 < p < 1$, we have $px + (1 - p)y \in C$. A function f on C with values in $[-\infty, \infty]$ is called *convex* iff, for all such x, y, and p,

$$f(px + (1 - p)y) \leq pf(x) + (1 - p)f(y).$$

If f is convex on a convex set C with closure \overline{C}, and \overline{f} is defined on \overline{C} by $\overline{f} = f$ on C and $\overline{f} = +\infty$ on $\overline{C} \backslash C$, then it is easily seen that \overline{f} is convex on \overline{C}.

We have $Ef(X) = f(EX)$ whenever f is a linear function and EX is finite. For a nonlinear function we can expect only a one-way inequality, and then only if the function has suitable properties. For example, if X is real-valued and $E|X| < \infty$, $(EX)^2 \leq E(X^2)$ by the Cauchy-Bunyakovsky-Schwarz inequality (5.1.4) applied to $X \cdot 1$. So for $f(x) = x^2$, we have $f(EX) \leq Ef(X)$. This f is a convex function, and the inequality extends to general convex functions as follows:

10.2.6. Jensen's Inequality Let C be a non-empty Borel measurable convex set in \mathbb{R}^k, f a convex function defined on C, and X a random variable with values in C such that $E|X| < \infty$ and $f(X)$ is a random variable. Then $EX \in C$, $Ef(X)$ is defined, with $-\infty < Ef(X) \leq +\infty$, and $Ef(X) \geq f(EX)$.

Remark. If C is open, or X has its values in the interior U of C, which is convex, then since f is continuous on U (Theorem 6.3.4), the fact that $f(X)$ is measurable follows from the other hypotheses. On the other hand, for example, let U be the open disk $\{\langle x, y \rangle : x^2 + y^2 < 1\}$. Let C be the union of U with an arbitrary subset T of the circle $x^2 + y^2 = 1$. Let f be defined as any convex, bounded function on U, for example $f \equiv 0$ or $f(x, y) = (x^2 + y^2)^{1/2}$, so $f \leq 1$ on U. Let f be defined as any function on T with $f(z) \geq 1$ for all $z \in T$. Then f is convex but need not be continuous or even Borel measurable, so that, even if T is the whole unit circle, if X has values in T with positive probability, $f(X)$ may not be measurable.

Proof. It is easily seen that the closure \overline{C} is convex, so it is the intersection of all closed half-spaces including it (Theorem 6.2.9). Such half-spaces are of the form $\{x : h(x) \geq t\}$, where h is a nonzero linear function on \mathbb{R}^k. Then $Eh(X) = h(EX)$. It is clear that EX is in each half-space including \overline{C}, so $EX \in \overline{C}$. If EX is on the boundary of C, then by choosing a support hyperplane for C at EX (Theorem 6.2.7), we have a linear $h \neq 0$ and $t \in \mathbb{R}$ with $h(x) \geq t$ for all $x \in C$, so that $h(X) \geq t$, and $Eh(X) = h(EX) = t$. But then $h(X) = t$ a.s. Then we can replace \mathbb{R}^k by the hyperplane $h^{-1}\{t\}$, and C by $C \cap h^{-1}\{t\}$, reducing the dimension k by 1. Continuing in that way, if $k = 1$, then C is an interval, and if EX is on its boundary, then X is a constant a.s. and the theorem holds. So it can be assumed from here on that EX is in the interior U of C, which is then non-empty and the same as the interior of the closure \overline{C} by Proposition 6.2.10.

Let $D := \{\langle x, y \rangle \in \mathbb{R}^{k+1} : x \in C, y \geq f(x)\}$. From the definition of convex function (6.3.1), D is a convex set. Clearly $V := \langle EX, f(EX) \rangle$ is on the boundary of D. By Theorem 6.2.7, D has a support hyperplane H at V, say $a_0 y + g(x) = c$, where g is a linear form on \mathbb{R}^k, $g(x) \equiv \sum_{1 \leq j \leq k} c_j x_j$. If $a_0 = 0$, then $g \neq 0$. Since EX is in U, an open set, there are points x in C with $g(x) > g(EX) = c$ and other points x with $g(x) < c$. Taking y large enough, there are points $\langle x, y \rangle$ in D with $g(x) > c$ and others with $g(x) < c$, contradicting the fact that H is a support hyperplane of D (for example, if $k = 1$, the line $g(x) = c$ is vertical and splits D, having points of D on both sides). So $a_0 \neq 0$. Dividing by a_0, we can assume $a_0 = 1$. Then D is included in the closed half-space $\{\langle x, y \rangle : y \geq c - g(x)\}$. It follows that $f(x) \geq c - g(x)$ for all $x \in C$, where $f(EX) = c - g(EX)$. Thus

$$f(X) - f(EX) \geq g(EX) - g(X).$$

Since g is linear, the right side is integrable and has integral 0. By assumption, $f(X)$ is measurable. Thus $f^-(X)$ has a finite integral and $Ef(X)$ is defined, possibly as $+\infty$. Taking E of both sides gives $Ef(X) \geq f(EX)$. $\qquad\square$

10.2.7. Conditional Jensen's Inequality Let (Ω, \mathcal{A}, P) be a probability space, and f a random variable on Ω with values in an open, convex set C in \mathbb{R}^k. Let g be a real-valued convex function defined on C. If $|f|$ and $g \circ f$ are integrable, and \mathcal{C} is any sub-σ-algebra of \mathcal{A}, then a.s. $E(f \mid \mathcal{C}) \in C$ and

$$E(g(f) \mid \mathcal{C}) \geq g(E(f \mid \mathcal{C})).$$

Proof. As C is open in a complete metric space, it is a Polish space (Theorem 2.5.4). Thus there exist conditional distributions $P_{f \mid \mathcal{C}}(\cdot, \cdot)$ (Theorem 10.2.2). We can write the conditional expectations as integrals with respect to the conditional distributions (Theorem 10.2.5):

$$E(f \mid \mathcal{C})(x) = \int_C y \, P_{f \mid \mathcal{C}}(dy, x) \qquad \text{a.s., and}$$

$$E(g \circ f \mid \mathcal{C})(x) = \int_C g(y) P_{f \mid \mathcal{C}}(dy, x) \qquad \text{a.s.}$$

Then by the unconditional Jensen inequality (10.2.6), applied to the law $P_{f \mid \mathcal{C}}(\cdot, x)$ for almost all x, we get $g(E(f \mid \mathcal{C})(x)) \leq E(g \circ f \mid \mathcal{C})(x)$. $\qquad\square$

It follows easily from Theorem 10.2.2 that Theorem 10.2.1 applies if T is a Polish space:

10.2.8. Proposition *Let (S, \mathcal{D}) be a measurable space and T a Polish space with Borel σ-algebra \mathcal{B}. Let $Y(x, y) := y$ from $\omega := S \times T$ to T. Let \mathcal{A} be the product σ-algebra on Ω and \mathcal{C} the sub-σ-algebra $\{D \times T : D \in \mathcal{D}\}$. Let P be any probability measure on \mathcal{A}. Then there is a conditional distribution $P_{Y|\mathcal{C}}(\cdot, \cdot)$, where $P_x := P_{Y|\mathcal{C}}(\cdot, (x, y))$ doesn't depend on y. In other words, conditional distributions P_x exist.*

Proof. The conditions of Theorem 10.2.2 hold, so there is a conditional distribution $P_{Y|\mathcal{C}}$ on $\mathcal{B} \times \Omega$. For each $B \in \mathcal{B}, (x, y) \mapsto P_{Y|\mathcal{C}}(B, (x, y))$ is \mathcal{C}-measurable, so it doesn't depend on y. $\qquad\square$

One example of the properties of conditional expectation is as follows. Given a closed linear subspace F of a Hilbert space H, for each z in H there is a unique x in F such that $z - x$ is orthogonal to F (Theorem 5.3.8). The function taking z to x is easily seen to be linear and is called the orthogonal projection from H onto F. For square-integrable functions, conditional expectation is such an orthogonal projection:

10.2.9. Theorem *Let (S, \mathcal{B}, P) be any probability space and \mathcal{C} any sub-σ-algebra of \mathcal{B}. Then $F := L^2(S, \mathcal{C}, P)$ is a closed linear subspace of the Hilbert space $H := L^2(S, \mathcal{B}, P)$, and on H, conditional expectation given \mathcal{C} is the orthogonal projection from H onto F.*

Proof. By Theorems 5.2.1 and 5.3.1, L^2 spaces are Hilbert spaces and F is a closed linear subspace of H. For any $X \in H$, X is in \mathcal{L}^1 by the Cauchy-Bunyakovsky-Schwarz inequality applied to X and 1. Then $Y := E(X|\mathcal{C})$ is in \mathcal{L}^2 by the conditional Jensen inequality (10.2.7) for $g(t) = t^2$. So $Y \in F$. For any $V \in F$, $E|V(X - Y)|$ is finite, also by the Cauchy-Bunyakovsky-Schwarz inequality, and $E(V(X - Y)) = EE(V(X - Y)|\mathcal{C}) = E(VE(X - Y|\mathcal{C}))$ (by Theorem 10.1.9) $= 0$, so $X - Y$ is orthogonal to F, and $X = Y + (X - Y)$ is the orthogonal decomposition of X as in Theorem 5.3.8. $\qquad\square$

Example (The Borel Paradox). Let S^2 be the unit sphere $\{(x, y, z): x^2 + y^2 + z^2 = 1\}$. On S^2 we have spherical coordinates θ, ϕ with $-\pi < \theta \leq \pi, -\pi/2 \leq \phi \leq \pi/2$,

$$(x, y, z) = (\cos \phi \cos \theta, \cos \phi \sin \theta, \sin \phi).$$

Let P be the uniform probability on S^2, equal to surface area/(4π), where

$$dP(\theta, \phi) = \cos \phi \, d\phi \, d\theta/(4\pi) = (\cos \phi \, d\phi/2)(d\theta/(2\pi)),$$

a product measure, but P is rotationally invariant and so doesn't depend on the choice of coordinates. What is the conditional distribution of P on a great circle K?

(a) The conditional distributions of θ given ϕ (on parallels of latitude) can all be taken to be the uniform distribution $d\theta/(2\pi)$, $-\pi < \theta \leq \pi$; for example, for $\phi = 0$ (the equator, a great circle).

(b) The conditional distributions of ϕ given θ (on meridians) can all be taken as $\cos \phi \, d\phi/2$, a nonuniform distribution on halves of great circles. Two such halves with values of θ differing by π form a great circle.

Thus, the conditional distribution of P on a great circle K, a set with $P(K) = 0$, is not uniquely determined and does depend on the choice of coordinates. In each of (a) and (b), the conditional distribution of a coordinate is uniquely determined up to equality for *almost* all values of the other coordinate (Theorem 10.2.2).

Problems

1. Let P be a probability measure on \mathbb{R}^2 having a density $f(x, y)$ with respect to Lebesgue measure λ^2. Let \mathcal{A} be the Borel σ-algebra in \mathbb{R}^2 and \mathcal{C} the smallest σ-algebra for which x is measurable. Find the conditional densities of y for P given x if $f(x, y) = 3x^2 + 3y^2$ for $0 \leq y \leq x \leq 1$ and 0 elsewhere.

2. Show that for $1 \leq p < \infty$, any $f \in \mathcal{L}^p(X, \mathcal{S}, P)$ and σ-algebra $\mathcal{C} \subset \mathcal{A}$, $|E(f \mid \mathcal{C})|^p \leq E(|f|^p \mid \mathcal{C})$.

3. (a) Let $f > 0$ be a function such that $\log f$ and f are both in \mathcal{L}^1. Show that for any sub-σ-algebra \mathcal{C}, $E(\log f \mid \mathcal{C}) \leq \log E(f \mid \mathcal{C})$.
 (b) Give an example of a random variable $f > 0$ in \mathcal{L}^1 such that $\log f$ is not in \mathcal{L}^1. *Hint:* Let $f \leq 1$.

4. Give an example of a real-valued convex function f on the open interval $(0, 1)$ such that $\int_0^1 f(x) \, dx = +\infty$. *Hint:* Let $f(x) = 1/x$.

5. Let (X, \mathcal{A}, P) be a probability space and \mathcal{C} the collection of sets A in \mathcal{A} such that $P(A) = 0$ or 1. Show that \mathcal{C} is a σ-algebra and find a regular conditional probability $P(\cdot \mid \mathcal{C})(\cdot)$.

6. Find a probability space (X, \mathcal{A}, P) and a sub-σ-algebra $\mathcal{C} \subset \mathcal{A}$ for which there is no regular conditional probability $P(\cdot \mid \mathcal{C})(\cdot)$. *Hint:* Let $X = [0, 1]$ with Borel σ-algebra \mathcal{C} and $P = $ Lebesgue measure. Let \mathcal{C} be a set with inner measure 0 and outer measure 1 (Theorem 3.4.4). Let \mathcal{A} be generated by \mathcal{C} and C, with $P((A \cap C) \cup (B \backslash C)) = (P(A) + P(B))/2$ for

all $A, B \in \mathcal{C}$. *Hint:* By the uniqueness in Theorem 10.2.2, $P(\cdot \mid \mathcal{C})(x) = \delta_x$ on \mathcal{C}, and thus on \mathcal{A}, for P-almost all x.

7. Let f be a real-valued function on an open interval (a, b) such that f is not convex. Show that for some random variable X with $a < X < b$, $f(EX) > Ef(X)$.

8. Show that any open convex set C in \mathbb{R}^k is an intersection of open half-spaces $\{x: g(x) > t\}$ for linear (affine) functions g and $t \in \mathbb{R}$. *Hints:* See the proof of Jensen's inequality 10.2.6. Consider $C = \mathbb{R}^k$ as the intersection of an empty collection.

9. Show that if C is any convex set in \mathbb{R}^1, f is any convex function on C and X is any random variable with values in C, then $f(X)$ is always measurable (a random variable).

10. Let X and Y be two i.i.d. random variables with $X > 0$ a.s. and $EX < \infty$. Show that $E(Y/X) > 1$, unless $X = c$ a.s. for some constant c. Give an example where $E(Y/X) = +\infty$. (This may seem paradoxical: X and Y, being identically distributed, are in some sense of the same size, but $E(Y/X) > 1$ would suggest that Y tends to be larger than X, while of course $E(X/Y) > 1$ also.)

11. Prove or disprove: For any set X, σ-algebras $\mathcal{A} \subset \mathcal{B}$ of subsets of X, probability measures P and Q on \mathcal{B}, and $0 < t < 1$, $P(\cdot \mid \mathcal{A})$ and $Q(\cdot \mid \mathcal{A})$ can be chosen so that for each $B \in \mathcal{B}$, $(tP + (1-t)Q)(B \mid \mathcal{A}) = tP(B \mid \mathcal{A}) + (1-t)Q(B \mid \mathcal{A})$ almost everywhere for $P + Q$.

12. Prove the arithmetic-geometric mean inequality (5.1.6) $(x_1 x_2 \cdots x_n)^{1/n} \leq (x_1 + \cdots + x_n)/n$, for any $x_j \geq 0$, from Jensen's inequality. *Hint:* Treat any $x_j = 0$ separately; if all $x_j > 0$, use the variables $y_j = \log(x_j)$.

13. Let $X := (X_1, \ldots, X_k)$ have a normal distribution $N(m, C)$ on \mathbb{R}^k. Let \mathcal{A} be the smallest σ-algebra for which X_2, \ldots, X_k are measurable. Show that $E(X_1 \mid \mathcal{A}) = c_1 + \sum_{2 \leq j \leq k} c_j X_j$ for some constants c_1, \ldots, c_k. *Hint:* First show you can assume $m = 0$ (then c_1 will be 0). Let U be the linear space of random variables spanned by X_2, \ldots, X_k. Let Y be the orthogonal projection of X_1 into U. Then $V := X_1 - Y$ is orthogonal to X_2, \ldots, X_k: show that it is independent of (X_2, \ldots, X_k), using the joint characteristic function $E \exp(i(t_1, \ldots, t_k) \cdot (V, X_2, \ldots, X_k))$ and facts in §9.5.

14. Call a function f on a set C *strictly* convex if $f(tx + (1-t)y) < tf(x) + (1-t)f(y)$, for any $x \neq y$ in C and $0 < t < 1$. Show that then in Jensen's inequality (10.2.6), $Ef(X) > f(EX)$ unless $X = EX$ a.s.

10.3. Martingales

Given a set T, a measurable space (S, \mathcal{S}), and a probability space (Ω, \mathcal{B}, P), a *stochastic process* is a function $\langle t, \omega \rangle \mapsto X_t(\omega)$, $t \in T$, $\omega \in \Omega$, with values in S, such that for each $t \in T$, $X_t(\cdot)$ is measurable. Often, $S = \mathbb{R}$ and \mathcal{S} is the σ-algebra of Borel sets. Also, T is most often a subset of the real line and is considered as a set of times, so that X_t is the value of some quantity at time t.

Suppose (T, \leq) is linearly ordered and $\{\mathcal{B}_t\}_{t \in T}$ is a family of σ-algebras with $\mathcal{B}_t \subset \mathcal{B}_u \subset \mathcal{B}$ for $t \leq u$. Then $\{X_t, \mathcal{B}_t\}_{t \in T}$ is called a *martingale* iff $E|X_t| < \infty$ for all t and

$$X_t = E(X_u \mid \mathcal{B}_t) \qquad \text{whenever } t \leq u. \tag{10.3.1}$$

If " $=$ " is replaced by "\leq" or "\geq" in (10.3.1), and X_t is \mathcal{B}_t measurable for all t, then $\{X_t, \mathcal{B}_t\}$ is called a *submartingale* or *supermartingale* respectively.

In this book, T will most often be the set of positive integers $n = 1, 2, \ldots$. Then $\{X_n, \mathcal{B}_n\}_{n \geq 1}$ may be called a *martingale sequence* (or a submartingale or supermartingale sequence).

If we think of X_t as the fortune at time t of a gambler, then a martingale is a "fair" game in the sense that at any time t, no matter what the history up to the present (given by \mathcal{B}_t), the expected net gain or loss from further play to time u is 0. Likewise a submartingale is a game tending to favor the player, while a supermartingale is unfavorable.

Example. Let X_1, X_2, \ldots, be independent real random variables, $S_n := X_1 + \cdots + X_n$, and \mathcal{B}_n the smallest σ-algebra for which X_1, \ldots, X_n are measurable. Then for each n, X_{n+1} is independent of every event in \mathcal{B}_n: let \mathcal{C}_n be the collection of all events (whose indicator functions are) independent of X_{n+1}. Then \mathcal{C}_n contains all events $X_j^{-1}(A_j)$, for A_j Borel sets in \mathbb{R} and $j \leq n$, and intersections of such events, by definition of independence. Finite disjoint unions of sets in \mathcal{C}_n are in \mathcal{C}_n. Thus the algebra \mathcal{A} generated by the events $X_j^{-1}(A_j)$ is included in \mathcal{C}_n (by Propositions 3.2.2 and 3.2.3 and induction). Now clearly \mathcal{C}_n is a monotone class, so by Theorem 4.4.2 it includes the σ-algebra generated by \mathcal{A}, which is \mathcal{B}_n as desired.

Thus $\{S_n, \mathcal{B}_n\}$ is a martingale if for all n, $EX_n = 0$, a submartingale if $EX_n \geq 0$, and a supermartingale if $EX_n \leq 0$.

Iterated conditional expectations (10.1.3) and induction give:

10.3.2. Proposition *A sequence $\{X_n, \mathcal{B}_n\}$ is a martingale if and only if for all n, $\mathcal{B}_n \subset \mathcal{B}_{n+1}$, $E|X_n| < \infty$, and $X_n = E(X_{n+1} \mid \mathcal{B}_n)$, and likewise for*

*submartingale and supermartingale sequences with X_n measurable for \mathcal{B}_n
and "$=$" replaced by "\leq" or "\geq" respectively.*

The submartingale property is preserved by suitable functions:

10.3.3. Theorem *Let f be a convex function on an open interval U in \mathbb{R}
and $\{X_t, \mathcal{B}_t\}_{t \in T}$ a submartingale such that for all t, X_t has values in U and
$E|f(X_t)| < +\infty$. If either*

 (a) f is nondecreasing, or
 (b) $\{X_t, \mathcal{B}_t\}$ is a martingale,

then $\{f(X_t), \mathcal{B}_t\}_{t \in T}$ is a submartingale.

Proof. For $t \leq u$, in case (a), since $X_t \leq E(X_u \mid \mathcal{B}_t)$ and $f \uparrow$, $f(X_t) \leq
f(E(X_u \mid \mathcal{B}_t))$. In case (b), $f(X_t) = f(E(X_u \mid \mathcal{B}_t))$. In either case, $f(E(X_u \mid
\mathcal{B}_t)) \leq E(f(X_u) \mid \mathcal{B}_t)$ by conditional Jensen's inequality (10.2.7). □

Example. If $\{X_t, \mathcal{B}_t\}_{t \in T}$ is a martingale, then $\{|X_t|, \mathcal{B}_t\}_{t \in T}$ is a submartingale.

The next fact will help reduce the study of convergence of sub- and super-
martingales to that of martingales.

10.3.4. Theorem (Doob Decomposition) *For any submartingale sequence
$\{X_n, \mathcal{B}_n\}$, there exist random variables Y_n and Z_n with $X_n \equiv Y_n + Z_n$ for all
n, where $\{Y_n, \mathcal{B}_n\}$ is a martingale, $Z_1 \equiv 0$, Z_n is \mathcal{B}_{n-1} measurable for $n \geq 2$,
and $Z_n(\omega) \leq Z_{n+1}(\omega)$ for all n and almost all ω. With these properties, the
Y_n and Z_n are uniquely determined.*

Note. Such a sequence $\{Z_n\}$ is called an increasing process. We have for each
n, $E|Z_n| \leq E|X_n| + E|Y_n| < +\infty$. By the way, suppose $X_1 \leq X_2 \leq \cdots \leq
X_n \leq \cdots$, where X_n is measurable for \mathcal{B}_n. Then $\{X_n\}$ is a submartingale and,
in one sense, an increasing process, but the unique Doob decomposition of
$\{X_n\}$ is not $X_n = 0 + X_n$ (see the example after the proof) unless X_n is \mathcal{B}_{n-1}
measurable.

Proof. Let $D_1 := X_1$ and $D_n := X_n - X_{n-1}$, $n \geq 2$. Let $G_1 := 0$ and $G_n :=
E(D_n \mid \mathcal{B}_{n-1})$, $n \geq 2$. Then $G_n \geq 0$ for all n a.s. since $\{X_n\}$ is a submartingale.

Let $H_n := D_n - G_n$. Let

$$Z_n := \sum_{j=1}^n G_j, \qquad Y_n := X_n - Z_n = \sum_{j=1}^n H_j.$$

Then $\{Z_n\}$ is an increasing process. For each $n \geq 2$,

$$E(H_n \mid \mathcal{B}_{n-1}) = E(D_n \mid \mathcal{B}_{n-1}) - E(G_n \mid \mathcal{B}_{n-1}) = G_n - G_n = 0.$$

Thus $E(Y_{n+1} \mid \mathcal{B}_n) = E(Y_n \mid \mathcal{B}_n) = Y_n$ and $\{Y_n, \mathcal{B}_n\}$ is a martingale, so existence of Y_n and Z_n with the given properties is proved.

For the uniqueness, let $\{y_n\}$ and $\{z_n\}$ be other sequences with all the stated properties of $\{Y_n\}$ and $\{Z_n\}$. Then $z_1 = Z_1 \equiv 0$, so $y_1 = Y_1 = X_1$. To use induction, suppose that $z_j = Z_j$ and $y_j = Y_j$ for $j = 1, \ldots, n-1$. Then

$$
\begin{aligned}
z_n &= E(z_n \mid \mathcal{B}_{n-1}) && \text{since } z_n \text{ is } \mathcal{B}_{n-1} \text{ measurable} \\
&= E(X_n - y_n \mid \mathcal{B}_{n-1}) = E(X_n \mid \mathcal{B}_{n-1}) - y_{n-1} \\
&= E(X_n \mid \mathcal{B}_{n-1}) - Y_{n-1} = E(X_n - Y_n \mid \mathcal{B}_{n-1}) = E(Z_n \mid \mathcal{B}_{n-1}) \\
&= Z_n && \text{since } Z_n \text{ is also } \mathcal{B}_{n-1} \text{ measurable.}
\end{aligned}
$$

It follows that $Y_n = y_n$, and the uniqueness is proved. $\qquad\square$

Example. Let $X_1 \equiv 0$ and $X_2 = 2$ or 4 with probability $1/2$ each. Then $X_1 \leq X_2$. On the other hand, let $Z_2 = 3$ and $Y_2 = X_2 - Z_2$. Let $\mathcal{B}_1 := \{\emptyset, \Omega\}$, the trivial σ-algebra. Let $Y_1 = 0$. Then $X_n = Y_n + Z_n$, $n = 1, 2$, is the decomposition given by Theorem 10.3.4. $X_n = 0 + X_n$ is another such decomposition except that X_2 is not \mathcal{B}_1 measurable, showing that the \mathcal{B}_{n-1} measurability of Z_n is needed for the uniqueness of the decomposition.

A set $\{X_t\}_{t \in T}$ of random variables is called L^1-*bounded* iff $\sup_{t \in T} E|X_t| < +\infty$. The $\{X_t\}$ are called *uniformly integrable* if

$$\lim_{M \to \infty} \sup_{t \in T} E\left(|X_t| 1_{\{|X_t| > M\}}\right) = 0.$$

Example. The sequence $n 1_{[0,1/n]}$ is L^1-bounded but not uniformly integrable for Lebesgue measure on $[0, 1]$.

10.3.5. Theorem $\{X_t\}$ *is uniformly integrable if and only if it is L^1-bounded and, for every $\varepsilon > 0$, there is a $\delta > 0$ such that for each event A with $P(A) < \delta$, we have $E(|X_t| 1_A) < \varepsilon$ for all t.*

Proof. "If": Let $E|X_t| \leq K < \infty$ for all t. Given $\varepsilon > 0$, take δ as given. For $M > K/\delta$, $P(|X_t| > M) < \delta$ for all t, and then

$$E\big(|X_t|1_{\{|X_t|>M\}}\big) < \varepsilon, \qquad \text{as desired.}$$

Conversely, if $\{X_t\}$ is uniformly integrable, and M is such that $E|X_t|1_{\{|X_t|>M\}} < 1$ for all t, then $E|X_t| < M + 1$ for all t, so $\{X_t\}$ is L^1-bounded. Given $\varepsilon > 0$, if K is large enough so that for all t,

$$E\big(|X_t|1_{|X_t|>K}\big) < \varepsilon/2,$$

take $\delta < \varepsilon/(2K)$. Then whenever $P(A) < \delta$,

$$E|X_t|1_A \leq E|X_t|1_A 1_{|X_t|\leq K} + E|X_t|1_{|X_t|>K} < \frac{\varepsilon}{2} + \frac{\varepsilon}{2} = \varepsilon. \qquad \square$$

If $f \in \mathcal{L}^1$ and $|X_t| \leq f$ for all t, it is easily seen that $\{X_t\}$ is uniformly integrable, since for any M, $\{|X_t| > M\} \subset \{f > M\}$ for all t. But not all uniformly integrable families are bounded in absolute value by one function in L^1: for example, on $[0, 1]$ with Lebesgue measure λ consider the set of all functions of the form $n1_A$ where A is any set with $\lambda(A) = 1/n^2$, $n \geq 1$. The following, then, is an improvement on the dominated convergence theorem. Giving an "if and only if" condition, it cannot be further improved.

10.3.6. Theorem *Given X_n and X in \mathcal{L}^1, $E|X_n - X| \to 0$ as $n \to \infty$ if and only if both $X_n \to X$ in probability and $\{X_n\}_{n \geq 1}$ are uniformly integrable.*

Proof. "If": A subsequence $X_{n(k)} \to X$ a.s. (by Theorem 9.2.1), so by Fatou's lemma (4.3.3), setting $X_0 := X$, $\{X_n\}_{n \geq 0}$ are uniformly integrable. Given $\varepsilon > 0$, take $\delta > 0$ such that $P(A) < \delta$ implies $E(|X_n|1_A) < \varepsilon/4$ for all $n \geq 0$. Take n_0 large enough so that for all $n \geq n_0$, $P(|X_n - X| > \varepsilon/2) < \delta$. Then

$$E|X_n - X| \leq \varepsilon/2 + \varepsilon/2 = \varepsilon \qquad \text{as desired.}$$

Conversely, if $E|X_n - X| \to 0$, then $X_n \to X$ in probability: if $E|X_n - X| < \varepsilon^2$, then $P(|X_n - X| > \varepsilon) < \varepsilon$. Also, $E|X_n| \leq E|X| + E|X_n - X|$ and $X \in \mathcal{L}^1$ imply that $\{X_n\}$ is an L^1-bounded sequence, say $E|X_n| \leq K < \infty$ for all n. Given $\varepsilon > 0$, take $\gamma > 0$ such that $P(A) < \gamma$ implies $E|X|1_A < \varepsilon/2$. (Such a γ exists, or we could take $A(n)$ with $P(A(n)) < 1/2^n$, so $|X|1_{A(n)} \to 0$ a.s. by the Borel-Cantelli lemma (8.3.4), and $E|X|1_{A(n)} \geq \varepsilon/2$, contradicting dominated convergence.) Then take n_0 large enough so that $E|X_n - X| \leq \varepsilon/2$ for $n \geq n_0$. Take $\delta > 0$, $\delta < \gamma$, such that $P(B) < \delta$ implies $E|X_n|1_B < \varepsilon/2$

for $n < n_0$. (Clearly, a finite set of integrable functions is uniformly integrable.) Then for $n \geq n_0$, $E|X_n|1_B \leq \varepsilon/2 + \varepsilon/2 = \varepsilon$. ☐

Problems

1. Let $\{X_n, \mathcal{B}_n\}$ be a martingale, considered as the sequence of fortunes of one gambler after each play of a game. Suppose that another gambler, B, can place bets f_n, where f_n is a bounded \mathcal{B}_n-measurable function, on the $(n+1)$st outcome, so that B has the sequence of fortunes $Y_n := Y_1 + \sum_{1 \leq j < n}(X_{j+1} - X_j)f_j$, $n \geq 2$, where $Y_1 \in \mathcal{L}^1$ and Y_1 is \mathcal{B}_1 measurable. Show that $\{Y_n, \mathcal{B}_n\}$ is still a martingale. (Thus, no "system" of bets B tries, based on past outcomes, can produce any positive expected gain. Formerly, a "martingale" was such a system rather than the sequence $\{X_n\}$ or $\{Y_n\}$.)

2. Let X_1, X_2, \ldots, be i.i.d. with distribution $N(0, 1)$, $S_n := X_1 + \cdots + X_n$, and $Y_n := \exp(S_n - n/2)$. Show that $\{Y_n, \mathcal{B}_n\}$ is a martingale, where \mathcal{B}_n is the smallest σ-algebra for which X_1, \ldots, X_n are measurable. *Hint:* See Proposition 9.4.2.

3. Let Y_1, Y_2, \ldots, be i.i.d. with distribution $N(0, 4)$ and let $S_n := Y_1 + \cdots + Y_n$. Find constants c_n such that $\exp(S_n - c_n)$ is a martingale, with $c_1 = 0$. *Hint:* See Problem 2 above.

4. In Theorem 10.3.5, show that the "ε and δ" continuity condition implies the L^1-boundedness if the probability space is the unit interval with Lebesgue measure, but not for a probability space having atoms (as in §3.5).

5. Let X be any random variable with $0 < E|X| < \infty$ and $EX = 0$. Show that there exists a martingale $\{X_n\}_{n=0,1,2}$ such that $X_0 = 0$, $X_2 = X$, and $P(X_1 \neq 0) = 1$. *Hint:* Let \mathcal{B}_1 be generated by the one set $\{X < 0\}$.

6. Let X, Y, and Z be independent real random variables with $Ee^X < \infty$, $E|X| < \infty$, $EY = 0$, and $EZ^2 < \infty$. Show that $\{X, e^X + Y + Z^2\}$ is a submartingale where \mathcal{B}_1 is the smallest σ-algebra making X measurable, and find its Doob decomposition. *Hint:* See Problem 10.1.5.

7. Let $f_n := a_n 1_{[0,1/n]}$ on $[0,1]$ with Lebesgue measure. For what sequences $\{a_n\}$ is $\{f_n\}$ uniformly integrable?

8. Let $\{X_n\}$ be a submartingale. Show that EX_n is a nondecreasing sequence of numbers, and give an example to show that $X_n - EX_n$ is not necessarily a martingale. *Hint:* Consider the uniqueness in the Doob decomposition.

9. Give an example of a martingale $\{X_n\}_{n\geq 1}$ which converges in probability but not a.s. *Hints:* Let $\{u_n\}_{n\geq 1}$ be i.i.d. uniformly distributed in $[0, 1]$ (with distribution = Lebesgue measure). Let $X_1 \equiv 0$. If $X_n = 0$, let $X_{n+1} = -1$ if $0 \leq u_n < 1/(2n)$, $X_{n+1} = 1$ if $1/(2n) \leq u_n < 1/n$, and $X_{n+1} = 0$ otherwise. If $X_n = a \neq 0$, let $X_{n+1} = 0$ if $1/2^n \leq u_n \leq 1$. Otherwise let X_{n+1} have a suitable value depending on a.

10.4. Optional Stopping and Uniform Integrability

Given a probability space (Ω, \mathcal{B}, P), a set $T \subset \mathbb{R}$, and a family $\{\mathcal{B}_t\}_{t \in T}$ of sub-σ-algebras of \mathcal{B} such that $\mathcal{B}_t \subset \mathcal{B}_u$ whenever $t < u$, a random variable α on Ω with values in $T \cup \{+\infty\}$ is called a *stopping time* iff for each $t \in T$, $\{\omega: \alpha(\omega) \leq t\} \in \mathcal{B}_t$. In this chapter, T will be a countable set, such as the set \mathbb{N}^+ of all positive integers, the set of negative integers, or $\mathbb{N}^+ \cup \{\infty\}$. If for each $n \in T$, X_n is a random variable, then the family $\{X_n\}_{n \in T}$ or $\{X_n, \mathcal{B}_n\}_{n \in T}$ is said to be *adapted* iff for all n, X_n is measurable for \mathcal{B}_n. When a family $\{X_n\}$ is mentioned, and nothing to the contrary is indicated, it will be assumed to be adapted. On the other hand, for any random variables $X_n, n \in T$, let \mathcal{A}_n be the smallest σ-algebra for which all X_k are measurable for $k \leq n$. Then $\{X_n, \mathcal{A}_n\}_{n \in T}$ is adapted. If σ-algebras \mathcal{B}_n are not mentioned or specified, then one can take $\mathcal{B}_n = \mathcal{A}_n$. In general, T is considered as a set of times, and the events in \mathcal{B}_n are those such that we know by time n whether they have occurred or not. For example, let (X_n, \mathcal{B}_n) be a submartingale, where X_n is the price of some commodity at time n, and let $\alpha(\omega)$ be the time an individual takes action (such as buying or selling the commodity). In deciding whether to act by time n, the individual has information on what has happened only up to time n, so that α is a stopping time.

Examples. If $T = \{1, 2, \ldots\}$, $\{X_n\}$ are adapted, and $y \in \mathbb{R}$, then $\alpha(\omega) := \inf\{n: X_n(\omega) > y\}$, or $\alpha(\omega) := +\infty$ if $X_n(\omega) \leq y$ for all n, then α is a stopping time.

Any constant $\alpha(\cdot) \equiv c \in T \cup \{+\infty\}$ is a stopping time.

For any measurable function $f \geq 0$ and fixed n, $n + f(X_1, \ldots, X_n)$ is a stopping time.

Given a stopping time α, let \mathcal{B}_α be the collection of all events $A \in \mathcal{B}$ such that for all t, $A \cap \{\omega: \alpha(\omega) \leq t\} \in \mathcal{B}_t$. Here \mathcal{B}_α may be called the set of events up to time α. Clearly \mathcal{B}_α is a σ-algebra. The idea, as for fixed times, is that \mathcal{B}_α is the set of events such that it is known by time α whether they have occurred or not.

Suppose $T = \{1, 2, \ldots\}$ and let α be a stopping time with values in T. Let $X_\alpha(\omega) := X_{\alpha(\omega)}(\omega)$. Then α has countably many values, each on a measurable set, so X_α is a random variable (Lemma 4.2.4). In fact, it is easy to check that X_α is measurable for \mathcal{B}_α. If α is bounded, say $\alpha \le m$, and $X_n \in \mathcal{L}^1$ for all n, then also $X_\alpha \in \mathcal{L}^1$ since $|X_\alpha| \le \max(|X_1|, \ldots, |X_m|)$.

It turns out that the submartingale, supermartingale, and martingale properties are preserved under what is called "optional stopping," that is, evaluation at stopping times, if the stopping times are bounded.

Example. Let X_1, X_2, \ldots, be independent variables with $P(X_n = 2^n) = P(X_n = -2^n) = 1/2$ for all n. Let $S_n := X_1 + \cdots + X_n$. Then $\{S_n, \mathcal{B}_n\}$ is a martingale, as in the example before Proposition 10.3.2. Let α be the least n such that $X_n > 0$. Then α is finite almost surely. Note that $S_\alpha = 2$ a.s. So although 1 and α are stopping times with $1 \le \alpha$, it is not true that $E(S_\alpha \mid \mathcal{B}_1) = S_1$ a.s. Thus the following theorem does not hold for some unbounded stopping times, and we can be glad that it holds for bounded ones.

10.4.1. Optional Stopping Theorem *Let $\{X_n, \mathcal{B}_n\}$ be any martingale sequence and let α and β be two stopping times for $\{\mathcal{B}_n\}$ with $\alpha \le \beta \le N$ for some $N < \infty$. Then $\{X_\alpha, X_\beta; \mathcal{B}_\alpha, \mathcal{B}_\beta\}$ is a martingale, so that $E(X_\beta \mid \mathcal{B}_\alpha) = X_\alpha$ a.s. The same holds if we replace "martingale" by "submartingale" or "supermartingale" and "=" by "\ge" or "\le" respectively. In other words, if $Y_1 := X_\alpha, Y_2 := X_\beta, S_1 := \mathcal{B}_\alpha, S_2 := \mathcal{B}_\beta, D := \{1, 2\}$, then $\{Y_j, S_j\}_{j \in D}$ is a martingale, submartingale, or supermartingale if $\{X_n, \mathcal{B}_n\}$ has the same property.*

Proof. If $A \in \mathcal{B}_\alpha$, then for all n,

$$A \cap \{\beta \le n\} = (A \cap \{\alpha \le n\}) \cap \{\beta \le n\} \in \mathcal{B}_n,$$

so $A \in \mathcal{B}_\beta$. Thus $\mathcal{B}_\alpha \subset \mathcal{B}_\beta$.

Let $\{X_n, \mathcal{B}_n\}$ be a submartingale. Given any $j = 1, \ldots, N$, and $A \in \mathcal{B}_\alpha$, let $A_j := A(j) := A \cap \{\alpha = j\}$, so that $A_j \in \mathcal{B}_j$. For $j \le k \le N$ let

$$A_{jk} := A(j, k) := A_j \cap \{\beta = k\},$$

$$U(j, k) := \bigcup_{i \ge k} A_{ji} = A_j \cap \{\beta \ge k\},$$

$$V(j, k) := U(j, k + 1) = A_j \cap \{\beta > k\}.$$

Then $V(j, k) \in \mathcal{B}_k$ since $\{\beta > k\} = \Omega \backslash \{\beta \le k\} \in \mathcal{B}_k$.

By definition of submartingale, $E(X_k 1_{V(j,k)}) \leq E(X_{k+1} 1_{V(j,k)})$. Thus

$$E\left(X_k 1_{U(j,k)}\right) \leq E\left(X_k 1_{A(j,k)}\right) + E\left(X_{k+1} 1_{V(j,k)}\right), \qquad \text{so that}$$
$$E\left(X_k 1_{U(j,k)}\right) - E\left(X_{k+1} 1_{U(j,k+1)}\right) \leq E\left(X_\beta 1_{A(j,k)}\right).$$

Summing from $k = j$ to N, we have a telescoping sum on the left, and $U(j, N + 1) = \varnothing$, so

$$E\left(X_j 1_{A(j)}\right) = E\left(X_j 1_{U(j,j)}\right) \leq E\left(X_\beta 1_{A(j)}\right).$$

Now summing from $j = 1$ to N gives $E(X_\alpha 1_A) \leq E(X_\beta 1_A)$, as desired.

The supermartingale case is symmetrical (replacing X_n by $-X_n$). In the martingale case, inequalities for expectations are replaced by equalities, completing the proof. □

In the martingale case of Theorem 10.4.1, suppose $\alpha \leq \beta \equiv N$. Then $E(X_N | \mathcal{B}_\alpha) = X_\alpha$, so $EX_\alpha = EX_N = EX_1$. Thus in a fair game one cannot improve (or worsen) one's expectations by optional stopping (based on information available up to the time of stopping) for a bounded stopping time. Suppose you try, for example, to "quit when you're ahead," letting α be a time when $X_\alpha - X_1$ is positive or at least equal to a given positive goal, if possible, but you can play only until a given finite time N. Then the probability that you don't reach your goal may be small, but your expected losses must exactly balance your expected gains, so your losses if you do lose may be quite large.

Given random variables X_1, \ldots, X_n and $M > 0$, an event is defined by $A(M, n) := \{\max_{1 \leq j \leq n} X_j \geq M\}$.

10.4.2. Theorem (Doob's Maximal Inequality) *For any submartingale sequence $\{X_n, \mathcal{B}_n\}$ any n, and $M > 0$, we have*

$$M P(A(M, n)) \leq E\left(X_n 1_{A(M,n)}\right) \leq E \max(X_n, 0).$$

Note. Doob's inequality improves neatly on Markov's inequality, $M P(X \geq M) \leq EX^+$, where $X^+ := \max(X, 0)$. In one sense, Markov's inequality is sharp, since it becomes an equation if $X = M$ with some probability and is 0 otherwise. On the other hand, Doob's inequality gives us the same upper bound for the probability of the union of the events $\{X_j \geq M\}$ for $j = 1, \ldots, n$ that Markov's inequality would give us for the one event $\{X_n \geq M\}$.

Proof. Let α be the least $j \leq n$ such that $X_j \geq M$, or if there is no such $j \leq n$, let $\alpha := n$. Then α is a stopping time and $\alpha \leq n$. Let $A := A(M, n)$. Then

$A \in \mathcal{B}_\alpha$ since for any $m \leq n$,

$$A \cap \{\alpha \leq m\} = \left\{\max_{j \leq m} X_j \geq M\right\} \in \mathcal{B}_m.$$

By the optional stopping theorem (10.4.1) with $\beta \equiv n$, $\{X_\alpha, X_n; \mathcal{B}_\alpha, \mathcal{B}_n\}$ is a submartingale. On A, $X_\alpha \geq M$, so

$$M P(A) \leq E(X_\alpha 1_A) \leq E(X_n 1_A) \leq E \max(X_n, 0). \qquad \square$$

A martingale sequence $\{X_n, \mathcal{B}_n\}$ will be called *right-closable* iff it can be extended to the linearly ordered set $\{1, 2, \ldots, \infty\}$, so that there is a random variable $X_\infty \in \mathcal{L}^1$ with $E(X_\infty \mid \mathcal{B}_n) = X_n$ for all n. Let \mathcal{B}_∞ be the smallest σ-algebra including \mathcal{B}_n for all n. Then letting $Y := E(X_\infty \mid \mathcal{B}_\infty)$, Y is \mathcal{B}_∞-measurable and by Theorem 10.1.3, $E(Y \mid \mathcal{B}_n) = X_n$ for all n. So we can assume that X_∞ is \mathcal{B}_∞-measurable. In that case, $\{X_n, \mathcal{B}_n\}_{1 \leq n \leq \infty}$ will be called a *right-closed* martingale. On the other hand, for any $Y \in \mathcal{L}^1$ and any increasing sequence of σ-algebras \mathcal{B}_n, setting $X_n := E(Y \mid \mathcal{B}_n)$ always gives a martingale sequence, also by Theorem 10.1.3. Here is a criterion for martingales to be of this form.

10.4.3. Theorem *A martingale sequence $\{X_n, \mathcal{B}_n\}$ is right-closable if and only if $\{X_n\}$ are uniformly integrable.*

Note. Once this equivalence is proved, the term "right-closable" is no longer needed: the right-closable martingales $\{X_n, \mathcal{B}_n\}_{1 \leq n < \infty}$ will be called uniformly integrable.

Proof. If $E(Y \mid \mathcal{B}_n) = X_n$ for all n, with $E|Y| < \infty$, then for any set $A \in \mathcal{B}_n$, $E(|X_n| 1_A) \leq E(|Y| 1_A)$, since $|\cdot|$ is a convex function, by conditional Jensen's inequality (10.2.7). Now $\{Y\}$ is uniformly integrable, meaning by definition that $E(|Y| 1_{|Y|>M}) \to 0$ as $M \to \infty$, by dominated convergence. It follows from our alternate characterization of uniform integrability (Theorem 10.3.5) that $E(|Y| 1_A) \to 0$ as $P(A) \to 0$. Applying this to sets $A := \{|X_n| \geq M\}$ gives that $\{X_n\}$ are uniformly integrable, since

$$\sup_n P(|X_n| \geq M) \leq \sup_n E|X_n|/M \leq E|Y|/M \to 0 \qquad \text{as } M \to \infty.$$

Conversely, suppose $\{X_n\}$ are uniformly integrable. Let $\mathcal{A} := \bigcup_{n \geq 1} \mathcal{B}_n$, an algebra (any increasing union of algebras is an algebra). For any $A \in \mathcal{B}_n$

let $\gamma(A) := E(X_n 1_A)$. Then for all $m \geq n$, by the martingale property,

$$\gamma(A) = E(X_m 1_A) = \lim_{k \to \infty} E(X_k 1_A).$$

Thus γ is well-defined on \mathcal{A} ($\gamma(A)$ does not depend on n for n large enough so that $A \in \mathcal{B}_n$). By the uniform integrability, for any $\varepsilon > 0$ there is a $\delta > 0$ such that whenever $A \in \mathcal{A}$ and $P(A) < \delta$, then $|\gamma(A)| < \varepsilon$. Here the following will help:

10.4.4. Lemma *Let \mathcal{A} be an algebra of sets and \mathcal{S} the σ-algebra generated by \mathcal{A}. Let μ be a finite, nonnegative measure on \mathcal{S} and γ a finitely additive, bounded, real-valued set function on \mathcal{A}. Suppose that as $\delta \downarrow 0$, $\sup\{|\gamma(A)|\colon A \in \mathcal{A}, \mu(A) < \delta\} \to 0$. Then γ extends to a countably additive signed measure on \mathcal{S} which is absolutely continuous with respect to μ, so that for some $f \in L^1(\mu)$,*

$$\gamma(B) = \int_B f \, d\mu \qquad \text{for all } B \in \mathcal{S}.$$

Proof. If $A_n \downarrow \emptyset$, $A_n \in \mathcal{A}$, then $\mu(A_n) \downarrow 0$ (by Theorem 3.1.1), so $\gamma(A_n) \to 0$. Thus γ is countably additive on \mathcal{A} (by 3.1.1 again). Being bounded, γ extends to a countably additive signed measure on \mathcal{S} (Theorem 5.6.3), which will still be called γ. Then there is the Jordan decomposition $\gamma = \gamma^+ - \gamma^-$ where γ^+ and γ^- are finite, nonnegative measures, with $\gamma^+(C) = \sup\{\gamma(B)\colon B \in \mathcal{S}, B \subset C\}$ for each $C \in \mathcal{S}$ (Theorem 5.6.1).

Given any $\varepsilon > 0$, take $\delta > 0$ such that if $\mu(A) < \delta$, $A \in \mathcal{A}$, then $|\gamma(A)| \leq \varepsilon$. Let \mathcal{M} be the collection of all sets B in \mathcal{S} such that $\mu(B) \geq \delta$ or $|\gamma(B)| \leq \varepsilon$ (or both). Then $\mathcal{A} \subset \mathcal{M}$. To show that \mathcal{M} is a monotone class, let $B_n \uparrow B$, $B_n \in \mathcal{M}$. If $|\gamma(B_n)| \leq \varepsilon$ for all n, then $|\gamma(B)| \leq \varepsilon$ by countable additivity (specifically, continuity of γ^+ and γ^- under monotone convergence). Otherwise, $\mu(B_n) \geq \delta$ for some n and then $\mu(B) \geq \delta$, so $B \in \mathcal{M}$. Next, suppose $B_n \downarrow B$, $B_n \in \mathcal{M}$. If $|\gamma(B_n)| \leq \varepsilon$ for all large enough n, then $|\gamma(B)| \leq \varepsilon$. Otherwise, for some subsequence $n(k)$, $\mu(B_{n(k)}) \geq \delta$ so $\mu(B) \geq \delta$. So again $B \in \mathcal{M}$, and \mathcal{M} is a monotone class. So $\mathcal{M} = \mathcal{S}$ (by Theorem 4.4.2).

Since $\mathcal{M} = \mathcal{S}$ holds for all $\varepsilon > 0$ and suitable $\delta = \delta(\varepsilon) > 0$, it follows that if $\mu(B) = 0$, then $\gamma(B) = 0$, and $\gamma^+(B) = 0$ so $\gamma^-(B) = 0$. Thus γ, γ^+, and γ^- are absolutely continuous with respect to μ. The Radon-Nikodym theorem (5.5.4 and Corollary 5.6.2) then gives the lemma. \square

Now to continue the proof of Theorem 10.4.3, using Lemma 10.4.4, let X_∞ be the Radon-Nikodym derivative $d\gamma/dP$. Then for any n and set $A \in \mathcal{B}_n$, $\int_A X_n \, dP = \gamma(A) = \int_A X_\infty \, dP$, so $E(X_\infty \mid \mathcal{B}_n) = X_n$ for all n. \square

Problems

1. A player throws a fair coin and wins \$1 each time it's heads, but loses \$1 each time it's tails. The player will stop when his or her net winnings reach \$1, or after n throws, whichever comes sooner. What is the player's probability of winning, and expected loss conditional on not winning, if (a) $n = 3$? (b) $n = 5$?

2. Let α be a stopping time and $\tau(n)$ a nondecreasing sequence of bounded stopping times, both for a martingale $\{X_n, \mathcal{B}_n\}_{n\geq 1}$. Show that α is also a stopping time for the martingale $\{X_{\tau(n)}, \mathcal{B}_{\tau(n)}\}_{n\geq 1}$, in the sense that $\{\alpha \leq \tau(n)\} \in \mathcal{B}_{\tau(n)}$ for each n.

3. Let $\{X_n, \mathcal{B}_n\}$ be an L^1-bounded martingale. Let \mathcal{T} be the set of all stopping times τ for it such that $\tau < +\infty$ a.s. Show that the set of all X_τ for $\tau \in \mathcal{T}$ is also L^1-bounded. *Hint:* Consider the stopping times $\min(\tau, n)$.

4. Let $\{X_n\}_{1\leq n\leq\infty}$ be a right-closed martingale. Suppose that $E|X_\infty|^p < \infty$ for some $p > 1$. Let $Y := \sup_n |X_n|$ and $0 < r < p$. Show that $EY^r < \infty$. *Hint:* Apply the maximal inequality to the submartingale $|X_n|^p$. Then apply Lemma 8.3.6 to Y^r.

5. In Problem 4, if $p = 1$ show that it can happen that $EY = +\infty$. *Hint:* Let the probability space be $(0, 1/2)$ with uniform distribution (twice Lebesgue measure). Let $X_\infty(t) = t^{-1}|\log t|^{-3/2}$ for $0 < t < 1/2$. Let \mathcal{B}_n be the σ-algebra generated by the intervals $[j/2^n, (j+1)/2^n), j = 0, 1, \ldots$. Use the fact that $Y \geq X_n$ on the interval $(1/2^{n+1}, 1/2^n)$. *Note:* This is another example of a uniformly integrable family not dominated by any one integrable function.

6. (Random walk.) Let X_1, X_2, \ldots, be i.i.d. random variables with $X_j = +1$ or -1 with probability $1/2$ each and $S_n := X_1 + \cdots + X_n$, giving a martingale (see the example just before Proposition 10.3.2). Let $\tau := \min\{n: S_n = 1\}$. Prove that $\tau < \infty$ a.s. *Hint:* Let $p = P(\tau < \infty)$. Consider the two possibilities for X_1 to show that $p = (p^2 + 1)/2$, so $p = 1$. (Note that $ES_\tau = 1 \neq ES_1 = 0$, so this is another example showing that Theorem 10.4.1 may fail for unbounded stopping times ($\alpha = 1 \leq \tau$).)

7. Show that if μ is a finite measure and $\sup\{|\gamma(A)|: A \in \mathcal{A}, \mu(A) < \delta\} \to 0$ as $\delta \downarrow 0$, as in Lemma 10.4.4, and γ is finite on each atom of μ, then γ is bounded. *Hint:* See §3.5.

8. (Optional sampling.) Let (X_n, \mathcal{B}_n) be a submartingale and let $\tau(1) \leq \tau(2) \leq \cdots$ be a nondecreasing sequence of bounded stopping times (so for each n, there is some constant $M_n < \infty$ with $\tau(n) \leq M_n$). Show that $(X_{\tau(n)}, \mathcal{B}_{\tau(n)})$ is a submartingale.

10.5. Convergence of Martingales and Submartingales

The proof that L^1-bounded martingales converge a.s. will be done first for right-closed martingales. Convergence shows that not only the index set $\{1, 2, \ldots\}$ but the martingale $\{X_1, X_2, \ldots\}$ is "closed" by adjoining X_∞.

As an example, for any random variables X_1, X_2, \ldots (not necessarily a martingale, submartingale, etc.), let \mathcal{B}_n be the smallest σ-algebra for which X_1, \ldots, X_n are measurable and \mathcal{B}_∞ the smallest σ-algebra including all the \mathcal{B}_n. Let A be any set in \mathcal{B}_∞. Then the conditional expectation $E(1_A \mid \mathcal{B}_n)$ is also called the conditional probability $P(A \mid \mathcal{B}_n) = P(A \mid X_1, \ldots, X_n)$. The following will imply that this sequence converges to 1_A a.s. as $n \to \infty$.

10.5.1. Theorem (Doob) *For any right-closed martingale sequence* $\{X_n, \mathcal{B}_n\}_{1 \le n \le \infty}$, X_n *converges to* X_∞ *a.s. as* $n \to \infty$.

Proof (C. W. Lamb). It can be assumed that the martingale is defined on a probability space (Ω, \mathcal{B}, P) with $\mathcal{B} = \mathcal{B}_\infty$. Let F be the set of all $Y \in \mathcal{L}^1(\Omega, \mathcal{B}, P)$ such that for some $n = n(Y) < \infty$, Y is \mathcal{B}_n measurable. To show that F is dense in \mathcal{L}^1, it suffices to approximate simple functions, and hence indicator functions 1_A, $A \in \mathcal{B}$, by elements of F. The collection of events A which can be approximated is a monotone class including the algebra $\mathcal{A} = \bigcup_{n < \infty} \mathcal{B}_n$ and hence equals \mathcal{B} by Theorem 4.4.2.

Given $\varepsilon > 0$, choose $Y_\infty \in F$ such that $E|X_\infty - Y_\infty| < \varepsilon^2$. Let $Y_n := E(Y_\infty \mid \mathcal{B}_n)$. Then $Y_n = Y_\infty$ for all $n \ge n(Y_\infty)$, so $Y_n \to Y_\infty$ a.s. Also, note that $\{X_n - Y_n, \mathcal{B}_n\}_{1 \le n \le \infty}$ is a martingale. Thus $\{|X_n - Y_n|, \mathcal{B}_n\}$ is a submartingale (Theorem 10.3.3(b) for $f(x) \equiv |x|$). Then by the maximal inequality (10.4.2) and monotone convergence (Theorem 4.3.2), $P(\sup_{n < \infty} |X_n - Y_n| > \varepsilon) \le \sup_{n < \infty} E|X_n - Y_n|/\varepsilon \le E|X_\infty - Y_\infty|/\varepsilon < \varepsilon$. Thus $P\{\limsup_{n \to \infty} X_n - Y_\infty > \varepsilon\} < \varepsilon$, $P\{\liminf_{n \to \infty} X_n - Y_\infty < -\varepsilon\} < \varepsilon$, and $P(|X_\infty - Y_\infty| > \varepsilon) < \varepsilon$, so $P(\limsup X_n - X_\infty > 2\varepsilon) < 2\varepsilon$ and $P(\liminf X_n - X_\infty < -2\varepsilon) < 2\varepsilon$. Letting $\varepsilon \downarrow 0$ gives $X_n \to X_\infty$ a.s. $\qquad \square$

10.5.2. Corollary *For any uniformly integrable martingale* $\{X_n, \mathcal{B}_n\}_{1 \le n < \infty}$, X_n *converges a.s. as* $n \to \infty$ *to some* X_∞ *with* $E(X_\infty \mid \mathcal{B}_n) = X_n$ *for all* n, *and* $E|X_n - X_\infty| \to 0$.

Proof. Apply Theorems 10.4.3, 10.5.1, and 10.3.6. $\qquad \square$

It is easy to see that any maximum or minimum of two stopping times is a stopping time. Let $a \wedge b := \min(a, b)$ and $a \vee b := \max(a, b)$. So, for any stopping time α and any fixed n, $\alpha \wedge n$ is a stopping time.

10.5.3. Lemma *For any martingale* $\{X_n, \mathcal{B}_n\}$ *and stopping time* α, $\{X_{\alpha \wedge n}, \mathcal{B}_{\alpha \wedge n}\}$ *is a martingale.*

Proof. Apply Theorem 10.4.1 (optional stopping) and Proposition 10.3.2.

\square

Lemma 10.5.3 says: Suppose a gambler is playing a fair game, so that if he continued indefinitely, his sequence X_n of fortunes at time n would be a martingale. The gambler, by some rule based only on information in hand when the decision must be made, decides to stop at time α. Then the sequence of the gambler's fortunes is $X_{\alpha \wedge n}$ at time n and is still a martingale.

Here is a main convergence theorem:

10.5.4. Martingale Convergence Theorem (Doob) *Let* $\{X_n, \mathcal{B}_n\}_{1 \le n < \infty}$ *be any* L^1-*bounded martingale sequence. Then* X_n *converges a.s. to some random variable X.*

Before proving the theorem, let's look at some examples. Let μ and ν be probability measures on the same σ-algebra \mathcal{S}. Let \mathcal{A}_n be an increasing sequence of σ-algebras whose union generates \mathcal{S}. For example, let \mathcal{S} be the Borel σ-algebra in $[0, 1)$. For $n = 1, 2, \ldots$, let \mathcal{A}_n be the algebra generated by the intervals $[(k-1)/2^n, k/2^n), k = 1, 2, \ldots, 2^n$. Suppose that ν is absolutely continuous with respect to μ on each \mathcal{A}_n, with Radon-Nikodym derivative f_n. Then the $\{f_n, \mathcal{A}_n\}$ form a martingale with respect to μ, which is always L^1-bounded. If ν is absolutely continuous with respect to μ on \mathcal{S}, with Radon-Nikodym derivative f, then by Theorem 10.5.1, f_n converges to f a.s. and in L^1. On the other hand, if ν is singular with respect to μ, Theorem 10.5.4 tells us that f_n still converges to something a.s. (actually, to 0: Problem 4 asks for a proof).

Proof. Given $M > 0$, let $\alpha(\omega)$ be the least n such that $|X_n(\omega)| \ge M$, or $+\infty$ if there is no such n. Then α is a stopping time. The proof will use the following fact.

10.5.5. Lemma $\{X_{\alpha \wedge n}, \mathcal{B}_{\alpha \wedge n}\}_{1 \le n < \infty}$ *is a uniformly integrable martingale sequence.*

Proof. First, $\{X_{\alpha \wedge n}\}$ is a martingale by Lemma 10.5.3. To show it's uniformly integrable, let $Z := |X_\alpha|$ if $\alpha < \infty$; otherwise, $Z := M$. Then $|X_{\alpha \wedge n}| \le Z$ for all n a.s. So it will be enough to prove that $EZ < \infty$. (For example, if the increments $X_{n+1} - X_n$ are uniformly bounded, say $|X_{n+1} - X_n| \le K < \infty$

a.s. for all n, then $Z \le M + K$.) Now $E(Z1_{\{\alpha=\infty\}}) \le M$ and

$$E\left(Z1_{\{\alpha<\infty\}}\right) = E\left(|X_\alpha|1_{\{\alpha<\infty\}}\right) = E\left(\lim_{n\to\infty} |X_{\alpha\wedge n}|1_{\{\alpha<\infty\}}\right)$$

$$\le \liminf_{n\to\infty} E\left(|X_{\alpha\wedge n}|1_{\{\alpha<\infty\}}\right) \qquad \text{(by Fatou's lemma)}$$

$$\le \sup_n E|X_{\alpha\wedge n}| \le \sup_n E|X_n| < \infty,$$

by optional stopping (10.4.1) for $\alpha \wedge n$ and $\beta := n$, and Theorem 10.3.3. So $E|Z| < \infty$. \square

Now to finish the proof of martingale convergence, on the set where $|X_n| < M$ for all n, $X_n \equiv X_{\alpha\wedge n}$, so X_n converges a.s. there by Lemma 10.5.5 and Corollary 10.5.2. By the maximal inequality (Theorem 10.4.2) and monotone convergence (Theorem 4.3.2) $P(|X_n| \ge M$ for some $n) \le \sup_n E|X_n|/M \to 0$ as $M \to \infty$. Thus X_n converges a.s. \square

Example (Polya's Urn Scheme). An urn contains one black and one red ball. Repeatedly, one ball is drawn at random from the urn and then replaced, along with an additional ball of the same color. What is the distribution after n draws? Let B_n be the number of black balls and R_n the number of red balls after n turns. Then $B_n/(B_n+R_n)$ forms a martingale (Problem 2 below) which is clearly uniformly integrable, so it converges a.s. and in L^1; its limit is not a constant.

To extend the convergence theorem to sub- and supermartingales, the next fact about the Doob decomposition (10.3.4) will be of use.

10.5.6. Lemma *If (X_n, \mathcal{B}_n) is an L^1-bounded submartingale, then in the Doob decomposition $X_n = Y_n + Z_n$, the variables Z_n of the increasing process are uniformly integrable, so Z_n and Y_n are L^1-bounded. If X_n are uniformly integrable, then so are Y_n as well as Z_n. The corresponding statements also hold for supermartingales.*

Proof. We have $Z_1 = 0 \le Z_2 \le Z_3 \le \cdots$, and $EZ_n = EX_n - EY_n = EX_n - EY_1 = EX_n - EX_1$. Thus $\sup_n EZ_n < \infty$ and by monotone convergence, $E \sup_n Z_n < \infty$. So the Z_n are dominated by an integrable function and hence uniformly integrable. The rest of the lemma is then clear. \square

10.5.7. Sub- and Supermartingale Convergence Theorem *Any L^1-bounded submartingale or supermartingale converges a.s.; if uniformly integrable, it also converges in L^1.*

Proof. For the Doob decomposition, $Z_n(\omega)$ is nondecreasing in n a.s. and converges a.s. up to an integrable limit by the proof of Lemma 10.5.6. Then $\{Y_n, \mathcal{B}_n\}$ is an L^1-bounded martingale, converging a.s. by Theorem 10.5.4. The uniformly integrable case follows from Corollary 10.5.2 for Y_n and monotone or dominated convergence for Z_n. □

For example, for any L^1-bounded martingale $\{X_n\}$, which thus converges a.s., the submartingale $\{|X_n|\}$ also converges a.s., just because $\{X_n\}$ does, or as a special case of submartingale convergence.

Finally, for nonnegative supermartingales the optional stopping (Theorem 10.4.1) can be improved in that the stopping times need only be finite, not necessarily bounded:

***10.5.8. Theorem** *Let (X_n, \mathcal{B}_n) be a supermartingale with $X_n(\omega) \geq 0$ for all n and ω. Let σ and τ be two stopping times with $\sigma \leq \tau$. Then $E(X_\tau 1_{\{\tau < \infty\}} | \mathcal{B}_\sigma) \leq X_\sigma 1_{\{\sigma < \infty\}}$.*

Proof. For any n, $EX_{\tau \wedge n} \leq EX_1$ by optional stopping (Theorem 10.4.1). Then by Fatou's lemma, $EX_\tau 1_{\{\tau < \infty\}} \leq EX_1 < \infty$. Likewise, $EX_\sigma 1_{\{\sigma < \infty\}} < \infty$. On the set in \mathcal{B}_σ where $\sigma = \infty$, also $\tau = \infty$, so both sides of the stated inequality are 0 and it holds.

For any $m = 1, 2, \ldots,$ and $A \in \mathcal{B}_\sigma$, it will be shown that

$$A^{(m)} := A \cap \{\sigma \leq m\} \in \mathcal{B}_{\sigma \wedge m}.$$

First, if $m \leq j$, then $A^{(m)} \cap \{\sigma \wedge m \leq j\} = A^{(m)} \in \mathcal{B}_m \subset \mathcal{B}_j$, while if $m > j$, $A^{(m)} \cap \{\sigma \wedge m \leq j\} = A \cap \{\sigma \leq j\} \in \mathcal{B}_j$ as desired. Thus for any $m \leq n$, applying optional stopping (10.4.1) to $\alpha := \sigma \wedge m \leq \beta := \tau \wedge n$,

$$E1_A X_\sigma 1_{\sigma \leq m} = E1_A X_{\sigma \wedge m} 1_{\sigma \leq m} \geq E1_A X_{\tau \wedge n} 1_{\sigma \leq m}$$
$$\geq E1_A X_{\tau \wedge n} 1_{\tau \leq n} 1_{\sigma \leq m}$$
$$= E1_A X_\tau 1_{\tau \leq n} 1_{\sigma \leq m}.$$

Letting $n \to \infty$, $1_{\tau \leq n}$ increases up to $1_{\tau < \infty}$. All the other factors being nonnegative, monotone convergence gives

$$E1_A X_\tau 1_{\sigma \leq m} 1_{\tau < \infty} \leq E1_A X_\sigma 1_{\sigma < \infty}.$$

Likewise, $1_{\sigma \leq m}$ increases up to $1_{\sigma < \infty}$, so monotone convergence again applies. Also, since $\sigma \leq \tau$, $1_{\sigma < \infty} 1_{\tau < \infty} = 1_{\tau < \infty}$, so we get

$$E1_A X_\tau 1_{\tau < \infty} \leq E1_A X_\sigma 1_{\sigma < \infty},$$

which implies the conclusion. □

Example. Let $X_1, X_2, \ldots,$ be independent real random variables with mean 1 and $T_n := X_1 X_2 \cdots X_n$. Let \mathcal{B}_n be the smallest σ-algebra making X_1, \ldots, X_n measurable. Then since X_{n+1} is independent of \mathcal{B}_n, $E(X_{n+1} \mid \mathcal{B}_n) = E X_{n+1} = 1$. Thus Theorem 10.1.9 implies $E(T_{n+1} \mid \mathcal{B}_n) = T_n$, so $\{T_n, \mathcal{B}_n\}$ is a martingale (Proposition 10.3.2). If X_j are identically distributed and $X_j > 0$ a.s., let $Y_j := \log X_j$. Then by Jensen's inequality (10.2.6) and Problem 14 of §10.2, $E Y_1 < 0$ unless $X_1 = 1$ a.s. If $-\infty < E Y_1 < 0$, then a.s. $(Y_1 + \cdots + Y_n)/n \to E Y_1$, so $Y_1 + \cdots + Y_n \to -\infty$ a.s. If $E Y_1 = -\infty$, then take M large enough so that $E \max(Y_j, -M) < 0$ and infer that $Y_1 + \cdots + Y_n \to -\infty$ in this case also. Thus $T_n \to 0$ a.s. Such $\{T_n\}$ give examples of L^1-bounded, positive martingales that do not converge in L^1, since $E T_n \equiv 1$.

Example. The martingale and stopping time in this example are best known by the name "double or nothing." Let D_n be independent random variables such that $P(D_n = 2^n) = P(D_n = -2^n) = 1/2$ for $n = 0, 1, 2, \ldots$. Let $X_n := \sum_{j=0}^{n} D_j$. Then X_n form a martingale since $E D_j = 0$, taking the σ-algebra for each n to be \mathcal{A}_n, the smallest σ-algebra for which D_0, \ldots, D_n or equivalently X_0, X_1, \ldots, X_n are measurable.

Let τ be the least n such that $X_n > 0$ or equivalently $D_n > 0$ since $-1 - 2 - \cdots - 2^{n-1} + 2^n = 1 > 0$ for each n. Then $P(\tau \geq n) = 1/2^n$, so τ is finite a.s., and since $X_\tau \equiv 1$ we have $E X_\tau = 1$. However, for each n, $E X_{\tau \wedge n} = 0$ by the optimal stopping theorem 10.4.1. In fact $X_{\tau \wedge n} = 1$ with probability $1 - 2^{-n-1}$ and $-2^{n+1} + 1$ with probability 2^{-n-1}. Also, $X_{\tau \wedge n}$ are L^1-bounded, and $(X_{\tau \wedge n}, \mathcal{A}_{\tau \wedge n})$ is a martingale by Lemma 10.5.3, while $X_{\tau \wedge n}$ converges a.s. to 1, illustrating Theorem 10.5.4. The variables $X_{\tau \wedge n}$ are not uniformly integrable, as can be checked directly, and also because they don't converge in expectation (Corollary 10.5.2). The original martingale $\{X_n\}$ is not L^1-bounded.

Problems

1. Let $X_1, X_2, \ldots,$ be i.i.d. random variables with $X_1 \geq 0$, $1 < E X_1 < \infty$, $T_n := X_1 X_2 \cdots X_n$, and $\mathcal{B}_n :=$ the smallest σ-algebra making X_1, \ldots, X_n measurable.
 (a) Show that $\{T_n, \mathcal{B}_n\}$ is a submartingale.
 (b) Find the Doob decomposition of T_n.
 (c) If $P(X_1 = 0) > 0$, show that $T_n \to 0$ a.s. although $\{T_n\}$ is not L^1-bounded.

2. In Polya's urn scheme (example just before Lemma 10.5.6) prove that $B_n/(B_n + R_n)$ is a martingale. Show that it converges a.s. to a nonconstant limit.

3. Prove or disprove: For every martingale $\{X_n, \mathcal{B}_n\}$ and stopping time α, $\{X_{\alpha \wedge n}, \mathcal{B}_n\}$ is a martingale. *Note:* $\{X_{\alpha \wedge n}, \mathcal{B}_{\alpha \wedge n}\}$ is a martingale by Lemma 10.5.3.

4. Let \mathcal{A}_n be an increasing sequence of σ-algebras whose union generates a σ-algebra \mathcal{S}. Let μ and ν be probability measures on \mathcal{S} such that ν is singular with respect to μ on \mathcal{S}, but absolutely continuous with respect to μ on each \mathcal{A}_n with Radon-Nikodym derivative f_n. Show that $f_n \to 0$ a.s. for μ. *Hint:* If $f_n \to f$ a.s., show by Fatou's lemma that $\int_A f \, d\mu \leq \nu(A)$ for every $A \in \mathcal{S}$.

5. Let X_1, X_2, \ldots, be independent real random variables with $EX_n = 0$ for all n and $S_n := X_1 + \cdots + X_n$. Let \mathcal{B}_n be the smallest σ-algebra for which X_1, \ldots, X_n are measurable. Then $\{S_n, \mathcal{B}_n\}$ is a martingale (see the example before Proposition 10.3.2).
 (a) Give an example where S_n converge a.s., but are not L^1-bounded.
 (b) If $|X_j| \leq 2$ a.s. for all j, and S_n converges a.s., show that S_n converges in L^2. *Hint:* Use the three-series theorem (9.7.3).

6. (Wald's identity.) Let X_1, X_2, \ldots, be i.i.d. real random variables with $E|X_1| < \infty$ and S_n and \mathcal{B}_n as in Problem 5. Then $\{S_n, \mathcal{B}_n\}$ is a submartingale or supermartingale, depending on the sign of EX_1. Let τ be a stopping time for $\{S_n\}$. If $E\tau < \infty$, show that $ES_\tau = E\tau E X_1$. *Hints:* $ES_\tau = \sum_{n \geq 1} P(\tau = n) \sum_{i=1}^n E(X_i \mid \tau = n) = \sum_{i \geq 1} E(X_i \mid \tau \geq i) \times P(\tau \geq i)$. Now $\{\tau \geq i\} = \{\tau > i - 1\} \in \mathcal{B}_{i-1}$, and X_i is independent of this event, while $\sum_{i \geq 1} P(\tau \geq i) = E\tau$. Justify the interchange of sums by considering $|X_i|$ in place of X_i.

7. Show that for any random variable Y, $E|Y|^p = \int_0^\infty pt^{p-1} P(|Y| \geq t) \, dt$. *Hint:* Unless you can justify integrating by parts, prove this first for simple functions Y, then take a limit.

8. Let X and Y be two nonnegative random variables such that for every $t > 0$, $P(Y \geq t) \leq t^{-1} \int_{Y \geq t} X \, dP$. For any $p > 1$, $\|f\|_p := (\int |f|^p \, dP)^{1/p}$, and $q^{-1} + p^{-1} = 1$, show that $\|Y\|_p \leq q\|X\|_p$. *Hints:* Use Problem 7, interchange integrals, and apply the Rogers-Hölder inequality (5.1.2).

9. Given a nonnegative submartingale $\{X_n, \mathcal{B}_n\}$, let $X_n^* := \max_{j \leq n} X_j$ and $X^* := \max_{j \geq 1} X_j$. Prove that for any $p > 1$ and $p^{-1} + q^{-1} = 1$, $\|X^*\|_p \leq q \cdot \sup_n \|X_n\|_p$. *Hints:* Use the previous problem and Theorem 10.4.2; first deal with finite n.

10. Let X_1, X_2, \ldots, be independent random variables with means $EX_j = 0$ and finite variances σ_j^2. Let $S_n = X_1 + \cdots + X_n$.
 (a) Show that $\{S_n^2\}$ is a submartingale.
 (b) Evaluate the Doob decomposition of $\{S_n^2\}$.
 (c) If $\sum_j \sigma_j^2 < \infty$, show that the martingale S_n converges a.s. (giving a martingale proof of part of the three-series theorem, §9.7).

11. In Problem 10, suppose that $\sigma_j^2 = \sigma^2$ for all j. Let T be a stopping time for $\{S_n\}$ with $ET < \infty$. Show that $ES_T^2 = \sigma^2 ET$. *Hint:* Prove this for $T \wedge n$ and let $n \to \infty$. If $T \leq n$, use induction on n. See also the hints to Problem 6.

12. Let X_j be independent with $P(X_j = 1) = P(X_j = -1) = 1/2$, so that S_n as in Problems 10 and 11 is a "simple random walk." For positive integers m and n let T_{mn} be the least k such that $S_k = -m$ or $S_k = n$. Evaluate ET_{mn}, using Problem 11. *Hint:* For $t = T_{mn}$ show that $ES_t = 0$ and find $\mathcal{L}(S_t)$.

13. (a) If $Y_n \to Y$ a.s. and Y_n are all independent of Y, show that Y is a constant a.s.
 (b) In the example just before Theorem 10.5.1, if A is a "tail event" in the sense that it is independent of X_1, \ldots, X_n for all n, show that $P(A) = 0$ or 1, giving a martingale proof of an improved form of the Kolmogorov 0–1 law (8.4.4).

*10.6. Reversed Martingales and Submartingales

One way to look at "reversed" martingales is to take a sequence $(X_n, \mathcal{B}_n)_{n \geq 1}$ where the σ-algebras \mathcal{B}_n decrease as n increases. Here, though, the way of setting up such sequences will be to take sequences indexed by the negative integers $n = -1, -2, \ldots$, in their usual ordering. Such a sequence will be written $(X_n, \mathcal{B}_n)_{n \leq -1}$. Then a martingale, submartingale, or supermartingale, for the definitions at the beginning of §10.3 applied to the index set of negative integers, will be called a *reversed* martingale, submartingale, or supermartingale, respectively. Clearly, a reversed martingale satisfies $X_n = E(X_{-1} \mid \mathcal{B}_n)$ for $n = -1, -2, \ldots$. In this sense it is like a right-closed martingale, being the set of conditional expectations of one random variable for a sequence of σ-algebras, and thus uniformly integrable by the proof of Theorem 10.4.3. But for the interesting cases of convergence of reversed martingales and submartingales, n will go to $-\infty$.

Example. Let X_1, X_2, \ldots, be i.i.d. with $E|X_1| < \infty$ and $S_n := X_1 + \cdots + X_n$. Let $Y_{-n} := S_n/n$ and let \mathcal{B}_{-n} be the smallest σ-algebra for which S_k

are measurable for all $k \geq n$. Then $\{Y_n, \mathcal{B}_n\}_{n \leq -1}$ is a reversed martingale, whose a.s. convergence is equivalent to the strong law of large numbers (see Problem 1 below).

10.6.1. Theorem *For any reversed martingale* $(X_n, \mathcal{B}_n)_{n \leq -1}$, X_n *converges a.s. and in* \mathcal{L}^1 *as* $n \to -\infty$ *to* $X_{-\infty} := E(X_{-1} \mid \mathcal{B}_{-\infty})$ *where* $\mathcal{B}_{-\infty} := \bigcap_n \mathcal{B}_n$.

Proof. First suppose $E X_{-1}^2 < \infty$. If the X_n are defined on a probability space (Ω, \mathcal{S}, P), then X_{-1} belongs to the Hilbert space $H := L^2(\Omega, \mathcal{S}, P)$. By Theorem 10.2.9, the other variables $X_n = E(X_{-1} \mid \mathcal{B}_n)$ also belong to H, and the operator $T_n(X) := E(X \mid \mathcal{B}_n)$ is the orthogonal projection of H onto the linear subspace $H_n := L^2(\Omega, \mathcal{B}_n, P)$. Let $D_n := X_n - X_{n-1} = E(X_{-1} \mid \mathcal{B}_n) - E(X_{-1} \mid \mathcal{B}_{n-1})$ for $n = -1, -2, \dots$. Then the D_n are all orthogonal to each other in H, with

$$\|X_{-1}\|^2 = \|D_{-1}\|^2 + \cdots + \|D_{n+1}\|^2 + \|X_n\|^2 \qquad \text{for } n = -2, -3, \dots,$$

since $X_n \in H_n$ and by the Parseval equality (Theorem 5.4.1). It follows that $\sum_{n \leq -1} \|D_n\|^2$ converges. Hence, the series $D_{-1} + D_{-2} + \cdots$ of orthogonal elements converges, by the Riesz-Fischer theorem (5.4.5) (with $x_n := \|D_n\|$, and $e_n := D_n / \|D_n\|$ if $\|D_n\| > 0$). Thus the sequence $X_n = X_{-1} - D_{-1} - \cdots - D_{1+n}$ converges in L^2 as $n \to -\infty$ to some Y. Let $H_{-\infty}$ be the intersection of all the H_n. To see that $H_{-\infty} = L^2(\Omega, \mathcal{B}_{-\infty}, P)$, the following will help:

10.6.2. Lemma *Let* $\{\mathcal{B}_n\}_{n \leq -1}$ *be* σ*-algebras, as in Theorem 10.6.1. If* B *is a measurable set such that for each* n, $P(A_n \triangle B) = 0$ *for some* $A_n \in \mathcal{B}_n$, *then* $P(A \triangle B) = 0$ *for some* $A \in \mathcal{B}_{-\infty}$.

Proof. Let $A := \limsup_{n \to -\infty} A_n := \bigcap_m \bigcup_{n \leq m} A_n$. Then since $C_m := \bigcup_{n \leq m} A_n$ decreases as $m \downarrow -\infty$ and $C_m \in \mathcal{B}_m$, $A \in \mathcal{B}_{-\infty}$. For each m, $P(C_m \backslash B) \leq \sum_{n \leq m} P(A_n \backslash B) = 0$ and $P(B \backslash C_m) \leq P(B \backslash A_m) = 0$. Then by monotone convergence, $P(A \triangle B) = 0$. \square

10.6.3. Lemma $H_{-\infty} = L^2(\Omega, \mathcal{B}_{-\infty}, P)$; *that is, if* $F \in L^2(\Omega, P)$, *the equivalence class* F *is in* $H_{-\infty}$ *if and only if* $g \in F$ *for some* $g \in \mathcal{L}^2(\Omega, \mathcal{B}_{-\infty}, P)$.

Proof. Clearly, each \mathcal{L}^2 function measurable for $\mathcal{B}_{-\infty}$ is in H_n for each n and so in $H_{-\infty}$. Conversely, if $f \in H_{-\infty}$, for each rational t, $\{f > t\}$ almost surely equals a set in \mathcal{B}_n for each n and so, by Lemma 10.6.2, almost surely equals a set $C_t \in \mathcal{B}_{-\infty}$. Let $h(\omega) := \sup\{t \in \mathbb{Q} : \omega \in C_t\}$, where $-\infty \leq t \leq \infty$. Then

for any $y \in \mathbb{R}$, $\{h > y\} = \bigcup \{C_t : t > y\}$, so h is $\mathcal{B}_{-\infty}$ measurable. Outside of $\bigcup_{t \in \mathbb{Q}} \{f > t\} \triangle C_t$, a set of probability 0, $h = f$, proving Lemma 10.6.3.
$\qquad \square$

Now to continue the proof of Theorem 10.6.1, X_n converges in L^2 as $n \to -\infty$ to a function Y in $\mathcal{L}^2(\Omega, \mathcal{B}_{-\infty}, P)$. For any set $T \in \mathcal{B}_{-\infty}$, $E(X_n 1_T) = E(X_{-1} 1_T)$ for all n, so these terms also equal $E(Y 1_T)$. It follows that $Y = E(X_{-1} \mid \mathcal{B}_{-\infty})$, by uniqueness of conditional expectation (Theorem 10.1.1). Now L^2 convergence implies L^1 convergence, since $E|f| = E|f| \cdot 1 \le (Ef^2)^{1/2}$ for any f by the Cauchy-Bunyakovsky-Schwarz inequality. Let $X_{-\infty} := Y$.

Now $\{X_n - X_{-\infty}\}_{n \le -1}$ is also a (reversed) martingale. Thus for any finite set F of values of n, with largest member m, and $\varepsilon > 0$, by the Doob maximal inequality (10.4.2),

$$ Pr\left\{ \max_{k \in F} X_k - X_{-\infty} > \varepsilon \right\} \le E(X_m - X_{-\infty})^+ / \varepsilon. $$

Then by monotone convergence, let F increase up to the set of all negative integers to get $Pr\{\sup_{k \le m} X_k - X_{-\infty} > \varepsilon\} \le E(X_m - X_{-\infty})^+ / \varepsilon$, which converges to 0 as $m \to -\infty$. There is a symmetric inequality for $\sup_{k \le m} X_{-\infty} - X_k$. Letting $\varepsilon \downarrow 0$, we find that X_n converges almost surely as $n \to -\infty$ to $X_{-\infty}$ in this case.

Now suppose that X_{-1} is not necessarily in \mathcal{L}^2. Given $\varepsilon > 0$, there is a $Y_{-1} \in \mathcal{L}^2$ with $E|X_{-1} - Y_{-1}| < \varepsilon^2$; for example, let $Y_{-1} = X_{-1}$ for $|X_{-1}| \le M$ and $Y_{-1} = 0$ otherwise for a large enough value of M. Let $Y_n := E(Y_{-1} \mid \mathcal{B}_n)$ for $n = -2, 3, \ldots$. Then $\{Y_n\}$ and $\{X_n - Y_n\}$ are both reversed martingales. The conclusion holds for Y_n. For $X_n - Y_n$, we have again by the maximal inequality

$$ Pr\left\{ \sup_n |X_n - Y_n| > \varepsilon \right\} \le E|X_{-1} - Y_{-1}| / \varepsilon < \varepsilon. $$

Thus $Pr(\limsup X_n - \liminf X_n > 2\varepsilon) < \varepsilon$. Letting $\varepsilon \downarrow 0$, we see that X_n converges a.s. as $n \to -\infty$. Since it is uniformly integrable, the sequence also converges in \mathcal{L}^1. The limit is again $E(X_{-1} \mid \mathcal{B}_{-\infty})$ by uniqueness of conditional expectations, proving Theorem 10.6.1.
$\qquad \square$

For any martingale $\{(Y_j, \mathcal{B}_j)\}_{1 \le j \le n}$ with only finitely many indices, let $X_{j-n-1} := Y_j$ for $j = 1, \ldots, n$ and $X_m := X_{-n}$ for $m \le -n$. Then $\{X_m\}_{m \le -1}$ is an example of a reversed martingale, which trivially converges as $m \to -\infty$.

So the essential difference between martingales and reversed martingales appears only for infinite sets of indices.

To treat reversed submartingales, much as for ordinary submartingale sequences, a Doob decomposition will be useful.

10.6.4. Theorem (Doob Decomposition and Convergence for Reversed Submartingales) *Let* $\{X_n, \mathcal{B}_n\}_{n \leq -1}$ *be a reversed submartingale, that is,* \mathcal{B}_n *are σ-algebras with* $\cdots \subset \mathcal{B}_{-2} \subset \mathcal{B}_{-1}$, X_n *is \mathcal{B}_n measurable and integrable for each n, and* $E(X_{n+1} \mid \mathcal{B}_n) \geq X_n, n = -2, -3, \ldots$. *Suppose that* $K := \inf_n E X_n > -\infty$. *Then there is a decomposition* $X_n \equiv Y_n + Z_n$ *where* Z_n *is measurable for* \mathcal{B}_{n-1}, $\{Y_n, \mathcal{B}_n\}$ *is a reversed martingale, and* $Z_{-1} \geq Z_{-2} \geq \cdots$ *almost surely, with* $Z_n \downarrow 0$ *a.s. as* $n \downarrow -\infty$. *Thus* X_n *converges almost surely and in* \mathcal{L}^1 *as* $n \to -\infty$.

Proof. Let $D_n := E(X_n \mid \mathcal{B}_{n-1}) - X_{n-1}$ for $n = -1, -2, \ldots$. By assumption, $D_n \geq 0$ a.s. Now $\sum_{n \leq -1} E D_n = \sum_{n \leq -1} E(X_n - X_{n-1}) = E X_{-1} - K$; note that $E X_n \downarrow K$ as $n \to -\infty$. Since the sum is finite, the series $\sum D_n$ converges by monotone convergence (Theorem 4.3.2) to an integrable function.

Let $Z_n := \sum_{k \leq n} D_k$ for each n. Then Z_n is nondecreasing as n increases almost surely, and $Z_n \downarrow 0$ a.s. as $n \to -\infty$. Let $Y_n := X_n - Z_n$. To show that Y_n is a (reversed) martingale, we have

$$E(Y_n \mid \mathcal{B}_{n-1}) = E(X_n - Z_n \mid \mathcal{B}_{n-1}) = D_n + X_{n-1} - \sum_{k \leq n} D_k$$

$$= X_{n-1} - \sum_{k < n} D_k = X_{n-1} - Z_{n-1} = Y_{n-1},$$

proving the decomposition as desired. Now Z_n converges in \mathcal{L}^1 by monotone or dominated convergence, and Y_n converges a.s. and in \mathcal{L}^1 by Theorem 10.6.1, so X_n converges likewise. \square

Problems

1. As in Problem 4 in §10.1, let $X_1, X_2, \ldots,$ be i.i.d. real random variables with $E|X_1| < \infty$. Let $S_n := X_1 + \cdots + X_n$. For $n = 1, 2, \ldots,$ let $T_{-n} := S_n / n$. Let \mathcal{B}_{-n} be the smallest σ-algebra for which S_k are measurable for all $k \geq n$. Show that $\{T_k, \mathcal{B}_k\}_{k \leq -1}$ is a reversed martingale and $\{|T_k|, \mathcal{B}_k\}_{k \leq -1}$ is a reversed submartingale.

2. (Continuation.) Let Y be a random variable independent of all the X_j, with $E|Y| < \infty$. Let $U_j := Y + T_j$ for $j = -1, -2, \ldots$. Let \mathcal{C}_k be the smallest σ-algebra for which U_j is measurable for all $j \leq k$. Show that

$\{U_k, C_k\}_{k \leq -1}$ is a reversed martingale and find its limit as $k \to -\infty$. Let $C_{-\infty} := \bigcap_{k \leq -1} C_k$. Is Y equal a.s. to a variable measurable for $C_{-\infty}$?

3. Show that any reversed supermartingale $\{X_n\}$ with $X_n \leq 0$ a.s. for all n is uniformly integrable.

4. Let $\{\mathcal{B}_n\}_{n \leq -1}$ be σ-algebras with $\mathcal{B}_{n-1} \subset \mathcal{B}_n$ for all $n \leq -1$. Let B be a measurable set such that for some $A_n \in \mathcal{B}_n$ for each n, $P(A_n \triangle B) \to 0$ as $n \to -\infty$. Show that for some $A \in \mathcal{B}_{-\infty}$, $P(A \triangle B) = 0$ (this improves on Lemma 10.6.2).

5. Show that in Theorem 10.6.4, the decomposition $X_n = Y_n + Z_n$ with the stated properties is unique. *Hint:* First show $Z_n - Z_{n-1}$ is unique.

6. Show however that if Z_n is taken to be only \mathcal{B}_n measurable rather than \mathcal{B}_{n-1} measurable, then the decomposition $X_n = Y_n + Z_n$ is not unique.

*10.7. Subadditive and Superadditive Ergodic Theorems

This section will prove an extension of the ergodic theorem (§8.4) to situations where, instead of partial sums of variables transformed by successive powers of a measure-preserving transformation, we have only one-sided inequalities relating such partial sums.

Let (X, \mathcal{A}, P) be a probability space and T a measure-preserving transformation of X onto itself. Then a sequence $\{f_m\}_{m \geq 1}$ of real-valued, integrable random variables on X is called *superadditive* iff for all positive integers m and n, $f_{m+n} \geq f_m + f_n \circ T^m$ a.s. The sequence is *subadditive* if the reverse inequality holds, and *additive* if both hold. Clearly, a sequence $\{f_n\}$ is subadditive if and only if $\{-f_n\}$ is superadditive, so that facts about subadditive sequences are equivalent to facts about superadditive sequences.

Let f be integrable for P and $S_n := \sum_{0 \leq j < n} f \circ T^j$, as in the ergodic theorem (8.4.1), where T^0 is defined as the identity, so $f \circ T^0 \equiv f$. Then it is easily seen that $\{S_n\}_{n \geq 1}$ is additive. Conversely, for any additive sequence $\{f_n\}$, let $f := f_1$. Then $S_n = f_n$ a.s. for all n, as can be seen by induction on n, so every additive sequence is of the form $\{S_n\}$.

Example. Given a law P on \mathbb{R}, take a countable product of copies of (\mathbb{R}, P) as in §8.2, and let X_1, X_2, \ldots, be the coordinates, so that the X_i are i.i.d. with law P. Let T be the shift transformation $T(\{X_n\}) := \{X_{n+1}\}$ as in §8.4. Then the partial sums $S_n := X_1 + \cdots + X_n$ form an additive sequence and $\{|S_n|\}_{n \geq 1}$ is subadditive.

Let $\mathcal{A}_{\text{inv}(T)}$ be the σ-algebra of *invariant* sets B in \mathcal{A} for which $B = T^{-1}(B)$. Let E_I denote conditional expectation with respect to $\mathcal{A}_{\text{inv}(T)}$. Recall (§8.4)

that if T is ergodic, as it is for shift transformations in the last example (by Theorem 8.4.5), the invariant sets will have probabilities 0 or 1, so that $E_I g = Eg$ for any integrable random variable g.

Here is the main convergence theorem for superadditive sequences:

10.7.1. Theorem (J. F. C. Kingman) *For any superadditive sequence $\{f_n\}$ of integrable random variables, f_n/n converges a.s. as $n \to \infty$ to $\gamma :=$ $\sup_n E_t(f_n)/n \le +\infty$. Here γ is integrable if and only if $\sup_n Ef_n/n < +\infty$, and if γ is integrable, then f_n/n converges to γ also in L^1.*

Proof (J. Neveu). Let $f_0 := 0$; this is consistent with superadditivity. Let $f^+ := \max(f, 0)$, $f^- := f^+ - f$ for any f. Let

$$F_n := \sum_{j=0}^{n-1} f_1 \circ T^j, \qquad H_n := \sum_{j=0}^{n-1} f_1^+ \circ T^j.$$

Then $\{F_n\}_{n \ge 1}$ is additive and $f_n \ge F_n$ for all n, by superadditivity and induction on n:

$$f_{n+1} \ge f_n + f_1 \circ T^n \ge F_n + f_1 \circ T^n = F_{n+1}.$$

Let $G_n := f_n - F_n$. Then $G_n \ge 0$ and $\{G_n\}$ is superadditive. Let $J_n := H_n - F_n = \sum_{0 \le j < n} f_1^- \circ T^j$. Then $\{H_n\}$ and $\{J_n\}$ are nonnegative and additive, with $f_n \equiv G_n + H_n - J_n$. For the additive sequence $\{J_n\}$, $E_I(J_n)/n = E_I f_1^-$ for all n, so that $\sup_n E_I(J_n/n) = \inf_n E_I(J_n/n)$ a.s. Thus if J_n/n converges a.s. to its $\gamma := \gamma(\{J_n\})$, then also $-J_n/n \to \gamma(\{-J_n\}) = -\gamma(\{J_n\})$ a.s., and likewise for $\{H_n\}$. Hence, if the theorem holds for $\{G_n\}$, $\{H_n\}$, and $\{J_n\}$, then since $\gamma(\{f_n\}) = \gamma(\{G_n\}) + \gamma(\{H_n\}) - \gamma(\{J_n\})$, it also holds for $\{f_n\}$. So we can assume $f_n \ge 0$.

The main part of the proof is based on the method of a proof of the Birkhoff ergodic theorem due to F. Riesz, beginning with the following:

10.7.2. Lemma (Riesz) *Let u_1, \ldots, u_n be any real numbers and $s_j := u_1 + \cdots + u_j$. Let $v_j := v(j, n) := \max_{k: j \le k \le n} s_k - s_j$. Then*

$$\sum_{j=0}^{n-1} u_{j+1} 1_{\{v(j,n) > 0\}} \ge 0.$$

Remarks. A term u_j is called an "m-leader" if $u_j + \cdots + u_{j+m-1} > 0$. So the lemma says that the sum of all u_j which are m-leaders (for any m with $j \le j + m - 1 \le n$) is nonnegative.

Proof. We have $v_j \geq 0$ for all j (consider $k = j$). For $j = 0, 1, \ldots, n - 1$, we have $v_j = (u_{j+1} + v_{j+1})^+$, as follows: $v_j = \max(0, u_{j+1}, u_{j+1} + u_{j+2}, \ldots, u_{j+1} + \cdots + u_n)$, and likewise $v_{j+1} = \max(0, u_{j+2}, \ldots, u_{j+2} + \cdots + u_n)$, or $v_n = 0$ if $j = n - 1$, so $v_j = \max(0, u_{j+1} + \max(0, u_{j+2}, \ldots, u_{j+2} + \cdots + u_n)) = \max(0, u_{j+1} + v_{j+1})$, as claimed. It follows that

$$v_j \leq v_{j+1} + u_{j+1} 1_{\{v(j,n) > 0\}},$$

if $v_j = 0 \leq v_{j+1} + 0$ or if $0 < v_j = v(j, n) = v_{j+1} + u_{j+1}$. Summing over j, with $v_j - v_{j+1}$ summing to $v_0 - v_n$, gives that the sum in the lemma is at least $v_0 - v_n = v_0 \geq 0$. □

10.7.3. Lemma (Maximal Inequality) *For a superadditive sequence $\{f_n\}$ with $f_n(x) \geq 0$ for all n and x, let g be any nonnegative, integrable function. Set*

$$v := \sup_{n \geq 0} \left(f_n - \sum_{0 \leq j < n} g \circ T^j \right).$$

Then

$$E_I\left(g 1_{\{v > 0\}}\right) \leq \sup_{n \geq 1} E_I(f_n)/n.$$

Proof. The Riesz lemma will be applied for each x to the sequence $u_j := f_i - f_{i-1} - g \circ T^{i-1}$. Then, with a telescoping sum, we have

$$v(j, n) = \max_{k: j \leq k \leq n} \left(f_k - f_j - \sum_{i=j}^{k-1} g \circ T^i \right), \qquad 0 \leq j < n.$$

The Riesz lemma gives

$$\sum_{j=0}^{n-1} (f_{j+1} - f_j - (g \circ T^j)) 1_{\{v(j,n) > 0\}} \geq 0, \qquad n = 1, 2, \ldots.$$

For each j, $f_{j+1} \geq f_j + (f_1 \circ T) \geq f_j$ since $f_1 \geq 0$, so the sequence $\{f_j\}$ is nondecreasing. Thus we have

$$f_n = f_n - f_0 = \sum_{j=0}^{n-1} f_{j+1} - f_j \geq \sum_{j=0}^{n-1} (f_{j+1} - f_j) 1_{\{v(j,n) > 0\}}$$

$$\geq \sum_{j=0}^{n-1} (g \circ T^j) 1_{\{v(j,n) > 0\}}.$$

For $j \leq k$, we have $f_k - f_j \geq f_{k-j} \circ T^j$. Thus $v(j, n) \geq v(0, n - j) \circ T^j$. Since $g \geq 0$, it follows that

$$f_n \geq \sum_{j=0}^{n-1} \left(g 1_{\{v(0,n-j)>0\}} \right) \circ T^j, \quad \text{and}$$

$$E_I(f_n) \geq \sum_{k=1}^{n} E_I \left(g 1_{\{v(0,k)>0\}} \right).$$

As $k \to \infty$, the events $\{v(0, k) > 0\}$ increase up to the event $\{v > 0\}$ in the statement of Lemma 10.7.3. Thus

$$\frac{1}{n} \sum_{k=1}^{n} 1_{\{v(0,k)>0\}} \uparrow 1_{\{v>0\}} \quad \text{as } n \to \infty.$$

Then the monotone convergence theorem for conditional expectations (10.1.7) and a supremum over n give the maximal inequality as stated. □

Let $\bar{f} := \limsup_{n\to\infty} f_n/n$. About \bar{f} we will prove:

10.7.4. Lemma *We have $\bar{f} \geq \bar{f} \circ T$ a.s. and for any measurable function h such that $h \geq h \circ T$ a.e., where T is a measure-preserving transformation of a finite measure space (X, \mathcal{A}, μ), we have $h \circ T = h$ a.e.*

Proof. First, for all n, $f_{n+1} \geq f_1 + f_n \circ T$, so

$$\frac{f_{n+1}}{n+1} \geq \frac{f_1}{n+1} + \frac{f_n \circ T}{n+1}.$$

Letting $n \to \infty$, $f_1/(n + 1) \to 0$ and $n/(n + 1) \to 1$, so the lim sup gives $\bar{f} \geq \bar{f} \circ T$, as stated. For the second half, if the statement fails, there is a rational q with $\mu(h \circ T < q < h) > 0$. Then since μ is finite, $\mu(h \circ T < q) > \mu(h < q)$, but these measures are equal since T is measure-preserving, a contradiction, proving Lemma 10.7.4. □

Now to continue the proof of Theorem 10.7.1, since the sequence of constants $\{n\}_{n\geq 1}$ is additive, the sequence $\{f_n + n\}_{n\geq 1}$ is superadditive. If γ' is the function defined for $\{f_n + n\}$ in the statement of Theorem 10.7.1, as γ is for $\{f_n\}$, then $\gamma' \equiv \gamma + 1$ and $(f_n + n)/n = (f_n/n) + 1$, so the statement of the theorem for $\{f_n + n\}$ is equivalent to that for $\{f_n\}$. So we can assume $f_n \geq n$ for all n. Then $f \geq 1$.

For $r = 1, 2, \ldots$, let $g_r := \min(r, \bar{f} - 1/r)$. Then each g_r is an invariant function ($g_r \circ T = g_r$ a.s.), nonnegative and integrable. Apply the maximal inequality (10.7.3) to $g = g_r$. Since $g_r < \bar{f}$ everywhere, we have $v > 0$

everywhere in Lemma 10.7.3, so it gives

$$g_r = E_I g_r \leq \sup_n E_I(f_n)/n = \gamma$$

as defined in the theorem (10.7.1). Letting $r \to \infty$ gives

$$(*) \quad \bar{f} \leq \gamma \qquad \text{a.s.}$$

To prove $\liminf f_n/n \geq \gamma$ a.s., first suppose $\{f_n\}$ is additive, and let $h :=$ f_1, so that

$$f_n = \sum_{j=0}^{n-1} h \circ T^j.$$

Take any a with $0 < a < \infty$. Note that

$$(**) \quad h \geq \min(h, a) = a - (a - h)^+, \qquad \text{so}$$

$$\liminf_{n \to \infty} f_n/n \geq a - \limsup_{n \to \infty} n^{-1} \sum_{0 \leq j < n} (a - h)^+ \circ T^j.$$

Applying $(*)$ to $(a-h)^+$, then noting that $\gamma = E_I(h)$ for an additive sequence, as at the beginning of the proof, gives

$$\liminf_{n \to \infty} f_n/n \geq a - E_I((a - h)^+) = E_I(\min(h, a))$$

by $(**)$. Letting $a \uparrow +\infty$ then gives that

$$(***) \quad f_n/n \to \gamma \qquad \text{a.s. for additive sequences.}$$

In the general case, $f_k \circ T^j \leq f_{k+j} - f_j$ for each j gives

$$(****) \quad \sum_{j=0}^{n-1} f_k \circ T^j \leq \sum_{j=0}^{n-1} (f_{k+j} - f_j) \leq k f_{n+k-1},$$

since f_1, \ldots, f_{k-1} are nonnegative and the sequence is nondecreasing, so $f_r \leq f_{n+k-1}$ for $r = n, \ldots, n + k - 1$. Now noting that $(n + k - 1)/n \to 1$ as $n \to \infty$ for fixed k, we get

$$\liminf_{n \to \infty} f_n/n = \liminf_{n \to \infty} f_{n+k-1}/(n + k - 1) = \liminf_{n \to \infty} f_{n+k-1}/n$$

$$\geq \frac{1}{k} \liminf_{n \to \infty} \frac{1}{n} \sum_{j=0}^{n-1} f_k \circ T^j \quad \text{by } (****)$$

$$= \frac{1}{k} E_I(f_k) \qquad \text{by } (***) \text{ with } h \text{ replaced by } f_k.$$

Taking the supremum over k then gives $\gamma \leq \liminf_{n \to \infty} f_n/n$, so that by $(*)$, $f_n/n \to \gamma$ a.s.

Now set $z_n := f_n/n$. Then we have $z_n \to \sup_m E_I z_m = \gamma$, and $E\gamma \geq EE_I z_m = E z_m$ for all m. By Fatou's lemma (4.3.3), $E\gamma \leq \liminf_{n\to\infty} E z_n$, so $E z_n \to E\gamma$. It follows that γ is integrable if and only if $\sup_n E z_n < \infty$, as stated. For L^1 convergence, $0 \leq (\gamma - z_n)^+ \leq \gamma$, so if $E\gamma < \infty$, then by dominated convergence (Theorem 4.3.5), $E(\gamma - z_n)^+ \to 0$. Thus since $E(\gamma - z_n) \to 0$, we also have $E(\gamma - z_n)^- \to 0$ and $E|\gamma - z_n| \to 0$, finishing the proof of Theorem 10.7.1. $\qquad\square$

Example (Random Products). Suppose M is a semigroup (often of matrices or linear operators), so that an associative product operation is defined. Suppose also that $\|\cdot\|$ is a *submultiplicative* function from M into the positive real numbers, that is, $\|AB\| \leq \|A\|\|B\|$ for all $A, B \in M$. Let A_1 be a measurable function from a probability space into M (for a σ-algebra on M such that the product is jointly measurable and $\|\cdot\|$ is measurable), let T be again a measure-preserving transformation of the probability space, and let $A_n := A_1 \circ T^{n-1}$ for $n = 1, 2, \ldots$. Let $f_n := \log \|A_n A_{n-1} \cdots A_1\|$. Then $\{f_n\}$ is subadditive.

In one case, let M be the set of all linear transformations of \mathbb{R}^k into itself, which can be represented by $k \times k$ matrices. If $\|\cdot\|$ is the usual norm on \mathbb{R}^k, and $A \in M$, let $\|A\|$ be the operator norm $\|A\| := \sup\{|Ax|/|x| : x \neq 0, x \in \mathbb{R}^k\}$. It is easily seen that $\|\cdot\|$ is submultiplicative.

If the A_j are i.i.d. elements of M, then without changing their joint distribution, we can assume that they are the coordinates for a countable Cartesian product of copies of M, with the product probability distribution and σ-algebra, and where T is the shift transformation $\{A_j\} \mapsto \{A_{j+1}\}$. Then the invariant σ-algebra is trivial, by the Kolmogorov 0–1 law (8.4.4), so that f_n/n converges a.s. to some constant. It turns out, however, that it may be nontrivial to evaluate the constant (see Furstenberg and Kesten, 1960; Furstenberg, 1963; Ledrappier, 1984).

Example (Lattice-Indexed Variables). The theory of subadditive sequences seems to have originated from work in percolation theory (Hammersley and Welsh, 1965). For example, let the plane be decomposed by a square lattice formed by parallel horizontal lines at distance 1 apart and vertical lines likewise. Suppose that for each line segment S forming a side of two squares in the lattice there is given a nonnegative random variable $X(S)$. For any two vertices u and v of the lattice, let $L(u, v)$ be the infimum of all sums of $X(S)$ over all sequences of segments S joining u to v. Then for any vertices u, v, and w, $L(u, w) \leq L(u, v) + L(v, w)$. So L is a kind of subadditive process indexed by points in the lattice. Restricted to vertices $(n, 0)$, $n = 1, 2, \ldots$, it gives a subadditive process as defined above.

Problems

1. Let $X_1, X_2, \ldots,$ be i.i.d. real random variables with $E|X_1| < \infty$ and $S_n := X_1 + \cdots + X_n$. Verify that $|S_n|$ is a subadditive sequence and find the limit of $|S_n|/n$ in terms of an integral for X_1.

2. Let (X, \mathcal{A}, P) be a probability space and T a measure-preserving transformation of X onto itself. Let Y be a real random variable on X and $Y_j := Y \circ T^j$ for $j = 1, 2, \ldots$. Let $S_0 := 0$ and for $n = 1, 2, \ldots,$ let $S_n := Y_1 + \cdots + Y_n$ and $f_n := \max\{S_k - S_j : 0 \le j \le k \le n\}$. Prove that $\{f_n\}$ is subadditive.

3. Let $(X, \|\cdot\|)$ be a normed linear space. A linear function A from X into itself is called an *operator* and is called *bounded* if $\|A\| := \sup\{\|Ax\| : \|x\| = 1\} < \infty$.
 Then $\|A\|$ is called the *operator norm* of A.
 (a) Prove that the operator norm is submultiplicative on the set of all bounded linear operators.
 (b) In \mathbb{R}^2, with its usual Euclidean norm, let $A_1, A_2, \ldots,$ be i.i.d. linear operators, each represented by the following matrices with probability $1/3$ each:

$$\begin{pmatrix} 1 & 0 \\ 0 & 1 \end{pmatrix}, \quad \begin{pmatrix} 2 & 0 \\ 0 & 2 \end{pmatrix}, \quad \text{and} \quad \begin{pmatrix} 1 & 1 \\ 0 & 1 \end{pmatrix}.$$

 Find the limiting constant in the random products example (near the end of the section).

4. Let $\{f_n\}$ be a subadditive sequence and $A := \{\sup_n f_n > 0\}$. Show that $\int_A f_1 \, dP \ge 0$. *Hint:* See the maximal ergodic lemma (8.4.2).

5. The subadditive ergodic theorem is often stated for an array $\{f_{m,n}\}_{0 \le m < n}$ of random variables whose joint distribution is the same as that of $\{f_{m+1,n+1}\}_{0 \le m < n}$ and which satisfies $f_{k,n} \le f_{k,m} + f_{m,n}$ whenever $k < m < n$. Show that if $\{f_n\}$ and T satisfy the hypotheses of Theorem 10.7.1, then $f_{m,n} := f_{n-m} \circ T^m$ satisfies those just stated.

6. Conversely, suppose $\{f_{m,n}\}_{0 \le m < n}$ satisfy the conditions of Problem 5. Let Ω be a Cartesian product $\Pi_{0 \le m < n} \mathbb{R}_{m,n}$ of copies of \mathbb{R}, with coordinates $\{x_{m,n}\}_{0 \le m < n}$, and let P be the law of $\{f_{m,n}\}_{0 \le m < n}$ on Ω. Show that $T : \{x_{m,n}\} \mapsto \{x_{m+1,n+1}\}$ is measure-preserving on Ω, that $f_n := x_{0,n}$ is a subadditive sequence, and that $x_{m,n} = f_{n-m} \circ T^m$ as in Problem 5.

7. In a theorem of T. Liggett (1985), the hypotheses of Problem 5 are weakened by assuming subadditivity $f_{k,n} \le f_{k,m} + f_{m,n}$ only for $0 = k < m < n$, and equal joint distributions only as follows:

(a) For each $m \geq 0$, $\{f_{m+1,m+k+1}\}_{k\geq 1}$ have the same joint distribution as $\{f_{m,m+k}\}_{k\geq 1}$, and

(b) For each $k \geq 1$, $\{f_{nk,(n+1)k}\}_{n\geq 1}$ have the same joint distribution as $\{f_{(n+1)k,(n+2)k}\}_{n\geq 1}$. Show that the hypotheses of Problem 5 imply (a) and (b).

8. Show that (a) implies that for each m and n with $0 \leq m < n$, $f_{m+1,n+1}$ has the same distribution as $f_{m,n}$.

9. Show that (a) and (b) do not imply that $\{f_{m+1,n+1}\}_{0\leq m<n}$ and $\{f_{m,n}\}_{0\leq m<n}$ have the same joint distribution. *Hint:* Let $f_{0,n}$ be i.i.d. with law $N(0, 1)$ and let $f_{m,m+k} = f_{0,k}$ for all m and k except $f_{1,3} = -f_{0,2}$.

Notes

§10.1 For a probability space (Ω, \mathcal{B}, P), Kolmogorov (1933, Chap. V, Sec. 4) defined the conditional expectation $E(X \mid Y)$ of X given Y, for two random variables X and Y on Ω with $E|X| < \infty$. This is the same as $E(X \mid \mathcal{B}_Y)$ where \mathcal{B}_Y is the smallest σ-algebra for which Y is measurable. Likewise, $E(X \mid Y) := E(X \mid Y_1, \ldots, Y_n) = E(X \mid \mathcal{B}_Y)$ was defined when $Y := (Y_1, \ldots, Y_n)$ is a random vector, or n-tuple of real random variables. Let (S, \mathcal{T}) be a measurable space and Y a measurable function from Ω to S. Halmos (1950, p. 209) defines conditional expectation given Y. While "conditional expectation given a function" might seem more special than given a σ-algebra, since one can always take the smallest σ-algebra for which Y is measurable, one can also go in the other direction by taking $S = \Omega$, Y the identity, and \mathcal{T} any sub-σ-algebra of \mathcal{B}. Doob (1953, pp. 17–18, 623) defines conditional expectation given a sub-σ-algebra and attributes it to Kolmogorov (1933), who gave the proof of existence using the Radon-Nikodym theorem, which applies generally.

Blackwell and Dubins (1963) proved the following converse to the dominated convergence theorem for conditional expectations (10.1.8): if $f_n \geq 0$ and $f_n \to f_0$ a.s., where $f_n \in \mathcal{L}^1(P)$ for all $n \geq 0$, and if $E \sup_n f_n = +\infty$, then on some probability space there are random variables $\{g_n\}_{n\geq 0}$ with the same joint distribution as $\{f_n\}_{n\geq 0}$ and a σ-algebra \mathcal{C} of measurable sets such that $P(E(g_n \mid \mathcal{C}) \to E(g_0 \mid \mathcal{C})) = 0$.

§10.2 Doob (1938), then Halmos (1941) apparently first undertook to prove existence of regular conditional probabilities, or somewhat more generally, what are called "quotient measures." Their statements, however, had insufficient hypotheses, requiring separability, in effect, but not completeness, in Theorem 10.2.2. Andersen and Jessen (1948a) gave a counterexample to a related statement on existence of measures on infinite product spaces. Dieudonné (1948, pp. 39–51) showed that regular conditional probabilities need not exist on arbitrary separable metric spaces but do for compact metric spaces; this fact is not far from the result for complete separable metric spaces, in view of Ulam's theorem.

"Jensen's inequality" was first proved for \mathbb{R}^1, if X is a simple function and f has a first derivative everywhere, by Hölder (1889). Then Jensen (1906, p. 186) proved it for arbitrary continuous convex functions. Hölder's name is generally associated with another famous inequality (5.1.2), so I don't suggest the name "Hölder-Jensen inequality" for 10.2.6. The proof of existence of regular conditional probabilities and

their use in proving conditional Jensen inequalities are based on the exposition in Bauer (1968, §10.3).

§10.3 The example just before 10.3.4 has a converse: for any nonnegative submartingale sequence $\{X_n, \mathcal{B}_n\}$ on a probability space (Ω, \mathcal{B}, P), there is a probability space (V, \mathcal{C}, Q) and a martingale sequence $\{Y_n, \mathcal{C}_n\}$ on it such that on \mathbb{R}^∞, the distributions of $\{X_n\}_{n \geq 1}$ and $\{|Y_n|\}_{n \geq 1}$ are the same; there are analogous results for nonnegative submartingales $\{X_t\}_{0 \leq t < \infty}$ (Gilat, 1977).

Random House (1987) gives four meanings for "martingale," two for parts of a horse's harness, one for part of a sailing ship's rigging, and one for a betting system such as double or nothing, which led to the mathematical meaning. Some meanings involve ways of branching. The etymology goes back to Middle French before 1600 and then has two possible branches: one through *martegalo*, Provençal for female resident of the town of Martigues, and another through Spanish *almártaga*, "harness," and Arabic *almarta'ah*, "the vein."

The mathematical theory of martingales began to develop strongly around 1940. Ville (1939, pp. 111–130) used the word "martingale" for a sequence of random variables. Doob (1940) undertook a rigorous and deeper theory.

Vitali (1904–05, 1907) was apparently the first to give forms of Theorems 10.3.5 and 10.3.6, for Lebesgue measure on bounded measurable sets in \mathbb{R}.

§10.4–5 Doob (1940) first proved most of the results of these sections on martingales (he then used the phrase "Property \mathcal{E}" rather than "martingale"). Earlier Lévy (1937) proved a.s. convergence for complete martingales such that X_∞ is the indicator function of an event. Doob (1953, p. 294) introduced and treated submartingales (and so, by symmetry, supermartingales), which he then called "semimartingales." Doob (1953, pp. 629–634) gives notes on martingales, mentioning Ville (1939), and extensively describing relations of Doob's own work with overlapping work of Andersen and Jessen (1946, 1948b). The proofs in §10.5 through Lemma 10.5.6 are as in Lamb (1973). Most other proofs count the number of "upcrossings" of any interval $[a, b]$, $X_{n(2j-1)}(\omega) \leq a < b \leq X_{n(2j)}(\omega)$, $j = 1, 2, \ldots, n(1) < n(2) < \cdots$. Lamb's proof seems to be shorter. Baez-Duarte (1968) and P.-A. Meyer (1972) give other interesting proofs without upcrossings. For further developments, see the book of Hall and Heyde (1980).

§10.6 Doob (1940, p. 458; 1953) apparently was the first to treat reversed martingales. The results of this section were given in Doob (1953, p. 329) with somewhat different proofs. In that book Doob apparently introduced reversed submartingales (as well as submartingales) for the first time.

§10.7 The subadditive ergodic theorem (10.7.1) is due to Kingman (1968). The proof given is from Neveu (1983). Most applications use subadditive rather than superadditive sequences. Riesz (1945, p. 225) proved his m-leaders lemma (10.7.2).

Some previous proofs of the subadditive ergodic theorem have applied the classical Birkhoff ergodic theorem (8.4.1 above) for additive sequences. Neveu's proof does not, and thus it proves the Birkhoff theorem at the same time. Kingman (1973, 1976) surveyed subadditive processes.

Liggett (1985) proved the conclusion of the subadditive ergodic theorem under weakened hypotheses (mentioned in Problem 7) and indicated the broader applications of this extended theorem. Leventhal (1988) gave new proofs of Liggett's theorem and of the Birkhoff theorem.

References

An asterisk identifies works I have found discussed in secondary sources but have not seen in the original.

Andersen, Erik Sparre, and Børge Jessen (1946). Some limit theorems on integrals in an abstract set. *Danske. Vid. Selsk. Mat.-Fys. Medd. 22*, no. 14. 29 pp.

———, and ——— (1948a). On the introduction of measures in infinite product sets. *Danske Vid. Selsk. Mat.-Fys. Medd. 25*, no. 4. 8 pp.

———, and ——— (1948b). Some limit theorems on set-functions. *Danske Vid. Selsk. Mat.-Fys. Medd. 25*, no. 5. 8 pp.

Baez-Duarte, Luis (1968). Another look at the martingale theorem. *J. Mathematical Analysis and Appls. 23*: 551–557.

Bauer, Heinz (1968). *Wahrscheinlichkeitstheorie und Grundzüge der Masstheorie*. De Gruyter, Berlin. Publ. in English as *Probability Theory and Elements of Measure Theory*. Transl. Lisa Rosenblatt. Holt, Rinehart & Winston, New York, 1972.

Blackwell, David, and Lester E. Dubins (1963). A converse to the dominated convergence theorem. *Illinois J. Math. 7*: 508–514.

Dieudonné, Jean (1948). Sur le théorème de Lebesgue-Nikodym (III). *Ann. Inst. Fourier* (Grenoble) (Ser. 2) *23*: 25–53.

Doob, Joseph L. (1938). Stochastic processes with an integral-valued parameter. *Trans. Amer. Math. Soc. 44*: 87–150.

——— (1940). Regularity properties of certain families of chance variables. *Trans. Amer. Math. Soc. 47*: 455–486.

——— (1953). *Stochastic Processes*. Wiley, New York.

Furstenberg, Harry (1963). Non-commuting random products. *Trans. Amer. Math. Soc. 108*: 377–428.

———, and Harry Kesten (1960). Products of random matrices. *Ann. Math. Statist. 31*: 457–469.

Gilat, David (1977). Every nonnegative submartingale is the absolute value of a martingale. *Ann. Probability 5*: 475–481.

Hall, Peter, and C. C. Heyde (1980). *Martingale Limit Theory and Its Application*. Academic Press, New York.

Halmos, Paul R. (1941). The decomposition of measures. *Duke Math. J. 8*: 386–392.

——— (1950). *Measure Theory*. Van Nostrand, Princeton.

Hammersley, J. M., and D. J. A. Welsh (1965). First-passage percolation, subadditive processes, stochastic networks, and generalized renewal theory. *Bernoulli-Bayes-Laplace Anniversary Volume*, ed. J. Neyman and L. M. Le Cam, pp. 61–110. Springer, New York.

Hölder, Otto (1889). Ueber einen Mittelwerthssatz. *Gesellschaft der Wissenschaften, Göttingen, Nachrichten (= Göttinger Nachrichten)*, pp. 38–47.

Jensen, J. L. W. V. (1906). Sur les fonctions convexes et les inégalites entre les valeurs moyennes. *Acta Math. 30*: 175–193.

Kingman, John F. C. (1968). The ergodic theory of subadditive stochastic processes. *J. Roy. Statist. Soc.* (Ser. B) *30*: 499–510.

——— (1973). Subadditive ergodic theory (With discussion). *Ann. Probability 1*: 883–899.

——— (1976). Subadditive processes. *École d'été de probabilités de St-Flour*, 1975. *Lecture Notes in Math.* (Springer) *539*: 167–223.

Kolmogoroff, Andrei N. [Kolmogorov, A.N.] (1933). *Grundbegriffe der Wahrscheinlich-keitsrechnung.* Springer, Berlin. 2d English ed., *Foundations of the Theory of Probability,* transl. ed. Nathan Morrison. Chelsea, New York, 1956.

Lamb, Charles W. (1973). A short proof of the martingale convergence theorem. *Proc. Amer. Math. Soc. 38:* 215–217.

Ledrappier, F. (1984). Quelques propriétés des exposants caractéristiques. *École d'été de probabilités de St-Flour,* 1982. *Lecture Notes in Math.* (Springer) *1097*: 305–396.

Levental, Shlomo (1988). A proof of Liggett's version of the subadditive ergodic theorem. *Proc. Amer. Math. Soc. 102*: 169–173.

Lévy, Paul (1937). *Théorie de l'addition des variables aléatoires.* 2d. ed. 1954. Gauthier-Villars, Paris.

Liggett, Thomas M. (1985). An improved subadditive ergodic theorem. *Ann. Probability 13*: 1279–1285.

Meyer, Paul-André (1972). *Martingales and Stochastic Integrals I. Lecture Notes in Math. 284* (Springer): Berlin.

Neveu, J. (1983). Courte démonstration du théorème ergodique sur-additif. *Ann. Inst. Henri Poincaré* (Ser. B) *19*: 87–90.

Random House Dictionary of the English Language (1987), 2d ed., unabridged, Random House, New York.

Riesz, F. (1945). Sur la théorie ergodique. *Comment. Math. Helvetici 17*: 221–239.

*Ville, Jean (1939). *Étude critique de la notion de collectif.* Gauthier-Villars, Paris.

*Vitali, Giuseppe (1904–05). Sulle funzioni integrali. *Atti Accad. Sci. Torino 40*: 1021–1034.

——— (1907). Sull'integrazione per serie. *Rend. Circolo Mat. Palermo 23*: 137–155.

11

Convergence of Laws on Separable
Metric Spaces

So far we have dealt with convergence of laws mainly on finite-dimensional Euclidean spaces \mathbb{R}^k, for the central limit theorem (§§9.3–9.5). Now we'll treat converging laws on more general, possibly infinite-dimensional spaces. Here are some cases where such spaces and laws can be helpful.

Let $x(t, \omega)$ be the position of a randomly moving particle at time t, where $\omega \in \Omega$, for some probability space (Ω, \mathcal{B}, P). For each ω, we then have a continuous function $t \mapsto x(t, \omega)$. Suppose we consider times t with $0 \le t \le 1$ and that x is real-valued (the particle is moving along a line, or we just consider one coordinate of its position). Then $x(\cdot, \omega)$ belongs to the space $C[0, 1]$ of continuous real-valued functions on $[0, 1]$. The space $C[0, 1]$ has a usual norm, the supremum norm $\|f\| := \sup\{|f(t)|: 0 \le t \le 1\}$. Then $C[0, 1]$ is a complete separable metric space for the metric d defined as usual by $d(f, g) := \|f - g\|$. It may be useful to approximate the process x, for example, by a process y_n such that for each ω and each $k = 1, \ldots, n$, $y_n(\cdot, \omega)$ is linear on the interval $[(k - 1)/n, k/n]$. Thus it may help to define y_n converging to x in law (or in probability or a.s.) in $C[0, 1]$.

A number of useful facts about the convergence of laws hold in metric spaces, so the convergence will be studied in such spaces. Often separability and/or completeness will be assumed.

11.1. Laws and Their Convergence

We recall some definitions from §§7.1 and 9.3. Let (S, \mathcal{T}) be a topological space. Then the σ-algebra \mathcal{B} generated by the \mathcal{T} (the smallest σ-algebra including \mathcal{T}) is called the *Borel* σ-algebra. Sets in it are called *Borel* sets. A measure defined on \mathcal{B} may be called a *Borel* measure. A *law* is a Borel probability measure; in other words, a probability measure defined on \mathcal{B}. The simplest examples of laws are the point masses δ_x defined by $\delta_x(A) := 1_A(x)$ for any (Borel) set A.

S or (S, \mathcal{T}) is called a *Polish* space iff \mathcal{T} is metrizable by some metric d for which S is separable and complete. Recall that a separable metric space (S, e) is Polish if and only if it is a countable intersection of open sets in its completion (Theorem 2.5.4).

Let $C_b(S) := C_b(S, \mathcal{T})$ be the space of all bounded continuous real-valued functions on S. Recall that a sequence P_n of laws is said to *converge* to a law P, written $P_n \to P$, iff $\int f \, dP_n \to \int f \, dP$ for every $f \in C_b(S)$ (§9.3). Note that for a metric space (S, d), this convergence depends only on the topology of S, not on the particular metric.

Recall that if random variables with values in a separable metric space converge a.s. or in probability, then their laws converge (Proposition 9.3.5). On the other hand, the sequence of laws δ_n (unit mass at n) does not converge on \mathbb{R} (consider $f(t) := \cos(\pi t)$).

Recall (§2.1) that for any topological space (S, \mathcal{T}) and any set $A \subset S$, \overline{A} denotes the closure of A, int A its interior (the largest open set included in A), and ∂A the boundary of A, $\partial A := \overline{A} \backslash \text{int } A$. (Note that ∂A is always closed.) For a measure μ defined on \mathcal{B}, a set A is called a *continuity set* iff $\mu(\partial A) = 0$.

For example, an interval $[a, b]$ in \mathbb{R} is a continuity set of a point mass δ_x if and only if x is not a or b. The next fact, giving several conditions equivalent to convergence of laws on a metric space, has been called the *portmanteau theorem*.

11.1.1. Theorem *For laws P_n and P on a metric space (S, d) the following are equivalent:*

(a) $P_n \to P$.
(b) *For all open U,* $\liminf_{n\to\infty} P_n(U) \geq P(U)$.
(c) *For all closed F,* $\limsup_{n\to\infty} P_n(F) \leq P(F)$.
(d) *For all continuity sets A of P,* $\lim_{n\to\infty} P_n(A) = P(A)$.

Example. Let P_n be the unit mass $\delta_{1/n}$ and $P = \delta_0$ on \mathbb{R}. Then $P_n \to P$. In (b), let U be the open half-line $(0, \infty)$ and in (c), let F be the closed half-line $(-\infty, 0]$. Then, even though the lim inf in (b) and the lim sup in (c) are actual limits ($P_n(U) \equiv 1$ and $P_n(F) \equiv 0$), the inequalities in (b) and (c) are strict. F and U are not continuity sets of P. This example can help to recall and visualize the different conditions in the portmanteau theorem.

Proof. (a) implies (b): let U be open, $F := S \backslash U$, $\varepsilon > 0$, and $d(x, F) := \inf_{y \in F} d(x, y)$ (as in 2.5.3 and Theorem 7.1.1). Let $f_m(x) := \min(1, md(x, F))$, so that $0 \leq f_m \uparrow 1_U$ (as in the proof of Lemma 9.3.2). Let

$F_m := f_m^{-1}(\{1\})$. Then for m large, $\liminf_{n\to\infty} P_n(U) \geq \lim_{n\to\infty} \int f_m \, dP_n = \int f_m \, dP \geq P(F_m) > P(U) - \varepsilon$, which implies (b).

(b) and (c) are clearly equivalent by taking complements. If they hold, then for any Borel set A,

$$P(\text{int } A) \leq \liminf_{n\to\infty} P_n(\text{int } A) \leq \liminf_{n\to\infty} P_n(A) \leq \limsup_{n\to\infty} P_n(A)$$

$$\leq \limsup_{n\to\infty} P_n(\overline{A}) \leq P(\overline{A}).$$

If A is a continuity set of P, then $P(\text{int } A) = P(\overline{A})$, so $P_n(A) \to P(A)$ and (d) holds.

Now assume (d). Let $f \in C_b(S)$. The sets $F_y := \{x \in S : f(x) = y\}$ are disjoint for different values of y, so $P(F_y) > 0$ for at most countably many values of y. By adding a constant to f, which does not affect convergence of $\int f \, dP_n$ to $\int f \, dP$, we can assume $P(F_0) = 0$. Then for $\varepsilon > 0$ and each integer k let

$$B_{k,\varepsilon} := \{x : k\varepsilon \leq f(x) < (k+1)\varepsilon\}.$$

Then $\partial B_{k,\varepsilon} \subset F_{k\varepsilon} \cup F_{(k+1)\varepsilon}$. So there exist arbitrarily small ε such that $B_{k,\varepsilon}$ are continuity sets of P for all integers k. Then for all k, $(P_n - P)(B_{k,\varepsilon}) \to 0$ as $n \to \infty$. Since f is bounded, for the given ε, $B_{k,\varepsilon}$ is non-empty for only finitely many values of k. Now

$$\int f \, dP - \varepsilon \leq \sum_k k\varepsilon P(B_{k,\varepsilon}) = \lim_{n\to\infty} \sum_k k\varepsilon P_n(B_{k,\varepsilon}) \leq \liminf_{n\to\infty} \int f \, dP_n$$

$$\leq \limsup_{n\to\infty} \int f \, dP_n \leq \lim_{n\to\infty} \sum_k (k+1)\varepsilon P_n(B_{k,\varepsilon})$$

$$= \sum_k (k+1)\varepsilon P(B_{k,\varepsilon}) \leq \int f \, dP + \varepsilon.$$

Letting $\varepsilon \downarrow 0$ completes the proof. $\qquad\square$

In the example $\delta_{1/n} \to \delta_0$ mentioned after Theorem 11.1.1, we will have distribution functions $F_n(t)$ converging to $F(t)$ for all t except $t = 0$, where F is discontinuous. This illustrates the next fact:

11.1.2. Helly-Bray Theorem *If P_n and P are laws on \mathbb{R} with distribution functions F_n and F respectively, $n = 1, 2, \ldots$, then $P_n \to P$ if and only if $F_n(t) \to F(t)$ at all t where F is continuous.*

Proof. F is continuous at t if and only if $(-\infty, t]$ is a continuity set for P. Thus the portmanteau theorem, (a) implies (d), gives the "only if" part, which was also proved directly earlier (Theorem 9.3.6).

Conversely, to prove "if," let U be any open set in \mathbb{R} and $\varepsilon > 0$. Then U is a countable union of open intervals (open intervals with rational endpoints form a base of the usual topology). So there is a finite union $V = \bigcup_{1 \le j \le m} (a_j, b_j) \subset U$ with $P(V) > P(U) - \varepsilon$. Here a_j and b_j can be chosen such that $[a_j, b_j] \subset U$ for all j. Then, taking a_j a little smaller and b_j a little larger, we can assume that none of these points are discontinuities of F or of F_n for any n, since such discontinuities form a countable set. Then

$$\liminf_{n \to \infty} P_n(U) \ge \lim_{n \to \infty} P_n(V) = P(V) > P(U) - \varepsilon,$$

so the fact that (b) implies (a) in the portmanteau theorem completes the proof. □

Recall that in a separable metric space, convergence in probability implies convergence in law (Proposition 9.3.5), but the converse is often not true. Here is a situation where the converse holds:

11.1.3. Proposition *If (S, d) is a metric space, $p \in S$, and X_n are random variables on a probability space (Ω, \mathcal{B}, P) with $\mathcal{L}(X_n) \to \delta_p$, then $X_n \to p$ in probability.*

Proof. Let $f_k(x) := \max(0, 1 - kd(x, p))$. Then $f_k \in C_b(S)$ and $P\{d(X_n, p) < 1/k\} \ge Ef_k(X_n) \to 1$ as $n \to \infty$ for all k. □

In Proposition 11.1.3, convergence in probability cannot always be replaced by almost sure convergence (this is Problem 1 below).

11.1.4. Proposition *For any topological space (S, \mathcal{T}) and measure μ on the Borel sets of S, the continuity sets of μ form an algebra.*

Proof. For any set $A \subset S$, $\partial A = \partial(S \setminus A)$, so A is a continuity set if and only if its complement is one. The empty set \emptyset and S are always continuity sets. For any set C, the closure \overline{C} may be written as $(C)^-$. Next, for any two sets A and B, $\mathrm{int}(A \cup B) \supset \mathrm{int}\, A \cup \mathrm{int}\, B$, and $(A \cup B)^- = \overline{A} \cup \overline{B}$, so

$$\partial(A \cup B) \subset (\overline{A} \cup \overline{B}) \setminus (\mathrm{int}\, A \cup \mathrm{int}\, B) \subset \partial A \cup \partial B.$$

Thus if A and B are continuity sets, so is $A \cup B$. □

For example, let A and B be the sets of rational and irrational numbers in \mathbb{R} respectively. Then in the last proof, we see that the inclusions int $A \cup$ int $B \subset$ int $(A \cup B)$ and $\partial(A \cup B) \subset \partial A \cup \partial B$ can be strict.

Problems

1. Give an example of real random variables X_n with $\mathcal{L}(X_n) \to \delta_0$ such that X_n does not converge to 0 a.s. ($X_n \to 0$ in probability by Proposition 11.1.3). *Hint:* See §9.2, Problem 1.

2. Show that for any countable set S with discrete topology, convergence of laws is equivalent to convergence in total variation (defined in §9.3).

3. Give an example where the continuity sets of P do not form a σ-algebra. *Hint:* For Lebesgue measure on $[0, 1]$, consider singletons.

4. If F_n and F are distribution functions on \mathbb{R}, F is continuous, and $F_n(t) \to F(t)$ for all t, show that $F_n(t) \to F(t)$ uniformly: $\sup_t |(F_n - F)(t)| \to 0$ as $n \to \infty$. *Hint:* Given $\varepsilon > 0$, find a finite set on which the values of F are dense in $[0, 1]$ within ε, and use monotonicity of the F_n and F.

5. Let $x(n)$ be any points in \mathbb{R} with $|x(n)| \to \infty$ as $n \to \infty$. Show that the laws $\delta_{x(n)}$ do not converge.

6. Give an example of laws $P_n \to P$ on \mathbb{R} with distribution functions $F_n(t) \to F(t)$ whenever F is continuous at t (by the Helly-Bray theorem) but where $F_n(t)$ does not converge to $F(t)$ for t in a dense set. *Hint:* Define a law P on \mathbb{R} such that $P(\{q\}) > 0$ for each rational q. Let $P_n(A) = P(A - 1/n)$ for any Borel set A.

7. For a law P on \mathbb{R}^2, the distribution function F is defined by $F(x, y) := P((-\infty, x] \times (-\infty, y])$. Let P_n be laws converging to a law P, with distribution functions F_n and F, respectively. Show that $F_n(x, y)$ converges to $F(x, y)$ for almost all (x, y) with respect to Lebesgue measure λ^2 on \mathbb{R}^2.

8. Show that in Problem 7, either $F_n(x, y)$ converges to $F(x, y)$ for all (x, y) in \mathbb{R}^2, or else convergence fails for an uncountable set of values of (x, y).

9. In the situation of Problems 7 and 8, suppose $P(\{(x, y)\}) = 0$ for each single point (x, y) in \mathbb{R}^2. Show that for some rotation of coordinates $x' = x \cos\theta - y \sin\theta$, $y' = x \sin\theta + y \cos\theta$, the distribution functions $G_n(x', y')$ of P_n in the (x', y') coordinates converge everywhere to the corresponding distribution function of P.

10. Call a set A in a topological space *full* iff int $A =$ int \overline{A}. Prove or disprove: In \mathbb{R}, the full sets form an algebra.

11. For any nondecreasing function F on \mathbb{R} and $t \in \mathbb{R}$ let $F(t^-) :=$ $\lim_{x \uparrow t} F(x)$. Suppose F_n and F are distribution functions such that for all $t \in \mathbb{R}$, $F_n(t) \to F(t)$ and $F_n(t^-) \to F(t^-)$. Show that then $\lim_{n \to \infty} \sup_t |(F_n - F)(t)| = 0$ (in other words, $F_n \to F$ uniformly on \mathbb{R}). *Hints:* The case where F is continuous is Problem 4. In this case, given $\varepsilon > 0$, find a finite set G such that $F(t)$ for t in G are dense in the range of F within ε and such that for any jump of height $F(t) - F(t^-) \geq \varepsilon$, we have $t \in G$. Show that if a nondecreasing function H is within ε of F at t and t^- for each $t \in G$, then $\sup_t |(F - H)(t)| \leq 4\varepsilon$.

11.2. Lipschitz Functions

Continuous functions can be defined on any topological space. Functions with derivatives of any order can be defined on a Euclidean space or more generally on a differentiable manifold. On a general metric space, the most natural forms of smoothness are Lipschitz conditions, as follows.

Let (S, d) be a metric space. Recall that for a real-valued function f on S, the Lipschitz seminorm is defined (as in §6.1) by $\|f\|_L := \sup_{x \neq y} |f(x) - f(y)|/d(x, y)$. Then $\|f\|_L = 0$ if and only if f is a constant function. Call the supremum norm $\|f\|_\infty := \sup_x |f(x)|$. Let $\|f\|_{BL} := \|f\|_L + \|f\|_\infty$ and $BL(S, d) := \{f \in \mathbb{R}^S : \|f\|_{BL} < \infty\}$. Here "$BL$" stands for "bounded Lipschitz," and $BL(S, d)$ is the set of all bounded, real-valued Lipschitz functions on S.

Examples. If $f(x) = \sin x$ and $g(x) := \cos x$, then f and g are bounded Lipschitz functions on \mathbb{R}. The functions $f(x) := 2x$ and $g(x) := |x|$ are Lipschitz but, of course, not bounded. The function $f(x) := x^2$ is not a Lipschitz function on \mathbb{R}.

Now $\|\cdot\|_{BL}$ not only is a norm but has a submultiplicative property:

11.2.1. Proposition *$BL(S, d)$ is a vector space, $\|\cdot\|_{BL}$ is a norm on it, and for any f and g in $BL(S, d)$, $fg \in BL(S, d)$ and $\|fg\|_{BL} \leq \|f\|_{BL}\|g\|_{BL}$.*

Proof. The vector space property is clear. Since on $BL(S, d)$, $\|\cdot\|_L$ is a seminorm and $\|\cdot\|_\infty$ is a norm, $\|\cdot\|_{BL}$ is a norm. Clearly $\|fg\|_\infty \leq \|f\|_\infty\|g\|_\infty$. For any x and y in S,

$$|f(x)g(x) - f(y)g(y)| \leq |f(x)(g(x) - g(y))| + |g(y)(f(x) - f(y))|$$
$$\leq (\|f\|_\infty\|g\|_L + \|g\|_\infty\|f\|_L)\, d(x, y).$$

Thus $fg \in BL(S, d)$ and the norm inequality for products follows. \square

Next, consider lattice operations $f \vee g := \max(f, g)$, $f \wedge g := \min(f, g)$. These operations, like addition, also preserve the Lipschitz property:

11.2.2. Proposition *For any* f_1, \ldots, f_n *and* $* = \vee$ *or* \wedge,

(a) $\|f_1 * \cdots * f_n\|_L \le \max_{1 \le i \le n} \|f_i\|_L$ *and*

(b) $\|f_1 * \cdots * f_n\|_{BL} \le 2 \max_{1 \le i \le n} \|f_i\|_{BL}$.

Proof. To prove (a) by induction on n, it is enough to treat $n = 2$. There if $(f \vee g)(x) \ge (f \vee g)(y)$ and $(f \vee g)(x) = f(x)$, we have $|(f \vee g)(x) - (f \vee g)(y)| = f(x) - (f \vee g)(y) \le f(x) - f(y)$. The three other possible cases are symmetrical, interchanging x and y and/or f and g, so

$$|(f \vee g)(x) - (f \vee g)(y)| \le \max(|f(x) - f(y)|, |g(x) - g(y)|).$$

This implies (a) for maxima. The case of minima is symmetrical. Inequality (a) for $\|\cdot\|_\infty$ in place of $\|\cdot\|_L$ is clear. So for any f_1, \ldots, f_n, there exist i and j such that

$$\|f_1 * \cdots * f_n\|_L \le \|f_i\|_L \quad \text{and} \quad \|f_1 * \cdots * f_n\|_\infty = \|f_j\|_\infty.$$

These facts imply (b). $\qquad\square$

If a function f from \mathbb{R} into itself has a derivative $f'(x)$ for all x, then it can be shown that f is a Lipschitz function if and only if f' is bounded (this is Problem 1 below). On the other hand, $x \vee -x = |x|$, where each of x and $-x$ is differentiable (in fact, C^∞) for all x, but $|x|$ is not differentiable at 0; still, it is a Lipschitz function.

We have an extension theorem for bounded Lipschitz functions:

11.2.3. Proposition *If* $A \subset S$ *and* $f \in BL(A, d)$, *then* f *can be extended to a function* $h \in BL(S, d)$ *with* $h = f$ *on* A *and* $\|h\|_{BL} = \|f\|_{BL}$.

Proof. By the Kirszbraun-McShane extension theorem (6.1.1) there is a function g on S with $g = f$ on A and $\|g\|_L = \|f\|_L$. Let $h := (g \wedge \|f\|_\infty) \vee (-\|f\|_\infty)$. Then $h = f$ on A, $\|h\|_L \le \|g\|_L$ by Proposition 11.2.2(a), twice, so $\|h\|_L \le \|f\|_L$, and $\|h\|_L = \|f\|_L$. Also, $\|h\|_\infty \le \|f\|_\infty$ so $\|h\|_\infty = \|f\|_\infty$. $\qquad\square$

Recall that since continuous functions on a compact space S are bounded, the space $C(S)$ of all continuous real functions on S coincides with $C_b(S)$. Such functions can be approximated by Lipschitz functions:

11.2.4. Theorem *If (S, d) is any compact metric space, then $BL(S, d)$ is dense in $C(S)$ for $\|\cdot\|_\infty$.*

Proof. $BL(S, d)$ is an algebra by Proposition 11.2.1 and contains the constants. It separates points by Proposition 11.2.3, taking A to be a two-point set and f any nonconstant function on it. Thus the Stone-Weierstrass theorem (2.4.11) applies. \square

Example. If $S = [0, 1]$, and f is any continuous real function on S, let $f_n(x) = f(x)$ for $x = j/n$, $j = 0, 1, \ldots, n$, and let f_n be linear on each interval $[(j - 1)/n, j/n]$, $j = 1, \ldots, n$. Then the f_n are bounded and Lipschitz, and $\| f_n - f \|_\infty \to 0$ as $n \to \infty$.

11.2.5. Corollary *For any compact metric space (S, d), $(C(S), \|\cdot\|_\infty)$ is a separable Banach space.*

Proof. For each $n = 1, 2, \ldots$, the sets $\{ f : \| f \|_{BL} \le n \}$ are uniformly bounded and equicontinuous, thus totally bounded (actually compact) and hence separable for $\|\cdot\|_\infty$, by the Arzelà-Ascoli theorem (2.4.7). The union of these sets is dense in $C(S)$ by Theorem 11.2.4. \square

Examples. If S is a compact set in a finite-dimensional Euclidean space \mathbb{R}^k, then the polynomials with rational coefficients form a countable dense set in $C(S)$, by the Weierstrass approximation theorem (Corollary 2.4.12) or its extension, the Stone-Weierstrass theorem (2.4.11).

Problems

1. Let f be a function from \mathbb{R} into \mathbb{R} such that a derivative $f'(x)$ exists for all x. Show that f is a Lipschitz function, $\| f \|_L < \infty$, if and only if f' is bounded.

2. Show that a polynomial on \mathbb{R} is a Lipschitz function if and only if its degree is 0 or 1 (it is constant or linear).

3. Show that $BL(S, d)$ is complete for $\|\cdot\|_{BL}$ and hence a Banach space.

4. Give examples for $S = R$ to show that the constant 2 in Proposition 11.2.2(b) cannot be replaced by any smaller constant in general.

5. Show that for $S = [0, 1]$ with usual metric d, $BL(S, d)$ is not separable for $\|\cdot\|_L$ and hence not for $\|\cdot\|_{BL}$. *Hint:* Consider $f_t(x) := |x - t|$.

6. Show that in Proposition 11.2.2(a) the inequality may be strict.

7. Give an alternate proof of Corollary 11.2.5 as follows. Let $\{x_n\}$ be a dense sequence in S. Map S into a countable Cartesian product of lines

by $x \mapsto \{d(x, x_n)\}_{1 \le n < \infty}$. Thus reduce to the case where S is a countable product of intervals with a product metric, as in Proposition 2.4.4. Consider polynomials in the coordinates (finitely many in each polynomial) and again apply the Stone-Weierstrass theorem (2.4.11).

8. (Fixed point theorem.) Let (S, d) be a complete metric space. Let F be a function from S into S such that for some $t < 1, d(F(x), F(y)) \le td(x, y)$ for all x and y in S. Let x_0 be any point in S and define a sequence $\{x_n\}$ recursively by $x_{n+1} = F(x_n)$ for $n = 0, 1, \ldots$.
 (a) Show that $\{x_n\}$ is a Cauchy sequence.
 (b) Let $x_n \to y$. Show that $F(y) = y$ (y is a "fixed point" of F) and is the unique point of S with this property.
 Note: If $S = \mathbb{R}$ with usual metric, then the assumption on F is that $\|F\|_L < 1$.

9. The problem is to solve a differential equation $dy/dx = G(x, y)$, where G is continuous in (x, y) and Lipschitz in y, so that $\|G(x, \cdot)\|_L \le K < \infty$ for all x in an interval $[a, b]$. For $a \le t \le u \le b$, let S be $C[t, u]$ with supremum distance and define a function F from S into itself by $F(f)(x) = \int_t^x G(y, f(y)) \, dy, t \le x \le u$.
 (a) Under what condition on t and u does this F satisfy the conditions of Problem 8?
 (b) Show that for any $c \in \mathbb{R}$ there is a unique function f, continuous for $a \le x \le b$, such that $y = f(x)$ satisfies the differential equation for $a < x < b$, with $f(a) = c$. *Hint:* Cover $[a, b]$ with finitely many intervals $[t, u]$.

10. Show that the differential equation $dy/dx = |y|^{1/2}$ has more than one solution with $y = 0$ when $x = 0$.

11. Let f be a continuous function from \mathbb{R} into \mathbb{R} such that for some $M < \infty, f'(x)$ exists with $|f'(x)| \le M$, for all but finitely many x. Show that $\|f\|_L \le M$.

12. Show that there is a continuous function f from $[0, 1]$ into itself with $|f'(x)| \le 1$ for Lebesgue almost all x, which is not a Lipschitz function. *Hint:* Consider the Cantor function of Proposition 4.2.1.

11.3. Metrics for Convergence of Laws

For any subset A of a set S, let A^c denote the complement $S \setminus A$. For (S, d) a metric space and $\varepsilon > 0$ let

$$A^\varepsilon := \{y \in S : d(x, y) < \varepsilon \text{ for some } x \in A\}.$$

Definition. For any two laws P and Q on S let

$$\rho(P, Q) := \inf\{\varepsilon > 0: P(A) \le Q(A^\varepsilon) + \varepsilon \quad \text{for all Borel sets } A\}.$$

Since $A^\varepsilon = (\overline{A})^\varepsilon$, we get an equivalent definition if A is required to be closed. This distance ρ between laws allows for a distance in S by comparing A and A^ε, and for a difference in probabilities by adding ε.

Examples. For two point masses δ_x and δ_y (where $\delta_x(A) := 1_A(x)$ for any x and set A), $\rho(\delta_x, \delta_y) = d(x, y)$ if $d(x, y) \le 1$ (to show that $\rho(\delta_x, \delta_y) \le d(x, y)$, consider the cases where $x \in A$ or not; to show that the bound is attained for $d(x, y) \le 1$, take $A = \{x\}$). If $d(x, y) > 1$, then $\rho(\delta_x, \delta_y) = 1$. In fact, note that $\rho(P, Q) \le 1$ for any laws P and Q. So $\rho(\delta_x, \delta_y) \equiv \min(d(x, y), 1)$.

11.3.1. Theorem *For any metric space (S, d), ρ is a metric on the set of all laws on S.*

Proof. Clearly $\rho(P, Q) \ge 0$ and $\rho(P, P) = 0$ for any P and Q.

If $\rho(P, Q) > \varepsilon$, then for some Borel set A, $P(A) > Q(A^\varepsilon) + \varepsilon$. Now $A^{\varepsilon c \varepsilon} \subset A^c$, that is, if $d(x, y) < \varepsilon$ for some $y \in A^{\varepsilon c}$, then $x \notin A$, since if $x \in A$, then $y \in A^\varepsilon$. So $Q(A^{\varepsilon c}) > P(A^c) + \varepsilon \ge P(A^{\varepsilon c \varepsilon}) + \varepsilon$. Thus $\rho(Q, P) \ge \rho(P, Q)$. Interchanging P and Q gives $\rho(P, Q) = \rho(Q, P)$.

If $\rho(P, Q) = 0$, then letting A be any closed set F and $\varepsilon = 1/n$, $n = 1, 2, \ldots$, note that the intersection of the $F^{1/n}$ is F. Thus $P(F) \le Q(F)$ and likewise $Q(F) \le P(F)$, so $P(F) = Q(F)$. Thus $P = Q$ by closed regularity (Theorem 7.1.3).

If M, P, and Q are laws on S, $\rho(M, P) < x$ and $\rho(P, Q) < y$, then for any Borel set A,

$$M(A) \le P(A^x) + x \le Q((A^x)^y) + y + x \le Q(A^{x+y}) + x + y,$$

so $\rho(M, Q) \le x + y$. Letting $x \downarrow \rho(M, P)$ and $y \downarrow \rho(P, Q)$ gives $\rho(M, Q) \le \rho(M, P) + \rho(P, Q)$. □

The metric ρ is called the *Prohorov metric*, or sometimes the *Lévy-Prohorov* metric. Now for any laws P and Q on S, let $\int f\, d(P - Q) := \int f\, dP - \int f\, dQ$ and

$$\beta(P, Q) := \sup\left\{\left|\int f\, d(P - Q)\right| : \|f\|_{BL} \le 1\right\}.$$

11.3.2. Proposition *For any metric space (S, d), β is a metric on the set of all laws on S.*

Proof. By Theorem 6.1.3, we need only check that if $\beta(P, Q) = 0$, then $P = Q$. For any closed set F, if $g(x) := d(x, F)$, then $\|g\|_L \leq 1$ (by 2.5.3). Thus for the functions $f_m := md(x, F) \wedge 1$ we have $\|f_m\|_{BL} \leq m + 1$. If U is the complement of F, then since $f_m \uparrow 1_U$, we get $P(U) = Q(U)$, $P(F) = Q(F)$, and $P = Q$, as in the last proof. $\qquad\square$

Next it will be shown that ρ and β each metrize convergence of laws on separable metric spaces. The metrizability of convergence of laws has consequences such as the following: suppose P_{nk} and P_n are laws such that $P_n \to P_0$ and for each n, $P_{nk} \to P_n$. Then there is a subsequence $P_{nk(n)}$ such that $P_{nk(n)} \to P_0$. This does not follow directly or easily from the definition of convergence of laws, in part because the space $C_b(S)$ is not separable in general.

11.3.3. Theorem *For any separable metric space (S, d) and laws P_n and P on S, the following are equivalent:*

(a) $P_n \to P$.
(b) $\int f\, dP_n \to \int f\, dP$ for all $f \in BL(S, d)$.
(c) $\beta(P_n, P) \to 0$.
(d) $\rho(P_n, P) \to 0$.

Proof. Clearly (a) implies (b). To prove that (b) implies (c), let T be the completion of S. Then each $f \in BL(S, d)$ extends uniquely to a function in $BL(T, d)$, giving a 1–1 linear mapping from $BL(S, d)$ onto $BL(T, d)$ which preserves $\|\cdot\|_{BL}$ norms. The laws P_n and P on S also define laws on T. Thus in this step it can be assumed that S is complete. Then by Ulam's theorem (7.1.4), for any $\varepsilon > 0$ take a compact $K \subset S$ with $P(K) > 1 - \varepsilon$. The set of functions $B := \{f : \|f\|_{BL} \leq 1\}$, restricted to K, forms a compact set of functions for $\|\cdot\|_\infty$ by the Arzelà-Ascoli theorem (2.4.7). Thus for some finite k there are $f_1, \ldots, f_k \in B$ such that for any $f \in B$, there is a $j \leq k$ with $\sup_{y \in K} |f(y) - f_j(y)| < \varepsilon$. Then

$$\sup\{|f(x) - f_j(x)| : x \in K^\varepsilon\} < 3\varepsilon,$$

since if $y \in K$ and $d(x, y) < \varepsilon$, then

$$|f(x) - f_j(x)| \leq |f(x) - f(y)| + |f(y) - f_j(y)| + |f_j(y) - f_j(x)|$$
$$\leq \|f\|_L\, d(x, y) + \varepsilon + \|f_j\|_L\, d(x, y) < 3\varepsilon.$$

Let $g(x) := 0 \vee (1 - d(x, K)/\varepsilon)$. Then $g \in BL(S, d)$ (using 2.5.3 and Proposition 11.2.2) and $1_K \leq g \leq 1_{K^\varepsilon}$. For n large enough, $P_n(K^\varepsilon) \geq \int g\, dP_n > 1 - 2\varepsilon$.

Thus for each $f \in B$ and f_j as above,

$$\left| \int f\, d(P_n - P) \right| \leq \int |f - f_j|\, d(P_n + P) + \left| \int f_j\, d(P_n - P) \right|$$

$$\leq 2(P_n + P)(S \backslash K^\varepsilon) + (3\varepsilon) \cdot 2 + \left| \int f_j\, d(P_n - P) \right|$$

$$\leq 12\varepsilon + \left| \int f_j\, d(P_n - P) \right| \leq 13\varepsilon$$

for each $j = 1, \ldots, k$ for n large enough by (b). Thus (c) follows.

Next, to show (c) implies (d): Given a Borel set A and $\varepsilon > 0$, let $f(x) := 0 \vee (1 - d(x, A)/\varepsilon)$. Then $\|f\|_{BL} \leq 1 + \varepsilon^{-1}$. For any laws P and Q on S, $Q(A) \leq \int f\, dQ \leq \int f\, dP + (1 + \varepsilon^{-1})\beta(P, Q) \leq P(A^\varepsilon) + (1 + \varepsilon^{-1})\beta(P, Q)$. Thus $\rho(P, Q) \leq \max(\varepsilon, (1 + \varepsilon^{-1})\beta(P, Q))$. Hence if $\beta(P, Q) \leq \varepsilon^2$, then $\rho(P, Q) \leq \varepsilon + \varepsilon^2$. Since $\rho(P, Q) \leq 1$, it follows that $\rho(P, Q) \leq 2\beta(P, Q)^{1/2}$ in all cases, so (d) follows.

To show that (d) implies (a), using the portmanteau theorem (11.1.1), let A be a continuity set of P and $\varepsilon > 0$. Then for $0 < \delta < \varepsilon$ and δ small enough, $P(A^\delta \backslash A) < \varepsilon$ and $P(A^{c\delta} \backslash A^c) < \varepsilon$. Then for n large enough, $P_n(A) \leq P(A^\delta) + \delta \leq P(A) + 2\varepsilon$ and $P_n(A^c) \leq P(A^{c\delta}) + \delta \leq P(A^c) + 2\varepsilon$, so $|(P_n - P)(A)| \leq 2\varepsilon$. Letting $\varepsilon \downarrow 0$, we have $P_n(A) \to P(A)$, so (a) follows. \square

11.3.4. Corollary *Let (S, T) be any topological space with a countable dense subset, suppose laws $P_n \to P$ on S, and let \mathcal{F} be any uniformly bounded, equicontinuous family of functions on S. Then $P_n \to P$ uniformly on \mathcal{F}, in other words*

$$\lim_{n \to \infty} \sup \left\{ \left| \int f\, d(P_n - P) \right| : f \in \mathcal{F} \right\} = 0.$$

Note. "Equicontinuous" means equicontinuous at each point. On a general topological space, as here, the notion "uniformly equicontinuous" is not even defined. An example of a uniformly bounded, equicontinuous family is the set \mathcal{F} of all functions f on a metric space with $\|f\|_{BL} \leq 1$. On the other hand, any finite set of bounded continuous functions is uniformly bounded and equicontinuous, whether or not the functions are Lipschitzian for a given metric.

Proof. Let $d(x, y) := \sup\{|f(x) - f(y)|: f \in \mathcal{F}\}$ for each $x, y \in S$. Then d is a pseudometric on S, that is, a metric except that possibly $d(x, y) = 0$ for some $x \neq y$. Let T be the set of equivalence classes of points of S for the relation $\{\langle x, y\rangle: d(x, y) = 0\}$. Then there is a natural map G of S onto T such that for a metric e on T, $d(x, y) = e(Gx, Gy)$ for all x and y. For each $f \in \mathcal{F}$, $|f(x) - f(y)| \leq d(x, y)$ for all x and y, so $\|f\|_L \leq 1$ for d. Also, if $d(x, y) = 0$, then $f(x) = f(y)$. So $f(x)$ depends only on Gx: for some function h on T, $f(x) = h(Gx)$ for all x, and $\|h\|_L \leq 1$ for e. Thus the set of such h remains equicontinuous as well as uniformly bounded. So we can assume that d is a metric.

Since \mathcal{F} is equicontinuous, d is jointly continuous on $S \times S$. Now (S, d) is separable. Defining Lipschitz norms with respect to d,

$$\|f\|_{BL} \leq 1 + \sup\{\|f\|_\infty: f \in \mathcal{F}\} < \infty$$

for all $f \in \mathcal{F}$. Thus the fact that (a) implies (c) in Theorem 11.3.3 gives the conclusion. $\qquad\qquad\square$

Now recall the Ky Fan metric α for random variables defined by $\alpha(X, Y) := \inf\{\varepsilon > 0: P(d(X, Y) > \varepsilon) \leq \varepsilon\}$, which metrizes convergence in probability (Theorem 9.2.2). The fact that convergence in probability implies convergence in law (Proposition 9.3.5) can be made more specific in terms of the Prohorov and Ky Fan metrics:

11.3.5. Theorem *For any separable metric space (S, d) and random variables X and Y on a probability space with values in S, $\rho(\mathcal{L}(X), \mathcal{L}(Y)) \leq \alpha(X, Y)$.*

Proof. Take any $\varepsilon > \alpha(X, Y)$. Then $P(d(X, Y) \geq \varepsilon) < \varepsilon$. For any Borel set A, if $X \in A$ and $d(X, Y) < \varepsilon$, then $Y \in A^\varepsilon$, so

$$\mathcal{L}(X)(A) = P(X \in A) \leq P(Y \in A^\varepsilon) + \varepsilon = \mathcal{L}(Y)(A^\varepsilon) + \varepsilon.$$

So $\rho(\mathcal{L}(X), \mathcal{L}(Y)) \leq \varepsilon$. Letting $\varepsilon \downarrow \alpha(X, Y)$ finishes the proof. $\qquad\square$

It will be shown later, in Corollary 11.6.4, that for any two laws P and Q on a complete separable metric space, there exist random variables X and Y with those laws such that $\rho(P, Q) = \alpha(X, Y)$. In this sense, Theorem 11.3.5 cannot be improved on.

Problems

1. Show that in the definition of $\beta(P, Q)$ the supremum can be restricted to just those f with $\|f\|_{BL} \leq 1$ such that both sup $f = \|f\|_\infty$ and inf $f = -\|f\|_\infty$.

2. For $a < b$ let $P_{a,b}$ be the uniform distribution over the interval $[a, b]$, having density $1_{[a,b]}/(b - a)$ with respect to Lebesgue measure. Evaluate: (a) $\rho(P_{0,1}, P_{0,2})$; (b) $\beta(P_{0,1}, P_{0,2})$; (c) $\rho(P_{0,1}, P_{1/2,3/2})$.

3. Give an alternate proof that (b) implies (a) in Theorem 11.3.3 using part of the proof of the portmanteau theorem (11.1.1).

4. Suppose F is closed in S, $x \geq 0$, $y > 0$, and $P(F) > Q(F^y) + x$. Prove that $\beta(P, Q) \geq 2xy/(2 + y)$. *Hint:* Show that there is a function f equal to 1 on F and -1 outside F^y with $\|f\|_L \leq 2/y$. Note that $\int f\, d(P - Q) = \int f + 1\, d(P - Q)$.

5. Let $g(x) := 2x^2/(2 + x)$, $x \geq 0$.
 (a) Show that for any two laws P and Q, $g(\rho(P, Q)) \leq \beta(P, Q)$. *Hint:* Show that g is increasing, and use Problem 4.
 (b) Infer that
 $$\rho(P, Q) \leq \left(\frac{3}{2}\beta(P, Q)\right)^{1/2}.$$
 (c) Show that the inequality in part (a) is sharp by finding, for any t with $0 \leq t \leq 1$, laws P and Q on \mathbb{R} with $\rho(P, Q) = t$ and $\beta(P, Q) = g(t)$. *Hint:* Let P be a point mass δ_0 and let Q be concentrated at 0 and one other point.

6. Consider point masses $P := \delta_p$ and $Q := \delta_q$. Show that as $d(p, q) \to 0$, $\rho(P, Q)/\beta(P, Q) \to 1$. On the other hand, for $\mu := (P + Q)/2$, $\mu_n := \mu + (P - Q)/n$, $d(p, q) = 1/n$, show that $\beta(\mu_n, \mu)/\rho(\mu_n, \mu)^2 \to 1$ as $n \to \infty$.

7. Define $\|f\|_L$ and $\|f\|_\infty$ for complex-valued functions just as for real-valued functions. Show that the metric $\beta'(P, Q)$ for probability measures, defined as β but with complex-valued functions, is actually equal to β. *Hint:* A complex function f can be multiplied by a complex constant z with $|z| = 1$ to make $\int zf\, d(P - Q)$ real.

8. *Lévy's metric.* Let P and Q be two laws on \mathbb{R} with distribution functions F and G, respectively. Let $\lambda(P, Q) := \inf\{\varepsilon > 0 : F(x - \varepsilon) - \varepsilon \leq G(x) \leq F(x + \varepsilon) + \varepsilon$ for all $x\}$.
 (a) Show that λ is a metric metrizing convergence of laws on \mathbb{R}. *Hint:* Use the Helly-Bray theorem (11.1.2).

(b) Show that $\lambda \le \rho$, but that there exist laws P_n and Q_n with $\lambda(P_n, Q_n) \to$ 0 while $\rho(P_n, Q_n)$ does not converge to 0.

11.4. Convergence of Empirical Measures

For any probability space (S, \mathcal{B}, μ) there is a probability space (Ω, P) on which there are independent random variables X_1, X_2, \dots, with values in S and $\mathcal{L}(X_j) = \mu$ for all j, since we can take Ω as a Cartesian product of a sequence of copies of S and X_j as coordinates (§8.2). The empirical measures μ_n are defined by

$$\mu_n(A)(\omega) := n^{-1} \sum_{j=1}^{n} \delta_{X_j(\omega)}(A), \qquad A \in \mathcal{B}, \quad \omega \in \Omega.$$

If S is any topological space and f a bounded, continuous, real-valued function on S, then $\int f \, d\mu_n = (f(X_1) + \cdots + f(X_n))/n$, which converges to $\int f \, d\mu$ almost surely as $n \to \infty$ by the strong law of large numbers (Theorem 8.3.5). But the set of probability 0 on which convergence fails may depend on f, and the space of bounded continuous functions in general is nonseparable, so the following fact is not immediate:

11.4.1. Theorem (Varadarajan) *Let (S, d) be a separable metric space and μ any law (Borel probability measure) on S. Then the empirical measures μ_n converge to μ almost surely:*

$$P(\{\omega: \mu_n(\cdot)(\omega) \to \mu\}) = 1.$$

Proof. The proof will use the fact (Theorem 2.8.2) that there is a totally bounded metric e on S for the d topology. For example, the real line \mathbb{R} of course is not totally bounded with its usual metric, but $x \mapsto \arctan x$ is a homeomorphism of \mathbb{R} to the bounded open interval $(-\pi/2, \pi/2)$, which is totally bounded. So we can assume that (S, d) is totally bounded. By Theorem 11.3.3, it is enough to show that almost surely $\int f(x) \, d\mu_n(x)(\omega) \to \int f \, d\mu$ for all $f \in BL(S, d)$. But $BL(S, d)$ is separable for the supremum norm $\|\cdot\|_\infty$ (not usually for $\|\cdot\|_{BL}$) by Corollary 11.2.5, since $BL(S, d)$ is naturally isometric to $BL(C, d)$ where C is the completion of S and is compact. Let $\{f_m\}$ be dense in $BL(S, d)$ for $\|\cdot\|_\infty$. Then by the strong law of large numbers (Theorem 8.3.5), almost surely we have convergence for $f = f_m$ for all m, and then for general $f \in BL(S, d)$ by considering f_m close to f. $\qquad\square$

On the real line \mathbb{R}, let μ be a probability measure with distribution function $F(t) := \mu((-\infty, t])$, $-\infty < t < \infty$. Then the empirical measures μ_n have distribution functions $F_n(t)(\omega) := \mu_n((-\infty, t])(\omega)$. Here F_n is called an *empirical distribution function* for F. Here is a classic limit theorem:

11.4.2. Theorem (Glivenko-Cantelli) *Let μ be any law on \mathbb{R} with distribution function F. Then almost surely $F_n(\cdot)(\omega) \to F$ uniformly on \mathbb{R} as $n \to \infty$.*

Proof. By Theorems 11.4.1 and 11.1.2, almost surely $F_n(t) \to F(t)$ for all t at which F is continuous. At each of the remaining at most countable set of values of t where F may have a jump (Theorem 7.2.5), we also have $F_n(t) \to F(t)$ by the strong law of large numbers (Theorem 8.3.5) applied to the random variables $1_{(-\infty,t]}(X_j)$. So almost surely $F_n(t) \to F(t)$ for all t.

To prove uniform convergence, the general case will be reduced to the case of Lebesgue measure λ on $[0, 1]$. Let G be its distribution function $G(x) = \max(0, \min(x, 1))$ for all x.

For any distribution function F on \mathbb{R} and $0 < t < 1$ let $X_F(t) := \inf\{x: F(x) \geq t\}$. Recall that for $\lambda =$ Lebesgue measure on $(0, 1)$, X_F is a random variable with distribution function F by Proposition 9.1.2. In other words, $\lambda \circ X_F^{-1} = P$ where F is the distribution function of P. For example, if F is the distribution function of $(2\delta_0 + \delta_2)/3$, then $X_F(t) = 0$ for $0 < t < 2/3$ and $X_F(t) = 2$ for $2/3 < t < 1$. The following fact will be useful:

11.4.3. Lemma *If G_n are empirical distribution functions for G, then for any distribution function F on \mathbb{R}, $G_n \circ F$ are empirical distribution functions for F.*

Proof. Let Y_j be i.i.d. (λ), so that

$$G_n(F(x)) = n^{-1} \sum_{j=1}^{n} 1_{Y_j \leq F(x)}.$$

Then $X_j := X_F(Y_j)$ are i.i.d., with the law having distribution function F by Proposition 9.1.2, which also yields that $X_j \leq x$ if and only if $Y_j \leq F(x)$, so the lemma follows. □

So continuing the proof of Theorem 11.4.2, to prove $F_n \to F$ uniformly a.s., it is enough to prove that $G_n \to G$ uniformly a.s. Given $\varepsilon > 0$, choose m large enough so that $1/m < \varepsilon/2$. Let $E := \{k/m: k = 0, 1, \ldots, m\}$. Then almost surely, G_n converges to G on the finite set E, so for almost all ω

there is an n large enough so that $|G_n - G|(x)(\omega) < \varepsilon/2$ for all $x \in E$. Then, for any $x \in [0, 1]$, take u and v in E with $u \le x \le v$ and $v - u = 1/m$. It follows that $G_n(x) \ge G_n(u) > u - \varepsilon/2 > x - \varepsilon$ and likewise $G_n(x) < x + \varepsilon$, so $|G_n - G|(x) < \varepsilon$ for all x. $\qquad\square$

Problems

1. Let P be the uniform distribution $\lambda/2$ on the interval $[0, 2]$. Suppose $n = 3$, $X_1 = 1.1$, $X_2 = 0.4$, and $X_3 = 1.7$. Sketch a graph of the distribution function F of P and the empirical distribution function F_3. Evaluate $\sup_x |F_n - F|(x)$.

2. Show that there are distribution functions H_n with $H_n(t) \to H(t)$ for all t such that
 (a) H is not a distribution function, or
 (b) H is a distribution function but H_n does not converge to H uniformly.
 Hint: Let $H_n(t) := 0$ for $t \le 0$, $\min(t^n, 1)$ for $t > 0$.

3. Finish the proof of the Glivenko-Cantelli theorem (11.4.2) without using Proposition 9.1.2 and Lemma 11.4.3. Instead use pointwise convergence and consider, for each t where F has a jump at t, the open interval $J_t := (-\infty, t)$. Show that almost surely $\mu_n(J_t) \to \mu_n(J_t)$ for all such t and use this to complete the proof.

4. For any separable metric space (S, d) and law μ on S, show that $\rho(\mu_n, \mu) \to 0$ and $\beta(\mu_n, \mu) \to 0$ a.s. and in probability as $n \to \infty$. *Hint:* The main problem is to show that $\rho(\mu_n, \mu)$ and $\beta(\mu_n, \mu)$ are measurable random variables. For the set \mathcal{P} of all laws on S with ρ topology (Theorem 11.3.3), show that (\mathcal{P}, ρ) is separable, ρ and β are measurable $\mathcal{P} \times \mathcal{P} \mapsto \mathbb{R}$ and $\omega \mapsto P_n(\cdot)(\omega)$ is measurable from Ω into \mathcal{P}.

5. Let P be a law on S.
 (a) If $P(F) = 1$ for some Borel set F, specifically a finite or countable set, show that for any other law Q, the infimum in the definition of $\rho(P, Q)$ can be taken over $A \subset F$.
 (b) Let $B(x, r) := \{y : d(x, y) < r\}$ for $x \in S$ and $r > 0$. For a fixed r let $f(x) := P(B(x, r))$. Show that f is upper semicontinuous: for each $t \in \mathbb{R}$, $\{y \in S : f(y) \ge t\}$ is closed.
 (c) Apply (a) to μ_n and (b) to μ to get another proof that $\rho(\mu_n, \mu)$ is measurable (Problem 4).

6. If F is a finite set $\{x_1, \ldots, x_n\}$ and P is a law with $P(F) = 1$, show that for any law Q the supremum $\sup \{|\int f \, d(P - Q)| : \|f\|_L \le 1\}$ can be

restricted to functions of the form $f(x) = \min_{1 \le i \le n}(c_i + d(x, x_i))$ for rational c_1, \ldots, c_n. *Hint:* Use Proposition 11.2.2.

7. Suppose F is a continuous distribution function on \mathbb{R} and let F_n be the corresponding empirical distribution functions. Find the supremum over all $t \in R$ of the variance of $F_n(t) - F(t)$. Let $f_n = O_P(a_n)$ mean that for any $\varepsilon > 0$, there is an $M < \infty$ such that $P(|f_n/a_n| > M) < \varepsilon$ for all n. Find the minimal c such that for all t, $F_n(t) - F(t) = O_P(n^c)$.

11.5. Tightness and Uniform Tightness

Recall the definitions of a tight probability measure and a uniformly tight set of laws (just before Theorem 9.3.3 above). This section will prove, first, that on any reasonably "measurable" metric space, all laws are tight. Then it will be shown that being uniformly tight is closely connected to compactness properties of sets of laws.

Definition. A separable metric space (S, d) is *universally measurable (u.m.)* iff for every law P on the completion T of S there are Borel sets A and B in T with $A \subset S \subset B$ and $P(A) = P(B)$, so that S is measurable for the (measure-theoretic) completion of P (as defined in §3.3).

Examples. If S is a Borel set in T, then clearly S is u.m., so most handy spaces S will be u.m. If S is not Lebesgue measurable in $[0, 1]$ (§3.4), it is not u.m. There exist u.m. sets which are not Borel: for example, some analytic sets $\{f(x): x \in S\}$, where f is continuous and S is a complete separable metric space (see §13.2).

11.5.1. Theorem *A separable metric space (S, d) is u.m. if and only if every law P on S is tight.*

Proof. If (S, d) is u.m., let P be any law on S. Then P defines a law \overline{P} on the completion T by $\overline{P}(A) = P(A \cap S)$ for every Borel set A in T (here $A \cap S$ is always a Borel set in S, as can be seen beginning with open sets, and so on). By Ulam's theorem (7.1.4) applied to \overline{P}, and universal measurability, there are compact sets K_n included in S with $\overline{P}(\bigcup_{n \ge 1} K_n) = 1$. Then $P(\bigcup_{n \ge 1} K_n) = 1$. Compactness of sets does not depend on which larger sets they are considered in, so the K_n are compact sets in S and P is tight.

Conversely, if every law on S is tight, let Q be any law on T. Define Q outer measure: $Q^*(A) := \inf\{Q(B): B \supset A\}$. If $Q^*(S) = 0$, then S is

measurable for the completion of Q. If $Q^*(S) > 0$, let $P(B) := Q^*(B)/Q^*(S)$ for each Borel set in S. Then P is a law on S by Theorem 3.3.6. Since this P is tight, take compact $K_n \subset S$ with $P(A) = 1$ where $A = \bigcup_{n \geq 1} K_n$ Then $Q(A) = Q^*(S)$, so S is u.m. $\qquad\square$

11.5.2. Corollary *The u.m. property of (S, d) depends on the metric d only through its topology.*

In fact, the u.m. property is preserved by any 1–1 (Borel) measurable function with measurable inverse (see the Notes), in other words, by Borel isomorphism. On the other hand, any two separable metric spaces which are Borel subsets of their completions (and so trivially u.m.) are Borel isomorphic if and only if they have the same cardinality, as will be shown in Theorem 13.1.1.

It was shown in Proposition 9.3.4 that on \mathbb{R}^k, every converging sequence of laws is uniformly tight. Here is an extension to general metric spaces, provided that the laws themselves are tight.

11.5.3. Theorem (Le Cam, 1957). *Let (S, d) be a metric space and suppose laws P_n converge to P_0 where P_n is tight for all $n \geq 0$. Then $\{P_n\}$ is uniformly tight. So if S is separable and u.m., every converging sequence of laws is uniformly tight.*

Proof. Each P_n, being tight, is concentrated in a countable union of compact sets K_{nm}. A compact set in a metric space is separable (since it's totally bounded). Thus the union of all the K_{nm} is separable. Let T be its closure, which is also separable. Any bounded continuous real function on T can be extended to all of S as a bounded continuous function by the Tietze extension theorem (2.6.4). Thus the laws P_n restricted to T still converge, and so they converge by Theorem 11.3.3 for the Prohorov metric, which is the same for these laws on S as on T.

Given $\varepsilon > 0$, first take a compact K with $P_0(K) > 1 - \varepsilon$. For each $n = 1, 2, \ldots$, let

$$a(n) := \max(1/n, \inf\{\delta > 0: P_n(K^\delta) > 1 - \varepsilon\})$$

(for K^δ as defined in 11.3). Then $P_n(K^{2a(n)}) > 1 - \varepsilon$. Since P_n converge to P_0 for the Prohorov metric (Theorem 11.3.3), $a(n) \to 0$ as $n \to \infty$. Take a compact $K_n \subset K^{2a(n)}$ with $P_n(K_n) > 1 - \varepsilon$. Let $L := K \cup \bigcup_{n=1}^{\infty} K_n$. A sequence $\{x_m\} \subset L$ has a convergent subsequence if $x_m \in K$ or a particular K_j for infinitely many m. Otherwise, there is a subsequence $x_{m(k)} \in K_{n(k)}$ with

$n(k) \to \infty$. Then for some $y_k \in K$, $d(x_{m(k)}, y_k) < 2a(n(k)) \to 0$ as $k \to \infty$. There is a subsequence $y_{k(j)} \to y \in K$. So $x_{m(k(j))} \to y \in L$. Thus $\{x_m\}$ always has a subsequence converging in L, so L is compact. Now $P_n(L) > 1 - \varepsilon$ for all n. □

The next fact relates uniform tightness to the two metrics for laws treated in §11.3, the Prohorov metric ρ and the dual-bounded-Lipschitz metric β.

11.5.4. Theorem *Let (S, d) be a complete separable metric space. Let $A be a set of laws on S. Then the following are equivalent:*

 (I) *A is uniformly tight.*
 (II) *Every sequence P_n in A has a subsequence $P_{n(k)} \to P$ for some law P on S.*
 (III) *For the metric β (or ρ) on the set of all laws on S, A has compact closure.*
 (IV) *A is totally bounded for β (or ρ).*

Remarks. Conditions (I) and (II) depend only on the topology of S, not on the specific metric. So (I) and (II) are equivalent in Polish spaces, such as $S = (0, 1)$. On the other hand, for S not complete, (IV) does depend on the metric: for example, the laws $\delta_{1/n}$ form a totally bounded set for β or ρ on $(0, 1)$, just as on $[0, 1]$ with usual metric, but (I) fails for them on $(0, 1)$, yet (I) and (IV) are equivalent on \mathbb{R}, which is the same topologically as $(0, 1)$.

Proof. The equivalence of (II) and (III) follows from the metrization theorem (11.3.3) and the equivalence of compactness and existence of convergent subsequences in metric spaces (Theorem 2.3.1). Clearly (III) implies (IV).

(I) implies (II): suppose K_n are compact and $P(K_n) > 1 - 1/n$ for all $P \in A$. Let $\{P_m\}$ be a sequence in A. For each n, the space of continuous functions $C(K_n)$ with supremum norm is separable (Corollary 11.2.5 above). The proof for $S = \mathbb{R}^k$ (Theorem 9.3.3) then applies to give a converging subsequence $P_{n(k)} \to P$. P is defined on the smallest σ-algebra making all $f \in C_b(S, d)$ measurable, which is the Borel σ-algebra since (S, d) is a metric space (Theorem 7.1.1). Thus P is a law as desired.

The last step will be to show that (IV) implies (I). If A is totally bounded for β, then it is also for ρ, since $\rho \leq 2\beta^{1/2}$, as shown in the proof of Theorem 11.3.3. So assume A is totally bounded for ρ. Given any $\varepsilon > 0$, take a finite set $B \subset A$ such that $A \subset B^{\varepsilon/2}$ for ρ. Each P in B is tight by Theorem 11.5.1 or 7.1.4, so there is a compact K_P with $P(K_P) > 1 - \varepsilon/2$. Let K_B be the union

of the sets K_P for P in B, a compact set with $P(K_B) > 1 - \varepsilon/2$ for all $P \in B$. Take a finite set $F := F(\varepsilon) \subset S$ such that $K_B \subset F^{\varepsilon/2}$, so $P(F^{\varepsilon/2}) > 1 - \varepsilon/2$ for all $P \in B$. Then $P(F^\varepsilon) > 1 - \varepsilon$ for all $P \in A$ by definition of ρ. Next, for any $\delta > 0$, let $\varepsilon(m) := \delta/2^m$, $m = 1, 2, \ldots$, and let K be the intersection of the closures of $F(\varepsilon(m))^{\varepsilon(m)}$. Then K is compact and $P(K) > 1 - \delta$ for all $P \in A$, so A is uniformly tight. $\qquad\square$

11.5.5. Corollary (Prohorov, 1956) If (S, d) is a complete separable metric space, then the set of all laws on S is complete for ρ and for β.

Proof. A Cauchy sequence (for either metric) is totally bounded, hence it has a convergent subsequence by Theorem 11.5.4 and so converges. $\qquad\square$

Problems

1. Let X_j be i.i.d. variables with distribution $N(0, 1)$. Let H be a Hilbert space with orthonormal basis $\{e_j\}_{j \geq 1}$. Let $X := \sum_j X_j e_j/j$. Find, for each $\varepsilon > 0$, a compact set $K \subset H$ such that $P(X \in K) > 1 - \varepsilon$.

2. Prove that in any metric space, a uniformly tight sequence of laws has a convergent subsequence.

3. Show that in a metric space S, the collection of all universally measurable subsets is a σ-algebra.

4. A point x in a set A in a topological space is called *isolated* iff $\{x\}$ is open in the relative topology of A. A compact set is called *perfect* iff it has no isolated points. A compact, perfect metric space has uncountably many points in any neighborhood of each of its points (Problem 10 in §2.3). Show that for any non-empty perfect compact metric space K there is a law P on the Borel sets of K with $P\{x\} = 0$ for all x and $P(U) > 0$ for every non-empty open set U. *Hint:* Find a sequence of laws for which a subsequence converges to such a P.

5. Define a specific, finite set F of laws on $[0, 1]$ such that for every law P on $[0, 1]$ there is a Q in F with $\rho(P, Q) < 0.1$, where ρ is Prohorov's metric.

6. Let $C_0(\mathbb{R})$ be the set of all continuous real functions f on \mathbb{R} such that $f(x) \to 0$ as $|x| \to \infty$. If μ_n and μ are finite measures on \mathbb{R}, say that $\mu_n \to \mu$ vaguely if $\int f \, d\mu_n \to \int f \, d\mu$ for all $f \in C_0(\mathbb{R})$. Show that every sequence of laws on \mathbb{R} has a subsequence converging vaguely to some measure μ on \mathbb{R} with $0 \leq \mu(\mathbb{R}) \leq 1$. *Hint:* See Problem 1(a) of §2.8.

Convergence of Laws on Separable Metric Spaces

7. Let H be an infinite-dimensional, separable Hilbert space with an orthonormal basis $\{e_n\}_{n\geq 1}$. Let $e(n) := e_n$ and $P_n := \delta_{e(n)}$. For any subsequence $P_{n(k)}$, find a bounded continuous f on H such that $\int f\, dP_{n(k)}$ does not converge.

8. Let (S, d) be a metric space. Let A be a uniformly tight set of laws on S. Show that A has compact closure for ρ or β. *Hint:* See Problem 2.

9. Call a set A in a topological space S *universally of measure* 0 iff for every measure μ on the Borel sets of A which is nonatomic (that is, $\mu\{x\} = 0$ for every point x), $\mu^*(A) = 0$. Assuming the continuum hypothesis, show that there is an uncountable set A universally of measure 0 in $[0, 1]$. *Hint:* Show that there are c nonatomic laws on the Borel σ-algebra of $[0, 1]$, using the equivalence theorem 1.4.1. In a transfinite recursion (1.3.2), let μ_α be the nonatomic laws. Alternately put a new point $x_\alpha \in A$ and put $B_\alpha \subset A^c$ where B_α is a set of first category disjoint from the countable set $\{x_\beta\}_{\beta \leq \alpha}$ with $\mu_\alpha(\bigcup_{\beta \leq \alpha} B_\beta) = 1$.

11.6. Strassen's Theorem: Nearby Variables with Nearby Laws

Strassen's theorem says that if two laws are close to each other in the Prohorov metric ρ, then there are random variables X and Y with these laws which are equally close (for the Ky Fan metric) in probability. In other words, the joint distribution of (X, Y) will be concentrated near the diagonal $X = Y$ (so that X and Y will be far from independent). The proof will be based on a finite combinatorial fact called a pairing theorem.

Given two sets X and Y, a *relation* is a subset $K \subset X \times Y$. Then xKy will mean $\langle x, y\rangle \in K$. For any $A \subset X$ let $A^K := \{y \in Y : xKy$ for some $x \in A\}$. A *K-pairing f* of A into B is a 1–1 function f from A into B such that $xKf(x)$ for all $x \in A$. For any finite set A, card(A) (the cardinality of A) means the number of elements in A.

11.6.1. Pairing Theorem (D. König-P. Hall) *Let X and Y be finite sets with a relation $K \subset X \times Y$ such that for any set $A \subset X$, card(A^K) \geq card(A). Then there exists a K-pairing f of X into Y, and so onto Y if* card(Y) $=$ card(X).

Note. The given sufficient condition for the existence of a K-pairing is also necessary, since f must be 1–1 from A into A^K for each A. The theorem has also been called the "marriage lemma": if X is a set of women, Y is a set of men, and the relation xKy means x and y are compatible, then the theorem gives the exact condition that each x can be married to a compatible y.

Proof. Let card$(X) = m$. The proof is by induction on m. For $m = 1$, the result is clear. Assume it holds for $1, \ldots, m-1$. Choose some $x \in X$. Then $x K y$ for some $y \in Y$. If there is a K-pairing f of $X \setminus \{x\}$ into $Y \setminus \{y\}$, let $f(x) = y$ to give the desired f. Otherwise, by induction assumption, there is a set $A \subset X \setminus \{x\}$ such that card$(A^K \setminus \{y\}) <$ card(A), so card$(A^K) =$ card$(A) > 0$, and there is a K-pairing f of A onto A^K. If the hypothesis holds for $X \setminus A$ and $Y \setminus A^K$ (restricting the relation K to the product of those sets), we can define the pairing on all of X. Otherwise there is a set $D \subset X \setminus A$ with card$(D^K \setminus A^K) <$ card(D). But then card$((A \cup D)^K) =$ card$(A^K) +$ card$(D^K \setminus A^K) <$ card$(A) +$ card$(D) =$ card$(A \cup D)$, a contradiction. \square

Let (S, d) be a separable metric space. $\mathcal{P}(S)$ will denote the set of all laws on S. Given $A \subset S$ and $\delta > 0$, recall that

$$A^\delta = \{x : d(x, y) < \delta \text{ for some } y \in A\} = \{x : d(x, A) < \delta\},$$

as in the definition of Prohorov's metric (§11.3); also, $d(\cdot, A)$ is as in (2.5.3). Then A^δ is open. A closed set $A^{\delta]}$ is defined by $A^{\delta]} := \{x : d(x, A) \leq \delta\}$.

On a Cartesian product $A \times B$, we have the coordinate projections $p_1(\langle x, y \rangle) := x$ and $p_2(\langle x, y \rangle) := y$. For a law μ on $A \times B$, its *marginals* are the laws $\mu \circ p_1^{-1}$ on A and $\mu \circ p_2^{-1}$ on B. For example, if μ is a product measure $P \times Q$, where P and Q are laws, then P and Q are the marginals. On the other hand, if P is a law on A, and μ is the "diagonal" law $P \circ D^{-1}$ on $A \times A$, where $D(x) := \langle x, x \rangle$, then both marginals of μ are equal to P.

Theorem 11.3.5 showed that the Prohorov distance between two laws is bounded above by the Ky Fan distance of any two random variables with those laws. On the other hand, two random variables with the same law could be independent and so not close in the Ky Fan metric. The next fact is a converse of Theorem 11.3.5 which will imply (for $\alpha = \beta$) that for any two laws, there are random variables with those laws, as close (or nearly so, for incomplete spaces) in the Ky Fan metric as the laws are in the Prohorov metric (Corollary 11.6.4).

11.6.2. Theorem *For any separable metric space (S, d), any two laws P and Q on S, $\alpha \geq 0$ and $\beta \geq 0$, the following are equivalent:*

(I) For any closed set $F \subset S$, $P(F) \leq Q(F^{\alpha]}) + \beta$.

(II) For any $a > \alpha$ there is a law $\mu \in \mathcal{P}(S \times S)$ with marginals P and Q such that $\mu\{\langle x, y \rangle : d(x, y) > a\} \leq \beta$.

If P and Q are tight, then μ can be chosen in (II) for $a = \alpha$.

Remarks. If $\alpha = \beta$, then (I) says that $\rho(P, Q) \leq \alpha$ where ρ is Prohorov's metric. All laws on S are tight if it is complete (Ulam's theorem, 7.1.4) or just universally measurable (Theorem 11.5.1). Problem 5 below is to show that for a general (nonmeasurable) set S we may not be able to take $a = \alpha$. Theorem 11.6.2 will be proved by way of the following more detailed fact.

11.6.3. Theorem *Assuming (I) of Theorem 11.6.2, for any $a > \alpha$ and $b > \beta$ there exist two nonnegative Borel measures η and γ on $S \times S$ such that:*

(1) $\eta + \gamma$ is a law on $S \times S$ having marginals P and Q on S.
(2) η is concentrated in the set of $\langle x, y \rangle$ such that $d(x, y) \leq a$.
(3) γ has total mass $\leq b$.
(4) η and γ are both finite sums of product measures.

Proof. Take $\varepsilon > 0$ with $\varepsilon < \min(a - \alpha, b - \beta)/5$. Some simpler cases will be treated first.

 Case A: There is an integer n and sets $M \subset S$, $N \subset S$ with $\mathrm{card}(M) = \mathrm{card}(N) = n$ such that $P\{x\} = Q\{y\} = 1/n < \varepsilon$ for all $x \in M$ and $y \in N$. Take an integer k with $n\beta < k < n(\beta + \varepsilon)$. Take sets U and V with k elements each, disjoint from all sets previously mentioned. Let $X := M \cup U$ and $Y := N \cup V$. For $x \in X$ and $y \in Y$ define $x K y$ iff: (a) $x \in U$, or (b) $y \in V$, or (c) $d(x, y) \leq \alpha$, $x \in M$ and $y \in N$. Given $A \subset X$ with $\mathrm{card}(A) = r$, to show $\mathrm{card}(A^K) \geq r$ it can be assumed that $A \subset M$, since otherwise $A^K = Y$. Then

$$r/n = P(A) \leq Q(A^{\alpha]}) + \beta \leq \beta + \mathrm{card}(A^K \cap N)/n, \quad \text{so}$$
$$r \leq n\beta + \mathrm{card}(A^K \cap N) < k + \mathrm{card}(A^K \cap N) = \mathrm{card}(A^K).$$

 So the pairing theorem (11.6.1) applies, giving a K-pairing f of X onto Y. Then $f(x) \in N$ for at least $n - k$ members x of M, forming a set T, and then $d(x, f(x)) \leq \alpha$. Let $g(x) = f(x)$ for such x and extend g to a 1–1 function from M onto N. Let

$$\mu := \sum_{x \in M} \delta_{\langle x, g(x) \rangle}/n, \quad \eta := \sum_{x \in T} \delta_{\langle x, g(x) \rangle}/n, \quad \gamma := \mu - \eta.$$

 Then $\mu \in \mathcal{P}(S \times S)$, μ has marginals P and Q, and as desired, η and γ have properties (1) to (4).
 Case B: Each of P and Q is concentrated in finitely many points and gives each a rational probability. Then for some positive integer n with $1/n < \varepsilon$, P and Q have values included in $\{j/n: j = 0, 1, \ldots, n\}$. Let

J be a set (disjoint from S) with n elements, say $J = \{1, 2, \ldots, n\}$. For $i, j \in J$ let $f(i, j) = \varepsilon$ for $i \neq j$, and 0 if $i = j$. On $S \times J$ let

$$e(\langle x, i \rangle, \langle y, j \rangle) := d(x, y) + f(i, j) \leq d(x, y) + \varepsilon.$$

Then e is a metric. Define a law $P_1 \in \mathcal{P}(S \times J)$ so that for each atom x of P with $P\{x\} = j/n$, $P_1(\langle x, i \rangle) = 1/n$ for $i = 1, \ldots, j$, and 0 for $i > j$. Then P_1 has marginal P on S. Let Q_1 be the analogous law on $S \times J$ with marginal Q on S.

Then the hypotheses of Case A hold for P_1 and Q_1, with $\alpha + \varepsilon$ in place of α. So by case A, there is a law $\mu_1 = \eta_1 + \gamma_1$ on $(S \times J) \times (S \times J)$ with marginals P_1 and Q_1 on $S \times J$ such that η_1 and γ_1 have properties (1) to (4) for μ_1. Let μ, η, and γ be the marginals of μ_1, η_1, and γ_1, respectively, on $S \times S$. Then η and γ have properties (1) to (4), since $\alpha + \varepsilon < a$.

Case C: This is the general case. Given $\varepsilon > 0$, let A be a maximal subset of S with $d(x, y) \geq \varepsilon$ for all $x \neq y$ in A. Then A is countable, say $A = \{x_j\}_{j \geq 1}$, possibly finite. Let

$$B_j := \{x \in S : d(x, x_j) < \varepsilon \leq d(x, x_i) \quad \text{for all } i < j\}.$$

Then the B_j are disjoint Borel sets with union S. Define a law P' by $P'\{x_j\} = p_j := P(B_j)$ for $j = 1, 2, \ldots$. Likewise define Q' from Q. Then for any Borel set A, $P'(A) \leq P(A^\varepsilon)$. First suppose $A \neq S$ and choose $x_0 \in S \backslash A$. For some integer n, let $P''\{x_j\} = [np_j]/n$ for all $j \geq 1$ where $[x]$ is the greatest integer $\leq x$. Define $P''\{x_0\} := 1 - \sum_{j \geq 1} P''\{x_j\}$. Define Q'' from Q likewise. Let n be chosen large enough so that $P''\{x_0\} < \varepsilon$ and $Q''\{x_0\} < \varepsilon$. Then for any closed F, since $(F^\varepsilon)^\alpha \subset F^{\varepsilon+\alpha}$, and the closed sets $F^{\delta]} \uparrow F^\varepsilon$ as $\delta \uparrow \varepsilon$,

$$P''(F) \leq P'(F) + \varepsilon \leq P(F^\varepsilon) + \varepsilon \leq Q(F^{\varepsilon+\alpha}) + \beta + \varepsilon$$
$$\leq Q'(F^{2\varepsilon+\alpha}) + \beta + \varepsilon \leq Q''(F^{2\varepsilon+\alpha}) + \beta + 2\varepsilon,$$

where the next-to-last inequality follows from $Q(C) \leq Q'(C^\varepsilon)$ for any Borel set C, which holds since if C intersects B_j, C^ε contains x_j. Now Case B applies to P'' and Q'', giving $\mu'' = \eta'' + \gamma''$ on $S \times S$ with marginals P'' and Q'', $\eta''(d > 3\varepsilon + \alpha) = 0$ and $\gamma''(S \times S) \leq \beta + 3\varepsilon$. These measures are concentrated in points $\langle x_i, x_j \rangle$, giving them masses which are multiples of $1/n$. If we change the definition of γ'' by including in it all the mass at $\langle x_0, x_j \rangle$ or $\langle x_i, x_0 \rangle$ for any i or j, then all the above holds except that now $\gamma''(S \times S) \leq \beta + 5\varepsilon$. Now in the exceptional case $S = A$, we perform the above construction for $A \cup \{x_o\}$ for some point

x_o not in A and then take image measures under the mapping leaving A fixed and taking x_o to x_1.

Let $P_i(C) := P(C|B_i) := P(C \cap B_i)/P(B_i)$ for any Borel set C, or $P_i := 0$ if $P(B_i) = 0$, for $i = 1, 2, \ldots$. Likewise define Q_j from Q for $j = 1, 2, \ldots$. Let

$$\eta'' = \sum_{i,j} \frac{k(i,j)}{n} \delta_{\langle x_i, x_j \rangle}, \qquad \eta := \sum_{i,j} \frac{k(i,j)}{n}(P_i \times Q_j)$$

where $k(i,j)$ are nonnegative integers and the sums are over $i, j \geq 1$ with $d(x_i, x_j) \leq 3\varepsilon + \alpha$. Let η have marginals u and v. Then it can be checked that $u \leq P$ and likewise $v \leq Q$. Let $t := (P - u)(S) = (Q - v)(S) \leq \beta + 5\varepsilon$. If $t = 0$, let $\gamma := 0$, otherwise let $\gamma := (P - u) \times (Q - v)/t$. Then clearly $\eta + \gamma$ is a law with marginals P and Q, and (1) to (4) hold, proving Theorem 11.6.3. $\qquad \square$

Proof of Theorem 11.6.2. To show that (II) implies (I), for μ satisfying (II) and any closed set F,

$$P(F) = \mu(F \times S) \leq \beta + \mu\{\langle x, y \rangle : x \in F, d(x, y) \leq a\}$$
$$\leq \beta + \mu\{\langle x, y \rangle : y \in F^a\} = \beta + Q(F^a).$$

Let $a = a(n) := \alpha + 1/n$ for $n = 1, 2, \ldots$. Then $\bigcap_n F^{a(n)} = F^{\alpha]}$ and (I) holds.

For the converse, to show (I) implies (II), first suppose P and Q are tight. For $n = 1, 2, \ldots$, by Theorem 11.6.3 there are laws μ_n on $S \times S$ with marginals P and Q for all n such that

$$\mu_n\{d > \alpha + 1/n\} < \beta + 1/n, \qquad n = 1, 2, \ldots.$$

If K and L are compact, $P(K) > 1 - \varepsilon$ and $Q(L) > 1 - \varepsilon$, then $\mu_n((S \times S) \setminus (K \times L)) \leq 2\varepsilon$ for all n, so $\{\mu_n\}$ are uniformly tight and by Theorem 11.5.4 have a subsequence convergent to some μ, which has marginals P and Q. Then by the portmanteau theorem (11.1.1), for each n, $\mu(d > \alpha + 1/n) \leq \beta$, so $\mu(d > \alpha) \leq \beta$, proving the theorem in that case.

Now in the general case, let \overline{S} be the completion of S. For each Borel set C in \overline{S} let $\overline{P}(C) := P(C \cap S)$. Then \overline{P} is a law on \overline{S}. Likewise define \overline{Q} on \overline{S}. By Ulam's theorem (7.1.4), \overline{P} and \overline{Q} are tight. For any closed set F in \overline{S},

$$\overline{P}(F) = P(F \cap S) \leq Q((F \cap S)^{\alpha]} + \beta) \leq \overline{Q}(F^{\alpha]}) + \beta.$$

So by the tight case, there is a law v on $\overline{S} \times \overline{S}$ with marginals \overline{P} and \overline{Q} satisfying $v(d > \alpha) \leq \beta$. As in the proof of Theorem 11.6.3, Case C, given

$0 < \varepsilon < (a-\alpha)/2$, let \overline{S} be the union of disjoint Borel sets \overline{B}_j with $d(x,y) \le \varepsilon$ for all $x, y \in \overline{B}_j$ for each j. Let $B_j := \overline{B}_j \cap S$. Let $c_{jk} := \nu(\overline{B}_j \times \overline{B}_k)$, $p_j := \overline{P(\overline{B}_j)} = P(B_j)$, and $q_k := \overline{Q(\overline{B}_k)} = Q(B_k)$ for $j, k = 1, 2, \dots$. Then $p_j \equiv \sum_k c_{jk}$ and $q_k \equiv \sum_j c_{jk}$. Define P_j and Q_k as before. Let $\mu := \sum_{j,k} c_{jk}(P_j \times Q_k)$. Let $\overline{\mu}(D) := \mu(D \cap (S \times S))$ for Borel sets $D \subset \overline{S} \times \overline{S}$. Then μ has marginals P and Q, $\overline{\mu}(\overline{B}_j \times \overline{B}_k) = c_{jk}$ for all j and k, and

$$\mu(d > \alpha + 2\varepsilon) = \overline{\mu}(d > \alpha + 2\varepsilon)$$
$$\le \sum_{j,k} \{c_{jk} : d(y,z) > \alpha \quad \text{for all } y \in \overline{B}_j \text{ and } z \in \overline{B}_k\}$$
$$\le \nu(d > \alpha) \le \beta, \quad \text{proving Theorem 11.6.2.} \qquad \square$$

Now the Ky Fan metric α for random variables, which metrizes convergence in probability (Theorem 9.2.2), relates to the Prohorov metric for laws as follows:

11.6.4. Corollary *For any separable metric space (S, d), laws P and Q on S, and $\varepsilon > 0$, or $\varepsilon = 0$ if P and Q are tight, there is a probability space (Ω, μ) and random variables X, Y on Ω with*

$$\alpha(X, Y) \le \rho(P, Q) + \varepsilon, \qquad \mathcal{L}(X) = P, \text{ and } \mathcal{L}(Y) = Q.$$
$$\text{So } \rho(P, Q) = \inf\{\alpha(X, Y) : \mathcal{L}(X) = P, \mathcal{L}(Y) = Q\}.$$

Proof. For the first part, apply Theorem 11.6.2 with $\alpha = \beta = \rho(P, Q)$ and $\alpha < a < \alpha + \varepsilon$ in general, or $a = \alpha$ if P and Q are tight. Since always $\rho(\mathcal{L}(X), \mathcal{L}(Y)) \le \alpha(X, Y)$ (Theorem 11.3.5), letting $a \downarrow \alpha(X, Y)$ gives the conclusion. $\qquad \square$

Now recall that the dual-bounded-Lipschitz metric β and the Prohorov metric ρ metrize the same topology on laws (Theorem 11.3.3). More specifically, in the proof of Theorem 11.3.3, (c) implies (d), it was shown that $\rho \le 2\beta^{1/2}$. Here is a specific inequality in the other direction:

11.6.5. Corollary *For any separable metric space S and laws P and Q on S, $\beta(P, Q) \le 2\rho(P, Q)$.*

Proof. Given $\varepsilon > 0$, take random variables X and Y from Corollary 11.6.4. Let $A := \{\langle x, y \rangle \in S \times S : d(x, y) \le \rho(P, Q) + \varepsilon\}$. Take any $f \in BL(S, d)$.

Then

$$\left|\int f \, d(P - Q)\right| = |Ef(X) - Ef(Y)| \le E|f(X) - f(Y)|$$

$$\le [\|f\|_L E1_A(X, Y) + 2\|f\|_\infty](\rho(P, Q) + \varepsilon)$$

$$\le 2\|f\|_{BL}(\rho(P, Q) + \varepsilon).$$

Letting $\varepsilon \downarrow 0$ finishes the proof. □

It follows that ρ and β define the same uniform structure (uniformity) on the set of all laws on S, to be treated in the next section.

Problems

1. Let $n_{ij} = 0$ or 1 for $i = 1, \dots, k$ and $j = 1, \dots, m$, so that $\{n_{ij}\}$ forms a rectangular $k \times m$ matrix of zeroes and ones. Call a pair $\langle A, B \rangle$, where $A \subset \{1, \dots, k\}$ and $B \subset \{1, \dots, m\}$, a *cover* if whenever $n_{ij} = 1$, either $i \in A$ or $j \in B$. Let c be the smallest sum of the cardinalities $\text{card}(A) + \text{card}(B)$ for all covers $\langle A, B \rangle$. Let d be the largest cardinality of a set $D \subset \{1, \dots, k\} \times \{1, \dots, m\}$ such that $n_{ij} = 1$ for all $\langle i, j \rangle \in D$ and for any $\langle i, j \rangle \ne \langle r, s \rangle$ in D, both $i \ne r$ and $j \ne s$. Prove that $c = d$. *Hint:* Use Theorem 11.6.1.

2. Show that Corollary 11.6.5 is sharp, in \mathbb{R}, in the sense that for $0 \le t \le 1$, $\sup\{\beta(P, Q): \rho(P, Q) = t\} = 2t$. *Hint:* Let P and Q each have atoms of size t at points far apart, with P and Q otherwise the same. (In the other direction, see §11.3, Problem 5.)

3. (a) Show that for any compact set K and $\delta > 0$, if $y \in K^{\delta]}$, then for some $x \in K, d(x, y) \le \delta$.
 (b) Show that in an infinite-dimensional Hilbert space, there is a closed set F with discrete relative topology and $y \in F^{1]}$ with $d(x, y) > 1$ for all $x \in F$. *Hint:* Let $F = \{a_n e_n\}_{n \ge 1}$ for e_n orthonormal and some a_n.

4. In \mathbb{R}, given a probability distribution function F, recall the random variable X_F defined by $X_F(t) := \inf\{x: F(x) \ge t\}$ defined for $0 < t < 1$, with respect to Lebesgue measure λ on $(0, 1)$ (Proposition 9.1.2). Give an example of two laws P and Q with distribution functions F and G, respectively, such that $\alpha(X_F, X_G) > \rho(P, Q)$, so that in Corollary 11.6.4 we cannot take $X = X_F$ and $Y = X_G$ (for small enough ε).

5. Let E be a set in $[0, 1]$ with Lebesgue inner measure 0 and outer measure 1 (Theorem 3.4.4), $\lambda_*(E) = 0$ and $\lambda^*(E) = 1$. Let $T := \{x + 1: x \in [0, 1] \backslash E\}$ and $S := E \cup T$. Define laws P and Q on S by $P(A) := \lambda^*(A \cap E)$ and

$Q(A) := \lambda^*(A \cap T)$ whenever A is a Borel set as a subset of S (not, in general, a Borel set in \mathbb{R}), by Theorem 3.3.6. Show that condition (I) in Theorem 11.6.2 holds for $\alpha = 1$ and $\beta = 0$, while condition (II) does not hold for $a = 1$ and $\beta = 0$, showing that taking $a = \alpha$ can fail without tightness.

6. In Theorem 11.6.3, suppose $\gamma(S \times S) > 0$. Let γ' be the product measure $\gamma' = (\gamma \circ p_1^{-1}) \times (\gamma \circ p_2^{-1})/\gamma(S \times S)$. Prove, or disprove, which of properties (1) to (4) are preserved when γ is replaced by γ'.

7. Recall the notion of conditional distribution $P_{\cdot|\cdot}(\cdot, \cdot)$ as defined in §10.2. Let (X, \mathcal{A}) and (Y, \mathcal{B}) be two measurable spaces. Let P be a probability measure on the product σ-algebra in $X \times Y$. Let $h(x, y) \equiv y$. Let \mathcal{C} be the smallest σ-algebra of subsets of $X \times Y$ for which x is measurable, so $\mathcal{C} = \{A \times Y: A \in \mathcal{A}\}$. Suppose P is a finite sum of product measures. Show that there exists a conditional distribution $P_{h|\mathcal{C}}(\cdot, \cdot)$ on $\mathcal{B} \times (X \times Y)$, where $P_{h|\mathcal{C}}(\cdot, (x, y))$ does not depend on y.

*11.7. A Uniformity for Laws and Almost Surely Converging Realizations of Converging Laws

First, instead of converging sequences of laws, sequences of pairs of laws will be considered. This amounts to treating uniformities (uniform structures, as defined in §2.7) rather than topologies on spaces of laws. Then it will be shown that for a converging sequence of laws, on a separable metric space, there exist random variables with those laws converging almost surely.

11.7.1. Theorem *For any separable metric space (S, d) and sequences of laws P_n and Q_n on S, the following are equivalent as $n \to \infty$:*

(a) $\beta(P_n, Q_n) \to 0.$
(b) $\rho(P_n, Q_n) \to 0.$
(c) There exist on some probability space random variables X_n and Y_n with values in S such that X_n has law P_n and Y_n has law Q_n for all n, with $d(X_n, Y_n) \to 0$ in probability.
(d) The same as (c) with "in probability" replaced by "almost surely."

Remark. There is a bounded continuous function f on \mathcal{R} with $f(n) = 0$ and $f(n + 1/n) = 1$ for all $n = 1, 2, \ldots$, by Theorem 2.6.4. Letting $P_n = \delta_n$ and $Q_n = \delta_{n+(1/n)}$ shows that conditions (a) and (b) in Theorem 11.7.1 are *not* equivalent to $\int f \, d(P_n - Q_n) \to 0$ for all bounded continuous f, although they are if the Q_n are all the same by Theorem 11.3.3.

Proof. (b) implies (a) by Corollary 11.6.5 and (a) implies (b) since $\rho \leq 2\beta^{1/2}$ as shown in the proof that 11.3.3(c) implies 11.3.3(d). Clearly (d) implies (c).

(c) implies (a): Let $\|f\|_{BL} \leq 1$. Given $\varepsilon > 0$, let A_n be the event that $d(X_n, Y_n) > \varepsilon$. Then

$$\left| \int f \, d(P_n - Q_n) \right| = |E(f(X_n) - f(Y_n))| \leq 2\Pr(A_n) + \varepsilon.$$

Thus $\beta(P_n, Q_n) \leq 3\varepsilon$ for n large enough, so (a) holds.

So it will be enough to prove that (b) implies (d). Theorem 11.6.3, with $\alpha = \beta = \alpha_n = \rho(P_n, Q_n)$ and $a = b = \alpha_n + 1/n$, gives measures $\mu_n = \eta_n$ and $\gamma = \gamma_n$ on $S \times S$. Let $t_n = \gamma_n(S \times S)$. Let T be the Cartesian product of countably many spaces $S_n \times S_n$ where each S_n is a copy of S. On $I := [0, 1)$ take Lebesgue measure λ. Let $\Omega := I \times T$. For each $x \in I$, let \Pr_x be the probability measure on T which is the Cartesian product, by Theorem 8.2.2, of the laws $\mu_n/(1 - t_n)$ for those n such that $x < 1 - t_n$ and of γ_n/t_n for n with $x \geq 1 - t_n$. Let A be any measurable subset of Ω. A law \Pr on Ω will be defined by $\Pr(A) = \int_I \int_T 1_A(x, y) \, d\Pr_x(y) \, dx = \int_I \Pr_x(A_x) \, d\lambda(x)$, where $A_x := \{y \in T : \langle x, y \rangle \in A\}$. To show \Pr is well-defined, let's first show that $g_A(x) := \Pr_x(A_x)$ is measurable in x. Let \mathcal{M} be the collection of all sets $C \subset \Omega$ such that g_C is measurable. Let \mathcal{R} be the semiring of finite-dimensional rectangles in Ω, as in Proposition 8.2.1, so in this case each set B in \mathcal{R} is a product of a Borel set in I, Borel sets in $S_n \times S_n$ for finitely many n, and all of $S_n \times S_n$ for other values of n. There are only finitely many possibilities for \Pr_x on a finite product, and for B_x given B, each on a Borel set of x in I. It follows that $\mathcal{R} \subset \mathcal{M}$. Next, each set A in the algebra \mathcal{A} generated by \mathcal{R} is a finite disjoint union $A = \cup_i B(i)$ of sets in \mathcal{R}, by Proposition 8.2.1. Then $g_A = \sum_i g_{B(i)}$, so g_A is measurable and $\mathcal{A} \subset \mathcal{M}$. Now monotone convergence of sets $C(n)$ in Ω implies monotone convergence of the functions $g_{C(n)}$, which preserves measurability, so \mathcal{M} is a monotone class and by Theorem 4.4.2, \mathcal{M} includes the product σ-algebra \mathcal{S} on Ω, which is generated by \mathcal{A}, or by \mathcal{R}. Clearly $\Pr(\Omega) = 1$ and $\Pr(\emptyset) = 0$. \Pr is countably additive since if sets A_j in \mathcal{S} are disjoint, then for each $x \in I$, the sets $(A_j)_x$ are disjoint in T, with union A_x, their \Pr_x probabilities add up to $\Pr_x(A_x)$, and we can apply dominated or monotone convergence. So \Pr is a probability measure on Ω.

Its marginal on $S_n \times S_n$, found by integrating the marginal of \Pr_x with respect to $d\lambda(x)$ on I, is $(1 - t_n)\mu_n/(1 - t_n) + t_n\gamma_n/t_n = \mu_n + \gamma_n$, for each n (replacing $0/0$ by 1), which has marginals P_n and Q_n on S_n as desired. The coordinates (X_n, Y_n) in $S_n \times S_n$ satisfy $d(X_n, Y_n) \leq \alpha_n + 1/n$ almost surely

for those n such that $x < 1 - t_n$, which holds for n large enough for all x. Thus $d(X_n, Y_n) \to 0$ almost surely. \square

Next, if the Q_n are all the same in Theorem 11.7.1, it will be shown that we can take all the Y_n the same. This will not be a corollary of Theorem 11.7.1; the proof is different and harder.

11.7.2. Theorem *Let S be any separable metric space and P_n, $n = 0, 1, \ldots$, laws on S with P_n converging to P_0 as $n \to \infty$. Then on some probability space there exist random variables X_n, $n = 0, 1, \ldots$, with values in S, such that X_n has law P_n for all n and $X_n \to X_0$ almost surely.*

Proof. Recall that, as in Theorem 11.1.1, a set $A \subset S$ is called a *continuity set* of P_0 iff $P_0(\partial A) = 0$, where ∂A is the boundary of A, and that for a set A in a metric space (S, d), $\mathrm{diam}(A) := \sup\{d(x, y): x, y \in A\}$. The following will be useful:

11.7.3. Lemma *For any separable metric space (S, d), law P on S, and $\varepsilon > 0$, there are disjoint Borel continuity sets A_j of P with $S = \bigcup_{j \geq 1} A_j$ and $\mathrm{diam}(A_j) < \varepsilon$ for all j.*

Proof. For any $x \in S$ and $r > 0$, the ball $B(x, r) := \{y: d(x, y) < r\}$ is a continuity set of P unless $P\{y: d(x, y) = r\} > 0$, which can happen for at most countably many values of r. Let $\{x_j\}_{j \geq 1}$ be dense in S. For each j, $B(x_j, r)$ is a continuity set for some $r = \varepsilon_j$ with $\varepsilon/4 < \varepsilon_j < \varepsilon/2$. Let $A_1 := B(x_1, \varepsilon_1)$ and for $j > 1$ let

$$A_j := B(x_j, \varepsilon_j) \setminus \bigcup_{i < j} B(x_i, \varepsilon_i).$$

Since the continuity sets form an algebra (Proposition 11.1.4), the A_j are continuity sets and the rest follows. \square

Now to continue the proof of Theorem 11.7.2, let $P := P_0$. Let $\varepsilon := \varepsilon_m = 1/m^2$ for $m = 1, 2, \ldots$. Given such an ε, choose continuity sets $A_j = A_{jm}$ for each $j = 1, 2, \ldots$, as given by Lemma 11.7.3 for P and ε. Take $k(m)$ large enough so that $P(\bigcup_{i \leq k(m)} A_{im}) > 1 - \varepsilon$. We can assume that $P(A_{im}) > 0$ for each $i \leq k(m)$ (otherwise renumber the A_{im} and/or reduce $k(m)$). By Theorem 11.1.1, $P_n(A_{jm}) \to P(A_{jm})$ for each j and m as $n \to \infty$. So for some n_m large enough, $P_n(A_{jm}) > p_{jm} := (1 - \varepsilon_m)P(A_{jm})$ for all $j =$

$1, \ldots, k(m)$ and $n \geq n_m$. Replacing n_m by $\max\{n_i : i \leq m\}$, we can take the sequence $\{n_m\}_{m \geq 1}$ nondecreasing. We can also assume $n_m \uparrow \infty$ as $m \to \infty$.

Let $n_0 := 1$. For each $n = 1, 2, \ldots$, there is a unique $m = m(n)$ such that $n_m \leq n < n_{m+1}$. Let, for $m \geq 1$,

$$\eta_n(B) := \sum_{1 \leq j \leq k(m)} p_{jm} P_n(B | A_{jm})$$

for each Borel set $B \subset S$, and let $\alpha_n := P_n - \eta_n$. By choice of p_{jm} and n_m, $P_n \geq \eta_n$, so α_n is a nonnegative measure on S. The total mass of α_n is

$$t_n := 1 - \sum_{j \leq k(m)} p_{jm} = 1 - (1 - \varepsilon_m) \sum_{j \leq k(m)} P(A_{jm}) \leq 1 - (1 - \varepsilon_m)^2 < 2\varepsilon_m.$$

Then

$$\sum_{j \leq k(m)} P(A_{jm}) = (1 - t_n)/(1 - \varepsilon_m). \tag{11.7.4}$$

Also, $t_n \geq 1 - (1 - \varepsilon_m) = \varepsilon_m = 1/m^2 > 0$.

Now, the probability space Ω will be $I \times \Pi_{n \geq 0} S_n$ where each S_n is a copy of S. For each $t \in I := (0, 1]$, $x \in S_0$, and $n = 1, 2, \ldots$, define a law $\mu_n(x, t)(\cdot)$ on S_n by $\mu_n \equiv P_n$, if $n < n_1$; if $m = m(n) \geq 1$, let $\mu_n(t, x)(\cdot) := P_n(\cdot | A_{jm})$ if $t \geq \varepsilon_m$, $x \in A_{jm} = A_{jm(n)}$, and $j \leq k(m)$. Otherwise, if $x \notin \bigcup_{j \leq k(m)} A_{jm}$ or $t < \varepsilon_m$, let $\mu_n(t, x)(\cdot) := \alpha_n/t_n$. So μ_n is defined in each case as a probability law on the Borel sets of S. Let λ be Lebesgue measure on I. Given $(t, x) \in I \times S_0$, let $\mu_{t,x}$ on $V := \Pi_{n \geq 1} S_n$ be the product of the laws $\mu_n(t, x)(\cdot)$ for each n, which exists by Theorem 8.2.2. A probability measure Pr on Ω will be defined by $Pr(B) := \int_I \int_S \int_V 1_B(x, s, v) \, d\mu_{s,x}(v) \, dP(s) \, dx$. So (t, x) in $I \times S_0$ will have the product law $\lambda \times P$. To show Pr is well-defined, we need to show, as in the proof of Theorem 11.7.1, that the inner integral is a measurable function of (s, x). For finite-dimensional rectangles, there are again only finitely many possibilities for the product of the $\mu_n(t, x)$ for finitely many values of $n \geq 1$, each occurring on a measurable set of (t, x) in $I \times S_0$, so we have the desired measurability for such rectangles. We can pass from the semiring of rectangles (Proposition 8.2.1) to their generated algebra of finite disjoint unions of rectangles and then by monotone classes (Theorem 4.4.2) to the product σ-algebra, just as in the proof of Theorem 11.7.1. So we get a well-defined law Pr on Ω.

Let X_n be the nth coordinate on Ω, with values in S_n. Then $Pr \circ X_n^{-1} = P_n$ clearly for $n < n_1$, so suppose $n \geq n_1$ and $m = m(n) \geq 1$, $\varepsilon = \varepsilon_m$, and $k := k(m)$. Then the conditional distribution of X_n given X_0, found by integrating with respect to t in I, is $(1 - \varepsilon) P_n(\cdot | A_{jm}) + \varepsilon \alpha_n/t_n$ if $X_0 \in A_{jm}$ for some $j = 1, \ldots, k$; otherwise the conditional distribution of X_n given X_0 is α_n/t_n.

By (11.7.4), $P(S \backslash \bigcup_{j \leq k(m)} A_{jm}) = (t_n - \varepsilon)/(1 - \varepsilon)$. Then integrating with respect to X_0 for P, we get

$$Pr \circ X_n^{-1} = (t_n - \varepsilon)\alpha_n/(t_n(1 - \varepsilon)) + \sum_{1 \leq j \leq k} P(A_{jm})\{(1 - \varepsilon)P_n(\cdot | A_{jm})$$

$$+ \varepsilon\alpha_n/t_n\} = \eta_n + c\alpha_n$$

for a constant $c = c(n, \varepsilon)$ which must be 1 since we get a probability law (and which can be checked directly), so, as desired, the distribution of X_n is $\eta_n + \alpha_n = P_n$.

Since $\sum_m \varepsilon_m = \sum_m 1/m^2 < \infty$, the Borel-Cantelli lemma for P implies that for almost all $X_0 \in S_0$, for all m large enough, $X_0 \in A_{jm}$ for some $j \leq k(m)$. Also, for all $t \in I$, $t > \varepsilon_m$ for m large enough and so also for n large enough. When these things occur, the conditional distribution of X_n given X_0 is concentrated in A_{jm}, so $d(X_n, X_0) < \varepsilon_m$. Thus as $n \to \infty$, $X_n \to X_0$ a.s. □

Theorems 11.7.1 and 11.7.2 both spoke of "some probability space." There are sometimes advantages in using a fixed probability space, such as the unit interval $[0, 1]$ with Lebesgue measure λ. The next fact will be useful in replacing complete separable metric spaces by $[0, 1]$ as probability space:

11.7.5. Theorem *For any complete separable metric space (S, d) and law P on S, there is a Borel measurable function f from $[0, 1]$ into S with $\lambda \circ f^{-1} = P$.*

Proof. For each $m = 1, 2, \ldots$, let $S = \bigcup_n A_{mn}$ where the A_{mn} are disjoint, Borel, continuity sets of P, and diam$(A_{mn}) \leq 1/m$, by Lemma 11.7.3. Take $0 = t_0 \leq t_1 \leq \cdots$ such that $P(A_{1n}) = t_n - t_{n-1}$ for all n. Let k-tuples $(n(1), \ldots, n(k))$ of positive integers be denoted by (n, k). Let $B_{(n,k)} := \bigcap_{1 \leq j \leq k} A_{jn(j)}$. Let $I_{(n,1)} := [t_{n-1}, t_n)$ and for $k \geq 2$, recursively on k, define left-closed, right-open intervals $I_{(n,k)}$ forming a decomposition of each $I_{(n,k-1)}$ into disjoint intervals for $n(k) = 1, 2, \ldots$, and such that $\lambda(I_{(n,k)}) = P(B_{(n,k)})$ for all (n, k). Then $I_{(n,k)}$ is empty whenever $P(B_{(n,k)}) = 0$.

Define a sequence of functions f_k from $[0, 1)$ into S as follows. Whenever $I_{(n,k)} \neq \emptyset$, choose an $x_{(n,k)} \in B_{(n,k)}$ and let $f_k(t) := x_{(n,k)}$ for all $t \in I_{(n,k)}$. Then as $k \to \infty$, the Borel functions f_k converge uniformly to some Borel function f from $[0, 1)$ into S (f is Borel measurable by Theorem 4.2.2). Let $P_k := \lambda \circ f_k^{-1}$. Then $P_k \to \lambda \circ f^{-1}$.

Now $P_k(B_{(n,k)}) = P(B_{(n,k)})$ for all (n, k). Then for any Borel set $A \subset S$,

$$P_k(A) = \sum_{(n,k)} \left\{ \lambda\left(I_{(n,k)}\right) : x_{(n,k)} \in A \right\}$$

(where the sum is over all k-tuples (n, k) for the given k)

$$= \sum_{(n,k)} \left\{ P\left(B_{(n,k)}\right) : x_{(n,k)} \in A \right\} \leq P\left(A^{\varepsilon+1/k}\right)$$

for any $\varepsilon > 0$, where A^δ is as in the definition of the Prohorov metric ρ (§11.3). We obtain $\rho(P_k, P) \leq 1/k$, and by Theorem 11.3.3, $P_k \to P$, so $P = \lambda \circ f^{-1}$.
$\qquad\qquad\qquad\qquad\qquad\qquad\qquad\qquad\qquad\qquad\qquad\qquad\qquad\qquad\square$

Recall the Lévy metric: for laws P and Q on \mathbb{R} with distribution functions F and G,

$$\lambda(P, Q) := \inf\{\varepsilon > 0 : F(x - \varepsilon) - \varepsilon \leq G(x) \leq F(x + \varepsilon) + \varepsilon \quad \text{for all } x\}.$$

Then λ metrizes convergence of laws on \mathbb{R}, but λ defines a uniformity weaker than the Fortet-Mourier-Prohorov uniformity (defined by ρ or β), as shown by Problem 8 in §11.3.

Since convergence of laws was proved in the central limit theorem by way of pointwise convergence of characteristic functions (§§9.5, 9.8), it may be surprising that even stronger (uniform) closeness of characteristic functions doesn't imply that laws are close otherwise:

11.7.6. Proposition *There exist laws P_n and Q_n on \mathbb{R}, with $P_n([1, \infty)) = Q_n((-\infty, -1]) = 1$ for all n, and with characteristic functions f_n and g_n, respectively, such that $\sup_t |(f_n - g_n)(t)| \to 0$ as $n \to \infty$.*

Remark. For such a P_n and Q_n, clearly $\rho(P_n, Q_n) = 1$, so these laws are as far apart as possible for the Prohorov metric.

Proof. Let P_n have the density $1/(x \log n)$ on the interval $[1, n]$ and 0 elsewhere. Let Q_n be the image of P_n by the transformation $x \mapsto -x$, so $dQ_n(x) = dP_n(-x)$. Then for any t, and $y := xt$,

$$(f_n - g_n)(t) = (\log n)^{-1} \int_1^n e^{ixt} - e^{-ixt} \, dx/x$$

$$= 2i(\log n)^{-1} \int_t^{nt} (\sin y)/y \, dy.$$

Now, $\int_0^u (\sin y)/y \, dy$ is bounded uniformly in u, since the integrals from $m\pi$

to $(m + 1)\pi$ alternate in sign and decrease in absolute value as $m \to \infty$. The conclusion follows. □

Problems

1. Show that if S is the real line \mathbb{R} with usual metric, then Theorem 11.7.2 holds for the probability space $(0, 1)$ with Lebesgue measure. *Hint:* Let X_n be the inverse of the distribution function of P_n as defined in Proposition 9.1.2.

2. Do the same if S is any complete separable metric space. *Hint:* A countable product of complete separable metric spaces S_n, with product topology, is also metrizable as a complete separable metric space by Theorem 2.5.7.

3. Suppose $\{P_n\}$ and $\{Q_n\}$ are two uniformly tight sequences of laws on \mathbb{R} with characteristic functions f_n and g_n, respectively. Suppose for each $M < \infty$, $\sup_{|t| \le M} |(f_n - g_n)(t)| \to 0$ as $n \to \infty$. Show that $\rho(P_n, Q_n) \to 0$. *Hint:* Suppose not. Take subsequences.

4. If (S, d) and (T, e) are metric spaces, a function f from S into T is called an *isometry* iff $e(f(x), f(y)) = d(x, y)$ for all x and y in S. Suppose γ is a function such that for any metric space (S, d), $\gamma = \gamma_{S,d}$ is a metric on the set of all laws on S. Say that γ is *invariantly defined* if both
 (a) for any laws P and Q on S and isometry f of S into T, $\gamma(P \circ f^{-1}, Q \circ f^{-1}) = \gamma(P, Q)$, and
 (b) if two laws P and Q each give total mass 1 to some measurable subset A, then $\gamma_{A,d}(P, Q) = \gamma_{S,d}(P, Q)$.
 Show that the metrics ρ and β are both invariantly defined.

5. This begins a sequence of related problems. Let F be the set of all functions f on S of the form $f(x) \equiv \delta(1 - d(x, B)/\varepsilon) \vee 0$ where B is a bounded set, so that diam $B := \sup\{d(x, y): x, y \in B\} < \infty$, and where $0 < \varepsilon < 1$ and $0 < \delta \le \min(\varepsilon, 1/\text{diam } B)$. Let $d_F(P, Q) := \sup\{|\int f\, d(P - Q)|: f \in F\}$. Let $\overline{d}_F(P, Q) := \sup_{g,T} d_F(P \circ g^{-1}, Q \circ g^{-1})$ over all isometries g of S into other metric spaces T. Show that d_F and \overline{d}_F are metrics and metrize the usual convergence of laws on S. *Hint:* See the portmanteau theorem (11.1.1) and its proof. If $f \in F$ on T, consider $\|f \circ g\|_{BL}$.

6. Show that \overline{d}_F is invariantly defined.

7. Show that \overline{d}_F may define a different uniformity than that of β and ρ: for example, on the real line \mathbb{R}, if P_n is the average of point masses at $1, \dots, n$, then P_n converges to 0 for \overline{d}_F, as does Q_n, the average of point masses at $-1, \dots, -n$. So $\overline{d}_F(P_n, Q_n) \to 0$, although P_n and Q_n do not become close for ρ or β.

*11.8. Kantorovich-Rubinstein Theorems

Strassen's theorem (Corollary 11.6.4) said that for two laws P and Q at distance r in the Prohorov metric there are random variables having laws P and Q and at Ky Fan distance r, or a little more than r if the space is not complete. This section will prove a very similar fact but with other distances in place of those of Prohorov and Ky Fan.

For any two laws P and Q on S, let $M(P, Q)$ denote the set of all laws on $S \times S$ with marginals P and Q. One member of it is the product law $P \times Q$.

The distance for random variables will be $Ed(X, Y)$. This is not always finite, so random variables will be considered for which it will be finite.

11.8.1. Proposition *Let (S, d) be a separable metric space. Let X be a random variable with values in S. Then the following are equivalent.*

(a) $Ed(X, y) < \infty$ *for some y in S.*
(b) $Ed(X, z) < \infty$ *for every z in S.*

Proof. (a) implies (b) since $d(X, z) \leq d(X, y) + d(y, z)$. □

$\mathcal{P}_1(S)$ will denote the collection of all laws P on S for which $\int d(x, z) \, dP(x) < \infty$ for some (hence every) z in S. Note that if (S, d) is bounded, all laws on S belong to $\mathcal{P}_1(S)$. For example, if $S = \mathbb{R}$ and X is a real-valued random variable, we have $\mathcal{L}(X) \in \mathcal{P}_1(S)$ if and only if $E|X| < \infty$.

For any two laws P and Q in $\mathcal{P}_1(S)$, the *Wasserstein* or *Monge-Wasserstein distance* W will be defined by

$$W(P, Q) := \inf \left\{ \int d(x, y) \, d\mu(x, y) : \mu \in M(P, Q) \right\}.$$

Since for any $z \in S$, $d(x, y) \leq d(x, z) + d(y, z)$, $W(P, Q)$ is finite for any P and Q in $\mathcal{P}_1(S)$. This W is related to what are called transportation problems. Let P be the distribution of points at which a good is produced and Q the distribution of points where it is consumed. Let $d(x, y)$ be the cost of transporting the good from x to y. Then $W(P, Q)$ is the smallest cost at which the total production can be transported to the consumers.

Now recall the Lipschitz seminorm for suitable real-valued functions f on S:

$$\|f\|_L := \sup\{|f(x) - f(y)|/d(x, y): x \neq y \text{ in } S\},$$

treated in §6.1 and §11.2. For two laws P and Q on S, let

$$\gamma(P, Q) := \|P - Q\|_L^* := \sup \left\{ \left| \int f \, d(P - Q) \right| : \|f\|_L \le 1 \right\}.$$

For example, in \mathbb{R} let $P_n := ((n-1)\delta_0 + \delta_n)/n$. and $P := \delta_0$. Then $P_n \to P$ for the previously defined convergence of laws, and so, equivalently, for the ρ and β metrics, but $\gamma(P_n, P) \equiv 1$ and $W(P_n, P) \equiv 1$.

Here is the main theorem of this section:

11.8.2. Theorem (Kantorovich-Rubinstein) *For any separable metric space (S, d) and any two laws P and Q in $\mathcal{P}_1(S)$,*

$$W(P, Q) = \gamma(P, Q).$$

If P and Q are tight, for example if S is complete, then there exists a law Pr *in $M(P, Q)$ such that $\int d(x, y) \, d\mathrm{Pr}(x, y) = W(P, Q)$, so that the infimum in the definition of $W(P, Q)$ is attained.*

Proof. Let $\mu \in M(P, Q)$ and $\|f\|_L \le 1$. Then

$$\left| \int f \, d(P - Q) \right| = \left| \int f(x) - f(y) \, d\mu(x, y) \right| \le \int d(x, y) \, d\mu(x, y).$$

Taking the supremum over f, and the infimum over μ, gives $\gamma(P, Q) \le W(P, Q)$. Next, the following will be useful.

11.8.3. Lemma *Both W and γ are pseudometrics on $\mathcal{P}_1(S)$.*

Remark. Actually, W and γ are both metrics. After Theorem 11.8.2 is proved, it will be enough to show that one of the two equal pseudometrics is a metric.

Proof. Clearly, for any P and Q, $W(P, Q) = W(Q, P) \ge 0$ and $\gamma(P, Q) = \gamma(Q, P) \ge 0$. In fact, γ is clearly a pseudometric.

Let $\varepsilon > 0$. Then S is a countable union of disjoint Borel sets $S_n, n = 1, 2, \dots$, such that $d(x, y) < \varepsilon$ for any n, $x \in S_n$, and $y \in S_n$, as in the proof of Theorem 11.6.3, Case C. Let μ in $M(P, Q)$ be such that $\int d(x, y) \, d\mu(x, y) < W(P, Q) + \varepsilon$. For any Borel set $A \subset S$ let $\mu_{mn}(A) := \mu((A \cap S_m) \times S_n)$ and $\nu_{mn}(A) := \mu(S_m \times (A \cap S_n))$. Let $\mu' := \sum_{m,n} (\mu_{mn} \times \nu_{mn})/\mu(S_m \times S_n)$, where $0/0$ is replaced by 0. Then μ' is a law on $S \times S$ such that $\mu'(S_m \times S_n) = \mu(S_m \times S_n)$ for all m and n, while μ' is a product measure on $S_m \times S_n$, $\mu' \in M(P, Q)$, and

$$\int d(x, y) \, d\mu'(x, y) < W(P, Q) + 3\varepsilon.$$

For any other law T on S, there is likewise a law $\nu \in M(Q, T)$ which is a sum of a sequence of product measures such that

$$\int d(y, z)\, d\nu(y, z) < W(Q, T) + 3\varepsilon.$$

Now, given $y \in S$, the conditional distribution of x given y for μ' (as defined early in §10.2) has countably many possible values, each on a Borel set S_m. Likewise, the conditional distribution of z given y for ν has countably many possible values, each on a Borel set B_i. Define a law of (x, y, z) in $S \times S \times S$ such that (y, z) has distribution ν and (x, y) has distribution μ', where y has distribution Q and for each y, x and z are conditionally independent. Then the set of possible conditional distributions of $\langle x, z \rangle$ given y, which are product laws, is again countable and each occurs on a Borel set $S_m \cap B_i$. Thus a law M is well-defined on $S \times S \times S$, with marginals P, Q, and T on S, and satisfying

$$\int d(x, z)\, dM(x, y, z) \le W(P, Q) + W(Q, T) + 6\varepsilon.$$

Letting $\varepsilon \downarrow 0$ and considering the $\langle x, z \rangle$ marginals in $M(P, T)$ gives $W(P, T) \le W(P, Q) + W(Q, T)$. So W is a pseudometric. □

11.8.4. Lemma *For any $P \in \mathcal{P}_1(S)$ there are laws P_n, with finite sets F_n such that $P_n(F_n) = 1$ for all n, such that $W(P_n, P) \to 0$ and $\gamma(P_n, P) \to 0$ as $n \to \infty$.*

Proof. For each n, there are disjoint Borel sets $S_{nk}, k = 1, 2, \ldots$, with $\bigcup_k S_{nk} = S$, such that $d(x, y) < 1/n$ for all $x, y \in S_{nk}$ for each k. Choose a point x_{nk} in S_{nk} for each n and k such that S_{nk} is non-empty. Let $f_n(x) := x_{nk}$ for each $x \in S_{nk}$ and each k. Also let $f_{nk}(x) := x_{nj}$ for each $x \in S_{nj}$ for $j = 1, \ldots, k$ and $f_{nk}(x) := x_{n1}$ for each $x \in S(n, j) := S_{nj}, j > k$. (It can be assumed that S_{n1} is non-empty.)

Now $\sum_{j>k} \int_{S(n,j)} d(x_{n1}, x)\, dP(x) < 1/n$ for $k \ge k_n$ large enough by dominated convergence. Then $\int d(f_{nk}(x), x)\, dP(x) < 2/n$. Let $P_n := P \circ f_{nk}^{-1}$. Then for the distribution μ_n of $\langle x, f_{nk}(x) \rangle$ on $S \times S$, where x has distribution P on X, we have $\mu_n \in M(P, P_n)$ and $\int d(x, y)\, d\mu_n(x, y) < 2/n$. So $W(P, P_n) \le 2/n$. Since $\gamma \le W$, as shown at the beginning of the proof of Theorem 11.8.2, Lemma 11.8.4 follows. □

Now, to prove Theorem 11.8.2, first suppose S is compact. For any continuous real function h on $S \times S$ and laws P and Q on S let

$$m_{P,Q}(h) := \sup\left\{\int f\, dP + \int g\, dQ \colon f(x) + g(y) < h(x, y) \text{ for all } x, y\right\}.$$

Let $C(S \times S)$ be the space of all continuous real functions on $S \times S$ with the usual product topology.

11.8.5. Lemma *For any compact metric space (S, d), laws P and Q on S and $h \in C(S \times S)$,*

$$m_{P,Q}(h) = \inf\left\{ \int h \, d\mu \colon \mu \in M(P, Q) \right\}.$$

Proof. For any $\mu \in M(P, Q)$, clearly $m_{P,Q}(h) \le \int h \, d\mu$. So in the stated equation, "\le" holds. For the converse inequality, let L be the linear space of all functions $\varphi(x, y) = f(x) + g(y)$ for any two continuous real functions f and g on S, and let

$$r(\varphi) := \int f \, dP + \int g \, dQ.$$

Then r is well-defined since if $f(x) + g(y) \equiv k(x) + j(y)$, then $f(x) - k(x) \equiv j(y) - g(y)$, which must be some constant c, so $k = f - c$ and $j = g + c$, giving $\int f - k \, dP + \int g - j \, dQ = 0$.

For any $h \in C(S \times S)$ let

$$U := U_h := \{ f \in C(S \times S) \colon f(x, y) < h(x, y) \quad \text{for all } x, y \}.$$

Then U is a convex set, open for the supremum norm (since S is compact). Now r is a linear form on L, not identically 0, and bounded above on the non-empty convex set $U \cap L$ since $f(x) + g(y) < h(x, y)$ for all x, y implies $r(\varphi) \le \sup(f) + \sup(g) \le \sup(h) < +\infty$. So by one form of the Hahn-Banach theorem (6.2.11), r can be extended to a linear form ρ on $C(S \times S)$ with $\sup_{u \in U} \rho(u) = \sup_{v \in U \cap L} r(v)$.

Suppose $u \in C(S \times S)$ and $u(x, y) \ge 0$ for all x, y. Then for any $c \ge 0$, $h - 1 - cu \in U$, so $\rho(h - 1 - cu)$ is bounded above as $c \to +\infty$, implying $\rho(u) \ge 0$. Also, for any $f \in C(S \times S)$, $|\rho(f)| \le \rho(1) \sup|f|$, so $\rho(\cdot)$ is bounded. By the Riesz representation theorem (7.4.1) there is a nonnegative, finite measure ρ on $S \times S$ such that $\rho(f) = \int f \, d\rho$ for all $f \in C(S \times S)$. Since $\rho = r$ on L, we have for any f and $g \in C(S)$ that $\int f(x) \, d\rho(x, y) = \int f \, dP$ and $\int g(y) \, d\rho(x, y) = \int g \, dQ$. Then since S is a compact metric space, $\rho \in M(P, Q)$ by uniqueness in the Riesz representation theorem (7.4.1) on S and the image measure theorem (4.1.11). Now

$$m_{P,Q}(h) = \sup_{u \in U \cap L} r(u) = \sup_U \rho = \int h \, d\rho,$$

so $m_{P,Q}(h) \ge \inf\{\int h \, d\mu \colon \mu \in M(P, Q)\}$, proving Lemma 11.8.5. $\quad\square$

11.8.6. Lemma *If S is a compact metric space and $h \in C(S \times S)$, the following are equivalent:*

(a) *There is a set $J \subset C(S)$ such that for all laws P and Q on S,*

$$m_{P,Q}(h) = \sup_{j \in J} \left| \int j \, d(P - Q) \right|;$$

(b) *h is a pseudometric on S.*
If h is a pseudometric, we can take

$$J := J_h := \{ j \in C(S) : |j(x) - j(y)| \le h(x, y) \quad \text{for all } x, y \}.$$

Proof. If J exists, take for any x and y, $P = \delta_x$ and $Q = \delta_y$. Then $M(P, Q)$ contains only $\delta_{\langle x,y \rangle}$, so by Lemma 11.8.5

$$h(x, y) = \sup_{j \in J} |j(x) - j(y)|,$$

which is a pseudometric.

Conversely, if h is a pseudometric and $f(x) + g(y) < h(x, y)$ for all x and y, let $j(x) := \inf_y(h(x, y) - g(y))$. Then $f \le j \le -g$, and for all x and x',

$$j(x) - j(x') \le \sup_y (h(x, y) - h(x', y)) \le h(x, x'),$$

so $j \in J_h$ and $\int f \, dP + \int g \, dQ \le \int j \, d(P - Q)$. Hence

$$m_{P,Q}(h) \le \sup_{j \in J} \left| \int j \, d(P - Q) \right|.$$

The converse inequality always holds, since for any $j \in J$, we can let $f = j$, $g = -j - \delta$, $\delta \downarrow 0$, in the definition of $m_{P,Q}$. □

Now for a compact metric space (S, d), letting $h = d$ in Lemma 11.8.5 gives $m_{P,Q}(d) = W(P, Q)$. Then by Lemma 11.8.6, $W(P, Q) = \|P - Q\|_L^*$, finishing the proof of Theorem 11.8.2 for S compact. Now in the general case, given any two laws P and Q on S, take P_n with finite supports converging to P from Lemma 11.8.4 and likewise Q_n converging to Q. For each n, $W(P_n, Q_n) = \|P_n - Q_n\|_L^*$, as just shown. From the properties of pseudometrics (Lemma 11.8.3) and the convergence in Lemma 11.8.4, it follows that $W(P, Q) = \|P - Q\|_L^*$, proving Theorem 11.8.2. □

Problems

1. Show that for laws P and Q on \mathbb{R}^1 with distribution functions F and G respectively, $W(P, Q) = \int_{-\infty}^{\infty} |F - G|(x)\, dx$. *Hint:* Let $h := 2 1_{\{F > G\}} - 1$, so that $h = 1$ when $F > G$ and $h = -1$ when $F \le G$. Let H be an indefinite integral of h, so that $\|H\|_L = 1$. Show that for any function J with $\|J\|_L \le 1$,

$$\left| \int J\, d(P - Q) \right| = \left| \int (F - G)(x) J'(x)\, dx \right| \le \int |F - G|(x)\, dx$$

and that this bound is attained for $J = H$.

2. For the inverse X_F of a distribution function F as defined at Proposition 9.1.2, show that $\int |F - G|(x)\, dx = \int_0^1 |X_F - X_G|(t)\, dt$. *Hint:* This is the area between the two graphs. (Thus, the Monge-Wasserstein distances between all laws on \mathbb{R}^1 can be attained simultaneously by random variables with those laws on one probability space, namely, the unit interval with Lebesgue measure.)

3. Show that there exist three laws α, β, and γ on \mathbb{R}^2 such that there is no law P on \mathbb{R}^6 with coordinates x, y, and z in \mathbb{R}^2 for which x, y, and z have laws α, β, and γ, respectively, and $E|x - y| = W(\alpha, \beta)$, $E|y - z| = W(\beta, \gamma)$, and $E|x - z| = W(\alpha, \gamma)$. *Hint:* Let a, b, and c be the vertices of an equilateral triangle of side 1. Let $\alpha := (\delta_a + \delta_b)/2$, $\beta = (\delta_a + \delta_c)/2$, and $\gamma = (\delta_b + \delta_c)/2$. (So the simultaneous attainment of Monge-Wasserstein distances by random variables for all laws, as in \mathbb{R}^1 in Problem 2, no longer is possible in \mathbb{R}^2.)

4. Let P be a law on a separable normed vector space $(S, \|\cdot\|)$, $x \in S$, and P_x the translate of P by x, so that $P_x(A) := P(A - x)$ for all Borel sets A, where $A - x := \{a - x : a \in A\}$. Show that $W(P, P_x) = \|x\|$.

5. Show that W is a metric on $\mathcal{P}_1(S)$. *Hint:* Consider the metric β.

6. For a separable metric space (S, d) and $1 \le r < \infty$ let $\mathcal{P}_r(S)$ be the set of all laws P on S such that $\int d(x, y)^r\, dP(x) < \infty$ for some $y \in S$. Show that the same is true for all $y \in S$.

7. Let $W_r(P, Q) := (\inf\{Ed(X, Y)^r : \mathcal{L}(X) = P, \mathcal{L}(Y) = Q\})^{1/r}$ for any $1 \le r < \infty$ and laws P and Q in $\mathcal{P}_r(S)$. Show that for any law P in $\mathcal{P}_r(S)$ there are finite sets $F_n \subset S$ and laws P_n with $P_n(F_n) = 1$ such that $W_r(P_n, P) \to 0$ as $n \to \infty$, as in Lemma 11.8.4 for $r = 1$.

8. Let S be a Polish space. Let μ and ν be laws on $S \times S$. For $\langle x, y \rangle \in S \times S$ let $p_1(x, y) := x$ and $p_2(x, y) := y$. Suppose that $\mu \circ p_2^{-1} = \nu \circ p_1^{-1}$ as laws on S. Show that there is a law τ on $S \times S \times S$ such that, for $f(x, y, z) := \langle x, y \rangle$

and $g(x, y, z) := \langle y, z \rangle$, we have $\tau \circ f^{-1} = \mu$ and $\tau \circ g^{-1} = \nu$. *Hint:* See the proof of Lemma 11.8.3.

9. Show that W_r is a metric on $\mathcal{P}_r(S)$ for any Polish space S. *Hint:* Use Problem 8 and §10.2.

*11.9. U-Statistics

A central limit theorem will be proved for some variables which are averages of symmetric functions of i.i.d. variables. Let X_1, X_2, \ldots, be i.i.d. variables with some law Q, taking values in some measurable space (S, \mathcal{A}), where most often $S = \mathbb{R}$ with its Borel σ-algebra. Let f be a measurable function on the Cartesian product S^m of m copies of S, so we can evaluate $f(X_1, \ldots, X_m)$. Suppose f and Q are such that $E|f(X_1, \ldots, X_m)| < \infty$. Then set $g(Q) := Ef(X_1, \ldots, X_m)$. Now suppose given X_1, \ldots, X_n for some $n \geq m$. In statistics, we assume given only the data or observations X_1, \ldots, X_n. Q is unknown except for the information provided about it by X_1, \ldots, X_n and possibly some other regularity conditions, in this case the fact that $g(Q)$ is finite. A *statistic* is any measurable function T_n of X_1, \ldots, X_n. A statistic T_n is called an *unbiased estimator* of a function $g(Q)$ iff $ET_n(X_1, \ldots, X_n) = g(Q)$ whenever X_1, \ldots, X_n are i.i.d. Q. A sequence $\{T_n\}$ of statistics, where $T_n = T_n(X_1, \ldots, X_n)$, is called a *consistent* sequence of estimators of $g(Q)$ iff T_n converges to $g(Q)$ in probability as $n \to \infty$ whenever X_1, X_2, \ldots, are i.i.d. with distribution Q.

If $g(Q) = Ef(X_1, \ldots, X_m)$ as above, one way to get a consistent sequence of unbiased estimators of $g(Q)$ is to set $Y_1 = f(X_1, \ldots, X_m)$, $Y_2 = f(X_{m+1}, \ldots, X_{2m})$, and so on, so that Y_1, Y_2, \ldots, are i.i.d. variables with $EY_i = g(Q)$ and by the strong law of large numbers, almost surely $\overline{Y} = \overline{Y}^{(k)} = (Y_1 + \cdots + Y_k)/k \to g(Q)$ as $k \to \infty$. We can let $T_n = \overline{Y}^{(k)}$ for $km \leq n < (k+1)m$. This method gives about n/m values of Y_i for n values of X_j. It turns out to be relatively inefficient. U-statistics will make more use of the information in X_1, \ldots, X_n to get a better approximation to $g(Q)$ for a given n.

A function f of m variables x_1, \ldots, x_m will be called *symmetric* if $f(x_1, \ldots, x_m) \equiv f(x_{\pi(1)}, \ldots, x_{\pi(m)})$ for every permutation π of $\{1, \ldots, m\}$, in other words any 1–1 function π from $\{1, \ldots, m\}$ onto itself. There are $m!$ such permutations. For $m = 2$, f is symmetric if and only if $f(x, y) = f(y, x)$ for all x and y. For any function f of m variables, the *symmetrization* is defined by

$$f^{(s)}(x_1, \ldots, x_m) := m!^{-1} \sum_\pi f\left(x_{\pi(1)}, \ldots, x_{\pi(m)}\right),$$

where the sum is over all $m!$ permutations π of $\{1, \ldots, m\}$. Then clearly $f^{(s)}$ is symmetric, and f is symmetric if and only if $f \equiv f_s$. For example, if $f(x, y) \equiv x + 3y$, then $f^{(s)}(x, y) \equiv 2x + 2y$.

Given S and f as above, and $n \geq m$, the nth U-statistic $U_n = U_n(f)$ is defined by

$$U_n := \frac{(n-m)!}{n!} \sum_{i_1, \ldots, i_m \neq}^{n} f\left(X_{i_1}, \ldots, X_{i_m}\right),$$

where the sum notation means that the sum runs over all ordered m-tuples of distinct integers i_1, \ldots, i_m from 1 to n. There are exactly $n!/(n-m)!$ such ordered m-tuples (there are n ways to choose i_1, then $n-1$ ways to choose i_2, \ldots, and $n - m + 1$ ways to choose i_m). So U_n is an average of terms, each of which is f evaluated at m variables i.i.d. with distribution Q, so that each term has expectation $g(Q)$, and $EU_n = g(Q)$ for all n.

For each set $\{i_1, \ldots, i_m\}$ of distinct integers from 1 to n, there are $m!$ possible orderings, each of which appears in the average. So without changing U_n, f can be replaced by its symmetrization $f^{(s)}$, and U_n can then be written as an average of fewer terms:

$$U_n = \frac{1}{\binom{n}{m}} \sum_{1 \leq i_1 < i_2 < \cdots i_m \leq n} f^{(s)}\left(X_{i_1}, \ldots, X_{i_m}\right).$$

Examples. For $m = 1$, a U-statistic U_n is simply the average of n i.i.d. variables $f(X_i)$, such as the sample mean \overline{X} for $f(x) \equiv x$. Next, for $n \geq 2$, let s^2 be the sample variance:

$$s^2 := \frac{1}{n-1} \sum_{i=1}^{n} (X_i - \overline{X})^2.$$

Let $f(x, y) := (x - y)^2/2$. Then s^2 will be shown to be the U_n statistic for this f. In view of common factors on both sides, this is equivalent to saying that $(\sum_i X_i^2) - n\overline{X}^2 = n^{-1} \sum_{i<j} (X_i - X_j)^2$. To check this, note that the coefficient of X_k^2 on each side is $1 - 1/n$ for $k = 1, \ldots, n$, and the coefficient of $X_i X_j$, $i < j$, is $-2/n$.

In the rest of this section we assume f is symmetric, $f = f^{(s)}$.

If h is a random variable defined on S^k for some k, let $E_Q h := Eh(X_1, \ldots, X_k)$ for X_1, \ldots, X_k i.i.d. with law Q. Thus $g(Q) = E_Q f$. Now suppose that $E_Q f^2 < \infty$. The problem at hand is to evaluate and analyze the variances of the U-statistics U_n in order to prove their asymptotic normality as $n \to \infty$.

For any integrable real random variable V, and random variable X taking values in some measurable space (M, \mathcal{B}), the conditional expectation $E(V|X)$ is defined as a random variable $v(X)$ which is a measurable function of X and such that for any $A \in \mathcal{B}$, $E1_A(X)V = E(1_A v)(X)$. Then $v(X)$ is the conditional expectation of V for the smallest σ-algebra for which X is measurable (by Theorem 4.2.8).

11.9.1. Proposition *Let X and Y be independent random variables with values in A and B, respectively, where (A, \mathcal{S}) and (B, \mathcal{T}) are measurable spaces. Let f be real-valued and measurable on $A \times B$ and $E|f(X, Y)| < \infty$. Let Y have a law μ on B. Then*

 (a) $f_c(X) := E(f(X, Y)|X) = \int f(X, y)\, d\mu(y)$.
 (b) If X, Y, and Z are jointly independent and Z also has values in B and the same law μ, and $Ef^2(X, Y) < \infty$, then $E(f(X, Y)f(X, Z)) = Ef_c^2(X)$.

Note. This fact will be applied when $A = S^k$, $B = S^{m-k}$, $X = (X_{i(1)}, \ldots, X_{i(k)})$, and $Y = (X_{i(k+1)}, \ldots, X_{i(m)})$ as in the definition of U-statistics, where also $Z = (X_{j(k+1)}, \ldots, X_{j(m)})$, $i(r) = j(r)$ for $r = 1, \ldots, k$, and the sets $\{i(k+1), \ldots, i(m)\}$ and $\{j(k+1), \ldots, j(m)\}$ are disjoint. Here the indices $i(r), r = 1, \ldots, m$ are not necessarily in increasing order, nor are $j(1), \ldots, j(m)$.

Proof. Part (a) follows directly from the definitions and the Tonelli-Fubini theorem. For (b), write

$$E(f(X, Y)f(X, Z)) = E(\{f(X, Y) - f_c(X) + f_c(X)\}\{f(X, Z) - f_c(X) + f_c(X)\}).$$

Then clearly $E(\{f(X, Y) - f_c(X)\}\{-f_c(X) + f_c(X)\}) = 0$. Also, $Ef_c^2(X) < \infty$ by the conditional Jensen inequality (10.2.7), and $E(f_c(X)\{f(X, Z) - f_c(X)\}) = 0$ by the Tonelli-Fubini theorem, since if we integrate first with respect to Z, $f(X, Z)$ is replaced by $f_c(X)$ from part (a). Likewise, $E(\{f(X, Y) - f_c(X)\}f(X, Z)) = 0$, integrating first with respect to Y. Then (b) follows. ☐

For any function $h(x_1, \ldots, x_m)$ let $V_Q(h)$ be the variance of $h(X_1, \ldots, X_m)$ when X_1, \ldots, X_m are i.i.d. (Q). Then $V_Q(f)$ is defined and finite when $E_Q f^2 < \infty$.

The next lemma involves hypergeometric probabilities: for any nonnegative integers k, m, n, and N, let $h(k, m, n, N)$ be the probability that for a given set F with N elements, and subset G with n elements, if another subset H with m elements is chosen at random, with equal probability for each possible H, then the intersection $G \cap H$ will contain exactly k elements. Let G^c denote the complement $F \backslash G$. The numbers of elements in each of the four sets $G \cap H$, $G \cap H^c$, $G^c \cap H$, and $G^c \cap H^c$ are often written in what is called a 2×2 contingency table, as follows:

	G	G^c	Total
H	k	$m - k$	m
H^c	$n - k$	$N - m - n + k$	$N - m$
Total	n	$N - n$	N

By counting the numbers of possible sets, one evaluates

$$h(k, m, n, N) = \frac{\binom{n}{k}\binom{N-n}{m-k}}{\binom{N}{m}} = \frac{n!(N-n)!m!(N-m)!}{k!(n-k)!(m-k)!(N-m-n+k)!N!}$$

Here $h(k, m, n, N) = 0$ unless $0 \le k \le m \le N$, $k \le n \le N$, and $N - m - n + k \ge 0$.

Let $f_k(X_1, \ldots, X_k) := E(f(X_1, \ldots, X_m)|X_1, \ldots, X_k)$. So f_k is the function f_c of Proposition 11.9.1 with $X = (X_1, \ldots, X_k)$ and $Y = (X_{k+1}, \ldots, X_m)$. Here is one step in the analysis of the variance of $V_Q(U_n)$:

11.9.2. Lemma $V_Q(U_n) = \sum_{k=1}^{m} h(k, m, m, n) V_Q(f_k)$.

Proof. The variance of U_n is the average of the covariances of $f_G := f(\{X_i\}_{i \in G})$ with $f_H := f(\{X_j\}_{j \in H})$ where G and H independently run over the collection of all subsets of $\{1, \ldots, n\}$ with m elements. This is true since, for one thing, $Ef_G = E_Q f$ for all G. If G and H have k elements in common, $k = 1, \ldots, m$, the covariance is $V_Q(f_k)$ by Proposition 11.9.1(b). If G and H have empty intersection, then f_G and f_H are independent and their covariance is 0. So by definition of hypergeometric probabilities the conclusion follows. \square

Recall that for any random variable Y, $\sigma(Y) := (\text{Var } Y)^{1/2}$. Here is a theorem on asymptotic normality of U-statistics:

11.9.3. Theorem *Whenever* $f(X_1, \ldots, X_m)$ *is a random variable with* $E_Q f^2 < \infty$, *we have as* $n \to \infty$

(a) $Z_n := n^{1/2}(U_n - g(Q)) \underset{\mathcal{L}}{\to} N(0, m^2 V_Q(f_1))$, *where* $N(0, 0) := \delta_0$.

(b) *If* $V_Q(f_1) > 0$, *then* $(U_n - g(Q))/\sigma(U_n) \underset{\mathcal{L}}{\to} N(0, 1)$.

Proof. First, $E_Q f_k^2$ and so $V_Q(f_k)$ are finite for $k = m$ by assumption, and then for $k = 1, \ldots, m - 1$ by the conditional Jensen inequality (10.2.7).

Let $Y_n := mn^{-1/2} \sum_{1 \le j \le n} (f_1(X_j) - g(Q))$. Then Y_n is $n^{-1/2}$ times a sum of n i.i.d. variables with mean 0 and variance $m^2 V_Q(f_1)$. So by the central limit theorem, $Y_n \underset{\mathcal{L}}{\to} N(0, m^2 V_Q(f_1))$ as $n \to \infty$.

The following lemma will be applied here to $S = \mathbb{R}$, $V_n = Z_n$ and $W_n = Y_n$. It is a slight extension of the fact that convergence in probability implies convergence in law (Proposition 9.3.5).

11.9.4. Lemma *Let* (S, d) *be any separable metric space. Let* μ *be a law on S and V_n and W_n random variables with values in S such that $W_n \underset{\mathcal{L}}{\to} \mu$ and $d(V_n, W_n) \to 0$ in probability as $n \to \infty$. Then $V_n \underset{\mathcal{L}}{\to} \mu$ as $n \to \infty$.*

Proof. By Theorem 11.7.1 for Prohorov's metric ρ, $\rho(\mathcal{L}(V_n), \mathcal{L}(W_n)) \to 0$ as $n \to \infty$. Then by Theorem 11.3.3, $\rho(\mathcal{L}(W_n), \mu) \to 0$ and so $\rho(\mathcal{L}(V_n), \mu) \to 0$, implying that $V_n \underset{\mathcal{L}}{\to} \mu$. □

Since $E(U_n - Y_n)^2 \to 0$ implies $|U_n - Y_n| \to 0$ in probability, the next lemma will finish the proof of Theorem 11.9.3(a). (Recall that $a_n = O(b_n)$ means that a_n/b_n is bounded as $n \to \infty$).

11.9.5. Lemma *For each m, with Z_n as in Theorem 11.9.3 and Y_n as in its proof,*

$$E((Z_n - Y_n)^2) = O(1/n) \to 0 \quad \text{as } n \to \infty.$$

Proof. We have

$$E((Z_n - Y_n)^2) = EZ_n^2 + EY_n^2 - 2E(Z_n Y_n).$$

Clearly, $EY_n^2 = m^2 \text{Var}(f_1)$ for all n. By Lemma 11.9.2,

$$EZ_n^2 = n \, \text{Var}(U_n) = \sum_{1 \le k \le m} nh(k, m, m, n) V_Q(f_k). \quad \text{Now}$$

$$h(1, m, m, n) = \binom{m}{1}\binom{n-m}{m-1} \Big/ \binom{n}{m} = \frac{m^2(n-m)!^2}{(n-2m+1)!n!}$$

$$= \frac{m^2(n-m)(n-m-1)\cdots(n-2m+2)}{n(n-1)\cdots(n-m+1)}$$

and as $n \to \infty$, $(n-m-j)/(n-1-j) = 1 + O(1/n)$ for $j = 0, 1, \ldots,$ $m-2$, so

$$nh(1, m, m, n)V_Q(f_1) = m^2 V_Q(f_1) + O(1/n).$$

For $k \geq 2$,

$$nh(k, m, m, n) = n\binom{m}{k}\binom{n-m}{m-k} \Big/ \binom{n}{m} = \frac{nm!^2(n-m)!^2}{k!(m-k)!^2(n-2m+k)!n!}.$$

This equals a factor $C(m, k)$ not depending on n times

$$\frac{(n-m)(n-m-1)\cdots(n-2m+k+1)}{(n-1)(n-2)\cdots(n-m+1)},$$

where there are $m-k$ factors in the numerator and $m-1$ in the denominator. So this quantity is of order $O(n^{1-k})$ as $n \to \infty$. Since m is fixed and finite, it follows that $EZ_n^2 = m^2 V_Q(f_1) + O(1/n)$ as $n \to \infty$.

Next, it will be shown that Y_n is the orthogonal projection of Z_n into the linear subspace of square-integrable random variables spanned by random variables $h(X_j)$, $j = 1, \ldots, n$, which are functions of the individual variables. Since Y_n is of this form, it remains to show that $Z_n - Y_n$ is orthogonal to Y_n, in other words $EY_n(Z_n - Y_n) = 0$, or $E(Y_n Z_n) = EY_n^2$. To do this, we have

$$E(Y_n Z_n) = m\binom{n}{m}^{-1}\sum_{i=1}^{n}\sum_{1 \leq i_1 < \cdots < i_m \leq n} E(\{f_1(X_i) - g(Q)\}$$

$$\times \{(f(X_{i_1}, \ldots, X_{i_m}) - g(Q)\})$$

and the summands are 0 by independence unless i is one of the i_j, in which case the term is $V_Q(f_1)$, as in the proof of Proposition 11.9.1, with $f = f - f_1 + f_1$. The number of ways to choose $i_1 < \cdots < i_m$, one of which is i, is $\binom{n-1}{m-1}$ for each i, so the effect of the outer $\sum_{i=1}^{n}$ is to multiply the expression by n. Since

$$mn\binom{n-1}{m-1} \Big/ \binom{n}{m} = m^2,$$

$E(Y_n Z_n) = m^2 V_Q(f_1) = EY_n^2$ as claimed. Thus, $E((Y_n - Z_n)^2) = \text{Var}(Z_n) - \text{Var}(Y_n) = O(1/n) \to 0$ as $n \to \infty$, proving Lemma 11.9.5 and so, with Lemma 11.9.4, also Theorem 11.9.3(a). Now if $V_Q(f_1) > 0$, we

have $\text{Var}(Z_n) = \text{Var}(Y_n) + O(1/n) = m^2 V_Q(f_1) + o(1/n)$ as $n \to \infty$, and $(U_n - EU_n)/\sigma(U_n) = Z_n/\sigma(Z_n)$, so part (b) follows. $\qquad\square$

Problems

1. Let X_1, X_2, \ldots, be i.i.d. with finite variance and an unknown mean μ. Let Y be an unbiased estimator of μ with $Y = c_1 X_1 + \cdots + c_n X_n$. Show that $E(Y - \mu)^2$ is minimized for $Y = \overline{X}$ ($c_j \equiv 1/n$), which is a U-statistic. *Suggestion:* First show that $\sum c_j = 1$, then apply the Lagrange multiplier technique, setting partial derivatives of $E(Y - \mu)^2 + \lambda(\sum_{j=1}^n c_j - 1)$ with respect to each c_j equal to 0.

2. The usual sample variance s^2 is a U-statistic of order 2. For $n = 4$, $(X_1 + X_2 - X_3 - X_4)^2/4$ is another possible estimator of variance. Show that it is an unbiased estimator of the variance of X_1 but that its variance is larger than that of s^2, at least when the i.i.d. X_i have a normal distribution. *Hint:* If X has a standard normal $N(0, 1)$ distribution, then $EX^3 = 0$ and $EX^4 = 3$.

3. Suppose that Y has a Poisson distribution, $P(Y = k) = e^{-\lambda}\lambda^k/k!, k = 0, 1, \ldots$ Suppose Y can be observed only when $Y \geq 1$, and we have no other information about λ. The problem is to estimate $e^{-\lambda}$. Let V be an unbiased estimator of $e^{-\lambda}$, so that $V(k)$ is defined for $k = 1, 2, \ldots$, with $\sum_{k \geq 1} e^{-\lambda}\lambda^k V(k)/(k!(1 - e^{-\lambda})) = e^{-\lambda}$, for all $\lambda > 0$. Solve, if possible, for the $V(k)$ and comment on the properties and usefulness of the resulting estimator.

4. For a given $n \geq 2$, consider the family of binomial distributions, for the number K of successes in n independent trials with probability p of success on each trial. Thus any estimator (statistic) is a function of $K = 0, 1, \ldots, n$. Then K/n is the usual unbiased estimator of p. Show that for $n = 2$ there is a unique unbiased estimator of p^2. For what value(s) of K are its values remarkable?

5. Let Q run through the collection of all laws for which $E_Q f^2 < \infty$. For the U-statistic U_n, show that
 (a) U_n is the unique unbiased estimator T of $g(Q)$ which is symmetric in $X_1, \ldots, X_n, n \geq m$. *Hint:* If $V = T - U_n, E_Q V = 0$. If $Q\{x_i\} = p_i, i = 1, \ldots, r, E_Q V$ is a symmetric polynomial $S(p_1, \ldots, p_r)$, homogeneous of degree n and 0 for $p_i \geq 0$ (the restriction $\sum_i p_i = 1$ can be removed). Show by induction on r that $S \equiv 0$, then that $V \equiv 0$.
 (b) U_n is the unbiased estimator of $g(Q)$ with smallest variance. *Hint:* For an unbiased estimator T with symmetrization W, $W = U_n$ by part (a).

6. For i.i.d. variables X_1, X_2, \ldots, let \mathcal{A}_n be the smallest σ-algebra for which X_1, \ldots, X_n are measurable and \mathcal{B}_n the smallest σ-algebra for which X_{n+1}, X_{n+2}, \ldots, are measurable. Let \mathcal{S}_n be the σ-algebra of sets in \mathcal{A}_n whose indicator functions are symmetric functions of X_1, \ldots, X_n. Let \mathcal{C}_n be the σ-algebra generated by \mathcal{S}_n and \mathcal{B}_n. Show that for any sequence of U-statistics $\{U_n, n \geq m\}$ generated as above by a random variable $f(X_1, \ldots, X_m)$ with f symmetric and $E_Q|f| < \infty$, $\{U_n, \mathcal{C}_n\}_{n \geq m}$ is a reversed martingale.

Notes

§11.1 Convergence of laws on \mathbb{R}, at least in the special case of the central limit theorem, might be said to go back to de Moivre and Laplace (see the notes to §9.4). More generally on \mathbb{R}, Helly (1911–1912) and Bray (1918–1919) proved forms of the Helly-Bray theorem (11.1.2), according to Hildebrandt (1966).

Apparently A. D. Alexandroff (1940) was the first to develop a general theory of convergence of laws, or "weak convergence" of finite measures, on rather general spaces—in fact, spaces even more general than topological spaces, since only countable unions of open sets were required to be open (I don't know if this additional generality has found real uses). Alexandroff (1940, p. 308) gives the definition without any earlier references.

The portmanteau theorem (11.1.1) is apparently due to Alexandroff (1940, p. 312, stated; 1943, Theorem 2, p. 180). He describes the situation as follows:

> The charges are as if swimming in the space striving to the limit distribution. . . .The charges moving in the space may enter in the limit the closed set F getting onto its boundary, but come out of it they cannot *only* in the limit, for to this end they have to get away from the closed set, which cannot occur in the limit, but always only at a certain stage of the motion and so before the passage to the limit.

On A. D. Alexandroff, see also the notes to §9.3, and P. S. Alexandroff et al. (1973).

The word "portmanteau," which earlier meant a large traveling bag, later acquired a figurative meaning of carrying several purposes or virtues.

§11.2 R. Lipschitz in 1864 defined conditions $|f(x) - f(y)| \leq K|x - y|$ (see the notes to §6.1). Lipschitz (1876, p. 150) treated such conditions for functions on \mathbb{R}^k, in connection with coefficients and solutions of differential equations. Liouville (1837, p. 19; 1838, p. 565) had used the method of "successive approximation" of solutions, which Picard (1893) developed in greater generality, under Lipschitz conditions; Ince (1926, p. 76) gives further history and references. Lipschitz (1876, p. 157) also mentioned conditions $|f(x) - f(y)| \leq K|x - y|^\delta$, where $0 < \delta < 1$. For any metric d, d^δ is also a metric, so that such conditions can also be viewed as Lipschitz conditions. Hölder (1882) made good use of the conditions with $\delta < 1$, which thus are often called *Hölder conditions*.

Proposition 11.2.1 shows that $BL(S, d)$ is what is called a Banach algebra. It has been studied from that point of view (Sherbert, 1964).

The use of the space $BL(S, d)$ and its properties, at least for applications in probability, apparently began with Fortet and Mourier (1953).

§11.3 Prohorov (1956) defined his metric ρ and proved it metrized convergence of laws on complete separable metric spaces. Fortet and Mourier (1953) defined the metric β. Ranga Rao (1962) proved that convergence of laws is uniform over uniformly bounded, equicontinuous classes of functions (Corollary 11.3.4) and thus implies β convergence. Dudley (1966, 1968) proved further results in the section.

§11.4 Glivenko (1933) first proved Theorem 11.4.2 when F is continuous. Cantelli (1933) extended it to a general distribution function (see summaries in *Jahrbuch, 59*: 1166). Glivenko, who lived from 1897 to 1940, also worked in mathematical logic, foundations of mathematics, functions of real variables, and algebra (there are eleven index entries for his name in *Istoriia*, 1970). On Cantelli, see also the notes to §8.3 (Borel-Cantelli lemma). Varadarajan (1958) first proved Theorem 11.4.1.

§11.5 In measure theory on locally compact spaces, a measure μ was called *regular* if $\mu(A) = \sup\{\mu(K): K \subset A, K \text{ compact}\}$ for all A in the domain of μ (e.g., Halmos, 1950, p. 224). The earliest use I know of the word "tight" is by Le Cam (1957) about sets of measures here called uniformly tight. Szpilrajn (1937, 2.5(v)) proved that if (S, d) is a separable metric space and there is a 1–1, Borel measurable function f from S onto another metric space T, with Borel measurable inverse, then S is u.m. if and only if T is. This fact improves on Corollary 11.5.2, where f and f^{-1} are continuous. For a proof in English see Shortt (1984). Szpilrajn's paper was in Polish. I thank Rae Shortt for the reference to it. Szpilrajn later changed his name to, and published under the name, Marczewski.

Theorem 11.5.3 is due to Le Cam (1957, Theorem 4, p. 222). In Theorem 11.5.4, for (II) to imply (I), the hypothesis that S be Polish (separable and complete for some metrization) cannot be weakened very much. The implication fails for all non-Polish separable metrizable spaces which are Borel sets in their completions, such as the space of all rational numbers, according to D. Preiss (1973). Recall that a separable metric space is Polish (complete for some other metric with the same topology) if and only if it is a G_δ (a countable intersection of open sets) in its completion (Theorem 2.5.4).

§11.6 P. Hall (1935) gave the first explicit statement and proof of the pairing theorem (11.6.1). It follows easily from an earlier theorem of D. König (1931, 1950); see M. Hall (1958).

Strassen (1965) proved the main theorem (11.6.2) for S complete, using a form of the Hahn-Banach theorem (separation of convex sets, as in §6.2 above). The proof and extension above, with $b > \beta$ in place of β in Theorem 11.6.2(II), was in Dudley (1968). The replacement of $b > \beta$ by β, and Theorem 11.6.3, had not been published before the first (1989) edition of this book to my knowledge. The theorem can also be viewed in terms of linear programming (Schay, 1974).

§11.7 Persi Diaconis (personal communication) asked about uniformities consistent with the topology of convergence of laws and how to characterize the ρ and β uniformity among them. Theorem 11.7.1 was published in the first edition of this book (1989) for the first time as far as I know.

Theorem 11.7.2 is due to Skorohod (1956) for complete separable metric spaces and Dudley (1968, Theorem 3) for separable metric spaces. Wichura (1970) and Dudley (1985) gave extensions to nonseparable metric spaces if the limit law P has separable

support. Dudley (1966) treated some uniformities for convergence of laws on topological spaces.

For any two complete, separable metric spaces X and Y and uncountable Borel sets $A \subset X$ and $B \subset Y$, there is a 1-1 Borel function f from A onto B with Borel inverse, as will be shown in §13.1 below. Using this somewhat deep theorem, one could prove Theorem 11.7.5 by first getting $P = \mu \circ f^{-1}$ for some law μ on $[0, 1]$, then $\mu = \lambda \circ X_F^{-1}$, where F is the distribution function of μ and X_F is given by Proposition 9.1.2, so $P = \lambda \circ (f \circ X_f)^{-1}$. Theorem 11.7.5 was proved in Halmos and von Neumann (1942, Theorem 2). Rokhlin (1949, §2, no. 7) showed that f can be taken to be 1-1. Dyson (1953) gave examples of laws on \mathbb{R} which are far apart but whose characteristic functions are uniformly close, as shown in Proposition 11.7.6.

§11.8 Monge (1781) posed the following "transportation problem": given two domains A and B in space, of equal volume $V(A) = V(B)$, how to decompose A and B into arbitrarily small parts and match each part A_i of A with a part B_i of B having equal volume so as to minimize the sum $\sum_i V(A_i)d(A_i, B_i)$ where $d(A_i, B_i)$ is the distance from A_i to B_i (becoming well defined as the sizes of A_i and B_i approach 0). The value of the minimum is seen to be $cW(P, Q)$ where P and Q are the uniform probability distributions on A and B, respectively, and $c = V(A)$. Such problems relate to mathematical economics, as mentioned in the text.

L. V. Kantorovich in 1939 invented linear programming—the maximization or minimization of a linear function of several variables subject to several simultaneous linear constraints. See, for example, Kantorovich (1940), which also uses a separation theorem for convex sets. On Kantorovich's work in mathematics see Akilov et al. (1962), Vulikh et al. (1972), and Aganbegyan et al. (1987). Kantorovich shared the Nobel Memorial Prize in Economic Science in 1975 with T. C. Koopmans for "contributions to the theory of optimum allocation of resources"—see Scarf, 1975. Kantorovich lived from 1912 to 1986 (*New York Times*, 1986).

The metric for laws on \mathbb{R} having finite first moments, $W(P, Q) := \inf\{E|X - Y|: \mathcal{L}(X) = P, \mathcal{L}(Y) = Q\}$, sometimes called the Gini index, was studied by Gini around 1914, as reviewed by Rachev (1984). Dall'Aglio (1956, p. 42) proved the representation of $W(P, Q)$ in terms of distribution functions after Salvemini (1943) had done so for discrete distributions.

Kantorovich and Rubinstein (Rubinshtein) (1958) first proved their theorem (only for compact metric spaces, but there in a somewhat more general form). No translation into English has been published as far as I know. The current proof is based on ideas of J. Neveu (I learned Lemmas 11.8.5 and 11.8.6 from him in 1974). The topic is related to infinite-dimensional linear programming (Rubinshtein, 1970, esp. pp. 182–183, and Vershik, 1970). The insufficient proof in Dudley (1976, Lecture 20) was completed by de Acosta (1982, pp. 368–372). Meanwhile Szulga (1978, 1982) and Fernique (1981) gave other proofs. Rachev (1984) surveyed this problem and related ones.

I have benefited from correspondence and discussions on this topic with A. de Acosta, J. Neveu, W. Philipp, S. T. Rachev, B. Simon, and V. M. Zolotarev.

§11.9 Hoeffding (1948) proved the main result (Theorem 11.9.3) of this section. Earlier, Halmos (1946) had shown optimal properties of U-statistics as unbiased estimators, and von Mises (1947) had some results on asymptotic distributions overlapping with Hoeffding's.

References

de Acosta, Alejandro (1982). Invariance principles in probability for triangular arrays of B-valued random vectors and some applications. *Ann. Probability 10*: 346–373.

Aganbegyan, A. G., A. D. Aleksandrov, D. K. Faddeev, M. K. Gavurin, S. S. Kuteladze, V. L. Makarov, Yu. G. Reshetnyak, I. V. Romanovskii, G. Sh. Rubinshtein, and S. L. Sobolev (1987). Leonid Vital'evich Kantorovich. *Russian Math. Surveys 42*, no. 2: 225–232. Transl. from *Uspekhi Mat. Nauk 42*, no. 2: 177–182.

Akilov, G. P., B. Z. Vulikh, M. K. Gavurin, V. A. Zalgaller, I. P. Natanson, A. G. Pinsker, and D. K. Faddeev (1962). Leonid Vital'evich Kantorovich (In honor of his fiftieth birthday). *Russian Math. Surveys 17*, no. 4: 115–128. Transl. from *Uspekhi Mat. Nauk 17*, no. 4: 201–215.

Alexandroff, Alexander Danilovich [Aleksandrov, A. D.] (1940, 1941, 1943). Additive set-functions in abstract spaces. *Mat. Sbornik* (N.S.) *8*: 307–348; *9*: 563–628; *13*: 169–238 (in English).

Alexandroff [Aleksandrov], P. S., N. V. Efimov, V. A. Zalgaller, and A. V. Pogorelov (1973). Aleksandr Danilovich Aleksandrov (on his sixtieth birthday). *Russian Math. Surveys 28*, no. 2: 225–230. Transl. from *Uspekhi Mat. Nauk 28*, no. 2: 249–253.

Bray, H. E. (1918–1919). Elementary properties of the Stieltjes integral. *Ann. Math.* (Ser. 2) *20*: 177–186.

*Cantelli, Francesco Paolo (1933). Sulla determinazione empirica della leggi di probabilità. *Giorn. Ist. Ital. Attuari, 4*: 421–424.

Dall'Aglio, Giorgio (1956). Sugli estremi dei momenti delle funzioni di ripartizione doppia. *Ann. Scuola Norm. Sup. Pisa* (Ser. 3) *10*: 35–74.

Dudley, R. M. (1966). Convergence of Baire measures. *Studia Math. 27*: 251–268.

———— (1968). Distances of probability measures and random variables. *Ann. Math. Statist. 39*: 1563–1572.

———— (1976). *Probabilities and Metrics*. Lecture Notes Ser. No. 45, Matematisk Institut, Aarhus Universitet.

———— (1985). An extended Wichura theorem, definitions of Donsker class, and weighted empirical distributions. *Probability in Banach Spaces V* (Proc. Conf. Medford, 1984). Ed. A. Beck et al. *Lecture Notes in Math.* (Springer) *1153*: 141–178.

Dyson, Freeman J. (1953). Fourier transforms of distribution functions. *Canad. J. Math. 5*: 554–558.

Fernique, Xavier (1981). Sur le théorème de Kantorovitch-Rubinstein dans les espaces polonais. *Séminaire de Probabilités XV. Lecture Notes in Math.* (Springer) *850*: 6–10.

Fortet, Robert, and Edith Mourier (1953). Convergence de la répartition empirique vers la répartition théorique. *Ann. Sci. Ecole Norm. Sup. 70*: 266–285.

*Glivenko, Valery Ivanovich (1933). Sulla determinazione empirica della leggi di probabilità. *Giorn. Ist. Ital. Attuari 4*: 92–99.

Hall, Marshall, Jr. (1958). A survey of combinatorial analysis. In *Some Aspects of Analysis and Probability*, pp. 35–104. Wiley, New York.

Hall, Philip (1935). On representatives of subsets. *J. London Math. Soc. 10*: 26–30.

Halmos, Paul (1946). The theory of unbiased estimation. *Ann. Math. Statist. 17*: 34–43.

————— (1950). *Measure Theory.* Van Nostrand, Princeton.

—————, and von Neumann, J. (1942). Operator methods in classical mechanics, II. *Ann. Math. 43*: 332–350.

∗Helly, E. (1911–1912). Über lineare Funktionaloperatoren. *Sitzungsber. Nat. Kais. Akad. Wiss. (Wien) Math. Nat. Kl.* IIa *121*: 265–297.

Hildebrandt, T. H. (1966). Linear continuous functionals on the space (*BV*) with weak topologies. *Proc. Amer. Math. Soc. 17*: 658–664.

Hoeffding, Wassily (1948). A class of statistics with asymptotically normal distribution. *Ann. Math. Statist. 19*: 293–325.

∗Hölder, Otto Ludwig (1882). Beiträge zur Potentialtheorie. Dissertation, Tübingen.

Ince, Edward Lindsay (1926). *Ordinary Differential Equations.* Repr. Dover, New York, 1956.

Istoriia otechestvennoi matematiki (in Russian) (1970). Vol. 4, bk. 2. Acad. Sci. USSR and Ukrain. SSR, Kiev.

Jahrbuch über die Fortschritte der Mathematik, 59 (1933).

Kantorovich, Leonid Vasilevich (1940). A new method of solving some classes of extremal problems. *Comptes Rendus (Doklady) Acad. Sci. USSR 28*: 211–214.

—————, and G. Sh. Rubinshtein (1958). On a space of completely additive functions. *Vestnik Leningrad Univ. 13*, no. 7, *Ser. Mat. Astron. Phys.* 2: 52–59 (in Russian).

∗König, Dénes (1931). Graphs and matrices. (In Hungarian.) *Mat-Fiz. Lapok 38*: 116–119.

————— (1950). *Theorie der endlichen und unendlichen Graphen.* Chelsea, New York.

Le Cam, Lucien (1957). Convergence in distribution of stochastic processes. *Univ. of Calif. Publs. in Statistics* 2, no. 11: 207–236 (Univ. of Calif. Press, Berkeley).

Liouville, Joseph (1837). Second Mémoire: Sur le développement des fonctions ou parties de fonctions en séries dont les divers termes sont assujettis à satisfaire à une même équation différentielle du second ordre contenant un parametre variable. *J. de Math. Pures et Appliquées* (Ser. 1) 2: 16–35.

————— (1838). Premier mémoire. Sur la théorie des équations différentielles linéaires et sur le développement des fonctions en séries. *J. de Math. Pures et Appliquées* (Ser. 1) 3: 561–614.

Lipschitz, Rudolf Otto Sigismund (1876). Sur la possibilité d'intégrer complètement un système donné d'équations différentielles. *Bull. Sci. Math.* (Ser. 1) *10*: 149–159.

von Mises, R. (1947). On the asymptotic distribution of differentiable statistical functionals. *Ann. Math. Statist. 18*: 309–348.

∗Monge, Gaspard (1781). Mémoire sur la théorie des déblais et des remblais. In *Histoire de l'Académie des Sciences de Paris, avec les Mémoires de mathématique et de physique pour la même année*, pp. 666–704.

New York Times [unsigned] (1986). Leonid V. Kantorovich dies; Won Nobel Economics Prize. April 11, 1986, p. D18.

Parthasarathy, K. R. (1967). *Probability Measures on Metric Spaces.* Academic Press, New York.

Picard, Émile (1893). Sur l'application des méthodes d'approximation successives a l'étude de certaines équations différentielles ordinaires. *J. de Math. Pures et Appliquées* (Ser. 4) *9*: 217–271.

Preiss, David (1973). Metric spaces in which Prohorov's theorem is not valid. *Z. Wahrscheinlichkeitsth. verw. Geb. 27*: 109–116.

Prohorov, Yurii Vasilevich (1956). Convergence of random processes and limit theorems in probability theory. *Theor. Probability Appls. 1*: 157–214.

Rachev, Svetlozar T. (1984). The Monge-Kantorovich mass transference problem and its stochastic applications. *Theor. Probability Appls. 29*: 647–676.

Ranga Rao, R. (1962). Relations between weak and uniform convergence of measures with applications. *Ann. Math. Statist. 33*: 659–680.

Rokhlin, Vladimir Abramovich (1949). On the fundamental ideas of measure theory. *Mat. Sbornik 25*: 107–150. (In Russian); *Amer. Math. Soc. Translations 71* (1952): 1–54.

Rubinshtein, Gennadii Shlemovich (1970). Duality in mathematical programming and certain problems in convex analysis. *Russian Mathematical Surveys 25*: no. 5: 171–201.

*Salvemini, T. (1943). Sull calcolo degli indici di concordanza tra due caratteri quantitativi. In *Atti della VI Riunione della Soc. Ital. di Statistica*. Rome.

Scarf, Herbert E. (1975). The 1975 Nobel Prize in Economics: Resource allocation. *Science 190*: 649, 710, 712.

Schay, Geza, (1974). Nearest random variables with given distributions. *Ann. Probability 2*: 163–166.

Sherbert, Donald S. (1964). The structure of ideals and point derivations in Banach algebras of Lipschitz functions. *Trans. Amer. Math. Soc. 111*: 240–272.

Shortt, Rae Michael (1984). Universally measurable spaces: An invariance theorem and diverse characterizations. *Fund. Math. 121*: 169–176.

Skorohod, Anatolii Vladimirovich (1956). Limit theorems for stochastic processes. *Theor. Probability Appls. 1*: 261–290.

Strassen, V. (1965). The existence of probability measures with given marginals. *Ann. Math. Statist. 36*: 423–439.

*Szpilrajn, E. (1937). On absolutely measurable sets and functions (in Polish). *C. R. Soc. Sci. Lett. Varsovie Cl. III 30:* 39–68.

Szulga, Alicja (1978). On the Wasserstein metric. *Trans. Eighth Prague Conf. Information Theory, Statistical Decision Functions Random Processes*, vol. B, pp. 267–273. Reidel, Dordrecht.

———(1982). On minimal metrics in the space of random variables. *Theory Probability Appls. 27*: 424–430.

Varadarajan, Veeravalli S. (1958). On the convergence of sample probability distributions. *Sankhyā, 19*: 23–26.

Vershik, A. M. (1970). Some remarks on infinite-dimensional linear programming problems. *Russian Mathematical Surveys 25*, no. 5: 117–124.

Vulikh, B. Z., M. K. Gavurin, A. N. Kolmogorov, Yu. V. Linnik, V. L. Makarov, B. S. Mityagin, A. G. Pinsker, G. S. Rubinshtein, and D. K. Faddev (1972). Leonid Vitalevich Kantorovich (On his 60th birthday). *Russian Math. Surveys 27*, no. 3: 193–201. Transl. from *Uspekhi Mat. Nauk 27*, no. 3: 221–227.

Wichura, M. J. (1970). On the construction of almost uniformly convergent random variables with given weakly convergent image laws. *Ann. Math. Statist. 41*: 284–291.

12

Stochastic Processes

Stochastic processes have been treated so far mainly in connection with martingales, although a general definition was given: a stochastic process is a function X of two variables t and ω, $t \in T$, $\omega \in \Omega$, where (Ω, \mathcal{B}, P) is a probability space and for each t, $X(t, \cdot)$ is measurable on Ω. Taking T to be the set of positive integers, any sequence of random variables is a stochastic process. In much of the more classical theory of processes, T is a subset of the real line. But by the 1950s, if not before, it began to be realized that there are highly irregular random processes, useful in representing or approximating "noise," for example, which are in a sense defined over the line but which do not have values at points t. Instead, "integrals" $W(f) = \int W(t) f(t) \, dt$ are defined only if f has some smoothness and/or other regularity properties. Thus an appropriate index set T for the process may be a set of functions on \mathbb{R} rather than a subset of \mathbb{R}. Such processes are also useful where we may have random functions not only changing in time but defined also on space, so that T may be a set of smooth functions of space as well as, or instead of, time variables. At any rate, the beginnings of the theory of stochastic processes, and a basic existence theorem, hold for an arbitrary index set T without any structure. Then what is from many points of view the single most important process, the "Wiener process" or "Brownian motion," will be defined and studied. If X_i are i.i.d. variables with mean 0 and variance 1 and $S_n = X_1 + \cdots + X_n$, $\limsup_{n \to \infty} S_n / (2n \log \log n)^{1/2} = 1$ (the law of the iterated logarithm) will be proved in §12.5, using Brownian motion.

12.1. Existence of Processes and Brownian Motion

Given a stochastic process $x \colon \langle t, \omega \rangle \mapsto x(t, \omega)$, $t \in T$, $\omega \in \Omega$, the function $x(t, \cdot)$ on Ω will often be written as x_t, so $x_t(\omega) := x(t, \omega)$. Let \mathbb{R}^T denote the set of all real functions on T.

Suppose for each finite set $F \subset T$, we are given a law P_F on \mathbb{R}^F. Is there a process x with $\mathcal{L}(\{x_t\}_{t \in F}) = P_F$ for each F? If so, the P_F would need to

be *consistent*: whenever $F \subset G$, and f_{GF} is the natural projection of \mathbb{R}^G onto \mathbb{R}^F, $P_G \circ f_{GF}^{-1} = P_F$. This condition will also be proved to be sufficient, taking $\Omega = \mathbb{R}^T$, with the natural evaluation $x(t, f) := f(t)$ as the stochastic process.

Only the case of real-valued functions will be needed in this chapter after Theorem 12.1.2 and its proof, so on first reading, it may be best just to consider \mathbb{R} instead of more general measurable spaces.

A measurable space (X, \mathcal{S}), where \mathcal{S} is a σ-algebra of subsets of X, will be called *standard* iff there is a topology on X for which it is a compact metric space and \mathcal{S} is the σ-algebra of Borel sets. If (X, \mathcal{S}) is standard, (Y, \mathcal{U}) is another measurable space, and f is a 1–1 measurable function from X onto Y with measurable inverse, then (Y, \mathcal{U}) is clearly also standard. For example, any countable set X with the σ-algebra \mathcal{S} of all its subsets is standard, being either finite or equivalent as a measurable space to the compact metric space $\{0\} \cup \{1/n : n = 1, 2, \ldots\}$ in \mathbb{R}. Theorem 13.1.1 will show that every separable metric space Y which is a Borel subset of its completion is standard (with its Borel σ-algebra). So any Borel set in \mathbb{R}^k with Borel σ-algebra is standard. For now, a simpler case will be proved:

12.1.1. Proposition $(\mathbb{R}, \mathcal{B})$ *is standard, where* \mathcal{B} *is the* σ*-algebra of Borel sets in* \mathbb{R}.

Proof. Write \mathbb{R} as the disjoint union of $\{0\}$ and the set of all half-open intervals $(n, n+1]$ and $[-n-1, -n)$ for $n \in \mathbb{N}$. Write $[0, 1]$ as the disjoint union of $\{0\}$ and the half-open intervals $(1/(m+1), 1/m]$ for $m = 1, 2, \ldots$. Thus there is a 1–1 Borel measurable function taking $[0, 1]$ onto \mathbb{R}. \square

A measurable space (X, \mathcal{S}) will be called *universally measurable* (u.m.) iff \mathcal{S} is the Borel σ-algebra for a metric d on X for which (X, d) is a separable metric space which is a universally measurable subset of its completion (measurable for the completed measure of every probability law on the Borel σ-algebra of the completion), as defined in §11.5. Recall that a separable metric space is u.m. if and only if every law on it is tight (Theorem 11.5.1). Clearly, any standard measurable space is u.m.

Let T be any set, and suppose that for each $t \in T$, (S_t, \mathcal{B}_t) is a measurable space. Suppose that for each finite $F \subset T$, a law P_F is given on the Cartesian product $S_F := \Pi_{t \in F} S_t$, with the product σ-algebra of the \mathcal{B}_t. Then whenever $F \subset G$, we have the natural projection f_{GF} from S_G onto S_F, and the given laws are *consistent* if $P_G \circ f_{GF}^{-1} = P_F$ whenever $F \subset G \subset T$ and G is finite.

Again, on first reading, suppose $S_t = \mathbb{R}$ with Borel σ-algebra for all t. Especially in that case, where each law P_F is defined on a finite-dimensional vector space \mathbb{R}^F, the P_F will be called *finite-dimensional distributions*. One way to define consistent finite-dimensional distributions is to assign a law P_t on each S_t and to let the coordinates be independent, in other words, to define each P_F as the product measure of the P_t for $t \in F$. In that case, and for T countable, the next theorem would follow from Theorem 8.2.2.

12.1.2. Theorem (Kolmogorov) *For any set T and u.m. spaces $(S_t, \mathcal{B}_t)_{t \in T}$, and any consistent family of laws $\{P_F : F$ finite, $F \subset T\}$ (with P_F on S_F), there is a probability measure P_T on S_T with $P_T \circ f_{TF}^{-1} = P_F$ for all finite $F \subset T$.*

Proof. We may as well assume that each S_t is a separable metric space which is a u.m. subset of its completion, with Borel σ-algebra. Let \mathcal{A} be the collection of all subsets of S_T of the form $A = f_{TF}^{-1}(B)$ for Borel sets $B \subset S_F$ and F finite. Recall that on a Cartesian product of two, or finitely many, separable metric spaces S_t, the Borel σ-algebra of the product topology equals the product σ-algebra of the Borel σ-algebras in each S_t (Proposition 4.1.7). To see that \mathcal{A} is an algebra, first, it contains the complement of such a set A in it since $A^c = f_{TF}^{-1}(B^c)$. For unions, suppose also $D = f_{TG}^{-1}(C)$ for a finite $G \subset T$ and Borel set $C \subset S_G$. Let $H := F \cup G$. Then $A \cup D = f_{TH}^{-1}(f_{HF}^{-1}(B) \cup f_{HG}^{-1}(C)) \in \mathcal{A}$, so \mathcal{A} is an algebra.

For each $A = f_{TF}^{-1}(B) \in \mathcal{A}$ with B measurable in S_F, let $P_T(A) := P_F(B)$. To see that P_T is well-defined, suppose also $A = D$ as above. Then

$$A = (f_{HG} \circ f_{TH})^{-1}(C) = f_{TH}^{-1}(f_{HG}^{-1}(C)) = f_{TH}^{-1}(f_{HF}^{-1}(B)).$$

Since f_{TH} is onto S_H, this implies $f_{HG}^{-1}(C) = f_{HF}^{-1}(B))$, and by consistency

$$P_F(B) = P_H(f_{HF}^{-1}(B)) = P_H(f_{HG}^{-1}(C)) = P_G(C)$$

as desired. To prove that P_T has a countably additive extension to the σ-algebra generated by \mathcal{A}, it will be enough to prove that P_T is countably additive on \mathcal{A} (Theorem 3.1.4).

First an easy proof will be given if the (S_t, \mathcal{B}_t) are all standard measurable spaces, such as \mathbb{R} with its Borel σ-algebra. So we can assume that the S_t are all compact metric spaces with Borel σ-algebras. Thus S_T is compact by Tychonoff's theorem (2.2.8). For each $A = f_{TF}^{-1}(B) \in \mathcal{A}$, we have by Theorem 7.1.3 that

$$P_T(A) = P_F(B) = \sup\{P_F(K): K \text{ compact}, K \subset B\}.$$

For each compact $K \subset B$, $L := f_{TF}^{-1}(K)$ is a closed subset of the compact space S_T and hence compact, and $L \subset A$. So $P_T(A) = \sup\{P_T(L): L \text{ compact}, L \in \mathcal{A}, \text{ and } L \subset A\}$. Thus P_T is countably additive on \mathcal{A} by Lemma 10.2.4, finishing the proof for standard spaces S_t.

Now for general u.m. spaces S_t, if P_T is not countably additive, then there exist sets $W_n \in \mathcal{A}$ and $\varepsilon > 0$ with $W_n \downarrow \emptyset$ but $P_T(W_n) > \varepsilon$ for all n, by Theorem 3.1.1. Then for some finite $F(n) \subset T$ and Borel sets $B_n \subset S_{F(n)}$, where we may assume $F(1) \subset F(2) \subset \cdots$, $W_n = f_{TF(n)}^{-1}(B_n)$. If $m \le n$, let $f_{nm} := f_{n,m} := f_{F(n)F(m)}$. Then

$$f_{TF(n)}^{-1}(B_n) \subset f_{TF(m)}^{-1}(B_m) = \left(f_{nm} \circ f_{TF(n)}\right)^{-1}(B_m) = f_{TF(n)}^{-1}\left(f_{nm}^{-1}(B_m)\right).$$

Thus $B_n \subset f_{nm}^{-1}(B_m)$.

Let $P_n := P_{F(n)}$ for each n. Thus $P_T(W_n) = P_n(B_n)$ for each n. By Theorem 7.1.3 and Theorem 11.5.1 for universally measurable spaces, take compact $K_n \subset B_n$ with $P_n(B_n \setminus K_n) < \varepsilon/3^n$. Let

$$C_n := \bigcap_{m=1}^{n} f_{nm}^{-1}(K_m).$$

Then for all n, $C_n \subset B_n$, and C_n is compact since K_n is compact and $\bigcap_{m < n} f_{nm}^{-1}(K_m)$ is closed. Now

$$B_n \setminus C_n = \bigcup_{m=1}^{n} B_n \setminus f_{nm}^{-1}(K_m) \subset \bigcup_{m=1}^{n} f_{nm}^{-1}(B_m \setminus K_m), \quad \text{so}$$

$$P_n(B_n \setminus C_n) \le \sum_{m=1}^{n} \varepsilon/3^m < \varepsilon/2, \quad \text{and } P_n(C_n) \ge \varepsilon - \varepsilon/2 = \varepsilon/2.$$

For $n \le r$,

$$f_{rn}^{-1}(C_n) = f_{rn}^{-1}\left(\bigcap_{m=1}^{n} f_{nm}^{-1}(K_m)\right) = \bigcap_{m=1}^{n} f_{rm}^{-1}(K_m) \supset C_r,$$

so $f_{rn}(C_r) \subset C_n$. Thus $f_{rj}(C_r) = f_{nj}(f_{rn}(C_r)) \subset f_{nj}(C_n)$ if $j \le n \le r$. Since $C_r \ne \emptyset$, $f_{rj}(C_r) \ne \emptyset$. Let

$$D_n := \bigcap_{r \ge n} f_{rn}(C_r) = \bigcap_{r > n} f_{rn}(C_r).$$

Then D_n, being the intersection of a decreasing sequence of non-empty compact sets, is also compact and non-empty. Choose $x_1 \in D_1$. Theorem 2.2.12(b)

implies that

$$f_{n+1,n}(D_{n+1}) = f_{n+1,n}\left(\bigcap_{r>n} f_{r,n+1}(C_r)\right) = \bigcap_{r>n} f_{r,n}(C_r) = D_n.$$

So $x_n \in D_n$ can be chosen recursively with $f_{n+1,n}(x_{n+1}) = x_n$ for all n. Then $f_{mn}(x_m) = x_n$ for $m > n$. Since $x_n \in D_n \subset S_{F(n)}$, let x_n have coordinates $x_n(u)$ for $u \in F(n)$. Let $x(u) := x_n(u)$ if $u \in F(n)$ for any n. Then $x(u)$ is well-defined for $u \in \bigcup_n F(n)$. For other u let $x(u)$ be any point of S_u. Then $x \in S_T$ and for all n, $f_{TF(n)}(x) = x_n$. Since $D_n \subset C_n \subset K_n \subset B_n$, we have $x \in f_{TF(n)}^{-1}(B_n) = W_n$ for all n, a contradiction. □

The normal laws $N(m, C)$ defined in and after Theorem 9.5.7 are also often called *Gaussian*. A stochastic process x is called *Gaussian* iff for every finite $F \subset T$, $\mathcal{L}(\{x_t\}_{t\in F})$ is Gaussian. These laws are determined by the mean function $m(t) := Ex_t$ from T into \mathbb{R} and the covariance function $C(s, t) := Ex_s x_t - m(s)m(t)$ from $T \times T$ into \mathbb{R}. An example of a possible covariance function C is any function which is nonnegative when $s = t$ and 0 when $s \neq t$, which will mean that x_s and x_t are independent. Here is an existence theorem.

12.1.3. Theorem *Let T be any set, m any function from T into \mathbb{R}, and C any function from $T \times T$ into \mathbb{R} such that $C(s, t) = C(t, s)$ for all $s, t \in T$ and for any finite $F \subset T$, $\{C(s, t)\}_{s,t \in F}$ is a nonnegative definite matrix. Then there exists a Gaussian process $\{x_t\}_{t \in T}$ with mean function m and covariance function C.*

Proof. If $\{y_t\}_{t\in T}$ is a Gaussian process with mean 0 and covariance C, then $x_t := y_t + m(t)$ is Gaussian with mean m and covariance C. So we can take $m \equiv 0$.

For any finite $F \subset T$, let C_F be the restriction of C to $F \times F$. Consider the normal law $N(0, C_F)$ on \mathbb{R}^F (Theorem 9.5.7). If $G \subset F$, and t is a linear function on \mathbb{R}^G, or equivalently a point of \mathbb{R}^G with the usual inner product, then $t \circ f_{FG}$ on \mathbb{R}^F is the linear form with the coordinates of t on G and 0 on $F\backslash G$. Thus we have equality of inner products $(C_F(t \circ f_{FG}), t \circ f_{FG}) = (C_G(t), t)$. Then $N(0, C_F) \circ f_{FG}^{-1} = N(0, C_G)$ since each has the characteristic function $\exp(-(C_G(t), t)/2)$ (by Theorem 9.5.7; also, this is a special case of Proposition 9.5.12). So the laws $N(0, C_F)$ for all finite $F \subset T$ form a consistent family and Kolmogorov's theorem (12.1.2) applies. □

Definition. Let $(H, (\cdot, \cdot))$ be a real inner product space. An *isonormal* process on H is a Gaussian process L with mean $EL(f) = 0$ and covariance $EL(f)L(g) = (f, g)$ for all $f, g \in H$.

12.1.4. Theorem *For any real inner product space H, there exists an isonormal process L on H. Any such L is linear, that is, for any $f, g \in H$ and $c \in \mathbb{R}$, we have $L(cf + g) = cL(f) + L(g)$ a.s.*

Proof. The nonnegative definite property follows from the definition of inner product, so an isonormal process L exists by Theorem 12.1.3. For the linearity, if we expand $E([L(cf + g) - cL(f) - L(g)]^2)$ and apply the properties of L we get

$$\|cf + g\|^2 - 2c(cf + g, f) + c^2\|f\|^2 - 2(cf + g, g) + 2c(f, g) + \|g\|^2$$
$$= \|f\|^2(c^2 - 2c^2 + c^2) + (f, g)[2c - 2c - 2c + 2c] + \|g\|^2[1 - 2 + 1] = 0,$$

so L is linear as stated. $\qquad\square$

Here is an informal introduction to Brownian motion. Microscopic particles suspended in a liquid are constantly colliding with molecules of the liquid and changing direction. In a mathematical idealization of this process, the particle changes direction at every time. So, if $s < t < u$ and X_t is the position of the particle at time t, then the increments $X_u - X_t$ and $X_t - X_s$ are independent. We will be concerned here only with the one-dimensional process, where X_t has real values, being for example the x coordinate of a three-dimensional motion. Also, the process will be homogeneous in time, so that the law of $X_t - X_s$ depends only on the difference $t - s$ for $0 \le s < t$. The process will start at 0 at time 0, so $X_0 = 0$, and have mean $EX_t = 0$ for all $t \ge 0$. Let $\sigma^2(t) := EX_t^2$. Now assuming that all increments $X_{tj} - X_{t(j-1)}$ are independent for $j = 1, \ldots, n$, we get $\sigma^2(nt) = n\sigma^2(t)$, $\sigma^2(t/m) = \sigma^2(t)/m$, and $\sigma^2(qt) = q\sigma^2(t)$ for any integers m and $n > 0$ and rational $q \ge 0$. Also, since X_t is a sum of arbitrarily many small i.i.d. increments, under mild conditions (as in Lindeberg's theorem, 9.6.1), X_t would have a normal distribution. At any rate, it will be assumed that $\mathcal{L}(X_t) = N(0, \sigma^2(t))$, and that any finite set of X_t's has a jointly normal distribution. The process will be normalized by setting $\sigma^2(1) = 1$, so $\sigma^2(q) = q$ for all rational $q \ge 0$. For any $s < t, \sigma^2(t) = \sigma^2(s) + \sigma^2(t - s) \ge \sigma^2(s)$, so $\sigma^2(\cdot)$ is nondecreasing, and $\sigma^2(t) = t$ for all $t \ge 0$. Then for $0 < s < t$, $EX_s(X_t - X_s) = 0$ by independence, so $EX_sX_t = EX_s(X_s + (X_t - X_s)) = EX_s^2 = s = \min(s, t)$. The

means $EX_t = 0$ and covariances EX_sX_t uniquely determine the laws of a Gaussian process, since the law $N(0, C)$ is uniquely determined according to Theorem 9.5.1 by its characteristic function $\exp(-(Ct, t)/2)$ (given in Theorems 9.5.6 and 9.5.7) and thus by the covariance C. In this case the process will be called Brownian motion, to be defined more formally a little later.

The next step is to look at the continuity of the Brownian motion process as a function of t. A process x defined for t in a topological space (which for Brownian motion will be the half-line $[0, \infty)$ or sometimes subintervals of it) will be called *sample-continuous* iff for all ω the function $t \mapsto x_t(\omega)$ is continuous. A process x is called *continuous in probability* iff whenever $t(n) \to t$, $x_{t(n)} \to x_t$ in probability.

For example, let Ω be the unit interval with Lebesgue measure and let $x(t, \omega) := 1_{\{t=\omega\}}$. This process is not sample-continuous; on the contrary, $t \mapsto x(t, \omega)$ is always discontinuous at $t = \omega$. On the other hand, for each t, $x(t, \omega) = 0$ a.s. So if we let $y(t, \omega) \equiv 0$, then y is sample-continuous and has the same finite-dimensional distributions (the P_F for F finite, as in Theorem 12.1.2) as x does. Thus sample continuity is not determined by these finite-dimensional distributions. So it is not true that every Brownian motion, as defined so far, is sample-continuous, but it will be shown that there exists one which is.

If there exists a process x with given P_F for F finite such that $t \mapsto x(t, \omega)$ is continuous for almost all ω, then we could assume it is sample-continuous (for all ω) without changing the laws P_F, since on a set of 0 probability where this function may not be continuous, we can change $x(t, \omega)$ to make it, say, identically 0, a continuous function.

Here is a formal definition of Brownian motion process:

Definition. A *Brownian motion* or *Wiener process* is a sample-continuous Gaussian process x on $T = [0, \infty)$ with mean 0 and covariance $Ex_sx_t = \min(s, t)$.

A process closely related to the Brownian motion is the *Brownian bridge*, defined as a sample-continuous Gaussian process y on $T = [0, 1]$ with mean 0 and $Ey_sy_t = s(1 - t)$ for $0 \le s \le t \le 1$. So $y_0 = y_1 = 0$ a.s.; the two ends of the "bridge" are at the same level (and that is about as far as the analogy with a bridge goes). The Brownian bridge turns up as a limit in a central limit theorem for empirical distribution functions, as follows. Recall (§11.4) that for any law on \mathbb{R}, with distribution function F, there are empirical distribution

functions F_n which converge to F uniformly on \mathbb{R} as $n \to \infty$ (the Glivenko-Cantelli theorem, 11.4.2). This is a type of strong law of large numbers. For ordinary random variables X_1, X_2, \ldots, i.i.d. with $EX_1^2 < \infty$ we have $S_n/n - EX_1 \to 0$ a.s. To understand the rate of convergence, the next step is the central limit theorem (9.5.6), which says that $n^{1/2}(S_n/n - EX_1)$ converges in law to a normal distribution. Likewise, for empirical distribution functions F_n, it's natural to consider $n^{1/2}(F_n - F)$. For any fixed x, the law of $n^{1/2}(F_n - F)(x)$ converges as $n \to \infty$ to $N(0, F(x)(1 - F(x)))$. In fact, the joint distribution of these quantities for any finite set of values of x converges to a normal distribution. Recall that, as in Lemma 11.4.3, the study of empirical distribution functions can be reduced to studying those for the uniform distribution function G on $[0, 1]$, namely, $G(x) := \max(0, \min(x, 1))$ for all x. For its empirical distribution functions G_n, let $\alpha_n(t) := n^{1/2}(G_n(t)-t)$, $0 \le t \le 1$, noting that $G(t) = t$ there. Since $G_n - G$ is the average of n independent copies of $G_1 - G$, the covariance of $\alpha_n(s)$ and $\alpha_n(t)$ for any s and t is the same for all n. It is easily computed for $n = 1$, where it equals $s(1 - t)$ for $0 \le s \le t \le 1$. So the Brownian bridge process is a Gaussian process with the same covariance as all the α_n. It is the limit to which the α_n converge in law, which in this book is seen only for finitely many values of t at a time.

Since by definition a Brownian motion or bridge must be sample-continuous, it is not obvious that these processes exist, but they do:

12.1.5. Theorem (Wiener) *There exist a Brownian motion x and Brownian bridge y (which are sample-continuous).*

Proof. If x is a Brownian motion and $y_t := x_t - tx_1$, $0 \le t \le 1$, it is straightforward to check that y is a Brownian bridge. So it is enough to prove the theorem for Brownian motion. Here bounds for normal distributions will be useful:

12.1.6. Lemma *For any $c > 0$,*

(a) $N(0, 1)([c, \infty)) \le (2\pi)^{-1/2}c^{-1} \exp(-c^2/2)$, *and the two sides are asymptotic (their ratio converges to 1) as $c \to +\infty$.*
(b) $N(0, 1)([c, \infty)) \le \exp(-c^2/2)$.

Remarks. The more refined bound in (a) is the more accurate of the two for large c, while the simpler bound (b) is adequate for many theoretical purposes, also for large c.

Proof. For (a),

$$N(0, 1)([c, \infty)) = (2\pi)^{-1/2} \int_c^\infty \exp(-x^2/2)\, dx$$

$$\leq (2\pi)^{-1/2} \int_c^\infty \frac{x}{c} \exp(-x^2/2)\, dx$$

$$= (2\pi)^{-1/2} c^{-1} \exp(-c^2/2).$$

The two sides are asymptotic by L'Hospital's rule. Now (b) follows from (a) for $c \geq (2\pi)^{-1/2}$. For $0 < c < (2\pi)^{-1/2}$, $N(0, 1)([c, \infty)) < N(0, 1)([0, \infty)) = 1/2 < \exp(-1/(4\pi)) \leq \exp(-c^2/2)$, proving (b) in that case also. □

Let H be the Hilbert space $L^2([0, \infty), \lambda)$, where λ is Lebesgue measure. Let L be isonormal on H (by Theorem 12.1.4). For $0 \leq t < \infty$ let $x_t := L(1_{[0,t]})$. Then $Ex_t = 0$ and for $0 \leq s < \infty$, $Ex_s x_t = (1_{[0,s]}, 1_{[0,t]}) = \min(s, t)$. So x has the means and covariances of Brownian motion, which uniquely determine the joint distribution of x_t for t in any finite set. From the linearity of L (Theorem 12.1.4) it follows that for $0 \leq s \leq t$, $\mathcal{L}(x_t - x_s) = N(0, t - s)$. The next problem is to find a process z such that for each t, $z_t = x_t$ a.s., so that z also has mean 0 and covariance $\min(s, t)$, and such that z is sample-continuous. This will be done first for $0 \leq t \leq 1$.

For each $n = 1, 2, \ldots$, let V_n be the set of all random variables equal to 0 or $x(j/2^n, \cdot) - x((j-1)/2^n, \cdot)$ for some $j = 1, \ldots, 2^n$. Then $\mathcal{L}(v) = N(0, 2^{-n})$ for all $v \in V_n$ except $v = 0$. Thus by Lemma 12.1.6,

$$Pr\{\sup\{|v|: v \in V_n\} \geq n^{-2}\} \leq 2^n N(0, 2^{-n})\{x: |x| \geq n^{-2}\}$$

$$\leq 2^{n+1} N(0, 1)(2^{n/2} n^{-2}, \infty) \leq p_n := 2^{n+1} \exp(-2^n/(2n^4)).$$

For n large, $\exp(-2^n/2n^4) < 4^{-n}$, so $\sum_n p_n < +\infty$. Then by the Borel-Cantelli lemma, for almost all ω there is an $n_o(\omega)$ such that for all $n \geq n_o(\omega)$, for all $v \in V_n$, $|v| \leq n^{-2}$.

For each $t \in [0, 1]$, take a binary expansion $t = \sum_{j \geq 1} t_j/2^j$ with $t_j = 0$ or 1. For t a dyadic rational, we choose the expansion with $t_j = 1$ for j large (except for $t = 0$). Let

$$t(n) := \sum_{j=1}^n t_j/2^j.$$

Then for each $n \geq 2$, $x_{t(n)} - x_{t(n-1)} \in V_n$. Since $\sum_n n^{-2} < \infty$, whenever $n_o(\omega)$ exists the limit $z_t := \lim_{n \to \infty} x_{t(n)}$ is defined for all $t \in [0, 1]$. Otherwise, set

448 *Stochastic Processes*

$z_t := 0$. Since $t \mapsto x_t$ is continuous in probability, for each t we have $z_t = x_t$ a.s., as desired (*here*, the set of probability 0 may depend on t). To prove z is sample-continuous on $[0, 1]$, suppose $n \geq n_o(\omega)$ and $|s - t| \leq 2^{-n}$. Then for $s(n) = j/2^n$ and $t(n) = k/2^n$, we have $|j - k| = 0$ or 1, so $|z_s - z_t| \leq |z_{s(n)} - z_{t(n)}| + |z_s - z_{s(n)}| + |z_t - z_{t(n)}| \leq n^{-2} + 2 \sum_{m > n} m^{-2} \to 0$ as $n \to \infty$. Thus z is sample-continuous, as desired. The same proof on $[k, k+1]$ for each positive integer k gives sample continuity for Brownian motion on $[0, \infty)$.

\square

Problems

1. Show that in the Kolmogorov existence theorem (12.1.2) the universally measurable spaces S_t can be replaced by any Hausdorff spaces S_t on which all laws are regular (as defined in §7.1).

2. In $[0, 1]$, let λ^* be outer measure for Lebesgue measure λ and let A_n be nonmeasurable sets with $A_n \downarrow \emptyset$ and $\lambda^*(A_n) = 1$ for all n, as in §3.4, Problem 2. For each n, let P_n be the probability measure (*hint:* see Theorem 3.3.6) defined by $P_n(B \cap A_n) := \lambda^*(B \cap A_n)$ for each Borel set B. For each $x \in A_n$ let

$$f_n(x) := (x, x, \ldots, x) \in B_n := \Pi_{j=1}^n A_j.$$

Let $Q_n := P_n \circ f_n^{-1}$ on B_n. Show that the Q_n define a consistent family of finite-dimensional distributions on $B_\infty := \Pi_{j=1}^\infty A_j$, but that there is no probability measure on B_∞ which has the given finite-dimensional distributions Q_n. (Thus the regularity assumption in Problem 1 cannot simply be removed.)

3. Let x_t be a Brownian motion. Show that for any $\varepsilon > 0$ and almost all ω there is an $M(\omega) < \infty$ such that for all s and t in $[0, 1]$, $|x_s - x_t| \leq M(\omega)|s - t|^{0.5-\varepsilon}$. *Hint:* First, in the proof of sample-continuity (12.1.5), replace $1/n^2$ by $1/2^{n(1-\varepsilon)/2}$. Note that these numbers form a geometric series in n, and that if the conclusion holds for each ω for $|s - t|$ small enough, then it holds for all s and t in $[0, 1]$ with perhaps a larger $M(\omega)$.

4. Note that for $0 \leq s \leq t \leq u \leq v$, $x_t - x_s$ and $x_v - x_u$ are independent. Show that for any Brownian motion x_t, $\lim_{n \to \infty} \sum_{j=1}^n (x_{j/n} - x_{(j-1)/n})^2 = 1$ in probability. Deduce that Problem 3 does not hold for $0.5 + \varepsilon$ in place of $0.5 - \varepsilon$. *Hint:* $EX^4 = 3\sigma^4$ if $\mathcal{L}(X) = N(0, \sigma^2)$ (Proposition 9.4.2).

5. In the proof of Proposition 12.1.1, a 1–1 measurable function f from \mathbb{R} onto $[0, 1]$ was constructed by representing each space as a union of countably many disjoint intervals and letting f be monotone from each interval in \mathbb{R} to an interval in $[0, 1]$. Show that this cannot be done with only finitely many intervals or half-lines.

6. Prove Lemma 12.1.6(b) for $0 < c < (2\pi)^{-1/2}$ using derivatives.

7. Let $M(a) := \int_a^\infty \exp(-x^2/2)\,dx / \exp(-a^2/2)$ for $a \geq 0$. This is called *Mills' ratio*. Let $f(a) := 2((a^2 + 4)^{1/2} + a)^{-1}$ and $g(a) := 2((a^2 + 2)^{1/2} + a)^{-1}$. Show that $f(a) \leq M(a) \leq g(a)$ for all $a \geq 0$. *Hints:* Show that $f' \geq af - 1$, $M' = aM - 1$, and $g' \leq ag - 1$, and that all three functions are bounded above by $1/a$. Consider $(M - f)'$ and show that if $M - f$ were ever negative it would go to $-\infty$, and likewise for $(g - M)'$.

8. Let F be the distribution function of the uniform distribution on the interval $[3, 7]$. Find the limit as $n \to \infty$ of the law on \mathbb{R}^2 of $n^{1/2}(F_n(4) - F(4), F_n(6) - F(6))$.

9. Let X_t be any stochastic process such that for some $p \geq 1$ and each $t \geq 0$, $E|X_s - X_t|^p \to 0$ as $s \to t$. Show that $t \mapsto EX_t$ is continuous.

10. Let X_t be a Gaussian process defined for $0 \leq t \leq 1$ such that for some $K < \infty$ and p with $0 < p \leq 2$, $E(X_t - X_s)^2 \leq K|s - t|^p$ for all s and t in $[0, 1]$. Show that the process can be taken to be sample-continuous. *Note: EX_t is not assumed to be 0 or constant.*

11. Let $T \subset \mathbb{R}$. Show that continuity in probability for processes defined on T depends only on the finite-dimensional distributions (unlike sample-continuity); specifically, if x and y are two processes such that $\mathcal{L}(\{x_t\}_{t \in F}) = \mathcal{L}(\{y_t\}_{t \in F})$ for every finite set $F \subset T$, then x is continuous in probability if and only if y is.

12. Recall the Poisson distribution P_λ on \mathbb{N} with parameter $\lambda \geq 0$, such that $P_\lambda(k) = e^{-\lambda}\lambda^k/k!$ for $k = 0, 1, \dots$. If X has distribution P_λ, then $EX = \lambda$ and $\text{var}(X) = \lambda$. Show that there exists a process p_t, $t \geq 0$, such that for any $0 \leq s < t$, the increment $p_t - p_s$ has distribution P_{t-s} and such that p has independent increments; that is, the increments $p_t - p_s$ for any finite set of disjoint intervals $(s, t]$ are independent. Then show that the process $v_t := p_t - t$ has the means and covariances of Brownian motion, $Ev_t \equiv 0$ and $Ev_s v_t = \min(s, t)$ for any s and $t \geq 0$, but that there is no process z with the same finite-dimensional joint distributions as v such that z is sample-continuous.

13. Define a 1–1 measurable function specifically to carry out the proof of Proposition 12.1.1.

12.2. The Strong Markov Property of Brownian Motion

In studying the Brownian motion process x_t, one can ask what are the distributions of, for example, $\sup_{0 \le t \le a} |x_t|$ or $\sup_{0 \le t \le a} x_t$ for $a > 0$. Specifically, to find the probability of the event $\{\sup_{0 \le t \le a} x_t \ge b\}$, where $b > 0$, note that on this event, there is a least time τ for which $x_\tau \ge b$, where $\tau \le a$ and by continuity $x_\tau = b$. Then consider the increment $x_a - x_\tau$. Since the process has independent normal increments with mean 0, it seems plausible (and will be shown in the proof of 12.3.1) that $P(x_a - x_\tau \ge 0 \mid \tau \le a) = 1/2$, and then

$$P\left(\sup_{0 \le t \le a} x_t \ge b\right) = P(\tau \le a) = 2P(x_a \ge b) = 2N(0, a)([b, \infty))$$

$$= 2N(0, 1)\left(-\infty, -b/a^{1/2}\right) = 2\Phi\left(-b/a^{1/2}\right)$$

where Φ is the standard normal distribution function, solving the problem. To fill in the details and justify this type of argument, it is useful to know that the Brownian motion makes a "fresh start" at "Markov times" such as τ; in other words, the process $(x_{\tau+t} - x_\tau)_{t \ge 0}$ has the same law on the space of continuous functions as the original Brownian motion process $(x_t)_{t \ge 0}$. This property is called the strong Markov property and will be proved, as the main fact in this section, in Theorem 12.2.7.

Let $x(t, \omega)$, $t \in T$, $\omega \in \Omega$, be any real-valued stochastic process, on a probability space (Ω, \mathcal{S}, P). On the space \mathbb{R}^T of all real functions on T, define the evaluation or coordinate function $e_t(f) := f(t)$ for all $t \in T$. The process x defines a function $X: \omega \mapsto x(\cdot, \omega)$ from Ω into \mathbb{R}^T, which is measurable for the smallest σ-algebra \mathcal{B}^T on \mathbb{R}^T such that e_t is measurable for each $t \in T$. Let P_T be the image measure $P \circ X^{-1}$ on \mathbb{R}^T. Then we have a process $Y: Y(t, f) := f(t)$, $t \in T$, $f \in \mathbb{R}^T$, such that for the probability space $(\mathbb{R}^T, \mathcal{B}^T, P_T)$, and any finite set $F \subset T$, the joint distributions of x and Y on \mathbb{R}^F are equal:

$$\mathcal{L}(\{Y(t, \cdot)\}_{t \in F}) = \mathcal{L}(\{x(t, \cdot)\}_{t \in F}).$$

Here P_T on \mathcal{B}^T may also be called the *law* $\mathcal{L}(\{x_t\}_{t \in T})$ of the process. For any subset S of \mathbb{R}^T which contains each function $x(\cdot, \omega)$, $\omega \in \Omega$, we can also restrict P_T to S.

Let $T := [0, \infty)$ Let $C(T)$ denote the set of all continuous real-valued functions on T. Then for the Brownian motion process x, the set $S \subset \mathbb{R}^T$ can be taken as $C(T)$ by Theorem 12.1.5.

It will be useful to define a topology and metric on $C(T)$. For each $n = 1$, $2, \ldots$, define a seminorm $\|\cdot\|_n$ on $C(T)$ by

$$\|f\|_n := \sup\{|f(x)|: 0 \le x \le n\}.$$

Thus for each n, $d_n(f, g) := \|f - g\|_n$ gives a pseudometric on $C(T)$. Let $h(t) := t/(1 + t)$, $t \geq 0$, and let d be the metric on $C(T)$ defined from the pseudometrics d_n as in Proposition 2.4.4, that is, $d(f, g) := \sum_n h(d_n(f, g))/2^n$. Then a sequence $\{f_m\}$ in $C(T)$ converges for d if and only if it converges for each d_n, that is, it converges uniformly on compact subsets of T. Also, a sequence $\{g_j\}$ is a Cauchy sequence for d if and only if it is one for d_n for each n. Then it converges uniformly on $[0, n]$ to some function $f_n \in C[0, n]$ for each n, with $f_k \equiv f_n$ on $[0, k]$ for $k \leq n$. Thus for some $f \in C(T)$, g_j converges to f uniformly on $[0, n]$ for each n, so $d(g_j, f) \to 0$, proving $(C(T), d)$ is complete (this is related to Theorem 2.5.7). As each $C[0, n]$ is separable (Corollary 11.2.5), it follows that $C(T)$ is separable for d. By the way, a real vector space, complete for a metric defined as here from a sequence of seminorms, is called a *Fréchet space*.

For example, if $f_n(t) = 0$ for $0 \leq t \leq n$ and f_n is any continuous function for $t \geq n$, then $f_n \to 0$ in $C(T)$ no matter how large f_n may be for $t > n$.

Let $x(t, \omega)$, $t \in T$, $\omega \in \Omega$ be a real-valued stochastic process where (T, \mathcal{U}) is a measurable space and (Ω, \mathcal{C}, P) a probability space. Then the process is called *measurable* iff it is jointly measurable, that is, measurable for the product σ-algebra $\mathcal{U} \otimes \mathcal{C}$. The following will be useful:

12.2.1. Proposition *If (T, e) is a separable metric space and x is a sample-continuous process, then x is measurable, for the Borel σ-algebra on T.*

Proof. Let $\{t_k\}_{1 \leq k < \infty}$ be dense in T. For each n and $k = 1, 2, \ldots$, let $U_{nk} := \{t : e(t, t_k) < 1/n\}$, and $V_{nk} := U_{nk} \setminus \bigcup_{j < k} U_{nj}$. Then for each n, the V_{nk} are disjoint and their union is T. Let $x_n(t, \omega) := x(t_k, \omega)$ for each $t \in V_{nk}$, $k = 1, 2, \ldots$. Then clearly each x_n is jointly measurable and for all t and ω, $x_n(t, \omega) \to x(t, \omega)$ as $n \to \infty$. Thus x is measurable (by Theorem 4.2.2). $\qquad\square$

It will be helpful to relate the Borel σ-algebra $\mathcal{B} := \mathcal{B}_{(d)}$ on $C(T)$ for d to the smallest σ-algebra \mathcal{S}_T for which the evaluation e_t (recall $e_t(f) := f(t)$) is measurable for each $t \in T$. (Here $\mathcal{S}_T = \{A \cap C(T) : A \in \mathcal{B}^T\}$ for the product σ-algebra \mathcal{B}^T on \mathbb{R}^T.)

12.2.2. Proposition *On $C(T)$, $T := [0, \infty)$, the σ-algebras \mathcal{B} and \mathcal{S}_T are equal. Any law P on \mathcal{B} is uniquely determined by the set of finite-dimensional distributions $\{\mathcal{L}(\{f(t)\}_{t \in G}) : G \text{ finite}, G \subset T\}$.*

Proof. Each evaluation e_t is continuous, thus Borel measurable, so $\mathcal{S}_T \subset \mathcal{B}$. Conversely, for each $n = 1, 2, \ldots$, the supremum norm on $[0, n]$ equals a

supremum over the countable set Q_n of all rational numbers in $[0, n]$. Each closed ball

$$\{f: \|f - g\|_n \le c\}, \qquad g \in C(T), \qquad c \ge 0,$$

is the intersection of the sets $\{f: |f(t) - g(t)| \le c\}$ for $t \in Q_n$. Thus such closed balls are in \mathcal{S}_T, so the function G_n defined by $G_n(f) := d_n(f, g)$ is \mathcal{S}_T measurable for each n and each $g \in C(T)$. It follows that $G(f) := d(f, g)$ is also measurable. Thus each open ball $\{f: d(f, g) < r\}$, $g \in C(T)$, $r > 0$, is in \mathcal{S}_T. Since $(C(T), d)$ is separable, by Proposition 2.1.4 each open set for d is in \mathcal{S}_T, hence each Borel set, so $\mathcal{B} = \mathcal{S}_T$.

Now as in the proof of Kolmogorov's theorem (12.1.2), the collection of events determined by the finite-dimensional joint distributions is an algebra which generates \mathcal{S}_T. Two probability measures that agree on an algebra also agree on the σ-algebra it generates, by the monotone class theorem (4.4.2). □

For any probability space (Ω, \mathcal{A}, P), two sub-σ-algebras $\mathcal{B} \subset \mathcal{A}$ and $\mathcal{C} \subset \mathcal{A}$ will be called *independent* iff $P(B \cap C) = P(B)P(C)$ for all $B \in \mathcal{B}$ and $C \in \mathcal{C}$. For a set \mathcal{Y} of random variables, let $\sigma(\mathcal{Y})$ be the smallest σ-algebra for which all $Y \in \mathcal{Y}$ are measurable. Then \mathcal{Y} will be said to be *independent* of a σ-algebra \mathcal{C} iff $\sigma(\mathcal{Y})$ is. If \mathcal{Z} is another set of random variables, then \mathcal{Y} and \mathcal{Z} will be called *independent* iff $\sigma(\mathcal{Y})$ and $\sigma(\mathcal{Z})$ are. It is equivalent, via monotone classes, that any finite subset of \mathcal{Y} is independent of any finite subset of \mathcal{Z}.

If $\{x_t\}_{t \ge 0}$ is a Brownian motion process and $0 \le t_0 < t_1 < t_2 < \cdots < t_n$, the increments $x(t_j) - x(t_{j-1})$, $j = 1, \ldots, n$, are independent, since they have a jointly normal distribution with zero convariances. Specifically, their covariance matrix has $t_j - t_{j-1}$ as its jth diagonal term and is 0 off the diagonal. Thus, for each $t \ge 0$, the set of all x_u, $0 \le u \le t$, is independent of the set of all increments $x_{t+h} - x_t$, $h \ge 0$, by Theorem 9.5.14.

It will be useful that independence is preserved under some limits:

12.2.3. Proposition *Let (Ω, \mathcal{F}, P) be a probability space and X_n, $n = 0$, $1, \ldots$, random variables on Ω with values in a separable metric space S. Let \mathcal{D} be a sub-σ-algebra of \mathcal{F} and for each $n \ge 1$ let \mathcal{A}_n be a σ-algebra for which X_1, \ldots, X_n are measurable. Suppose that $\mathcal{A}_1 \subset \mathcal{A}_2 \subset \cdots$ and each \mathcal{A}_n is independent of \mathcal{D}. Let $X_n \to X_0$ in probability. Then X_0 is also independent of \mathcal{D}.*

Proof. Taking a subsequence, we can assume that $X_n \to X_0$ a.s. by Theorem 9.2.1. Let $\mathcal{A} := \bigcup_n \mathcal{A}_n$, an algebra. Let $\mathcal{J} := \{C \in \mathcal{F} : P(C \cap B) = P(C)P(B)$ for all $B \in \mathcal{D}\}$. Then \mathcal{J} is a monotone class including \mathcal{A}, so by Theorem 4.4.2, it includes the σ-algebra \mathcal{S} generated by \mathcal{A}. Since X_0 is equal a.s. to a measurable function for \mathcal{S} (by Theorem 4.2.2), it is independent of \mathcal{D}. $\qquad\square$

Given a probability space (Ω, \mathcal{B}, P), a collection $\{\mathcal{B}_t\}_{t \geq 0}$ of σ-algebras with $\mathcal{B}_t \subset \mathcal{B}_u \subset \mathcal{B}$ for $0 \leq t \leq u$ is called a *filtration*.

Definition. Let (Ω, \mathcal{B}, P) be a probability space, let $\{x_t\}_{0 \leq t < \infty}$ be a sample-continuous Brownian motion process on Ω, and let $\{\mathcal{B}_t\}_{0 \leq t < \infty}$ be a filtration. Then $(x_t, \mathcal{B}_t)_{t \geq 0}$ will be called a *Brownian motion* iff

(1) x_t is \mathcal{B}_t measurable for $0 \leq t < \infty$, and
(2) for any t, the process $\{x_u - x_t\}_{u \geq t}$ is independent of $\mathcal{B}_{t+} := \bigcap_{v > t} \mathcal{B}_v$.

12.2.4. Proposition *If x_t is a sample-continuous Brownian motion, and \mathcal{F}_t is the smallest σ-algebra for which all the x_s are measurable for $0 \leq s \leq t$, then $(x_t, \mathcal{F}_t)_{t \geq 0}$ is a Brownian motion.*

Proof. This would be clear if in (2) we had \mathcal{B}_t rather than \mathcal{B}_{t+}. For each $h > 0$, $\{x_{u+h} - x_{t+h}\}_{u \geq t}$ is independent of $\mathcal{B}_{t+h} \supset \mathcal{B}_{t+}$. Then letting $h \downarrow 0$ through some sequence, and applying sample continuity, Proposition 12.2.3 applies to $\mathcal{S} = C([t, \infty))$ and the result follows. $\qquad\square$

The \mathcal{F}_t in Proposition 12.2.4 are the smallest possible σ-algebras giving a Brownian motion. If for example X is a random variable independent of $\{x_t\}_{t \geq 0}$ and \mathcal{C}_t is the smallest σ-algebra including \mathcal{F}_t for which X is measurable, then $(x_t, \mathcal{C}_t)_{t \geq 0}$ is also a Brownian motion.

If $(x_t, \mathcal{B}_t)_{t \geq 0}$ is a Brownian motion, then a a function τ from Ω into $[0, +\infty]$ will be called a *stopping time* for $\{\mathcal{B}_t\}_{t \geq 0}$ if $\{\tau \leq t\} \in \mathcal{B}_t$ for all $t \geq 0$, or a *Markov time* if $\{\tau < t\} \in \mathcal{B}_t$ for all $t > 0$.

An example of a stopping time is a *hitting time* $h_a := \inf\{t : x_t = a\}$ for $a \in \mathbb{R}$, as will be shown in the proof of 12.3.1.

Let $(x_t, \mathcal{B}_t)_{0 \leq t < \infty}$ be any Brownian motion. If τ is a stopping time for $\{\mathcal{B}_t\}_{t \geq 0}$ let

$$\mathcal{B}_\tau := \{A \in \mathcal{B} : A \cap \{\tau \leq t\} \in \mathcal{B}_t \text{ for all } t \geq 0\}.$$

If instead τ is a Markov time let

$$\mathcal{B}_{\tau+} := \{A \in \mathcal{B}: A \cap \{\tau < t\} \in \mathcal{B}_t \text{ for all } t > 0\}.$$

Then \mathcal{B}_τ and $\mathcal{B}_{\tau+}$ are clearly σ-algebras. Note that $\Omega \in \mathcal{B}_\tau$ if and only if τ is a stopping time, and $\Omega \in \mathcal{B}_{\tau+}$ if and only if τ is a Markov time.

12.2.5. Theorem (a) *If τ is a stopping time it is \mathcal{B}_τ measurable.*
 (b) *If τ is a Markov time it is $\mathcal{B}_{\tau+}$ measurable.*
 (c) *If τ is a stopping time then it is also a Markov time and $\mathcal{B}_\tau \subset \mathcal{B}_{\tau+}$.*

Proof. (a) For any $s \geq 0$ and $t \geq 0$,

$$\{\tau \leq s\} \cap \{\tau \leq t\} = \{\tau \leq \min(s, t)\} \in \mathcal{B}_{\min(s,t)} \subset \mathcal{B}_t,$$

so $\{\tau \leq s\} \in \mathcal{B}_\tau$ for all $s \geq 0$. Then (a) follows since the closed intervals $[0, s]$ for $0 \leq s < \infty$ generate the Borel σ-algebra of the extended half-line $[0, \infty]$. Likewise, (b) follows via sets $\{\tau < s\}, \{\tau < t\}, 0 < s, t < \infty$.

For (c), first, the event $\{\tau < t\}$ is the union of events $T_n := \{\tau \leq t - 1/n\}$ for $n = 1, 2, \ldots$, where $T_n = \emptyset \in \mathcal{B}_t$ for $t < 1/n$ and $T_n \in \mathcal{B}_{t-1/n} \subset \mathcal{B}_t$ otherwise. Taking a union over n, τ is a Markov time. Let $A \in \mathcal{B}_\tau$. Then likewise $A \cap T_n \in \mathcal{B}_t$ for each n; another union gives $A \in \mathcal{B}_{\tau+}$. □

A filtration $\{\mathcal{B}_t\}_{t \geq 0}$ is called *right-continuous* iff for each $t \geq 0$,

$$\mathcal{B}_t = \mathcal{B}_{t+} := \bigcap_{v > t} \mathcal{B}_v.$$

12.2.6. Proposition *For a right-continuous filtration $\{\mathcal{B}_t\}_{t \geq 0}$, τ is a Markov time if and only if it is a stopping time, and then $\mathcal{B}_\tau = \mathcal{B}_{\tau+}$.*

Proof. A stopping time is always a Markov time by Theorem 12.2.5(c). Conversely, let τ be a Markov time. For each $t \geq 0$,

$$\{\tau \leq t\} = \bigcap_{n=1}^{\infty} \{\tau < t + 1/n\}.$$

Since the intersection is decreasing it can be restricted to $n \geq n_0$ for any n_0. It follows that $\{\tau \leq t\} \in \mathcal{B}_{t+} = \mathcal{B}_t$, so τ is a stopping time. We have $\mathcal{B}_\tau \subset \mathcal{B}_{\tau+}$ always by Theorem 12.2.5(c). Conversely let $A \in \mathcal{B}_{\tau+}$. Then for $0 \leq t < u$, $A \cap \{\tau \leq t\} = A \cap \{\tau < u\} \cap \{\tau \leq t\} \in \mathcal{B}_u$, so $A \cap \{\tau \leq t\} \in \mathcal{B}_{t+} = \mathcal{B}_t$. Thus $A \in \mathcal{B}_\tau$. □

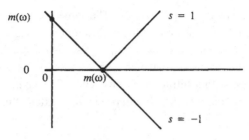

Figure 12.2

It is immediate from the definition of Brownian motion $(x_t, \mathcal{B}_t)_{t \geq 0}$ that $(x_t, \mathcal{B}_{t+})_{t \geq 0}$ is also a Brownian motion. Clearly the filtration $\{\mathcal{B}_{t+}\}_{t \geq 0}$ is right-continuous. As defined here, some but not all filtrations are right-continuous. Thus not all Markov times need be stopping times. At some points it may be reassuring to know that alternately one could take Brownian motions with right-continuous filtrations and apply Proposition 12.2.6.

The σ-algebra $\mathcal{B}_{\tau+}$ gives information about what happens "immediately after" τ, as in Lemma 12.2.8 below and the next example.

Example. Define a process z_t as follows. Let $m(\omega)$ be a random time uniformly distributed in $[0, 1]$. Let $s = s(\cdot)$ be a random variable independent of $m(\cdot)$ such that $P(s = 1) = P(s = -1) = 1/2$. Let $z_t := m(\omega) - t$ if $s = -1$ and $z_t := |m(\omega) - t|$ if $s = 1$ (Figure 12.2). Let \mathcal{C}_t be the smallest σ-algebra for which z_u is measurable for $0 \leq u \leq t$. Then m is a stopping time for the σ-algebras \mathcal{C}_t, and $s(\cdot)$ is measurable for \mathcal{C}_{m+} but not for \mathcal{C}_m, showing a difference between the two σ-algebras for the z_t process. The value of s can't be found from the values of z_t for $t \leq m(\omega)$, but it shows itself immediately after $m(\omega)$.

Let (Ω, \mathcal{A}, P) be a probability space and $v(t, \omega), 0 \leq t < \infty, \omega \in \Omega$, a sample-continuous real-valued stochastic process. Let $V: \omega \mapsto v(\cdot, \omega)$ from Ω into $C([0, \infty))$. Let $\mathcal{L}(v) := \mathcal{L}(V) := P \circ V^{-1}$ on the Borel σ-algebra $\mathcal{B} = \mathcal{S}_{[0,\infty)}$ as shown in Proposition 12.2.2.

If $A \in \mathcal{A}$ and $P(A) > 0$, let $P_{|A}(E) := P(E \mid A) := P(E \cap A)/P(A)$ for $E \in \mathcal{A}$. Then $P_{|A}$ is a probability measure on \mathcal{A}. Let $\mathcal{L}(v \mid A) := P_{|A} \circ V^{-1}$. If \mathcal{D} is a sub-σ-algebra of \mathcal{A}, then $v(\cdot, \cdot)$ will be said to be *independent of* \mathcal{D} iff $\{v(t, \cdot)\}_{t \geq 0}$ is, which is equivalent to $\mathcal{L}(v \mid A) = \mathcal{L}(v)$ for each $A \in \mathcal{D}$ with $P(A) > 0$.

For any fixed $t > 0$ and Brownian motion x, clearly $\{x_{t+h} - x_t\}_{h \geq 0}$ has the same finite-dimensional laws (for h restricted to a finite set) as $x = \{x_h\}_{h \geq 0}$,

and so the same law on $(C([0, \infty)), \mathcal{B})$. The same holds for t replaced by a Markov time τ taking just finitely many different (nonnegative real) values $t_i := t(i)$, since for each i, the law of $\{x_{t(i)+h} - x_{t(i)}\}_{h \geq 0}$ conditional on $\{\tau = t(i)\} \in \mathcal{B}_{t(i)+}$ is the same as that of Brownian motion, by the definition of Brownian motion with a filtration. In this sense, Brownian motion makes a fresh start at such Markov times. It will be useful to extend this fact to general Markov times.

12.2.7. Theorem (Strong Markov Property of Brownian Motion) *Let* $(x_t, \mathcal{B}_t)_{t \geq 0}$ *be any Brownian motion and* τ *a Markov time for* $\{\mathcal{B}_t\}_{t \geq 0}$ *with* $P(\tau < \infty) > 0$. *Then* $\omega \mapsto x_\tau := x(\tau(\omega), \omega)$ *is* $\mathcal{B}_{\tau+}$ *measurable on the set* $\{\tau < \infty\}$. *For* $h \geq 0$ *let*

$$W(h, \omega) := W_h(\omega) := x(\tau(\omega) + h, \omega) - x(\tau(\omega), \omega) \quad \text{if } \tau(\omega) < \infty;$$

undefined, otherwise. Then for any $C \in \mathcal{B}_{\tau+}$ *with* $P(C) > 0$ *on which* $\tau < \infty$, *the conditional law* $\mathcal{L}(W \mid C) = \mathcal{L}(x)$, *where* $x = \{x_t\}_{t \geq 0}$ *is Brownian. In other words, for* $F := \{\tau < \infty\} \in \mathcal{B}_{\tau+}$, *the process* W *on the probability space* $(F, P_{\mid F})$ *is a Brownian motion and is independent of the* σ-*algebra* $\mathcal{B}^F_{\tau+} := \{B \cap F : B \in \mathcal{B}_{\tau+}\}$.

Proof. First, some lemmas will be useful.

12.2.8. Lemma *If* τ *is a constant* c, *then* $\mathcal{B}_{\tau+} = \mathcal{B}_{c+} := \bigcap_{t > c} \mathcal{B}_t$.

Proof. First, note that a constant is a stopping time and (so) a Markov time. If $A \in \mathcal{B}_{\tau+}$, it follows from the definition of $\mathcal{B}_{\tau+}$ that $A \in \mathcal{B}_t$ for all $t > c$, so $A \in \mathcal{B}_{c+}$. Conversely if $A \in \mathcal{B}_{c+}$, then $A_t := A \cap \{\tau < t\} = A \in \mathcal{B}_t$ if $t > c$, while otherwise $A_t = \emptyset \in \mathcal{B}_t$. So $A \in \mathcal{B}_{\tau+}$. □

The next fact seems intuitive since if $u \leq v$ and we know whether or not an event has happened by time u, then we certainly know it by time v.

12.2.9. Lemma *If* u *and* v *are Markov times and* $u \leq v$, *then* $\mathcal{B}_{u+} \subset \mathcal{B}_{v+}$.

Proof. If $A \in \mathcal{B}_{u+}$, then for any $t > 0$,

$$A \cap \{v < t\} = (A \cap \{u < t\}) \cap \{v < t\} \in \mathcal{B}_t. □$$

To continue the proof of the strong Markov property (Theorem 12.2.7), let $\tau_n(\omega) := \tau(n)(\omega) := k/2^n$ iff $(k-1)/2^n \leq \tau(\omega) < k/2^n, k = 1, 2, \ldots,$

$n = 1, 2, \ldots$, or $\tau_n = +\infty$ iff $\tau = +\infty$. Then the τ_n are stopping times since for $k/2^n \le t < (k+1)/2^n$,

$$\{\tau_n \le t\} = \{\tau < k/2^n\} \in \mathcal{B}_t,$$

while for $t < 1/2^n$, $\{\tau_n \le t\} = \emptyset \in \mathcal{B}_t$.

We have $\tau_n \downarrow \tau$, with $\tau_n > \tau$ for all n if $\tau < \infty$. To show that x_τ is $\mathcal{B}_{\tau+}$ measurable, it is enough (by Theorem 4.1.6) to show that for each real y and $t > 0$, $\{x_\tau > y\} \cap \{\tau < t\} \in \mathcal{B}_t$. This event is equivalent to: for some $m = 1, 2, \ldots$, and all $n \ge m$, $\tau(n) < t$ and $x_{\tau(n)} \ge y + (1/m)$. Now $\tau(n)$ has just countably many possible values $t(n, k)$, and each event $\{\tau(n) = t(n, k) < t\} \cap \{x_{t(n,k)} \ge y + (1/m)\}$ is in \mathcal{B}_t, where for $t(n, k) < t$, $\{\tau(n) = t(n, k)\} \in \mathcal{B}_{\tau(n)}$ by Theorem 12.2.5(a), so x_τ is $\mathcal{B}_{\tau+}$ measurable.

Let $W_n(h, \omega) := x(\tau_n(\omega) + h, \omega) - x(\tau_n, \omega)$ if $\tau(\omega) < \infty$; undefined otherwise. Then for $\tau < \infty$, $W_n(h, \omega) \to W(h, \omega)$ for all $h \ge 0$ and ω by sample continuity. Because each function $x(\cdot, \omega)$ is uniformly continuous on compact sets, the convergence of W_n to W is uniform on compact sets, as metrized by d (before Proposition 12.2.1). Thus $d(W_n(\cdot, \omega), W(\cdot, \omega)) \to 0$ as $n \to \infty$ for any $\omega \in F$. Note that W is (jointly) measurable on $\{\tau < \infty\}$ by Proposition 12.2.1.

If τ is a constant, the conclusion holds by Lemma 12.2.8 and the definition of Brownian motion with a filtration.

Next let $c := c(n, k) := k/2^n$. For fixed k, n and thus c, if $A \in \mathcal{B}_{c+}$ and $P(A \cap \{\tau_n = c\}) > 0$, then $\mathcal{L}(W_n \mid A \cap \{\tau_n = c\}) = \mathcal{L}(x)$ by the case when τ is constant since $\{\tau_n = c\} \cap A \in \mathcal{B}_{c+}$. If $C \in \mathcal{B}^F_{\tau(n)+}$ and $P(C) > 0$, let $C_k := C \cap \{\tau_n = c(n, k)\}$. Then $C_k \in \mathcal{B}_{c(n,k)+}$ and $C = \bigcup_k C_k$, a disjoint union. Since W_n is independent of each C_k and has the same conditional law, namely $\mathcal{L}(x)$, on each, the same is true of their union C. So W_n is independent of any $C \in \mathcal{B}^F_{\tau+}$ by Lemma 12.2.9. For any such C with $P(C) > 0$ and any bounded continuous real-valued function f on $C[0, \infty)$,

$$E(f(W_n) \mid C) = \int_C f(W_n)\, dP/P(C) = Ef(x).$$

Since W_n converge to W for d, $Ef(W \mid C) = Ef(x)$, so $\mathcal{L}(W \mid C) = \mathcal{L}(x)$ on $C[0, \infty)$ by Lemma 9.3.2. This law doesn't depend on C, so W on $(F, P_{|F})$ is independent of $\mathcal{B}^F_{\tau+}$. Theorem 12.2.7 is proved. □

Problems

In problems 1–3 and 5–6, take Brownian motion with minimal σ-algebras \mathcal{F}_t as in Proposition 12.2.4.

1. Given a Brownian motion, define a random variable $\tau \geq 0$ on its probability space which is not a Markov time and such that $x_{\tau+h} - x_\tau$ does not have the same law as x_h for some $h > 0$. *Hint:* Let $\tau := \inf\{t: x_{t+1} - x_t \geq 0\}$, $h = 1$.

2. For a Brownian motion $\{x_t\}_{t \geq 0}$, let $s(\omega) := \inf\{t: x_t > 1\}$ and $\tau(\omega) := \inf\{u > s(\omega): x_u < 0\}$. Show that τ is a Markov time.

3. Give an example of a Markov time τ for Brownian motion such that if $s(\omega) = n$ for $n \leq \tau(\omega) < n + 1, n = 0, 1, \ldots$, then s is not a Markov time.

4. If ρ and τ are two Markov times for a filtration $\{\mathcal{B}_t\}_{t \geq 0}$, show that $\rho + \tau$ is also a Markov time for $\{\mathcal{B}_t\}_{t \geq 0}$.

5. Give an example of two stopping times ρ and τ for Brownian motion with $\tau < \rho < \infty$ a.s. such that $x_\rho - x_\tau$ is not independent of $\mathcal{B}_{\tau+}$. *Hint:* Let $\tau \equiv 1, \rho := \inf\{t > 1: x_t = 0\}$.

6. Show that for $b > 0$ and $c > 0$, $c + |x_b|$ is a Markov time for Brownian motion if and only if $c \geq b$.

7. Let X_1, X_2, \ldots, be i.i.d. real random variables and $S_n := X_1 + \cdots + X_n$. Let \mathcal{B}_n be the smallest σ-algebra for which X_1, \ldots, X_n are measurable. Let $k := k(\cdot)$ be a positive integer-valued stopping time for $\{\mathcal{B}_n\}_{n \geq 0}$, meaning that $\{\omega: k(\omega) \leq n\} \in \mathcal{B}_n$ for all $n = 1, 2, \ldots$. Show that $\{S_{n+k} - S_k\}_{n \geq 1}$ is independent of $\mathcal{B}_k := \{A: A \cap \{k \leq n\} \in \mathcal{B}_n \text{ for all } n = 1, 2, \ldots\}$ and has the same distribution as the sequence $\{S_n\}_{n \geq 1}$.

8. Show that the law of Brownian motion, on the Borel σ-algebra in $C[0, \infty)$ for the metric d metrizing uniform convergence on compact intervals, is unique. (This fact has been assumed in the text.) *Hint:* Use Proposition 12.2.2.

9. (a) If τ_n are stopping times for $n = 1, 2, \ldots$, for a filtration $\{\mathcal{B}_t\}_{t \geq 0}$, and $\tau_n \uparrow \tau$ as $n \to \infty$, show that τ is also a stopping time for $\{\mathcal{B}_t\}_{t \geq 0}$.

 (b) Do the same for Markov times. *Hint:* Show that $\{\tau < t\} = \bigcup_{q \in \mathbb{Q}, q < t} \bigcap_n \{\tau_n < q\}$ for $t > 0$.

10. Show that for the z_t process and stopping time m in the example of Figure 12.2, $z_{m+h} - z_m$ is independent of \mathcal{B}_m but not of \mathcal{B}_{m+} (in contrast to the strong Markov property of Brownian motion).

11. Show that if τ is a Markov time for a filtration $\{\mathcal{B}_t\}_{t \geq 0}$, then it is a stopping time for $\{\mathcal{B}_{t+}\}_{t \geq 0}$.

12. Let $s = \pm 1$ with probability $1/2$ each and $X_t(\omega) := s(\omega) 1_{(1,2]}(t)$. Let $\{\mathcal{B}_t\}$ be the smallest σ-algebra making X_u measurable for $0 \leq u \leq t$. Let

$\tau := 2 - s$. Show that τ is a Markov time but not a stopping time for $\{\mathcal{B}_t\}_{t \geq 0}$.

12.3. Reflection Principles, The Brownian Bridge, and Laws of Suprema

Let X_n be independent and equal $+1$ or -1 with probability $1/2$ each ("coin tossing" random variables) and $S_n := X_1 + \cdots + X_n$. The S_n are said to define a simple *random walk*. Note that n and S_n have the same parity (both even or both odd). Let $0 < k < n$ where k and n have different parity. Let A_{kn} be the event that $S_j \geq k$ for some $j \leq n$. On A_{kn}, let $m(\omega)$ be the least such j. Then since $n - m(\omega)$ is odd, $S_n - S_{m(\omega)}$ cannot be 0. For each $j < n$, we have the conditional probability $P(S_n - S_{m(\omega)} > 0 \,|\, m(\omega) = j) = 1/2$. Since the events $\{m(\omega) = j\}$ for $j = 1, \ldots, n$ are disjoint with union A_{kn}, we have $P(S_n - S_{m(\omega)} > 0 \,|\, A_{kn}) = 1/2$. Thus $P(S_n > k) = P(S_n \geq k) = P(A_{kn})/2$, so $P(A_{kn}) = 2P(S_n > k)$. This fact is known as the "reflection principle of Désiré André." The idea is that starting at time $m(\omega)$, for any set of possible paths, reflection in the line $y = k$ preserves probabilities (see Figure 12.3A).

In this section, the distributions of suprema of Brownian motion and the Brownian bridge and their absolute values over some finite intervals will be calculated by reflection methods. First comes an extension of André's reflection principle to Brownian motion:

12.3.1. Reflection Principle Let $(x_t, \mathcal{B}_t)_{t \geq 0}$ be a Brownian motion and $b, c > 0$. Then

$$P(\sup\{x_t : t \leq b\} \geq c) = 2P(x_b \geq c) = 2N(0, b)([c, \infty)).$$

Proof. Let $\tau := \tau(\omega) := \inf\{t : x_t = c\}$. Then the hitting time τ is a stopping time (possibly infinite), since by sample continuity, for $t < \infty$, $\{\tau \leq t\} = \bigcap_{q < c} \bigcup_{r < t} \{x_r > q\}$, where r and q are restricted to be rational. Also, $\sup\{x_t : t \leq b\} \geq c$ if and only if $\tau \leq b$. Let $B := \{\tau \leq b\}$. Then $P(B) > 0$, $B \in \mathcal{B}_\tau \subset \mathcal{B}_{\tau+}$ by Proposition 12.2.5, and $\tau < \infty$ on B, so the strong Markov property (Theorem 12.2.7) applies. Let W_h be the Brownian process $x_{\tau+h} - x_\tau$,

Figure 12.3A

defined on the event $\{\tau < \infty\}$. Then $\{x_b \geq c\} \subset B$ and

$$P(x_b \geq c \mid B) = P(x_b > c \mid B) = P(W_{b-\tau} > 0 \mid B).$$

Let $s := b - \tau$ on B. We have $P(s = 0) \leq P(x_b = c) = 0$, so $s > 0$ a.s. on B. Now $\omega \mapsto W_{s(\omega)}(\omega)$ is measurable by Proposition 12.2.1, since W is sample-continuous.

Let \mathcal{A} be the smallest σ-algebra of subsets of B for which all W_t for $t \geq 0$, restricted to $\omega \in B$, are measurable. Let $\mathcal{B}'_{\tau+} := \{C \cap B : C \in \mathcal{B}_{\tau+}\}$. For any event $D \subset B$ let $P_B(D) := P(D)/P(B)$. Let B' and B'' be two copies of B with different σ-algebras $(B', \mathcal{B}'_{\tau+})$ and (B'', \mathcal{A}). Let u and v be the identity functions from B into B' and B'' respectively. Then for the given σ-algebras, u and v are independent for P_B by the strong Markov property, so the law of (u, v) on $B' \times B''$ is the product $\mu \times \rho$ of their two laws. We can write $s = s(u)$ and $W_t(\omega) = W_t(v)$ for $t \geq 0$. Then by the Tonelli-Fubini theorem,

$$P\big(W_{s(\omega)} > 0 \mid B\big) = \int \rho\big(\{v : W_{s(u)}(v) > 0\}\big)\, d\mu(u) = \int \frac{1}{2}\, d\mu(u) = \frac{1}{2}.$$

Thus $P(x_b \geq c \mid B) = 1/2$ and $P(x_b \geq c) = P(B)/2$. \square

The Brownian bridge process y_t, $0 \leq t \leq 1$, is a Gaussian process with mean 0 and covariance $E y_s y_t = s(1 - t)$ for $0 \leq s \leq t \leq 1$. The distribution of the supremum of this process and of its absolute value will be evaluated in Propositions 12.3.3 and 12.3.4. These distributions can be applied in statistics as follows.

As noted just before Theorem 12.1.5, y_t arises as a limit of processes $n^{1/2}(G_n(t) - G(t))$ where G_n is an empirical distribution function for the uniform distribution function G on $0 \leq t \leq 1$, namely, $G(t) := \max(0, \min(t, 1))$ for all real t. If F is any continuous distribution function on \mathbb{R}, then $n^{1/2}(G_n(F(s)) - F(s))$ will have the same distribution for its supremum, or supremum of its absolute value, as for G the uniform distribution on $[0, 1]$, since F takes \mathbb{R} onto $(0, 1)$. (F may or may not have 0 or 1 in its range, but $G_n(0) - 0 = 0 = G_n(1) - 1$ a.s. in any case.) Also, $G_n \circ F$ have the same distributions as empirical distribution functions F_n for F, as shown in Lemma 11.4.3. So for n large enough, given n independent observations from an otherwise unknown distribution function F, and so given F_n, one can test the hypothesis that F is a given distribution function H by taking $n^{1/2}(F_n - H)$, finding its supremum or the supremum of its absolute value (these are called "Kolmogorov-Smirnov statistics"), and seeing whether there is a small probability (say, ≤ 0.05) for as large or larger a value in case of the Brownian bridge, in which case we would reject the hypothesis $F = H$.

As noted in the proof of sample continuity (Theorem 12.1.5), if x is a Brownian process, then $y_t := x_t - t x_1, 0 \le t \le 1$, gives a Brownian bridge. It is sometimes helpful to have another way to get from x_t to y_t: the idea is that the distribution of y_t is the conditional distribution of x_t given that $x_1 = 0$. Since the latter is only one event and has probability 0, such a conditional distribution is not well-defined, but the idea can be carried out by a limit process as follows. For any $\varepsilon > 0$, $P(|x_1| < \varepsilon) > 0$, so the conditional distribution $\mathcal{L}(\{x_t\}_{0 \le t \le 1} | |x_1| < \varepsilon)$ is defined on $C[0, 1]$. As usual, on $C[0, 1]$ we have the supremum norm and the metric and topology it defines, for which one can define convergence of laws.

12.3.2. Proposition *As $\varepsilon \downarrow 0$, $\mathcal{L}(x | |x_1| < \varepsilon) \to \mathcal{L}(y)$ on $C[0, 1]$.*

Proof. Let $y_t := x_t - t x_1$, as just recalled. Then for all t in $[0, 1]$, $E x_1 y_t = 0$. By Proposition 9.5.12, all these variables are jointly Gaussian with mean 0. By Theorem 9.5.14, this implies that the process y is independent of x_1.

Let F be the function from $C[0, 1] \times \mathbb{R}$ into $C[0, 1]$ defined by $F(g, u)(t) := g(t) + ut, 0 \le t \le 1$. Then F is jointly continuous and hence Borel measurable. Since $x_t = y_t + t x_1$, $\mathcal{L}(x)$ on $C[0, 1]$ is the image measure $(\mathcal{L}(y) \times N(0, 1)) \circ F^{-1}$. Let N_ε be the conditional distribution of x_1 given $|x_1| < \varepsilon$ and let u_ε be a random variable independent of the y process with law N_ε; in other words, we take the law $\mathcal{L}(y) \times N_\varepsilon$ on $C[0, 1] \times \mathbb{R}$, with coordinates (y, u_ε). Then

$$\mathcal{L}(x \mid |x_1| < \varepsilon) = \mathcal{L}(F(y, u_\varepsilon)).$$

Thus for any bounded continuous real function G on $C[0, 1]$,

$$E(G(x) \mid |x_1| < \varepsilon) = E(G(F(y, u_\varepsilon))) \to E(G(F(y, 0))) = E(G(y))$$

as $\varepsilon \downarrow 0$, since $u_\varepsilon \to 0$ and G and F are continuous. □

In view of Proposition 12.3.2, the Brownian bridge y is sometimes called "tied down Brownian motion," "Brownian motion tied down at 0 and 1," or "pinned Brownian motion." By the way, for a constant c, $x_t + c$ is called "Brownian motion starting at c" (at $t = 0$).

The next fact is a reflection principle for the Brownian bridge. What is the probability that y_t ever reaches the height $b > 0$? Heuristically, it is the conditional probability that x_t does (for $t \le 1$) given that $x_1 = 0$. By reflection, the probability of hitting b, then returning to 0 at time 1, equals the probability of hitting b, then reaching $2b$ at time 1. But by continuity, if $x_1 = 2b$, then x_t must have passed through b somewhere earlier, so the conditional probability

is just the ratio of the $N(0, 1)$ density at $2b$ to the density at 0, namely, $\exp(-2b^2)$. A detailed proof will be given:

12.3.3. Proposition *If y is a Brownian bridge and $b \in \mathbb{R}$, then $P(y_t = b$ for some $t \in [0, 1]) = \exp(-2b^2)$.*

Proof. Let $Q_b := P(y_t = b$ for some $t \in [0, 1])$. Then $Q_0 = 1$ and $Q_{-b} = Q_b$, since $\mathcal{L}(-y) = \mathcal{L}(y)$ on $C[0, 1]$. So we can assume $b > 0$. Let $x_t := y_t + tx_1$, where x_1 has law $N(0, 1)$ and is independent of y. Then x is a Brownian motion. For any $\varepsilon > 0$ let

$$P_{b,\varepsilon} = P(x_t \geq b \text{ for some } t \in [0, 1] \mid |x_1| < \varepsilon).$$

Since ε will be converging to 0, we can assume $0 < \varepsilon < b$. Then $P_{b-\varepsilon,\varepsilon} \geq Q_b \geq P_{b+\varepsilon,\varepsilon}$.

Let $\tau := \inf\{t \colon x_t = b\}$. Then as shown in the proof of 12.3.1, τ is a stopping time. Let $B := \{\tau < 1\}$. By the sample continuity and intermediate value theorem, $P(B) > 0$. For each measurable event A, let $P_b(A) := P_{|B}(A) := P(A \mid B)$. Then with respect to P_b, the process $W_s := x_{\tau+s} - x_\tau$ has the law of Brownian motion and is independent of $\mathcal{B}_{\tau+}^B := \{D \cap B \colon D \in \mathcal{B}_{\tau+}\}$, by the strong Markov property (Theorem 12.2.7). Note that $x_\tau \equiv b$ on B by sample continuity. Let $s := 1 - \tau$. Then s is a $\mathcal{B}_{\tau+}^B$ measurable random variable by Theorem 12.2.5, so it is independent of W for P_b. As noted in the proof of 12.3.1, $\omega \mapsto W_{s(\omega)}(\omega)$ is measurable. Now for $0 < \varepsilon < b$ we have

$$\begin{aligned} P_{b,\varepsilon} &= P\{\tau < 1 \text{ and } |W_s + b| < \varepsilon\}/P(|x_1| < \varepsilon) \\ &= P_b\{|W_s + b| < \varepsilon\}P(\tau < 1)/P(|x_1| < \varepsilon) \end{aligned}$$

(since for any event A and $B := \{\tau < 1\}$, $P(A \cap B) = P(A \mid B)P(B)$).

Just as in the proof of 12.3.1, we can write $s = s(u)$ and write $W_t(\omega), t \geq 0$, as $W_t(v), t \geq 0$, where u and v are independent for P_b. So we can write $W_{s(\omega)}(\omega) \equiv W_{s(u)}(v)$.

Replacing $W_t(v)$ for all $t \geq 0$ by $-W_t(v)$ does not change its distribution, which remains that of a Brownian motion. By independence, the joint distribution of $(s(u), \{W_t(v)\}_{t \geq 0})$ for P_b (a product measure) equals that of $(s(u), \{-W_t(v)\}_{t \geq 0})$. So the distributions of W_s and $-W_s$ for P_b are equal. Thus

$$P_b(|W_s + b| < \varepsilon) = P_b(|-W_s - b| < \varepsilon) = P_b(|W_s - b| < \varepsilon).$$

(For this reflection, see Figure 12.3B.)

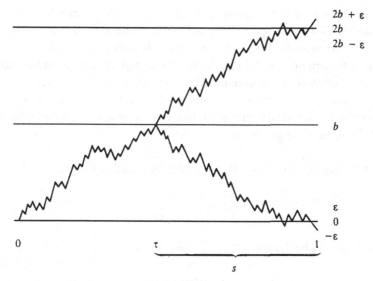

Figure 12.3B

By definition of W_t, and since $x_\tau = b$, $W_s - b = x_1 - 2b$ whenever $\tau < 1$. By sample continuity, and since $0 < \varepsilon < b$, $|x_1 - 2b| < \varepsilon$ implies $x_1 > b$, so $\tau < 1$. It follows that

$$P_b(|W_s + b| < \varepsilon) = P(|x_1 - 2b| < \varepsilon)/P(\tau < 1).$$

Thus by the last equation for $P_{b,\varepsilon}$,

$$P_{b,\varepsilon} = P(|x_1 - 2b| < \varepsilon)/P(|x_1| < \varepsilon) \to \exp(-2b^2) \qquad \text{as } \varepsilon \downarrow 0.$$

Then for any $\delta > \varepsilon > 0$,

$$Q_b \le P_{b-\varepsilon,\varepsilon} \le P_{b-\delta,\varepsilon} < \exp(-2(b-\delta)^2) + \delta$$

for ε small enough, and likewise

$$Q_b \ge P_{b+\varepsilon,\varepsilon} \ge P_{b+\delta,\varepsilon} > \exp(-2(b+\delta)^2) - \delta.$$

Letting $\delta \downarrow 0$ gives $Q_b = \exp(-2b^2)$. $\qquad \square$

We next turn to another problem: to find the distribution of the supremum of $|y_t|$, for $0 \le t \le 1$. This is more difficult. The probability that $|y_t|$ reaches b is the probability that y_t reaches b or $-b$, or twice the probability that y_t reaches b minus the probability that it reaches both b and $-b$. Now in view of Proposition 12.3.2, consider the probability that x_t reaches b, then $-b$, then $|x_1| < \varepsilon$. This equals the probability that x_t reaches b, then $3b$, then

$|x_1 - 4b| < \varepsilon$. This last inequality implies, for $\varepsilon < b$, that x_t must have reached b and $3b$, so the latter probability is just the probability that $|x_1 - 4b| < \varepsilon$. The process may also first reach $-b$, then b, then return to near 0 at 1. As we repeatedly write the probability of a union as a sum of probabilities minus the probability of an intersection, the intersection involves more and more visits back and forth between b and $-b$ before returning to 0 at 1. After reflection, these paths become paths reaching larger and larger multiples of b at 1. Eventually we get the following series:

12.3.4. Proposition *For a Brownian bridge y and any $b > 0$,*

$$P(\sup\{|y_t|: 0 \le t \le 1\} \ge b) = 2 \sum_{n \ge 1} (-1)^{n-1} \exp(-2n^2 b^2).$$

Proof. Let A_n be the event that for some t_j with $0 < t_1 < \cdots < t_n < 1$, $y(t_j, \omega) = (-1)^{j-1} b$ for $j = 1, \ldots, n$. Let $P_n := P(A_n)$, $s := s(\omega) := \inf\{t: y_t = b\}$, $s' := s'(\omega) := \inf\{t: y_t = -b\}$, where s or s' is defined as $+\infty$ if the corresponding set of t is empty. Let $Q_n := P(A_n \text{ and } s < s')$. Then $Q_n = P_n - Q_{n+1}$, since Q_{n+1} is unchanged when y is replaced by $-y$. By Proposition 12.3.3, $P_1 = \exp(-2b^2)$. By another reflection, as indicated just above, one can see that $P_2 = \exp(-8b^2)$ and, iterating the process, we have $P_n = \exp(-2n^2 b^2)$. Then $Q_n \to 0$ as $n \to \infty$ and

$$P(\sup\{|y_t|: 0 \le t \le 1\} \ge b) = 2Q_1 = 2(P_1 - P_2 + P_3 - \cdots),$$

which gives the desired sum. □

Remark. Although the sum has infinitely many terms, it is both alternating and quite rapidly converging, so that it can be computed rather easily to any desired accuracy, except for small values of b.

Next, the joint distribution of $\sup_t y_t$ and $\inf_t y_t$ will be found. If $a > 0$ and $b > 0$, let

$$\Phi(a, b) := P(-a < y_t < b \quad \text{for } 0 \le t \le 1),$$

$$\Pi(a, b) := P\left(\inf_t y_t \le -a \quad \text{and} \quad \sup_t y_t \ge b \right).$$

Then Proposition 12.3.3 implies

$$\Phi(a, b) = 1 - \exp(-2a^2) - \exp(-2b^2) + \Pi(a, b).$$

12.3.5. Proposition *For any $a > 0$ and $b > 0$,*

$$\Phi(a, b) = \sum_{m=-\infty}^{\infty} \exp(-2m^2(a + b)^2) - \exp(-2[(m + 1)a + mb]^2).$$

Proof. Let A_n be the event that for some $t_1 < t_2 < \cdots < t_n < 1$, $y(t_j) = -a$ for j odd and $y(t_j) = b$ for j even. Let B_n be defined as A_n with "odd" and "even" interchanged. Let $u := u(\omega) := \inf\{t : y_t = -a\}$ and $v := v(\omega) := \inf\{t : y_t = b\}$, with $u = +\infty$ or $v = +\infty$ respectively if the corresponding set of t is empty. Let $P_n := P(A_n)$, $Q_n := P(A_n$ and $u < v)$, $R_n := P(B_n)$, and $S_n := P(B_n$ and $v < u)$. Then $Q_n = P_n - S_{n+1}$ and $S_n = R_n - Q_{n+1}$. By the reflection method, as in the last two proofs,

$$P_{2n} = R_{2n} = \exp(-2n^2(a + b)^2),$$

$$P_{2n+1} = \exp(-2[(n + 1)a + nb]^2), \quad \text{and}$$

$$R_{2n+1} = \exp(-2[na + (n + 1)b]^2). \quad \text{Thus}$$

$$\Phi(a, b) = 1 - Q_1 - S_1 = 1 + \sum_{m \geq 1}(-1)^m (P_m + R_m)$$

$$= \sum_{m=-\infty}^{\infty} \exp(-2m^2(a + b)^2) - \sum_{m=-\infty}^{\infty} \exp(-2[(m + 1)a + mb]^2).$$

\square

Remark. Proposition 12.3.5 gives Propositions 12.3.4 (for $a = b$) and 12.3.3 (let $a \to \infty$) as corollaries.

12.3.6. Proposition *If y is a Brownian bridge, then for all $x > 0$,*

$$P(\sup y - \inf y \geq x) = 2\sum_{m \geq 1}(4m^2 x^2 - 1)\exp(-2m^2 x^2).$$

Proof. If the series for $\Phi(a, u)$ in Proposition 12.3.5 is differentiated termwise with respect to u, we get the series $h(a, u) := \sum_{m=-\infty}^{\infty} -4m^2(a + u) \cdot \exp(-2m^2(a + u)^2) + 4m[(m + 1)a + mu]\exp(-2[(m + 1)a + mu]^2)$. It will be shown that this series converges uniformly and absolutely for $a + u \geq \varepsilon, a > 0$, and $u > 0$ for any $\varepsilon > 0$. Let $v := a + u$. Then $v \cdot \exp(-2m^2 v^2)$ has a relative maximum for $v > 0$ only at $v = 1/(2|m|) < \varepsilon$ for $|m| > 1/(2\varepsilon)$, so its value for $v \geq \varepsilon$ is dominated by its value at $v = \varepsilon$. We have $\sum_m m^2 \exp(-2m^2\varepsilon^2) < \infty$. Also, $|a + mv|\exp(-2(a + mv)^2)$ is maximized with respect to $a \geq 0$ either for $a = 0$, which reduces to the case

just treated, or, by differentiation, if $m \geq 0$, when $1/2 = a + mv \geq m\varepsilon$, which is impossible for m large ($m > 1/(2\varepsilon)$). Or, for $m < 0$, we get instead $1/2 = |(m+1)v - u| \geq |m+1|\varepsilon$, which likewise cannot happen for $|m| > 1 + 1/(2\varepsilon)$. So the uniform absolute convergence holds.

Also, as $a \downarrow 0$ or $b \downarrow 0$ while the other stays constant, $\Phi(a, b) \to 0$ since

$$\Phi(b, a) = \Phi(a, b) \leq \Phi(a, +\infty) = 1 - \exp(-2a^2)$$

by Proposition 12.3.3. Thus for all $a > 0$, since the uniform absolute convergence justifies termwise integration,

$$\Phi(a, b) = \int_0^b h(a, u)\, du. \tag{12.3.7}$$

Let $\xi := \inf_{0 \leq t \leq 1} y_t$ and $\eta := \sup_{0 \leq t \leq 1} y_t$, so $\Phi(a, b) \equiv P(\xi > -a$ and $\eta < b)$. By 12.3.3, η has the density $4b \cdot \exp(-2b^2)$ on the half-line $b > 0$. Let $g(a, u) := h(a, u)/(4u \cdot \exp(-2u^2))$ for any $a > 0$ and $u > 0$. Some conditional probabilities will be useful:

For each $a > 0$, $\quad P(\xi > -a \mid \eta) = g(a, \eta)$ for almost all η. \quad (12.3.8)

To prove this, first, for a given value of a, let B be an event in the smallest σ-algebra for which η is measurable, so that for some Borel set $C \subset \mathbb{R}$, $B = \eta^{-1}(C)$. We need to prove

$$P(B \cap \{\xi > -a\}) = E(1_B g(a, \eta)) = E(1_C(\eta) g(a, \eta)) \tag{12.3.9}$$

where the latter equality is clear. The first equation in (12.3.9) holds for $C = [0, b)$ for any $b > 0$ by (12.3.7). Thus it holds for $C = [c, b)$ for $c \leq b$ by differences, and so for any finite disjoint union of such intervals by addition. Such unions form a ring, by Propositions 3.2.1 and 3.2.3. The collection of all C for which (12.3.9) holds is a monotone class and contains the complement of any set in it, so it is a σ-algebra by Proposition 3.2.5 and Theorem 4.4.2, and thus contains all Borel sets, proving (12.3.9) and thus (12.3.8) for each $a > 0$. It follows that

$$P(\xi \leq -a \mid \eta) = 1 - g(a, \eta) \quad \text{for almost all } \eta.$$

For any $x > 0$, $P(\eta - \xi \geq x \mid \eta) = 1$ a.s. if $\eta \geq x$, so

$$P(\eta - \xi \geq x) = P(\eta \geq x) + P(\eta < x \text{ and } \xi \leq \eta - x).$$

We can assume (12.3.8) holds for all $a > 0$ and $\eta > 0$ (not only almost all η), which makes it plausible for variable $a = x - \eta$; here is a proof. For each $n = 1, 2, \ldots$, decompose the interval $0 \leq \eta < x$ into n equal intervals

$I_j := [\eta_{j-1}, \eta_j), j = 1, \ldots, n$ where $\eta_j := \eta(j) := jx/n$ for $j = 0, 1, \ldots, n$.
Define the events $A_x := \{\eta < x \text{ and } \xi \leq \eta - x\}$,

$$D_n := \{\eta \in I_j \text{ and } \xi \leq \eta_j - x \text{ for some } j\}, \quad \text{and}$$

$$E_n := \{\eta \in I_j \text{ and } \xi \leq \eta_{j-1} - x \text{ for some } j\}.$$

Then clearly $E_n \subset A_x \subset D_n$ for each n. We have for $j = 1, \ldots, n$

$$P(D_n) = \sum_{j=1}^{n} P(\eta \in I_j \text{ and } \xi \leq \eta_j - x)$$

$$= \sum_{j=1}^{n} \int_{\eta(j-1)}^{\eta(j)} (1 - g(x - \eta_j, u)) 4u \cdot \exp(-2u^2) \, du, \quad \text{and}$$

$$P(E_n) = \sum_{j=1}^{n} P(\eta \in I_j \text{ and } \xi \leq \eta_{j-1} - x)$$

$$= \sum_{j=1}^{n} \int_{\eta(j-1)}^{\eta(j)} (1 - g(x - \eta_{j-1}, u)) 4u \cdot \exp(-2u^2) \, du.$$

Letting $n \to \infty$ and noting that $g(t, u)$ is bounded by 1 by (12.3.8) and continuous in t for $t > 0$ (since $h(t, u)$ is), we get

$$P(A_x) = \int_0^x 4u(1 - g(x - u, u)) \exp(-2u^2) \, du.$$

Thus using Proposition 12.3.3,

$$P(\eta - \xi \geq x) = \exp(-2x^2) + \int_0^x 4u \cdot \exp(-2u^2) - h(x - u, u) \, du$$

$$= 1 + 8 \sum_{m=1}^{\infty} m^2 x^2 e^{-2m^2 x^2} - \sum_{m=-\infty}^{\infty} m \left[e^{-2m^2 x^2} - e^{-2(m+1)^2 x^2} \right]$$

$$= 1 - 1 + \sum_{m=1}^{\infty} [8m^2 x^2 + (m - 1) - (m + 1)] \exp(-2m^2 x^2)$$

$$= \sum_{m=1}^{\infty} (8m^2 x^2 - 2) \exp(-2m^2 x^2). \qquad \square$$

Problems

1. Calculate $P(\sup_t |y_t| > b)$ for $b = 2$ and 3.

2. Calculate $P(\sup_s y_s - \inf_t y_t > x)$ for $x = 2$ and 4.

3. Let C be the usual unit circle in the plane. On C let P be the normalized arc length measure $d\theta/(2\pi)$. Let \mathcal{A} be the set of all sub-arcs of C. Show that there exists a Gaussian process G with $T = \mathcal{A}$ having mean 0 such that

$$EG(A)G(B) = P(A \cap B) - P(A)P(B)$$

 for any two arcs A and B. Show that the laws of G are preserved by rotations of C. Show that G can be defined from a Brownian bridge $\{y_t\}_{0 \le t \le 1}$ where for each arc A of the form $\{\theta : \phi < \theta \le \eta\}$ for $0 \le \phi \le \eta \le 2\pi$ we have $G(A) = y_{\eta/2\pi} - y_{\phi/2\pi}$, or for the complementary arc A^c, $G(A^c) = -G(A)$. Metrizing \mathcal{A} by $d(A, B) := P(A \triangle B)$ where \triangle is the symmetric difference, show that G can be taken to be sample-continuous. Show that then $\sup_{A \in \mathcal{A}} G(A)$ has the law of $(\sup - \inf)y$ given in Proposition 12.3.6.

4. For each real $v \ne 0$ find the distribution function and density function of the random variable (hitting time) $h_v = \inf\{t : x_t = v\}$ for a Brownian motion x_t. (Show that h_v is finite a.s.) Show that the density $f(t)$ is asymptotic to c/t^p as $t \to \infty$ for some constants c and p and evaluate c and p. *Hint:* Use 12.3.1 and differentiate with respect to b.

5. For the Brownian motion x and Brownian bridge y, show that the process $\{x_t\}_{t \ge 0}$ has the same law as $(1+t)y_{g(t)}$ where $g(t) := t/(1+t)$.

6. For a Brownian process $x : x_t(\omega) := x(t)(\omega)$ and constants a and b, find $V(a, b) := P(x_t \ge at + b \text{ for some } t \ge 0)$. *Hints:* Clearly $V(a, b) = 1$ for $b \le 0$. Let $b > 0$. To show $V(a, b) = 1$ for $a < 0$, consider large t. For $a = 0$, use Problem 4. So take $a > 0$ and $b > 0$. For $c > 0$, note that $\{x(c^2t)/c\}_{t \ge 0}$ has the same law as $\{x_t\}_{t \ge 0}$. Thus $V(a, b)$ is a function of ab. In case $a = b$, apply Problem 5 and a fact from this section. Thus find $V(a, b) \equiv e^{-2ab}$.

7. Show in detail that Proposition 12.3.3 follows from Proposition 12.3.5.

8. Let $\rho \le \tau$ be two bounded Markov times for a Brownian motion (x_t, \mathcal{B}_t), so that $\tau \le M < \infty$ a.s. for some constant M. Show that $Ex_\tau^2 < \infty$ and $E(x_\tau \mid \mathcal{B}_{\rho+}) = x_\rho$ a.s. *Hint:* See Theorem 10.4.1. Approximate ρ and τ from above by discrete-valued stopping times as in the proof of Lemma 12.2.9.

9. Building on the result of Problem 8, let $c_n < \infty$ and let $\tau(n)$ be Markov times for a Brownian motion (x_t, \mathcal{B}_t), with $\tau(n) \le c_n$ and $\tau(n) \le \tau(n+1)$ for all n and ω.
 (a) Show that $\{x_{\tau(n)}, \mathcal{B}_{\tau(n)+}\}_{n \ge 1}$ is a martingale.
 (b) Show that $E(x_{\tau(n+1)} - x_{\tau(n)})x_{\tau(n)} = 0$ for all n.

10. Show that $\xi := \inf_{0 \le t \le 1} y_t$ and $\eta := \sup_{0 \le t \le 1} y_t$ have a joint density $f(v, u)$, so that for any Borel set $B \subset \mathbb{R}^2$, $P((\xi, \eta) \in B) = \int \int_B f(v, u) \, dv \, du$. *Hint:* Show that for any $\delta > 0$, the series in Proposition 12.3.5 and its termwise partial derivatives $\partial/\partial a$, $\partial/\partial b$ and $\partial^2/\partial a \partial b$ converge absolutely and uniformly for $a + b \ge \delta > 0$, $a > 0$ and $b > 0$.

11. For a Brownian motion x_t, let $X := \inf_{0 \le t \le 1} x_t$ and $Y := \sup_{0 \le t \le 1} x_t$. For $a > 0$ and $b > 0$ evaluate $P(X \ge -a \text{ and } Y \le b)$. *Hint:* This is similar to Proposition 12.3.5 except that the process does not return to 0 at time 1.

12. For a fixed $v \ne 0$ and the hitting time h_v defined in Problem 4, let $u(t, \omega) := x_t(\omega)$ for $t \ne h_v$ and $u(h_v, \omega) := 0$. Show that the process u has the same finite-dimensional joint distributions as Brownian motion x_t but u is not sample-continuous.

12.4. Laws of Brownian Motion at Markov Times: Skorohod Imbedding

The main fact in this section, Theorem 12.4.2, will show that for any random variable X with mean 0 and finite variance, there is a Markov time τ for Brownian motion such that $\mathcal{L}(X) = \mathcal{L}(x_\tau)$. This fact, and its extension to sequences of partial sums of i.i.d. variables (Theorem 12.4.5), will then be useful in the next section, on the almost sure behavior of the partial sums as $n \to \infty$.

The equations $Ex_t = 0$ and $Ex_t^2 = t$ will be extended to $Ex_\tau = 0$ and $Ex_\tau^2 = E\tau$ for any Markov time τ with $E\tau < \infty$. Imagine a player observing a Brownian motion who can stop at any Markov time τ and then is paid x_τ (or has to pay $-x_\tau$ if $x_\tau < 0$). If τ has finite expectation, then the average gain will be 0, as with bounded stopping times for martingales (Theorem 10.4.1 and the remarks after it). On the other hand let $\tau := \inf\{t: x_t = 1\}$. Then $\tau < \infty$ a.s. by 12.3.1 as $b \to \infty$ but $Ex_\tau = 1$, and so $E\tau = +\infty$, as will follow from the next theorem. (The player can make money only by waiting, on average, an extremely long time.)

12.4.1. Theorem *Let $(x_t, \mathcal{A}_t)_{t \ge 0}$ be a Brownian motion and τ a Markov time for it with $E\tau < \infty$. Then $Ex_\tau = 0$ and $Ex_\tau^2 = E\tau$.*

Proof. We have that x_τ is measurable as shown after Lemma 12.2.9. First suppose τ is a stopping time and has only finitely many values $0 \le t_1 < \cdots < t_n < \infty$. For $n = 1$ the results are clear. Let $t(j) := t_j$. It follows from the definition of Brownian motion $(x_t, \mathcal{A}_t)_{t \ge 0}$ that $(x_{t(j)}, \mathcal{A}_{t(j)})_{1 \le j \le n}$ is a martingale

sequence: for $0 \le s < t$, $x_t - x_s$ is independent of x_s, so $E(x_t - x_s \mid x_s) = 0$. Thus we have by optional stopping (Theorem 10.4.1) that $Ex_\tau = Ex_{t(n)} = 0$. For Ex_τ^2 induction on n will be used. Let $\alpha := \min(\tau, t_{n-1})$, a stopping time with $n - 1$ values. Then

$$Ex_\tau^2 = Ex_\alpha^2 - Ex_{t_{n-1}}^2 1_{\tau=t_n} + E\left[x_{t_{n-1}} + \left(x_{t_n} - x_{t_{n-1}}\right)\right]^2 1_{\tau=t_n}.$$

Now $\{\tau = t_n\} = \{\tau > t_{n-1}\} \in \mathcal{A}_{t(n-1)}$, and $x_{t(n)} - x_{t(n-1)}$ has mean 0 and is independent of $\mathcal{A}_{t(n-1)}$ by definition of Brownian motion, again, so it is independent of, thus orthogonal to, the random variable $x_{t(n-1)} 1_{\tau=t(n)}$. So, by the induction hypothesis on α,

$$Ex_\tau^2 = E\alpha + E\left(x_{t(n)} - x_{t(n-1)}\right)^2 1_{\tau=t(n)}$$
$$= E\alpha + (t_n - t_{n-1})P(\tau = t_n) = E\tau.$$

Next suppose τ is bounded, $\tau < M < \infty$ a.s. Then there are simple functions $\tau_m \downarrow \tau$ with $\tau_1 \le M$, as in the proof of Theorem 12.2.7 after Lemma 12.2.9, and these τ_m are stopping times (as they may not be if $\tau_m \uparrow \tau$). Let $\tau(m) := \tau_m$. As $m \to \infty$, $x_{\tau(m)} \to x_\tau$ a.s. by sample continuity. As just shown, $Ex_{\tau(m)}^2 = E\tau_m$, which converges to $E\tau$ by monotone convergence (or dominated convergence).

For each m and each of the finitely many possible values s of τ_m, on $\{\tau_m = s\}$, which is in $\mathcal{A}_{\tau(m)+}$, $x_M - x_{\tau(m)}$ is conditionally independent of $\mathcal{A}_{\tau(m)+}$ by the strong Markov property (Theorem 12.2.7) and so has conditional distribution $N(0, M - s)$. Thus $E(x_M \mid \mathcal{A}_{\tau(m)+}) = x_{\tau(m)}$.

Then by conditional Jensen's inequality (10.2.7), for each m, $x_{\tau(m)}^2 \le E(x_M^2 \mid \mathcal{A}_{\tau(m)+})$. It follows that the $x_{\tau(m)}^2$ are uniformly integrable (as in the proof of Theorem 10.4.3). Thus $E(x_{\tau(m)}^2 - x_\tau^2) \to 0$ by Theorem 10.3.6, so $Ex_\tau^2 = E\tau$. Also, $Ex_\tau = \lim_{m\to\infty} Ex_{\tau(m)} = 0$.

If γ is another Markov time with $\gamma \le \tau \le M < \infty$, then there are also simple stopping times $\gamma(m) := \gamma_m \downarrow \gamma$ with $\gamma_m \le \tau_m$. Since $x_{\gamma(m)}^2$ and $x_{\tau(m)}^2$ are uniformly integrable as just shown, so are the $x_{\gamma(m)}x_{\tau(m)}$ by the Cauchy-Bunyakovsky-Schwarz inequality. We have $E(x_{\gamma(m)} - x_{\tau(m)})x_{\gamma(m)} = 0$ for all m since for m fixed, $\{x_{\gamma(m)}, x_{\tau(m)}\}$ is a martingale by optional stopping (10.4.1), and then we can apply Theorem 10.1.9. So letting $m \to \infty$ gives $E(x_\tau - x_\gamma)x_\gamma = 0$.

Now take any Markov time β with $E\beta < \infty$. Let $\beta(n) := \min(\beta, n)$. Then the $\beta(n)$ are Markov times, $\beta(n) \uparrow \beta$, and $Ex_{\beta(n)}^2 = E\beta(n) \uparrow E\beta$. Since $x_{\beta(n)} \to x_\beta$, Fatou's lemma (4.3.3) implies $Ex_\beta^2 \le E\beta$. Also, for $m \le n$, we have $E(x_{\beta(n)} - x_{\beta(m)})x_{\beta(m)} = 0$ from the previous part of the proof. Thus

$$E\left(\left(x_{\beta(n)} - x_{\beta(m)}\right)^2\right) = E(\beta(n) - \beta(m)),$$

which goes to 0 uniformly in $n \geq m$ as $m \to \infty$. Again by Fatou's lemma (4.3.3), $E((x_\beta - x_{\beta(m)})^2) \leq E\beta - E\beta(m) \to 0$ as $m \to \infty$, so

$$Ex_\beta = 0, \quad Ex_\beta^2 = \lim_{m \to \infty} Ex_{\beta(m)}^2 = \lim_{m \to \infty} E(\beta(m)) = E\beta. \qquad \square$$

The possible laws of $X = x_\tau$ where τ is a Markov time with $E\tau < \infty$ were just proved to satisfy $EX = 0$ and $EX^2 < \infty$. The next theorem will show that these are the *only* restrictions on the law of X.

12.4.2. Theorem (Skorohod Imbedding) Let $(x_t, \mathcal{A}_t)_{t \geq 0}$ be a Brownian motion. Let μ be any law on \mathbb{R} with $\int x \, d\mu = 0$ and $\int x^2 \, d\mu < \infty$. Then there is a Markov time τ for $\{\mathcal{A}_t\}_{\tau \geq 0}$ with $\mathcal{L}(x_\tau) = \mu$ and $E\tau < +\infty$, so $E\tau = \int x^2 \, d\mu$.

Remarks. (i) We can take \mathcal{A}_t to be the minimal σ-algebras \mathcal{F}_t of Proposition 12.2.4. (ii) If μ is concentrated in the two points -1 and 1, for example, then the condition $\int x \, d\mu = 0$ implies $\mu\{-1\} = \mu\{1\} = 1/2$. Then τ can be the least t such that $x_t = \pm 1$. Other examples where μ has two-point support will be treated in Case I of the proof.

Proof. Let X be a random variable with law μ, so $EX = 0$ and $EX^2 < \infty$. The conclusions $E\tau = Ex_\tau^2 = EX^2$ follow from the others by Theorem 12.4.1.

 Case I. Suppose μ is concentrated in two points $-a$ and b, where $a > 0$ and $b > 0$, so that $\mu\{-a\} = b/(a + b)$ and $\mu\{b\} = a/(a + b)$. Let $\tau := \inf\{t : x_t = -a \text{ or } b\}$. Since $x_n - x_{n-1}$ are independent for $n = 1, 2, \ldots$, with distribution $N(0, 1)$, there is a $\gamma > 0$ such that for all n,

$$P(|x_j - x_{j-1}| \leq a + b \text{ for } j = 1, \ldots, n\}) \leq (1 - \gamma)^n.$$

If $|x_{j+1} - x_j| > a + b$, then either one of x_j and x_{j+1} is less than $-a$, or one is larger than b. Thus $\tau < \infty$ a.s., and since $(1 - \gamma)^n \to 0$ geometrically, $E\tau < \infty$. τ is a stopping time since it is the minimum of the two hitting times h_{-a}, h_b. Hitting times were shown to be stopping times in the proof of 12.3.1. By Theorem 12.2.5, τ is also a Markov time. Since x_τ has only the two values $-a$ and b, and $Ex_\tau = 0$ by Theorem 12.4.1, its distribution is uniquely determined, so $\mathcal{L}(X_\tau) = \mathcal{L}(X)$, as desired.

 Case II. X is simple, that is, it has just finitely many possible values ($\mu(F) = 1$ for some finite set F).

12.4.3. Lemma *For any simple random variable X with $EX = 0$ there is a probability space (Ω, \mathcal{B}, P), an $n < \infty$, a martingale $\{X_j, \mathcal{B}_j\}_{0 \le j \le n}$ with $X_0 \equiv 0$, $\mathcal{L}(X_n) = \mathcal{L}(X)$, and such that for each $j = 1, \dots, n-1$ and each value of X_j, X_{j+1} has at most two possible values.*

Proof. The probability space will be taken as $(\mathbb{R}, \mathcal{B}, \mu)$ with Borel σ-algebra \mathcal{B} where $\mu = \mathcal{L}(X)$. Let \mathcal{B}_0 be the trivial algebra $\{\varnothing, \mathbb{R}\}$. Let \mathcal{B}_1 be the algebra generated by $(-\infty, 0]$. Then \mathcal{B}_j will be an increasing sequence of finite algebras generated by left-open, right-closed intervals, where if A is an interval which is an atom of \mathcal{B}_j, and A is finite, $A = (a, b]$, then $(a, (a+b)/2]$ and $((a+b)/2, b]$ will be atoms of \mathcal{B}_{j+1}. For each $j \ge 1$, $(-\infty, -j], (-j, 1-j], (j-1, j]$, and (j, ∞) will be atoms of \mathcal{B}_{j+1}. Thus each \mathcal{B}_j will have 2^j atoms, each divided in two to form atoms of \mathcal{B}_{j+1}. Since X is simple, having only finitely many values, for j large enough, say $j = n$, each atom of \mathcal{B}_j will contain at most one value of X. Then let X_n be the identity on \mathbb{R}. X_n is equal a.s. to a \mathcal{B}_n-measurable function. Let $X_j := E(X_n \mid \mathcal{B}_j)$ for $j = 0, 1, \dots, n-1$. Then $\{X_j, \mathcal{B}_j\}_{0 \le j \le n}$ is a martingale with the desired properties. $\qquad\square$

Now to continue the proof of Theorem 12.4.2, if X_1 has only one possible value, it must be 0 and $X = 0$ a.s., so let $\tau = 0$. Otherwise X_1 has two possible values $-a$ and b with $a > 0$ and $b > 0$. Let τ_1 be the least t such that $x_t = -a$ or b. By Case I, $E\tau_1 < \infty$ and x at time τ_1 has the law of X_1. Let $x(t) := x_t$.

Inductively, suppose given Markov times ρ_j for the original Brownian motion such that $E\rho_j < \infty$ and $Y_j := x(\rho_j)$ has the law of X_j with $\tau_1 = \rho_1 \le \rho_2 \le \cdots \le \rho_j$.

Let $z_t := x(t + \rho_j) - Y_j$. Let $\rho(j) := \rho_j$ and $\mathcal{C}_0 := \mathcal{A}_{\rho(j)+}$. For $t > 0$ let \mathcal{C}_t be the smallest σ-algebra including \mathcal{C}_0 and making z_s measurable for $0 < s \le t$.

For each value z of Y_j, which is a value of X_j, there are at most two possible values of X_{j+1}, say $c = c(z)$, $d = d(z)$ with $c = d = z$ or $c < z < d$. Here $c(\cdot)$ and $d(\cdot)$ are (simple) random variables measurable for \mathcal{C}_0. Let $z := Y_j$. Let ζ be a stopping time for $\{\mathcal{C}_t\}_{t \ge 0}$, defined as the least t such that $z_t = c(z) - z$ or $d(z) - z$. By Case I, conditional on each value z of Y_j, $E\zeta < \infty$ and $E(z_\zeta \mid \mathcal{A}_{\rho(j)+}) = 0$. Let $\rho_{j+1} := \rho_j + \zeta$. This is a Markov time for $\{\mathcal{A}_t\}_{t \ge 0}$ by the following fact:

12.4.4. Lemma *Let ρ be a Markov time for a Brownian motion $\{x_t, \mathcal{A}_t\}_{t \ge 0}$ with $\rho < \infty$ a.s. Let $z_t := x_{\rho+t} - x_\rho$, $\mathcal{C}_0 := \mathcal{A}_{\rho+}$ and let \mathcal{C}_t be the smallest σ-algebra including \mathcal{C}_0 for which z_s are measurable for $0 \le s \le t$. Then*

$\{z_t, C_t\}_{t\geq 0}$ *is a Brownian motion. Let ζ be a Markov time for $\{C_t\}_{t\geq 0}$. Then $\rho + \zeta$ is a Markov time for $\{A_t\}_{t\geq 0}$.*

Proof. To show that $\{z_t, C_t\}_{t\geq 0}$ is a Brownian motion, z_t is C_t measurable for each t by definition. Next, $\{z_u\}_{u\geq 0}$ is a Brownian process independent of C_0 by the strong Markov property 12.2.7. For $t \geq 0$ let \mathcal{F}_t be the smallest σ-algebra for which z_s are all measurable for $0 \leq s \leq t$, and \mathcal{G}_t the smallest σ-algebra for which $z_u - z_t$ are measurable for all $u \geq t$. Then since C_0 is independent of the σ-algebra generated by \mathcal{F}_{t+} and \mathcal{G}_t, while \mathcal{F}_{t+} and \mathcal{G}_t are independent by Proposition 12.2.4, the three σ-algebras C_0, \mathcal{F}_{t+}, and \mathcal{G}_t are jointly independent, $P(A \cap B \cap C) = P(A)P(B)P(C)$ for any $A \in C_0$, $B \in \mathcal{F}_{t+}$ and $C \in \mathcal{G}_t$. Now C_{t+} is generated by C_0 and \mathcal{F}_{t+}. It will be shown that (not surprisingly) \mathcal{G}_t is independent of C_{t+}. For any two algebras \mathcal{A} and \mathcal{B}, the class of all sets $A \cap B$ for $A \in \mathcal{A}$ and $B \in \mathcal{B}$ is a semiring since for any $A, C \in \mathcal{A}$ and B, $D \in \mathcal{B}$,

$$(A \cap B)\backslash(C \cap D) = [(A\backslash C) \cap B] \cup [A \cap C \cap (B\backslash D)].$$

Thus the set \mathcal{D} of all finite disjoint unions $\bigcup_{i=1}^{n} A_i \cap B_i$ for $A_i \in C_0$ and $B_i \in \mathcal{F}_{t+}$ is an algebra by Proposition 3.2.3. The collection of sets $H \in C_{t+}$ such that $P(H \cap C) = P(H)P(C)$ for all $C \in \mathcal{G}_t$ is a monotone class including \mathcal{D} and thus is all of C_{t+} by Theorem 4.4.2. So $\{z_t, C_t\}_{t\geq 0}$ is a Brownian motion.

Now note that for any fixed $r > 0$, $\rho + r$ is a Markov time, since for each $t > 0$, $\{\rho + r < t\} = \{\rho < t - r\} \in \mathcal{A}_v \subset \mathcal{A}_t$ where $v := \max(0, t - r)$. Next, it will be shown that $C_r \subset \mathcal{A}_{(\rho+r)+}$. Let $B \in C_r$. To show that for any $t > 0$, $B \cap \{\rho + r < t\} \in \mathcal{A}_t$, it will be enough by Lemma 12.2.9 to consider B of the form $\{z_s \in C\} = \{x_{\rho+s} - x_\rho \in C\}$ for Borel sets $C \subset \mathbb{R}$ and $0 \leq s \leq r$. If $t \leq r$, we get the empty set, so we can assume $u := t - r > 0$. Now, it will be shown that there exist stopping times $\rho(n)$, each with only rational values, such that (a) $\rho(n) \downarrow \rho$ as $n \to \infty$, and (b) $\rho < u$ implies $\rho(n) < u$ for all n. For (a), apply the proof of Theorem 12.2.7 after Lemma 12.2.9 to get stopping times $\zeta(n) \downarrow \rho$ with rational values. Then let $u_k \uparrow u$ be rational with $u_0 := 0 < u_1 < \cdots < u_n < \cdots$. Let $\xi := u_k$ for $u_{k-1} \leq \rho < u_k$ for each k and $\xi := \zeta(1)$ for $\rho \geq u$. Then ξ is a stopping time since $\zeta(1) > \rho$ and $\rho(n) := \min(\xi, \zeta(n))$ is a stopping time satisfying (a) and (b).

For each n, $x_{\rho(n)+s} - x_{\rho(n)}$ is measurable, and its restriction to $\{\rho + r < t\}$ is measurable for \mathcal{A}_t. Since a limit of measurable functions is measurable (Theorem 4.2.2), $z_s = x_{\rho+s} - x_\rho$ restricted to $\{\rho + r < t\}$ is measurable for \mathcal{A}_t, and $C_r \subset \mathcal{A}_{(\rho+r)+}$, as desired.

Take any $t > 0$. For each rational $r > 0$, $\{\zeta < r\} \in C_r \subset \mathcal{A}_{(\rho+r)+}$. Let $A_r := \{\zeta < r\} \cap \{\rho < t - r\}$. Then by definition of $\mathcal{A}_{(\rho+r)+}$, $A_r \in \mathcal{A}_t$. The

event $\{\rho + \zeta < t\} = \{\zeta < t - \rho\}$ is the union over all rational $r > 0$ of $\{\zeta < r < t - \rho\} = A_r$, so $\{\rho + \zeta < t\} \in \mathcal{A}_t$, proving Lemma 12.4.4. $\quad\square$

Now to return to the proof of Theorem 12.4.2, $Y_{j+1} = x(\rho_{j+1})$ has the law of X_{j+1}, with $E\rho_{j+1} < \infty$, so the induction can continue. Then for $j = n$, we get a Markov time $\tau = \rho_n$ such that $E\tau < \infty$ and $\mathcal{L}(X_\tau) = \mathcal{L}(X)$ as desired, finishing the proof for Case II.

Now for the general case, where X is not necessarily simple, recall the σ-algebras \mathcal{B}_j in \mathbb{R} as in Lemma 12.4.3. Let X_∞ be the identity on \mathbb{R}, with the law $\mu = \mathcal{L}(X)$ on the Borel σ-algebra \mathcal{B}_∞. Let $X_j := E(X_\infty \mid \mathcal{B}_j)$ for $j = 0, 1, 2, \ldots$. Then $(X_j, \mathcal{B}_j)_{0 \le j \le \infty}$ is a right-closed martingale, so $X_j \to X_\infty$ a.s. as $j \to \infty$ by Theorem 10.5.1. Define the $\rho_j := \rho(j)$ and Y_j also as in Case II. Then for each $j = 0, 1, \ldots, Y_{j+1} - Y_j = z_\zeta$, $E(z_\zeta \mid \mathcal{A}_{\rho(j)+}) = 0$ as shown in Case II, and Y_j is measurable for $\mathcal{A}_{\rho(j)+}$ by Theorem 12.2.7. It follows that $E(Y_{j+1} \mid \mathcal{A}_{\rho(j)+}) = Y_j$ and $(Y_j, \mathcal{A}_{\rho(j)+})_{0 \le j < \infty}$ is a martingale by Proposition 10.3.2. Thus for $1 \le j < k$, $E((Y_k - Y_j)Y_j \mid \mathcal{A}_{\rho(j)+}) = Y_j^2 - Y_j^2 = 0$ by Theorem 10.1.9. So $E(Y_j(Y_k - Y_j)) = 0$, and

$$EY_k^2 = E((Y_j + Y_k - Y_j)^2) = EY_j^2 + E((Y_k - Y_j)^2) \ge EY_j^2.$$

Also, $EY_k^2 = EX_k^2 = E\rho_k$ and $EX_k^2 \le EX^2$ for all k by the conditional Jensen inequality (10.2.7) for $f(x) := x^2$. So the nondecreasing sequence of Markov times ρ_j has a finite limit τ a.s., with $E\tau \le EX^2$, where for any $t > 0$, $\{\tau < t\} = \bigcup\{\bigcap_{n \ge 1}\{\rho_n < q\}: q \in \mathbb{Q}, q < t\} \in \mathcal{A}_t$, so τ is a Markov time. The Y_k converge a.s. to x_τ, just as the X_k converge to X_∞, so the equal laws of X_k and Y_k converge to the laws of X and of x_τ, which are thus equal, proving the Skorohod imbedding (Theorem 12.4.2). $\quad\square$

For example, let $\mu(-2) = \mu(2) = 1/4$ and $\mu(0) = 1/2$. To find a Markov time τ "as small as possible" with $\mathcal{L}(x_\tau) = \mu$, it would seem one should get the value $x_\tau = 0$ by stopping "right away," say to flip a coin; if it's heads, let $\tau = 0$, otherwise let $\tau = \sigma$, the least time t with $x_t = \pm 2$. Another way is first to reach ± 1 and then stop the next time the process reaches 0 or ± 2. Though it appears less efficient to reach 0 by first going to ± 1, in fact $E\tau = \int x^2 \, d\mu$ in both cases, since $E\tau < \infty$ (Theorem 12.4.1), see also Remark (i) after Theorem 12.4.2 and Problem 5 below.

Theorem 12.4.2 extends to sequences of partial sums of i.i.d. variables as follows. A partial sum S_n of i.i.d. variables X_i with mean 0 and finite variance will have the law of a variable $x_{T(n)}$ where $T(n)$ is a Markov time which is "asymptotically constant" in the sense that $T(n)/ET(n)$ converges

in probability to 1 as $n \to \infty$. Thus, $x_{T(n)}$, which has the same law as S_n, can be approximated by $x_{ET(n)}$, which has a normal law. The approximation will be shown in §12.5. It provides an improvement on the central limit theorem in regard to the behavior of the sequence S_1, S_2, \ldots.

12.4.5. Theorem *Under the conditions of Theorem 12.4.2, there are independent, identically distributed random variables $\tau(j) \geq 0$, $j = 1, 2, \ldots$, such that for $T(0) := 0$ and $T(n) := \tau(1) + \cdots + \tau(n)$, $n = 1, 2, \ldots$, each $T(n)$ is a Markov time, and $x_{T(j)} - x_{T(j-1)}$ are i.i.d. with law μ and $Ex_{T(j)}^2 = ET(j) = j \int x^2 \, d\mu < +\infty$ for $j = 1, 2, \ldots$.*

Proof. The $\tau(j)$ for $j \geq 1$ will be defined recursively. It will be shown using induction that for $1 \leq j \leq n$, $T(j)$ and $x_{T(j)}$ are all $\mathcal{A}_{T(n)+}$ measurable. For $j = n = 1$ we apply the Skorohod embedding 12.4.2. The Markov time $\tau(1) = T(1)$ is $\mathcal{A}_{T(1)+}$ measurable by Theorem 12.2.5(b), and $x_{T(1)}$ is $\mathcal{A}_{T(1)+}$ measurable by the strong Markov property Theorem 12.2.7.

For the recursion and induction step, given $T(n)$, let $z_u := x_{T(n)+u} - x_{T(n)}$. Then by the strong Markov property of Brownian motion (Theorem 12.2.7), z_u is a Brownian process independent of $\mathcal{A}_{T(n)+}$ and hence of $T(j)$ and $x_{T(j)}$ for $j \leq n$ by the induction hypothesis.

Let $\mathcal{F}_{z,u}$ be the smallest σ-algebra for which z_s are measurable for $0 \leq s \leq u$, as in Proposition 12.2.4. By the strong Markov property (Theorem 12.2.7), events in $\mathcal{F}_{z,u}$ are independent of events in $\mathcal{A}_{T(n)+}$, which includes $\mathcal{A}_{T(j)+}$ for $j < n$ by Lemma 12.2.9. As in the proof of Lemma 12.4.4, for each $t \geq 0$, let \mathcal{C}_t be generated by $\mathcal{C}_0 := \mathcal{A}_{T(n)+}$ and $\mathcal{F}_{z,t}$. Then $\{z_t, \mathcal{C}_t\}_{t \geq 0}$ is a Brownian motion. By the Skorohod imbedding (Theorem 12.4.2), and the first Remark after it, there is a Markov time $\tau(n + 1)$ for $\{\mathcal{F}_{z,u}\}_{u \geq 0}$, such that $\mathcal{L}(\tau(n + 1))$ is the same for all $n = 0, 1, 2, \ldots$, and $\mathcal{L}(z_{\tau(n+1)}) = \mu$, with $\tau(n + 1)$ and $z_{\tau(n+1)}$ independent of $\mathcal{A}_{T(n)+}$. Let $T(n+1) := T(n) + \tau(n+1)$. Then $T(n+1)$ is a Markov time for $(x_t, \mathcal{A}_t)_{t \geq 0}$ by Lemma 12.4.4. It is $\mathcal{A}_{T(n+1)+}$ measurable by Theorem 12.2.5(b), and $x_{T(n+1)}$ is by Theorem 12.2.7. Thus the recursive construction can continue and Theorem 12.4.5 is proved. \square

Problems

1. For any real c let $\tau := \inf\{t: x_t = c\}$ for a Brownian process x_t. Using the exact distribution of τ as found in Problem 4 of §12.3, verify that for $c \neq 0$, $E\tau = +\infty$ without using Theorem 12.4.1.

2. Let $\sigma := \inf\{t: x_t = 2 \text{ or } -1\}$. Find $E\sigma$. *Hint:* It is unnecessary to find the distribution of σ for this, but find that of x_σ.

3. Find the distribution of σ in Problem 2. *Hints:* See Problem 11 of §12.3. Also, note that for any constant c the processes cx_t and x_{c^2t} have the same distribution.

4. If x_t is a Wiener (Brownian) process, \mathcal{F}_t is the smallest σ-algebra for which x_s is measurable for $0 \le s \le t$, and $A \in \mathcal{F}_t$ for all $t > 0$, show that $P(A) = 0$ or 1. *Hints:* Use the fact that the process has independent increments, and compare the Kolmogorov 0–1 law (8.4.4).

5. Let $\mu(\{-1\}) = \mu(\{1\}) = 1/4$ and $\mu(\{0\}) = 1/2$. Consider the following stopping times: let η' be the least time such that $x_t = \pm 1/2$ and η the least $t > \eta'$ for which $x_t = -1, 0$, or 1. Let ρ be the least t for which $x_t = 1$ or $-1/3$. If $x_\rho = 1$, let $\tau = \rho$; otherwise, let τ be the least $t > \rho$ such that $x_t = 0$ or -1. Suppose that \mathcal{A}_0 contains an event A with $P(A) = 1/2$. Let $\xi = 0$ on A. On the complement of A let ξ be the least time such that $x_t = \pm 1$.

 Show that the three stopping times η, τ, and ξ satisfy Theorem 12.4.2 for μ and so have the same expectation. (In this sense, ξ has no advantage.)

6. For a Brownian motion $(x_t, \mathcal{A}_t)_{t \ge 0}$, let τ be the least t such that $x_t = 1$ or -1. Let σ be the least t such that $x_t = \pm 1/2$. Let ρ be the least $r > 0$ such that $x_r = 1/2$ or $-3/2$. Evaluate $E\tau$, $E\sigma$, and $E\rho$. Verify that $E\tau = E\sigma + E\rho$.

7. Let τ be a Markov time for some filtration $\{\mathcal{B}_t\}_{t \ge 0}$ and f a nondecreasing function from $[0, \infty]$ into itself with $f(t) \ge t$ for all t such that f is right-continuous, $f(t) = \lim_{u \downarrow t} f(u)$ for $0 \le t < \infty$. Show that $f(\tau)$ is also a Markov time for $\{\mathcal{B}_t\}_{t \ge 0}$. *Note:* f need not be 1–1.

12.5. Laws of the Iterated Logarithm

In case of i.i.d. random variables X_j with mean 0 and $EX_1^2 < \infty$, the strong law of large numbers, $S_n/n \to 0$ a.s., can be much improved: $S_n/n^\alpha \to 0$ a.s. for any $\alpha > 1/2$, as will follow from Theorem 12.5.1 below. On the other hand, from the central limit theorem, if $EX_1^2 > 0$, then S_n/\sqrt{n} does not converge to 0. On average, $|S_n|$ is of order of magnitude \sqrt{n}. If one runs along the sequence S_n, it turns out that from time to time $|S_n|$ has values of somewhat larger order of magnitude than \sqrt{n}. Actually, as will be shown, $|S_n|/(n \log \log n)^{1/2}$ is a.s. bounded but does not go to 0 a.s. The factor $(\log \log n)^{1/2}$ goes to ∞ quite slowly as $n \to \infty$. It follows that $|S_n|/(n^{1/2}(\log n)^\alpha) \to 0$ a.s. for any $\alpha > 0$ (Problem 1).

Let $u(t) := (2t \log \log t)^{1/2}$ for $t > e$. Facts such as the two theorems in this section are known as *laws of the iterated logarithm*, or *log log laws*.

12.5.1. Theorem (Hartman and Wintner) *Let* X_1, X_2, \ldots, *be independent, identically distributed random variables with* $EX_1 = 0$ *and* $EX_1^2 = 1$. *Let* $S_n := X_1 + \cdots + X_n$. *Then almost surely*

(a) $\limsup_{n \to \infty} S_n / u(n) = 1$ *and*
(b) $\liminf_{n \to \infty} S_n / u(n) = -1$.

Proof. Part (a) will imply (b), replacing X_j by $-X_j$. The following will be proved first:

12.5.2. Theorem (Khinchin) *For any sample-continuous Brownian process* x_t, *almost surely* $\limsup_{t \to \infty} x_t / u(t) = 1$.

Functions F and G are said to be *asymptotic* as $x \to \infty$, written $F \sim G$, iff $\lim_{x \to \infty} F(x)/G(x) = 1$.

12.5.3. Lemma *For any* ε *with* $0 < \varepsilon < 1$,

$$\limsup_{s \to \infty} \sup\{|x_t - x_s|: s \le t \le s(1 + \varepsilon)\}/u(s) \le 4\varepsilon^{1/2} \quad \text{a.s.}$$

Proof. Let $t_k := t(k) := (1 + \varepsilon)^k$ for $k = 1, 2, \ldots$ For $M > 0$, let $E_{k,M} := \{\omega: \sup\{|x_t - x_{t(k)}|: t_k \le t < t_{k+1}\} \ge M\}$. For each k, the process $y_h := x_{t(k)+h} - x_{t(k)}$ is a Brownian process, to which we can apply the reflection principle (12.3.1), and then the normal tail upper bound (Lemma 12.1.6(b)), giving $P(E_{k,M}) \le 2 \exp(-M^2/(2t_k\varepsilon))$.

Take any $\alpha > \varepsilon^{1/2}$ and let $M := M(k) := \alpha u(t_k)$. Then

$$P(E_{k,M(k)}) \le 2 \exp(-\alpha^2 (\log \log t_k)/\varepsilon) = 2/(k \log(1 + \varepsilon))^D,$$

where $D := \alpha^2/\varepsilon > 1$. Thus $\sum_k P(E_{k,M(k)}) < \infty$. So almost surely for k large enough, by the Borel–Cantelli lemma (8.3.4),

$$\sup\{|x_t - x_{t(k)}|: t_k \le t \le t_{k+1}\} < \alpha u(t_k). \tag{12.5.4}$$

Also for k large enough, $u(t_{k+1})/u(t_k) < 1 + \varepsilon < 2$. Now if $t_k \le s \le t \le t_{k+1}$ and (12.5.4) holds, then

$$|x_t - x_s|/u(s) \le (|x_t - x_{t(k)}| + |x_s - x_{t(k)}|)/u(t_k) \le 2\alpha.$$

If $t_k \le s \le t_{k+1} \le t \le s(1+\varepsilon)$, and (12.5.4) holds for k and for $k+1$ in place of k, then

$$|x_t - x_s|/u(s) \le \left(|x_t - x_{t_{k+1}}| + |x_{t_{k+1}} - x_{t_k}| + |x_s - x_{t_k}|\right)/u(t_k)$$
$$\le \alpha u(t_{k+1})/u(t_k) + 2\alpha < 4\alpha.$$

Letting $\alpha \downarrow \varepsilon^{1/2}$ proves the lemma. $\qquad\square$

Proof of Theorem 12.5.2. For any $t > e^e$ and $\delta > 0$ we have by Lemma 12.1.6(b)

$$P(x_t \ge (1+\delta)u(t)) = N(0,1)\left(\left[(1+\delta)(2\log\log t)^{1/2}, \infty\right]\right)$$
$$\le \exp(-(1+\delta)^2 \log\log t) = (\log t)^{-B},$$

where $B := (1+\delta)^2 > 1$. Then for the same $t_k := t(k)$ as in the last proof,

$$\limsup_{k\to\infty} x_{t(k)}/u(t_k) \le 1+\delta \quad \text{a.s.}$$

Applying Lemma 12.5.3 and letting $\varepsilon \downarrow 0$ and $\delta \downarrow 0$ give a.s.

$$\limsup_{t\to\infty} x_t/u(t) \le 1. \tag{12.5.5}$$

Now for a lower bound, let $0 < \delta < 1$. Let $\gamma := \delta/2$ and take any $T > 1$ large enough so that

$$(1-\gamma)(1-T^{-1})^{1/2} - (1+\delta)T^{-1/2} > 1-\delta. \tag{12.5.6}$$

For $j = 1, 2, \ldots$, let E_j be the event

$$E_j := \{x(T^j) - x(T^{j-1}) > (1-\gamma)u(T^j - T^{j-1})\}.$$

Then the E_j are independent and by the asymptotic statement in Lemma 12.1.6(a), with $\psi(c) := (2\pi)^{-1/2}c^{-1}\exp(-c^2/2)$,

$$P(E_j) = N(0, T^j - T^{j-1})([(1-\gamma)u(T^j - T^{j-1}), \infty))$$
$$= N(0,1)\left(\left[(1-\gamma)(2\log\log(T^j - T^{j-1}))^{1/2}, \infty\right)\right)$$
$$\sim \psi\left((1-\gamma)(2\log\log(T^j - T^{j-1}))^{1/2}\right)$$
$$= (2\pi)^{-1/2}(1-\gamma)^{-1}(2\log\log(T^j - T^{j-1}))^{-1/2}$$
$$\cdot \exp(-(1-\gamma)^2 \log\log(T^j - T^{j-1})).$$

Since $\log\log(T^j - T^{j-1}) \le \log(j \log T)$, we have for some constants C and $A := 1 - \gamma$,

$$\liminf_{j\to\infty} P(E_j)(j\log T)^A \ge C > 0.$$

Thus $\sum_j P(E_j)$ diverges. So by independence, a.s. E_j occurs for infinitely many j.

Now in (12.5.5), we can replace x_t with $-x_t$ by symmetry and hence with $|x_t|$. So almost surely for j large enough we will have $x(T^{j-1}) \geq -(1 + \delta)u(T^{j-1})$. Then if E_j occurs,

$$x(T^j) \geq (1 - \gamma)u(T^j - T^{j-1}) - (1 + \delta)u(T^{j-1}).$$

As $j \to \infty$, $\log\log(T^j - T^{j-1})$, $\log\log(T^j)$, and $\log\log(T^{j-1})$ are all asymptotic to each other. Thus

$$\lim_{j \to \infty} u(T^j - T^{j-1})/u(T^j) = (1 - T^{-1})^{1/2} \quad \text{and}$$

$$\lim_{j \to \infty} u(T^{j-1})/u(T^j) = T^{-1/2}.$$

So by choice of T (12.5.6), we have a.s. $x(T^j) \geq (1 - \delta)u(T^j)$ for infinitely many j. Letting $\delta \downarrow 0$, it follows that $\lim \sup_{t \to \infty} x_t/u(t) \geq 1$ a.s., so by (12.5.5) $\lim \sup_{t \to \infty} x_t/u(t) = 1$ a.s., proving Theorem 12.5.2. □

Proof of Theorem 12.5.1. Apply the Skorohod imbedding of sums (Theorem 12.4.5) with $\mu = \mathcal{L}(X_1)$. Then the sequence $\{x_{T(n)}\}_{n \geq 1}$ has the same distribution as $\{S_n\}_{n \geq 1}$. Since the $\tau(n)$ are i.i.d., with $E\tau(1) = 1$, we have by the strong law of large numbers that $T(n)/n \to 1$ a.s. Hence $u(T(n)) \sim u(n)$ a.s., and by Lemma 12.5.3 with $\varepsilon \downarrow 0$, a.s. $(x_{T(n)} - x_n)/u(n) \to 0$ as $n \to \infty$. So it is enough to prove $\lim \sup_{n \to \infty} x_n/u(n) = 1$ a.s. The lim sup is 1 a.s. by Theorem 12.5.2 and Lemma 12.5.3. □

Problems

1. Under the conditions of Theorem 12.5.1, and as a corollary of it, show that for any $\alpha > 0$, $\lim \sup_{n \to \infty} S_n/n^{1/2}(\log n)^\alpha = 0$ a.s.

2. Let G_n be i.i.d. variables with law $N(0, 1)$. Show that $\lim \sup_{n \to \infty} G_n/\sqrt{\log n} = \sqrt{2}$ a.s. *Hint:* Use Lemma 12.1.6 and the Borel-Cantelli lemma (8.3.4) for $\sqrt{2}$ replaced by $\sqrt{2} \pm \delta, \delta \downarrow 0$.

3. Let $t(k) := t_k := \exp(e^k)$. Find v_k such that $\lim \sup_{k \to \infty} x_{t(k)}/v_k = 1$ a.s., where x_t is a Brownian process. Is $v_k \sim u(t_k)$? *Hint:* $x_{t(k)} - x_{t(k-1)}$ are independent for $k \geq 2$, and $x_{t(k-1)}/\sqrt{t_k} \to 0$ a.s. as $k \to \infty$. Let $v_k := \sqrt{2t_k \log k}$ for $k \geq 2$ and apply Problem 2.

4. Let X_j be i.i.d. with law $N(0, 1)$. Show that $\sum_{n \geq 3} P(|S_n| > 2u(n)) = +\infty$. *Hint:* Use Lemma 12.1.6. (So, the log log law cannot be proved by direct application of the Borel-Cantelli lemma, even for normal variables.)

5. Find the smallest integer m such that for $t(m) := 10^m$ and $n = t(t(m))$, $\sqrt{\log\log n} \geq k$ for $k = 2, 5$, and 10. (Logarithms are to base e.)

6. Evaluate $\lim\sup_{n\to\infty} \sup_{n \leq t < n+1}(x_t - x_n)/\sqrt{\log n}$. *Hint:* See Problem 2 and 12.3.1.

7. Suppose that X_j are i.i.d. with $E|X_j|^\alpha = +\infty$ for some α with $0 < \alpha < 2$. Show that the log log law fails for such X_j, specifically that $\lim\sup_{n\to\infty} |S_n|/u(n) = +\infty$ a.s. *Hints:* Use Lemma 8.3.6. Show, using the Borel-Cantelli lemma, that $\lim\sup_{n\to\infty} |X_n|/u(n) = +\infty$ a.s. If X_n is large, either S_n or S_{n-1} is.

8. Prove the central limit theorem for i.i.d. real-valued random variables X_i with mean 0 and positive, finite variance, by way of the Skorohod imbedding (§12.4).

Notes

§12.1 Kolmogorov (1933a, Chap. 3, §4) proved his existence theorem (12.1.2) for stochastic processes with real values. The theorem also extends to rather general "projective limits" of probability spaces: Bochner (1955, pp. 118–120) and Frolík (1972). Recall that on arbitrary range spaces, product measures always exist (§8.2). Andersen and Jessen (1948) showed that some regularity assumption on the measure spaces is needed for existence of stochastic processes, correcting Doob (1938, pp. 90–93).

Doob (1953, p. 72) stated the existence of Gaussian processes with given nonnegative definite covariance (Theorem 12.1.3). It is essentially a corollary of existence of finite-dimensional normal laws with given nonnegative definite, symmetric covariance matrix (9.5.7) and the Kolmogorov existence theorem.

Robert Brown, a British botanist (who discovered the nuclei of cells, according to Thompson, 1959, p. 73), noticed in 1827 an irregular movement of microscopic organic or inorganic particles suspended in liquid (Brown, 1828). This movement was named "Brownian motion." Bachelier (1900) made a substantial beginning of a theory of random processes x_t with continuously varying t in general and Brownian motion in particular. Of Bachelier's work, Félix (1970) wrote: "...lack of clarity and precision, certain considerations of doubtful interest, and some errors in definition explain why, in spite of their originality, his studies exerted no real scientific influence."

Einstein (1905, 1906, 1926) explained Brownian motion in terms of the molecular theory of matter, finding that the distribution of position of a Brownian particle at time t, starting at x at time 0, was of the form $N(x, at)$ for a constant $a > 0$ depending on parameters of the particle and the liquid, not depending on events before time 0, so that the movement follows a Brownian motion in the sense of the Gaussian process defined in the text. Smoluchowski (1906) did related work, which Ulam (1957) reviewed.

Wiener (1923) proved existence of a sample-continuous Brownian process (Theorem 12.1.5), where specifically for any $\alpha < 1/2$, almost surely there is an $M(\omega)$ with $|X_s - X_t|(\omega) \leq M(\omega)|s - t|^\alpha$ for $0 \leq s \leq t \leq 1$. The simpler proof of sample continuity above is due to P. Lévy (1939, 1948).

For the Hilbert space $H = L^2([0, 1], \lambda)$, $\lambda =$ Lebesgue measure, Paley, Wiener, and Zygmund (1933) defined $L(f) := \int_0^1 f(t) dx_t$, where $f \in H$ and x_t is a Brownian motion process, first if f is of bounded variation, using integration by parts, then extending to all $f \in H$ using the fact that L is an isometry from H into $L^2(\Omega, P)$. This was apparently the first definition of a "stochastic integral" and thereby of an isonormal process (contrary to Dudley et al., 1972). Paley died in 1933 at age 26, in a skiing accident near Banff, while the paper was in press. In his brief career he published some 35 papers, several with co-authors: see Hardy (1934) and Poggendorff (1979). Another leading work is Paley and Wiener (1934), in Fourier analysis.

Wiener is known for his work not only in probability and analysis but in the field he called "cybernetics," out of which computer science, control theory, and communication theory might be said to have developed. Wiener (1953, 1956) wrote autobiographies. His work was collected: Wiener (1976–1986). Levinson (1966) is a biographical memoir, and Doob (1966) reviewed Wiener's work in probability. See also Browder, Spanier, and Gerstenhaber (eds.) (1966).

Nowadays it seems easier to obtain an isonormal process first and then to define the Brownian motion process from it. Kahane (1976, p. 558) attributes this approach to Kakutani (1944) but I could not find it there.

Itô (1944) substantially extended the stochastic integral by allowing suitable random integrands; see also McKean (1969).

Segal (1954, 1956, etc.) treated the isonormal process for a general abstract Hilbert space H, and from a slightly different point of view (finitely additive measures on the dual Hilbert space).

The diffusion of heat in k space dimensions is, as first shown by Fourier (1807; see Grattan-Guinness, 1972, pp. 109–111), subject (approximately) to the partial differential equation (heat equation)

$$c \frac{\partial f}{\partial t} = \sum_{j=1}^{k} \frac{\partial^2 f}{\partial x_j^2},$$

where c is a constant depending on the properties of the (homogeneous) medium of heat conduction. (On Fourier, see also Herivel (1975).) Laplace (1809) showed that a solution for $t > 0$ with a given initial value $g(x)$ when $t = 0$ is given for $c = k = 1$ by

$$f(x, t) = \frac{1}{\sqrt{\pi}} \int_{-\infty}^{\infty} g\left(x + 2zt^{1/2}\right) e^{-z^2} dz = \frac{1}{2\sqrt{\pi t}} \int_{-\infty}^{\infty} g(x - u) \exp(-u^2/(4t)) du,$$

where the last expression, found by a simple change of variables, can now be recognized as the convolution of g with $N(0, 2t)$. Thus a unit of heat, starting concentrated at the point x for $t = 0$, can be viewed as diffusing to an $N(x, 2t)$ density of heat at $t > 0$. Similar formulas hold for $k > 1$ and $c \neq 1$.

§12.2 The strong Markov property for Brownian motion had been used since the late 1930s, in effect, for special stopping times such as the hitting time of a point, but apparently Hunt (1956) gave the first general, rigorous statement and proof of the theorem. Itô and McKean (1965, pp. 22, 26) give another proof.

482 Stochastic Processes

§12.3 D. André (1887) treated a so-called ballot problem: if A received α votes and B received β votes, with $\alpha > \beta$, and the votes were cast in random order, what is the probability that A was always ahead? André used a symmetry argument which can be viewed as a reflection principle for random walk; see Feller (1968, p. 72). Bachelier (1939, pp. 29–31) and Lévy (1939, p. 293) stated the reflection principle (12.3.1) for Brownian motion, although a rigorous proof based on the strong Markov property was apparently not available until the work of Hunt (1956). Bachelier (1939, p. 32) briefly indicated the technique of repeated reflections. Propositions 12.3.3 and 12.3.4 emerged from work of Kolmogorov (1933b) and Smirnov (1939). Doob (1949) explicitly stated them and Proposition 12.3.5. Smirnov lived from 1900 to 1966; selected works were published in Smirnov (1970). Kac, Kiefer, and Wolfowitz (1955) proved Proposition 12.3.6, which Kuiper (1960) rediscovered and applied to get a rotationally invariant test for uniformity on a circle.

§12.4 Skorohod (1961) found his imbedding (Theorem 12.4.2), assuming that \mathcal{A}_0 contains sets of all probabilities between 0 and 1. Root (1969) showed that the imbedding still holds even if all events in \mathcal{A}_0 have probability 0 or 1. For further results along this line see Sheu (1986) and references there. Dubins (1968) extended the imbedding to martingales z_t (where Ez_t^2 may be infinite).

§12.5 Khinchin (1923, 1924) first discovered the law of the iterated logarithm for the special case of binomial variables X_j having just two values. Kolmogorov (1929) proved a log log law for certain individually bounded, independent, not necessarily identically distributed X_n, as follows: Let $EX_n := 0$, $s_n := (E\sum_{1\le j\le n} X_j^2)^{1/2}$. Assume $|X_n(\omega)| \le a_n < \infty$ a.s. where $a_n(\log\log s_n)^{1/2}/s_n \to 0$ and $s_n \to \infty$ as $n \to \infty$. Then $\limsup_{n\to\infty} S_n/u(s_n^2) = 1$ a.s.

Kolmogorov's proof was rather difficult; Stout (1974, p. 272) writes of the "Herculean effort required." Loève (1977, pp. 266–272) gave an exposition of it, stating the following lower exponential bound: "Let $c = \max_{k\le n}|X_k|/S_n$ Given $\gamma > 0$, if $c = c(\gamma)$ is sufficiently small and $\varepsilon = \varepsilon(\gamma)$ is sufficiently large, then $P(S_n/s_n > \varepsilon) > \exp(-\varepsilon^2(1+\gamma)/2)$." Tucker (1967, p. 132) agrees. But for large enough ε, since the variables are bounded, the left side is evidently 0. (In Kolmogorov's original proof, variables were chosen in a different, correct order.) Stout (1974, p. 262) gave a proof with a corrected hypothesis for the lower exponential bound: "Let $|X_i| \le cs_n$ a.s. for each $1 \le i \le n$ and [some] $n \ge 1$... there exist constants $\varepsilon(\gamma)$ and $\pi(\gamma)$ such that if $\varepsilon \ge \varepsilon(\gamma)$ and $\varepsilon c \le \pi(\gamma)$, then. ..." (If $|X_1| \le cs_1$, then $c \ge 1$, which may make it impossible to satisfy the other hypotheses, but things work out if n is large enough and c small enough.)

Khinchin (1933) proved his log log law for Brownian motion (Theorem 12.5.2).

Hartman and Wintner (1941) proved their theorem (12.5.1) using Kolmogorov's result and a delicate truncation argument. Strassen (1964) gave, along with an extended form of the theorem, the proof using Skorohod imbedding; Breiman (1968) gave another exposition. The proof is much shorter than the original Kolmogorov-Hartman-Wintner proof, and much shorter still if one does not include in the comparison the proof (§12.4) of Skorohod imbedding, which has independent interest (although no other use of it is made in this book). There is a fairly easy proof without Skorohod imbedding if $E|X_1|^{2+\delta} < \infty$ for some $\delta > 0$ (Feller, 1943; Pinsky, 1969). Kostka (1973) shows how

that method does not apply for $\delta = 0$. A. de Acosta (1983) gave a reasonably short proof of the Hartman-Wintner theorem without the Skorohod imbedding.

Strassen (1966) proved a converse of the Hartman-Wintner log log law: if X_n are i.i.d. and $\lim \sup_{n\to\infty} |S_n|/u(n) = 1$, then $EX_1 = 0$ and $EX_1^2 = 1$. The "one-sided" converse with S_n in place of $|S_n|$ was proved independently by Martikainen (1980), Rosalsky (1980), and Pruitt (1981, Theorem 10.1). All three used results of Kesten (1970) and Klass (1976, 1977).

Bingham (1986) gives a general survey of iterated logarithm laws, which have been extended in various directions.

References

de Acosta, Alejandro (1983). A new proof of the Hartman-Wintner law of the iterated logarithm. *Ann. Probability 11*: 270–276.

Andersen, Erik Sparre, and Børge Jessen (1948). On the introduction of measures in infinite product sets. *Danske Vid. Selsk. Mat.-Fys. Medd. 25*, no. 4, 8 pp.

André, Désiré (1887). Solution directe du problème résolu par M. Bertrand. *Comptes Rendus Acad. Sci. Paris 105*: 436–437.

*Bachelier, Louis Jean Baptiste Alphonse (1900). Théorie de la spéculation. *Ann. Ecole Norm. Sup.* (Ser. 3) *17*: 21–86.

——— (1910). Mouvement d'un point ou d'un système matériel soumis à l'action de forces dépendant du hasard. *Comptes Rendus Acad. Sci. Paris 151*: 852–855.

——— (1939). *Les nouvelles méthodes du calcul des probabilités.* Gauthier-Villars, Paris.

Bingham, N. H. (1986). Variants on the law of the iterated logarithm. *Bull. London Math. Soc. 18*: 433–467.

Bochner, Salomon (1955). *Harmonic Analysis and the Theory of Probability.* University of Calif. Press, Berkeley and Los Angeles.

Breiman, Leo (1968). *Probability.* Addison-Wesley, Reading, Mass.

Browder, Felix, E. H. Spanier, and M. Gerstenhaber (eds.) (1966). *Norbert Wiener, 1894–1964.* Amer. Math. Soc., Providence, R.I. Also in *Bull Amer. Math. Soc. 72*, no. 1, Part II.

Brown, Robert (1828). A Brief Description of Microscopical Observations made in the Months of June, July and August 1827, on the Particles contained in the Pollen of Plants; and on the General Existence of Active Molecules in Organic and Inorganic Bodies, London. German transl. in *Ann. Phys. 14* (1828): 294–313.

Doob, Joseph L. (1938). Stochastic processes with an integral-valued parameter. *Trans. Amer. Math. Soc. 44*: 87–150.

——— (1949). Heuristic approach to the Kolmogorov-Smirnov theorems. *Ann. Math. Statist. 20*: 393–403.

——— (1953). *Stochastic Processes.* Wiley, New York.

——— (1966). Wiener's work in probability theory. In Browder et al. (1966), pp. 69–71.

Dubins, Lester (1968). On a theorem of Skorohod. *Ann. Math. Statist. 39*: 2094–2097.

Dudley, R. M., Jacob Feldman, and Lucien Le Cam (1972). Some remarks concerning priorities, in connection with our paper "On seminorms and probabilities." *Ann. Math. 95*: 585.

Einstein, Albert (1905). On the movement of small particles suspended in a stationary liquid demanded by the molecular-kinetic theory of heat [in German]. *Ann. Phys.* (Ser. 4) *17*: 549–560. English transl. in Einstein (1926), pp. 1–18.

——— (1906). Zur Theorie der Brownschen Bewegung. *Ann. Phys.* (Ser. 4) *19*: 371–381. Transl. in Einstein (1926).

——— (1926). *Investigations on the theory of the Brownian movement.* Ed. with notes by R. Fürth. Transl. by A. D. Cowper, E. P. Dutton, New York.

Félix, Lucienne (1970). Bachelier, Louis. *Dictionary of Scientific Biography, 1*, pp. 366–367. Scribner's, New York.

Feller, Willy (1943). The general form of the so-called law of the iterated logarithm. *Trans. Amer. Math. Soc. 54*: 373–402.

——— (1968). *An Introduction to Probability Theory and Its Applications*, 1, 3d ed. Wiley, New York.

Fourier, Joseph B. J. (1822). *Théorie analytique de la chaleur.* Gauthier-Villars, Paris.

Frolík, Zdenek (1972). Projective limits of measure spaces. *Proc. Sixth Berkeley Symp. Math. Statist. Prob. 2*: 67–80. Univ. of Calif. Press, Berkeley and Los Angeles.

Grattan-Guinness, I[vor] (1972). *Joseph Fourier, 1768–1830: A survey of his life and work, based on a critical edition of his monograph on the propagation of heat, presented to the Institut de France in 1807.* M.I.T. Press, Cambridge, Mass.

Hardy, Godfrey H. (1934). Raymond Edward Alan Christopher Paley. *J. London Math. Soc. 9*: 76–80.

Hartman, Philip, and Aurel Wintner (1941). On the law of the iterated logarithm. *Amer. J. Math. 63*: 169–176.

Herivel, John (1975). *Joseph Fourier: The man and the physicist.* Clarendon Press, Oxford.

Hunt, Gilbert A. (1956), Some theorems concerning Brownian motion. *Trans. Amer. Math. Soc. 81*: 294–319.

Itô, Kiyosi (1944). Stochastic integral. *Proc. Imp. Acad. Tokyo 20*: 519–524.

———, and Henry P. McKean, Jr. (1965). *Diffusion Processes and Their Sample Paths.* Springer, New York.

Kac, Mark, Jack Kiefer, and Joseph Wolfowitz (1955). On tests of normality and other tests of goodness of fit based on distance methods. *Ann. Math. Statist. 26*: 189–211.

Kahane, Jean-Pierre (1976). Commentary on Paley et al. (1933). In Wiener (1976), *1*, pp. 558–563.

Kakutani, Shizuo (1944). On Brownian motions in *n*-space. *Proc. Imp. Acad. Tokyo 20*: 648–652.

Kesten, Harry (1970). The limit points of a normalized random walk. *Ann. Math. Statist. 41*: 1173–1205.

Khinchin, Alexander Yakovlevich (1923). Über dyadische Brüche. *Math. Z. 18*: 109–116.

——— (1924). Über einen Satz der Wahrscheinlichkeitsrechnung. *Fund. Math. 6*: 9–20.

——— (1933). *Asymptotische Gesetze der Wahrscheinlichkeitsrechnung.* Springer, Berlin; Chelsea, New York (1948).

Klass, Michael J. (1976, 1977). Toward a universal law of the iterated logarithm. Part I, *Z. Wahrscheinlichkeitsth. verw. Geb. 36*: 165–178, Part II, ibid. *39*: 151–165.

Kolmogorov, Andrei N. (1929). Über das Gesetz des iterierten Logarithmus. *Math. Ann. 101*: 126–135.

—————— (1933a). *Grundbegriffe der Wahrscheinlichkeitsrechnung*. Springer, Berlin. Transl. as *Foundations of Probability*. Chelsea, New York (1956).

—————— (1933b). Sulla determinazione empirica di una legge di distribuzione (in Italian). *Giorn. Ist. Ital. Attuar. 4*: 83–91. Russian transl. in Kolmogorov (1986), 134–141.

—————— (1986). *Probability Theory and Mathematical Statistics* (selected works; in Russian). Moscow, Nauka.

Kostka, D. G. (1973). On Khintchine's estimate for large deviations. *Ann. Probability 1*: 509–512.

Kuiper, Nicolaas H. (1960). Tests concerning random points on a circle. *Proc. Kon. Akad. Wetensch. A (Indag. Math. 22) 63*: 38–47.

Laplace, Pierre Simon de (1809). Mémoire sur divers points d'analyse. *J. Ecole Polytechnique*, cahier 15, tome 8, pp. 229–264; *Oeuvres, XIV*, pp. 178–214, esp. pp. 184–193.

Levinson, Norman (1966). Wiener's life. In Browder et al. (1966), pp. 1–32.

Lévy, Paul (1939). Sur certains processus stochastiques homogènes. *Compositio Math. 7*: 283–339.

—————— (1948). *Processus stochastiques et mouvement brownien*. Gauthier-Villars, Paris.

Loève, Michel (1977). *Probability Theory 1*. 4th ed. Springer, New York.

Martikainen, A. I. (1980). A converse to the law of the iterated logarithm for a random walk. *Theory Probability Appl. 25*: 361–362.

McKean, Henry P. (1969). *Stochastic Integrals*. Academic Press, New York.

Paley, Raymond Edward Alan Christopher, Norbert Wiener, and Antoni Zygmund (1933). Notes on random functions. *Math. Z. 37*: 647–668. Also in Wiener (1976), *1*, pp. 536–557.

——————, and N. Wiener (1934). *Fourier Transforms in the Complex Domain. Amer. Math. Soc. Colloq. Publs. 19*.

Pinsky, Mark (1969). An elementary derivation of Khintchine's estimate for large deviations. *Proc. Amer. Math. Soc. 22*: 288–290.

J. C. Poggendorffs biographisch-literarisches Handwörterbuch 7b Teil 6 (1979). Paley, R. E. A. C. Akademie-Verlag, Berlin.

Pruitt, William E. (1981). General one-sided laws of the iterated logarithm. *Ann. Probab. 9*: 1–48.

Root, David H. (1969). The existence of certain stopping times on Brownian motion. *Ann. Math. Statist. 40*: 715–718.

Rosalsky, Andrew (1980). On the converse to the iterated logarithm law. *Sankhyā* Ser. A *42*: 103–108.

Segal, Irving Ezra (1954). Abstract probability spaces and a theorem of Kolmogoroff. *Amer. J. Math. 76*: 721–732.

—————— (1956). Tensor algebras over Hilbert spaces, I. *Trans. Amer. Math. Soc. 81*: 106–134.

Sheu, Shey Shiung (1986). Representing a distribution by stopping a Brownian motion: Root's construction. *Bull. Austral. Math. Soc. 34*: 427–431.

Skorohod, Anatolii Vladimirovich (1961). *Studies in the Theory of Random Processes* [in Russian]. Univ. of Kiev. Transl. Addison-Wesley, Reading, Mass. (1965).

Smirnov, Nikolai Vasil'evich (1939). Estimation of the deviation between empirical

distribution curves of two independent samples [in Russian]. *Bull. Univ. Moscow* 2, no. 2, pp. 3–14. Repr. in Smirnov (1970), pp. 117–127, 267.

———— (1970). *Theory of probability and mathematical statistics: Selected works* [in Russian]. Nauka, Moscow.

Smoluchowski, Marian (1906). Zur kinetischen Theorie der Brownschen Molekularbewegung und der Suspensionen. *Ann. Phys.* (Ser. 4) *21*: 756–780.

Stout, William F. (1974). *Almost Sure Convergence.* Academic Press, New York.

Strassen, Volker (1964). An invariance principle for the law of the iterated logarithm. *Z. Wahrsch. verw. Geb. 3*: 211–226.

———— (1966). A converse to the law of the iterated logarithm. *Z. Wahrsch. verw. Geb. 4*: 265–268.

Thompson, D'Arcy (1959). *Growth and Form.* Cambridge Univ. Press. (1st ed. 1917.)

Tucker, Howard G. (1967). *A Graduate Course in Probability.* Academic Press, New York.

Ulam, Stanislaw (1957). Marian Smoluchowski and the theory of probabilities in physics. *Amer. J. Phys. 25*: 475–481.

Wiener, Norbert (1923). Differential space. *J. Math. Phys. M.I.T. 2*: 131–174. Also in Wiener (1976), *1*, pp. 455–498.

———— (1953). *Ex-Prodigy: My Childhood and Youth.* Simon & Schuster, New York. Repr. M.I.T. Press, Cambridge, Mass. (1964).

———— (1956). *I Am a Mathematician.* Doubleday, New York. Repr. M.I.T. Press, Cambridge, Mass. (1964).

———— (1976–1986). *Norbert Wiener: Collected Works with Commentaries.* Ed. Pesi Masani. 4 vols. M.I.T. Press, Cambridge, Mass.

13

Measurability

Borel Isomorphism and Analytic Sets

*13.1. Borel Isomorphism

Two measurable spaces (X, \mathcal{B}) and (Y, \mathcal{C}) are called *isomorphic* iff there is a one-to-one function f from X onto Y such that f and f^{-1} are measurable. Two metric spaces (X, d) and (Y, e) will be called *Borel-isomorphic*, written $X \sim Y$, iff they are isomorphic with their σ-algebras of Borel sets.

Clearly, Borel isomorphism comes somewhere between being homeomorphic topologically and being isomorphic as sets, which means having the same cardinality. The following main fact of this section shows that in many cases, surprisingly, Borel isomorphism is just equivalent to having the same cardinality:

13.1.1. Theorem *If X and Y are two separable metric spaces which are Borel subsets of their completions, then $X \sim Y$ if and only if X and Y have the same cardinality, which moreover is either finite, countable, or c (the cardinal of the continuum, that is, of $[0, 1]$).*

Remarks. In general, the continuum hypothesis, stating that no sets have cardinality uncountable but strictly less than c, is independent of the other axioms of set theory, including the axiom of choice (see the notes to Appendix A.3). For Borel sets in complete separable metric spaces, however, the continuum hypothesis follows from the axioms, by the theorem about to be proved. Examples of the isomorphism are $\mathbb{R} \sim \mathbb{R}^2$ and $\mathbb{R} \sim \mathbb{R}\backslash\mathbb{Q}$, the space of irrational numbers.

The proof will be based on several other facts. For any metric space S, let S^∞ be a countable product of copies of S, with the product topology, which is metrized in Proposition 2.4.4. If S is complete, then S^∞ with this metric is also complete by Theorem 2.5.7.

Let "2" denote the discrete space with two points $\{0, 1\}$, so that 2^∞ will be the compact metrizable space which is the countable product of copies

of $\{0, 1\}$ (and, by the way, is homeomorphic to the Cantor set treated in Proposition 3.4.1, via $f(\{x_n\}) = \sum_n 2x_n/3^n$). As usual, let $I := [0, 1]$.

13.1.2. Lemma *There are Borel sets $B \subset 2^\infty$ and $C \subset 2^\infty$ with $B \sim I$ and $C \sim I^\infty$.*

Proof. Let B be the set of all $\{x_n\} \in 2^\infty$ such that either $x_n = 1$ for all n or $x_n = 0$ for infinitely many n. Then B has countable complement in 2^∞, so it is a Borel set. Define f by $f(\{x_n\}) = \sum_n x_n/2^n$ (binary expansion). Then f takes 2^∞ onto I, and is 1–1 from B onto I. Since the series defining f converges uniformly and the finite partial sums are continuous, f is continuous from 2^∞ onto I and thus Borel measurable. On the other hand, it is clear that for the inverse of the restriction of f to B, each digit x_n is measurable on I. Each n-tuple of digits (x_1, \ldots, x_n) is measurable from I onto a finite set 2^n. For $x \in I$ let $g_n(x) = (x_1(x), \ldots, x_n(x), 0, 0, \ldots)$, which is measurable. Then g_n converge pointwise to f^{-1}, which is thus measurable (Theorem 4.2.2), so $B \sim I$. Then $B^\infty \sim I^\infty$ and $(2^\infty)^\infty \sim 2^\infty$, so $B^\infty \sim C$ for some Borel set $C \subset 2^\infty$. \square

13.1.3. Lemma *For any complete separable metric space Y and Borel subset X there are Borel sets $A \subset B \subset I^\infty$ with $X \sim A$ and $Y \sim B$.*

Proof. By Proposition 2.4.4 and the proof of Theorem 2.8.2, Y is homeomorphic to a subset B of I^∞. Then B is (the intersection of its closure with) a countable intersection of open sets (a G_δ) by Theorem 2.5.4 and thus a Borel set. The Borel isomorphism $Y \sim B$ then gives a Borel A with $A \sim X$ (since a Borel subset of a Borel set is a Borel set, as one can see beginning with A a relatively open set, and so on). \square

On the space \mathbb{N} of nonnegative integers we have, as usual, the discrete topology.

13.1.4. Lemma *For any (non-empty) complete separable metric space (X, d) there is a continuous function f from \mathbb{N}^∞ onto X.*

Example. Let $X = [0, 1]$ with usual metric. For any integer n let $L(n)$ be the last decimal digit of n, for example $L(317) = 7$. For any sequence $\{n_i\}_{i \geq 1}$ of nonnegative integers let $f(\{n_i\}) := \sum_i L(n_i)/10^i$, the decimal expansion with $L(n_i)$ as digits. Then f is continuous from \mathbb{N}^∞ onto $[0, 1]$.

Proof. For any subset A of X let $\text{diam}(A) := \sup\{d(x, y): x, y \in A\}$, called the diameter of A. We have $X = \bigcup_{n\geq 1} A_n$ where each A_n is a non-empty closed set with $\text{diam}(A_n) \leq 1$ (the A_n are not disjoint in general). Let $A(n) := A_n$. Recursively, for $k = 1, 2, \ldots$, there are non-empty closed sets $A(n_1, \ldots, n_k)$ with diameters at most $1/k$ such that for each k and $n_j \in \mathbb{N}$, $j = 1, \ldots, k$,

$$A(n_1, \ldots, n_k) = \bigcup_{i\in\mathbb{N}} A(n_1, \ldots, n_k, i).$$

Then for each $\{n_j\} \in \mathbb{N}^\infty$, $\{F_k\}_{k\geq 1} := \{A(n_1, \ldots, n_k)\}_{k\geq 1}$ is a decreasing sequence of non-empty closed sets. Choosing any sequence $x_k \in F_k$, we have a Cauchy sequence which is in F_m for $k \geq m$ and thus converges to an $x \in \bigcap_k F_k$. This x is unique since $\text{diam}(F_k) \downarrow 0$. Let $f(\{n_j\}_{j\geq 1}) = x$.

Since \mathbb{N} has discrete topology, if a sequence $\{z_n\}$ converges in \mathbb{N}^∞ to some y, then each coordinate z_{nk} converges in \mathbb{N} and thus is eventually equal to y_k, the kth coordinate of y. Once the first m coordinates of z_n have stabilized, $f(z_n)$ thereafter moves a distance at most $1/m$. Thus $f(z_n)$ converges, so f is continuous. It is clearly onto. □

13.1.5. Theorem *For any non-empty Borel set B in a complete separable metric space X, there is a continuous function f from \mathbb{N}^∞ onto B.*

Example. Let B be the set \mathbb{Q} of rational numbers, which is not complete for any metric metrizing its usual topology (relative topology as a subset of \mathbb{R}), by Corollary 2.5.6. Let $\mathbb{Q} = \{q_n\}_{n\geq 0}$. For any $n \in \mathbb{N}^\infty$ with coordinates $n(1), n(2), \ldots$, let $f(n) := q_{n(1)}$. Then f is continuous from \mathbb{N}^∞ onto \mathbb{Q}.

Proof. Let \mathcal{C} be the collection of all Borel sets in X which are the ranges of continuous functions on \mathbb{N}^∞. Then all closed sets, being complete, are in \mathcal{C} by Lemma 13.1.4.

Let $A_n \in \mathcal{C}$ for $n = 1, 2, \ldots$. Let f_n be continuous from \mathbb{N}^∞ onto A_n for each n. For $\{n_j\}_{j\geq 1} \in \mathbb{N}^\infty$ let $f(\{n_j\}) := f_{n(1)}(\{n_{j+1}\}_{j\geq 1})$ where $n(1) := n_1$. Then f takes \mathbb{N}^∞ onto $\bigcup_n A_n$, and f is continuous since the topology on \mathbb{N} is discrete (for the n_1 coordinate) and each f_n is continuous. So $\bigcup_n A_n \in \mathcal{C}$. Open sets in metric spaces are F_σ sets, that is, countable unions of closed sets (as in the proof of Theorem 7.1.3), so all open sets are in \mathcal{C}.

Let $F := \{\{\alpha_j\}_{j\geq 1} \in (\mathbb{N}^\infty)^\infty: f_1(\alpha_1) = f_j(\alpha_j) \text{ for all } j\}$. Then F is an intersection of closed sets and hence closed. Define g on F by $g(\alpha) := f_1(\alpha_1)$. Then g is continuous and takes F onto $\bigcap_j A_j$. By Lemma 13.1.4 there is a

continuous h from \mathbb{N}^∞ onto F, so $g \circ h$ is continuous from \mathbb{N}^∞ onto $\bigcap_j A_j$, which is thus in C.

Let \mathcal{D} be the collection of sets B such that both B and $X \backslash B$ are in C. Then all open sets are in \mathcal{D}. Any countable union of sets in \mathcal{D} is in \mathcal{D}. If $B \in \mathcal{D}$, then $X \backslash B \in \mathcal{D}$. Thus \mathcal{D} is a σ-algebra, and so equals the whole Borel σ-algebra, which thus also equals C. □

Definitions. A set S in a topological space is called *dense in itself* if for each x in S, every neighborhood of x contains points of S other than x. A compact set dense in itself is called *perfect*.

13.1.6. Lemma *For any separable metric space X, there is a countable set $C \subset X$ such that $X \backslash C$ is dense in itself.*

Proof. Let C be the set of all $y \in X$ such that some open neighborhood of y in X is countable. The collection of such open neighborhoods gives an open cover of C, which has a countable subcover ("Lindelöf's theorem"; specifically, by Proposition 2.1.4, we can take the neighborhoods all in some countable base of the topology of X). So C is countable, and by definition of C its complement is dense in itself. □

13.1.7. Theorem (Alexandroff-Hausdorff) *Every uncountable Borel set B in a complete separable metric space (X, e) includes a perfect set C which is homeomorphic to 2^∞.*

Example. The Cantor set C in $[0, 1]$ is the set of all sums $x = \sum_{i \ge 1} n_i / 3^i$ where $n_i = 0$ or 2 for all i. C is perfect and is homeomorphic to 2^∞ by the correspondence of x with $\{n_i / 2\}_{i \ge 1}$.

Proof. By Theorem 13.1.5, there is a continuous function f from a complete separable metric space (S, d) onto B. For each $y \in B$, choose one $x := x_y$ in S with $f(x) = y$. Let A be the set of all x_y. Using Lemma 13.1.6, by deleting a countable set, we can assume that A is dense in itself. Take any two different points x_0 and x_1 in A. Then since f is continuous and 1–1 on A, there are disjoint closed neighborhoods F_i of x_i in A such that the closures of the ranges $f(F_i)$ are disjoint. Likewise, each F_i includes two closed sets F_{i0} and F_{i1} with non-empty interiors (all in the relative topology of A) such that the closures of the ranges $f(F_{ij})$ are all disjoint. Continuing recursively, we get closed sets $F_{i(1)i(2)...i(m)}$ for $m = 1, 2, \ldots$, where $i(k) = 0$ or 1 for each k, and for

each m, the closures of the ranges of f on these sets are all disjoint. Also, the sets are chosen with $F_{i(1)i(2)...i(m)} \subset F_{i(1)...i(m-1)}$ for all m and $i(1), \ldots, i(m)$. We can also assume that $d(x, y) < 1/m$ for all $x, y \in F_{i(1)...i(m)}$ in all cases.

Then for each $i = \{i(k)\}_{k \geq 1} \in 2^\infty$, the intersection of all the $F_{i(1)...i(m)}$ consists of a unique point in S which will be called $g(i)$. This gives a function from 2^∞ into S. Clearly g is 1–1 and continuous. By the choices made, f is 1–1 on the range of g. Thus $f \circ g$ is 1–1 and continuous from 2^∞ into B. Since 2^∞ is compact, $f \circ g$ is a homeomorphism by Theorem 2.2.11. Now 2^∞ is perfect by definition of product topology, so the range of $f \circ g$ is perfect. □

13.1.8. Lemma *If A, B, and C are Borel subsets of a complete separable metric space S with $A \subset B \subset C$ and $A \sim C$, then $A \sim B$.*

Remark. Let $A \subset B \subset C$ be any sets and suppose there is a 1–1 function from A onto C. Then there is a 1–1 function from A onto B by the equivalence theorem (1.4.1). In this sense, Lemma 13.1.8 is the analogue for Polish spaces and measurable functions of the equivalence theorem for general functions.

Proof. Let $A_0 := A$ and $D_0 := C \backslash A$. Recursively, for $n = 0, 1, \ldots$, let f_n be a Borel isomorphism of A_n onto $A_n \cup D_n$ for disjoint Borel sets A_n and D_n, so f_0 exists. Let $A_{n+1} := f_n^{-1}(A_n)$, $D_{n+1} := f_n^{-1}(D_n)$. Then $A_n = A_{n+1} \cup D_{n+1}$ where A_{n+1} and D_{n+1} are disjoint Borel sets and $A_{n+1} \sim A_n$, $D_{n+1} \sim D_n$ via f_n. Let f_{n+1} be f_n restricted to A_{n+1}, a Borel isomorphism onto $A_{n+1} \cup D_{n+1}$. So the recursion can continue. Let $E := \bigcap_n A_n$. Then E, D_0, D_1, ... are disjoint Borel sets with $A = E \cup \bigcup_{n \geq 0} D_n$.

Let $F_0 := C \backslash B$ and $G_0 := B \backslash A$. Then the decomposition $D_0 = F_0 \cup G_0$ yields a decomposition $D_n = F_n \cup G_n$ into disjoint Borel sets with $F_n \sim F_{n+1}$ and $G_n \sim G_{n+1}$ for all n.

If X_n are disjoint Borel sets (in some X) and Y_n are disjoint Borel sets (in some Y) with $X_n \sim Y_n$ for all n, then (by Lemma 4.2.4) $\bigcup_n X_n \sim \bigcup_n Y_n$. So

$$C = E \cup \bigcup_{n \geq 0} D_n = E \cup \bigcup_{n \geq 0} F_n \cup \bigcup_{n \geq 0} G_n$$

$$\sim E \cup \bigcup_{n \geq 1} F_n \cup \bigcup_{n \geq 0} G_n = C \backslash F_0 = B,$$

so $A \sim C \sim B$. □

Proof of Theorem 13.1.1. Clearly, if $X \sim Y$, then X and Y have the same cardinality. Conversely, if X is countable, then its σ-algebra of Borel sets contains all subsets, so if Y has the same cardinality, then $X \sim Y$.

Suppose X is uncountable. Then by Theorem 13.1.7, X includes a set K homeomorphic to 2^∞. On the other hand, by Lemma 13.1.3, $X \sim H$ for some Borel set $H \subset I^\infty$, so by Lemma 13.1.2, $X \sim H \sim D$ for a Borel set $D \subset 2^\infty$. So for some Borel set A, $2^\infty \sim A \subset D \subset 2^\infty$. Then $X \sim D \sim 2^\infty$ by Lemma 13.1.8. So if Y is also uncountable, $Y \sim 2^\infty \sim X$. □

Problems

1. Show that $[0, 1]$ is not homeomorphic to 2^∞. *Hint:* A topological space X is called *connected* if it is not the union of two non-empty disjoint open sets.

2. A topological space X is called *totally disconnected* iff for every two points $x \neq y$ in X there are disjoint open sets U and V with $X = U \cup V$, $x \in U$ and $y \in V$. Show that 2^∞ is totally disconnected.

3. Let X and Y be two countably infinite Hausdorff topological spaces. Show that there is a 1–1 Borel measurable function with Borel inverse from X onto Y.

4. Find a specific countably infinite subset A of $[0, 1]$ and a Borel isomorphism of $[0, 1]$ onto $[0, 1] \backslash A$. *Hint:* See Proposition 12.1.1 and its proof.

5. Give a specific 1–1 Borel function, with Borel inverse, from $[0, 1]$ onto 2^∞. *Hint:* See Problem 4 and the proof of Lemma 13.1.2.

6. Give a specific 1–1 Borel function, with Borel inverse, from $[0, 1]$ onto the square $[0, 1] \times [0, 1]$. *Hint:* Use Problem 5.

7. Prove or disprove: For every separable metric space (X, d), among the countable sets C for which $X \backslash C$ is dense in itself, there is always (a) a largest C, (b) a smallest C. *Hint:* See §1.3.

8. Prove or disprove: Let $X := (-2, -1) \cup \mathbb{N}$ as a subset of \mathbb{R} with usual topology. For every Borel set B in a complete separable metric space, there is a continuous function from X onto B. *Hint:* Consider $B = 2^\infty$.

9. Let (S, d) be a separable metric space and (T, e) a metric space. Let f be a Borel measurable function from S into T. Assuming the continuum hypothesis, prove that the range $f[S]$ is separable. *Hints:* If not, show that $f[S]$ includes an uncountable closed set A, with cardinality c, and with discrete relative topology. Thus all subsets C of A are closed and all sets $f^{-1}(C)$ are Borel in S, yielding 2^c Borel sets in S, which is impossible as in §4.2, Problem 8.

10. Let Ω be the least uncountable ordinal, that is, $(\Omega, <)$ is an uncountable well-ordered set such that for each $\alpha \in \Omega$, $\{\beta \in \Omega : \beta < \alpha\}$ is countable. On Ω, take the interval topology, with a subbase given by all open intervals $\{\gamma : \gamma < \beta\}$ and $\{\gamma : \gamma > \alpha\}$ for α and β in Ω. Show that there is no 1–1 Borel measurable function f from Ω onto $[0, 1]$. *Hint:* If f is Borel measurable from Ω onto 2^{∞}, show by a method from §7.3, Problem 2 that, for some $n = \{n_j\} \in 2^{\infty}$, $f^{-1}\{n\}$ is uncountable.

13.2. Analytic Sets

So far in this book, measurable sets in metric spaces have generally been either Borel sets or sets measurable for the completion of some measure, such as Lebesgue measurable sets in the line. The Borel σ-algebra is generated by the topology of a space, so it does not depend on the particular metric for the topology. Recall that a topological space metrizable by a metric for which it is complete and separable is called a *Polish* space. Any Cartesian product of countably many Polish spaces with product topology is Polish, by Proposition 2.4.4 and Theorem 2.5.7. Theorem 13.1.5 showed that any Borel set in a Polish space is a continuous image of \mathbb{N}^{∞}, itself a Polish space. For example, if V is an open or closed set in \mathbb{R}^k, then V is a countable union of compact sets, and so is any continuous image of it, which is then in particular a Borel set. It turns out, surprisingly, that not every continuous image of \mathbb{N}^{∞} is a Borel set. The continuous or Borel measurable images of Borel sets are described as follows. Recall the direct image $f[A]$ of a set A by a function f, defined by $f[A] := \{f(x) : x \in A\}$.

13.2.1. Theorem *Let Y be a Polish space and A a non-empty subset of Y. Then the following six conditions are equivalent:*

(a) $A = f[\mathbb{N}^{\infty}]$ *for some continuous f.*
(a′) $A = f[\mathbb{N}^{\infty}]$ *for some Borel measurable f.*
(b) $A = f[X]$ *for some Polish space X and continuous f.*
(b′) $A = f[X]$ *for some Polish space X and Borel measurable f.*
(c) $A = f[B]$ *for some Borel set B in a Polish space X and f continuous from B into Y.*
(c′) $A = f[B]$ *for some Borel set B in a Polish space X and f Borel measurable from B into Y.*

Proof. Since any continuous function f is Borel measurable (by Theorem 4.1.6), (a) implies (a′), (b) implies (b′), and (c) implies (c′).

Since \mathbb{N}^{∞} is Polish (by Proposition 2.4.4 and Theorem 2.5.7), (a) implies (b) and (a$'$) implies (b$'$). Clearly (b) implies (c) and (b$'$) implies (c$'$). So the proof will be done if it is shown that (c$'$) implies (a). In (c$'$), choose $c \in A$ and let $f(x) := c$ for all $x \in X \backslash B$. So (b$'$) holds and we need to prove (b$'$) implies (a).

Now $X \times Y$ is a Polish space. Suppose the graph H of f is Borel measurable in $X \times Y$. For example, the graph of a continuous function is always closed. Let g be the projection $g(x, y) := y$ from $X \times Y$ onto Y. Then g is continuous. By Theorem 13.1.5, $H = h[\mathbb{N}^{\infty}]$ for some continuous h. Then

$$A = (g \circ h)[\mathbb{N}^{\infty}] = g[h[\mathbb{N}^{\infty}]],$$

where $g \circ h$ is continuous, proving (a). So we just need to prove:

13.2.2. Lemma *If X and Y are Polish spaces and f is a Borel measurable function from X into Y, then the graph of f is a Borel set in $X \times Y$.*

Proof. By the Borel isomorphism theorem (13.1.1), Y is Borel-isomorphic either to 2^{∞} or to some countable subset of 2^{∞}, so we can assume $Y = 2^{\infty}$. For each $n = 1, 2, \ldots$, let

$$T(n) := \{s = \{s_j\}_{1 \le j \le n} : s_j = 0 \text{ or } 1 \text{ for each } j\}.$$

For each $s \in T(n)$ let

$$C_s := \{u = \{u_j\}_{j \ge 1} \in 2^{\infty} : u_j = s_j \text{ for } j = 1, \ldots, n\}.$$

Let $B_n := \bigcup_{s \in T(n)} f^{-1}(C_s) \times C_s$. Clearly each C_s and $f^{-1}(C_s)$ is a Borel set and $T(n)$ is finite, so B_n is a Borel set. Let G be the intersection of all the $B_n, n = 1, 2, \ldots$, so G is a Borel set. For each $y \in 2^{\infty}$ and n, $y \in C_s$ for a unique $s = s(n, y) \in T(n)$. To show that G is the graph of f, we have $(x, y) \in B_n$ for all n if and only if $f(x) \in C_{s(n,y)}$ for all n, but this means $f(x) = y$, proving Lemma 13.2.2 and so also Theorem 13.2.1. \square

A set A which either is empty or satisfies the equivalent conditions in Theorem 13.2.1 will be called an *analytic* set. Clearly, any Borel set in a Polish space is an analytic set. Examples of non-Borel analytic sets are not trivial. Before showing that such sets exist, we can just note that the direct image does not preserve some properties. If f is a continuous function and U is an open set, then $f[U]$ is not necessarily open: let $f(x) := x^2$ and $U := (-1, 1)$. If K is compact, then $f[K]$ is compact, but if F is closed, then $f[F]$ is not

necessarily closed: let $F := \{(x, y): xy = 1\}$ in \mathbb{R}^2, and let f be the projection $f(x, y) := x$.

The construction of non-Borel analytic sets will be based on some sets called universal sets. Let X and Y be sets and let S be a subset of $X \times Y$. Then for each $y \in Y$, we have a *section* of S, which is a subset of X defined by $\{x \in X: (x, y) \in S\}$. Let C be a collection of subsets of X. Then S is called a *universal set* for C iff C is the set of all sections of S. If C is a countable collection $\{C_n\}_{n \geq 0}$, then a simple universal set for it in $X \times \mathbb{N}$ is $C := \{(x, n): x \in C_n, n \in \mathbb{N}\}$. If each set C_n is open in X, then C is open in $X \times \mathbb{N}$, with \mathbb{N} having the discrete topology. But if C is an uncountable collection, an uncountable discrete space Y would not be a separable metric space. The space \mathbb{N}^∞ is convenient as a factor space in defining a universal open set since \mathbb{N}^∞ is a Polish space, but as a product of discrete spaces it has some "discrete" properties. For example, \mathbb{N}^∞ is totally disconnected: for any two distinct points m and n of \mathbb{N}^∞, the space \mathbb{N}^∞ can be written as the union of two disjoint open sets U and V with $m \in U$ and $n \in V$. With \mathbb{N}^∞ there are universal open sets, as follows:

13.2.3. Proposition *For any second-countable topological space X, there is an open set U in $X \times \mathbb{N}^\infty$ which is universal for the topology of X (the collection of all open sets in X), and a closed set F in $X \times \mathbb{N}^\infty$ which is universal for the collection of all closed sets in X.*

Proof. Let $\{U_n\}_{n \geq 1}$ be a countable base for the topology of X, with $U_0 = \emptyset$. Let U be the union of $U_n \times \{\{n_j\}_{j \geq 0}: n_k = n\}$ over all n and k. Each of these sets is open, so their union U is open. Let $n(j) := n_j$, $j = 1, 2, \ldots$. For a given point $\{n_j\}$ of \mathbb{N}^∞, the corresponding section of U is the union of all the $U_{n(j)}$ for the given sequence $n(j)$, which by definition of base gives all open sets in X. Taking F as the complement of U then gives a closed set in $X \times \mathbb{N}^\infty$ which is universal for the closed sets of X. □

There exist Polish spaces X for which there is no Borel set in $X \times \mathbb{N}^\infty$ universal for the collection of all Borel sets in X (see Problem 8 below). Even though the class of analytic sets is larger than the class of Borel sets, it turns out to be possible to define universal analytic sets (which themselves will be analytic, but not Borel):

13.2.4. Theorem *For any Polish space X, there is an analytic set A in $X \times \mathbb{N}^\infty$ which is universal for the collection of all analytic sets in X.*

Proof. Let F be a closed set in $(X \times \mathbb{N}^\infty) \times \mathbb{N}^\infty$ which is universal for the collection of closed sets in the Polish space $X \times \mathbb{N}^\infty$, by Proposition 13.2.3. Let $f(x, \{m_j\}, \{n_k\}) := (x, \{n_k\})$ from $X \times \mathbb{N}^\infty \times \mathbb{N}^\infty$ to $X \times \mathbb{N}^\infty$. Let $A :=$ $f[F]$. Then A is analytic by definition (Theorem 13.2.1). For each $n :=$ $\{n_k\}_{k \geq 1} \in \mathbb{N}^\infty$, the section $\{x \in X : (x, n) \in A\}$ equals

$$\pi[\{(x, m) \in X \times \mathbb{N}^\infty : (x, m, n) \in F\}],$$

where π is the natural projection $\pi(x, m) := x$ from $X \times \mathbb{N}^\infty$ onto X. The set of all such sections in X equals the set of all $\pi[H]$ for all closed $H \subset X \times \mathbb{N}^\infty$, since F is a universal closed set. Now, the graph of any continuous function g from \mathbb{N}^∞ into X is a closed set H whose projection into X is the range of g. So by Theorem 13.2.1, A is a universal analytic set. □

Now analytic non-Borel sets can be shown to exist. Notably, unlike the proof of existence of Lebesgue nonmeasurable sets, the following proof does not use the axiom of choice; it gives a specific, if somewhat complicated, example of an analytic non-Borel set, and a set needs to be rather complicated not to be Borel.

13.2.5. Proposition *In any uncountable Polish space X there exists an analytic set A such that*

 (a) the complement of A is not an analytic set, and
 (b) A is not a Borel set.

Proof. Since (b) follows directly from (a), it will be enough to prove (a). By the Borel isomorphism theorem (13.1.1), we can assume $X = \mathbb{N}^\infty$. By Theorem 13.2.4, take an analytic set C in $\mathbb{N}^\infty \times \mathbb{N}^\infty$ which is universal for the analytic sets of X. Let D be the diagonal $\{(n, n) : n \in \mathbb{N}^\infty\}$ in $\mathbb{N}^\infty \times \mathbb{N}^\infty$. It will be shown that $C \cap D$ is analytic. Let f be continuous from \mathbb{N}^∞ onto C by Theorem 13.2.1. Then $f^{-1}(C \cap D) = f^{-1}(D)$ is closed and hence Polish, so $C \cap D = f[f^{-1}(C \cap D)]$ is analytic by Theorem 13.2.1. Let $A := \{n \in \mathbb{N}^\infty : (n, n) \in C \cap D\}$. Now $A = \pi[C \cap D]$ where $\pi(m, n) := m$ is the natural projection, so by composition of continuous functions and Theorem 13.2.1, A is analytic in X. Suppose its complement is analytic. Then it equals some section of the universal analytic set C, and for some $m \in \mathbb{N}^\infty$, $n \notin A$ if and only if $(n, m) \in C$. For $m = n$, this gives a contradiction. □

Although analytic sets need not be Borel, it turns out that they will always be measurable for the completion of any probability measure defined on the

Borel sets. In other words, analytic sets are universally measurable, as defined in §11.5. It seems that analytic non-Borel sets are the most accessible examples of universally measurable non-Borel sets. From the definition of "universally measurable," of course, all Borel sets are universally measurable.

13.2.6. Theorem *In any Polish space X, any analytic set is universally measurable.*

Proof. Let A be analytic, so $A = f[\mathbb{N}^\infty]$ where f is continuous, by Theorem 13.2.1. Let μ be any probability measure on the Borel σ-algebra of X. For any positive integers k and M let

$$N(k, M) := \{\{n_j\}_{j \geq 1} \in \mathbb{N}^\infty : n_k \leq M\}.$$

Let $\varepsilon > 0$. Since $f[N(1, M)] \uparrow f[\mathbb{N}^\infty]$ as $M \uparrow \infty$, by continuity of outer measures from below (Theorem 3.1.11) there is an M_1 such that

$$\mu^*(f[N(1, M_1)]) \geq \mu^*(A) - \varepsilon/2.$$

Likewise, applying Theorem 3.1.11 repeatedly, we get M_k for all k such that

$$\mu(f[F_k]^-) \geq \mu^*(f[F_k]) \geq \mu^*(A) - \sum_{1 \leq j \leq k} \varepsilon/2^j > \mu^*(A) - \varepsilon,$$

where $F_k := \bigcap_{1 \leq j \leq k} N(j, M_j)$. As $k \to \infty$, $F_k \downarrow C := \bigcap_{j \geq 1} N(j, M_j)$. Each F_j is closed, and C is compact by Tychonoff's theorem (2.2.8), since the finite sets $\{1, \ldots, M_j\}$ are compact. To show that $f[F_k]^- \downarrow f[C]$, Theorem 2.2.12 will be applied. For any open $U \supset C$, since C is compact, $C \subset V \subset U$ where V is a finite union of sets in the base of the product topology in the definition of product topology (§2.2). Each set in the base depends on only finitely many coordinates. Let J be the largest index of any coordinate in the definition of the sets in the finite subcover. Since the first through Jth coordinates of points of F_J are exactly those for points of C, we have $F_J \subset U$. Thus Theorem 2.2.12 does apply, and $\mu(f[C]) \geq \mu^*(A) - \varepsilon$. Since $f[C]$ is compact, and taking $\varepsilon = 1/n$, $n = 1, 2, \ldots$, we get a countable union of compact sets, which is a Borel set B, with $B \subset A$ and $\mu(B) = \mu^*(A)$. Thus $\mu^*(A \setminus B) = 0$ and A is measurable for μ (by Proposition 3.3.2). $\quad\square$

If $A(y)$ is non-empty for each y in a set C, the axiom of choice says there is a function g on C with $g(y) \in A(y)$ for all y. If each $A(y)$ is a measurable set, and the sets $A(y)$ depend on y in a measurable way, can we take g to be measurable? Here is an answer, sometimes called a "cross section" theorem.

The measurability assumption on $A(\cdot)$ will be that $A := \{(x, y): x \in A(y)\}$ is a measurable set in a product space. This measurable form of the axiom of choice will not depend on the ordinary axiom of choice.

13.2.7. Theorem *Let X and Y be Polish spaces and let A be an analytic subset of $X \times Y$. Let C be the projection of A into Y, $C := \{y \in Y: (x, y) \in A$ for some $x \in X\}$. Then there is a function g from the analytic set C into X such that $(g(y), y) \in A$ for all $y \in C$, and such that g is measurable from the σ-algebra of universally measurable sets of Y to the Borel sets of X.*

Notes. In applications usually A is a Borel set in $X \times Y$, and then C is usually, but not necessarily, a Borel set in Y.

The axiom of choice is equaivalent to the well-ordering principle (Theorem 1.5.1). In $[0, \infty)$, every non-empty *closed* set has a least member. This weaker form of well-ordering, in \mathbb{N}^∞, will help with "measurable choice."

Proof. The lexicographical ordering on \mathbb{N}^∞ is defined by $\{m_j\}_{j\geq 1} < \{n_j\}_{n\geq 1}$ iff for some i, $m_j = n_j$ for all $j < i$ and $m_i < n_i$. Then we have:

13.2.8. Lemma *Any non-empty closed subset F of \mathbb{N}^∞ has a smallest member.*

Proof. Let n_1 be the smallest first coordinate of any point of F and given n_1, \ldots, n_{k-1}, let n_k be the smallest kth coordinate of the points of F having first $k - 1$ coordinates n_1, \ldots, n_{k-1}. For each k, there is a point $m^{(k)}$ of F having first k coordinates equal to n_1, \ldots, n_k, and these $m^{(k)}$ converge to $\{n_j\}_{j\geq 1}$, which thus must be in F and is its smallest member. \square

Now continuing with the proof of Theorem 13.2.7, by Theorem 13.2.1, let f be continuous from \mathbb{N}^∞ onto A. Let $\pi_1(x, y) := x$, $\pi_2(x, y) := y$. Then $\gamma := \pi_2 \circ f$ is continuous from \mathbb{N}^∞ onto C. So C is analytic by Theorem 13.2.1 and universally measurable by Theorem 13.2.6. For each $y \in C$, $\gamma^{-1}(\{y\})$ is closed in \mathbb{N}^∞. Let $h(y)$ be its smallest member, by Lemma 13.2.8. If h is measurable for the universally measurable σ-algebra in Y (to the Borel σ-algebra on its range), then so is $g := \pi_1 \circ f \circ h$, which then has the desired properties. So we just have to prove the measurability of h.

The set $C(n)$ of all $y \in C$ such that $\gamma(m) = y$ for some $m = \{m_j\}_{j\geq 1} \in \mathbb{N}^\infty$ with $m_1 = n$ is clearly an analytic set. Let $h_1(y)$ be the smallest n such that $y \in C(n)$. Then $h_1(y) = n$ if and only if $y \in C(n) \backslash \bigcup_{j < n} C(j)$, so h_1 is measurable for the σ-algebra $\mathcal{S}(\mathcal{A})$ generated by the analytic sets. Recursively, given h_1, \ldots, h_k, let $h_{k+1}(y)$ be the least n such that $\gamma(m) = y$ for some m

with $m_j = h_j(y)$ for $j \leq k$ and $m_{k+1} = n$. Then each h_k is measurable, and so is $y \mapsto \{h_k(y)\}_{k \geq 1}$, proving Theorem 13.2.7. □

Problems

1. Let A_n be analytic sets in a Polish space S for each $n = 1, 2, \ldots$. Show that $\bigcup_n A_n$ and $\bigcap_n A_n$ are also analytic. *Hint:* See the proof of Theorem 13.1.5.

2. A set B is called an F_σ if it is a countable union of closed sets. Let B be an F_σ in \mathbb{R}^k and f a continuous function from B into a metric space S. Show that $f[B]$ is an F_σ in S.

3. A countable intersection of open sets is called a G_δ (set). Find an example of a G_δ set B in \mathbb{R} and a continuous function f from B into \mathbb{R} such that $f[B]$ is not a G_δ.

4. Let f be a continuous function from X into Y for some Hausdorff topological spaces X and Y. Show that the graph of f is closed in $X \times Y$ with the product topology.

5. Let f be a function from a compact Hausdorff space K into another such space L such that the graph of f is closed in $K \times L$. Show that f is continuous. *Hint:* Use Theorem 2.2.11.

6. Give an example of a function f from a compact metric space K into a metric space Y such that the graph of f is closed in $K \times Y$ but f is not continuous. *Hint:* $K = $ unit circle, $Y = [0, 2\pi)$.

7. Let (X, \mathcal{B}) be a measurable space. Show that for any set A in the product σ-algebra $\mathcal{B} \otimes \mathcal{B}$ in $X \times X$, $\{x : (x, x) \in A\}$ is in \mathcal{B}. *Hint:* "Rectangles, semiring, algebra, monotone class."

8. For any measurable space (X, \mathcal{B}), show that there is never a set A in the product σ-algebra $\mathcal{B} \otimes \mathcal{B}$ in $X \times X$ which is universal for \mathcal{B}. *Hint:* See Problem 7 and the end of the proof of Proposition 13.2.5. *Note:* For $X = \mathbb{N}^\infty$ with Borel σ-algebra, it follows that "analytic" can't be replaced by "Borel" in Theorem 13.2.4.

9. Let the space \mathbb{Z} of all integers have the discrete topology. Let $\mathbb{Z}^{\mathbb{N}}$ be the set of all functions from \mathbb{N} into \mathbb{Z}, with product topology. Then $\mathbb{Z}^{\mathbb{N}}$ is a Polish space. On $\mathbb{Z}^{\mathbb{N}}$, for $n \in \mathbb{N}$, let \mathcal{F}_n be the smallest σ-algebra for which the coordinates z_0, \ldots, z_n of $z = \{z_n\}_{n \geq 0} \in \mathbb{Z}^{\mathbb{N}}$ are measurable. A *stopping time* is a function τ from $\mathbb{Z}^{\mathbb{N}}$ into $\mathbb{N} \cup \{+\infty\}$ such that $\{\tau \leq n\} \in \mathcal{F}_n$ for all n. On $\mathbb{N} \cup \{+\infty\}$, put the topology such that $\{n\}$ is open for each $n < \infty$, and each neighborhood of ∞ contains all but at most finitely many n. On the set K of all functions from $\mathbb{Z}^{\mathbb{N}}$ into $\mathbb{N} \cup \{+\infty\}$, put the product

topology. Show that (a) K is compact; (b) the set of all stopping times, as a closed subset F of K, is also compact; (c) K is not metrizable, but F is. *Hint:* Each $t \in F$ is uniquely determined by its values on sequences $\{z_n\}$ such that for some m, $z_n = z_m$ for all $n \geq m$.

10. (Continuation.) Let T be the set of all stopping times τ such that for some $f \in \mathbb{Z}^{\mathbb{N}}$, $\tau(f) = +\infty$. Show that T is an analytic set. *Note:* T is not a Borel set.

Notes

§13.1 This section was based on Parthasarathy (1967, Chapter 1, §2), who refers to Kuratowski (1933; 1966, p. 489), who in turn gives several references. That an uncountable Borel set in a complete separable metric space has a perfect subset (Theorem 13.1.7) was proved independently by Alexandroff (1916) and Hausdorff (1916). For the other facts, some references are Lusin (1930, Chapter 2; see p. 114) and Kuratowski (1934).

§13.2 Lebesgue (1905, pp. 191–192, 195–196) thought he had proved that for every Borel set in the (x, y) plane, its projection into the x axis is a Borel set in the line. Suslin (1917) found an error in the proof, invented analytic sets, and found that there exists an analytic set in the line whose complement is not analytic and so is not a Borel set. Suslin at the time was a research student of N. Lusin. Sierpiński (1950, p. 28) writes: "Par hasard j'étais présent au moment où Michel Souslin communiqua à M. Lusin sa remarque et lui donna le manuscrit de son premier travail. C'est tout sérieusement que M. Lusin a traité le jeune étudiant qui lui déclarait avoir trouvé une faute dans un Mémoire d'un savant éminent...." From Yushkevich (1968, p. 575) we learn that Suslin, born Nov. 15, 1894, died only 25 years later in 1919. He was hardly able to publish more than the 1917 note, although a short statement of a problem on linearly ordered sets (Suslin, 1920) has also kept his name alive; see, for example, Mauldin and Ulam (1987, p. 285).

Lusin (1917), in a note just after Suslin's, pointed out among other things that every analytic set in the line is Lebesgue measurable, which is a main example of the fact that analytic sets in Polish spaces are universally measurable.

Paplauscas (1973) writes of Lusin (1883–1950): "During the period 1914–1924 Luzin, a brilliant lecturer and scientific organizer, was the center of a Moscow school of [real] function theory.... Such outstanding mathematicians as P. S. Alexandrov, A. Y. Khinchin, D. E. Menshov, M. Y. Suslin, A. N. Kolmogorov, N. K. Bari, and P. S. Novikov were his pupils." Lusin (1930) wrote a book on analytic sets. Lebesgue, in a preface, wrote that "l'origine de tous les problèmes dont il va s'agir ici est une grossière erreur de mon Mémoire.... Fructueuse erreur, que je fus bien inspiré de la commettre!"

Calling an analytic set an A set and its complement a CA set, if one projects a CA set in \mathbb{R}^2 onto a set in \mathbb{R}, one gets a set called a PCA set, which need not be either an A or a CA set. Alternating taking complements and projections gives a hierarchy of "projective" sets, treated by Sierpiński (1950).

Cohn (1980, Chapter 8) was a source for some of the facts and proofs in this section and gives further information on analytic sets. Dellacherie (1978) gives examples of analytic (non-Borel) sets useful in probability theory.

References

Alexandroff [Aleksandrov], P. (1916). Sur la puissance des ensembles mesurables B. *Comptes Rendus Acad. Sci. Paris 162*: 323–325.

Cohn, Donald Lee (1980). *Measure Theory*. Birkhäuser, Boston.

Dellacherie, Claude (1978). *Quelques exemples familiers, en probabilités, d'ensembles analytiques, non boréliens, Séminaire de Probabilités, Univ. Strasbourg* (1976–1977); *Lecture Notes in Math.* (Springer) *649*: 746–756.

Hausdorff, Felix (1916). Die Mächtigkeit der Borelschen Mengen. *Math. Ann. 77:* 430–437.

Kuratowski, Kazimierz [Casimir] (1933). *Topologie I.* Monografje Matematyczne, Warsaw and Lwow.

———— (1934). Sur une généralisation de la notion d'homéomorphie. *Fund. Math. 22:* 206–220.

———— (1966). *Topology*, vol. 1, transl. from 2d French ed. (1958) by J. Jaworowski. Academic Press, New York; PWN, Warsaw.

Lebesgue, Henri (1905). Sur les fonctions représentables analytiquement. *Journal de Mathématiques* (Ser. 6) *1*: 139–216.

Lusin [Luzin], Nikolai Nikolaevich (1917). Sur la classification de M. Baire. *Comptes Rendus Acad. Sci. Paris 164*: 91–94.

———— (1930). *Leçons sur les ensembles analytiques et leurs applications*, with a Note by W. Sierpiński and a preface by H. Lebesgue. Gauthier-Villars, Paris.

Mauldin, R. Daniel, and Stanislaw M. Ulam (1987). Mathematical problems and games. *Adv. Appl. Math. 8*: 281–344.

Paplauscas [Paplauskas], A. B. (1973). Luzin, Nikolai Nikolaievich. *Dictionary of Scientific Biography, 8*: 557–559.

Parthasarathy, K. R. [Kalyanapuram Rangachari] (1967). *Probability Measures on Metric Spaces*. Academic Press, New York.

Sierpiński, Wacław (1950). *Les ensembles projectifs et analytiques*. In *Mémorial des Sciences Mathématiques*, fasc. 112. Gauthier-Villars, Paris.

Souslin [Suslin], Mikhail Yakovlevich (1917). Sur une définition des ensembles mesurables B sans nombres transfinis. *Comptes Rendus Acad. Sci. Paris 164*: 88–91.

———— (1920). [Problem]. *Fund. Math. 1*: 223.

Yushkevich, A. P. (1968). *Istoriya Matematiki v Rossii do 1917 goda* [in Russian]. Nauka, Moscow.

A

Axiomatic Set Theory

Although more detailed and formal than the presentation in §1.1, this appendix does not claim to provide a complete, rigorous presentation of axiomatic set theory (there are several entire books devoted to the subject, some of them listed in the references). Although axioms for set theory will be stated in detail, some definitions, such as linear ordering and well-ordering, will be assumed to be known (from Chapter 1).

A.1. Mathematical Logic

Around 300 B.C., Euclid's geometry presented "a strictly logical deduction of theorems from a set of definitions, postulates and axioms" (Struik, 1948, p. 59). Euclid went a long way, although not all the way, to the modern ideal of the axiomatic method, where, when the proof of a theorem is written out in detail, it can be checked mechanically and precisely to ascertain that it is (or is not) a proof. From a modern point of view, perhaps the least strictly logical part of Euclid's system is his definitions—for example, "a point is that which has no extension," "a line is a length, without width . . ." As was noted in §1.1, a truly precise mathematical system, or 'formal system', begins with some basic undefined terms. Then other terms can be defined from the basic ones.

The most widely accepted formal systems, giving a foundation for modern mathematics, are based on propositional calculus and first-order predicate logic. Only a very brief introduction to these topics will be given here. For more details see, for example, Kleene (1967).

In propositional calculus, statements are combined into other statements using propositional connectives. Two basic propositional connectives will be the words "and" and "not." Several other commonly used connectives can be defined from these two. Let P and Q be two statements, such as "$x = y$," "$u \neq v$." Then P or Q is defined as not (not P and not Q).

This last definition illustrates a certain difficulty in being precise about mathematical expressions. In mathematical logic, quotation marks are used

as follows: any expression beginning and ending with quotation marks is a name of the expression without those quotation marks. Thus "A" means the letter A, and "$A(B)$" means the string of four characters: A, left parenthesis, B, right parenthesis. (While using ordinary double quotation marks (") in this way, single quotes (', ') can serve for other uses of quotation marks.) If the definition of "or" had been written with quotation marks, saying "P or Q" means "not (not P and not Q)", it would have been undesirably specific, applying only to the expressions inside the quotation marks, containing the actual letters "P" and "Q". Further, these letters are not even statements. What is meant is that P and Q should be variables ranging over statements, so that each of "P" and "Q" can be replaced by a statement. To provide a convenient and yet precise form for such definitions, the logician W. V. Quine invented a variant of quotation marks, called 'corners', \ulcorner and \urcorner. For example, one can say $\ulcorner P$ or $Q\urcorner$ means \ulcornernot (not P and not Q)\urcorner. Here, whenever P is replaced by one statement and Q by another, in both of the expressions inside corners, the result is taken as a definition. Only after such replacements are made is it actually correct to use ordinary quotation marks and say, for example, "$x = y$ or $u = v$" means "not ($x \neq y$ and $u \neq v$)".

A distinction needs to be made between statements within a formal system and statements about the system. Statements about the system may contain 'syntactic variables' such as "P" or "Q" that range over statements or other expressions in the system. Quine's corners are a way of forming complex syntactic variables from simpler ones.

Now to define other important connectives, $\ulcorner P \Rightarrow Q\urcorner$, which is read $\ulcorner P$ implies $Q\urcorner$ or \ulcornerif P, then $Q\urcorner$, is defined by \ulcornernot (P and not Q)\urcorner. Also, $\ulcorner P \Leftrightarrow Q\urcorner$ is defined by $\ulcorner(P \Rightarrow Q)$ and $(Q \Rightarrow P)\urcorner$.

Propositional calculus has its own axioms and rules of inference which are incorporated into those of other, more complex formal systems. For example, (the) one rule of inference is 'modus ponens':

$$\text{from } \ulcorner P\urcorner \quad \text{and} \quad \ulcorner P \Rightarrow Q\urcorner \text{ one can infer } \ulcorner Q\urcorner.$$

There seems to be no need to state the other propositional axioms or rules here, as they also correspond to very familiar modes of mathematical reasoning.

The next main stage on the way to axiomatic set theory is first-order predicate logic or predicate calculus. Here there are two logical symbols called quantifiers: "\forall", read 'for all', and "\exists", read 'there exist(s)'. Using the quantifiers "\forall" and "\exists", variables such as "x" and "y", the propositional connectives "not", "and" and others (definable from these two as above), the signs "$=$", "\neq", and parentheses, we can construct 'well-formed formulas', or 'wffs' for short. There is a precise syntax, or set of rules, on what are wffs. Among

other things, left and right parentheses must match up properly, as usual. For example,

$$\text{``}(\forall x)(\exists y)((y = x) \quad \text{and} \quad (y \neq z))\text{''}$$

is a wff. When a variable, such as "x" or "y" in this formula, is governed by a quantifier, it is called 'bound.' Otherwise, like "z" in that formula, it is called 'free.'

Again, predicate calculus has its own axioms and rules of inference which would need to be included in a full presentation of axiomatic set theory. Among them are 'substitution rules' on when one expression can be substituted for another, for example, to keep order on which variables are bound, and how, and which are free. These rules have caused some technical difficulties in the field (Kleene, 1967, p. 107; Church, 1956, pp. 289–290) but seem not to reflect problems that workers in analysis or probability need to notice.

A.2. Axioms for Set Theory

The most widely used system of axioms for set theory is called ZF, or Zermelo-Fraenkel, after Ernst Zermelo, who proposed a system of axioms in 1908, and Abraham Fraenkel, who improved the system. In ZF, the only objects are sets. There is just one basic relation between them, membership, written "\in". For example, if x and y are sets, then "$x \in y$" is read 'x is a member of y' or 'x belongs to y.' The negation of the relation is written "$x \notin y$", read 'x is not a member of y.' In a formal system of axiomatic set theory such as ZF, 'set' and '\in' begin as undefined notions, gaining content through the axioms. To have an exact definition of 'set variable'—that is, a variable ranging over sets—one could use only the expressions "x", "x'", "x''", "x'''", But here, less formally, other variables, such as "y" and "X", will be allowed.

In ZF, $\ulcorner\forall x\urcorner$ means \ulcornerfor all sets $x\urcorner$, and $\ulcorner\exists y\urcorner$ means \ulcornerthere exists a set y [such that...]\urcorner.

The syntax governing well-formed formulas includes rules on usage of "\in". Each "\in" must be immediately preceded and followed either by a set variable or by some other expression which is allowed to play the same role. One source for more details is Cohen, 1966, pp. 6–7. For example, "$(\forall x)(\exists y))((x \in y) \Rightarrow (x \in z))$" is a wff, while

$$\text{``}(\forall x)(\exists z)\forall y(x \in yz)\exists \in\text{''}$$

is not a wff for several reasons.

The axioms for ZF vary somewhat from one presentation to another. Here is one version, consisting of nine axioms (not counting the axiom of choice).

ZF1. Extensionality $\forall y \forall z((y = z)$ iff $\forall x(x \in y$ iff $x \in z))$.

ZF1 states that two sets are equal if and only if they have the same members.

ZF2. Empty set $(\exists y)(\forall x)x \notin y$.

ZF2 says that some set y has no members. By extensionality, there is exactly one such set. This empty set will be called \emptyset.

ZF3. Unordered pairs $\forall x \forall y \exists z \forall u((u \in z)$ iff $((u = x)$ or $(u = y)))$.

Again, the set z in ZF3, with just the members x and y, is unique by extensionality. It will be called $\{x, y\}$. Note that $\{x, y\} = \{y, x\}$ for any x and y. Here $\{x, y\}$ is called the unordered pair with members x and y. Setting $x = y$, let $\{x\}$ denote $\{x, x\}$. Then $\{x\}$ is a set with exactly one member, x. Here $\{x\}$ is read 'singleton x.'

The following is not just one axiom but an axiom schema: it describes how (infinitely many) wffs can be formed, all of which will be taken as axioms.

ZF4. Selection For every wff $\ulcorner P \urcorner$ containing just one free variable "x", and not containing variables "y" or "z", the following is an axiom: $\ulcorner \forall y \exists z \forall x(x \in z$ iff $(x \in y$ and $P))\urcorner$.

Note that corners (rather than quotes) are needed because the wff P varies. (Putting corners in place of quotes around "x", "y", and "z" would allow more formulas as axioms but would give equivalent axioms defining no additional sets.) The set z in ZF4 is unique by extensionality. It will be denoted $\ulcorner \{x \in y : P\} \urcorner$, read \ulcornerthe set of all x in y such that $P \urcorner$. ZF4 is often called the Aussonderung axiom, from a German word meaning 'selection': from the set y, we can select all members satisfying P to form a set. But the limitation in ZF4 to subsets of some set is needed to avoid contradictions. Without it, if we try to define, for example, $r = \{x : x \notin x\}$, then $r \in r$ implies $r \notin r$ and vice versa (Bertrand Russell's paradox).

The next axiom allows formation of the union u of any collection x of sets, where a collection of sets is, in set theory, just a set:

ZF5. Unions $\forall x \exists u \forall y((y \in u)$ iff $(\exists z)((y \in z)$ and $(z \in x)))$.

The set u is uniquely determined, by extensionality, once x is given. It is called the *union* of x and written $\bigcup x$. This simple notation is common in

set theory but is not so often used in real analysis or other fields. Another possible notation is $\bigcup_{y \in x} y$. We can also define the union of two sets: let $x \cup y = \bigcup \{x, y\} = \{z: z \in x \text{ or } z \in y\}$. (The first term $x \cup y$ is being defined, the middle term $\bigcup\{x, y\}$ is well-defined according to ZF3 and ZF5, and the last description, though in notation similar to that of ZF4, is not formally allowed by ZF4.)

From the unordered pair and union axioms, we can get for any x_1, x_2, x_3, and x_4 a set $\{x_1, x_2, x_3, x_4\}$ having the given four elements. Iterating this, one can get sets containing any given finite list of sets.

Definitions. For any wff $\ulcorner Q \urcorner$ containing a free variable $\ulcorner x \urcorner$, $\ulcorner (\forall x \in y)Q \urcorner$ means $\ulcorner \forall x((x \in y) \Rightarrow Q) \urcorner$. Then $\ulcorner y \subset z \urcorner$, read '$y$ is included in z,' means $\ulcorner (\forall x \in y)x \in z \urcorner$.

ZF6. Power set For any set x, there is a set z such that $y \in z$ iff $y \subset x$.

The set z in ZF6, unique as usual, is the set of all subsets of x. It is sometimes written 2^x. (If x has n members, 2^x has 2^n.) For any two sets x and y, their *intersection* is defined by $x \cap y = \{u \in x: u \in y\}$, a set by ZF4. Then x and y are said to be *disjoint* iff $x \cap y = \emptyset$. If x is any non-empty set, its intersection is defined by

$$\bigcap x = \bigcap_{v \in x} v = \{u: u \in v \text{ for all } v \in x\}.$$

This is a set by ZF4 (selection), since, for example, we can take $\{u \in y: \ldots\}$ where $y = \bigcup x$. The difference $x \backslash y$ is defined by $x \backslash y = \{u \in x: u \notin y\}$.

With the axioms so far, one cannot prove that there is an infinite set. The next axiom provides one.

ZF7. Infinity $\exists M(\emptyset \in M \text{ and } (\forall x \in M)x \cup \{x\} \in M)$.

A set M as in ZF7 contains as members $\emptyset, \{\emptyset\}, \{\emptyset, \{\emptyset\}\}, \ldots$. The standard method of constructing nonnegative integers in set theory is to define the above sets as equal to $0, 1, 2, \ldots$, respectively. The set \mathbb{N} of all nonnegative integers will be defined as a minimal set M satisfying the conditions of ZF7 and is shown to exist as follows.

A.2.1 Theorem *There exists a unique set \mathbb{N} satisfying the conditions on M in ZF7 such that for any other such set M, $\mathbb{N} \subset M$.*

Proof. Let M be any set satisfying ZF7. Let $H(M)$ be the set of all subsets of M which also satisfy the conditions. Then $H(M)$ is indeed a set by selection (ZF4), since it is a subset of 2^M. Let $N(M) = \bigcap H(M)$. Then $N(M) \subset M$. If $J \in H(M)$ and $K \in H(M)$, then $J \cap K \in H(M)$. It follows that $N(J) = N(M)$. Also, $N(M) \in H(M)$. Now let M' be any other set satisfying ZF7. Then so does $M \cap M'$. Hence $N(M) = N(M \cap M') = N(M')$. So $N(M)$ does not depend on M, and one can set $\mathbb{N} := N(M)$. Then $\mathbb{N} \subset M$, so a set as described in the theorem exists. Let K be any other such set. Then $K \subset \mathbb{N}$ and $\mathbb{N} \subset K$, so $\mathbb{N} = K$, finishing the proof. $\qquad\square$

Definition. An *ordered pair* is a set of the form $\{\{x\}, \{x, y\}\}$ for some sets x and y. For any ordered pair Q, its *first member* is the unique x such that $\{x\} \in Q$. Its *second member* is the unique y such that $Q = \{\{x\}, \{x, y\}\}$ where x is the first member. Let $\langle x, y \rangle$ denote $\{\{x\}, \{x, y\}\}$.

The conventional definition above is only one way of encoding the desired pair and its ordering purely by sets. By the way, note that $\{\{x\}, \{x, y\}\} = \{\{y, x\}, \{x\}\}$, and so forth.

Definition. For any two sets X and Y, their *Cartesian product* $X \times Y$ is defined as the set $\{\langle x, y \rangle : x \in X, y \in Y\}$.

$X \times Y$ is a set, by the selection axiom ZF4, as a subset of 2^U where $U = 2^{X \cup Y}$. A *relation* is any set of ordered pairs. For any relation E, the inverse relation is defined by $E^{-1} := \{\langle y, x \rangle : \langle x, y \rangle \in E\}$. This is also a set by ZF4, as it is a subset of 2^U.

Definition. A *function* is a set f, all of whose members are ordered pairs, such that for any x, if $\langle x, y \rangle \in f$ and $\langle x, z \rangle \in f$, then $y = z$. For any function f, we write $f(x) = y$ to mean $\langle x, y \rangle \in f$. The *domain* of f is defined as the set of all x for which $f(x)$ is defined, in other words, dom $f = \{x : \exists y \langle x, y \rangle \in f\}$. Here ZF4 applies, since we can replace $\{x : \ldots\}$ by $\{x \in \bigcup \bigcup f : \ldots\}$. The *range* of f is the set of its values: ran $f := \{y : f(x) = y$ for some $x\}$, which again is a subset of $\bigcup \bigcup f$. Here ":=" means, as always, "equals by definition."

Suppose we have a wff with two free variables x and y, which we call $\ulcorner E(x, y)\urcorner$, and for each x in a set t, $E(x, y)$ holds for at most one value of y. Then $\{x \in t : \exists y\, E(x, y)\}$ is a set by selection. To define a function f such that $(f(x) = y)$ is equivalent to $(E(x, y)$ and $x \in t)$, it would be enough if $\{y : (\exists x \in t) E(x, y)\}$ were a set. (*Note:* $(\exists x \in t) \ldots$ means $(\exists x)(x \in t$

and ...).) Since this is not otherwise clear, it will be assumed next, where the function may also depend on other variables. Like ZF4, this will not be a single axiom, but an axiom-schema. Here $E(u, v)$ denotes the result of putting u in place of x and v in place of y throughout $E(x, y)$. For example, if $E(x, y)$ is "$(\forall t)(t \in x$ or $t \notin y)$", then $E(u, v)$ is "$(\forall t)(t \in u$ or $t \notin v)$", and likewise for wffs containing more variables.

ZF8. Replacement or substitution Let $\ulcorner E(x, y, x(1), \ldots, x(k))\urcorner$ be a wff containing free variables "x", "y", "$x(1)$", "$x(2)$",..., "$x(k)$", where $k = 0, 1, \ldots$, and no other free variables, and not containing bound variables "t", "u", "v", "w", or "z". Then each statement of the following form is an axiom:

$\forall x(1) \ldots \forall x(k)(\forall x \forall y \forall z((E(x, y, x(1), \ldots, x(k)))$ and

$E(x, z, x(1), \ldots, x(k)) \Rightarrow y = z))) \Rightarrow \forall t \exists w \forall v(v \in w$ iff $(\exists u)(u \in t$ and

$E(u, v, x(1), \ldots, x(k))))).$

In one sense, y is a function of the $k + 1$ variables $x, x(1), \ldots, x(k)$. In another sense, the idea is that y is a function of x in which $x(1), \ldots, x(k)$ are parameters. For example, the relation E (once algebraic operations were defined) might be $y = ax^2 + bx + c$, where $k = 3$, $a = x_1$, $b = x_2$, and $c = x_3$. Axiom ZF8 is useful in constructing ordinals (see the next section, A.3) and certain very large sets. For example, if we take an infinite set M from ZF7, we can form 2^M, then 2^{2^M}, and so on. Suppose we would like to define a set X such that $\mathbb{N} \in X$ and for all $A \in X$ we have $2^A \in X$. It seems natural that a 'countable union' of sets we can write down should also be a set, but it does not follow from ZF1 through ZF7 that it is. It does follow if ZF8 is included. Such large sets are hardly ever needed in mathematics and are not used in this book. Very occasionally, when looking for a proof of a very general assertion but not succeeding, one can gain some insight by finding a counterexample among each very large sets.

One more axiom helps to simplify and clarify set theory by preventing sets from being mutually members of each other, or from having infinite chains of membership: $\cdots \in x(3) \in x(2) \in x(1)$, for example.

ZF9. Regularity For any set $x \neq \emptyset$, $(\exists y \in x) x \cap y = \emptyset$.

A.2.2. Theorem For all x and y, $x \notin x$ and if $x \in y$, then $y \notin x$.

Proof. By ZF9, $x \cap \{x\} = \emptyset$. If $x \in x$, then $x \in x \cap \{x\}$, a contradiction. So $x \notin x$. (This conclusion also follows from the second half of the theorem, for

$x = y$.) If $x \in y \in x$, then by ZF9, $\{x, y\}$ has a member disjoint from it, say x by symmetry. Then $y \in x \cap \{x, y\}$, a contradiction. □

The remaining axioms that are usually, though not always, assumed are the axiom of choice and the (generalized) continuum hypothesis, treated in §1.5.

Problems

1. For the given definition of ordered pair, find:
 (a) $\bigcup \langle x, y \rangle$; (b) $\bigcup \bigcup \langle x, x \rangle$.

2. If $x \in y \in z \in w$, prove $w \notin x$.

3. Prove that there is no function f defined on \mathbb{N} such that $f(n+1) \in f(n)$ for all $n \in \mathbb{N}$. *Hint:* Apply ZF9 to the range of f.

4. Prove that if A and B are sets, so is $A \times B$.

5. Prove that if E is a relation, then E^{-1} is also.

6. Prove in detail, using ZF8, that there exists a set X with $\mathbb{N} \in X$ and such that for each $B \in X$, we have $2^B \in X$.

A.3. Ordinals and Cardinals

Two partially ordered sets (X, \leq) and (Y, \leq) are said to be *order-isomorphic* iff there is a 1–1 function f from X onto Y such that for any $x \in X$ and $t \in X$, $x \leq t$ if and only if $f(x) \leq f(t)$. The function f is called an *order-isomorphism*. Being order-isomorphic is an equivalence relation (on any set of partially ordered sets). There are ways to choose well-ordered sets, called ordinals, such that every well-ordered set is order-isomorphic to one and only one ordinal. A simple definition is as follows. A set x is called *full* iff for all $y \in x$, $y \subset x$. An *ordinal* is a full set X linearly ordered by \in, so that $\{\langle y, z \rangle \in X \times X : y \in z\}$ is a linear ordering on X. (By Theorem A.2.2, this ordering must be strict.) For ordinals, $x < y$ will mean $x \in y$.

The empty set \emptyset is clearly an ordinal (fullness and linear ordering by \in hold vacuously). It is easy to check that if x is an ordinal, so is $x \cup \{x\}$. Let $\{a, b, c\}$ be the set whose members are just a, b, and c, and likewise for any finite number of elements. (Such sets exist, by the unordered pair and union axioms.) Thus the following are ordinals: $\emptyset, \{\emptyset\}, \{\emptyset, \{\emptyset\}\}, \{\emptyset, \{\emptyset\}, \{\emptyset, \{\emptyset\}\}\}, \dots$. Viewed as ordinals, these sets are called $0, 1, 2, 3, \dots$, so that each nonnegative

integer n is identified with an ordinal which is a set having n elements. To deal with infinite ordinals it will help to develop some more theory.

A.3.1. Theorem *Every ordinal z is well-ordered by \in.*

Proof. By definition, z is linearly ordered by \in. Let $A \subset z$, $A \neq \emptyset$. By ZF9, take $x \in A$ such that $x \cap A = \emptyset$. Then for all $y \in A$, $y \notin x$, so $x \in y$ or $x = y$. Then x is the desired smallest member of A. □

A.3.2. Proposition *If $u \in v$ and v is an ordinal, then u is an ordinal.*

Proof. Since v is full, $u \subset v$, so u is linearly ordered by \in. If $x \in u$, then $x \in v$, so $x \subset v$. If $x \not\subset u$ (x is not included in u), take $y \in x \backslash u$. Then $y \in v$ and $y \notin u$. Since v is linearly ordered by \in, either $y = u$ or $u \in y$. If $u = y$, then since $y \in x \in u$, this contradicts Theorem A.2.2. If $u \in y$, then $u \in y \in x \in u$, contradicting the axiom of regularity (ZF9) applied to $\{x, y, u\}$. Thus u is full. □

For any partially ordered set $(X, <)$, an *initial segment* is a set $A \subset X$ such that $x \in A$ whenever $x < y \in A$.

A.3.3. Proposition *For any ordinals x and y, $x \cap y$ is an initial segment of x.*

Proof. If $u < v \in x \cap y$, then $u \in v \in x$ and $v \in y$. Since x and y are full, $v \subset x \cap y$, so $u \in x \cap y$. □

A.3.4. Proposition *Any set of ordinals is linearly ordered by inclusion.*

Proof. If u and v are ordinals, $u \not\subset v$ and $v \not\subset u$, let s and t be the least members of $u \backslash v$ and $v \backslash u$ respectively. If $y \in s$, then since u is full, $y \in u$, and $y < s$, so $y \notin u \backslash v$ and $y \in u \cap v$. Thus $s \subset u \cap v$. Likewise $t \subset u \cap v$. Conversely, by Proposition A.3.3, if $x \in u \cap v$, then $x < s$; in other words, $x \in s$, so $u \cap v \subset s$. Then $s = u \cap v = t$, a contradiction. □

A.3.5. Proposition *Any set of ordinals is linearly ordered by \in.*

Proof. Suppose u and v are ordinals with $u \subset v$ and $u \neq v$. By Theorem A.3.1, let y be the least element of $v \backslash u$. Then, by Proposition A.3.2, and since v is linearly ordered by \in, $u \subset y$. For any $x \in y$, we have $x \in v$, and by definition of y, $x \in u$. So $u = y \in v$. □

A.3.6. Corollary *Any set S of ordinals is well-ordered by* \in.

Proof. Let $A \subset S$, $A \neq \emptyset$. By regularity (ZF9) take $x \in A$ with $x \cap A = \emptyset$. Then by Proposition A.3.5, x is the desired least element of A. □

A.3.7. Theorem *The union of any set S of ordinals is an ordinal.*

Proof. Recall that all members of ordinals are ordinals (Proposition A.3.2). Let U be the union and $x, y \in U$. Let $x \in u \in S$ and $y \in v \in S$. By Proposition A.3.4, we can take $u = v$. Thus by Proposition A.3.5, either $x \in y, x = y$, or $y \in x$. If $x \in y \in z \in U$, then $x \in z$, since z is full. So U is linearly ordered by \in. Since u is full, $x \subset u \subset U$, so U is full. □

Definition. A set x is a *Peano set* if it is an ordinal and is well-ordered by the relation \in^{-1} (so that every non-empty subset of it has a largest member for \in).

It will turn out that the Peano sets are the finite ordinals 0, 1, 2, ... shown above. First, here are some other facts:

A.3.8. Theorem *If x is a Peano set and* $y \in x$, *then y is also a Peano set.*

Proof. One sees that y is an ordinal by Proposition A.3.2. Since x is full, $y \subset x$, so y is also well-ordered by \in^{-1}. □

For each ordinal x define $x + 1 := x \cup \{x\}$. This is also an ordinal, as noted just after the definition of ordinal.

A.3.9. Proposition *If y is an ordinal and x a largest member of y for* \in, *then* $y = x + 1$.

Proof. First, $x \subset y$, since y is full, so $x + 1 \subset y$. Conversely, let $v \in y$, $v \neq x$. Then $v \in x$. So $y \subset x + 1$ and $y = x + 1$. □

A.3.10. Theorem *The empty set* \emptyset *is a Peano set, and for any Peano set x, $x + 1$ is also a Peano set.*

Proof. For \emptyset this is clear. Recall that $x + 1$ is an ordinal. If $A \subset x + 1$, either $x \in A$ and x is the largest element of A, or $A \subset x$. Then if $A \neq \emptyset$, it has a largest member, since x is a Peano set. □

A.3.11. Proposition *If x is a Peano set, $x + 1 \neq \emptyset$.*

Proof. This is clear, since $x \in x + 1$. $\qquad\square$

A.3.12. Proposition *For any ordinal x, $\bigcup(x + 1) = x$.*

Proof. If $y \in x$, then since $x \in x + 1$, $y \in \bigcup(x + 1)$. Conversely, if $t \in \bigcup(x + 1)$, either $t \in x$, or $(\exists y)t \in y \in x$, but then $t \in x$ since x is full. $\qquad\square$

A.3.13. Corollary *If x and y are Peano sets and $x + 1 = y + 1$, then $x = y$.*

A.3.14. Theorem (Principle of Mathematical Induction) If x is a set of ordinals containing only Peano sets, $\emptyset \in x$ and for each $y \in x$, we have $y + 1 \in x$, then x is the set of all Peano sets.

Proof. First, note that there is no set of all ordinals, and it is not clear *a priori* that there is a set of all Peano sets. Suppose that u is a Peano set and $u \notin x$. Then $u + 1$, as an ordinal, is well-ordered by \in (Theorem A.3.1). It has a smallest member z which is not in x. Then $z \neq \emptyset$, and z is a Peano set by Theorems A.3.8 and A.3.10. Thus z has a largest member v. Likewise, v is a Peano set. By definition of z, we have $v \in x$. By Proposition A.3.9, $z = v + 1$ so $z \in x$ by the hypothesis, a contradiction. $\qquad\square$

Now recall the set \mathbb{N} from Theorem A.2.1, the smallest set satisfying the axiom of infinity, ZF7.

A.3.15. Theorem \mathbb{N} *is the set of all Peano sets and is an ordinal.*

Proof. Let ω be the set of all Peano sets in \mathbb{N}. By Theorems A.2.1 and A.3.14, ω is the set of all Peano sets. By Theorem A.3.10, ω satisfies the conditions of ZF7, so by definition of \mathbb{N}, $\omega = \mathbb{N}$. It is a set of ordinals and hence linearly ordered by \in (Corollary A.3.6). It is full by Theorem A.3.8, so it is an ordinal. $\qquad\square$

Thus we have an infinite ordinal ω. We can then form further ordinals $\omega + 1, (\omega + 1) + 1 := \omega + 2, \omega + 3$, and so on. Although it will not be proved here (in fact, it depends on the replacement axiom ZF9), the union of the latter sequence of ordinals is an ordinal 2ω; likewise, there are ordinals $2\omega + 1, \ldots, 3\omega, \ldots, 4\omega, \ldots$, with a union ω^2; ordinals $2\omega^2, 3\omega^2, \ldots$, with union ω^3; ordinals $\omega^4, \omega^5, \ldots$, with union ω^ω. (*Note:* A^B often means the set

of all functions from B into A, but not for ordinals, as here.) There are a great many, in fact uncountably many, countably infinite ordinals, all having the same cardinality as ω. There is a succession of larger and more complicated ordinals. There is no end to the process, no largest ordinal A, for $A + 1$ would be an ordinal. (The set of all ordinals, if it existed, would be the largest ordinal; this is called the "Burali-Forti paradox.")

Next it will be shown that ordinals, as formally defined, do what they were informally supposed to do:

A.3.16. Theorem *Every well-ordered set $(X, <)$ is order-isomorphic to some ordinal.*

Proof. For each $x \in X$ let $J(x) := \{y \in X : y \leq x\}$ and $I(x) := \{y \in X : y < x\}$. Let $Y := \{x \in X : J(x)$ is order-isomorphic to some ordinal$\}$. If $x < y \in Y$, let f be an order-isomorphism of $J(y)$ to some ordinal B. Then $f \restriction J(x)$ is an order-isomorphism from $J(x)$, an initial segment of $J(y)$, to some initial segment of B, which is an ordinal. Thus $x \in Y$, so Y is an initial segment of X. Suppose there is an order-isomorphism h of $J(x)$ to some ordinal with $h \neq f \restriction J(x)$. Let t be the least element of X with $h(t) \neq f(t)$. But as each ordinal is the set of all smaller ordinals, $h(t) = \{h(u) : u < t\} = \{f(u) : u < t\} = f(t)$, a contradiction. So the order-isomorphisms agree on $J(x)$ and are unique there. Let f be the union of all the order-isomorphisms for all $y \in Y$. (Here f exists by the axiom of replacement, ZF8.) Then the range of f, being an initial segment of ordinals, is an ordinal B, and f is an order-isomorphism. If $Y = X$, we are done. If not, let v be the least element of $X \setminus Y$. But then $f \cup \{\langle v, B \rangle\}$ is an order-isomorphism from $J(v)$ to $B + 1$, so $v \in Y$, a contradiction. □

Cardinals. Recall that two sets are said to have the same cardinality iff there is a one-to-one function from one onto the other. Just as each well-ordered set is order-isomorphic to one and only one ordinal, cardinals will be sets such that for every set A, there is one and only one cardinal having the same cardinality as A. The following is the usual way to define cardinals: a *cardinal* is an ordinal which does not have the same cardinality as any of its members (smaller ordinals). The next fact shows that this definition succeeds, assuming the axiom of choice (AC) or equivalently the well-ordering principle (WO), as treated in §1.5.

A.3.17. Corollary *AC or WO implies that every set X has the same cardinality as some cardinal.*

Proof. By WO, take a well-ordering of X. By Theorem A.3.16, take an order-isomorphism h from X to some ordinal B. Let C be the least ordinal with the same cardinality as B. Let g be a 1–1 function from B onto C. Let $f(x) = g(h(x))$ for all $x \in X$. Then f is 1–1 from X onto the cardinal C.

\square

With the definition of cardinals as certain ordinals, given above, the fact that every set has the same cardinality as some cardinal implies WO and hence AC. So this method of dealing with cardinality depends crucially on AC. By Corollary A.3.6, any set of cardinals is well-ordered by \in. Infinite cardinals are generally denoted by Hebrew alephs with ordinal subscripts, as follows: $\aleph_0 := \omega$, the countably infinite cardinal. For each nonzero ordinal k, \aleph_k is the smallest cardinal larger than all \aleph_j for all $j < k$. The *continuum hypothesis* says that 2^{\aleph_0} has cardinal \aleph_1.

Via binary expansions, there is a function f from 2^ω onto the unit interval $[0, 1] := \{x \in \mathbb{R}: 0 \le x \le 1\}$, where f is 1–1 except on a countable set; as in §13.1, f can be chosen to be 1–1, so that $[0, 1]$ and 2^ω have the same cardinality. Their cardinal is usually called c, the cardinal of 'the continuum.' The *generalized continuum hypothesis* (GCH) states that for any infinite sets X and Y, if there are 1–1 functions from X into Y and from Y into 2^X, then Y has the same cardinality as either X or 2^X. Then GCH implies that for every infinite cardinal \aleph_k, its power set 2^{\aleph_k} has the same cardinality as \aleph_{k+1}. It is known that GCH implies AC and that AC can be neither proved nor disproved from the usual axioms of set theory (ZF1–9) (see the notes).

Problems

1. Show that for any ordinal y, the set $\{x: x \le y\}$ is compact for the interval topology (defined in §2.2, Problem 9).

2. An ordinal y is called the *successor* of an ordinal x iff $y = x + 1$; in other words, $y = x \cup \{x\}$. Any ordinal y equal to its union $\bigcup y$ is called a *limit ordinal*. Prove that every ordinal is either a limit ordinal or the successor of some ordinal, but not both.

A.4. From Sets to Numbers

Set theory serves as a foundation for mathematics by providing sets to represent various kinds of numbers. The first step in this process has been to represent the nonnegative integers 0, 1, 2, ..., by the sets \emptyset, $\{\emptyset\}$, $\{\emptyset, \{\emptyset\}\}$, ...,

setting $n + 1 := n \cup \{n\}$ as above. This 'successor' of n is also sometimes written n'. Each nonnegative integer n is then a set with n elements, namely $\{0, 1, \ldots, n-1\}$. Now, the arithmetic operations can be defined by simple recursion. First, addition will be defined, as a function from $\mathbb{N} \times \mathbb{N}$ onto \mathbb{N}. With $+(m, n)$ written in the usual form, $m + n$, a recursive definition is $0 + n := n$, and $m' + n := (m + n)'$. Thus for each fixed $n \in \mathbb{N}$, $m + n$ is defined for all m by simple recursion (Corollary 1.3.3). Multiplication is similarly defined by simple recursion, setting $0n := 0$ and $m'n = mn + n$.

To represent integers that may be negative, we first define an equivalence relation E on $\mathbb{N} \times \mathbb{N}$: $\langle m, n \rangle E \langle j, k \rangle$ iff $m + k = j + n$. (The idea is that $m - n = j - k$, and that the ordered pair $\langle m, n \rangle$ of nonnegative integers is one representative of the integer $m - n$.) Then the set \mathbb{Z} of all integers is defined as the set of all equivalence classes of ordered pairs of nonnegative integers for the given equivalence relation. Note that a nonnegative integer will be coded by a different set in \mathbb{Z} than it was in \mathbb{N} previously. It is then possible to define addition, subtraction, and multiplication from $\mathbb{Z} \times \mathbb{Z}$ onto \mathbb{Z} using the operations from \mathbb{N}. The details are not very important or interesting and will not be given here. Next, one can define another equivalence relation F between ordered pairs: $\langle u, v \rangle F \langle z, w \rangle$ iff $uw = vz$, where u, v, z, and w are all members of \mathbb{Z}, $v \neq 0$, and $w \neq 0$. Here the idea is that $u/v = z/w$. Let \mathbb{Q}, the set of rational numbers, be defined as the set of all equivalence classes for F of ordered pairs $\langle u, v \rangle \in \mathbb{Z} \times \mathbb{Z}$ with $v \neq 0$. The arithmetic operations in \mathbb{Q} can be defined from those in \mathbb{Z}. The usual relation $<$ can also be defined in \mathbb{Q}. Again, the details are unimportant. Now, real numbers will be defined in terms of what are called Dedekind cuts. A set $C \subset \mathbb{Q}$ is called a *cut* iff the following three conditions all hold:

(a) $\emptyset \neq C \neq \mathbb{Q}$.
(b) If $x < y \in C$, then $x \in C$.
(c) If $x \in C$, then $(\exists y \in C) y > x$.

The set \mathbb{R} of all real numbers is then defined as the set of all cuts. The idea is that a real number x is represented by the set of all rationals q such that $q < x$. Then the ordering \leq for real numbers can be defined simply by inclusion of cuts. Addition is defined by $C + D := \{q + r: q \in C, r \in D\}$. The other usual arithmetic operations on real numbers can also be defined in terms of cuts, but again, the details will not be given.

A set B of real numbers is called *bounded above* iff $(\exists x \in \mathbb{R})(\forall y \in B) y \leq x$. Such an x is called an *upper bound* for B. If for all $u < x$, u is not an

upper bound for B, then x is called the *supremum* or *least upper bound* of B, $x := \sup B$.

A.4.1. Theorem *Every non-empty set B of real numbers which is bounded above has a supremum.*

Proof. Using cuts, let $c = \bigcup B$. One can prove that c is a cut and that $c = \sup B$. □

Likewise, a set B of real numbers is said to be *bounded below* iff $(\exists x)(x < y$ for all $y \in B)$. A non-empty set B of real numbers which is bounded below has a greatest lower bound or infimum called inf B.

Real numbers can also be defined in a more classical way through infinite decimal expansions. There are then some difficulties. To define $0.352745536 \ldots + 1.647254463\ldots$, for example, one may have to look at all the digits of each summand to find the first digit of the sum. The definitions of addition and $<$ and the proof of Theorem A.4.1 all are simpler for cuts than they would be for decimal expansions. Once the real numbers have been developed, with their crucial properties such as Theorem A.4.1, the various representations of numbers by sets have served their purpose, and it is generally advantageous to visualize real numbers as points on a line.

Problems

1. Prove or disprove: If B is any non-empty set of real numbers which is bounded below, then inf $B = \bigcap B$. (Use the definition of real numbers by cuts.)

2. Prove in detail (referring to specific axioms of ZF, as well as to simple recursion, Corollary 1.3.3) that addition is a function from $\mathbb{N} \times \mathbb{N}$ into \mathbb{N}.

Notes

§A.2 There are other axioms and axiom-schemata equivalent to ZF9, called axioms of 'foundation': see Fraenkel et al. (1973, pp. 86–91). Another much-studied system of axioms for set theory is the von Neumann-Gödel-Bernays (GB) system. By contrast to ZF, in GB there are not only sets but objects called classes. All sets are classes. Some classes are not sets and are called *proper* classes. A class A is a set if and only if $A \in B$ for some class B, so "set" is defined in GB: "class" and "\in" are the primitive, undefined notions of GB. Axioms ZF1, ZF2, ZF3, ZF5, ZF6, ZF7, and ZF9 are also axioms (or theorems) in GB, and ZF1 is assumed for classes as well as sets. Instead of ZF4, GB has a set of finitely many (8 or 9) axioms which together are equivalent to the axiom-schema

that for any wff P with just one free variable x there is a class A such that $x \in A$ iff P. Instead of the axiom-schema ZF8, GB has an axiom:

$$\forall A(\forall \text{ set } t)(((\forall x \in t)\forall y \forall z((\langle x, y \rangle \in A \quad \text{and} \quad \langle x, z \rangle \in A \Rightarrow y = z)) \Rightarrow$$
$$(\exists \text{ set } v)(\forall y)(y \in v \text{ iff } (\exists x \in t)\langle x, y \rangle \in A))).$$

Thus GB has a finite set of axioms and no axiom-schemata. It allows large objects to be named, such as the 'universe' V, the class of all sets. (In general, proper classes are those which are too large to be sets.) Large classes arise in some work on category theory (homological algebra). For a while, GB seemed to provide a more convenient foundation for categories than ZF. But category theory on occasion goes beyond GB too, considering, along with the universe V, objects such as 2^V (see MacLane, 1971, for example).

The representation of ordered pairs by sets is due to Wiener (1914) and Kuratowski (1921). Wiener (1894–1964), better known for his work in other fields, wrote his paper at age 19 in the language of the 'Theory of Types,' developed by Bertrand Russell and Alfred North Whitehead, which is no longer in use. Kuratowski, who published his paper in the second volume of the Polish journal *Fundamenta Mathematicae*, was later chief editor of that journal until his death at age 84 (see *Fundamenta*, 1980).

The references contain several books on set theory. Bishop (1967) presents a quite different foundation for analysis, without any precise axiomatic system. On the relation of ZF and GB see Cohen (1966, pp. 73–78). Quine (1951) presents his 'corners' in his section on 'quasi-quotation,' pp. 33–37.

§A.3 The development of ordinals has been based on Kelley (1955), Appendix. He attributes the definition of ordinal to Raphael M. Robinson.

Dedekind (1888) and Peano (1889, 1891) gave a set of axioms for the nonnegative integers in terms of 0 and a successor operation. These axioms do hold for $0 := \emptyset$ and $x' := x \cup \{x\}$ according to A.3.10, 11, 13, 14, and 15 (see also Kleene, 1952, p. 20). Although I had not seen the term "Peano set" as such in the literature, it seemed useful to have such a term.

Gödel (1940) proved that GCH (and so CH) and AC are consistent with ZF1–ZF9 if ZF1–ZF9 are consistent. In other words, if a contradiction follows from GCH and ZF1–ZF9, then one also follows just from ZF1–ZF9. Set theorists strongly believe that ZF1–ZF9 are consistent. Paul Cohen (1963–1964, 1966) proved that CH is independent of ZF1–ZF9 and AC. He found a *model*, a collection of sets for which AC and ZF1–ZF9 hold, but in this case CH does not; specifically, the cardinality of 2^{\aleph_0} is \aleph_2. He also showed that AC is independent of ZF1–ZF9, by finding a model for ZF1–ZF9 in which AC fails, even for a Cartesian product of countably many sets A_i with two elements each, or in which \mathbb{R} cannot be well-ordered (Cohen, 1966, p. 138). Counterintuitively, in the model where each $A_i = \{x_i, y_i\} = \{y_i, x_i\}$ for some x_i, y_i, $A := \bigcup_{i=1}^\infty A_i$ is not countable: there is no function from \mathbb{N} onto A.

Consider the set S of all well-orderings of \mathbb{N}. Let S/I be the set of all equivalence classes of elements of S under order isomorphism. Then S/I is a set with cardinal \aleph_1. There is no known function from S/I onto $[0, 1]$ and indeed there cannot be, since CH is independent. In this sense, some set theorists say, CH is false in 'the real world.'

Nevertheless, it remains a useful, consistent, and rather popular assumption. On the other hand, since S/I is well-ordered, CH implies that $[0, 1]$ can be well-ordered, and so implies the axiom of choice for subsets of $[0, 1]$. This is an example of the fact, which Sierpiński (1947) proved, that GCH implies AC.

§A.4 Landau (1930) gave a detailed development of the real numbers from the positive integers with a successor operation satisfying Peano's axioms. J. van Heijenoort (1967) provides other literature on the foundations of mathematics.

In 'nonstandard analysis' there are real numbers which are infinitely small but not 0. See, for example, Nelson (1977) and Robinson (1979). Some facts in ordinary analysis apparently can be proved rather efficiently by nonstandard analysis techniques.

References

Bishop, Errett (1967). *Foundations of Constructive Analysis*. McGraw-Hill, New York.

Church, Alonzo (1956). *Introduction to Mathematical Logic, I.* Princeton University Press.

Cohen, Paul J. (1963–1964). The independence of the continuum hypothesis. *Proc. Nat. Acad. Sci. USA 50*: 1143–1148, *51*: 105–110.

——— (1966). *Set Theory and the Continuum Hypothesis*. Benjamin, New York.

Dedekind, Richard (1888). *Was sind und was sollen die Zahlen?* Braunschweig.

Fraenkel, Abraham A., Yehoshua Bar-Hillel, and Azriel Levy (1973). *Foundations of Set Theory*. 2d ed. North-Holland, Amsterdam.

Fundamenta Mathematicae 109 (1980). Obituary of K. Kuratowski. pp. i–ii.

Gödel, Kurt (1940). *The Consistency of the Continuum Hypothesis*. Ann. Math. Studies, no. 3. Princeton University Press.

van Heijenoort, Jean (1967). *From Frege to Gödel: A Source Book in Mathematical Logic, 1879–1931.* Harvard University Press.

Kelley, John L. (1955). *General Topology*. Van Nostrand, Princeton.

Kleene, Stephen C. (1952). *Introduction to Metamathematics*. Van Nostrand, Princeton.

——— (1967). *Mathematical Logic*. Wiley, New York.

Krivine, Jean-Louis (1971). *Introduction to Axiomatic Set Theory*. Reidel, Dordrecht.

Kuratowski, Kazimierz (1921). Sur la notion d'ordre dans la théorie des ensembles. *Fund. Math. 2*: 161–171.

Landau, Edmund (1930). *Foundations of Analysis*. Transl. Chelsea, New York, 1951.

MacLane, Saunders (1971). *Categories for the Working Mathematician*. Springer, New York.

Nelson, Edward (1977). Internal set theory: A new approach to nonstandard analysis. *Bull. Amer. Math. Soc. 83*: 1165–1198.

Peano, Giuseppe (1889). *Arithmetices principia, nova methodo exposita*. Turin. Also in *Opere scelte*, II, pp. 20–55 (Edizioni Cremonese, Rome, 1958); transl. in van Heijenoort (1967), pp. 85–97.

——— (1891). Sul concetto di numero. *Rivista di Matematica 1*: 87–102, 256–267.

Quine, Willard Van Orman (1951). *Mathematical Logic*. 2d ed. Harvard University Press.

Robinson, Abraham (1979). *Selected Papers of Abraham Robinson*. Vol. II, *Nonstandard*

Analysis and Philosophy. Edited and with introductions by W. A. J. Luxemburg and S. Körner. Yale University Press.

Sierpiński, Wacław (1934, 1957). *Hypothèse du continu*. Monografje Matematyczne, Warsaw. 2d ed., "Containing as Appendix a number of research papers, 1934–1956." Chelsea, New York.

———— (1947). L'hypothèse généralisée du continu et l'axiome du choix. *Fundamenta Mathematicae 34*: 1–5. Repr. in Sierpiński (1957).

Struik, Dirk J. (1948). *A Concise History of Mathematics*. 2d ed. Dover, New York.

Suppes, Patrick (1960). *Axiomatic Set Theory*. Van Nostrand, Princeton.

Wiener, Norbert (1914). A simplification of the logic of relations. *Proc. Camb. Philos. Soc. 17*: 387–390.

B

Complex Numbers, Vector Spaces, and Taylor's Theorem with Remainder

Complex numbers will be defined here, starting from real numbers. (Real numbers were defined, as Dedekind cuts, in Section 1.1 and Appendix A.4.) The field \mathbb{C} of complex numbers is defined as the plane \mathbb{R}^2 with the usual addition

$$(x, y) + (u, v) := (x + u, y + v)$$

and with the *complex multiplication*

$$(x, y)(u, v) := (xu - yv, xv + yu).$$

The real line \mathbb{R} is considered as a subset of \mathbb{C}, with the real number x corresponding to $(x, 0)$ in \mathbb{C}: let $x := (x, 0)$. The addition and multiplication on $\mathbb{R} \times \mathbb{R}$ then agree with those on $\mathbb{C} \times \mathbb{C}$ restricted to $\mathbb{R} \times \mathbb{R}$.

The complex number i is defined as $(0, 1)$. Thus $i^2 = (-1, 0) = -1$, and $(x, y) = x + iy$; complex numbers are usually written in the latter form. Their addition and multiplication then are determined by those for real numbers together with $i^2 = -1$.

For any complex number $z = x + iy$, the complex *conjugate* is defined by $\bar{z} := x - iy$. The absolute value is defined by $|z| := (x^2 + y^2)^{1/2}$. For any complex $z \neq 0$, let

$$1/z := \bar{z}/|z|^2 = (x/(x^2 + y^2), -y/(x^2 + y^2)).$$

Then $z(1/z) = 1$. The division of a complex number w by $z \neq 0$ is defined by $w/z := w(1/z)$.

If $z = x + iy$ where x and y are real, then x is called the *real part* of z, $x = Re(z)$, and y is called the *imaginary part* of z, $y = Im(z)$. Numbers iy for real y are called *purely imaginary*.

Now vector spaces will be defined over a field K, which for this book will just be either $K = \mathbb{R}$ or $K = \mathbb{C}$. A *vector space* V over K is given by an "addition" $(v, w) \mapsto v + w$ from $V \times V$ into V and a "scalar multiplication" $(c, v) \mapsto cv$ from $K \times V$ into V, with the following properties: $(V, +)$ is

an abelian group, meaning that for all u, v, and w in V, $u + v = v + u$, $u + (v + w) = (u + v) + w$, there is a $0 \in V$ such that $u + 0 = u$ for all u, and for each $v \in V$ there is some $-v \in V$ with $v + (-v) = 0$. Also, the scalar multiplication is required to satisfy, for any $v \in V$ and $a, b \in K$, $(ab)v = a(bv)$, $1 \cdot v = v$, $(-1) \cdot v = -v$, and $(a + b)v = av + bv$. It follows that $0v = 0$. A vector space is also sometimes called a *linear space*.

A real-valued function f on an open interval (a, b) is said to be C^n if the derivatives $f^{(j)}$ exist for $j = 0, 1, \ldots, n$, where $f^{(0)} \equiv f$ and $f^{(1)} \equiv f'$, and $f^{(n)}$ is continuous (the lower order derivatives are automatically continuous), throughout the interval (a, b). Suppose that $0 \in (a, b)$. Then a partial Taylor series expansion of f around 0 is defined, and we have the remainder

$$R_n(f, x) := f(x) - \sum_{j=0}^{n-1} f^{(j)}(0) x^j / j! \tag{B.1}$$

for $a < x < b$. We then have

B.2. Taylor's theorem with integral remainder *If f is C^n on (a, b) and $a < 0 < b$, then*

$$R_n(f, x) = \int_0^x f^{(n)}(t)(x - t)^{(n-1)} \, dt / (n - 1)!. \tag{B.3}$$

Proof. For $n = 1$, (B.3) follows from the fundamental theorem of calculus. The general case will be proved by induction on n. If $n \geq 2$ the induction hypothesis gives

$$R_{n-1}(f, x) = -\int_0^x f^{(n-1)}(t) \, d_t (x - t)^{n-1} / (n - 1)!,$$

where d_t indicates that t is the variable of integration, with x held fixed. Integration by parts gives

$$\frac{f^{(n-1)}(0) x^{n-1}}{(n - 1)!} + \int_0^x f^{(n)}(t)(x - t)^{n-1} \, dt / (n - 1)!,$$

which implies (B.3). □

B.4. Corollary *Under the same conditions,*

$$|R_n(f, x)| \leq \sup_{0 \leq t \leq 1} \left| f^{(n)}(tx) \right| |x|^n / n!.$$

Now let's consider possibly complex-valued functions. A series $\sum_{n=1}^{\infty} z_n$ with complex terms z_n is said to be *absolutely convergent* if $\sum_n |z_n| < \infty$. An absolutely convergent series is easily seen to be convergent in \mathbb{C} with respect to the usual metric $d(z, w) := |z - w|$. If a function f on an interval $(a - r, a + r)$ has a convergent Taylor series

$$f(x) = \sum_{n=0}^{\infty} f^{(n)}(a)(x - a)^n / n!,$$

then $f(z)$ is defined for all complex z in the disk $|z - a| < r$ by the same series

$$f(z) := \sum_{n=0}^{\infty} f^{(n)}(a)(z - a)^a / n!,$$

which is absolutely convergent for each such z. Specifically, the complex *exponential function* is defined by $e^z := \sum_{n=0}^{\infty} z^n / n!$, which is absolutely convergent for all complex z. For any real t, from the known Taylor series of the sine and cosine functions we see that $e^{it} = \cos t + i \sin t$, so $|e^{it}| = 1$. The multiplication rule $e^{w+z} = e^w e^z$ still holds for any complex w, z by the binomial theorem and absolute convergence. It is easily seen that $|e^{x+iy}| = e^x$ for any real x, y. It follows that for $w, z \in \mathbb{C}$, $e^w = e^z$ if and only if $z = w + 2m\pi i$ for some $m \in \mathbb{Z}$.

Any $z \in \mathbb{C}$ can be written, uniquely unless $z = 0$, as $z = re^{i\theta}$ where $r \geq 0, \theta \in \mathbb{R}$ and $-\pi < \theta \leq \pi$. Then the *principal branch of the logarithm* is defined by $\mathrm{plog}(z) := \log r + i\theta$ for $r > 0$, undefined for $r = 0$, where $\log r$ is the usual real-valued logarithm for $r > 0$. The function plog is analytic for $|\theta| < \pi$ but discontinuous when $|\theta| = \pi$. We have $\mathrm{plog}(1 + z) = z - \frac{1}{2}z^2 + \frac{1}{3}z^3 - \cdots$ for $|z| < 1$. For $w, z \in \mathbb{C}$ we have $w = e^z$ if and only if $z = \mathrm{plog}(w) + 2m\pi i$ for some $m \in \mathbb{Z}$.

The complex derivative is defined, if it exists, by $f'(z) := \lim_{w\to 0} (f(z + w) - f(z))/w$. A function f from an open subset U of the complex plane \mathbb{C} into \mathbb{C} is called *holomorphic* or *analytic* iff the derivative $f'(z) \in \mathbb{C}$ exists for all $z \in U$. If f is holomorphic on U, then its derivatives of all orders exist on U; for every $w \in U$ there is a largest r, $0 < r \leq +\infty$, such that $|z - w| < r$ implies $z \in U$, and for any such w and z the Taylor series around w for $f(z)$ converges:

$$f(z) = \sum_{n=0}^{\infty} \frac{f^{(n)}(w)}{n!}(z - w)^n$$

(Ahlfors, 1979, pp. 38, 119, 179). Now in (B.1), (B.2), (B.3), and (B.4) the real variable x can be replaced by a complex variable z if in (B.2) f is holomorphic

on an open disk $D := \{z: |z| < r\}$, $z \in D$, and the variable of integration t in (B.3) and (B.4) is replaced by a complex variable w, integrating along the line segment from 0 to z. Marsden (1973, p. 83) gives the fundamental theorem of calculus in the complex case.

Corollary B.4 is most useful for $|x|$ or $|z|$ small. For larger $|z|$, still with $z \in D$, the corollary may not be so helpful, as will be seen in the following four examples.

Example (a). Let $f(z) := 1/(1-z) = 1 + z + z^2 + \cdots$ for $|z| < 1$ (geometric series). Then $f^{(n)}(z) = n!/(1-z)^{n+1}$. Thus for $0 < x < 1$, Corollary B.4 gives only $|R_n(f, x)| \le x^n/(1-x)^{n+1}$, which approaches 0 as $n \to \infty$ only for $x < 1/2$.

Thus convergence of infinite Taylor series in the largest open disk in which a function is holomorphic, which is proved from Cauchy integral formulas (Ahlfors, 1979, pp. 119, 179), does not follow from Corollary B.4 and not easily from (B.3).

Example (b). Let $f(z) = \mathrm{plog}(1+z)$. Then $f^{(n)}(z) = (-1)^{n-1}(n-1)!/(1+z)^n$ for $n = 1, 2, \ldots$. Corollary B.4 gives $|R_n(f, z)| \le |z|^n/(n(1-|z|)^n)$, which converges to 0 as $n \to \infty$ only for $|z| \le 1/2$.

Example (c). Recall that $\binom{x}{k} := x(x-1)\cdots(x-k+1)/k!$ for $x \in \mathbb{R}$ and $k = 1, 2, \ldots$, while $\binom{x}{0} := 1$. The function g defined by $g(t) := (1-t)^{1/2}$ for real t with $|t| < 1$ has the Taylor series $f(t) := \sum_{n=0}^{\infty} \binom{1/2}{n}(-t)^n$. Comparing factors in the numerator and denominator, we see that $|\binom{1/2}{n}| < 1$ for all $n = 0, 1, \ldots$. Thus for $|t| \le r < 1$ the series $f(t)$ converges absolutely and uniformly and defines $f(t)$. By Theorem B.2 it follows that $f(t) = (1-t)^{1/2}$ at least for $|t| < 1/2$. Also, the double series

$$f(t)^2 = \sum_{j,m=0}^{\infty} \binom{1/2}{j}\binom{1/2}{m}(-t)^{j+m}$$

converges absolutely and uniformly for $|t| \le r$ and so can be rearranged as $\sum_{k=0}^{\infty} C_k t^k$, where

$$C_k := (-1)^k \sum_{i=0}^{k} \binom{1/2}{i}\binom{1/2}{k-i}.$$

From $f(t) = g(t) = (1-t)^{1/2}$ for $|t| < 1/2$ it follows that $C_0 = 1$, $C_1 = -1$, and $C_k = 0$ for $k \ge 2$. Thus $f(t) = g(t) = (1-t)^{1/2}$ for $|t| < 1$ (by continuity

it cannot be $-(1 - t)^{1/2}$), and the series $f(t)$ converges to $g(t)$ uniformly for $|t| \le r$.

Another elementary proof of convergence of binomial series, for $(1 + x)^a$ where a is any real number, is given by Courant (1937), Appendix to Chapter VI, Section 3.

Example (d). Let $f(x) := e^{-1/x^2}$ for $x \ne 0$ and $f(0) := 0$. Then it is easily checked that f is C^{∞}, that is, it has continuous derivatives of all orders. Its Taylor series around 0 has all coefficients 0, so the series converges everywhere, but not to $f(x)$ except for $x = 0$.

Another fact on Taylor series with remainders will be stated for functions of a real variable. Recall that $f = o(g)$ (as $t \to 0$) means $f/g \to 0$ and $f = O(g)$ means f/g is bounded.

B.5. Theorem *If $n \ge 1$ and f is a real-valued function on an interval containing 0 such that the nth derivative $f^{(n)}(0)$ is defined and finite, then $R_{n+1}(f, t) = o(t^n)$ as $t \to 0$.*

Proof. By assumption, the derivatives $f^{(j)}(t)$, $j = 1, \ldots, n - 1$, exist for t in some neighborhood of 0. Subtracting a polynomial from f, we can assume that $f^{(j)}(0) = 0$ for $j = 0, 1, \ldots, n$. If $h'(t) = O(k(t))$ for some function h and nondecreasing, nonnegative function k as $t \to 0$, and $h(0) = 0$, then by the mean value theorem, $h(t) = O(tk(t))$ as $t \to 0$. Likewise if $h'(t) = o(k(t))$ we have $h(t) = o(tk(t))$. By definition of derivative, $f^{(n-1)}(t) = o(t)$. Then by iteration we get $f(t) = o(t^n)$. □

References

Ahlfors, Lars Valerian (1979). *Complex Analysis.* 3d ed. McGraw-Hill, New York.

Courant, R. (1937). *Differential and Integral Calculus*, Vol. I, 2d ed., transl. by E. J. McShane. Interscience, New York.

Marsden, Jerrold E. (1973). *Basic Complex Analysis.* W. H. Freeman, San Francisco.

C

The Problem of Measure

Assuming the axiom of choice, it is known, as shown in §3.4, that there exist nonmeasurable sets for Lebesgue measure. There is still the question whether Lebesgue measure could be extended to all subsets of \mathbb{R} as a countably additive measure. More generally, the "problem of measure" asks whether there is a measure μ on all subsets of an uncountable set X, with $\mu(X) = 1$ and $\mu\{x\} = 0$ for each point $x \in X$. This appendix will prove a partial answer, which will be used in Appendix E.

C.1. Theorem (Banach and Kuratowski) Assuming the continuum hypothesis, there is no measure μ defined on all subsets of $I := [0, 1]$ with $\mu(I) = 1$ and $\mu\{x\} = 0$ for each $x \in I$.

Proof. The proof will be based on the following:

C.2. Lemma *Assuming the continuum hypothesis, there exist subsets A_{ij} of I for i and $j = 1, 2, \ldots$, with the following properties:*

(a) *For each i, the sets A_{ij} for different j are disjoint and their union is I.*
(b) *For any sequence $k(i)$ of positive integers,*

$$\bigcap_i \bigcup_{1 \le j \le k(i)} A_{ij} \text{ is at most countable.}$$

Proof. For any two sequences $\{n_i\}$ and $\{k_i\}$ of positive integers, $\{n_i\} \le \{k_i\}$ will mean $n_i \le k_i$ for all i. The following will be proved first.

C.3. Lemma *Assuming the continuum hypothesis, there is a set \mathcal{F} of sequences of integers, where \mathcal{F} has cardinality c (the cardinality of I), such that for every sequence $\{m_j\}$ (in \mathcal{F} or not), the set of all sequences $\{n_j\}$ in \mathcal{F} with $\{n_j\} \le \{m_j\}$ is at most countable.*

526

Proof. Let \mathcal{S} be the set of all sequences of positive integers. Then \mathcal{S} has cardinality c (§13.1). By the continuum hypothesis, there is a well-ordering \leqq on \mathcal{S} such that for each $y \in \mathcal{S}$, $\{x : x \leqq y\}$ is countable. For each $\alpha \in \mathcal{S}$, let f_α be a function from \mathbb{N} onto $\{x : x \leqq \alpha\}$. Define a sequence $g_\alpha := \{g_\alpha(n)\}_{n \geq 0}$ of positive integers by $g_\alpha(n) := f_\alpha(n)(n) + 1$ for $n = 0, 1, \dots$. (Then g_α is called a "diagonal" sequence.) It is not true that $g_\alpha \leq x$ for any of the countably many $x \leqq \alpha$. Let \mathcal{F} be the set of all sequences $g_\alpha, \alpha \in \mathcal{S}$. Then if $g_\alpha \leq x$, it must not be the case that $x \leqq \alpha$, so $\alpha \leqq x$, and the set of such α is countable. Now \mathcal{F} is uncountable, since each g_α is the sequence y for some $y \in \mathcal{S}$, and if the set of such y were countable, they would have a supremum $\beta \in \mathcal{S}$ for \leqq, but the sequence g_β is different from all the sequences $y \leqq \beta$, a contradiction. So by the continuum hypothesis, \mathcal{F} has cardinality c, proving Lemma C.3. $\qquad\square$

Now to prove Lemma C.2, let h be a 1–1 function from $[0, 1]$ onto \mathcal{F}. Thus each $h(x)$ is a sequence $\{h(x)_n\}_{n \geq 0}$. Define sets A_{ij} by $x \in A_{ij}$ iff $j = h(x)_i$. Then for a fixed value of i, the sets A_{ij} for different j are clearly disjoint. Their union over all j gives the whole interval $[0, 1]$.

Let $\{k_i\} := \{k(i)\}$ be any sequence of positive integers. Let $x \in \bigcap_i \bigcup_{j \leq k(i)} A_{ij}$. Then by definition of A_{ij}, we have $h(x)_i \leq k(i)$ for all i, so $h(x) \leq \{k_i\}$. By Lemma C.3, there are only countably many sequences in \mathcal{F} which are $\leq \{k_i\}$, and since h is 1–1, there are only countably many such $x \in [0, 1]$, finishing the proof of Lemma C.2. $\qquad\square$

Now to prove Theorem C.1, choose $k(i)$ for each $i \geq 1$ such that $\mu(B_i) < 1/2^{i+1}$ where $B_i := \bigcup_{j > k(i)} A_{ij}$. By Lemma C.2, the intersection of the complements of all the B_i is countable, so it has μ measure 0. Thus $1 = \mu\left(\bigcup_i B_i\right) < \sum_{i \geq 1} 1/2^{i+1} = 1/2$, a contradiction. $\qquad\square$

Notes

This appendix is based on the paper of Banach and Kuratowski (1929). For more information on the problem of measure ("measurable cardinals," "real-valued measurable cardinals," etc.), see, for example, Jech (1978).

References

Banach, Stefan, and Casimir [Kazimierz] Kuratowski (1929). Sur une généralisation du problème de la mesure. *Fund. Math. 14*: 127–131.

Jech, Thomas (1978). *Set Theory*. Academic Press, New York.

D

Rearranging Sums of Nonnegative Terms

It is a basic fact of analysis that sums of nonnegative terms can be rearranged (for example, Stromberg, 1981, p. 61). It will be proved here for completeness.

D.1. Lemma *Suppose $a_{mn} \geq 0$ for all $m, n \in \mathbb{N}$. Let $k \mapsto (m(k), n(k))$ be any 1–1 function from \mathbb{N} onto $\mathbb{N} \times \mathbb{N}$. Then*

$$\sum_{m=0}^{\infty} \sum_{n=0}^{\infty} a_{mn} = \sum_{n=0}^{\infty} \sum_{m=0}^{\infty} a_{mn} = \sum_{k=0}^{\infty} a_{m(k),n(k)} = S$$

$$:= \sup \left\{ \sum_{\langle m,n \rangle \in F} a_{mn} : F \subset \mathbb{N} \times \mathbb{N}, \ F \text{ finite} \right\}$$

(whether S is finite or $+\infty$).

Proof. For finite partial sums, we have

$$S_{MN} := \sum_{m=0}^{M} \sum_{n=0}^{N} a_{mn} = \sum_{n=0}^{N} \sum_{m=0}^{M} a_{mn} \leq S,$$

$$T_K := \sum_{k=0}^{K} a_{m(k),n(k)} \leq S$$

for any positive integers M, N, and K. Thus we can replace the upper limits M, N, and K in each sum by $+\infty$ (first in the inner, then the outer sums) and they remain less than or equal to S. On the other hand, for any finite $F \subset \mathbb{N} \times \mathbb{N}$ there are some finite K and M with

$$S_F := \sum_{\langle m,n \rangle \in F} a_{mn} \leq T_K \leq S_{MM}.$$

Hence $S_F \leq T_\infty$, $S_F \leq \lim_{M \to \infty} \lim_{N \to \infty} S_{MN} := U$, and $S_F \leq \lim_{N \to \infty} \lim_{M \to \infty} S_{MN} := V$. Thus $S \leq T_\infty$, $S \leq U$, and $S \leq V$, so $S = T_\infty = U = V$.

\square

Reference

Stromberg, Karl R. (1981). *An Introduction to Classical Real Analysis*. Wadsworth, Belmont, Calif.

E

Pathologies of Compact Nonmetric Spaces

Recall that a topological space is called *locally compact* iff every point in it is contained in some open set with compact closure. For some time, locally compact spaces were considered the most natural spaces on which to study general measures and integrals. The regularity extension, treated in §7.3, appeared as a primary advantage of measures on compact or locally compact spaces. Local compactness also fits well with group structures. A *topological group* is a group G with a Hausdorff topology \mathcal{T} for which the group operations $\langle g, h \rangle \mapsto gh$ and $g \mapsto g^{-1}$ are continuous. A locally compact Hausdorff topological group has left and right *Haar measures*, which are Radon measures (finite on compact sets) invariant under all left or right translations $g \mapsto hg$ or $g \mapsto gh$ respectively. The theory of measures on locally compact spaces, particularly groups, occupied the last three of the twelve chapters of the classic text of Halmos (1950). Bourbaki (1952–1969) put even more emphasis on locally compact spaces. The main purpose of this appendix is to indicate why locally compact spaces have had less attention in this book than in those of Halmos and Bourbaki.

Another class of spaces for measure theory is complete separable metric spaces, or topological spaces which can be metrized to be separable and complete (Polish spaces). Among such spaces, for example, are separable, infinite-dimensional Banach spaces, none of which is locally compact. Yet many of the advantages of local compactness persist in such spaces, at least for finite measures such as probability measures, because of Ulam's theorem (7.1.4).

If measure theory is to be done primarily in locally compact spaces, as in Bourbaki's text, then a fairly general (completely regular, Hausdorff) topological space can be taken as a subset of a compact Hausdorff space (Theorem 2.8.3). Other structures, such as algebraic operations (addition, scalar multiplication), however, may not extend continuously to the compactification, and the general compactification of Tychonoff-Čech often also loses other properties, such as metrizability (see Problems 5 and 7 of §2.8).

Perhaps more serious, less widely recognized difficulties afflict measure theory on nonmetrizable compact Hausdorff spaces. For example, a sequence of continuous functions f_n from the unit interval [0, 1] into a compact Hausdorff space may converge everywhere to a function f which is quite nonmeasurable (for the Borel σ-algebra on the range space), as shown in Proposition 4.2.3. The main fact in this appendix has to do with construction of stochastic processes, as in Kolmogorov's theorem (12.1.2). Let T be any set and \mathbb{R}^T the set of all real-valued functions on T. When a probability measure is given on \mathbb{R}^T by Kolmogorov's theorem, it is defined only on the smallest σ-algebra for which the coordinate evaluations e_t, $e_t(f) := f(t)$, are measurable for each $t \in T$. A measurable set then depends only on countably many coordinates. Thus if T is uncountable, not even singletons $\{f\}$ are measurable, and the supremum of f over T is not a measurable function of f. So there is a need to extend the "law" of the process to a larger σ-algebra. This can be done, for example, for the Wiener process (Brownian motion) via sample continuity, defining the law on the space of continuous functions, by Theorem 12.1.5. More generally, for suitable processes with $T \subset \mathbb{R}$ one can define their laws on sets of functions continuous from the right with limits from the left, as in Theorem E.6 below.

The regularity extension of §7.3 can be applied to real-valued processes:

E.1. Construction For an arbitrary real-valued stochastic process $x(t, \omega)$, $t \in T$, $\omega \in \Omega$, we can take $x(\cdot, \cdot)$ as having values in the compact metrizable space $\overline{\mathbb{R}} := [-\infty, \infty]$. With its Borel σ-algebra, $\overline{\mathbb{R}}$ is a standard measurable space (§12.1). Kolmogorov's theorem (12.1.2) gives a law P_x of x on the compact product space $\overline{\mathbb{R}}^T$, so that for any finite set $F = \{t_1, \ldots, t_k\} \subset T$ and Borel sets $A_i \subset \mathbb{R}$ or $\overline{\mathbb{R}}$, $i = 1, \ldots, k$,

$$P\{x(t_i, \cdot) \in A_i \text{ for } i = 1, \ldots, k\} = P_x\{f \in \overline{\mathbb{R}}^T: f(t_i) \in A_i \text{ for } i = 1, \ldots, k\}.$$

P_x is defined on the smallest σ-algebra \mathcal{C} for which all the coordinate projections π_t on $\overline{\mathbb{R}}^T$ are measurable, where $\pi_t(f) := f(t)$ for each $f \in \overline{\mathbb{R}}^T$. As noted in the example following Theorem 7.1.1, \mathcal{C} equals the Baire σ-algebra in the compact Hausdorff space $\overline{\mathbb{R}}^T$. Thus by the regularity extension (§7.3), P_x extends to a regular Borel measure $P_{(x)}$ on $\overline{\mathbb{R}}^T$.

Note, however, that this construction makes no use whatever of any structure on T, such as an ordering, topology, or σ-algebra. But suppose that \mathcal{S} is a σ-algebra of subsets of T, (Ω, \mathcal{A}, P) is a probability space, and x is a measurable stochastic process, that is, a jointly measurable function from $\Omega \times T$ into \mathbb{R}.

Is the (seemingly natural) function $(f, t) \mapsto f(t)$ measurable from $\overline{\mathbb{R}}^T \times T$ into \mathbb{R}, for some completion of the product σ-algebra? It will be shown in Proposition E.2 that the answer is negative, even for an isonormal process (as defined in §12.1). This shows that compactification and the regularity extension fail to produce the desired results. On the other hand, if $T = [0, 1]$ and the process x has left and right limits everywhere, then the collection of functions with such limits in a Borel set and the regularity extension is more successful (Theorem E.6).

Let H be a separable, infinite-dimensional Hilbert space, with an orthonormal basis $\{e_n\}_{n \geq 1}$. Then for each $y \in H$, $y = \sum_n y_n e_n$ for some real numbers y_n with $\sum_n y_n^2 < \infty$. Let G_n be independent, identically distributed random variables with standard normal distribution $N(0, 1)$, defined on a probability space (Ω, \mathcal{A}, P). Specifically, let Ω be a countable Cartesian product of real lines with coordinates ω_n, where P is the product of standard normal laws $N(0, 1)$ on \mathbb{R} for each n. The isonormal process, as in §12.1, can be defined by $L(y)(\omega) := \sum_n y_n \omega_n$. For each $y \in H$, the series converges a.s. by the three-series theorem (9.7.3). Let $L(y)(\omega) := 0$ if the series does not converge. The finite partial sums of the series defining L are clearly jointly measurable on $H \times \Omega$, with the Borel σ-algebra on H. The set on which a sequence (or series) of measurable functions converges is measurable, as shown in the proof of Theorem 4.2.5. Thus L as defined is jointly measurable.

The stochastic process L on H defines, by Construction E.1, a probability measure P_L on the Baire σ-algebra $\overline{\mathbb{R}}^H$ and its regularity extension $P_{(L)}$, defined and regular as a measure on the σ-algebra of Borel sets in $\overline{\mathbb{R}}^H$. $P_{(L)}$ then has a completion \overline{P}_L (Proposition 3.3.3), so that whenever $A \subset B$ and $P_{(L)}(B) = 0$, then $\overline{P}_L(A) = 0$.

The function $(f, t) \mapsto f(t)$ will not be jointly measurable for a product σ-algebra on $\overline{\mathbb{R}}^H \times H$, even if the σ-algebra on $\overline{\mathbb{R}}^H$ is the collection of all its subsets, so long as at least one set $A \subset H$ is not in the σ-algebra considered on H (let $f = 1_A$). So to have a chance for joint measurability, the product σ-algebra of the domain of \overline{P}_L and the Borel σ-algebra of \mathcal{B} of H must be extended in some way. First, \mathcal{B} can be enlarged. There is the σ-algebra \mathcal{U} of universally measurable sets, which are measurable for the completions of all laws on \mathcal{B}. (This is strictly larger than \mathcal{B}, containing for example all analytic sets, as shown in §13.2.) Although no probability measure on H is given, if we choose one, say μ, defined on \mathcal{B} and also other sets so as to be complete, its σ-algebra $\mathcal{M}(\mu)$ of measurable sets will include \mathcal{U}. Also, we will then have a product measure $\overline{P}_L \times \mu$ on $\overline{\mathbb{R}}^H \times H$ for which we can take a completion. Then if the evaluation $(f, t) \mapsto f(t)$ is to be measurable, \overline{P}_L– almost all f should be measurable for $\mathcal{M}(\mu)$ (and this remains true, by the

way, if the usual completion of the product measure is extended by the method of Bledsoe and Morse (1955), treated in Problem 14 at the end of §4.4). For any σ-algebra \mathcal{S} on H, let $\mathcal{L}^0(H, \overline{\mathbb{R}}, \mathcal{S})$ be the collection of all functions from H into $\overline{\mathbb{R}}$, measurable for \mathcal{S}. Let $\mathcal{S}^0(H, \overline{\mathbb{R}}, \mu) := \mathcal{S}^0(H, \overline{\mathbb{R}}, \mathcal{M}(\mu))$.

Recall the outer measure, defined by $\mu^*(A) := \inf\{\mu(B): B \supset A, B$ measurable$\}$, and the inner measure, defined by $\mu_*(A) := \sup\{\mu(B): B \subset A, B$ measurable$\}$. Now the main fact in this appendix can be stated:

E.2. Proposition *Let P_L be the law of the isonormal process and \overline{P}_L its completed regularity extension on the compact space $\overline{\mathbb{R}}^H$. Then, assuming the continuum hypothesis, the spaces of Borel measurable or universally measurable functions on H have inner measure 0 for \overline{P}_L:*

$$(\overline{P}_L)_*(\mathcal{L}^0(H, \overline{\mathbb{R}}, \mathcal{B})) = (\overline{P}_L)_*(\mathcal{L}^0(H, \overline{\mathbb{R}}, \mathcal{U})) = 0.$$

In fact, there exists a complete law μ on H such that

$$(\overline{P}_L)_*(\mathcal{L}^0(H, \overline{\mathbb{R}}, \mu)) = 0.$$

Thus $(f, t) \mapsto f(t)$ is not measurable for the completion of the product measure $\overline{P}_L \times \mu$.

Proof. Let μ be the measure on H which is the distribution of $\sum_n y_n e_n$ where the y_n are independent random variables with $\mathcal{L}(y_n) = N(0, n^{-3/2})$. (Since the sum of the variances converges, this does give a law on H.) Let \mathcal{V} be the set of all functions V defined on H such that for each $f \in H$, $V_f := V(f)$ is a non-empty open interval in \mathbb{R} with rational endpoints. For any measurable set $A \subset \Omega$, finite set $F \subset H$, and $V \in \mathcal{V}$, let

$$A_{F,V} := \{z \in A: L(f)(z) \in V_f \text{ for all } f \in F\}.$$

The proof of Proposition E.2 is based mainly on the following:

E.3. Lemma *Assuming the continuum hypothesis, for every measurable set $A \subset \Omega$ with $P(A) > 0$ there is a set $S \subset H$ with $\mu^*(S) = 1$ such that for every finite set $F \subset S$ and any $V \in \mathcal{V}$,*

$$P(A_{F,V}) > 0. \qquad (*)$$

Proof. Note that for the conclusion to hold, F and so S must be linearly independent (for finite linear combinations). By the continuum hypothesis, take a well-ordered set $(J, <)$ where J has cardinality c (the cardinality of \mathbb{R}) and such that for each $\alpha \in J$, $\{\beta \in J: \beta < \alpha\}$ is countable. Since H is

separable, it has a countable base for its open sets (Proposition 2.1.4). The collection of all subsets of the countable base thus has cardinality c. Thus there are exactly c open sets in H and exactly c closed sets. Let the collection of all closed sets Y with $\mu(Y) > 0$ be indexed as $\{Y_\alpha : \alpha \in J\}$. The set $S :=$ $\{s_\alpha : \alpha \in J\}$ will be defined recursively.

For $n = 1, 2, \ldots$, let \mathcal{C}_n be the collection of all Cartesian products $\Pi_{1 \le k \le n}[a_k, a_k + 1/2^n) \times \Pi_{k > n}\mathbb{R} \subset \Omega$ where each a_k may have any of the values $-n, -n + 1/2^n, -n + 2/2^n, \ldots, n - 1/2^n$. Let $\mathcal{F}_n := \mathcal{C}_n \cup \{D_n\}$ where D_n is the complement of the union of \mathcal{C}_n. Let \mathcal{A}_n be the algebra generated by \mathcal{F}_n (or \mathcal{C}_n). Then, \mathcal{A}_n is an increasing sequence of finite algebras whose union generates the Borel σ-algebra of Ω. For each set $A \in \mathcal{F}_n$ we have $P(A) > 0$, and for any Borel set $B \subset \Omega$, $P(B \mid \mathcal{A}_n) = P(B \cap A)/P(A)$ on each $A \in \mathcal{F}_n$. By the martingale convergence theorem (10.5.4), $P(B \mid \mathcal{A}_n) \to 1$ at almost all points of B. The points of B where this occurs will be called *density points* of B (for \mathcal{A}_n). For any $z \in \Omega$ let $E^n(z)$ be the unique set in \mathcal{F}_n to which z belongs. Then z is a density point of B if and only if $z \in B$ and

$$\lim_{n \to \infty} P(E^n(z) \cap B)/P(E^n(z)) = 1.$$

A sequence $\{r_n\}$ of real numbers will be called *recurrent* iff for every non-empty open interval $U \subset \mathbb{R}$, $r_n \in U$ for infinitely many n. For any finite N, this property does not depend on r_1, \ldots, r_N, clearly. Let $A(k)$ be the set of integers $n = 2^k, \ldots, 2^{k+1} - 1, k = 0, 1, \ldots$. For $n = 1, 2, \ldots$, let $k(n)$ be the unique $k = 0, 1, \ldots$, such that $n \in A(k)$. To continue the proof of Lemma E.3, two more lemmas will help:

E.4. Lemma *Let* $y_n := (-1)^{k(n)}\omega_n/n^{3/4}$ *for* $n = 1, 2, \ldots$. *Then almost surely for* P, *the partial sums* $S_n := S(n) := \sum_{1 \le j \le n} y_j \omega_j$ *form a recurrent sequence.*

Proof. Let $Z_k := (-1)^k \sum_{j \in A(k)} \omega_j^2/j^{3/4}$. Then $EZ_k = (-1)^k \sum_{j \in A(k)} j^{-3/4}$, which is asymptotic as $k \to \infty$ to $(-1)^k 2^{(k+8)/4}(2^{1/4} - 1)$, by comparison with integrals of $x^{-3/4}$ from 2^k to 2^{k+1} and from $2^k - 1$ to $2^{k+1} - 1$. Since $E\omega_j^4 = 3$ (by Proposition 9.4.2c), the variance of Z_k is $2\sum_{j \in A(k)} j^{-3/2}$, which is asymptotic to a constant times $2^{-k/2}$ as $k \to \infty$. Thus by the Chebyshev inequality and Borel-Cantelli lemma, almost surely $|Z_k - EZ_k| < 1$ for all large enough k. Then Z_{k+1}/Z_k converges to $-2^{1/4} < -1$, and $Z_{2k+1} + Z_{2k} \to -\infty$, while $Z_{2k} + Z_{2k-1} \to +\infty$. It follows that $S(2^{2k} - 1) \to -\infty$ while $S(2^{2k+1} - 1) \to +\infty$ as $k \to \infty$. On the other hand, the individual terms $y_j \omega_j \to 0$ a.s. as $j \to \infty$, in other words, $\omega_j^2/j^{3/4} \to 0$, as follows again from $E\omega_j^4 \equiv 3$, the Chebyshev inequality, and the Borel-Cantelli lemma. So

as $k \to \infty$, in going from $S(4^k)$ to $S(2^{2k+1})$ in smaller and smaller steps, the S_n become more and more dense in \mathbb{R}, proving Lemma E.4. □

The set \mathcal{R} of all $\omega := \{\omega_j\}_{j \geq 1}$ for which $S(n)$ are recurrent is measurable (taking intervals with rational endpoints), with $P(\mathcal{R}) = 1$. Let \mathcal{M} be the set of all ω such that $\sum_j \omega_j^2 / j^{3/2} < \infty$ and such that for all N large enough, $\max_{j \leq N} |\omega_j| \leq N$. Since $\sum_N N P(|\omega_1| > N) < \infty$ by Lemma 12.1.6, $P(\mathcal{M}) = 1$. For each finite set $F \subset H$ and $V \in \mathcal{V}$, let

$$T_{F,V} := \{y \in H : \text{for some density point } \omega \text{ of } A_{F,V} \cap \mathcal{R} \cap \mathcal{M},$$
$$y_n = (-1)^{k(n)} \omega_n n^{-3/4} \text{ for all } n \geq m \text{ for some } m = m(y)\}.$$

E.5. Lemma $T_{F,V}$ *is measurable and if* $P(A_{F,V}) > 0$, *then* $\mu(T_{F,V}) = 1$.

Proof. The 1–1 measurable function $\omega \mapsto \sum_n (-1)^{k(n)} \omega_n n^{-3/4} e_n$, defined almost everywhere on Ω into H, is measure-preserving, taking P into μ. The set of all density points of $A_{F,V} \cap \mathcal{R} \cap \mathcal{M}$ is almost of all $A_{F,V}$ for P, so it has probability $P(A_{F,V})$. If in the definition of $T_{F,V}$ we fix $m = 1$, we get a measurable set with probability $P(A_{F,V})$. Suppose there we also replace the set of density points of $A_{F,V} \cap \mathcal{R} \cap \mathcal{M}$ by a Borel set included in it, with the same probability. Then for each fixed m in the definition of $T_{F,V}$, we get a set $C_{F,V,m}$ which is analytic by Theorem 13.2.1, and so universally measurable by Theorem 13.2.6. These sets increase with m. Thus their union $C_{F,V}$ is independent of y_1, \ldots, y_n for each n. On the other hand, $C_{F,V}$ is a function of the sequence $\{y_j\}_{j \geq 1}$ of variables independent for μ. We have $\mu(C_{F,V}) > 0$, so by the Kolmogorov 0–1 law (8.4.4), $\mu(C_{F,V}) = 1$. Since $C_{F,V} \subset T_{F,V}$, $\mu(T_{F,V}) = 1$ where $T_{F,V}$ is measurable since, by assumption, μ is complete, proving Lemma E.5. □

Now continuing the proof of Lemma E.3, to specify S, given $S_\alpha := \{s_\beta : \beta < \alpha\}$ such that $P(A_{F,V}) > 0$ for any finite subset F of S_α and $V \in \mathcal{V}$, let $T_\alpha := \bigcap_{F,V} T_{F,V}$ where F runs over all finite subsets of S_α and V over \mathcal{V}. Then $\mu(T_\alpha) = 1$ by Lemma E.5 since S_α is countable and the sets of possible F and V_f for $f \in F$ are countable. Thus we can and do choose $s_\alpha \in T_\alpha \cap Y_\alpha$. So the recursive definition of S is complete. Since S intersects each Y_α, and any measure on the Borel sets is closed regular (Theorem 7.1.3), $\mu^*(S) = 1$.

Now, $(*)$ in Lemma E.3 depends on only finitely many elements of S at a time. Any finite subset of S is included in S_α for some $\alpha \in J$, so it will be enough to show that S_α satisfies $(*)$ for each α. This will be done recursively, assuming that S_β satisfies $(*)$ for all $\beta < \alpha$. Also, if there is no largest $\gamma < \alpha$, then any finite subset of S_α is included in S_β for some $\beta < \alpha$, for which the

conclusion holds by induction assumption. So we can assume that there is a largest $\beta < \alpha$. Then $S_\alpha = S_\beta \cup \{s_\beta\}$.

Let F be a finite subset of S_α. Then (∗) holds if $F \subset S_\beta$, so to prove (∗) we can assume $F = G \cup \{t\}$ for a finite subset G of S_β and $t = s_\beta$. We need to prove $P(A_{F,V}) > 0$ for any $V \in \mathcal{V}$, given that $P(A_{G,V}) > 0$.

Since $t = s_\beta \in T_{G,V}$, there is a density point z of $A_{G,V} \cap \mathcal{R} \cap \mathcal{M}$ such that for some K,

$$t_j = (-1)^{k(j)} z_j j^{-3/4} \qquad \text{for all } j \geq K, \text{ and for all } N \text{ large enough,}$$
$$|z_j| \leq N \quad \text{for } j = 1, \ldots, N. \tag{\dagger}$$

Let $V_t = (a, b)$ and $\varepsilon := (b - a)/3 > 0$. There is an $M > K$ such that $P\{|\sum_{j \geq M} t_j \omega_j| \geq \varepsilon/2\} < 1/2$, since $\sum_{j \geq M} t_j \omega_j$ is a normal random variable with mean 0 and variance $\sum_{j \geq M} t_j^2 \to 0$ as $M \to \infty$.

Recall $E^n(z)$, defined in the proof of Lemma E.3, second paragraph. By independence, for all $N \geq M$,

$$P\left(E^N(z) \cap \left\{ \left| \sum_{j > N} t_j \omega_j \right| < \varepsilon/2 \right\} \right) > P(E^N(z))/2.$$

On the other hand, since z is a density point of $A_{G,V}$,

$$P(A_{G,V} \cap E^N(z)) > P(E^N(z))/2$$

for all large enough N. Then for

$$D_N := A_{G,V} \cap E^N(z) \cap \left\{ \left| \sum_{j > N} t_j \omega_j \right| < \varepsilon/2 \right\}, \quad P(D_N) > 0. \tag{\ddagger}$$

For any $\omega := \{\omega_j\}_{j \geq 1}$ in D_N,

$$\left| L(t) - \sum_{1 \leq j \leq N} t_j z_j \right| \leq S_1 + S_2 \quad \text{where}$$

$$S_1 := \left| \sum_{j > N} t_j \omega_j \right| < \varepsilon/2 \quad \text{and}$$

$$S_2 := \left| \sum_{1 \leq j \leq N} t_j (z_j - \omega_j) \right| \leq \|t\| \left(\sum_{1 \leq j \leq N} (z_j - \omega_j)^2 \right)^{1/2}$$

by the Cauchy-Bunyakovsky-Schwarz inequality. Now $\|t\|$ is fixed and $\omega \in E^N(z)$, with $|z_j| \leq N$ from (†), implies $|\omega_j - z_j| < 1/2^N$ for $j = 1, \ldots, N$,

so $S_2 \leq \|t\| N^{1/2}/2^N < \varepsilon/2$ for $N \geq N_o \geq M$, where N_o doesn't depend on ω. For such an N and any $\omega \in D_N$,

$$\left| L(t)(\omega) - \sum_{1 \leq j \leq N} t_j z_j \right| < \varepsilon.$$

Now the sequence $\sum_{1 \leq j \leq N} t_j z_j$ is recurrent, since t and z are related by (†), and $z \in \mathcal{R}$. So there is an $N \geq N_o$ such that $\sum_{1 \leq j \leq N} t_j z_j$ is in the middle third of V_t, which is the interval $(a + \varepsilon, a + 2\varepsilon)$. For such an N, we get that

$$P(A_{F,V}) \geq P(D_N \cap \{L(t) \in (a, b)\}) = P(D_N) > 0,$$

proving Lemma E.3. □

Continuing the proof of Proposition E.2, let C be a compact subset of \mathbb{R}^H, $\overline{P}_L(C) > 0$, and $C \subset \mathcal{L}^0(H, \mathbb{R}, \mu)$. Then $C \subset C_1$ for a compact Baire set C_1 with $\overline{P}_L(C) = \overline{P}_L(C_1) = P_L(C_1)$, as follows: by regularity, $\overline{P}_L(C) = \inf\{\overline{P}_L(U): U \text{ open}, U \supset C\}$. For $n = 1, 2, \ldots,$ take open U_n with $C \subset U_n$ and $\overline{P}_L(U_n \setminus C) < 1/n$. By Theorem 2.6.2 and Urysohn's lemma (2.6.3) there are continuous f_n with $0 \leq f_n \leq 1$, $f_n = 0$ on C and $f_n = 1$ outside U_n. Then $D_n := f_n^{-1}[0, 1/2]$ are compact Baire sets, whose intersection is a compact Baire set C_1, as desired.

Whether $f \in C_1$ depends only on $f(y)$ for y in a countable set $Y \subset H$, as shown in the example after Theorem 7.1.1. Let η be the function from Ω into \mathbb{R}^Y defined by $\eta(\omega)(y) := L(y)(\omega)$. This function is defined, for each y, for P-almost all ω, and since Y is countable, it is defined into \mathbb{R}^Y for P-almost all ω. Where defined, η is P-measurable. Now $C_1 = \pi^{-1}(C_2)$ where C_2 is a Borel set in \mathbb{R}^Y and π is the natural projection of \mathbb{R}^H onto \mathbb{R}^Y. From the definitions, $P(\eta^{-1}(C_2)) = P_L(C_1)$. Applying Lemma E.3 to $A = \eta^{-1}(C_2)$ gives a set $S \subset H$ with $\mu^*(S) = 1$ such that $P(\eta^{-1}(C_2)_{F,V}) > 0$ for every finite $F \subset S$ and $V \in \mathcal{V}$. Equivalently,

$$P_L\{\varphi \in C_1: \varphi(f) \in V_f \text{ for all } f \in F\} > 0.$$

By choice of C_1, we then have

$$\overline{P}_L\{g \in C: g(f) \in V_f \text{ for all } f \in F\} > 0.$$

Since C is compact for pointwise convergence, it follows that all functions from S into $\overline{\mathbb{R}}$ are restrictions of functions in C.

For any μ-measurable set $E \subset H$ let $\nu(E \cap S) := \mu^*(E \cap S)$. Then ν is a countably additive probability measure by Theorem 3.3.6. For each point $p \in S$, $\nu(\{p\}) \leq \mu(\{p\}) = 0$. So, by the continuum hypothesis again and Theorem C.1, ν is not defined on all subsets of S. So not every function from

S into $\overline{\mathbb{R}}$ can be extended to a μ-measurable function from H into $\overline{\mathbb{R}}$. This contradicts $C \subset \mathcal{L}^0(H, \overline{\mathbb{R}}, \mu)$. So the \overline{P}_L inner measure of $\mathcal{L}^0(H, \overline{\mathbb{R}}, \mu)$ is 0. □

Now it will be noted what the regularity extension *can* do for suitable processes. Let $E[0, 1]$ be the set of all real-valued functions on $[0, 1]$ such that for each $t \in [0, 1]$, the left limit $f(t^-) := \lim_{u \uparrow t} f(u)$ (except for $t = 0$) and the right limit $f(t^+) := \lim_{u \downarrow t}$ (except for $t = 1$) exist and are finite. A smaller space, which has been much studied, is the subspace $D[0, 1]$ of functions $f \in E[0, 1]$ such that $f(t) = f(t^+)$ for $0 \le t < 1$. Functions in $D[0, 1]$ are continuous from the right, with limits from the left, as are cumulative distribution functions F of probability laws P on \mathbb{R}, $F(x) := P((-\infty, x])$. If a process has sample functions in $E[0, 1]$, then the pathology as in Proposition E.2 does not occur:

E.6. Theorem (E. Nelson) *Let $I := [0, 1]$. Then $E[0, 1]$ is a Borel set in $\overline{\mathbb{R}}^I$, in fact a $K_{\sigma\delta}$, a countable intersection of countable unions of compact sets in $\mathbb{R}^I \subset \overline{\mathbb{R}}^I$. Each function in $E[0, 1]$ is Borel measurable on I. Let x be a stochastic process on $T = I$ and a probability space Ω such that for almost all $\omega \in \Omega$, the function $t \mapsto x_t(\omega)$ is in $E[0, 1]$. Then for the regularity extension \overline{P}_x of the law of x on $\overline{\mathbb{R}}^T$, $\overline{P}_x(E[0, 1]) = 1$. Thus the collections of Borel measurable functions, universally measurable functions, or functions measurable for a fixed measure on $[0, 1]$, each have inner measure 1.*

Proof. If u or v is $\pm\infty$, let $|u - v| := 0$ if $u = v$, otherwise $|u - v| := +\infty$. To show $E[0, 1]$, is a $K_{\sigma\delta}$, for $k, n = 1, 2, \ldots$, let

$$U_{nk} := \left\{ f \in \overline{\mathbb{R}}^I : \exists x_j, 0 \le x_1 < x_2 < \cdots < x_n \le 1, |f(x_j) - f(x_{j-1})| \right.$$
$$\left. > \frac{1}{k}, j = 2, \ldots, n \right\},$$

$V_n := \{ f \in \overline{\mathbb{R}}^I : \exists x, 0 \le x \le 1, |f(x)| > n \}$, both open sets. Let $W_{nk} := U_{nk} \cup V_n$. A function in $E[0, 1]$ must be bounded. We then have $E[0, 1] = \bigcap_{k=1}^{\infty} \bigcup_{n=1}^{\infty} W_{nk}^c$, a $K_{\sigma\delta}$ since each V_n^c and so W_{nk}^c is compact. For each k, $W_{1k} \supset W_{2k} \supset \cdots$, so $W_{1k}^c \subset W_{2c}^c \subset \cdots$.

If $f \in E[0, 1]$, then f is continuous except at most on a countable set, so it is Borel measurable.

If for some k and $\delta > 0$, $\overline{P}_x(W_{nk}^c) < 1 - \delta$ for all n, then there are Baire sets $B_n \supset W_{nk}^c$ with $P_x(B_n) = \overline{P}_x(W_{nk}^c)$. Let $B := \bigcup_n B_n$, a Baire set including

all W^c_{nk} with $P_x(B) \le 1 - \delta$. Then B only depends on the coordinates in a countable set T, and $B^c \subset W_{nk}$ for all n, contradicting the fact that x_t has sample functions in $E[0, 1]$. The rest follows. See also Nelson (1959, Theorem 3.4). □

Remarks. The collections of measurable functions which have inner measure 1 by Theorem E.6 as just stated all have inner measure 0 for the isonormal process L on a Hilbert space, as shown in Proposition E.2. A good many stochastic processes defined on [0, 1] can be taken to have their sample functions $t \mapsto x_t(\omega)$ in $D[0, 1]$ and so in $E[0, 1]$. One example is empirical distribution functions

$$F_n(t) := \sum_{1 \le j \le n} 1_{[0,t]}(X_j)$$

where the X_j are random variables with values in [0, 1], or specifically, i.i.d. variables with a distribution function F. Then the normalized functions $n^{1/2}(F_n - F)$ are also in $E[0, 1]$ (if F is the uniform distribution function, $F(t) = t$ for $0 \le t \le 1$, then these functions converge in law to the Brownian bridge as $n \to \infty$, as mentioned in §12.1). Other examples are Markov processes satisfying some usual regularity conditions; see, for example, Blumenthal and Getoor (1968, pp. 45–46).

On the other hand, $D[0, 1]$ is a highly nonmeasurable subset of $\overline{\mathbb{R}}^I$, as the following shows. Let $M_- := \{1_{[x,1]}: 0 \le x \le 1\}$ and $M_+ := \{1_{(x,1]}: 0 \le x < 1\}$. Then $M := M_- \cup M_+$ is compact in $I^I \subset \overline{\mathbb{R}}^I$, and $D[0, 1] \cap M = M_-$. But for any nonatomic regular Borel measure γ on I^I, M_- and M_+ both have inner measure 0, since a compact subset of either is countable. Thus if $\gamma(M) > 0$, M_- and M_+ are nonmeasurable for γ. (Note: M_- is the set of possible functions $F_1(\cdot)$.) Thus, the range of the map $x \mapsto 1_{[x,1]}$ from [0, 1] into I^I is not Borel or universally measurable in I^I, although the map is measurable with the Borel σ-algebra on I^I, since the inverse image of any open set in I^I is a union of intervals which are either nondegenerate or $\{0\}$; such a union is Borel, as the union of an open set and a countable set. Recall that by contrast $x \mapsto 1_{\{x\}}$ is nonmeasurable from I into I^I with Borel σ-algebra, or, measurable only for the σ-algebra of *all* subsets of the domain I, as arbitrary functions are (Proposition 4.2.3); its range is Borel, of the form $K \setminus \{0\}$ for a compact set K.

Notes

Proposition E.2 and its proof are from Dudley (1972, 1973); Vaclav Fabian gave very helpful advice toward the correction (1973). Proposition E.2 solves a problem posed by

Kakutani (1943) and Doob (1947). Bledsoe and Morse (1955) defined their extended product measure, which gives measure 0 to every set for which the iterated integrals of its indicator function are 0 in either order. Nelson (1959) proved positive results like Theorem E.6 more generally, showing that various classes of functions are $K_{\sigma\delta}$'s. For example, it suffices if $t \mapsto x_t(\omega)$ is continuous at t for almost all (t, ω) for some measure on T. Proposition E.2 shows, then, that such results do not extend further, to processes like the isonormal, where $t \mapsto x_t$ is continuous in probability but not for fixed ω. Tjur (1980, 10.9.4) gave a statement of nonmeasurability of M_- in the last paragraph. See also Dudley (1990).

References

Bledsoe, Woodrow W., and Anthony P. Morse (1955). Product measures. *Trans. Amer. Math. Soc. 79*: 173–215.

Blumenthal, R. M., and R. K. Getoor (1968). *Markov Processes and Potential Theory*. Academic Press, New York.

Bourbaki, N. (1952–1969). *Intégration*, Chaps. 1–4 (1952), 2d ed. (1965); Chap. 5 (1956), 2d ed. (1965); Chap. 6 (1959); Chaps. 7–8 (1963); Chap. 9 (1969). Hermann, Paris.

Doob, Joseph L. (1947). Probability in function space. *Bull. Amer. Math. Soc. 53*: 15–30.

Dudley, R. M. (1972, 1973). A counterexample on measurable processes. *Proc. 6th Berkeley Symp. Math. Statist. Prob.* (Univ. of Calif. Press) 2: 57–66; Correction, *Ann. Probability 1*: 191–192.

———— (1990). Nonmetric compact spaces and nonmeasurable processes. *Proc. Amer. Math. Soc. 108*: 1001–1005.

Halmos, Paul R. (1950). *Measure Theory*. Van Nostrand, Princeton.

Kakutani, Shizuo (1943). Notes on infinite product spaces, II. *Proc. Imp. Acad. Tokyo 19*: 184–188.

Nelson, Edward (1959). Regular probability measures on function space. *Ann. Math. 69*: 630–643.

Tjur, Tue (1980). *Probability Based on Radon Measures*. Wiley, New York.

Author Index

Morse, A. P., 149, 533, 540
Mourier, E., 433–434

Nagy, B. Sz., 246
Namioka, I., 218
Nathan, H., 22, 148
Nelson, E., 519, 540
Neugebauer, O., 183
Neumann, J. von, 112, 184, 246, 274, 275, 435, 517
Neveu, J., 382, 435
Nikodym, O. M., 184
Novikoff, A., 273, 276
Novikov, P. S., 500

Ohmann, D., 219
Ornstein, D. S., 277
Osgood, W. F., 79
Ott, E., 248
Ottaviani, G., 332
Oxtoby, J. C., 245

Paley, R. E. A. C., 481
Paplauskas [Paplauscas], A. B., 247, 500
Parseval, M.-A., 184
Parthasarathy, K. R., 500
Peano, G., 518
Pearson, Egon S., 331
Pearson, Karl, 331, 332
Picard, E., 433
Pinsky, M., 482
Plancherel, M., 277
Poisson, S. D., 332
Polya, G., 183, 219
Pontryagin, L. S., 77
Preiss, D., 434
Prokhorov, Yu. V., 434
Pruitt, W., 483
Pythagoras, 183

Quine, W. V., 504, 518

Rachev, S., 435
Radon, J., 111, 184
Ranga Rao, R., 434
Ravetz, J. R., 247
Reich, K., 78
Rényi, A., 276
Révész, P., 275
Riesz, F., 183–184, 219, 246, 330, 382
Roberts, A. W., 219

Robinson, R. M., 518
Rockafellar, R. T., 219
Rogers, L. J., 182–183
Rokhlin, V. A., 435
Root, D. H., 482
Rootselaar, B. van, 76
Rosalsky, A., 483
Rosenthal, Arthur, 185, 277
Rubin, H. and J. E., 22
Rubinstein, G. Sh., 435
Rüdenberg, L., 183
Rudin, W., 149, 246
Rudolph, D. J., 281
Russell, Bertrand, 22, 518

Saks, S., 148, 247
Salvemini, T., 435
Savage, I. R., 276
Savage, L. J., 112, 274, 278
Scarf, H. E., 435
Schaerf, H. M., 247
Schauder, J., 219
Schay, G., 434
Schmidt, E., 183–184
Schneider, I., 331
Schröder, E., 22
Schwartz, Jacob T., 78, 219, 277
Schwartz, Laurent, 138
Schwarz, H. A., 182
Schweder, T., 184, 332
Segal, I. E., 112, 481
Seidel, P. L. von, 77
Seneta, E., 276
Sherbert, D. S., 433
Sheu, S. S., 482
Sheynin, O. B., 276, 332
Shiryayev, A. N., 274
Shortt, R. M., 148, 434
Siegmund-Schultze, R., 76, 184
Sierpiński, W., 148, 149, 500, 519
Skorohod, A. V., 434, 482
Smirnov, N. V., 274, 482
Smith, H. L., 76
Smoluchowski, M., 480
Solovay, R. M., 112
Spanier, E. H., 481
Stanley, R. P., 183
Steinhaus, H., 183, 219
Stieltjes, T. J., 112
Stigler, S. M., 149, 331

Subject Index

First-order predicate logic, 504–505
Formal system, 503–505
Foundation, axioms of, 517
Fourier series, 171–173, 184, 240–243, 246–247, 314
Fourier transform, 298–310, 315, 331–332
 inversion, 303–305
Free variable (logic), 505
Fubini theorem, 137, 149
Full set, 510
Function, 5–6, 508
Functional analysis, 152–221
Fundamenta Mathematicae, 518
Fundamental theorem of calculus, 228–229

Game, fair or favorable, 353, 365
Gamma density and function, 286
Gaussian probability densities, 138, 299, 331
Gaussian stochastic process, 443
Generalized continuum hypothesis, 515, 518
Generated σ-algebra, 86
Generated σ-ring, 118
Geometry of numbers, 203, 218
Gini index, 435
Glivenko-Cantelli theorem, 400, 434
Gram-Schmidt orthonormalization, 168, 184
Graph (function), 5
Group, topological, 69, 219, 530

Haar measure, 530
Hahn decomposition, 178–179, 185
Hahn-Banach theorem, 191, 201–203, 218, 219
Half-open intervals, 25
Half-space, 200
Hamel basis, 168–169, 214
Hausdorff maximal principle, 19–22
Hausdorff space, 30, 76
Heat equation, 138, 481
Heine-Borel theorems, 76
Helly-Bray theorem, 387, 434
Hereditary collection of sets, 105
Hewitt-Savage 0–1 law, 272
Hilbert spaces, 160–174
Hitting time, 453
Hölder condition, 56, 433
Hölder inequality, 154, 157, 182–183, 208
Holomorphic function, 523
Homeomorphism, 40
Homological algebra, 518
Hypergeometric probabilities, 429
Hyperplane, 199

i.i.d. (independent, identically distributed), 261
Image measure, 121
Imaginary numbers, 521
In probability, convergence, 261, 287–291, 295
Includes, 3
Indecomposable law, 314
Independence (joint, pairwise), 252–253, 452
 of a σ-algebra, 340, 452
Indicator function, 6, 32, 253
Indiscrete topology, 39
Induction, 12–13, 513
Inductive set, 12
Inequalities for integrals, 152–157
Inf, infimum, 8, 34–35
Infinite sums of independent variables, 320–325
Infinitely divisible laws, 319, 326–329
Infinitely often, 262
Infinity, axiom of, 507
Initial segment, 13
Inner measure, 105, 533
Inner product, 160–162
Integrable function, 121
Integral: *see* Lebesgue, Riemann, Bochner, Pettis
Interior, 30
Intermediate value theorem, 43
Intersection, 5–7, 507
Interval topology, 42
Intervals, 24–25
Into (function), 5
Invariant:
 means, 274
 metric, 71
 set, 267, 374
Inverse:
 image, 27
 relation, 9, 508
Isolated point, 48
Isometry, 58, 169
Isonormal process, 444, 481, 532–533
Iterated integrals, 134–142, 343–344
Iterated logarithm, 476–480, 482–483

Jensen inequality, 348–349, 381
Jointly measurable, 139
Jointly normal, 312, 315
Jordan decomposition, 179

Notation Index

Symbols assignable to Greek or Latin letters:

\forall	for all, 7, 504
c_0	sequences $\to 0$, 202
\mathbb{C}	complex numbers, 153
$C(\cdot)$	set of continuous functions, 51
$C_b(\cdot)$	set of bounded continuous functions, 53, 158
d_{\sup}	supremum distance, 51, 52
\exists	(there) exists, 7, 504
δ_x	point mass at x, 292
E	expectation $\int \cdot \, dP$, 251, 306
\in	member of, 3, 505
\notin	not a member of, 4, 505
\varnothing	empty set, 4, 506
F_σ	countable union of closed sets, 60
G_δ	countable intersection of open sets, 60
\liminf	$\liminf x_n = \sup_m \inf_{n \geq m} x_n$, 129, 131
$\limsup_{n\to\infty}$	$\limsup x_n = \inf_m \sup_{n \geq m} x_n$, 129
$\limsup_{y\to x}$	44
ℓ^1	space of summable sequences, 72
$\mathcal{L}(\cdot)$	law of random variable, 282
λ	Lebesgue measure, 98
ℓ^p	p-summable sequences, 162
\mathcal{L}^0	set of all measurable functions, 119, 288
L^0	equivalence classes of them, 288
\mathcal{L}^p	p-integrable functions, 153, 157
L^p	equivalence classes of them, 158
\mathcal{L}^∞	essentially bounded functions, 155
L^∞	equivalence classes of them, 158
$N(m, C)$	normal law on \mathbb{R}^k, 307, 309
$N(m, \sigma^2)$	normal law on \mathbb{R}, 299
\mathbb{N}	set of nonnegative integers, 7, 507
\mathbb{Q}	set of rational numbers, 7, 58, 516
\mathbb{R}	set of real numbers, 7–8, 58, 516
\mathbb{R}^2	plane, 8
\mathbb{R}^k	Euclidean space, 38–39
\sup	supremum, 8, 34–35, 517
\mathbb{Z}	set of all integers $\ldots -1, 0, 1, \ldots$, 7, 516

554

Symbols not assignable to letters, alphabetized by defining words: